Interdisciplinary Biotechnological Advances

This series is an authoritative resource that explains recent advances and emerging directions in biotechnology, reflecting the forefront of research clearly and reliably, without excessive hype. Each volume is written by authors with excellent reputations and acknowledged expertise in the topic under discussion. The volumes span the entire field from an interdisciplinary perspective, covering everything from biotechnology principles and methods to applications in areas including genetic engineering, transgenic plants and animals, environmental problems, genomics, proteomics, diagnosis of disease, gene therapy, and biomedicine. The significance of these applications for the achievement of UN Sustainable Development Goals is highlighted. The series will be highly relevant for Master's and PhD students in Biotechnology, Nanochemistry, Biochemical Engineering, and Microbiology, medical students, academic and industrial researchers, agricultural scientists, farmers, clinicians, industry personnel, and entrepreneurs.

Satarupa Dey • Sayan Bhattacharya
Editors

Biotechnological Interventions in the Removal of Emerging Pollutants

Springer

Editors
Satarupa Dey
Department of Botany
Shyampur Siddheswari Mahavidyalaya
Howrah, West Bengal, India

Sayan Bhattacharya
School of Ecology and Environment Studies
Nalanda University
Rajgir, Bihar, India

ISSN 2730-7069 ISSN 2730-7077 (electronic)
Interdisciplinary Biotechnological Advances
ISBN 978-981-97-9921-3 ISBN 978-981-97-9922-0 (eBook)
https://doi.org/10.1007/978-981-97-9922-0

This Springer imprint is published by the registered company Springer Nature Singapore Pte Ltd.
The registered company address is: 152 Beach Road, #21-01/04 Gateway East, Singapore 189721, Singapore

We dedicate this book to our parents, family, and teachers for their guidance and motivation

Preface

Emergent pollutants are broadly categorized as a range of inorganic and organic compounds, which have detrimental effects on the well-being of living things. Different kinds of pollutants are produced by different human activities, and consequently, these pollutants have impacted public health and degraded biodiversity in different levels. The examples of emergent pollutants are pharmaceuticals, endocrine disruptors, heavy metals, antibiotics, personal care items produced by various sectors, etc. The majority of these emerging pollutants are released into surface water, where they contaminate ground water, soil, sediments, and seas. Both inorganic and biological treatments can be used to remove or degrade these pollutants from terrestrial and aquatic environments. Numerous microorganisms can be utilized effectively in bioremediation of emergent contaminants, through the processes of removal, breakdown, detoxification, and immobilization of contaminants. Simultaneously, enzyme biotechnology has also evolved as cost-effective low-energy requiring and eco-friendly technology, which can be used for remediation of various pollutants. However, bioremediation is not always adequate to treat emergent pollutants completely. Rather, a combination of both physical and chemical treatment, especially the applications of nanotechnology, can be effective in sustainable removal of the pollutants.

This book comprises 26 chapters, which address different perspectives related to emergent pollutants, the health hazard caused by them, and their bioremediation mechanisms. This book not only deals with the various health issues caused by emergent pollutants and their regulatory approaches, it also highlights the latest advancements in bioremediation techniques, exploring how microorganisms can break down various pollutants and examines the future potential of bioremediation to reduce global pollution. It will be a valuable resource for policymakers, educators, researchers, scientists, as well as undergraduate and graduate students in agriculture, forestry, ecology, soil science, and environmental sciences.

We extend our heartfelt gratitude to all the contributing authors and reviewers whose invaluable support has been instrumental in shaping this book. We sincerely thank the authors for their outstanding efforts in writing the informative chapters. Our appreciation also goes to the reviewers for their crucial inputs that have significantly improved the chapters. We would like to convey our sincere thanks to the entire team of Springer Nature for giving us the opportunity to edit this book. We also express our gratitude to our mentors, teachers, and students for their motivation. Their support has provided us with the assurance and strength to complete this project.

Finally, we thank our colleagues, family, and friends for their encouragement and support throughout the preparation of this book.

Howrah, West Bengal, India Satarupa Dey
Rajgir, Bihar, India Sayan Bhattacharya

Acknowledgment

We thank and express our sincere gratitude to all the authors and reviewers for their valuable contributions to this book. We appreciate their timely submissions and necessary corrections. We express our gratitude to the publishing team of Springer Nature for providing us the opportunity to edit this book. We convey our thanks to our colleagues for giving us the confidence and strength to accomplish this task. Finally, we would like to convey our gratitude to our family members for their constant support and encouragement.

Howrah, West Bengal, India — Satarupa Dey
Rajgir, Bihar, India — Sayan Bhattacharya

Contents

Chapter 1
An Overview and Impact of Emerging Pollutants on Human Health and Environment

Vishal Das, Kongkana Goswami, Sangeeta Biswas, Sosanka P. Sandilya, and Pobi Gogoi

Abstract Emerging pollutants (EPs) have been identified and quantified in living things and various environmental substances in recent years. Basically, chemicals that are either man-made or naturally occurring and have the potential to enter the environment and have a negative influence on both the ecosystem and human health are referred to as emerging pollutants. However, the harmful effects of environmental exposure on the general population are largely unknown and less covered. They are not currently being monitored and analyzed in the environment, although they contribute a major threat to the ecosystem and to human health, as well as causing environmental harm. Perfluorinated chemicals, by-products of water treatment, gasoline additives, pesticides, pharmaceuticals and personal care products, nanomaterials, human and veterinary medications, and UV filters, are some of the most common developing toxins. Numerous pollutants have been also reported and considered to increase the risk of cancer. Moreover, the EPs, also referred to as micropollutants, cause major harm to the environment and to human health and have made them a significant concern for the entire world's population in recent years. The existence of EP in the environment, its toxicological effects on health, and future perspectives regarding potential removal and treatment options including practical considerations, recent novel processes, new avenues, and solution strate-

V. Das (✉) · K. Goswami
Department of Medical Laboratory Technology, School of Health Sciences, The Assam Kaziranga University, Jorhat, Assam, India
e-mail: vishal@kazirangauniversity.in

S. Biswas
Department of Zoology, Royal School of Life Sciences, The Assam Royal Global University, Guwahati, Assam, India

S. P. Sandilya
Department of Microbiology, School of Health Sciences, The Assam Kaziranga University, Jorhat, Assam, India

P. Gogoi
Department of Zoology, DCB Girls' College, Jorhat, Assam, India

S. Dey, S. Bhattacharya (eds.), *Biotechnological Interventions in the Removal of Emerging Pollutants*, Interdisciplinary Biotechnological Advances,
https://doi.org/10.1007/978-981-97-9922-0_1

gies will be comprehensively discussed in this book chapter. Well, EPs are constantly being reported all over the world and considered as a major threat to the mankind and environment that needs to be addressed immediately on a global scale. The current study summarizes the overall impact and effect on emerging pollutants on human health and environment.

Keywords Emerging pollutants (EPs) · Human health · Toxicological effects

1.1 Introduction

The term "emerging pollutants" describes compounds that have just recently been discovered or have gained attention because of their environmental presence and their effects on ecosystems and human health. Numerous sources, including as industrial operations, farming practices, medications, personal care items, and more, can produce these contaminants (Gavrilescu et al. 2015). Research and regulatory efforts have focused on the effects of new contaminants on human health and the environment, as this is an increasingly pressing concern. Emerging pollutants are defined as naturally occurring or man-made substances that are not typically seen in the environment, but have the potential to do so and may have negative effects on the environment and human health that are either suspected or confirmed. It consists of items that are regularly used in homes, businesses, and other human endeavors (such as personal care and medications, plasticizers and gasoline additives, and degradation products and surfactants) (Fig. 1.1) (Arvaniti and Stasinakis 2015; Vasilachi et al. 2021).

These chemicals are being used in greater quantities in industry, transportation, agriculture, and urbanization, which are causing an increase in their hazardous waste and nonbiodegradable material levels in the environment. Furthermore, there is a lack of comprehensive and reliable epidemiological data regarding human exposure, serum and tissue concentrations, hazards to human and ecological health, and their behavior and fate in the global environment (Gavrilescu et al. 2015).

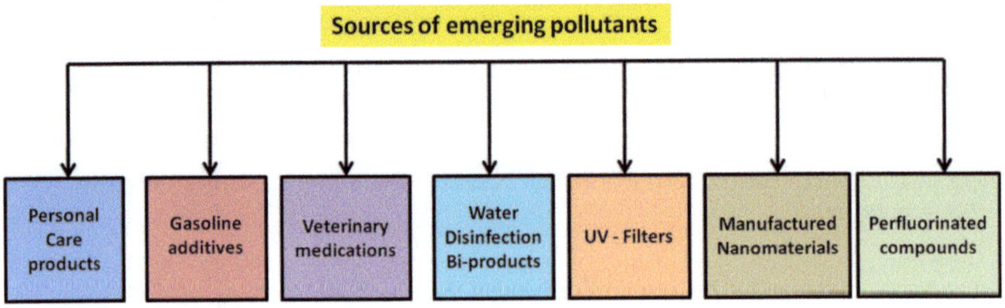

Fig. 1.1 Some of the emerging pollutants that have gained attention because of their environmental presence and their effects on ecosystems and human health

Environmental pollution ranks among the most difficult issues that face people on a daily basis. Concerning developments occurred in the quality of the environment as a result of industrialization and urbanization. Different forms of pollutants, such as persistent organic (pharmaceuticals, endocrine disrupting agents, personal care products, pesticides) and inorganic (e.g., heavy metals) are severe concerns globally that effects both the plants and animals (Bunke et al. 2019). "EPs" are chemicals that have the ability to bioaccumulate, persist in the environment, and pose a threat to human health and safety. Some of the effects of these substances include inappropriate growth, decreased fertility and reproductive health, delayed neurological development, deterioration of wildlife species, degradation of aquatic ecosystems, and possibly even immune system damage. It is crucial to emphasize that most toxins that are emerging are not newly created or have only recently entered the environment (Rodriguez-Narvaez et al. 2017). Emerging pollutants are coming from nowhere; rather, they are already here in detectable quantities, posing an irreversible threat to both human health and the ecosystem. It was previously present in extremely low concentrations, but as public knowledge of its adverse impact on the environment grows, the term "emerging pollutant" has been employed by the researchers.

This chapter summarizes and emphasizes the information on EPs and its impact on environment, a topic that has gotten a lot of attention lately. A study on emerging pollutants or emerging contaminants was examined in this work, paying particular attention to the materials' occurrence, environmental fate, and toxicity evaluation. Emerging pollutants are the threat to human health and the environment, and as a result, the study offers helpful insights to know and study about its impact on human and environment based on some thorough reports and comprehensive studies.

1.2 Perfluorinated Compounds

PFCs are aliphatic compounds with one or more carbon (C) atoms on which fluorine (F) atoms have completely replaced all hydrogen (H) atoms. The CnF2n+1− moiety is connected to a variety of functional groups in PFCs. Sulfonic acid (SO3H), carboxylic acid (−COOH), and sulfonamides (−SO2NH2) are the three main functional groups (). Produced since the late 1940s, perfluorinated compounds (PFCs) consist of a hydrophobic alkyl chain that is fully fluorinated and connected to a hydrophilic end group. The most important representative PFCs are perfluorooctanesulfonate (PFOS), perfluorooctanoic acid (PFOA), and their salts. PFCs pile up in the environment and are biomagnified by the food chain since they are incredibly stable and do not biodegrade (Morikawa et al. 2006). Compared to other environmental pollutants, PFCs are ubiquitous and can be found in blood samples from both industry employees and the general public at comparatively high concentrations. More significantly, their presence in breast milk and umbilical cord blood suggests that fetal development may be the beginning of a lifetime exposure to PFCs (Lee et al. 2013). PFC exposure is attributed to the existence of these species in blood, food, consumer products, indoor chemicals, and occupational exposure, per a study on PFCs (Wu et al. 2015).

Numerous studies have demonstrated that long-chain PFCs—that is, PFOA and PFOS—have longer elimination half-lives and are more bioaccumulative than short-chain PFCs, such as CnF2n+1COOH, where $n \geq 7$, and CnF2n+1−SO3H, where $n \geq 6$ (D'Hollander et al. 2010). The major manufacturer of PFOA and PFOS, the 3M Company, phased out production of these chemicals in 2000 due to widespread concern about the potential health and environmental effects of long-chain PFCs. This concern also facilitated regulatory initiatives in many countries to reduce the environmental release of these PFCs (Olsen et al. 2009).

Numerous toxicological consequences, such as altered thyroid hormone, hepatotoxicity, changes in glucose and cholesterol, toxicity to the reproductive and developmental systems, and carcinogenic effects, have been linked to PFAS exposure in epidemiological and experimental investigations (Rodriguez-Jorquera et al. 2019; Lau et al. 2007). The most prevalent malignancy among Chinese women is breast cancer. In the world, China is the country that accounts for 9.6% of all breast cancer fatalities and 12.2% of new diagnoses. Breast cancer risk factors include early menarche, late menopause, postmenopausal obesity, smoking, alcohol consumption, and excessive fat intake, in addition to genetic variables including mutations in BRCA-1/2. Fan L et al. in the year 2014 reported that less than one-third of cases are explained by known risk factors, and environmental exposure and lifestyle choices can significantly affect a person's risk of developing breast cancer (Fan et al. 2014). Numerous epidemiological studies have been conducted in Europe and the United States to evaluate the dangers of PFAS exposure and breast cancer (Fan et al. 2014). According to a recent study, perfluoro-n-tridecanoic acid (PFTrDA) was adversely related with breast cancer in Chinese women, but the plasma concentrations of Perfluorooctanoic acid (PFOA) and perfluoro-n-decanoic acid (PFDA) were positively associated with breast cancer. Breast cancer that was ER-, PR-, and HER2-positive was more likely to be linked to PFOA (Li et al. 2022). In their investigation, Seyyedsalehi MS et al. found a link between kidney cancer and total PFAS exposure, as well as a link between high PFAS concentrations and testicular cancer (Seyyedsalehi and Boffetta 2023). Many studies on human infertility, lactation, and semen quality have examined the relationship between the blood level of PFCs and reproductive failure in the general population. Fei et al. looked at the possibility that PFC exposure and related hormone disruptors may make infertility more likely (Fei et al. 2009). A study conducted by Joensen et al. suggested that, with the exception of primipara, PFOS may reduce the women's ability to lactate. Because multiparous women had previously breastfed and because excretion could lower PFOS serum levels, the connection between the PFOS serum level and multiparous women was not as strong. Joensen et al. looked at a possible relationship between testicular function, semen quality, and perfluoroalkyl acids (PFAAs) (Joensen et al. 2009).

The possible links between the levels of PFOA and PFOS in the serum and the prevalence of thyroid disease have been highlighted by other research. In order to determine whether there was a significant correlation between the serum and tissue concentrations of PFCs, Pirali et al. measured the PFOS and PFOA levels among 28 participants who underwent thyroid surgery for benign diseases (7 for Graves' disease and 15 for multinodular goiters) and malignant thyroid diseases (5 for papillary carcinoma and 1 for follicular carcinoma) (Pirali et al. 2009). The Northeastern states with the highest prevalence of thyroid cancer were the main locations where the higher PFAS levels were found. These findings support closer monitoring of PFAS exposure among local populations and provide guidance for future research on the possible consequences of exposure on the thyroid gland (Alsen et al. 2003).

1.3 Gasoline Additives

Chemical compounds or substances known as gasoline additives are added to gasoline in order to improve performance, lower emissions, increase fuel efficiency, and prevent engine deposits. These additives have multiple uses and are specifically designed to meet the needs of contemporary combustion engines.

From the 1920s to the late 1980s, lead-based antiknock additive mixes were widely employed to raise the octane ratings of gasoline. These mixtures contained the volatile organic compounds 1,2-dichloroethane (DCA) and 1,2-dibromoethane (ethylene dibromide; EDB). With federal maximum contaminant levels (MCLs) of 0.05 and 5 μg/L, respectively, EDB and DCA are likely human carcinogens. In the United States, lead is no longer used in gasoline, although EDB and DCA are still present in the environment (Falta et al. 2005).

In groundwater-dependent public drinking water systems in the United States, EDB and DCA are two of the most frequently found contaminants (EPA-815-R-03-006; U.S. EPA 2003). Generally, discharges from industries (DCA) or agriculture (EDB) are blamed for groundwater pollution by these chemicals. However, more than 80% of the EDB was consumed as an additive for gasoline in the 1970s. Often utilized as a solvent, DCA is a common industrial chemical feedstock (Falta et al. 2005).

In many nations, using leaded gasoline is still standard procedure. Tetraethyl lead (Pb (C2H5)4) is the most widely used lead compound, and it is used to increase the gasoline's octane level. Greater thermodynamic efficiency is achieved by using engines with higher compression ratios, which are made possible by higher octane fuel. Manufacturers are able to obtain more power and fuel efficiency out of lighter, smaller engines thanks to this increased efficiency. However, using this gasoline results in a lot of undesired emissions of particulates, carbon monoxide, and lead. For these reasons, a number of nations have worked to alter the usage of leaded gasoline in an effort to lessen or completely prevent the unintended pollution emissions that come with the rising use of gasoline (Bravo et al. 2006).

This substance is present in gasoline and is a member of the aromatic family. It is also utilized as a solvent. Exposure to toluene, whether acute (short-term) or chronic (long-term), can be hazardous to humans and animals. Following inhalation, symptoms include headaches, nausea, exhaustion, headaches, dizziness, and upper respiratory tract irritation (ATSDR 2017) (WHO 2016).

There are over 500 components in gasoline, including methyl tert-butyl ether (MTBE), 1,3-butadiene, and benzene, all of which are known or suspected carcinogens. Due to their high volatility and solubility, methyl tertiary butyl ether (MTBE) and ethyl tertiary butyl ether (ETBE) rapidly disperse in the environment and have low biodegradability. Consequently, their presence in surface waters is widespread even at low quantities of μg/L (Apetroaei et al. 2020). The most popular oxygenated bunker, MTBE, is being sporadically employed as a new unleaded gasoline additive, especially in developing areas (Mennear 1997). Tetraethyl lead as an antiknock species can be replaced by MTBE, which also improves the octane additive used in gasoline to increase its burning efficiency and reduce carbon monoxide and other dangerous chemicals, such benzene and ozone, in vehicle exhaust (McGregor 2006). According to research on animals, MTBE can damage the kidney, liver, immune system, and central nervous system in addition to causing cancer of the testicles, uterus, and kidneys (Mennear 1997). In an analytical study, Joseph and Weiner found a correlation between the levels of MTBE in the air and the following symptoms: skin rash, palpitations, allergy, otitis media, cough, headache, throat stimulation, hypersensitive rhinitis, upper respiratory transmissible illness, sicchasia, dizziness, wheezing, anxiety, and insomnia (Joseph and Weiner 2002).

Strong alkylating chemicals like chloroethane are utilized as industrial refrigerant, topical anesthetics, polymers, pigments, and gasoline additives (Fishbein 1979). Naturally occurring and extremely lipophilic, bromoethane finds usage in both industrial and pharmacological settings. Its ability to cross the blood–brain barrier is noteworthy (Klassen 1996). According to research conducted by the National Toxicology Program (NTP), uterine tumor incidence increased in B6C3F1 mice exposed to these chemicals for several weeks of inhalation (NTP 1989a, b). Aoyama et al. (2005) recently investigated the direct estrogenic effects of bromoethane in human Ishikawa cells and in ovariectomized B6C3F1 mice. For a duration of 3 days, mice were administered 0, 100, 500, or 1000 mg kg^{-1} of bromoethane.

1.4 Personal Care Products and UV Filters

Personal care products (PCPs) are a broad category of everyday home chemicals used for cleaning, cosmetics, and wellness. Products that easily expose people include those for hair and skin care, baby care, UV blocking lotions, facial cleansers, insect repellents, perfumes, fragrances, soap, detergents, shampoos, conditioners, toothpaste, and more.

These consist of, among other things, UV filters, scents, disinfectants, and insect repellents. Due to their existence and detrimental effects on aquatic ecosystems, particularly in relation to endocrine disruption and reproductive diseases, several of them are regarded as chemicals of growing concern. Studies done in 2017 has reported that, 72 PCPs that have been identified as emerging contaminants in 30 different nations back then, with concentrations ranging from 0.029 ng/L to 7.811×10^6 ng/L. The most often reported categories were sunscreens, antiseptics, and fragrances. As anticipated, wastewater treatment plant effluents included 64 PCPs total—more than twice as much as those identified in surface water and groundwater combined—than any other source. This finding provides proof of the anthropological input of PCPs to water bodies (Montes-Grajales et al. 2017).

Oxybenzone is widely used as a photo-stabilizer to reduce color and odor changes in sunscreen preparations at doses up to 6% and primarily as a short-wave UVA light absorber (320–340 nm) at concentrations up to 290–320 nm. According to the reviewed data, oxybenzone is present worldwide in human urine, serum, and breast milk in addition to being present in water, soil, sediments, sludge, and biota. It is not as effective as avobenzone, titanium dioxide, and/or zinc oxide as a sunscreen active in preventing UVA rays. This substance is a known contact and photocontact allergen that can cause contact urticaria and, to a lesser extent, contact-mediated anaphylaxis in people. It has also been connected to Hirschsprung's illness (DiNardo et al. 2018).

PCPs are present in human bodies at all stages of life, including intrauterine development. Inhalation, cutaneous contact, ingestion, and absorption are the direct routes of exposure; product use and environmental contamination are the indirect pathways (Heudorf et al. 2007). The estimated daily cutaneous route of exposure to titanium dioxide (TiO2) is between 2.8 and 21.4 mg/person/day, with toothpaste and sunscreen being major contributors to this exposure. In addition, it was calculated that the toothpaste caused ingestion of 0.15–3.9 mg TiO2/day. An estimated 35% of produced TiO2 is utilized in PCPs as UV shielding agents or to stop yellowing of the product (Wu and Hicks 2019).

Triclosan is the most common biocide ingredient in consumer liquid hand soaps. It is an antibacterial and antifungal agent that has been used as a biocide in personal hygiene products since the 1960s. It was originally used in the medical field in 1972 as a 1% component of surgical scrub and as toothpaste for dental care in Europe in 1985. Under the Federal Insecticide, Fungicide, and Rodenticide Act, triclosan has also been approved by the USEPA for use as an antimicrobial pesticide in bacteriostat and fungicide/fungistat applications (Jones et al. 2000).

A class of p-hydroxybenzoic acid ester derivatives (methyl, ethyl, propyl, isopropyl butyl, isobutyl, or benzyl) are known as parabens. Since parabens and their chlorinated derivatives are often detected in the aquatic compartment at low ng L^{-1} levels, they could be regarded as emerging pollutants. Because of their antibacterial properties as well as their durability at various pH and temperature ranges, parabens are frequently found in cosmetic products (Haman et al. 2015).

Due to their ability to scatter UV radiation with wavelengths ranging from 290 to 400 nm, sunscreens are a popular choice for protection against UV radiation risks, early skin aging, and skin cancer. Among the organic sunscreens that absorb new UV photons are benzophenone-3 (BP-3), 2-ethylhexyl 4-methoxycinnamate (OMC), 2-ethylhexyl 4-dimethylaminobenzoate (OD-PABA), homosalate (HMS), 3-(4-methylbenzylidene)camphor (4-MBC), and 3-benzylidene camphor (3-BC). Moreover, swimming, laundry, and bathing can introduce highly generated lipophilic sunscreens into the aquatic environment. Numerous investigations conducted on laboratory animals, both in vitro and in vivo, have demonstrated the endocrine disruptive effects of sunscreen, encompassing disturbances to the hypothalamic-pituitary-thyroid axis (HPT) as well as reproductive and developmental functions. Through the food chain, human and animal exposure patterns may overlap (Lei et al. 2015). The study conducted by Johnson et al. revealed that formaldehyde releasers, cyclosiloxanes, and parabens were the most commonly used compounds. Additional chemicals of concern (CoCs) detected by gas chromatography time-of-flight mass spectrometry (GCxGC-TOFMS) included pollutants, perfumes, solvents, and preservatives. Their findings supplement research on health disparities caused by chemical exposures from several sources, increase awareness of potentially harmful compounds in PCPs, and aid in estimating disparities in chemical exposure.

Because UV filters are found worldwide in sediments, seas, and biota, they have recently gained attention as pollutants. These substances are widely utilized in commercial items to stop photodegradation and in personal care products to stop UV spectrum infiltration. These pollutants are being found in the aquatic environment more frequently as consumption rises, most likely as a result of being rinsed off after application and then seeping into the ecosystem (Coleman et al. 2023). In the cosmetics fields, octyl-methoxycinnamate (OMC) is widely used as a UV-B filter. Although OMC was first created to lessen the harmful photobiological impacts of UV radiation, its safety has come under scrutiny after certain research revealed harmful effects on the environment (Lorigo et al. 2024).

1.5 Water Disinfection By-Products

A type of secondary pollutant called disinfection by-products (DBPs) is often found in the urban water cycle and has been shown to have (eco) toxicological effects on aquatic systems as well as human health. As a result, DBPs are already commonplace environmental pollutants that should raise alarm. The chemical disinfectants that are most frequently used in public water sources are chlorine, ozone, chloramine (chlorine + ammonia), and chlorine dioxide. According to recent research, relatively modest iodine intakes raise thyroid stimulating hormone levels (Robison et al. 1998).

Analytical chemists found in 1974 that consumed waters contained amounts of up to ~160 μg/L of trihalomethanes (THM4; chloroform, bromodichloromethane, dibromochloromethane, and bromoform), which form as byproducts of chlorine interactions with natural organic matter (NOM) (Rook 1974).

The list of harmful health effects linked to chemical disinfectants keeps getting longer. As of right now, the only significant risk that seems to be connected to drinking water chlorination is a higher risk of bladder cancer in men. It is crucial to identify the substances generating this effect in order to define alternatives that lower or eliminate the risk, not just to satiate scientific curiosity (Bellar et al. 1974). By the 1940s, widespread outbreaks of cholera, typhoid, and other waterborne illnesses had been mostly contained in the industrialized world because to chlorine cleaning of municipal drinking supplies, which had begun shortly after 1900. Ever since, epidemiological research has indicated links between drinking chlorinated tap water with high THM4 levels and unfavorable health consequences, such as bladder cancer (Costet et al. 2011).

Several halogenated organic compounds, such as haloacetic acids (HAAs), haloacetonitriles (HANs), and trihalomethanes (THMs), are included in DBPs. Of the previously listed DBPs, THMs are the most prevalent, followed by HAAs, which can be detected in finished water at quantities of around μg/L, and HANs, which can be found in finished water at values as low as ng/L and as high as μg/L. Chloroform, bromodichloromethane (BDCM), chlorodibromomethane (CDBM), and bromoform are examples of volatile compounds that belong to the THMs group. On the other hand, HANs, such as dichloroacetonitrile (DBAN) and monochloroacetic acid (MCAA), and HAAs, such as dichloroacetic acid (DCAA), trichloroacetic acid (TCAA), monobromoacetic acid (MBAA), and dibromoacetic acid (MBAA), are composed of volatile compounds (Kwarciak-Kozlowska 2020). Exposure to HOCs has been linked to a wide range of health consequences in both humans and wildlife, including neurological, immunological, endocrine, behavioral, and carcinogenic impacts. Furthermore, current research suggests that type 2 diabetes and obesity are influenced by exposure to HOCs (Kodavanti et al. 2023).

1.6 Veterinary Medications

Veterinarian medicines are one of the newly discovered toxins that have attracted a lot of interest because they have been found in water supplies and may be harmful to aquatic biota, which includes humans who use these water sources for purification. Anthelmintics make up an important portion of the animal pharmaceutical market. They are used to treat helminthic illnesses, or infections brought on by parasitic worms, and are administered to a variety of significant veterinary animals in aquaculture and agriculture (Horvat et al. 2012). Due to their widespread use as biocides, fungicides in agriculture, and antifungal agents in veterinary medicine, azole compounds have the potential to leave significant environmental residues due to their extensive use. It is believed that azoles may have an impact on aquatic species' endocrine systems (Peng et al. 2012). The pollution of soil, surface water, and groundwater caused major concerns because antibiotics, even at low concentrations, encourage the emergence and growth of antibiotic-resistant microbes because of their constant exposure to these substances. The authors report a rise in the development of harmful bacterial strains that are resistant, which can have detrimental effects on human health. A study by Vidhamaly et al. underline the necessity for additional thorough research with a solid methodology in order to better inform policy and put policies in place that ensure the quality of veterinary pharmaceuticals in supply chains. They suggested that transdisciplinary study is needed to understand the mechanism and effects of (substandard and fraudulent) SF veterinary products on human and animal health, agricultural output, their economies, and antimicrobial resistance (AMR) (Vidhamaly et al. 2022). Recent reports have linked the bicyclic nonsteroidal anti-inflammatory drug ketoprofen—which is frequently used in both human and veterinary medicine—to environmental contaminants that pose a threat to ecological health. Because of its sporadic mixture, enantiomers, and transformation products—which have ecotoxicological impacts on a variety of taxa, including plants, microbes, and vertebrates—it is becoming a more serious danger. Moreover, ketoprofen endangers the ecosystem's ability to operate by being bioaccumulated and biomagnified throughout the food chain (Tyumina et al. 2023).

1.7 Manufactured Nanomaterials

Nanomaterials (NMs) are ultrafine particles with a size range of 1–100 nm; however, size alone does not provide a valid criterion for NM classification. Materials may be incidental, man-made, or natural. Neural networks (NMs) display a variety of structural dimensions, such as zero-, one-, two-, and three-dimensional configurations (Cheriyamundath and Vavilala 2021). Because of their unique features and their applicability in a wide range of consumer and commercial applications, nanomaterials have sparked questions regarding possible environmental effects. Nanomaterials' small size may increase their bioavailability, or the ease with which living things can absorb them. Potentially negative consequences for ecological systems and creatures at various trophic levels may result from this. Because of their greater mobility, they may have a role in the environmental dispersion of nanomaterials. Additionally, they have the capacity to interact with other environmental contaminants in ways that have unclear antagonistic or synergistic effects (El-Kalliny et al. 2023).

According to the reports, NPs have the ability to damage DNA and trigger cell death since they can be found inside the cell in a variety of places, including the cytoplasm, nucleus, lipid vesicles, and cell membrane. An essential stage prior to cellular absorption is the interaction of NPs with cell membranes. It is believed that adhesive contact caused by steric interactions, electrostatic charges, Van der Waals forces, or interfacial tension effects is the mechanism of NP uptake by cells. These could consist of inflammation, oxidative stress, toxicity, and DNA damage. The transfer of NPs across tissue barriers into the bloodstream, where they can circulate and ultimately lodge in other organs, causes secondary NP exposure. These might include organ damage in the liver, spleen, or kidneys, as well as toxicity at the location of NP deposition (El-Kalliny et al. 2023).

Car exhaust is the primary source of atmospheric NPs. While gasoline engines release NPs between 20 and 60 nm, diesel engines release NPs between 20 and 130 nm. It has been discovered that CNTs and fibers are released as byproducts during the burning of gas and diesel. Since diesel-generated particles account for over 90% of all carbon nanoparticles (NPs) in the atmosphere, automobile pollution is a primary source of nanoparticulate contamination in metropolitan environments (Jeevanandam et al. 2018). In their study, Stefani et al. (2005) noted that lead, glass, respirable asbestos fibers, and other hazardous particles from everyday items could be present in the released particulates near the site of building destruction. The benthic organisms in freshwater were adversely impacted by the NPs aggregates that settled to the bottom and gradually accumulated in the silt. While NPs may build up in the marine ecology between warm and cold currents (Hyseni 2016), it has been reported that TiO2 NP exposure has been linked to negative effects on aquatic organisms, including "trout," according to certain study findings (Federici et al. 2007). Another nanoparticle Amorphous silicon dioxide particles (A-SiO2) inhalation exposure in animal has been reported partially linked to reversible inflammation in the respiratory system and alterations in blood parameters; however, these studies do not show evidence of lung fibrosis progression (Shin et al. 2017). Carbon nanotubes (CNTs), which can absorb nanomaterials by skin contact or oral ingestion through the gastrointestinal tract, have been the subject of several studies reporting possible effects on both aquatic and terrestrial creatures. According to tests on mammals, CNTs are both cytotoxic and genotoxic to various cell types, including macrophages. Exposure to CNTs can cause chromosomal aberrations, necrosis, apoptosis, and the release of reactive oxygen species (ROS), as well as the expression of inflammatory cytokines like IL-8 (Girardello et al. 2015) (Table 1.1).

1.8 Impacts of Emerging Pollutants on Human and Environment

Human health is seriously and increasingly at risk from pollution. It is now thought to be the primary environmental cause of health issues worldwide, causing over nine million preventable deaths every year. It causes the economy to suffer enormous losses, undermines efforts for economic improvement, and causes the Sustainable Development Goals (SDGs) to take longer to achieve. Pollution poses a threat to both the long-term survival of human populations and the integrity of the earth's support systems, much as climate change and the depletion of freshwater resources worldwide. The persistent challenge is the emerging pollutant. It is known that some EPs may affect aquatic ecosystems and creatures biochemically. Daphnia fish have been found to be susceptible to four distinct classes of EPs (antibiotics, antineoplastics, cardiac medications, and sex hormones) when these medicinal compounds are tested for aquatic toxicity; on the other hand, estrogens and antibiotics have been identified as the most toxic products for both public health and marine life (Rodriguez-Narvaez et al. 2017).

Chemicals are created and used by industries in over 100,000 distinct ways, making them an essential element of daily life. The employment and economic growth of nations are significantly impacted by the global chemical market. On the other hand, a lot of chemicals are released into the environment, and chemical contamination is becoming commonplace in both land and water. These environmental contaminants affect human health and the environment in a wide range of acute and

Table 1.1 Emerging pollutants, its sources, and their effects on human health and environment

S. no.	Emerging pollutants	Effect on human health and environment	Chemicals and sources	References
1	Perfluorinated compounds	Altered thyroid hormone, hepatotoxicity, changes in glucose and cholesterol, toxicity to the reproductive and developmental systems, and carcinogenic effects	Perfluoroalkyl acids	Rodriguez-Jorquera et al. (2019), Lau et al. (2007)
		Breast cancer, kidney cancer, and testicular cancer	PFAS	Fan et al. (2014), Seyyedsalehi and Boffetta (2023)
		Increased pregnancy losses and disruption in sex hormone homeostasis and sexual maturation	PFOA & PFOS	Potera (2009)
		Reduces women's ability to lactate	PFOS	Joensen et al. (2009)
2	Gasoline additives	Damages kidney, liver, immune system, and central nervous system in addition to causing cancer of the testicles, uterus, and kidneys	MTBE	Mennear (1997)
		Uterine tumor incidence.	Bromoethane	NTP (1989a, b)
		Following inhalation, symptoms include headaches, nausea, exhaustion, headaches, dizziness, and upper respiratory tract irritation	Toluene	ATSDR (2017)
		Suspected carcinogen	ETBE	Apetroaei et al. (2020)
3	Veterinary medications	Frequently used in both human and veterinary medicine has reported to have ecotoxicological impacts on a variety of taxa, including plants, microbes, and vertebrates	Ketoprofen	Tyumina et al. (2023)
		Impact on aquatic species' endocrine systems	Azole	Peng et al. (2012)
4	Water disinfection by-products	Widespread outbreaks of cholera, typhoid, and other waterborne illnesses had been mostly contained in the industrialized world because of chlorine cleaning of municipal drinking supplies.	Water chlorination	Costet et al. (2011)
		Linked to a wide range of health consequences in both humans and wildlife, including neurological, immunological, endocrine, behavioral, and carcinogenic impacts. Type 2 diabetes and obesity are influenced by exposure to HOCs	Halogenated organic compounds	Kodavanti et al. (2023)

5	Personal care products and UV filters	Oxybenzone is present worldwide in human urine, serum, and breast milk in addition to being present in water, soil, sediments, sludge, and biota. This substance is a known contact and photocontact allergen that can cause contact urticaria and, to a lesser extent, contact-mediated anaphylaxis in people. It has also been connected to Hirschsprung's illness	Oxybenzone	DiNardo et al. (2018)
		Harmful effects on the environment	Octyl-methoxycinnamate (OMC) is widely used as a UV-B filter	Lorigo et al. (2024)
		Early skin aging, and skin cancer.	Among the organic sunscreens that absorb new UV photons are benzophenone-3 (BP-3), 2-ethylhexyl 4-methoxycinnamate (OMC), 2-ethylhexyl 4-dimethylaminobenzoate (OD-PABA), homosalate (HMS), 3-(4-methylbenzylidene)camphor (4-MBC), and 3-benzylidene camphor (3-BC)	Lei et al. (2015)
6.	Manufactured nanomaterials	Negative effects on aquatic organisms	TiO2	Federici et al. (2007)
		Partially linked to reversible inflammation in the respiratory system and alterations in blood parameters	A-SiO2	Shin et al. (2017)
		CNTs are both cytotoxic and genotoxic to various cell types, including macrophages. Exposure to CNTs can cause chromosomal aberrations, necrosis, apoptosis, and the release of reactive oxygen species (ROS), as well as the expression of inflammatory cytokines like IL-8	Carbon nanotubes (CNTs)	Girardello et al. (2015)

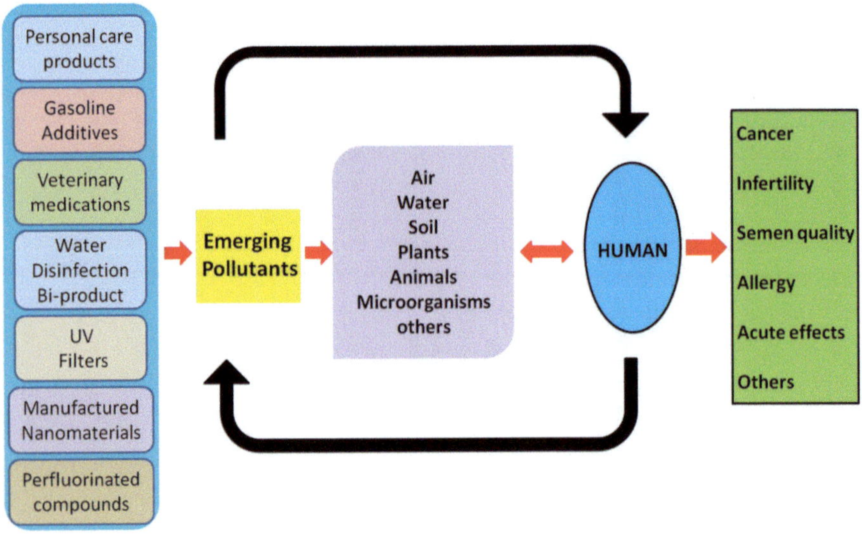

Fig. 1.2 Schematic diagram displaying the impact of some emerging pollutants on human and environment. It is also showcasing the interaction among air, water, soil, animals, microorganisms, and EPs that have harmful effects on humans upon exposure

long-term ways (e.g., immunotoxicity, neurological diseases, malignancies, endocrine disruption, etc.) (Sanchez and Egea 2018).

Chemicals and substances that have lately been determined to pose a threat to ecosystems, the environment, and human health are known as emerging pollutants, or EPs (Fig. 1.2). Furthermore, a large number of these newly discovered toxins are not covered by international or national regulations, which increases the risk to our way of life. Numerous substances, such as pharmaceuticals, steroids, endocrine disruptors, hormones, industrial additives, chemicals, microbeads, and microplastics are among the contaminants (Krishnakumar et al. 2022). Some of the emerging pollutants are discussed in this chapter that has gained attention lately.

Numerous investigations conducted on laboratory animals, both in vitro and in vivo, have demonstrated the endocrine disruptive effects of sunscreen, encompassing disturbances to the hypothalamic-pituitary-thyroid axis (HPT) as well as reproductive and developmental functions (Krause et al. 2012). Long-term exposure to particulate matter has been linked to a number of new concerns, including diabetes, prenatal mortality, neurodevelopment, and cognitive functions. Ultrafine particulate matter, secondary inorganic aerosols, organic and black carbons, and other newly discovered air pollutants have detrimental impacts on human health. Black carbon has been shown to raise mortality from all causes, as well as from cardiovascular and cardiopulmonary diseases (Yadav et al. 2021).

1.9 Conclusion and Future Recommendations

Through a variety of known and unknown routes, humans and the ecosystem are exposed to a wide range of new pollutants. Our chapter explores that these pollutants consistently create new and pressing problems for the air, water, soil, and ecosystems—especially for human health. Furthermore, novel chemical outputs proliferate and typically surpass the capabilities of risk assessment techniques, safety remediation strategies, monitoring methodologies, and existing preventative measures. The developing pollutants are suspected to be carcinogenic, teratogenic, and mutagenic to humans and other animals, according to ongoing research and studies. On the one hand, there isn't enough substantial data to establish a link between newly discovered pollutants and harmful impacts on human health and environment. However, we cannot discount the negative consequences suggested by animal tests, even though the exposure levels in animal trials could not accurately reflect human exposure levels and long-term chronic exposure is rarely used in animal models. We still don't fully understand the long-term impacts of EPs, though. The limited and sparse nature of EPs' fate and detrimental effects on aquatic life and human health demands for more research and understanding.

Therefore, in order to facilitate integrated research to reduce pollution inputs while maximizing available resources, future research endeavors should focus on the pollutants that have the most effects on both human health and environment.

References

Alsen M, Leung AM, van Gerwen M (2003) Per- and polyfluoroalkyl substances (PFAS) in community water systems (CWS) and the risk of thyroid cancer: an ecological study. Toxics 11(9):786. https://doi.org/10.3390/toxics11090786. PMID: 37755796; PMCID: PMC10537801

Aoyama H, Couse JF, Hewitt SC, Haseman JK, He H, Zheng X et al (2005) Upregulation of estrogen receptor expression in the uterus of ovariectomized B6C3F1 mice and Ishikawa cells treated with bromoethane. Toxicol Appl Pharmacol 209(3):226–235

Apetroaei MR, Avram ER, Atodiresei DV, Apetroaei GM (2020) Emerging pollutants-a potential threat to the marine environment. Sci Bull Nav Acad 23(1):171–178

Arvaniti OS, Stasinakis AS (2015) Review on the occurrence, fate and removal of perfluorinated compounds during wastewater treatment. Sci Total Environ 524–525:81–92

ATSDR (2017) Toxicological profile for toluene. Public Health Service, U.S. Department of Health and Human Services, Atlanta

Bellar TA, Lichtenberg JJ, Kroner RC (1974) The occurrence of organohalides in chlorinated drinking waters. J Am Water Works Assoc 66(12):703–706

Bravo AH, Sanchez AP, Sosa ER, Keener TC, Lu M (2006) The potential impact of gasoline additives on air quality in developing countries. Clean Techn Environ Policy 8:174–181

Bunke D et al (2019) Developments in society and implications for emerging pollutants in the aquatic environment. Environ Sci Eur 31(1):1–17

Cheriyamundath S, Vavilala SL (2021) Nanotechnology-based wastewater treatment. Water Environ J 35:123–132

Coleman C, Glinski D, Hankins C, Raimondo S, Henderson W (2023) Understanding the behavior of common ultraviolet filtering compounds under simulated environmental conditions. In: 2023 SETAC North America Annual Meeting, Louisville, KY, November 12–16, 2023

Costet N, Villanueva CM, Jaakkola JJK, Kogevinas M, Cantor KP, King WD, Lynch CF, Nieuwenhuijsen MJ, Cordier S (2011) Water disinfection by-products and bladder cancer: is there a European specificity? A pooled and meta-analysis of European case-control studies. Occup Environ Med 68(5):379–385

D'Hollander W, de Voogt P, De Coen W, Bervoets L (2010) Perfluorinated substances in human food and other sources of human exposure. Rev Environ Contam Toxicol 208:179–215

DiNardo JC, Downs CA (2018) Dermatological and environmental toxicological impact of the sunscreen ingredient oxybenzone/benzophenone-3. J Cosmet Dermatol 17(1):15–19

El-Kalliny AS, Abdel-Wahed MS, El-Zahhar AA, Hamza IA, Gad-Allah TA (2023) Nanomaterials: a review of emerging contaminants with potential health or environmental impact. Discov Nano 18(1):68

Falta RW, Bulsara N, Henderson JK, Mayer RA (2005) Leaded-gasoline additives still contaminate groundwater. Environ Sci Technol 39(18):379A–384A

Fan L, Strasser-Weippl K, Li JJ, St Louis J, Finkelstein DM, Yu KD et al (2014) Breast cancer in China. Lancet Oncol 15:e279–e289

Federici G, Shaw BJ, Handy RD (2007) Toxicity of titanium dioxide nanoparticles to rainbow trout (Oncorhynchus mykiss): Gill injury, oxidative stress, and other physiological effects. Aquat Toxicol 84(4):415–430

Fei C, McLaughlin JK, Lipworth L, Olsen J (2009) Maternal levels of perfluorinated chemicals and subfecundity. Hum Reprod 24(5):1200–1205. https://doi.org/10.1093/humrep/den490

Fishbein L (1979) Potential halogenated industrial carcinogenic and mutagenic chemicals. II. Halogenated saturated hydrocarbons. Sci Total Environ 11:163–195

Gavrilescu M, Demnerova K, Aamand J, Agathos S, Fava F (2015) Emerging pollutants in the environment present and future challenges in biomonitoring, ecological risks and bioremediation. New Biotechnol 32:147–156

Girardello R, Tasselli S, Baranzini N, Valvassori R, de Eguileor M, Grimaldi A (2015) Effects of carbon nanotube environmental dispersion on an aquatic invertebrate, Hirudo medicinalis. *PLoS One* 10(12):e0144361

Haman C, Dauchy X, Rosin C, Munoz JF (2015) Occurrence, fate and behavior of parabens in aquatic environments: a review. Water Res 68:1–11

Heudorf U, Mersch-Sundermann V, Angerer J (2007) Phthalates: toxicology and exposure. Int J Hyg Environ Health 210:623–634

Horvat AJ, Petrović M, Balbić S, Pavlović DM, Ašperger D, Pelko S, Mance AD, Kaštelan-Macan M (2012) Analysis, occurrence and fate of anthelmintics and their transformation products in the environment. TrAC Trends Anal Chem 31:61–84

Hyseni S (2016) Toxicological effects of nanomaterials on aqueous and terrestrial ecosystems. Cent Dev Strateg 2016(1):1–2

Jeevanandam J, Barhoum A, Chan YS, Dufresne A, Danquah MK (2018) Review on nanomaterials in medicine and their applications. Beilstein J Nanotechnol 9:1050–1074

Joensen UN, Bossi R, Leffers H, Jensen AA, Skakkebæk NE, Jørgensen N (2009) Do perfluoroalkyl compounds impair human semen quality? Environ Health Perspect 117(6):923–927. https://doi.org/10.1289/ehp.0800517

Jones RD, Jampani HB, Newman JL, Lee AS (2000) Triclosan: a review of effectiveness and safety in health care settings. Am J Infect Control 28(2):184–196

Joseph PM, Weiner MG (2002) Visits to physicians after the oxygenation of gasoline in Philadelphia. Arch Environ Health 57(2):137–154

Klassen CD (ed) (1996) Casarett and Doull's toxicology: the basic science of poisons, 5th edn. McGraw-Hill, New York

Kodavanti PRS, Costa LG, Aschner M (2023) Perspective on halogenated organic compounds. Adv Neurotoxicol 10:1–25

Krause M, Klit A, Jensen MB et al (2012) Sunscreens: are they beneficial for health? An overview of endocrine disrupting properties of UV-filters. Int J Androl 35(3):424–436

Krishnakumar S, Singh DSH, Godson PS et al (2022) Emerging pollutants: impact on environment, management, and challenges. Environ Sci Pollut Res 29:72309–72311

Kwarciak-Kozlowska A (2020) Methods used for the removal of disinfection by-products from water. In: Prasad MNV (ed) Disinfection by-products in drinking water: detection and treatment, 1st edn. Butterworth-Heinemann, Oxford, pp 1–22

Lau C, Anitole K, Hodes C, Lai D, Pfahles-Hutchens A, Seed J (2007) Perfluoroalkyl acids: a review of monitoring and toxicological findings. Toxicol Sci 99:366–394

Lee YJ, Kim MK, Bae J, Yang JH (2013) Concentrations of perfluoroalkyl compounds in maternal and umbilical cord sera and birth outcomes in Korea. Chemosphere 90:1603–1609

Lei M, Zhang L, Lei J, Zong L, Li J, Wu Z, Wang Z (2015) Overview of emerging contaminants and associated human health effects. Biomed Res Int 2015:1–12

Li X, Song F, Liu X, Shan A, Huang Y, Yang Z, Li H, Yang Q, Yu Y, Zheng H, Cao XC, Chen D, Chen KX, Chen X, Tang NJ (2022) Perfluoroalkyl substances (PFASs) as risk factors for breast cancer: a case-control study in Chinese population. Environ Health 21(1):83

Lorigo M, Quintaneiro C, Breitenfeld L, Cairrao E (2024) Effects associated with exposure to the emerging contaminant octyl-methoxycinnamate (a UV-B filter) in the aquatic environment: a review. J Toxicol Environ Health B Crit Rev 27(2):55–72

McGregor D (2006) Methyl tertiary-butyl ether: studies for potential human health hazards. Crit Rev Toxicol 36(4):319–358

Mennear JH (1997) Carcinogenicity studies on MTBE: critical review and interpretation. Risk Anal 17(6):673–681

Montes-Grajales D, Fennix-Agudelo M, Miranda-Castro W (2017) Occurrence of personal care products as emerging chemicals of concern in water resources: A review. Sci Total Environ 595:601–614

Morikawa A, Kamei N, Harada K, Inoue K, Yoshinaga T, Saito N et al (2006) The bioconcentration factor of perfluorooctane sulfonate is significantly larger than that of perfluorooctanoate in wild turtles (Trachemys scripta elegans and Chinemys reevesii): an Ai river ecological study in Japan. Ecotoxicol Environ Saf 65:14–21

National Toxicology Program (1989a) Toxicology and carcinogenesis studies of chloroethane (ethyl chloride) in F344/N rats and B6C3F1 mice (inhalation studies) (NTP TR No. 346, NIH Publication No. 75-00-3). National Institute of Environmental Health Sciences, Research Triangle Park, NC

National Toxicology Program (1989b) Toxicology and carcinogenesis studies of bromoethane (ethyl bromide) in F344/N rats and B6C3F1 mice (inhalation studies) (NTP TR No. 363, NIH Publication No. 74-96-4). National Institute of Environmental Health Sciences, Research Triangle Park, NC

Olsen GW, Chang SC, Noker PE, Gorman GS, Ehresman DJ, Lieder PH et al (2009) A comparison of the pharmacokinetics of perfluorobutanesulfonate (PFBS) in rats, monkeys, and humans. Toxicology 256:65–74

Peng X, Huang Q, Zhang K, Yu Y, Wang Z, Wang C (2012) Distribution, behavior, and fate of azole antifungals during mechanical, biological, and chemical treatments in sewage treatment plants in China. Sci Total Environ 426:311–317

Pirali B, Negri S, Chytiris S et al (2009) Perfluorooctane sulfonate and perfluorooctanoic acid in surgical thyroid specimens of patients with thyroid diseases. Thyroid 19(12):1407–1412

Potera C (2009) Reproductive toxicology: study associates PFOS and PFOA with impaired fertility. Environ Health Perspect 117(4):A148. PMCID: PMC2679623

Robison LM, Sylvester PW, Lang JP, Bull RJ (1998) Comparison of the effects of iodine and iodide on thyroid function in humans. J Toxicol Environ Health 55(2):93–106

Rodriguez-Jorquera IA, Colli-Dula RC, Kroll K, Jayasinghe BS, Parachu Marco MV, Silva-Sanchez C et al (2019) Blood transcriptomics analysis of fish exposed to perfluoro alkyls substances: assessment of a non-lethal sampling technique for advancing aquatic toxicology research. Environ Sci Technol 53:1441–1452

Rodriguez-Narvaez OM, Peralta-Hernandez JM, Goonetilleke A, Bandala ER (2017) Treatment technologies for emerging contaminants in water. A review. Chem Eng J 323:361–380

Rook JJ (1974) Formation of haloforms during chlorination of natural water. J Water Treat Exam 23(2):234–243

Sanchez W, Egea E (2018) Health and environmental risks associated with emerging pollutants and novel green processes. Environ Sci Pollut Res 25:6085–6086

Seyyedsalehi MS, Boffetta P (2023) Per- and poly-fluoroalkyl substances (PFAS) exposure and risk of kidney, liver, and testicular cancers: a systematic review and meta-analysis. Med Lav 114(5):e2023040

Shin JH, Jeon K, Kim JK et al (2017) Subacute inhalation toxicity study of synthetic amorphous silica nanoparticles in Sprague-Dawley rats. Inhal Toxicol 29(10):567–576

Stefani D, Wardman D, Lambert T (2005) The implosion of the Calgary General Hospital: ambient air quality issues. J Air Waste Manage Assoc 55(1):52–59

Tyumina E, Subbotina M, Polygalov M, Tyan S, Ivshina I (2023) Ketoprofen as an emerging contaminant: Occurrence, ecotoxicity, and (bio)removal. Front Microbiol 14:1200108

U.S. EPA (2003) Occurrence estimation methodology and occurrence findings report for the six-year review of existing National Primary Drinking Water Regulations; EPA-815-R-03-006. U.S. EPA, Office of Water, Washington, DC. www.epa.gov/safewater/standard/review/pdfs/support_6yr_occurancemethods_final.pdf

Vasilachi IC, Asiminicesei DM, Fertu DI, Gavrilescu M (2021) Occurrence and fate of emerging pollutants in water environment and options for their removal. Water 13(2):181

Vidhamaly V, Bellingham K, Newton PN et al (2022) The quality of veterinary medicines and their implications for One Health. BMJ Glob Health 7:e008564

WHO (2016) WHO expert consultation: available evidence for the future update of the WHO global air quality guidelines (AQGs). Meeting report, Bonn, Germany, 29 September – 1 October

Wu F, Hicks AL (2019) Estimating human exposure to titanium dioxide from personal care products through a social survey approach. Integr Environ Assess Manag 16:10–16

Wu XM, Bennett DH, Calafat AM et al (2015) Serum concentrations of perfluorinated compounds (PFC) among selected populations of children and adults in California. Environ Res 136:264–273

Yadav D, Rangabhashiyam S, Verma P, Singh P, Devi P, Kumar P et al (2021) Environmental and health impacts of contaminants of emerging concerns: Recent treatment challenges and approaches. Chemosphere 272:129492

Chapter 2
Evaluating Regulatory Approaches to Emerging Pollutants

Sangeeta Biswas, Pobi Gogoi, Matthew Wilkinson, Papori Deka, Vishal Das, Kongkana Goswami, and Sosanka P. Sandilya

Abstract This book chapter examines the evolving landscape of policies associated with emerging pollutants on a global scale. In light of the present scenario, where numerous unassessed emerging pollutants are widespread in the ecosystem, understanding existing policies and their shortcomings is imperative. By conducting an inclusive assessment of existing policies, case studies, and their validity, we divulge the convolutions and challenges in managing emerging pollutants. The chapter underscores the need for adaptable policies that consider the pertinacious nature of emerging pollutants and their varied nascencies. By evaluating the successes and shortcomings of existing policies, we offer insights into potential strategies for more effective regulatory frameworks. For instance, collaborative networks like NORMAN can be valuable for global regulatory bodies such as Environmental Protection Agency (EPA), European Chemicals Agency (ECHA), World Health Organization (WHO), United Nations Environment Programme (UNEP), Basel Convention, Stockholm Convention on Persistent Organic Pollutants, and Rotterdam

S. Biswas (✉)
Department of Zoology, Royal School of Life Sciences, The Assam Royal Global University, Guwahati, Assam, India
e-mail: sbiswas1@rgu.ac

P. Gogoi
Department of Zoology, DCB Girls' College, Jorhat, Assam, India

M. Wilkinson
University of New South Wales, Sydney, NSW, Australia

P. Deka
Department of Zoology, Handique Girls' College, Guwahati, Assam, India

V. Das · K. Goswami
Department of Medical Laboratory Technology, School of Health Sciences, The Assam Kaziranga University, Jorhat, Assam, India

S. P. Sandilya
Department of Microbiology, School of Health Sciences, The Assam Kaziranga University, Jorhat, Assam, India

S. Dey, S. Bhattacharya (eds.), *Biotechnological Interventions in the Removal of Emerging Pollutants*, Interdisciplinary Biotechnological Advances,
https://doi.org/10.1007/978-981-97-9922-0_2

Convention on Prior Informed Consent for monitoring, assessment, and regulation of ECs. This chapter adds to the ongoing discourse on environmental policy development and its crucial significance in mitigating escalating risks posed by emerging pollutants.

Keywords Emerging pollutants · Regulations · Global perspective · Challenges · Environmental governance

2.1 Introduction

The mounting pace of urbanization, industrial development, and human anthropogenic activities has led to the deterioration of the environment and, indeed, encumbered the same with toxic pollutants, generally termed emerging pollutants. Although these chemical substances have been in unvarying use for a very long time, they have gained more attention and are in focus lately because of the emergence of many groundbreaking analytical techniques that have revealed the existence of these pollutants even at the level of parts per trillion in all environmental metrics like air, water, and soil (Puri et al. 2023). Emerging contaminants (ECs) are a diverse collection of synthetic compounds, their derivatives, and byproducts that are widely used in agricultural practices, pharmaceuticals, and personal care products (PPCPs), mass-produced in various industries, emitted from automobiles, released as manufacturing waste, among other activities (Sanganyado 2022). Emerging contaminants also include synthetic as well as natural hormones, such as xenoestrogens and phytoestrogens, respectively, which usually end up in the environment (Puri et al. 2023). A common term called "contaminants of emerging concern" (CECs) emerged a decade ago for the chemicals that are presently not under regulation or not submitted to a routine monitoring organization, but may be under investigation for impending regulation (Dulio et al. 2018). These emerging pollutants often have severe impacts on the endocrine system, reproductive health, and embryonic development of humans and wildlife. These have been well documented, along with their roles in mutagenesis and carcinogenesis (Barroso et al. 2019; Ramírez-Malule et al. 2020).

Currently, about 275 million chemical substances are registered with the Chemical Abstracts Service (CAS), and it covers references of more than 59 million from the early 1800s to the present (CAS 2024). Nearly 700 new chemical substances per year are added to the US Toxic Substances Control Act (TSCA) on record, whereas the European Chemicals Agency (ECA 2021) has annually listed around 1700 chemicals under Registration, Evaluation, Authorization, and Restriction of Chemicals (REACH) since 2009 (Muir et al. 2023). A recently reported 50-year bibliometric study stated that out of the 19,776 CAS Registry Numbers (CASRNs) under consideration for "analytical study" and "pollutant," a data set of only 9.251×10^6 total counts of the CASRNs was found related to biological studies over a 55-year period (Muir et al. 2023).

Apprehensions have risen in recent decades since these chemicals are ubiquitous in nature, present at detectable levels in the environment and biological metrics, biologically active even at low concentrations, and have detrimental health effects on living organisms. A major concern is the occurrence of antibiotics in the aquatic environment, as it may result in the development of antimicrobial resistance genes (ARGs) in the bacterial population of the harboring environment and may cause several untreatable diseases (Biswas and Vellanki 2021). As stated by a recent review report, the development of antimicrobial resistance may lead to the deaths of 10 million individuals per year by 2050 (Chaturvedi et al. 2021). It is noteworthy to remark that, recently in the year 2021 at the 5th session of the United Nations Environment Assembly meeting (UNEA 5), environmental pollution in consort with biodiversity loss and climate challenge was categorized as priority global challenges (UNEA 2021). Subsequent to all the rising concerns, several environmental regulatory bodies were formed, different policies and acts were made, and international agreements were signed. For instance, in 2005, following a call by the European Commission, NORMAN was funded (Dulio et al. 2018). Presently, NORMAN is an autonomous and decidedly trustworthy open network of reference research centers, laboratories, associated establishments, private companies, and even NGOs for the monitoring of emerging contaminants. It, as an autonomous body with more than 70 members from 20 different countries across the globe, plays a very crucial role at the interface between the scientific community and policymakers (Dulio et al. 2018). Several other agencies include the Environmental Protection Agency (EPA), the European Chemicals Agency (ECHA), the World Health Organization (WHO), the United Nations Environment Programme (UNEP), etc.

In view of the fact that emerging pollutants are gradually loaded into the ecosystem due to anthropogenic activities, and many of them are not even evaluated for their ecotoxicity, understanding current regulatory policies, their shortcomings, and their challenges is crucial. Therefore, this review summarizes the existing regulatory policies and the evolving regulatory frameworks in recent times with reference to emerging pollutants on a global scale.

2.2 Understanding Emerging Pollutants

Pollutants are a broad term used for harmful substances that are released into the environment, such as heavy metals, oil, and toxins. Their effects are understood and long-standing regulations are in place to regulate them, whereas emerging pollutants are chemicals or contaminants that have only recently been recognized as a threat, are relatively new, are under the radar, and are often unregulated, with little or no monitoring and established limits for their harmful effects. "Emerging contaminants," now called contaminants of emerging concern, emerging pollutants, EPs, or new pollutants, is a loose term that has been in use since the mid-1990s by the US EPA and others, to describe a diverse group of ubiquitously present naturally occurring or chemically synthesized substances that are not subject to regulations, not monitored commonly, were only "discovered" relatively recently (often as a result of improved analytical chemistry detection levels), having long-distance mobility, and harmful to environment and health at relevant concentrations (Ma et al. 2018; Puri et al. 2023; The OW/ORD Emerging Contaminants Workgroup 2008; Xu and Liu 2023).

The extensive roster of emerging contaminants includes some of the following such as herbicides like glyphosate and atrazine; various pharmaceuticals like diclofenac, ibuprofen, various antibiotics, and hormones; both legal and illegal stimulants, including caffeine, cocaine, and amphetamines; preservative agents like parabens and triclosan; personal care items, for instance, sunscreens and UV filters; detergents and cleansers; industrial compounds and chemicals such as bisphenol A and chlorinated solvents; additives and materials used in food packaging, for example, phthalates and plasticizers; complex organic compounds like polycyclic aromatic hydrocarbons, polychlorinated biphenyls, and brominated dioxins; brominated/chlorinated flame retardants, perfluorinated compounds, and perfluoroalkyl substances; perfluorinated surfactants; endocrine disruptors; brominated/chlorinated flame retardants; polybrominated diphenyl ethers; PFAS (per-and polyfluoroalkyl substances), GenX, benzotriazoles; antibiotic-resistant bacteria such as *Escherichia coli* strains that produce extended-spectrum β-lactamase; other chemicals like alkylphenols, various metalloids, radionuclides, and rare earth elements; novel entities like nanomaterials and microplastics; potential agents of bioterrorism and sabotage; as well as indoor pollutants and various pathogens (Aguilar-Aguilar et al. 2023; Dong et al. 2023; Goot 2019; Morin-Crini et al. 2022).

In 2022, Morin-Crini et al. reviewed worldwide cases of water pollution by emerging contaminants and proposed a classification of emerging pollutants regardless of their effects on humans and the environment. These classes of emerging contaminants are pharmaceuticals (e.g., antibiotics, anti-inflammatory, hormones, psychiatric drugs, illicit drugs, veterinary medicine), personal care products (e.g., UV filters, disinfectants, biocides, cosmetic ingredients: bisphenols, parabens, triclosan), industrial substances (e.g., surfactants, alkylphenols, perfluorinated compounds, nanoparticles), pesticides (e.g., fungicides: triazoles, herbicides, insecticides: neonicotinoids, bactericides, rodenticides, nematicides), biological agents (e.g., pathogenic bacteria, multidrug resistant, microbes, antifungal resistance, antibiotic resistance genes), unintentional persistent organic pollutants (e.g., dioxins, brominated dioxins, halogenated polycyclic aromatic hydrocarbon, polychlorinated naphthalenes, and environmentally persistent free radicals), and other substances (rare earth elements, metalloids: Se, radionuclides, food packaging end containers, corrosion inhibitors, microplastics). However, Arman et al. (2021) use wider criteria to categorize emerging contaminants: (a) a compound that isn't necessarily new; (b) a compound that has been present in the environment for a while but whose presence has only recently been discovered and whose significance is starting to be understood; and (c) a well-known compound whose potential harm to people and the environment has only recently come to light.

These new pollutants have the potential to enter the environment through various pathways, such as wastewater discharges, agricultural runoff, atmospheric deposition, and improper disposal (Aguilar-Aguilar et al. 2023). Once these emerging pollutants are released into the environment, they can persist and accumulate in different compartments, such as water bodies and soil. The presence of emerging pollutants in the environment raises concerns due to their potential adverse effects.

Many of these emerging pollutants are not currently subject to routine monitoring programs, possibly due to their low concentrations (μg L^{-1}, ng L^{-1}) and undetectable contamination (Vargas-Berrones et al. 2020). Nonetheless, they are persistent, bioaccumulative, and hazardous, posing significant health and environmental risks, including disruption of homeostasis, development, behavior, embryo development, metabolism, reproductive system, and steroid levels; and they can cause cardiovascular issues, aberrant neuronal activity, neurotoxicity, reproductive and developmental toxicity, immune system deficiencies, carcinogenicity, gene expression aberrations, internal abrasion, gastrointestinal blockage, pathological stress, and oxidative stress (Cardoso 2022; Mahmud et al. 2022; Vargas-Berrones et al. 2020; Xu and Liu 2023).

2.3 Historical and Current Policy Context

Regulatory frameworks emerged during the late 1900s and early 2000s on controlled use of EC, mainly because of the issues related to their damaging effects. To protect human health and the environment, the EPA was founded in 1970 by Richard Nixon, former US president, and ever since, the regulatory powers and responsibilities of the agency have been a topic of intense argument (Ramírez-Malule et al. 2020; Britannica 2024). In the beginning, the EPA was charged with the management of the *Clean Air Act* (1970), the Federal Environmental Pesticide Control Act (1972), and the Clean Water Act (1972). By the mid-1990s, the EPA had imposed 12 major laws and continued to reinforce laws governing the quality of air, water, and noxious substances (Britannica 2024). Between 1999 and 2000, an important nationwide survey was carried out by the US Geological Survey and detected the presence of PPCPs in 80% of the 139 streams sampled along the country (Kolpin et al. 2002). Following the changes made in 1996 to the Safe Drinking Water Act, the EPA was asked to periodically list the contaminants that are not yet regulated but are suspected to be present in the water (Ramírez-Malule et al. 2020). Under the Clean Air Act (CAA), the Environmental Protection Agency (EPA) of the USA sets and enforces air pollution control standards nationwide (EPA US). In Europe, the Water Framework was implemented in 2000 and set some guiding principles for the sustainable use of aquatic bodies in Europe and to protect drinking water sources (EU 2000). Since the early 1970s, the EU has taken various actions for improving air quality including the New Air Quality Directive in the year 2008 (Directive 2008/50/EC) on ambient air quality and cleaner air for Europe (Marco and Bo 2013; UNION 2008). Some legislations for public participation were made; for instance, in 1998, the United Nations Economic Commission for Europe (UNECE) Convention or Aarhus Convention (UNECE 1998) was adopted in the Conference on the Environment for Europe process and entered into force in 2001. The convention grants access to data, public involvement in decision-making, and access to justice in matters related to the environment (Marco and Bo 2013). The information on Airbase is maintained by the European Environment Agency (EEA), and it publishes reports on air quality from time to time and the latest Air Quality Report was published by the EEA (EEA 2011, 2012). The constitution of the United Nations Educational, Scientific, and Cultural Organization (UNESCO) was taken in London in 1945, and later in 1946 it came into force. Ever since, UNESCO has launched the International Initiative on Water Quality (IIWQ), targeting to meet global water quality challenges (Ramírez-Malule et al. 2020).

Under the framework of IIWQ, a program was initiated by UNESCO called the Project on Emerging Pollutants in Water and Wastewater (2014–2018). In 2004, the call for NORMAN by the EU Commission was to create "a permanent network of reference laboratories and related organizations dealing with emerging environmental substances," and the same was funded in 2005. Over the past 10 years, as an independent body, NORMAN has endorsed a permanent network of reference labs, various types of research centers, NGOs, and even private establishments to work on emerging contaminants (Dulio et al. 2018). Today, NORMAN has around 70 members from 20 different countries around the globe. There are three pillars on which NORMAN relies: the first pillar is EMPODAT, a database of storage and management systems; the second pillar is the EMPODAT ECOTOX module, which is basically a platform for organized collection and assessment of the relevance and trustworthiness of ecotoxicity studies, the third pillar is nontarget screening (NTS) for identification of suspects and unknown ECs (Dulio et al. 2018). For instance, in the Republic of Korea, water policy started in the 1960s, mainly in the form of comprehensive water resource development projects (Lee 2019). The settlement and advancement of the comprehensive water resource development plan happened later in the 1970s and 1980s, respectively. While the application of environment-friendly water management plans started during the 1990s, a paradigm shift toward sustainable water management happened in the 2000s (Lee 2019). Since 2010, policies have been made focusing on combating climate change and also on progressing water management systems. A very recent amendment was made on May 28, 2018, when the national congress decided to restructure the government organization, which is at the central level, and the Ministry of Environment was given control over water management (Lee 2019). Hence, the policies were made at a very early time, and amendments are done in the same direction for sustainable development.

2.4 Methods for Detecting and Assessing Emerging Pollutants

The protection of the environment and human health from the threat of emerging pollutants encompasses a wide range of activities, starting from monitoring and assessing the pollutant sources to managing them under sustainable conditions. Emerging pollutants, if not monitored properly, may become unmanageable with increased and rapid industrialization, urbanization, and the widespread use of chemicals in agriculture and consumer products. To effectively confront these issues, a comprehensive monitoring program with data collection must be established.

Monitoring and data collection allow for the early detection of pollutants and the necessary information to evaluate the possible risks associated with them. These data sets are essential for regulatory bodies to establish and implement standards for the release of pollutants into the environment. This allows authorities to bind industries and entities under regulations to protect the environment.

Environmental monitoring stations play a crucial role in determining air quality, suspended particulate matter, volatile organic compounds, and nitrogen dioxide. However, a study by Langenbach and Wilson (2021) suggested that emerging pollutants like polyfluoroalkyl substances (PFAS) are becoming increasingly prevalent in air samples, emphasizing the need for expanded monitoring beyond conventional pollutants. Water quality monitoring requires the evaluation of parameters such as chemical oxygen demand (COD), heavy metals, and emerging contaminants like pharmaceutical and personal care products (PPCPs) (Bilela et al. 2023).

The detection and assessment of emerging pollutants include numerous methods ranging from traditional to advanced technologies, depending on the nature of the pollutants. Some conventional methods with a general reference to the field of environmental monitoring are analytical chemistry techniques, like high resolution mass spectrometry (HRMS) and high-performance liquid chromatography (HPLC), which are employed to identify and quantify pollutants in water as well as in environmental samples (Castro et al. 2021; Khalid et al. 2018; Yusuf et al. 2021).

Effect-based monitoring assays are increasingly used to assess chemical pollution and water quality, with potential applications beyond detecting pollutants. Studies have used estrogenicity measurement to evaluate wastewater treatment efficiency, percentage removal of chemical waste, and ecotoxicological risks. Implementing effect-based monitoring addresses the limitations of chemical analytical strategies, as they are inadequate to quantify decreasing concentrations of pollutants in aquatic ecosystems (Yusuf et al. 2021). Biomonitoring techniques involve bioindicators, biosensors, and biological assays. Bioindicators are used to assess environmental conditions by monitoring living organisms for physiological or behavioral changes due to exposure to emerging pollutants. Biosensors offer a rapid and sensitive approach by utilizing biological components to recognize and respond to specific pollutants. Biological assays are important tools that provide a complete understanding of the impacts of emerging pollutants on the ecosystem and its biotic components (Di Paolo et al. 2016; Gavrilaş et al. 2022; Yusa et al. 2012).

Remote sensing provides a cost-effective and long-term observation method with unprecedented frequency and spatial coverage for monitoring and assessing emerging pollutants. Algal blooms may be used as indicators of water quality and the presence of emerging contaminants, and it makes it possible to identify and track them. To gather data on changes in water color, variations in backscattered microwave radiation, and thermal data—which can provide information on the existence and distribution of emerging pollutants—remote sensing platforms, including ground-based, spaceborne, airborne, and UAV-based sensors, can be utilized. The use of several multi sensor data types, including optical and radar/LiDAR data, can enhance the precision of monitoring and assessing emerging contaminants. Artificial intelligence (AI)-based image processing algorithms may be used to evaluate and monitor emerging contaminants by analyzing remote sensing data and forecasting their presence and behavior (Rolim et al. 2023).

Environmental DNA analysis is an advanced technique that employs the detection and analysis of genetic material shed by organisms into the environment. This technique offers a valuable insight for environmental monitoring, allowing us to track biodiversity and assess the effect of pollutants on the ecosystem in an efficient manner (Nagarajan et al. 2022). Additionally, to predict the existence of analogous pollutants in environmental samples, machine learning algorithms like artificial neural network (ANN) may be trained on documented characteristics of recently identified pollutants (Rolim et al. 2023).

2.5 Challenges in Policy Implementation

The successful implementation of policies is an intricate endeavor that involves translating policy intentions into concrete actions, yet a number of challenges hinder effective enforcement. These challenges comprise economic, social, and political constraints that create complex barriers that policymakers must understand and manage thoughtfully to ensure their initiatives achieve the desired impact.

Adding to the complexity, emerging pollutants pose significant health risks, especially for vulnerable groups like fetuses, children, pregnant women, and the often overlooked poor. The highest rates of environmentally linked illness and mortality burden are observed in low- and middle-income countries, where impoverished populations, often associated with ethnic and religious minorities, experience an unequal impact (Prüss-Üstün et al. 2016; Sexton 2012). Moreover, these communities face additional threats from new potentially polluting activities in their vicinity, making environmental problems worse and putting them at higher risk of pollution (Haller et al. 2020).

The economic implication of pollution along with emerging pollutants incurs direct cleanup expenses ranging from tens of thousands to hundreds of millions of dollars annually across various countries. However, sufficient attention is often not given to additional expenses like impacts on agriculture, ecosystems, and biodiversity. Globally, pollution causes a 15–25% loss in agricultural productivity, with economies like Ghana experiencing a substantial 40% decline (Aragón and Rud 2016; Saha et al. 2017). In countries like China, annual economic losses total around USD 20 billion due to decreased crop yields and contaminated food (Zhou et al. 2020). The economic threat extends beyond air, water, and soil to broader pollution concerns. Diseases arising from emerging pollutants, often underestimated, lead to a decrease in the labor force and overall economic productivity.

Barriers to effective enforcement of emerging pollutants policies persist globally, despite increasing awareness of associated environmental and public health risks. The complexity of identifying and regulating emerging pollutants poses significant challenges. For instance, in India, the widespread use of pesticides in agriculture contaminates soil and water resources, endangering ecosystems and public health (Anju et al. 2010). Similarly, in the USA, the presence of per- and polyfluoroalkyl substances (PFAS) in drinking water raises health concerns due to their persistence (Domingo and Nadal 2019). Financial constraints further impede enforcement efforts, particularly in the Global South, where limited resources hinder investment in regulatory infrastructure and monitoring capabilities.

Weak enforcement mechanisms and limited government oversight have allowed these activities to persist, posing serious risks to human health and the environment. Weak enforcement in developing countries exacerbates environmental pollution and public health risks. In many regions of the Global South, regulatory agencies lack the resources, technical expertise, and political will to effectively enforce environmental regulations. This leads to widespread pollution and environmental degradation, with communities disproportionately bearing the burden of pollution-related health problems (Xiao 2020). In Kenya, for example, inadequate enforcement of e-waste management regulations has resulted in significant environmental and health hazards. The country faces challenges in managing the influx of electronic waste (e-waste), which often contains hazardous substances such as lead, mercury, and brominated flame retardants. Improper handling and disposal of e-waste leads to chemical runoff, contaminating water sources and soil. This contamination not only affects aquatic ecosystems but also poses risks to human health through the consumption of contaminated water and crops grown in contaminated soil. Moreover, informal e-waste recycling activities, often conducted without proper safety measures, expose workers to hazardous chemicals and pollutants, resulting in adverse health effects such as respiratory problems, neurological disorders, and skin diseases (Curtis et al. 2006). Power dynamics involving large global conglomerates exacerbate resourcing challenges in the Global South. In Sikkim, India, pharmaceutical companies, often operating as subsidiaries of multinational corporations, have been implicated in releasing hazardous waste, including expired drugs and chemical byproducts, into local water bodies at alarming rates (GroundXero 2019). With 29 operational pharmaceutical factories, Sikkim acts as a "lucrative backyard" for the Indian capital (McDuie Ra and Chettri 2018). Overall, governments in the Global South have desires to continue to attract investment that sit at odds with enforcing stringent limitations and checks on the externalities of this investment.

2.6 Case Studies

Policy responses to emerging pollutants are telling of the above challenges and offer insights into successful policy models and barriers to successful policy making and enforcement of policies aimed at reducing, treating, and diverting emerging pollutants.

South Korea's journey in water management unfolds as a nuanced exploration, particularly concerning the intricate challenges posed by emerging pollutants. Commencing with the development of river basins under Japanese rule, the nation's postwar recovery prompted the establishment of the River Basin Law in 1961. The surge in industrialization during the 1970s led to heightened water demand, necessitating the formulation of the Long-term Comprehensive

Water Resource Development Plan. However, the subsequent decades witnessed water quality deterioration attributable to economic growth and heavy metal utilization. The democratic regime's strategic emphasis on environment-friendly river basin restoration prompted the revision of the Long-term Comprehensive Water Resource Development Plan in 1996, accompanied by the enactment of the Han River Law in 1999, specifically directed at enhancing water quality. Amid persistent challenges, South Korea has emerged as a global frontrunner in water management, notably addressing emerging pollutants through initiatives exemplified by the Water Vision 2020. The 2017 Environmental Performance Review underscores the nation's commitment to green growth, highlighting advancements in policy frameworks, emission trading schemes, and environmental technology. Recognized by the OECD, Korea's transition toward sustainable water management is underscored by technological innovation, particularly in the realms of smart water management and inventive solutions for pollution abatement and water conservation. The nation's leadership in water-related innovation, epitomized by patented technologies for rainwater collection, storage, and seawater desalination, encapsulates a pioneering approach to mitigating the impact of emerging pollutants on water resources (Choi et al. 2017; OECD 2021).

Denmark, on the other hand, has emerged as a trailblazer in environmental regulation, prioritizing the prevention and management of emerging contaminants. The nation's proactive stance involves implementing precautionary measures even in the absence of conclusive scientific evidence, aiming to mitigate potential risks before they escalate. Denmark's stringent regulations often surpass the minimum standards set by the European Union, advocating for stricter controls on substances exhibiting endocrine-disrupting properties. Investment in research and development enables Denmark to detect and regulate emerging contaminants at an early stage. Through public engagement efforts, awareness is raised, fostering consumer demand for safer products and incentivizing market innovation to reduce the use of hazardous substances. The substitution principle is actively promoted, encouraging the replacement of harmful chemicals with safer alternatives, which is particularly crucial in addressing emerging contaminants. Denmark also engages in international collaboration and participates in global agreements to address chemical risks. The Danish Environmental Protection Agency (EPA) conducts comprehensive monitoring of chemical substances to identify emerging contaminants and assess associated risks, guiding regulatory actions. Denmark is recognized for its regulatory innovation, exemplified by the development of a publicly accessible chemical database by the Danish EPA. Moreover, Denmark influences global regulations through active participation in forums such as the European Chemicals Agency (ECHA), contributing expertise and experiences to shape broader regulatory frameworks. Continual refinement of regulations based on the latest scientific knowledge ensures Denmark effectively addresses new and emerging risks. This proactive, stringent, and innovative approach underscores Denmark's leadership in environmental protection and chemical safety (OECD 2019).

In Sweden and certain states in the USA, pharmaceutical take-back programs have mitigated the improper disposal of medications by providing designated drop-off locations in pharmacies. The roots of such initiatives can be traced back to the late twentieth century, when concerns about pharmaceutical pollution began to emerge. The 1988 discovery of pharmaceutical residues in surface waters in Sweden prompted early efforts to address the issue. Subsequent research highlighted the environmental and health risks posed by pharmaceutical waste, leading to the development of take-back programs as a preventive measure. Today, these programs continue to evolve with advancements in technology and policy frameworks, demonstrating the adaptability and effectiveness of regulatory responses to emerging pollutants (Rogowska and Zimmermann 2022). In the USA, significant progress has also been made in regulating hazardous air pollutants (HAPs), including many emerging pollutants. Beginning in the mid-twentieth century, growing public awareness of air pollution spurred legislative action. The Clean Air Act of 1970 established the framework for controlling air pollution, laying the groundwork for subsequent regulatory measures targeting specific pollutants. The Clean Air Act Amendments of 1990 represented a watershed moment in HAP regulation, introducing stricter emission standards and enforcement mechanisms. Since then, the Environmental Protection Agency (EPA) has worked to reduce HAP emissions from various industrial sources, achieving significant improvements in air quality and public health (U.S. EPA 2020).

Australia has been proactive in addressing emerging pollutants, particularly microplastics. The Australian government's initiatives to reduce plastic waste trace back to the 1970s and 1980s, when awareness of marine pollution began to grow. In response to mounting concerns, Australia implemented regulations to restrict the use of plastic microbeads in personal care products, building on earlier efforts to manage plastic pollution. The introduction of microbead bans in the early twenty-first century marked a significant milestone in Australia's environmental policy, reflecting a commitment to sustainable resource management and pollution prevention (Australian Government 2020).

As per Lapworth et al. (2012), cited in “Sustainable Water and Wastewater Management: Challenges, Innovations, and Lessons from Global Case Studies” by Amarachi et al. (2023), the enhancement of the water treatment facility in Cranbrook, British Columbia, Canada, involving the incorporation of a new storage pond, disinfection facilities, and an aerated lagoon, resulted in elevated water quality and offered tertiary treatment to a population of approximately 19,319. Furthermore, the utilization of reclaimed water contributed to improved agricultural production within the municipality. This successful initiative underscores the manifold environmental, social, and economic benefits associated with the implementation of green infrastructure in water and wastewater treatment.

Despite the above successful cases, emerging pollutants continue to challenge governance structures and enforcement regimes.

In India, where industrial growth has surged in recent decades, regulatory efforts have struggled to keep pace with emerging pollutants. The historical trajectory of environmental regulation in India reflects a complex interplay of economic development, environmental degradation, and policy responses. Rapid industrialization since the mid-twentieth century has led to widespread pollution, including contamination from industrial chemicals like perfluorinated compounds (PFCs). While India has made strides in environmental governance, enforcement challenges persist due to limited resources, institutional capacity, and regulatory fragmentation. Efforts to address PFC contamination are hampered by bureaucratic hurdles, legal complexities, and competing priorities, highlighting the need for integrated approaches to pollution control (Lei et al. 2015).

Nitrate pollution from agricultural runoff remains a significant concern in the USA. The historical roots of agricultural pollution can be traced back to the post-World War II era, when intensive farming practices expanded dramatically. The widespread adoption of synthetic fertilizers and pesticides revolutionized agriculture but also led to environmental problems, including nitrate contamination of water bodies. Efforts to regulate agricultural pollution gained momentum in the latter half of the twentieth century, culminating in the implementation of BMPs and regulatory measures to protect water quality. Despite these efforts, nitrate pollution persists, underscoring the complexities of agricultural pollution and the challenges of regulatory enforcement in rural areas (Miller and Smith 2020).

The examination of the regulatory failures concerning emerging pollutants (EPs) in Latin America by Vargas-Berrones et al. (2020) highlights a troubling absence of information and legislative structures. Latin American nations, especially those with fewer protected areas, confront concentrations and compound numbers surpassing those governed by established legislation. Scarce studies on EPs in wastewater, marine biota, and freshwater underscore the inadequacy of research in the region. The lack of regulations facilitates heightened EP discharge, presenting significant risks to environmental and public health. The aquatic environment, vital for biodiversity and ecosystem services, is notably susceptible. Comparisons with other regions underscore the scale of the problem, with Latin American wastewater and surface water concentrations far exceeding those in the USA, Europe, Asia, and South Africa. Challenges extend to inefficient water treatment plants lacking suitable EP removal processes. The imperative for regulatory intervention is stressed for countries such as Cuba, El Salvador, Haiti, Honduras, Nicaragua, Panama, Paraguay, Peru, and the Dominican Republic, where limited resources impede comprehensive monitoring and quantification.

These case studies underscore the importance of learning from past experiences to inform present and future policy responses to emerging pollutants. By understanding the evolution of environmental policies, policymakers can develop more proactive strategies based on the precautionary principle, which involve continuous monitoring, evidence-based decision-making, and concerted efforts among nations to develop comprehensive regulations that address contemporary challenges concerning emerging pollutants. Effective policy implementation requires technical expertise as well as political will, public engagement, and international cooperation to achieve meaningful and lasting change.

2.7 Future Trends and Recommendations

As the understanding of emerging pollutants evolves and new challenges arise, future regulatory approaches are expected to undergo significant transformations to effectively address emerging environmental and public health threats. Anticipated policy trends and innovations offer promising avenues for policymakers to enhance regulatory frameworks and ensure proactive management of emerging pollutants. One prominent trend is the adoption of a holistic and risk-based approach to regulatory decision-making (EPA 2020). Traditional regulatory frameworks often focus on individual pollutants or sources of pollution in isolation. However, emerging pollutants frequently exhibit complex interactions and synergistic effects with other pollutants, making a singular focus on individual compounds insufficient. Future regulatory frameworks are anticipated to consider the cumulative impacts of multiple pollutants and assess risks comprehensively across various environmental media, such as air, water, soil, and biota (Zhang et al. 2023). This integrated approach will enable regulators to prioritize pollutants based on their relative risks and allocate resources more effectively to mitigate the most significant threats to environmental and human health.

Innovations in monitoring and assessment technologies offer promising opportunities to enhance regulatory effectiveness (Krauss et al. 2010). Advances in sensor technologies, remote sensing, and analytical techniques enable real-time monitoring of pollutant concentrations and sources, providing policymakers with timely and actionable data for decision-making. Integrating these technologies into regulatory monitoring programs can improve the early detection of emerging pollutants, facilitate adaptive management strategies, and enhance public awareness and engagement. Future regulatory approaches are likely to emphasize proactive pollution prevention and pollution control strategies rather than reactive remediation measures (Hernandez 2015). Prevention-focused strategies, such as green chemistry, sustainable product design, and pollution prevention planning, aim to minimize the generation and release of pollutants at the source. By incentivizing cleaner production practices and promoting the use of safer alternatives, regulators can reduce their reliance on end-of-pipe treatment technologies and achieve more sustainable environmental outcomes in the long term. Additionally, promoting circular economy principles, such as product stewardship and extended producer responsibility, can incentivize product manufacturers and consumers to take greater responsibility for the entire lifecycle of products, including their disposal and recycling, thereby reducing the environmental footprint of consumer goods and minimizing the generation of emerging pollutants.

Future regulatory approaches should prioritize community engagement, transparency, and inclusivity in decision-making processes (Berry et al. 2019). Recognizing the diverse perspectives and knowledge of stakeholders, including affected communities, indigenous peoples, and marginalized groups, is essential for developing effective and equitable regulatory solutions to emerging pollutants. Meaningful engagement with stakeholders throughout the policy development and implementation phases can enhance public trust, foster collaboration, and ensure that regulatory decisions reflect the values and priorities of the communities most affected by pollution.

Recommendations for Policymakers

- Embrace an integrated, risk-based approach to regulatory decision-making that considers the cumulative impacts of multiple pollutants and prioritizes resources based on relative risks.
- Invest in the development and deployment of innovative monitoring and assessment technologies to enhance the early detection and characterization of emerging pollutants.
- Promote pollution prevention and pollution control strategies that prioritize cleaner production practices, sustainable product design, and circular economy principles.
- Enhance community engagement, transparency, and inclusivity in regulatory processes to ensure that regulatory decisions reflect the needs and priorities of affected communities.
- Foster collaboration and knowledge-sharing among governments, industries, academia, and civil society to leverage collective expertise and resources in addressing emerging environmental challenges.

Future regulatory approaches to emerging pollutants require a proactive, adaptive, and collaborative approach that leverages innovations in technology and policy to safeguard human health and the environment. By embracing holistic, risk-based strategies, investing in innovative solutions, and fostering inclusive and transparent decision-making processes, policymakers can effectively address emerging environmental challenges and ensure sustainable development for future generations.

2.8 Conclusion

Agriculture, urbanization, and industrial activities have led to the widespread presence of emerging pollutants, posing diverse health and environmental risks. Emerging pollutants encompass a wide range of substances, from pharmaceuticals to industrial compounds, entering the environment through various pathways and persisting with significant health impacts. Since the late 1900s, regulatory bodies like the EPA in the USA and initiatives in Europe and Asia have aimed to address emerging pollutants. Organizations like NORMAN have facilitated global collaboration in monitoring and regulating these pollutants. Despite regulatory efforts, challenges persist in effectively implementing policies due to economic, social, and political constraints. Vulnerable populations, particularly in low- and middle-income countries, bear unequal burdens of pollution. Comprehensive monitoring and assessment strategies are crucial for early intervention and risk evaluation. Advanced technologies, including remote sensing and environmental DNA analysis, offer diverse approaches to identifying and quantifying emerging pollutants. Case studies from various countries highlight different approaches to addressing emerging pollutants, showcasing both successes and challenges in policy implementation. Anticipated policy trends include adopting holistic and risk-based approaches to regulatory decision-making, enhancing monitoring technologies, promoting pollution prevention strategies, and fostering community engagement.

References

Aguilar-Aguilar A, de León-Martínez LD, Forgionny A, Soto NYA, Mendoza SR, Zárate-Guzmán AI (2023) A systematic review on the current situation of emerging pollutants in Mexico: a perspective on policies, regulation, detection, and elimination in water and wastewater. Sci Total Environ 905:167426

Amarachi AS, Obumneme AE, Agbai SO, Abuka CO, Nwaogazie FO, Ajamu PS, Ayinde AT, Ugwunna S (2023) Sustainable water and wastewater management: challenges, innovations, and lessons from global case studies. Path Sci 9(9):5014–5026

Anju A, Ravi SP, Bechan S (2010) Water pollution with special reference to pesticide contamination in India. J Water Resour Protect 2(5):432–448

Aragón FM, Rud JP (2016) Polluting industries and agricultural productivity: evidence from mining in Ghana. Econ J 126(597):1980–2011

Arman NZ, Salmiati S, Aris A, Salim MR, Nazifa TH, Muhamad MS, Marpongahtun M (2021) A review on emerging pollutants in the water environment: existences, health effects and treatment processes. Water 13(22):3258

Australian Government (2020) Reducing plastic pollution. Retrieved from https://www.environment.gov.au/protection/waste-resource-recovery/plastics-and- packaging/reducing-plastic-pollution. Accessed 10 Mar 2024

Barroso PJ, Santos JL, Martín J, Aparicio I, Alonso E (2019) Emerging contaminants in the atmosphere: analysis, occurrence and future challenges. Crit Rev Environ Sci Technol 49(2):104–171

Berry LH, Koski J, Verkuijl C, Strambo C, Piggot G (2019) Making space: how public participation shapes environmental decision-making. Stockholm Environment Institute, Stockholm

Bilela LL, Matijošytė I, Krutkevičius J, Alexandrino DA, Safarik I, Burlakovs J, Gaudêncio SP, Carvalho MF (2023) Impact of per-and polyfluorinated alkyl substances (PFAS) on the marine environment: raising awareness, challenges, legislation, and mitigation approaches under the one health concept. Mar Pollut Bull 194:115309

Biswas P, Vellanki BP (2021) Occurrence of emerging contaminants in highly anthropogenically influenced river Yamuna in India. Sci Total Environ 782:146741

Britannica T, Editors of Encyclopaedia (2024) Environmental protection agency. Encyclopedia Britannica. https://www.britannica.com/topic/Environmental-Protection-Agency. Accessed 10 Mar 2024

Cardoso PG (2022) Pharmaceuticals contamination: problematic and threats for the aquatic system. In: Life below water. Springer International Publishing, Cham, pp 801–809

CAS (Chemical Abstracts Service) (2024) CAS databases. https://www.cas.org/support/documentation/cas-databases. Accessed 10 Mar 2024

Castro G, Ramil M, Cela R, Rodríguez I (2021) Identification and determination of emerging pollutants in sewage sludge driven by UPLC-QTOF-MS data mining. Sci Total Environ 778:146256

Chaturvedi P, Shukla P, Giri BS, Chowdhary P, Chandra R, Gupta P, Pandey A (2021) Prevalence and hazardous impact of pharmaceutical and personal care products and antibiotics in environment: a review on emerging contaminants. Environ Res 194:110664

Choi IC, Shin HJ, Nguyen TT, Tenhunen J (2017) Water policy reforms in South Korea: a historical review and ongoing challenges for sustainable water governance and management. Water 9(9):717

Curtis L, Rea W, Smith-Willis P, Fenyves E, Pan Y (2006) Adverse health effects of outdoor air pollutants. Environ Int 32(6):815–830

Di Paolo C, Ottermanns R, Keiter S, Ait-Aissa S, Bluhm K, Brack W, Breitholtz M, Buchinger S, Carere M, Chalon C, Cousin X (2016) Bioassay battery interlaboratory investigation of emerging contaminants in spiked water extracts–towards the implementation of bioanalytical monitoring tools in water quality assessment and monitoring. Water Res 104:473–484

Domingo JL, Nadal M (2019) Human exposure to per-and polyfluoroalkyl substances (PFAS) through drinking water: a review of the recent scientific literature. Environ Res 177:108648

Dong M, He L, Jiang M, Zhu Y, Wang J, Gustave W, Wang S, Deng Y, Zhang X, Wang Z (2023) Biochar for the removal of emerging pollutants from aquatic systems: a review. Int J Environ Res Public Health 20(3):1679

Dulio V, van Bavel B, Brorström-Lundén E, Harmsen J, Hollender J, Schlabach M, Slobodnik J, Thomas K, Koschorreck J (2018) Emerging pollutants in the EU: 10 years of NORMAN in support of environmental policies and regulations. Environ Sci Eur 30:1–13.16

EEA (European Environment Agency) (2011) Air quality in Europe. European Environment Agency technical report no 12/2011. European Environment Agency, 84. https://www.eea.europa.eu/publications/air-quality-in-europe-2011. Accessed 10 Mar 2024

EEA (European Environment Agency) (2012) Air quality in Europe. European Environment Agency technical report no 4/2012. European Environment Agency, 84. https://www.eea.europa.eu/publications/air-quality-in-europe-2012. Accessed 10 Mar 2024

Environmental Protection Agency (EPA) (2020) Integrated risk information system. https://www.epa.gov/iris. Accessed 10 Mar 2024

European Chemicals Agency (ECHA) (2021) Emerging pollutants. https://echa.europa.eu/hot-topics/emerging-pollutants. Accessed 10 Mar 2024

European Union (2000) Directive 2000/60/EC of the European Parliament and of the Council of 23 October 2000 establishing a framework for Community action in the field of water policy. n. 60 327, Off J Eur Union L, Luxembourg, 2000. https://eur-lex.europa.eu/eli/dir/2000/60/oj. Accessed 10 Mar 2024

Gavrilaş S, Ursachi CŞ, Perţa-Crişan S, Munteanu FD (2022) Recent trends in biosensors for environmental quality monitoring. *Sensors* 22(4):1513

Goot M (2019) Emerging thoughts: a principled framework for regulating GenX as an emerging contaminant. NCL Rev 98:629

Ground Xero (2019) Alembic Pharmaceuticals Sikkim unit discharging highly toxic wastewater effluent in Tista River. (April 27) https://www.groundxero.in/2019/04/27/alembic-pharmaceuticals-sikkim-unit-discharging-highly-toxic-wastewater-effluent-in-tista-river/. Accessed 5 Mar 2024

Haller H, Flores-Carmenate G, Jonsson A (2020) Governance for sustainable remediation of polluted soil in developing countries. In: Sustainability concept in developing countries. IntechOpen, London

Hernandez RA (2015) Prevention and control of air pollution in China: a research agenda for science and technology studies. SAPI EN S Surveys Perspect Integr Environ Soc 8(1)

Khalid NK, Devadasan D, Aravind UK, Aravindakumar CT (2018) Screening and quantification of emerging contaminants in Periyar River, Kerala (India) by using high- resolution mass spectrometry (LC-Q-ToF-MS). Environ Monit Assess 190:1–12

Kolpin DW, Furlong ET, Meyer MT, Thurman EM, Zaugg SD, Barber LB, Buxton HT (2002) Pharmaceuticals, hormones, and other organic wastewater contaminants in US streams, 1999–2000: a national reconnaissance. Environ Sci Technol 36(6):1202–1211

Krauss M, Singer H, Hollender J (2010) LC–high resolution MS in environmental analysis: from target screening to the identification of unknowns. Anal Bioanal Chem 397:943–951

Langenbach B, Wilson M (2021) Per-and polyfluoroalkyl substances (PFAS): significance and considerations within the regulatory framework of the USA. Int J Environ Res Public Health 18(21):11142

Lapworth DJ, Baran N, Stuart ME, Ward RS (2012) Emerging organic contaminants in groundwater: a review of sources, fate and occurrence. Environ Pollut 163:287–303

Lee N (2019) Water policy and institutions in the Republic of Korea, ADBI working paper 985. Asian Development Bank Institute, Tokyo

Lei M, Zhang L, Lei J, Zong L, Li J, Wu Z, Wang Z (2015) Overview of emerging contaminants and associated human health effects. BioMed Res Int 2015:404796

Ma B, Arnold WA, Hozalski RM (2018) The relative roles of sorption and biodegradation in the removal of contaminants of emerging concern (CECs) in GAC-sand biofilters. Water Res 146:67–76

Mahmud A, Wasif MM, Roy H, Mehnaz F, Ahmed T, Pervez MN, Naddeo V, Islam MS (2022) Aquatic microplastic pollution control strategies: sustainable degradation techniques, resource recovery, and recommendations for Bangladesh. Water 14(23):3968

Marco G, Bo X (2013) Air quality legislation and standards in the European union: background, status and public participation. Adv Clim Chang Res 4(1):50–59

McDuie Ra D, Chettri M (2018) Himalayan boom town: rural–urban transformations in Namchi, Sikkim. Dev Chang 49(6):1471–1494

Miller GT, Smith TG (2020) Environmental science: working with the earth, 12th edn. Cengage Learning, California

Morin-Crini N, Lichtfouse E, Liu G, Balaram V, Ribeiro ARL, Lu Z, Stock F, Carmona E, Teixeira MR, Picos-Corrales LA, Moreno-Piraján JC (2022) Worldwide cases of water pollution by emerging contaminants: a review. Environ Chem Lett 20(4):2311–2338

Muir DC, Getzinger GJ, McBride M, Ferguson PL (2023) How many chemicals in commerce have been analyzed in environmental media? A 50 year bibliometric analysis. Environ Sci Technol 57(25):9119–9129

Nagarajan RP, Bedwell M, Holmes AE, Sanches T, Acuña S, Baerwald M, Barnes MA, Blankenship S, Connon RE, Deiner K, Gille D (2022) Environmental DNA methods for ecological monitoring and biodiversity assessment in estuaries. Estuar Coasts 45(7):2254–2273

OECD (2019) OECD environmental performance reviews: Denmark 2019. OECD Environmental Performance Reviews OECD Publishing. https://doi.org/10.1787/1eeec492-en. Accessed 10 Mar 2024

OECD (2021) 25 years of ambitious environmental reform. https://www.oecd.org/country/korea/thematic-focus/25-years-of-ambitious-environmental-reform-16cde12d/. Accessed 10 Mar 2024

OW/ORD Emerging Contaminants Workgroup (2008, June 03). White paper aquatic life criteria for contaminants of emerging concern. Part I: General challenges and recommendations [Report]. United States Environmental Protection Agency, Office of Water/Office of Research and Development. https://www.epa.gov/sites/default/files/2015-08/documents/white_paper_aquatic_life_criteria_for_contaminants_of_emerging_concern_part_i_general_challenges_and_recommendations_1.pdf. Accessed 10 Mar 2024

Prüss-Üstün A, Wolf J, Corvalán C, Bos R, Neira M (2016) Preventing disease through healthy environments: a global assessment of the burden of disease from environmental risks. World Health Organization, Geneva
Puri M, Gandhi K, Kumar MS (2023) Emerging environmental contaminants: a global perspective on policies and regulations. J Environ Manag 332:117344
Ramírez-Malule H, Quinones-Murillo DH, Manotas-Duque D (2020) Emerging contaminants as global environmental hazards. A bibliometric analysis. Emerg Contaminants 6:179–193
Rogowska J, Zimmermann A (2022) Household pharmaceutical waste disposal as a global problem—a review. Int J Environ Res Public Health 19(23):15798
Rolim SBA, Veettil BK, Vieiro AP, Kessler AB, Gonzatti C (2023) Remote sensing for mapping algal blooms in freshwater lakes: a review. Environ Sci Pollut Res 30(8):19602–19616
Saha JK, Selladurai R, Coumar MV, Dotaniya ML, Kundu S, Patra AK, Saha JK, Selladurai R, Coumar MV, Dotaniya ML, Kundu S (2017) Impacts of soil pollution and their assessment. In: Soil pollution-an emerging threat to agriculture. Springer, Singapore, pp 37–73
Sanganyado E (2022) Policies and regulations for the emerging pollutants in freshwater ecosystems: challenges and opportunities. In: Emerging freshwater pollutants. Elsevier, Amsterdam, pp 361–372
Sexton K (2012) Cumulative risk assessment: an overview of methodological approaches for evaluating combined health effects from exposure to multiple environmental stressors. Int J Environ Res Public Health 9(2):370–390
U.S. Environmental Protection Agency (U.S. EPA) (2020) Hazardous air pollutants (HAPs). Retrieved from https://www.epa.gov/haps. Accessed 10 Mar 2024
UNION P (2008) Directive 2008/50/EC of the European Parliament and of the Council of 21 May 2008 on ambient air quality and cleaner air for Europe. Official Journal of the European Union
United Nations Economic Commission for Europe (UNECE) (1998) Convention on access to information, public participation in decision-making and access to justice in environmental matters. 25 pp. https://unece.org/DAM/env/pp/documents/cep43e.pdf. Accessed 10 Mar 2024
United Nations Environment Assembly (2021) For People and Planet: The United Nations Environment Programme Strategy for 2022–2025 to Tackle Climate Change, Loss of Nature and Pollution, Fifth Session of the UN Environment Assembly (UNEA-5). Nairobi, Kenya. https://sustainabledevelopment.un.org/index.php?page=view&type=30022&nr=2802&menu=3170. Accessed 10 Mar 2024
Vargas-Berrones K, Bernal-Jácome L, de León-Martínez LD, Flores-Ramírez R (2020) Emerging pollutants (EPs) in Latin América: a critical review of under-studied EPs, case of study-nonylphenol. Sci Total Environ 726:138493
Xiao WQ (2020) The road to racial justice: resolving the disproportionate health burden placed on communities of color by highway pollution. Colum Hum Rts L Rev 52:911
Xu J, Liu J (2023) Managing the risks of new pollutants in China: the perspective of policy integration. Environ Health 1(6):360–366
Yusa V, Ye X, Calafat AM (2012) Methods for the determination of biomarkers of exposure to emerging pollutants in human specimens. TrAC Trends Anal Chem 38:129–142
Yusuf A, O'Flynn D, White B, Holland L, Parle-McDermott A, Lawler J, McCloughlin T, Harold D, Huerta B, Regan F (2021) Monitoring of emerging contaminants of concern in the aquatic environment: a review of studies showing the application of effect-based measures. Anal Methods 13(43):5120–5143
Zhang S, Yin Q, Wang S, Yu X, Feng M (2023) Integrated risk assessment framework for transformation products of emerging contaminants: what we know and what we should know. Front Environ Sci Engineer 17(7):91
Zhou Z, Liu J, Zeng H, Zhang T, Chen X (2020) How does soil pollution risk perception affect farmers' pro-environmental behavior? The role of income level. J Environ Manag 270:110806

Chapter 3
Toxicity and Health Impact of Emergent Pollutants and Sustainable Management as Bioremediation

Bhavna Nigam, Dheeraj Rathore and Indra Jeet Chaudhary

Abstract Pesticides are utilized to control pests and increase agricultural production. These chemical compounds are employed to repel, mitigate, or eradicate harmful pests that pose a threat to plant and humans. While pesticides provide benefits to crops, their usage also has significant adverse effects on the environment. They impact both intended targets and unintended species, including humans. The detrimental effects of pesticides, including their parent compounds, degradation products, and metabolites, can harm ecosystems, the environment, and human health. To safeguard public health, the World Health Organization (WHO) categorizes pesticides based on their harmful effects. Minimizing pesticide usage and understanding their classification can help reduce their impact on both the environment and human health. Currently, bioremediation strategies are being employed to combat pesticide pollution, offering an inexpensive and environment-friendly approach. This chapter covers various aspects of pesticides, including their global distribution, environmental impacts, human health concerns, and sustainable management through bioremediation.

Keyword Pesticides · Bioremediation · Environmental hazards · Toxicity · Health effects · Eco-sustainable technology

---

B. Nigam · D. Rathore
School of Environment & Sustainable Development, Central University of Gujarat, Vadodara, Gujarat, India

I. J. Chaudhary (✉)
Department of Environmental Science, Savitribai Phule Pune University (University of Pune), Pune, Maharashtra, India

S. Dey, S. Bhattacharya (eds.), *Biotechnological Interventions in the Removal of Emerging Pollutants*, Interdisciplinary Biotechnological Advances,
https://doi.org/10.1007/978-981-97-9922-0_3

3.1 Introduction

Agriculture plays a crucial role in the Indian economy, serving as its mainstay. Approximately 65–70% of the population in India relies on agriculture for their livelihood (Vaishampayan et al. 2019). A large part of agricultural land is depleting every year due to the rapid growth in human population, establishment of industries and urban encroachments which creates food scarcity. The agricultural landscape in India has undergone significant changes following the advent of the Green Revolution in the 1960s. Major part of population is being engaged with agriculture in India and for gaining self-sufficiency in food grains and in other agricultural commodities, pesticides and other chemicals were introduced on very large scale for high yielding crop varieties (Tudi et al. 2021). Application and consumption of pesticides are increasing due to increasing demand of agro product in regional climate changes (Strassemeyer et al. 2017).

The term pesticide covers a wide range of compounds including insecticides, fungicides, herbicides, rodenticides, molluscicides, nematicides, plant growth regulators, and others. Pesticides contamination from a wide variety of compounds may result from manufacturing, improper storage, handling, and disposal of pesticides, as well as from agricultural processes. Pesticides have been employed for many years to prevent the transmission of diseases carried by pests such as mosquitoes and fleas in animals and humans. They also contribute to food production by eliminating insects and other pests found in agricultural areas, and they help protect the environment by controlling the growth of molds, algae, and weeds (Zhang et al. 2018). Pesticides are biologically active substances used to prevent, eradicate, and manage pests by interfering with their metabolic activities (Lamichhane 2017). It issued globally and mostly in agrarian economy countries like India to fulfill growing populations' food demand.

3.2 Types of Pesticides and Target Organism

Pesticides are categorized into various types based on the organisms they target. Biocides are broad-spectrum pesticides designed to kill a wide range of living organisms. More specific substances are named according to their intended target. Herbicides are specifically formulated to destroy plants, insecticides target insects, and fungicides are used to combat fungi. There are also specific pesticides such as acaricides (targeting mites and ticks), rodenticides (used against rodents), and nematicides (designed to control nematodes) (Osman et al. 2011).

3.2.1 Herbicides

Herbicides are commonly referred to as weed killers and are utilized to eliminate unwanted plants. Selective herbicides are designed to target specific plants while minimizing damage to desired crops. Some herbicides interfere with the growth of weeds by imitating plant hormones. Nonselective herbicides, on the other hand, are not specific and will kill any plant they come into contact with. They are often employed to clear waste grounds, railway tracks, embankments, and industrial sites. In contrast, smaller quantities of herbicides are used in forestry, pasture systems, and the management of wildlife habitats (Aksakal et al. 2013).

3.2.2 Fungicides

Fungicides are substances, either chemical compounds or biological agents, used to kill or inhibit the growth of fungi or fungal spores. Fungi are small organisms that derive nourishment from living or nonliving sources. Fungicides find applications in agriculture to combat fungal infections in animals and humans. It is worth noting that chemicals used to control oomycetes, which are not fungi but employ similar mechanisms to infect plants, are also referred to as fungicides (Toda et al. 2021). Fungicides can be categorized as contact, translaminar, or systemic. Contact fungicides protect the areas of the plant where the spray has been applied but do not penetrate plant tissue. Translaminar fungicides redistribute within the sprayed leaf, reaching the unsprayed lower surface. Systemic fungicides are taken up by the plant and transported through the xylem vessels to protect new leaf growth and upper plant parts for a limited period (Rathour et al. 2022).

3.2.3 Insecticides

Insecticides are substances specifically employed to combat insects. They encompass ovicides and larvicides, which target the eggs and larvae of insects respectively. Insecticides find widespread applications in various sectors such as agriculture, industries, households, and medicine. The major factor behind increasing agriculture productivity in the twentieth century is because use of insecticides (Bowman and Zilberman 2013). Almost all the insecticides have potential to significantly alter ecosystems; many are concentrated in the food chain and toxic to humans (Landrigan et al. 2020).

3.2.4 *Acaricides*

Acaricides are substances that are utilized to kill members of the arachnid order Acari, which includes mites and ticks. These acaricides are employed in both agriculture and medicine, although the level of selective toxicity may vary between the two fields (Guo et al. 2021).

3.2.5 *Nematicides*

Nematicides are a specific type of pesticide that is utilized to eradicate plant parasitic nematodes. These nematicides are characterized by their broad-spectrum toxic properties and often possess high volatility or other properties that facilitate their migration through the soil (Gill and Garg 2014).

3.2.6 *Rodenticides*

Rodenticides are a specific category of pest control chemicals designed to eliminate rodents. Some rodenticides are formulated as single-feed baits, meaning that a single dose is potent enough to be lethal (Tosh et al. 2011). Rodents are challenging to control using poisons due to their scavenging behavior. Therefore, an effective rodenticide needs to be unappealing and odorless in lethal concentrations and exhibits delayed effects (Pelfrène 2010).

On the basis of chemical nature, pesticides are classified into five groups which are synthetic pyrethroids, organochlorine, carbamate, organophosphate, formamidine, and avermectin (Mandal 2019). Organochlorine, organophosphate, and carbamates are of major group in concern because of their toxicity and persistence in the environment. All these pesticides are widely used worldwide for controlling agriculture and household pests. Organophosphate represents largest group of chemical insecticides used in plant protection throughout the world (Ghorab and Khalil 2015). Organophosphate represents about half of the total insecticide used with annual application over 75 million pounds in the United States (Carvalho 2017). Seventy-five percent accounted by insecticides of the total market.

3.3 Sources of Pesticides in the Environment

The rapidly growing population in world needs increase in the food production; for this, one of main strategies to enhance crop productivity is effective pest management because more than 45% of annual food production is lost due to pest infestation (Dubey et al. 2011). So the application of a wide variety of pesticides becomes necessary in agricultural fields to combat pests. But improper and excessive misuse of pesticides in agriculture can heighten the risk of environmental contamination by causing the dispersion of pesticides into nontarget areas. The extensive utilization of chemicals in agriculture has resulted in the accumulation of numerous hazardous compounds in the air, water, and soil, leading to environmental pollution (Sharma et al. 2019). Uncontrolled use of pesticides in agricultural areas attributes greatly to environmental and health hazards. The usage of pesticides in India is relatively low compared to other regions. There has been a widespread contamination of food commodities with pesticides residues, basically due to nonjudicious usage. India unfortunately falls into the category of countries where the production and use of certain chlorinated pesticides, such as DDT and lindane, are still ongoing (Sharma et al. 2014). A survey conducted in Punjab and Haryana, focusing on pesticide application in agricultural fields by farmers, revealed various health issues including cancer, stillbirth, infertility, and kidney failure (Abhilash and Singh 2009). Globally, it has been reported that over 2500 pesticides are currently in use (Stehle and Schulz 2015). Pesticide poisoning leads to approximately 1 million deaths and chronic diseases worldwide each year (Boedeker et al. 2020). The US National Academy of Sciences has stated that the DDT metabolite DDE causes thinning of eggshells and that the decline in the bald eagle population in the United States was primarily due to exposure to DDT and its metabolites (Aktar et al. 2009).

Recent evidence suggests that pesticides have the potential to harm the immune system and act as hormone mimics, disrupting the endocrine system in both humans and animals, leading to various disorders. The United Nations reports that approximately 1 to 5 million cases of pesticide poisoning occur annually, resulting in several thousand fatalities among agricultural workers (Patel et al. 2012). The toxicity of pesticides is influenced by several factors, including the level of exposure, concentration of the pesticide, and duration of exposure. In India, the first report of poisoning due to pesticides was reported in Kerala in the year 1958, where over 100 people died after consuming wheat flour contaminated with parathion (Kumar et al. 2016). Due to those farmers of villages noted serious health problems to their cattle population due to water contamination. In children and women, the signs of serious ailments were seen (Mrema et al. 2017). In a study conducted on 356 workers in four HCH manufacturing units in India, it was found that 21% of the workers experienced neurological symptoms, and the intensity of exposure was linked to these symptoms (Hashim 2015). Another study collected data on reproductive toxicity from 1106 couples, specifically focusing on males involved in pesticide spraying (including OC, OP, and carbamates) in cotton fields (Hashim 2015). Early investigations into health effects such as liver function, immune function, neurologic impairment, and reproductive effects yielded inconclusive results. However, an increased mortality rate due to cardiovascular and respiratory diseases was observed, potentially related to both the psychosocial consequences of the accident and chemical contamination. Additionally, a higher number of diabetes cases were found. Pesticides have the potential to contaminate soil, water, turf, and other vegetation.

Insecticides are generally the most acutely toxic class of pesticides, but herbicides can also pose risks to nontarget organisms.

Pesticide residues can be detected in soil, air, surface water, and groundwater in various countries, and the use of pesticides in urban areas contributes to this issue. Runoff from treated plants and soil can transport pesticides to surface water, leading to water contamination. A series of extensive studies conducted by the US Geological Survey (USGS) in the early to mid-1990s on major river basins across the country revealed alarming findings. More than 90% of water and fish samples collected from streams contained one or, in many cases, multiple pesticides (Aktar et al. 2009). The USGS also found that concentrations of insecticides in urban streams commonly exceeded guidelines for protection of aquatic life (Phillips and Bode 2004). According to the USGS, urban streams tend to contain more detected pesticides compared to agricultural streams (Phillips and Bode 2004). Groundwater pollution resulting from pesticide contamination is a global issue. The USGS has reported the presence of at least 143 different pesticides and 21 transformation products in groundwater, representing various chemical classes (Aktar et al. 2009). In a survey conducted in India, it was found that 58% of drinking water samples from hand pumps and wells around Bhopal were contaminated with organochlorine pesticides above the EPA standards (Sarkar 2019). Excessive soil treatment with pesticides can lead to a decline in populations of beneficial soil microorganisms. The indiscriminate use of chemicals may initially provide short-term benefits, but over time, the reduced presence of beneficial soil organisms can negatively impact nutrient retention (Gyawali 2018). Pesticide sprays can have direct impacts on nontarget vegetation or can drift and volatilize from the intended area, leading to contamination of air, soil, and nontarget plants. Even with ground equipment, some pesticide drift can occur during every application (Gyawali 2018). It has been observed that nearly all investigated pesticides have been detected in rain, air, fog, or snow at different times throughout the country (Phillips and Bode 2004). Many sites sampled nationwide have shown the presence of multiple pesticides in the air. Certain herbicides in ester formulation have been found to volatilize from treated plants, releasing vapors that can cause severe damage to neighboring plants (Nieder et al. 2018).

3.4 Global Scenario of Pesticides Usage

At present, the global consumption of pesticides amounts to approximately 2 million tones. Herbicides account for 47.5% of the total usage, followed by insecticides at 29.5%, fungicides at 17.5%, and other pesticides at 5.5% (De et al. 2014). The United States, Brazil, China, Argentina, Russia, Canada, France, Australia, India, and Italy (Fig. 3.1) are the top ten pesticide-consuming nations in the world (World atlas 2018). Furthermore, it is predicted that global pesticide usage will rise to 3.5 million tons by 2020 (Zhang 2018).

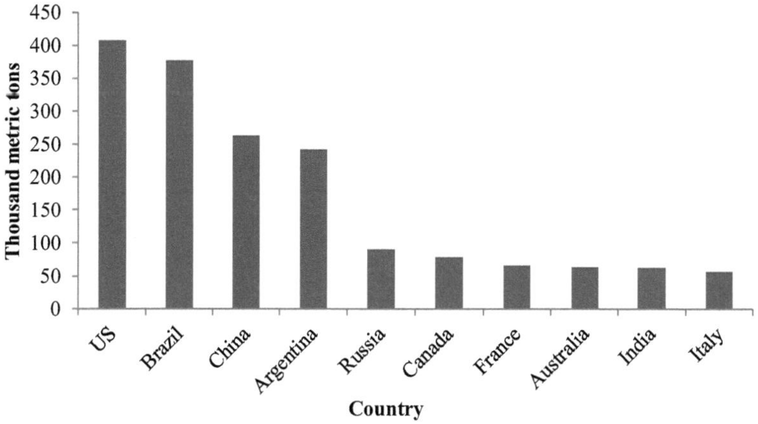

Fig. 3.1 Global pesticide consumption

Although use of chemicals to control pest dates back to 2500 BC, in the mid-1940s, the production and use of synthetic organic pesticides increased rapidly and by 1991, approximately 23,400 pesticide products were registered with the US Environmental Protection Agency (EPA). In India, use of pesticides began in 1948 when BHC (benzene hexachloride) for locust control and DDT (Dichloro-diphenyltrichloroethane) was imported for malaria control. The production of pesticide in India started with manufacturing plant for DDT and BHC near Calcutta in the year 1952 (Yadav et al. 2015). By the year 1958, India produced over 5000 metric tons pesticides including insecticides like BHC and DDT, and by the mid-nineties, approximately 145 registered pesticides were in use and later on production was increased to approximately by 85,5000 metric tons. Presently, India is the second largest manufacturer of pesticides in Asia after China and ranks 12th globally (Bhardwaj and Sharma 2013).

In India, pesticides are mainly used in public health and agriculture sector for controlling pests and diseases that affect man. The worldwide consumption of pesticides is about two million tons per year, in which 45% in Europe, 25% is consumed in the United States alone, and 25% in the rest of the world, in which India's share is only 3.75% (Kumar et al. 2019). Approximately 77% of pesticide consumption is attributed to the agriculture sector, while industrial, commercial, and government organizations account for 12%, and private households account for the remaining 11% (Grube et al. 2011). In India, insecticides make up the largest percentage share of pesticide classes used, accounting for 76%. This is followed by fungicides at 13%, herbicides at 11%, and other pesticides at 1% (Sharma et al. 2019). The use of herbicides and fungicides is comparatively less. The main use of pesticides in India is for the cotton crops (45%) followed by paddy then wheat crops (Mahajan et al. 2013). According to the data provided by the Directorate of Plant Protection, Quarantine, and Storage, Government of India, from 2005–06 to 2009–10, sulfur, a fungicide, was the most widely consumed pesticide in India,

Table 3.1 Example of most consumed pesticides in India (during 2005–06 to 2009–10)

S. no.	Pesticide (technical grade)	Quantity consumed (metric tons)
1	Sulfur (fungicide)	16,424
2	Endosulfan (insecticide)	15,537
3	Mancozeb (fungicide)	11,067
4	Phorate (insecticide)	10,763
5	Methyl parathion (insecticide)	8408
6	Monocrotophos (insecticide)	8209
7	Cypermethrin (insecticide)	**7309**
8	Isoproturon (herbicide)	7163
9	Chlorpyrifos (insecticide)	7163
10	Malathion (insecticide)	7103
11	Carbendazim (fungicide)	6767
12	Butachlor (herbicide)	6750
13	Quinalphos (insecticide)	6329
14	Copper oxychloride	6055
15	Dichlorvos (insecticide)	5833

Source—Directorate of Plant Protection, Quarantine and Storage, Govt of India

totaling 16,424 metric tons. This was followed by endosulfan, as indicated in Table 3.1.

3.5 Production and Usage of Pesticide in India

Pesticide production in India began in 1952 with the establishment of a plant for the production of benzene hexachloride (BHC) in Rishra near Calcutta. This was followed by the establishment of two units for manufacturing DDT (1,1,1-trichloro-2,2-bis(4-chlorophenyl) ethane) by Hindustan Insecticides Ltd. (Geetha and Fulekar 2008). The Indian pesticide industry is currently the fourth largest industry in the world and the leading producer of pesticides among Asian countries. In the year 2005–06, India produced 82,000 MT of pesticides, ranking second in Asia (behind China) and twelfth in the world. The annual production of pesticides in India is around 90,000 MT (Boricha and Fulekar 2009). Over the years, there has been a steady growth in the production of technical-grade pesticides in India, increasing from 5000 MT in 1958 to 102,240 MT in 1998. The value of pesticide demand in India was estimated to be around Rs. 22 billion (USD 0.5 billion) in 1996–97, accounting for approximately 2% of the total world market. The usage pattern of pesticides in India differs from the global trend. In India, around 76% of pesticide use is insecticides, compared to 44% globally (Geetha and Fulekar 2008). The usage of herbicides and fungicides is comparatively lower. The major use of pesticides in India is for cotton crops (45%), followed by paddy and wheat crops (Fig. 3.2).

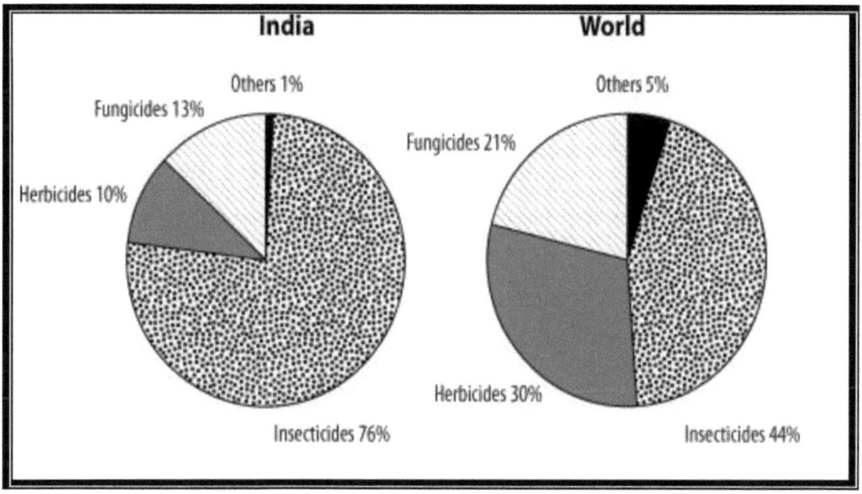

Fig. 3.2 Consumption pattern of pesticide (Bhat et al. 2019)

3.6 Use of Pesticides in Crop Protection

Pesticides are mixture or substances used for preventing, repelling, destroying or mitigating plants and animal's pests in agriculture as well as domestic field. According to census of India 2001, in Gujarat, approximately 97% of workers in rural areas are engaged in agricultural activities. The main cash crops of the state are cotton, groundnut, tobacco, cumin, rice and other vegetables. The total estimated consumption of technical grade pesticides is approx. 3642 metric tons in the year 2001–02 as per data available from department of agriculture, government of Gujarat.

Pest infestations are responsible for depleting approximately 45% of annual food production (Abhilash and Singh 2009). To protect crop plants from these pests, millions of tons of pesticides are released into the environment each year. India accounts for approximately 3% of the total pesticides used globally, and this usage is increasing at a rate of 2 –5% per annum. In India, about 67% of pesticides consumption is in the agriculture and horticulture sectors, with insecticides accounting for 75% of the total. The main types of insecticides used include organochlorines (40%), organophosphates (30%), carbamates (15%), synthetic pyrethroids (10%), and others (5%). The remaining 25% of pesticide usage in India is comprised of fungicides (10%), herbicides (7%), and others (8%). According to the Directorate of Plant Protection, Quarantine, and Storage, Government of India (2010), Uttar Pradesh is the state with the highest pesticide consumption during the period of 2005 to 2010, followed by Punjab, Haryana, Maharashtra, Rajasthan, Gujarat, and Tamil Nadu (Table 3.2).

During the period of 2005–06 to 2009–10, Uttar Pradesh was the state that consumed the highest number of pesticides, followed by Punjab, with Gujarat ranking

Table 3.2 Pesticide-consuming states in India during 2005–10

S. no.	State	Total pesticides consumed
1	Uttar Pradesh	39,948
2	Punjab	29,235
3	Haryana	21,908
4	Maharashtra	16,480
5	Rajasthan	15,239
6	Gujarat	13,430
7	Tamil Nadu	12,851
Total	All India	210,600

Source: http://www.indiaforsafefood.in/farminginindia.html

6th (Table 3.2). In terms of pesticide consumption across agricultural crops in India, cotton accounted for 45%, followed by rice (22%), vegetables (9%), plantation crops (7%), pulses (4%), wheat (4%), and others (9%) (Subramanyam and Hagstrum 2012). In Gujarat, major cash crops include cotton, groundnut, tobacco, sugarcane, cumin, rice, and vegetables such as green chilies, which commonly require the use of pesticides for pest protection. According to the 2001 Census of India, approximately 97% of rural workers are engaged in agriculture-related activities and use pesticides to protect crops and increase production. The Department of Agriculture, Government of Gujarat, reported an estimated consumption of approximately 3643 metric tons of technical-grade pesticides during the 2001–02 period.

3.7 Effect of Pesticide on Environment

Pesticides primarily enter the environment during planning and application. Application can happen through various procedures, contingent upon variables, for example, the controlled pest, the formulation type, and the application timing. Pesticides can be applied to crops or to the soil in agriculture. Boom sprayers, tunnel sprayers, and aerial application are all common ways to apply liquid sprays to crops. Systemic pesticides are another option. Pesticides can be applied to soils as granules, sprayed onto the soil surface where they may be incorporated into the soil top layer or injected as a fumigant. Before planting, pesticides are sometimes applied to seeds (Long and Krupke 2016).

Pesticides can enter target organisms, become degraded, or migrate to groundwater following application. They can also contaminate surface water bodies, volatilize into the air, or affect nontarget organisms through ingestion. The behavior and fate of pesticides are influenced by their physical and chemical properties, soil characteristics, site conditions, and management practices (Pérez-Lucas et al. 2019).

The process of adsorption is influenced by both the chemical properties of pesticides and the type of soil. Pesticide volatility refers to their tendency to transform into a gas, and higher volatility (higher vapor pressure) leads to greater loss to the atmosphere. Volatility can be influenced by humidity and temperature, and pesticide vapors can persist in soil, plants, or surface water for extended periods after application. Chemicals can be transported over long distances through the atmosphere, potentially resulting in surface water pollution through subsequent atmospheric deposition. Pesticide degradation plays a significant role in determining the behavior and fate of these compounds in the environment. Microorganisms, photodecomposition, and various physical and chemical reactions contribute to the degradation process, breaking down pesticides into different chemical forms (Gangola et al. 2022).

The following site conditions that can influence pesticide behavior in the environment such as topographical circumstances, geology and groundwater. The shallow groundwater is a condition, where the groundwater table high, or saturated soil and filters a lesser volume of pesticides and contamination due to their geological conditions. As to land conditions, the presence of sinkholes, wells, and exceptionally porous materials, for example, stores, rock works with groundwater pollution. The presence of waste trenches, streams, lakes, and ponds expands the likelihood that precipitation or water system spillover will taint surface water.

Flat landscapes, areas with closed drainage systems that direct water toward the center of a basin, and subsurface areas are particularly susceptible to groundwater contamination based on their topography. Climate conditions such as heavy rainfall or extensive irrigation can cause significant amounts of water to infiltrate the soil, potentially reaching the groundwater. Rainfall can also facilitate the transport of pesticides to nearby rivers, surface waters, lakes, and seas, leading to the contamination of these water bodies. Moreover, these chemicals can be carried over long distances to remote locations through various water channels and pathways. In terms of pesticide container handling, proper storage and disposal have an influence on environmental contamination (ANR 2014).

Pesticides that come into contact with nontarget organisms can undergo biotransformation through various enzymatic reactions in the liver, such as hydrolysis, oxidation, reduction, or conjugation. Biotransformation is the organism's way of detoxifying and eliminating foreign substances, but it can also produce metabolites that are more harmful than the original compound, known as bioactivation. An example of bioactivation is the transformation of DDT, which is relatively nontoxic to birds, into DDE, a compound that disrupts calcium metabolism and leads to eggshell thinning (Schweitzer and Noblet 2018).

In a study conducted by Semalulu et al. (2005) in the Ugandan catchment of Lake Victoria, the presence of banned organochlorine pesticides (such as DDT, endosulfan, lindane, and dieldrin) was detected in the air, suggesting that these pesticides may still be used in the Lake Victoria basin.

The United Kingdom aims to ensure pesticide sustainability by reducing the risks and impacts of pesticide use on human health and the environment. This involves promoting the development and implementation of alternative pest management methods and strategies (Ngowi et al. 2007).

3.8 Effect of Pesticides on Human Health

Employees and residents, especially those in rural areas, often have direct exposure to pesticides, which increases their risk of toxicity. Suicide by pesticides is popular in so many Latin American countries and Asian, as per World Health Organization (WHO) data (WHO 2014).

Pesticides are frequently largely uncontrollable and freely accessible, especially in low- and middle-income nations (Sarchiapone et al. 2011). The first epidemiological observations of pesticide-related suicides emerged in the early 1990s. Many organizations and governments are currently concerned about pesticide-related homicides and suicides because suicide and depression clearly relate with high pesticide exposure. This issue has inspired and continues to encourage several more studies into when or where and why pesticide exposure occurs; researchers have also discovered ways to resolve this severe social problem (Freire and Koifman 2013).

Pesticides enter the human body either directly or indirectly. Humans can be exposed to pesticides when they are applied to crops, resulting in adverse effects on the mouth, respiratory tract, eyes, and skin. These effects can manifest as vomiting, headaches, sneezing, skin rashes, and irritation. The toxicity of pesticides to humans depends on the duration and concentration of exposure. The body typically eliminates pesticides through excretion, including biliary, urinary, and secretory gland excretion. Prolonged consumption of pesticide-contaminated fruits and vegetables can lead to an accumulation of toxins in the body organs, potentially causing chronic diseases such as neurotoxicity, necrosis, cancer, reproductive disorders, asthma, diabetes, and cardiac disease, among others. Although quaternary nitrogen compounds like paraquat have been associated with neurodegenerative diseases like Parkinson's, the precise molecular mechanisms remain unknown. Similarly, the carbamate pesticide group inhibits the action of acetylcholinesterase (AChE), which serves as a biomarker for neurotoxicity. Pesticides, including organophosphorus compounds such as malathion and parathion, have been associated with various health issues, including cancer.

Breast cancer is one of the most common types of cancer, and it has been linked to the effects of organophosphorus compounds on cellular growth and proliferation (Calaf 2021). Similarly, organophosphorus compounds are implicated in asthma by affecting autoinhibitory M2 muscarinic receptors on parasympathetic neurons that regulate airway smooth muscle (Calaf 2021). These compounds can also interfere with endocrine hormone action, disrupt hormone release patterns, and imitate hormones, leading to decreased fertility and the development of genital tract anomalies in both males and females. Studies have shown that organophosphorus compounds reduce paraoxonase activity and increase the risk of coronary artery disease (Kabir et al. 2015). Hunger and malnutrition remain significant challenges in several African countries. The use of fertilizers and pesticides in agricultural lands can result in the transfer of these substances into crops, directly or indirectly impacting human health (Kumari and Sharma 2008).

Pesticides have been linked to endocrine disruption, chronic neurotoxicity, immune system disruption, carcinogenesis, genotoxicity, and mutagenicity (Oluwole and Checke 2019). Pesticide exposure has been linked to diabetes, asthma, Parkinson's disease, cancer, and leukemia (Kim 2017). Kumari and Sharma (2008) and Damalas (2009) conducted research on farmers' perceptions of the effects on the environment of pesticide use as well as strategies used in the Himalaya western mountains.

3.9 Bioremediation as Sustainable Management

Bioremediation is defined as the use of microorganisms for destroying or immobilize waste materials (Abatenh et al. 2017) without further disruption to the local environment. Bioremediation generally occurs due to microorganisms used the pollutants as a sole carbon and nitrogen source. So the degradation of pollutants is accompanied by microbial activities. The activity occurred by microbes in soil is affected by physicochemical and environmental conditions such as pH value, temperature, moisture content, oxygen, and so on. The degradation of pollutants is influenced by the availability of essential nutrients such as nitrogen, phosphorus, and carbon. Bioremediation, a process based on the activities of aerobic or anaerobic heterotrophic microorganisms, is commonly used to address soil contamination issues, particularly in industrialized areas worldwide. The control and optimization of bioremediation processes involve a complex system that takes into account various factors (Kensa 2011). These factors include the viability of microbial populations capable of degrading pollutants, the accessibility of contaminants for microbial degradation, and environmental conditions that support microbial growth. Bioremediation process promotes the microbial metabolism of contaminants by adjusting the water, air, and nutrient supply in the soil (Bhatt et al. 2021). If the bioremediation is performed properly, then it can be very a cost-effective technique. On-site remediation techniques are often favored due to their low cost and high level of public acceptance. These methods are generally considered cost-effective and can be implemented with relative ease. In this technique, it involves utilization of naturally occurring microorganisms to detoxify or degrade persistent pollutants which are hazardous to environment as well as human health. Bioremediation may use indigenous microorganisms to the polluted area or may isolate from elsewhere and brought to contaminated sites. Living microorganisms play a crucial role in the transformation of pollutants through various processes, including bioremediation, bioaccumulation, biomineralization, biodegradation, biotransformation, and cometabolism (Zhang et al. 2020; Chugh et al. 2022). These reactions facilitate the breakdown and conversion of pollutants by microorganisms in the environment. By actions of various microorganisms, biodegradation of pesticides take place. In remediation of pesticide, waste physicochemical methods are not sufficient and effective many conventional methods for treating synthetic pyrethroids contaminated sites includes chemical treatment, recycling, incineration, pyrolysis and landfills are not effective,efficient and costly which can lead to formation of toxic intermediates (Riser-Roberts 2020; Raheem et al. 2018). Thus, bioremediation seems to be a reliable technique that utilizes microorganism for the removal of pesticides from contaminated area. Bioremediation is very cost-effective and easy to use technique compare to other approaches. It does not pose any threaten of secondary pollution to the environment (Sharma 2020; Medfu Tarekegn et al. 2020).

Bioremediation process is relatively slow process and requires weeks to months for effective clean up but on other hand it seems good alternative to conventional technologies (Medfu Tarekegn et al. 2020). It can be effective only when environmental conditions are favorable for microbial growth and microbial activities. It often involves manipulation of environmental parameters to allow microbial growth and degradation rate proceeds faster. Some essential factors which affect bioremediation process are pH, temperature, moisture content, organic matter, oxygen, microbe, and level of nutrients and cosubstrates.

3.10 Conclusion

Pesticide use has increased dramatically in recent years, causing ecological harm, mainly contaminates the soil and water. Pesticides are available in a variety of forms, but the most commonly used pesticides are organophosphates, carbamate, organochlorine, and pyrethroids, which pose human and environmental concerns. Some pesticides, particularly organochlorides, are so persistent in the environment that they do not degrade. Consequently, these long-term environmental effects pose hazards to both nontarget organisms and human beings. Pesticide misuse and management, as well as pesticide behavior in the environment, contribute to environmental pollution, which includes water pollution, soil pollution, air pollution, and food contamination. The scientific community has been dedicated to developing innovative approaches to reduce pesticide pollution. Numerous bioremediation approaches and servers are used in environmentally sustainable management strategies to fix pesticide issues or identify new green solutions. Bioremediation approaches are also inexpensive and environment-friendly. Utilizing pesticide-degrading microorganisms in an environment-friendly manner is an effective approach to manage pesticide pollutants. Further research is needed to screen and identify efficient enzymes and microbial strains to minimize the risks of pesticides to the environment and human health.

References

Abatenh E, Gizaw B, Tsegaye Z, Wassie M (2017) The role of microorganisms in bioremediation-a review. Open J Environ Biol 2(1):038–046

Abhilash PC, Singh N (2009) Pesticide use and application: an Indian scenario. J Hazard Mater 165(1–3):1–2

Agriculture and Natural Resources (ANR) (2014) Water quality: Controlling Nonpoint Source (NPS) pollution. Pesticide management to protect water quality understanding pesticides and how they affect water quality 1999. Available from http://www.aces.edu/pubs/docs/A/ANR-0790/WQ4.5.1.pdf

Aksakal O, Erturk FA, Sunar S, Bozari S, Agar G (2013) Assessment of genotoxic effects of 2, 4-dichlorophenoxyacetic acid on maize by using RAPD analysis. Ind Crop Prod 1(42):552–557

Aktar W, Sengupta D, Chowdhury A (2009) Impact of pesticides use in agriculture: their benefits and hazards. Interdiscip Toxicol 2(1):1

Bhardwaj T, Sharma JP (2013) Impact of pesticides application in agricultural industry: an Indian scenario. Int J Agri Food Sci Technol 4(8):817–822

Bhat RA, Beigh BA, Mir SA, Dar SA, Dervash MA, Rashid A, Lone R (2019) Biopesticide techniques to remediate pesticides in polluted ecosystems. In Handbook of research on the adverse effects of pesticide pollution in aquatic ecosystems, IGI Global pp. 387–407

Bhatt P, Verma A, Gangola S, Bhandari G, Chen S (2021) Microbial glycoconjugates in organic pollutant bioremediation: recent advances and applications. Microb Cell Factories 20(1):1–8

Boedeker W, Watts M, Clausing P, Marquez E (2020) The global distribution of acute unintentional pesticide poisoning: estimations based on a systematic review. BMC Public Health 20(1):1–9

Boricha H, Fulekar MH (2009) Pseudomonas plecoglossicida as a novel organism for the bioremediation of cypermethrin. Biol Med 1(4):1

Bowman MS, Zilberman D (2013) Economic factors affecting diversified farming systems. Ecol Soc 18(1)

Calaf GM (2021) Role of organophosphorous pesticides and acetylcholine in breast carcinogenesis. Semin Can Biol 76:206–217. https://doi.org/10.1016/j.semcancer.2021.03.016

Carvalho FP (2017) Pesticides, environment, and food safety. Food Energy Secur 6(2):48–60

Chugh M, Kumar L, Bhardwaj D, Bharadvaja N (2022) Bioaccumulation and detoxification of heavy metals: an insight into the mechanism. In: Development in wastewater treatment research and processes. Elsevier, Amsterdam, pp 243–264

Damalas CA (2009) Understanding benefits and risks of pesticides use. Sci Res Essay 4(10):945–949

De A, Bose R, Kumar A, Mozumdar S (2014) Targeted delivery of pesticides using biodegradable polymeric nanoparticles, New Delhi, Springer India 10:978–81

Dubey NK, Ravindra Shukla RS, Ashok Kumar AK, Priyanka Singh PS, Bhanu Prakash BP (2011) Global scenario on the application of natural products in integrated pest management programmes. In: Natural products in plant pest managemen. CABI, Wallingford, pp 1–20

Freire C, Koifman S (2013) Pesticides, depression and suicide: a systematic review of the epidemiological evidence. Int J Hyg Environ Health 216(4):445–460

Gangola S, Bhatt P, Kumar AJ, Bhandari G, Joshi S, Punetha A, Bhatt K, Rene ER (2022) Biotechnological tools to elucidate the mechanism of pesticide degradation in the environment. Chemosphere 296:133916

Geetha M, Fulekar MH (2008) Bioremediation of pesticides in surface soil treatment unit using microbial consortia. Afr J Environ Sci Technol 2(2):036–045

Ghorab MA, Khalil MS (2015) Toxicological effects of organophosphates pesticides. Int J Environ Monitor Anal 3(4):218–220

Gill HK, Garg H, 2014. Pesticide: environmental impacts and management strategies. Pesticides-toxic aspects. 8(2014):187

Grube A, Donaldson D, Kiely T, Wu L (2011) Pesticides industry sales and usage. US EPA, Washington, DC

Guo L, Fan XY, Qiao X, Montell C, Huang J (2021) An octopamine receptor confers selective toxicity of amitraz on honeybees and Varroa mites. elife 12(10):e68268

Gyawali K (2018) Pesticide uses and its effects on public health and environment. J Health Promot 6:28–36

Hashim M (2015) Pesticides and drinking water. J Adv Botany Zool 3(1):1–5

Kabir ER, Rahman MS, Rahman I (2015) A review on endocrine disruptors and their possible impacts on human health. Environ Toxicol Pharmacol 40:241–258. https://doi.org/10.1016/j.etap.2015.06.009

Kensa VM (2011) Bioremediation-an overview. I Control Pollut 27(2):161–168

Kim KH, Kabir E, Jahan SA (2017) Exposure to pesticides and the associated human health effects. Science of the total environment, 575:525–535

Kumar S, Kaushik G, Villarreal-Chiu JF (2016) Scenario of organophosphate pollution and toxicity in India: a review. Environ Sci Pollut Res 23:9480–9491

Kumar A, Thakur A, Sharma V, Koundal S (2019) Pesticide residues in animal feed: status, safety, and scope. J Anim Feed Sci Technol 7:73–80

Kumari S, Sharma HR (2008) Farmer's perception on environmental effects of pesticides use, climate change and strategies used in mountain of Western Himalaya. Int J Agric Sci Res 1:57–68

Lamichhane JR (2017) Pesticide use and risk reduction in European farming systems with IPM: an introduction to the special issue. Crop Prot 1(97):1–6

Landrigan PJ, Stegeman JJ, Fleming LE, Allemand D, Anderson DM, Backer LC, Brucker-Davis F, Chevalier N, Corra L, Czerucka D, Bottein MY (2020) Human health and ocean pollution. Ann Glob Health 86(1)

Long EY, Krupke CH (2016) Non-cultivated plants present a season-long route of pesticide exposure for honey bees. Nat Commun 7(1):11629

Mahajan G, Chauhan BS, Gill MS (2013) Dry-seeded rice culture in Punjab State of India: lessons learned from farmers. Field Crop Res 20(144):89–99

Mandal SK (2019) Impact of pest control chemicals on biological activity of biocontrol agents. In The souvenir of Nafional Seminar on Agro-Chemical Inputs and Its Extension Approaches Towards Food-Security and Bio-Safety: prospects and Challanges (AEFS-2019), pp 80–84. Organized by: State Agricultural Management and Extension Training Institute – West Bengal.

Medfu Tarekegn M, Zewdu Salilih F, Ishetu AI (2020) Microbes used as a tool for bioremediation of heavy metal from the environment. Cogent Food Agric 6(1):1783174

Mrema EJ, Ngowi AV, Kishinhi SS, Mamuya SH (2017) Pesticide exposure and health problems among female horticulture workers in Tanzania. Environ Health Insights 17(11):1178630217715237

Ngowi AVF, Mbise TJ, Ijani ASM, London L, Ayayi OC (2007) Pesticides use by small-holder farmers in vegetable production in Northern Tanzania. Crop Prot 26(11):1617–1624

Nieder R, Benbi DK, Reichl FX, Nieder R, Benbi DK, Reichl FX (2018) Health risks associated with pesticides in soils. In: Soil components and human health. Springer, Dordrecht, pp 503–573

Oluwole O, Checke RA (2019) Health and environmental impacts of pesticides use practices a case study of farmers in Ekiti State, Nigeria. Int J Agric Sustain 7(3):153–163

Osman KA, Al-Humaid AI, Al-Rehiayani SM, Al-Redhaiman KN (2011) Estimated daily intake of pesticide residues exposure by vegetables grown in greenhouses in Al-Qassim region, Saudi Arabia. Food Control 22(6):947–953

Patel V, Ramasundarahettige C, Vijayakumar L, Thakur JS, Gajalakshmi V, Gururaj G, Suraweera W, Jha P (2012) Suicide mortality in India: a nationally representative survey. Lancet 379(9834):2343–2351

Pelfrène AF (2010) Rodenticides. In: Hayes' handbook of pesticide toxicology. Academic Press, London, pp 2153–2217

Pérez-Lucas G, Vela N, El Aatik A, Navarro S (2019) Environmental risk of groundwater pollution by pesticide leaching through the soil profile. In: Pesticides-use and misuse and their impact in the environment, vol 17, pp 1–28. IntechOpen

Phillips PJ, Bode RW (2004) Pesticides in surface water runoff in south-eastern New York State, USA: seasonal and storm flow effects on concentrations. Pest Manag Sci Formerly Pest Sci 60(6):531–543

Raheem A, Sikarwar VS, He J, Dastyar W, Dionysiou DD, Wang W, Zhao M (2018) Opportunities and challenges in sustainable treatment and resource reuse of sewage sludge: a review. Chem Eng J 337:616–641

Rathour SS, Kumar A, Maurya MK, Stephen AJ, Paul A (2022) Study on consumer preference of fungicide with special reference to custodia (fungicide) in Bijnore district of Uttar Pradesh

Riser-Roberts E (2020) Remediation of petroleum contaminated soils: biological, physical, and chemical processes. CRC Press, Boca Raton

Sarchiapone M, Mandelli L, Iosue M, Andrisano C, Roy A (2011) Controlling access to suicide means. Int J Environ Res Public Health 8:4550–4562

Sarkar L (2019) Emancipation of effluent from fertilizers and chemical pesticides causing water contamination at Rishipur Village, West Bengal, India. Int J Res Soc Sci 9(8):62–68

Schweitzer L, Noblet J (2018) Water contamination and pollution. In: Green chemistry. Elsevier, pp 261– 290. https://doi.org/10.1016/B978-0-12-809270-5.00011-X

Semalulu O, Hecky RE, Muir D (2005) Agricultural chemicals and metal contaminants in the Ugandan catchment of Lake Victoria. Water quality and quantity synthesis final report, LVEMP, pp 162–177

Sharma I (2020) Bioremediation techniques for polluted environment: concept, advantages, limitations, and prospects. In: Trace metals in the environment-new approaches and recent advances. IntechOpen. https://doi.org/10.5772/intechopen.90453

Sharma BM, Bharat GK, Tayal S, Nizzetto L, Čupr P, Larssen T (2014) Environment and human exposure to persistent organic pollutants (POPs) in India: a systematic review of recent and historical data. Environ Int 66:48–64

Sharma A, Kumar V, Shahzad B, Tanveer M, Sidhu GP, Handa N, Kohli SK, Yadav P, Bali AS, Parihar RD, Dar OI (2019) Worldwide pesticide usage and its impacts on ecosystem. SN Appl Sci 1:1–6

Stehle S, Schulz R (2015) Agricultural insecticides threaten surface waters at the global scale. Proc Natl Acad Sci 112(18):5750–5755

Strassemeyer J, Daehmlow D, Dominic AR, Lorenz S, Golla B (2017) SYNOPS-WEB, an online tool for environmental risk assessment to evaluate pesticide strategies on field level. Crop Prot 97:28–44

Subramanyam B, Hagstrum DW (2012) Alternatives to pesticides in stored-product IPM. Springer Science & Business Media New York. pages, 437, ISBN 1461543533, 9781461543534

Toda M, Beer KD, Kuivila KM, Chiller TM, Jackson BR (2021) Trends in agricultural triazole fungicide use in the United States, 1992–2016 and possible implications for antifungal-resistant fungi in human disease. Environ Health Perspect 129(5):055001

Tosh DG, McDonald RA, Bearhop S, Lllewellyn NR, Fee S, Sharp EA, Barnett EA, Shore RF (2011) Does small mammal prey guild affect the exposure of predators to anticoagulant rodenticides? Environ Pollut 159(10):3106–3112

Tudi M, Daniel Ruan H, Wang L, Lyu J, Sadler R, Connell D, Chu C, Phung DT (2021) Agriculture development, pesticide application and its impact on the environment. Int J Environ Res Public Health 18(3):1112. https://doi.org/10.3390/ijerph18031112

Vaishampayan MR, Nile UV, Patil DY (2019) Agriculture geography a study of cropping pattern in Western Satpuda region. Lulu com

WHO World Health Organization. Suicide prevention (SUPRE). http://www.who.int/mental_health/prevention/suicide/suicideprevent/en/. Accessed 16 Aug 2014

Worldatlas (2018) https://www.worldatlas.com/articles/top-pesticide-consuming countries-of-the-world.html

Yadav IC, Devi NL, Syed JH, Cheng Z, Li J, Zhang G, Jones KC (2015) Current status of persistent organic pesticides residues in air, water, and soil, and their possible effect on neighboring countries: a comprehensive review of India. Sci Total Environ 1(511):123–137

Zhang W (2018) Global pesticide use: profile, trend, cost/benefit and more. Proc Int Acad Ecol Environ Sci 8(1):1–27

Zhang C, Sun Y, Hu R, Huang J, Huang X, Li Y, Yin Y, Chen Z (2018) A comparison of the effects of agricultural pesticide uses on peripheral nerve conduction in China. Sci Rep 8(1):9621

Zhang H, Yuan X, Xiong T, Wang H, Jiang L (2020) Bioremediation of co-contaminated soil with heavy metals and pesticides: Influence factors, mechanisms and evaluation methods. Chem Eng J 398:125657

Chapter 4
Toxicity and Health Impacts of Emerging Pollutants

Deeptimayee Pal and Sujit Sen

Abstract Emerging pollutants (EPs) are a complex problem due to a wide range of environmental pollutants, leading to potential toxicity and health effects. This study investigates the intricate terrain of such pollutants, including pharmaceuticals and personal care products (PPCPs), endocrine-disrupting chemicals (EDCs), surfactants, flame retardants, plasticizers, and novel nanomaterials, which may endanger ecosystems and human welfare. EPs include a range of toxicological impacts, among which PPCPs are particularly noteworthy owing to their extensive use and subsequent release into aquatic environments. Per- and poly-fluoroalkyl substances (PFAS), recognized by their persistent nature and ability to accumulate in biological systems, are increasingly being identified as having a significant impact on human health, as supported by growing scientific research. Research findings indicate possible associations between exposure to PFAS and various health concerns, including delays in development, suppression of the immune system, and cancer. The potential of pollutants, such as pesticides and plasticizers, to disrupt the endocrine system substantially threatens human well-being, leading to detrimental effects on reproductive health, developmental abnormalities, and disturbances in hormonal equilibrium, underscoring the need for thorough risk evaluations and regulatory interventions. Nanomaterials, a relatively recent addition to developing contaminants, provide distinct and unprecedented concerns. The long-term health impacts of manufactured nanoparticles cause worry due to their ability to overcome biological barriers and accumulate in tissues. Adopting a comprehensive strategy is essential to effectively tackle the intricate nature of these pollutants, preserve ecosystems, and ensure the well-being of individuals by mitigating the potentially harmful consequences of persistent environmental pollution.

Keywords Emerging pollutants · Health impact · Toxicity · Risk assessment · Environmental analysis

D. Pal (✉) · S. Sen
Department of Chemical Engineering, National Institute of Technology Rourkela, Rourkela, Odisha, India

S. Dey, S. Bhattacharya (eds.), *Biotechnological Interventions in the Removal of Emerging Pollutants*, Interdisciplinary Biotechnological Advances,
https://doi.org/10.1007/978-981-97-9922-0_4

4.1 Introduction

In recent decades, the globe has faced the negative consequences of unregulated growth in many human activities such as industry, transportation, agriculture, and urbanization. The rising living standards and increased consumer demand have aggravated air pollution with substances such as CO2, greenhouse gases, NOx, SO2, and particulate matter. Water pollution has also increased due to various chemicals, nutrients, leachates, and oil spills, among other contaminants. Soil pollution has been exacerbated by the improper disposal of hazardous wastes, pesticide use, and sludge, the widespread use of disposable or nonbiodegradable materials, and inadequate waste management facilities. Emerging pollutants (EPs) have become a significant concern for the world's population due to their potential environmental and human health risks (Thomaidis et al. 2013). Chemical compounds known as EPs are prevalent in soil and water and have recently been acknowledged as major water contaminants. EPs include a wide range of synthetic chemicals used worldwide, including endocrine-disrupting compounds (EDCs), perfluorinated compounds, water disinfection byproducts, petroleum additives, UV filters, manufactured nanomaterials, pharmaceuticals, and personal care products, which are crucial in advancing contemporary civilization (Lapworth et al. 2012). Due to their growing use in industry, transportation, agriculture, and urban development, these chemicals are being introduced into the environment in higher quantities as hazardous wastes and nonbiodegradable materials.

Between 1930 and 2000, worldwide output of synthetic chemicals surged from 1 million to 400 million tons per year. In 2013, EUROSTAT statistics revealed that from 2002 to 2011, more than 50% of chemical manufacturing consisted of ecologically hazardous substances. More than 70% of these compounds have a substantial environmental effect. The US Geological Survey defined EP as any manufactured or naturally occurring substance, organism, or bacterium that is not typically found in the environment but can potentially adversely affect the ecosystem and human health. The pollutants are often present in trace amounts ranging from parts per trillion to parts per billion (Sivaranjanee and Kumar 2021). EPs are chemicals that can continue in the natural world, bioaccumulate, and potentially lead to significant damage, including abnormal growth, reduced fertility, neurodevelopmental delays, and harm to wildlife and aquatic ecosystems, possibly affecting the human immune system (Rodriguez-Narvaez et al. 2017). Most EPs are not novel or newly introduced pollutants in the environment. EPs are established pollutants with recently discovered adverse effects or mechanisms. Thus, the word "emerging" pertains to both the contamination and the emerging difficulties. EPs are sometimes called "pollutants of emerging concern" or "chemicals of emerging concern." From a more macro perspective, the following characteristics may be used to classify EPs: (i) not a novel compound, (ii) a compound that has been present in the environment for a long time but has only recently been identified, and its relevance is now being acknowledged, and (iii) a compound that has been known for a long time, but its possible adverse impacts on humans and the environment have only recently been comprehended (Arman et al. 2021).

4.2 Occurrence and Fate of EPs

EPs may enter the environment via several transportation and distribution systems. The physicochemical characteristics of chemicals dictate how they behave in the environment. The primary sources of ecologically significant developing pollutants are mainly wastewater treatment plant (WWTP) effluents and, secondarily, terrestrial run-offs from various surfaces. EPs have been detected in both surface and groundwater. Concentrations of pollutants in surface waters are often lower than those found at the discharge points of wastewater and sewage treatment facilities due to dilution and natural attenuation processes. Since surface water receives effluent directly from WWTPs and has a shorter residence time than groundwater, it contains higher quantities of EPs. Concentrations of groundwater could rise if the aquifer is located close to places where pollution is being produced. Some investigations have shown that the presence of certain PPCPs in surface water fluctuates significantly. The diversity is likely caused by the frequency of use and doses in various places, as well as the efficiency of the WWTPs system (la Farré et al. 2008).

Veterinary medications are intentionally released into the environment when liquid manure is applied to agricultural fields to treat and prevent illnesses in farming. Veterinary medicines and their metabolites may easily pollute soil and groundwater. Veterinary medications may reach groundwater by leaching or run-off from animal waste, with soil sorption potentially slowing their movement. Pharmaceuticals from human use enter water systems via excretion as either unchanged chemicals or metabolites via wastewater treatment plants. If drugs and their human metabolites can travel through WWTPs, they may infiltrate rivers or streams. They may access groundwater via leaching. Pharmaceuticals may enter surface waterways by run-off from farms where digested sludge has been applied. Personal care products (PCPs), such as scents, are released via shower drainage. They may then pass over wastewater treatment plants and into the environment. Nonylphenol and nonylphenol ethoxylates are often found in wastewater treatment plants because of their widespread usage in households and industries. Another significant category of developing contaminants is PFCs (Jahan et al. 2008). This class comprises PFOS and PFOA and several more structurally related chemicals. There is increasing concern about this group of substances due to research showing its toxicity, persistence, and ability to accumulate in living organisms. Perfluorocarbons and perfluorosulfonic acids are very stable and may be found in the environment, particularly in water. Both classes of chemicals have been produced for more than 50 years and have been extensively dispersed by water currents, leading to their presence in the deep sea and even in the Arctic (Ju et al. 2008).

The alteration and fate of these substances in environmental and biological systems are vital to this matter. Many commonly used PFCs are volatile and may

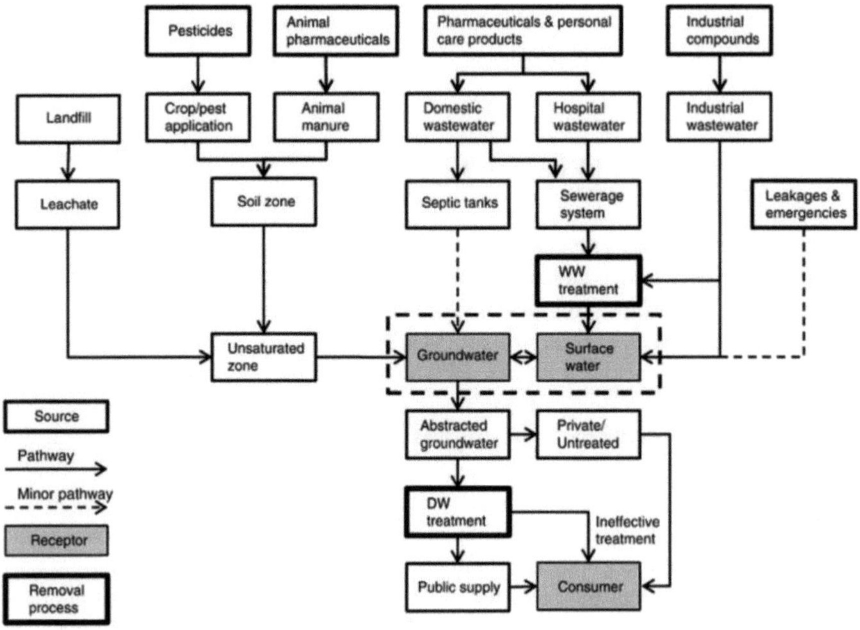

Fig. 4.1 Sources and routes of EPs in the aquatic environment. (Adapted from Stuart et al. (2012))

undergo metabolism or breakdown, resulting in the creation of their persistent sulfonate and carboxylic acid forms. Critical parts of this subject are still mostly unknown. It is essential to outline the fundamental metabolic and physical processes to comprehend the worldwide transportation of PFCs and their dispersion in the environment (la Farré et al. 2008). The sources and possible routes of EPs in the environment are illustrated in Fig. 4.1.

4.3 Toxicity and Health Impacts of EPs

EPs, a variety of compounds identified as environmental contaminants, present potential dangers to ecosystems and human health because of their different toxicity levels. EPs' toxicity may impact aquatic creatures' health and even penetrate the food chain, posing risks to human well-being. Long-term exposure to PPCPs and EDCs may disturb the hormonal equilibrium, perhaps leading to antibiotic resistance. This can impact aquatic life and human health, including developmental and reproductive problems. PFAS, recognized for their long-lasting nature, have been associated with harmful health impacts such as immune system suppression and cancer. Flame retardants, recognized for their ability to harm the nervous system and affect hormone function, underscore the many health risks linked to new types of pollutants. The toxicity and health impacts of these EPs are discussed briefly below.

4.3.1 PPCPs

Pharmaceutically active chemicals are EPs that are raising increasing concern due to their presence in water bodies. Different environmental matrices, globally have shown varying amounts of pharmaceutical chemicals. They endure in the environment because of their usage and disposal procedures (Wilkinson et al. 2017). The primary cause of these contaminants in aquatic bodies is the direct discharge of treated and untreated wastewater. Terrestrial run-off from farms, zoos, and aquariums is a secondary source. Furthermore, these substances are handled and used in varying ways, so a significant source in one location may not have the same importance in another. The various sources of PPCPs and their fate in the environment are depicted in Fig. 4.2. Waterways transport contaminants and their ability to dissolve in fats leads to bioaccumulation. Lipophilicity is not the only criterion for comprehending bioaccumulation. Recent research has elucidated other mechanisms, such as metabolism, inhalation exposure, and animal homeostasis, to enhance comprehension of the buildup of pollutants in marine organisms (Ebele et al. 2017).

Conventional WWTPs are ineffective at fully eliminating pharmaceutical chemicals. This aroused alarm over the potential hazards to aquatic species and people. Aquatic creatures have shown potential harmful effects via bioaccumulation research. Various pollutants were quantified in various fish organs, including gills, muscles, blood plasma, liver, and brain. Several investigations have detected antibiotics in algae, bivalves, and aquatic invertebrates. Simultaneously, antibiotic

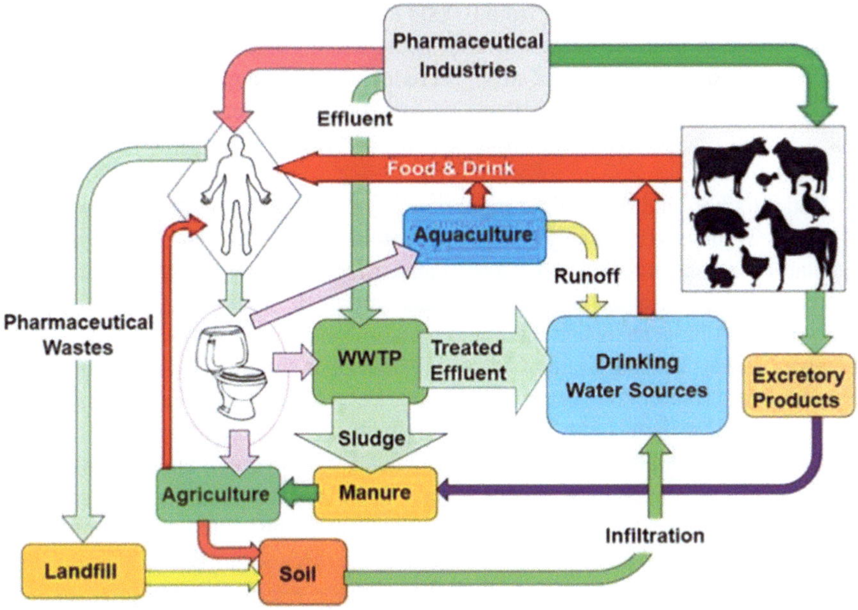

Fig. 4.2 Occurrence and fate of PPCPs (Majumdar and Pal 2020)

resistance poses a significant hazard to human health and has been a focus of concern in recent decades. Many investigations have been conducted on medicinal substances' destiny and harmful effects (Khan et al. 2020).

Pharmaceuticals are designed to elicit a biological reaction among particular beings, but exposure to ambient levels may also trigger a biological response in other organisms. Due to the extended presence of pharmaceuticals at low levels in aquatic environments and their lipophilic nature, their adverse consequences are more likely to be persistent rather than immediate. Both in vitro and in vivo tests are used to assess the hazardous potential of pharmacological substances, with the latter being more often utilized. Typical objectives for toxicity testing included growth, death, and reproduction. Recently, a significant focus has been investigating possible developmental, molecular, histological, and behavioral consequences. Microalgae, aquatic crustaceans, mollusks, worms, and fish are more extensively studied than amphibians and other vertebrates (Gunnarsson et al. 2019).

Diclofenac exposure at a dose of 5 μg/L resulted in gill changes and kidney lesions in rainbow trout. The fall in the vulture population in Pakistan and India was attributed to severe kidney failure in vultures that fed on diclofenac-treated dead cattle. There is no substantial evidence of reproductive defects produced by pain relievers at quantities relevant to the environment. One research found that diclofenac delayed Zebrafish hatching at 1000–2000 μg/L concentrations. The ecotoxicological potential of antibiotics is in its early stages. The observed effects generally include growth suppression in cyanobacteria and green algae and the growth of antibiotic resistance. Erythromycin at concentrations ranging from 0.0002 to 200 μg/L caused an increase in gill histopathological index, whereas oxytetracycline at concentrations ranging from 0.0004 to 400 μg/L led to an increase in pathological index in gilthead seabream after acute and chronic exposure, respectively. These complexes resulted in the strongest reduction of algal growth compared to the antibiotics alone. Disregarding these connections might lead to underestimating hazards and should be included in environmental risk assessment (Rodrigues et al. 2019).

Beta-blockers and blood lipid regulators have comparable impacts on nontarget creatures as they exert on target ones. Atorvastatin, a widely prescribed blood lipid regulator at a concentration of 1.2 μg/L, caused depletion of lipid, carbohydrate, and protein reserves. It also inhibited the transcription of crucial enzymes related to mitochondrial biogenesis and fatty acid metabolism in the digestive gland of a key bivalve species. The heart rate of Daphnia magna was significantly impacted by the beta-blocker propranolol (Jeong et al. 2018). Beta-2 receptors in fish reproductive, cardiac, and liver tissues may lead to ecotoxic effects from beta-blockers, similar to how they affect people. Data is mostly provided for propranolol and atenolol. Beta-blockers are considered to have little impact on aquatic creatures due to their strong tendency to bind to sediments, which limits their availability in the environment.

Neurotoxicity and behavioral abnormalities are often seen as side effects of psychiatric medications. Fluoxetine and carbamazepine are mostly explored for their harmful effects. Despite environmental concerns, some pharmacological groups, such as anticancer chemicals and X-ray contrast media, have unknown harmful effects on nontarget creatures. Nevertheless, they impact preserved biochemical pathways in lesser creatures similarly to humans (Toolaram et al. 2014). Contaminants at concentrations ranging from 0.2 to 123 μg/L affected visual acuity and caused mutagenic effects on tadpoles. Algae and cyanobacteria showed growth suppression when exposed to 5-fluorouracil at 0.13 mg/L and 1.20 mg/L, respectively. Endocrine disruptors are well-recognized for their ability to interfere with the reproductive system. They can bio-concentrate effectively due to their capacity to pass through gills easily and their attraction to sex steroid-binding proteins in aquatic species. Yet, only a few of these compounds have been thoroughly studied for their hazardous effects. Recent research revealed that the over-the-counter usage of ibuprofen may cause a reproductive condition in adult men known as compensated hypogonadism (Kristensen et al. 2018). This data might be used as a foundation for researching the impact of environmental exposures to toxins on human health. Antibiotics played a crucial role in transforming medical care by fighting infections formerly thought to be deadly. Antibiotics have been used since the 1940s and have become crucial in human healthcare, ranging from organ transplants to treating common infectious ailments. Without antibiotics, contemporary medicine might face significant challenges. Antibiotics have gained attention as environmental pollutants due to their frequent presence and persistence in various environmental sources globally, their ability to enter food chains and water systems, and their potential to damage nontarget creatures in relevant amounts. Residues were found in vegetables, milk, aquatic goods, and meat. Around 50% of persons aged 15 to 24 in the UK have taken antibiotics not prescribed for them. Seventy percent of antibiotics used in animals in the United States are crucial for human health. Global antimicrobial usage in meat production is projected to increase by 67% from 2010 to 2030 (Van Boeckel et al. 2015). The excessive use of antibiotics has compromised their efficiency since bacteria have developed resistance, even against powerful medicines like carbapenems, posing a threat to medicine and public health. Antibiotic resistance is said to result in 700,000 deaths annually worldwide and is expected to reach 10 million by 2050. The World Health Organization (WHO) has released its first list of antibiotic-resistant priority pathogens, including twelve bacterial species that present the most significant risk to human health and need urgent development of new medicines.

Antibiotic resistance results from evolutionary adaptation via Darwinian natural selection when microorganisms build strong mechanisms to survive exposure to harmful substances. While some microorganisms, plants, and fungi naturally create antibiotics, little evidence indicates that this plays a major role in developing antibiotic resistance in microbes within their natural habitat. New research suggests that bacterial persistence may promote and speed up the development of resistant mutants from the persister community via chromosomal alterations when exposed to antibiotics by offering a larger pool of viable cells for mutation. Horizontal development significantly contributes to the present antibiotic resistance dilemma. Resistance mutation is a fundamental process contributing to resistance, among others. The mutations modify the target protein by deactivating the antibiotic-binding site while keeping the protein's cellular function intact (Windels et al. 2019).

Coresistance is a method in which multiple resistance genes are positioned adjacent to one another on mobile genetic components, facilitating their simultaneous transmission by horizontal gene transfer. Heavy metal contamination in the environment may lead to the development of antibiotic resistance. Additionally, nonantibiotic medications may promote the development of multiple resistance to antibiotics via mutations. Recent investigations have shown the fundamental processes of multiple antibiotic resistance caused by nonantibiotics. Antidepressant fluoxetine induced antibiotic resistance in Escherichia coli by causing mutations in DNA-binding transcriptional regulators through reactive oxygen species (ROS). This led to increased expression of drug efflux systems that facilitated antibiotic removal via the AcrAB-TolC pump. This process also reduced the presence of outer membrane porin OmpF, preventing antibiotics from entering the cell. The research found that exposure to the antiepileptic medication carbamazepine increases the spread of antibiotic resistance in E. coli by horizontal gene transfer by activating pili-forming genes on the RP4 plasmid (Wang et al. 2019).

4.3.2 EDCs

The US Environmental Protection Agency (EPA) has defined an EDC as an agent that disrupts the synthesis, secretion, transport, binding, or elimination of natural hormones in the body. These hormones maintain homeostasis, reproduction, development, and/or behavior. The sources of exposure to EDCs are often varied and extensively dispersed across the global environment and culture (Kabir et al. 2015). However, the situation is inconsistent and cannot be readily anticipated because of the substantial variation in the consumption of these chemicals across different nations. The situation grows further intricate as some chemicals have been prohibited in certain nations while others remain unaffected. On a global scale, it is evident that EDCs mostly consist of steroid hormones, synthetic steroids, polychlorinated dibenzo dioxins, biphenyls, and similar substances. Previous research has clearly shown that these chemicals have significant hazardous effects on both humans and wildlife (Diamanti-Kandarakis et al. 2009). However, prior research has not extensively addressed endocrine disruptors such as alkylphenol ethoxylate, gonadotropin chemicals, and pesticides. Endocrine disruptors may be derived from natural sources or manufactured in industrial settings, depending on their wide range of uses.

Exposure to EDCs can happen through various means, such as drinking contaminated water, breathing contaminated air, consuming contaminated food, or coming into contact with contaminated soil. These soils are commonly exposed to pesticides, plasticizers, alkylphenols, and flame retardants in agriculture, industry, and household applications. Several endocrine disruptors can enter the aquatic environment, making water a potential reservoir of EDCs. Possible causes include inadequate elimination of pollutants during sewage treatment procedures, soil run-off, or uncontrolled release of pollutants into aquatic bodies. Certain water treatment systems, such as flocculation, sedimentation, filtration, and chlorination, may not possess the capability to completely eradicate or entirely remove certain contaminants from water. The current scenario is a cause for concern since it is evident that the release of EDCs into water bodies, such as lakes and streams, may result in harmful consequences for the ecosystem despite their low quantities.

The nervous system is a vital bodily system that is crucial in maintaining all body components' synchronization and optimal functioning. However, it is susceptible to the influence of EDCs, which many processes may generate. EDCs may directly impact endocrine glands, leading to changes in hormonal balance and function in the body. In contrast, the substances have the potential to first impact the central nervous system (CNS), such as neuroendocrine disruptors, which may subsequently affect the endocrine system. The impact on any of these factors could lead to neuroendocrine disruption, resulting in changes in metabolic rate, indirect behavioral effects, alterations in sexual differentiation within the brain that may affect both sexually dimorphic reproductive and nonreproductive neural processes, and potential neuroteratogenic effects. Numerous research and literature about humans and animals have been conducted to examine the impact of exposure to EDCs on human behavior, learning, memory, attention, sensory function, and neurological development. Neuroendocrine disruptors that have the potential to impact individuals either directly, indirectly, or both are various substances (Mellanen et al. 1996). The various health impacts of EDCs are illustrated in Fig. 4.3.

The three isotypes of the peroxisome proliferator-activated receptor (PPAR), alpha, beta, and gamma, are crucial for regulating cellular differentiation programs

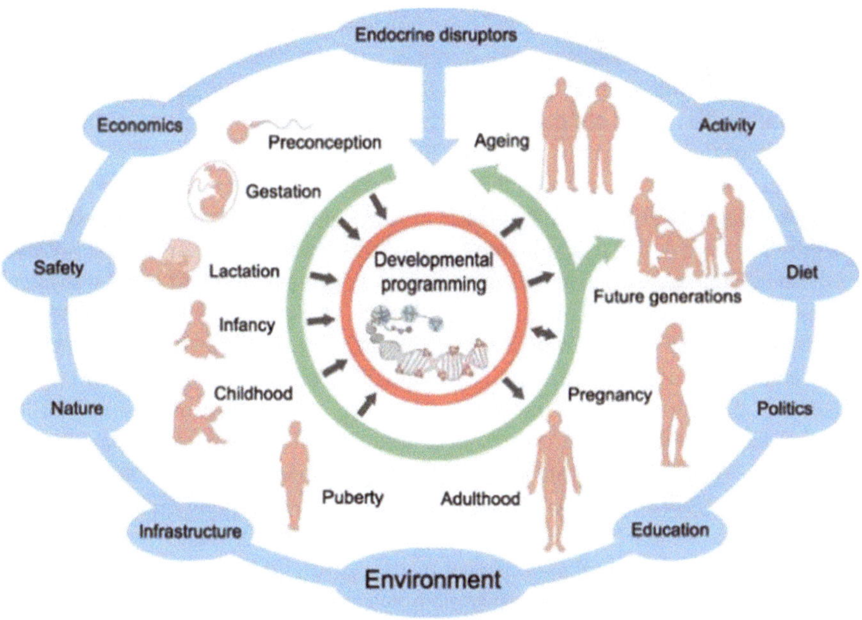

Fig. 4.3 Effect of EDCs on health (Sargis and Simmons 2019)

and lipid and carbohydrate metabolism via transcriptional regulation. The metabolic abnormalities are mostly caused by PPAR alpha, found in the liver and often linked to endocrine disruptors (Casals-Casas et al. 2008). Homeostatic activities, including development, metabolism, energy balance, stress response, and reproduction, are regulated by neuroendocrine systems, which bridge the central nervous and peripheral endocrine systems. Disruptions to the hypothalamic-pituitary-thyroid system's neuroendocrine homeostasis may impact metabolism and energy balance, and there is evidence that endocrine-disrupting drugs can cause such disruptions. For instance, when exposed to the thyrotropin-releasing hormone, PCBs lower thyroxin and thyroid-stimulating hormone (TSH), leading to hypothalamic and pituitary dysregulation. Another potential side effect of EDC exposure is a decrease in adult reproductive success due to inappropriate hypothalamus programming caused by these hormone-active chemicals (Gore 2010).

Possible female reproductive system disorders linked to EDC include polycystic ovarian syndrome, premature ovarian failure, and early puberty. Some undesirable side effects of EDCs in women include an increased risk of breast cancer and endometrial development. The progeny of these ladies had to battle serious illnesses, including vaginal cancer, as a result. The average age of menarche has dropped from about 16 or 17 to around 13 years old throughout the last century. Insulin resistance, metabolic syndrome, breast cancer, and other reproductive system malignancies may all have an increased incidence among girls and young women who reach puberty at a younger age. Research has shown that around 1% of women under the age of 40 have primary ovarian failure. Researchers have looked at several potential explanations, and EDC is one of them. Period abnormalities, such as longer periods, may lower fecundability (the capacity to conceive during a menstrual cycle), and EDC has the potential to disrupt the hormonal control of the menstrual cycle. This endocrine condition is more common in females. Anovulation, hyperandrogenism, and an increased risk of obesity, insulin resistance, and metabolic abnormalities are symptoms of the condition (Costa et al. 2014).

Sperm abnormalities, hypospadias, and ectopic testes are among the sexual organ problems that may affect men. Exposure to EDCs during the prenatal period, when the growing fetus is still in a vulnerable sexual differentiation stage, is a likely cause of these disorders. Several studies have shown a decrease in sperm quality in many nations since the 1930s. This loss is evident in sperm count, fraction of normal sperm, and semen volume, which might impact fertility. Other surveys contradict this declining trend in sperm quality in humans. There is a large range in sperm quality, both nationally and internationally, even if sperm quality is not falling everywhere. Based on our current understanding of how the testis grows and works, it stands to reason that substances with endocrine-like properties may influence sperm quality (Bay et al. 2006). However, the link between sperm quality and exposure to EDCs has not been investigated in any study so far. Fertility rates among males have fallen in several nations over the last several decades. People in the UK, France, and Denmark also observed something similar. Experiments on humans and animals have demonstrated that pesticides and PCBs, among other chemicals, may reduce fertility; however, whether or not this has anything to do with endocrine disruption is still debatable.

Reproductive problems, developmental delays, metabolic abnormalities, elevated cancer risk, and long-term impacts on children's health are just some of the human health issues that may result from exposure to EDCs. In children, EDCs have been linked to cognitive deficits, behavioral issues, neurodevelopmental abnormalities, hormone imbalances, decreased sperm quality, and infertility (Bay et al. 2006). There is evidence that some EDCs raise the chance of developing cancers of the breast, prostate, and testicles. Diabetes, obesity, and thyroid dysfunction are metabolic illnesses that may develop when EDCs disrupt the endocrine system (Diamanti-Kandarakis et al. 2009). EDCs exposed to pregnant women's bodies raise the risk of obesity, diabetes, and cardiovascular disease in their children. Exposure time, length, and individual susceptibility variables influence how EDCs affect human health (Costa et al. 2014).

4.3.3 Flame Retardants

Flame retardants (FRs) are classified as environmentally hazardous substances (EDs) and include a wide range of chemical compounds, including bromine, chlorine, and phosphorus. The use of these compounds in various commercial materials since the 1970s has mostly been aimed at mitigating their flammability (Zota et al. 2018). Although the usage of FRs is gradually eliminated, their persistent existence in long-lasting consumer goods, food, and indoor dust indicates that people will still encounter them because these exposures will persist in accumulating in living organisms. As a result, individuals around the globe will continue to be exposed to these toxins for many decades. Brominated flame retardants (BFRs) are a prominent category within the realm of flame retardants. BFRs, or biofloc reactors, are widely used industrial chemicals that continue to be generated in significant quantities on a worldwide scale. Approximately 21% of the overall flame retardant output is attributed to these particular flame retardants, a significant proportion of which have been recognized for their persistent nature, bioaccumulative properties, and hazardous nature (Waaijers and Parsons 2016).

These compounds find widespread use in many industries, such as electronics, clothing, children's toys, carpets, beds, television sets, automobiles, and numerous items. Consequently, several biofloc receptors (BFRs) may be detected in measurable quantities in animals and people (de Wit et al. 2010). Polybrominated diphenyl ethers (PBDEs), hexabromocyclododecane (HBCD), and tetrabromobisphenol A (TBBPA) are the most often used basic flame retardants (BFRs). PBDEs are classified as additive BFRs, meaning they are added to the polymer before, during, or after polymerization without doing so by chemical binding. These PBDEs are known to have a greater likelihood of leaching than reactive BFRs. These compounds are widely distributed worldwide in the environment and animals due to their lipophilicity and stable structure. HBCD, which has 16 distinct stereoisomers exhibiting diverse biological properties, finds use as a flame retardant in polystyrene foams in various industrial applications (Tomy et al. 2004). The objective is to halt the removal of these compounds by granting exemptions for manufacture (as permitted for the indicated parties) and for specified applications in buildings. The brominated phenolic chemical TBBPA has been used as a flame retardant in many materials such as plastic, cloth, and paper. These compounds can serve as additives or reactive flame retardants (BFRs). The primary purpose of their chemical manufacture is to be used as flame retardants in printed circuit boards. It is worth noting that over 80% of TBBPA is produced for this purpose, accounting for around 25% of the worldwide flame retardant industry (Abdallah et al. 2016).

The extensive use of HBCD and TBBPA in industrial settings has resulted in the pollution of the environment by these substances. Additionally, these compounds have been identified in many biological samples, including human blood, plasma, urine, and breast milk (Bjermo et al. 2017). Multiple studies have shown an upward trend in the concentrations of HBCD and TBBPA in the environment, as well as in people and animals. The detrimental health consequences of these substances may be attributed to their persistence, bioaccumulation, and biomagnification in human populations (Kim et al. 2014). Several studies have shown that brominated flame retardants have the potential to function as endocrine disruptors.

Furthermore, several research conducted on animals have also linked exposure to BFRs to detrimental health consequences. Exposure to HBCD has the potential to impact the liver hormone system and be associated with neurobehavioral changes. Similarly, TBBPA has been demonstrated to have consequences for thyroid hormones, neurological function, and reproductive processes. Furthermore, the presence of HBCD and TBBPA might lead to the development of cardiovascular diseases.

In terms of human health, there has been research that points to the negative impacts of HBCD. For example, a study by Zhang et al. concluded that the cytotoxicty in the Hep G2 human liver cancer cell line was caused by HBCD, with the effect being higher for γ-HBCD than β-HBCD and α-HBCD. As Fery et al. had previously discovered in Hep G2 cells, HBCD is a shared activator of the human pregnane-X receptor (PXR). Furthermore, HBCD can disrupt T4 metabolism by inducing enzymes like UGT via PXR and the constitutive androstane receptor. Hinkson et al. examined how HBCD affected the lytic function and ATP concentrations associated with human natural killer (NK) cells in the immunologic system (Hinkson and Whalen 2009). Lysing cancer cells, virally infected cells, and cells coated with antibodies is something that NK cells are trained to do. In this work, NK cells were exposed to varying concentrations of HBCD at several intervals after being obtained from the peripheral blood of healthy volunteer donors. According to the data, there was a significant decline in lytic function and ATP levels after HBCD exposure. This leads Hinkson et al. to believe that HBCD may raise cancer risk in those who are exposed to it.

According to a study examining the effects of HBCD on the endocrine system, higher dust concentrations may cause males to have lower levels of sex hormone-binding globulin because they raise free androgens. According to Al-Mousa and

Michelangeli, HBCD is a very hazardous chemical that promotes cell death via the reliant caspase pathway; it is also the most deadly BFR in the research (LC_{50}, 3 μM). When the microsomal sarcoplasmic-endoplasmic reticulum Ca2 + -ATPase (SERCA) is inhibited, leading to a rise in intracellular calcium ion concentrations, mitochondrial dysfunction is caused by either the release of cytochrome c or the depolarization of the membrane, activating this pathway. Furthermore, our findings demonstrated that HBCD induced cell death in SH-SY5Y by inhibiting SERCA, ATP binding, and SERCA's E2 to E1 transition. The researchers deduced that the E2 to E1 transition phase, phosphorylation, and ATP binding are all impacted by HBCD on the Ca2 + -ATPase (Al-Mousa and Michelangeli 2012).

4.3.4 PFCs

Perfluorinated compounds (PFCs) are chemical substances of a hydrophobic alkyl chain completely fluorinated and connected to a hydrophilic end group. They have been manufactured since the late 1940s. Perfluorooctanesulfonate (PFOS), perfluorooctanoic acid (PFOA), and their salts are critical types of PFCs used in various products. PFCs have exceptional resistance to heat, light, and chemicals, and microbial processes do not readily break them down (Miralles-Marco and Harrad 2015). PFCs are widely transported via several environmental pathways, originating from direct sources, such as the manufacturing, use, and disposal of consumer goods containing these chemicals, and indirect sources, such as the degradation of volatile and neutral PFC precursors (Pal and Sen 2023a). PFOA and PFOS have been found in surface water, oceans, animals, drinking water, human blood, and breast milk. The research found that PFCs are bioaccumulative, present in blood, food, indoor substances, consumer items, and occupational settings, contributing to PFC exposure (Ahrens 2011). However, there is little and restricted evidence about the negative impacts on individuals exposed to the environment. Some of the adverse effects of PFCs have been listed below.

PFOS and PFOA have been studied in laboratory animals to determine their potential to cause cancer (Pal and Sen 2023b). Yet, there is little evidence of the possible carcinogenic effects of PFCs in the general population. Research has extensively examined the correlation between serum PFC levels and reproductive issues in the general population, focusing on infertility, lactation, and semen quality in humans. Fei et al. investigated the possibility that PFC exposure and related hormone disruptors may make infertility worse (Fei et al. 2009). Previous research has highlighted the possible links between the levels of PFOA and PFOS in the blood and the presence of thyroid illness. Research has also examined the possible links between the levels of PFOA and PFOS in the blood and metabolic disorders. Furthermore, Fei et al. investigated the relationship between parental PFC levels and nursing time using the Danish National Birth Cohort as a basis. The results suggested that, except primipara, PFOS may reduce the women's ability to lactate (Fei et al. 2010).

Joensen et al. explored a possible relationship between perfluoroalkyl acids (PFAAs), testicular function, and semen quality. The authors evaluated semen quality in 105 Danish men from the general community by classifying PFAA levels into 10 degrees and looking at the serum concentrations of reproductive hormones under each PFAA degree. The findings suggested, albeit not statistically significant, that lower-than-normal sperm counts were correlated with greater PFAA concentrations (Joensen et al. 2009). According to this research, there may be some reproductive dysfunction brought on by PFC exposure; however, it is unknown how much disruption is caused by exposure levels. Shi conducted research where zebrafish eggs were exposed to PFOS to examine its developmental toxicity. Zebrafish embryos were subjected to concentrations of 0.1, 0.5, 1, 3, and 5 mg/L of the chemical at the four-hour post fertilization stage. Delayed hatching, lower hatching rates, and larval survivability were seen when embryos were exposed to 1, 3, and 5 mg/L PFOS for 132 hours. The fry revealed extensive developmental abnormalities following exposure to PFOS concentrations of 1 mg/L or more. Growth was markedly decreased in the groups treated with 3 mg/L and 5 mg/L of PFOS. The developmental deformity was tested for apoptosis mediation using flow cytometry measurement of DNA content, acridine orange staining, and terminal deoxynucleotidyl transferase dUTP nick-end labeling assay. These methods showed more apoptotic cells in the embryos treated with PFOS than in the control embryos. Zebrafish embryos were shown to be an appropriate model for assessing the developmental toxicity of PFCs. The gene-expression profiles in the embryo helped uncover potential toxicity pathways to development (Shi et al. 2008).

4.3.5 Surfactants

Surfactants are found in various forms, from ancient Babylonian ash-oil soap formulas to modern soaps, cleansers, and detergents. Detergents have become essential components of human existence, crucial in maintaining cleanliness and tidiness. The industrial sector extensively uses surfactants found in paints, textile softeners, antistatic agents, metal processing, and oil drilling, among other things. Certain surfactants possess antibacterial characteristics that make them effective as biocides (van Ginkel 1996). Industries globally release surfactants, often surface-active agents, into their wastewater treatment plants. After being used, surfactants are released into water bodies, where their prolonged presence may result in the buildup of potentially dangerous or harmful compounds, leading to significant environmental issues. Surfactants are often found in untreated effluents, and certain surfactants may reach levels between 0.4 and 40 mg/L, posing toxicity risks to aquatic life (Abel 1974).

The effects of surfactants on the world of microbes differ depending on the species and pollution level. Bacteria break down surfactants, a topic that will be covered in the biodegradation of surfactants. Surfactants harm several bacteria, including phosphate-solubilizing *Acinetobacter junii*, autotrophic ammonia-oxidizing Nitrosomonas and Nitrosospira strains (Brandt et al. 2001), and bioluminescent *Vibrio fischeri*. Even in bacteria that use SDS as their sole carbon source, you may witness surfactant-based membrane rupture, DNA damage, and nutritional deprivation. The effects of LAS on soil populations of actinomycetes, fungus, and bacteria were shown to be statistically significant for the former two (Asok and Jisha 2012). Algal development, motility, and photosynthetic ability are most affected by surfactants. The level of toxicity is determined by the kind and concentration of surfactants and the type of algae. At a concentration of 0.1 mg/mL, SDS leads to the absence of zygospore development or the generation of defective/abnormal spores. Algae's photosynthetic capacity declined exponentially as surfactant concentration increased and algal biomass fell, with cationic surfactants showing the most inhibitory effect.

It seems that surfactants considerably change soil physics, soil chemistry, and soil biology, with sorption processes playing a prominent role, even at low concentrations. Although surfactants mostly depress or destroy plant roots, they have a lesser inhibitory effect on wheat seedling shoots. When bean plants are cultivated in detergent water, their photosynthetic activity and chlorophyll concentration rate are reduced (Jovanic et al. 2010). Soil dehydrogenase activity was reduced, while acid and alkaline phosphatase and arylsulfatase activities were enhanced by applying the surfactant with anionic LAS properties. A study conducted in the late 1960s found that surfactant toxicity may be traced back to the destruction of chemoreceptors in the fish *Ictalurus natalis* when exposed to detergent levels as low as 0.5 ppm, which is much lower than the sublethal doses. Typically, dosages over 0.1 mg/L are necessary for anionic and nonionic surfactants to cause chronic and sublethal toxicity to aquatic organisms. Surfactants have detrimental biological impacts on aquatic creatures, particularly at very high concentrations (Romanelli et al. 2004).

Fish organs such as the gills, liver, kidneys, spleen, and intestines display the visible effects of surfactant exposure. Fishes become more susceptible to microbial attack when exposed to sublethal concentrations of surfactants, which cause gill epithelial hyperplasia, oxidative stress, and mucus layer damage (Rosety-Rodríguez et al. 2009). Conversely, exposure to high surfactant concentrations disrupts the gill epithelium, leading to asphyxiation or osmoregulatory failure. While SDS has been shown to have less of an impact on crustaceans, there have been reports of negative impacts on mussel suspension feeding and filter-feeding behaviors in bivalves. Dermatitis and aphthous ulcers are the principal outcomes of anionic surfactants targeting the stratum corneum of the delicate skin's membrane bilayer in humans (Marrakchi and Maibach 2006). Therefore, an acute hazardous impact from anionic surfactants is quite unlikely. Still, a chronic effect is more plausible, given that the average human dose is around 5 mg/person, found in water, detergent, toothpaste, and food.

In most cases, surfactants aid in the breakdown of polycyclic hydrocarbons; however, SDS prevented PAH degradation due to its preference as a growth substrate. This indicates that additional hydrocarbons might be bioaccumulated in water bodies due to the presence of this detergent. In addition, some reports indicate that the biodegradation of polycyclic aromatic hydrocarbons in soil polluted with weathered creosote is inhibited when SDS and *Pseudomonas aeruginosa* UG2 biosurfactants are added (Deschênes et al. 1996). Due to their ability to make hydrophobic pesticides seem more soluble, surfactants may significantly influence the pesticide destiny at effluent-irrigated locations. Worldwide, efforts have been made to mitigate the aftereffects of surfactant toxicity. Some examples of these efforts include the push for environmentally friendly surfactants, stringent rules on using phosphate-free surfactants and treating wastewater before disposal. According to a case study conducted in Ludhiana, using liquid detergents instead of powder forms reduces the toxicity of surfactants. To further our efforts to lessen the toxicity of surfactants, there are some intriguing proposals for the manual manufacture of both hard and soft water liquid detergents and laundry detergents. Another potential strategy is using natural, biodegradable soaps and soapnuts that are not harmful (Rebello et al. 2014).

4.3.6 Engineered Nanomaterials

Engineered nanomaterials are defined as having a particle size ranging from 1 to 100 nm. Examples include amorphous silicon dioxide, carbon nanotubes, and titanium dioxide. These materials are classified as EPs (Dreher 2004). Engineered nanomaterials are commonly utilized in various industrial applications. Nanoscale materials may exhibit unique physical and chemical characteristics, including surface, size, and quantum effects, leading to unpredictable biohazard consequences. Nanomaterials have an atomic interface accounting for 15–50% of the total surface area. This structure gives nanomaterials a high air, water, and soil adsorption capacity, allowing them to adsorb toxic gases, heavy metals, and biologically active substances. Nanomaterials possess unique qualities, including catalytic behavior, exceptional toughness, and strength, which render them resistant to destruction by chemical and biological means. Nanomaterials endure extended migration, transformation processes, and intricate chemical interactions in the environment as they adsorb diverse compounds on their surfaces. As a consequence, new contaminants are formed (Lei et al. 2015).

Manufactured nanomaterials have shown apparent carcinogenic properties, including reproductive and developmental toxicity in animal testing and in vitro experiments at higher concentrations (Schilling et al. 2010). Scientists have proposed that nanoparticles might be carcinogenic, irrespective of their chemical composition. Various epidemiological studies conducted in multiple periods and locations have assessed the correlation between exposure to titanium dioxide and human lung cancer. However, none of this research has shown statistically significant findings. No significant research has been published on this sector's carbon nanotubes. Pelucchi et al. conducted a meta-analysis of 28 cohorts and 15 case controls to examine the relationship between silica exposure levels and lung cancer risk (Pelucchi et al. 2006). Their findings did not provide substantial evidence to support the carcinogenicity of silicon dioxide. Biological safety concerns related to nanomaterials have garnered global attention. Currently, the processes causing toxicity in humans and animals are yet unidentified. Consequently, it is essential to prioritize the environmental safety of produced nanomaterials and enhance research on human health impacts.

4.3.7 Plasticizers

Phthalates are a class of plasticizers that enhance the flexibility and durability of various polyvinyl chloride (PVC) goods. These products include various applications, such as construction materials, food packaging, clothes, toys, children's items, blood bags, intravenous fluid bags, infusion sets, and certain medical equipment (Huber et al. 1996). Additionally, these compounds find use in solvents, lubricating lubricants, fixatives, detergents, and other items, including cosmetics and wood finishes . Phthalates lack covalent bonding with the polymeric matrix and undergo leaching from polyvinyl chloride (PVC) upon exposure to lipophilic chemicals. These pollutants are widely present in several sources, such as food, indoor air, soils, and sediments Furthermore, they are emitted directly into the environment during the manufacturing and use processes and after the disposal of PVC and other phthalate items. Phthalates can accumulate in invertebrates, fish, and plants, but they do not undergo biomagnification due to higher species' efficient metabolism and excretion of phthalates (Giuliani et al. 2020).

Phthalates, classified as EPs, have been identified as substances that manifest toxicological impacts on human well-being. Multiple pathways exist to enter pathogens into the human body, including eating, inhalation, cutaneous absorption, and intravenous tubing. Reproductive diseases, including infertility, diminished sperm quality, and hormonal abnormalities, have been linked to the presence of phthalates. Additionally, it is worth noting that these factors may exert developmental impacts on children, including neurodevelopmental abnormalities, cognitive impairments, and behavioral challenges. There exists a correlation between phthalates and an increased susceptibility to certain types of cancer, including breast, prostate, and testicular cancer (López-Carrillo et al. 2010). Phthalates have the potential to interfere with the functioning of the endocrine system, hence resulting in the development of metabolic illnesses, including obesity, diabetes, and thyroid dysfunction (Botelho et al. 2009). Further investigation on the toxicity and ramifications of phthalates is necessary to enhance comprehension of their influence on human well-being and formulate approaches to mitigate exposure.

4.4 Treatment Technologies for the Degradation of EPs

Various treatment methods are employed to eliminate EPs from water and wastewater, considering the wide range of these pollutants. Some commonly used treatment methods are adsorption, membrane filtration, advanced oxidation processes, ozonation, phytoremediation, biological and electrochemical methods (Ahmed et al. 2021). The selection of treatment technique is often based on the unique attributes of the developing contaminants, water quality, and treatment objectives. Hybrid systems combine several technologies and are used more often to improve efficiency in eliminating a more comprehensive range of EPs (Dhangar and Kumar 2020). Various adsorbents are often used to remove multiple pollutants via adsorption. Phase-shifting technologies function by transferring pollutants from one state (such as water) to another (such as solid) and have been shown to eliminate contaminants of emerging concern, such as ibuprofen, norfloxacin, ciprofloxacin, and paracetamol. Research and development of photocatalysts for visible light and phase-shifting technologies have significantly risen in addressing EPs. Solar photocatalysis may effectively handle various EP sources in water and soil (Cemre Birben and Bekbolet 2019).

4.5 Challenges and Opportunities

Monitoring and assessing EPs provide particular challenges. One problem is that these toxins are only sometimes monitored, although they can impact human and ecological health. The antibiotic resistance of aquatic pathogens has a growing impact, highlighting the increasing need for monitoring and assessing EPs. Advancements in technology and the use of novel compounds have improved the monitoring and evaluation of EPs to remove or contain pollutants. Here are some of the main obstacles and possibilities encountered in monitoring and assessing EPs.

EPs generating chemicals and methods undergo fast production, disposal, and use changes. Understanding the routes and transformation products of the EPs is crucial. The monitoring frequency for EPs varies according to their characteristics and the medium via which they are transported, such as biological, water, or air (Aalizadeh et al. 2019). Modeling methods for validating, adjusting, and comprehending transportation processes are essential and beneficial. However, these strategies are limited by computational and data needs at bigger scales (Brunner and Simmons 2012). Data availability and validation are crucial in addressing this difficulty. As EPs arise, risk assessment techniques related to monitoring and detection technologies and conceptual and methodological knowledge must continuously progress. For any evaluation system to sustain monitoring and real utilization results, it must be consistently informed by the outcomes of its implementation. Due to the worldwide nature of EPs, it is necessary to establish international norms and regulations to standardize the monitoring and exchange of research and technology related to EPs.

Comprehensive information on diverse kinds of pollutants is crucial for risk assessment studies, including EPs like EDCs, since these pollutants may combine to produce synergistic harmful effects on exposed animals (Noguera-Oviedo and Aga 2016). The risk of exposure to EPs is determined by their chemical properties and use. Phthalates provide a challenge for exposure assessment owing to their widespread use and presence in the environment. The endocrine-disrupting phenomenon is a synergistic impact caused by many EDCs. This synergistic impact necessitates using multiresidue approaches that identify and separate the many EDCs in the EP mix for risk assessment (Jiménez-Díaz et al. 2015). Various analytical techniques have identified EDC's existence in urban areas' water supply. Exposure to EDCs throughout different biological stages in animals and humans, such as gestation and breastfeeding, requires more investigation. More studies on EPs, explicitly focusing on EDCs, will be necessary as human activities become more varied, leading to greater resource use and waste generation (Yadav et al. 2021).

4.6 Limitations in the Management of EPs

The wide variety of EPs, their mechanisms of action, and the unknown transformation products significantly burden managing and establishing policies to control and eliminate EPs. Optimization tactics are necessary to provide a security monitoring and removal system capable of handling high loads. This includes selecting adsorbents and flocculants, membranes, dosing processes, system configurations, and mixing conditions (Qu et al. 2013). Therefore, it is crucial to systematize information and provide effective methods for simulating the behavior of EPs in water, soil, and air. The absence of comparison between various EP removal strategies, both conventional and emergent, hinders the ability to identify cost-effective and contaminant-specific treatments for re-emerging pollutants. Specific site factors may also be crucial in determining the constraints of EP treatment techniques (Krauss et al. 2019). The relative analysis should include the site-specific limits regarding the safe disposal or legal limit standard for EPs.

In addition to families, hospitals, and enterprises, alternative wastewater sources arise from new activities such as hydraulic fracturing. Recent research has focused on several methods for EP removal; algal technologies exhibit exceptional removal effectiveness for heavy metals and eliminate harmful nutrients. Ozonation and powdered activated charcoal effectively remove iohexol, iomeprol, and iopromide with an efficiency of 97% and diatrizoate with an efficiency of 90%. Various (bio)transformation byproducts exhibit distinct toxicological characteristics, which may be evaluated using modern modeling techniques. These advanced oxidation processes are economically restricted due to their high capital and operational expenses. Recent shifts in public perception of wastewater have resulted in three main changes: viewing wastewater as a water resource rather than hazardous waste for water reuse, considering wastewater as a source of health information through wastewater-based epidemiology, and implementing treatment processes to extract valuable components like precious metals or produce goods like biofuel for a circular economy (Villarín and Merel 2020). Despite being sophisticated, cost limits hinder their true impact in specific industries. Their significance remains for future studies to reduce the cost of EP treatment methods.

4.7 Conclusions and Future Perspectives

Humans and the environment are exposed to various EPs via numerous known and undiscovered pathways. These pollutants provide ongoing and severe threats to soil, water, air, and ecosystems, especially human health. Additionally, EPs sometimes surpass the capabilities of safety measures, risk assessment methodologies, monitoring techniques, and existing preventive measures. To evaluate the impact of pollutant exposure on human health under present circumstances, it is essential to consider several interactive factors. Removing EPs relies on upcoming technology and research to develop comprehensive degradation techniques. Current research and studies indicate that the EPs may have mutagenic, teratogenic, and carcinogenic effects on humans and other animals. Insufficient and extensive research has not shown causal links between EPs and negative consequences on the human body. Nevertheless, the negative consequences shown in animal trials should not be neglected. Extensive investigations with enhanced methodologies are required to provide more compelling and unambiguous results.

The themes and ideas highlighted need more investigation in the bibliographic evaluation should be investigated. The impact of EDCs and antibiotics on personal care items as EPs deserves more investigation. Organic micropollutants, pesticides, and PFCs in water and wastewater must be evaluated. Global monitoring of EPs is infrequent, necessitating research to analyze their current levels, movement, and possible changes to comprehend and control their movement within the hydrological system. Identifying potential tracers and indicators of EPs is essential for analyzing their dangers and determining economic feasibility. Understanding the characteristics of EPs at their origin and their appropriate disposal is crucial in reducing the threat of contaminating surface and subterranean water. Exploring comprehensive solutions for removing related EPs and studying the impact of additional EPs, affinity, and synergistic behavior are crucial topics that need more investigation.

There is a crucial need to improve the efficacy of treatment technologies to remove EPs from industrial and other kinds of wastewater. Various treatment procedures, including physical–chemical, biological, and advanced oxidation processes, are accessible. These procedures have constraints, including economic feasibility, environmental friendliness, the increasing number of discovered EPs, and the potential for intermediate compounds to be more harmful than the original compounds. Utilizing biochar as an adsorbent for removing EPs is a cost-effective and environmentally benign approach worth considering for large-scale operations, although more study may be necessary. Biological processes may be environmentally beneficial and economically viable, but they have constraints such as climatic conditions, pollutant load, and the type of pollutant microbial acclimations. Approximately 3000 chemicals are evaluated and used as pharmaceutical components, such as antibiotics, antidiabetic agents, contraceptives, antidepressants, analgesics, beta-blockers, lipid regulators, and medicines for erectile dysfunction. Only a limited portion of these EPs and their influence on the aquatic environment has been thoroughly researched. The widespread use of medications and personal care items has increased their concentration in metropolitan regions' surface water, groundwater, wastewater, and stormwater run-off.

Research has been conducted on the dangers of EPs on the environment and human health. Yet, there needs to be more comprehension of the enduring impacts of EPs. There needs to be more and more information on the effects of EPs on human health and aquatic life, highlighting the need for more research and understanding. Because EPs are rapidly evolving in the environment, it is challenging to establish cost-effective and sustainable programs for detecting, assessing risks, and removing all these components. EPs have become a difficulty in sustainable water management due to the exacerbation of water source challenges by climate change and population increase. The main issue with these pollutants is their small size, which makes them resistant to removal using traditional water treatment methods. WWTPs are crucial in removing contaminants before they are released into the river system. However, certain substances exhibit high persistence in water and are resistant to fast removal or biodegradation. Pollutant residues will disperse in the atmosphere and pollute drinking water sources. Future research should focus on contaminants with the most significant effect on human health and the aquatic environment. This will enable coordinated research efforts to reduce pollution using the best resources possible.

References

Aalizadeh R, Nika M-C, Thomaidis NS (2019) Development and application of retention time prediction models in the suspect and non-target screening of emerging contaminants. J Hazard Mater 363:277–285. https://doi.org/10.1016/j.jhazmat.2018.09.047

Abdallah MA-E, Bressi M, Oluseyi T, Harrad S (2016) Hexabromocyclododecane and tetrabromobisphenol-A in indoor dust from France, Kazakhstan and Nigeria: implications for human exposure. Emerg Contam 2(2):73–79. https://doi.org/10.1016/j.emcon.2016.03.006

Abel PD (1974) Toxicity of synthetic detergents to fish and aquatic invertebrates. J Fish Biol 6(3):279–298. https://doi.org/10.1111/j.1095-8649.1974.tb04545.x

Ahmed SF, Mofijur M, Nuzhat S, Chowdhury AT, Rafa N, Uddin MA, Inayat A, Mahlia TMI, Ong HC, Chia WY, Show PL (2021) Recent developments in physical, biological, chemical, and hybrid treatment techniques for removing emerging contaminants from wastewater. J Hazard Mater 416:125912. https://doi.org/10.1016/j.jhazmat.2021.125912

Ahrens L (2011) Polyfluoroalkyl compounds in the aquatic environment: a review of their occurrence and fate. J Environ Monit 13(1):20–31. https://doi.org/10.1039/C0EM00373E

Al-Mousa F, Michelangeli F (2012) Some commonly used brominated flame retardants cause Ca2+-ATPase inhibition, beta-amyloid peptide release and apoptosis in SH-SY5Y neuronal cells. PLoS One 7(4):e33059. https://doi.org/10.1371/journal.pone.0033059

Arman NZ, Salmiati S, Aris A, Salim MR, Nazifa TH, Muhamad MS, Marpongahtun M (2021) A review on emerging pollutants in the water environment: existences, health effects and treatment processes. Water 13(22):3258. https://doi.org/10.3390/w13223258

Asok AK, Jisha MS (2012) Assessment of soil microbial toxicity on acute exposure of the anionic surfactant linear alkylbenzene sulphonate. J Environ Sci Technol 5(5):354–363. https://doi.org/10.3923/jest.2012.354.363

Bay K, Asklund C, Skakkebaek NE, Andersson A-M (2006) Testicular dysgenesis syndrome: possible role of endocrine disrupters. Best Pract Res Clin Endocrinol Metab 20(1):77–90. https://doi.org/10.1016/j.beem.2005.09.004

Bjermo H, Aune M, Cantillana T, Glynn A, Lind PM, Ridefelt P, Darnerud PO (2017) Serum levels of brominated flame retardants (BFRs: PBDE, HBCD) and influence of dietary factors in a population-based study on Swedish adults. Chemosphere 167:485–491. https://doi.org/10.1016/j.chemosphere.2016.10.008

Botelho GGK, Golin M, Bufalo AC, Morais RN, Dalsenter PR, Martino-Andrade AJ (2009) Reproductive effects of Di(2-ethylhexyl)phthalate in immature male rats and its relation to cholesterol, testosterone, and thyroxin levels. Arch Environ Contam Toxicol 57(4):777–784. https://doi.org/10.1007/s00244-009-9317-8

Brandt KK, Hesselsøe M, Roslev P, Henriksen K, Sørensen J (2001) Toxic effects of linear alkylbenzene sulfonate on metabolic activity, growth rate, and microcolony formation of Nitrosomonas and Nitrosospira strains. Appl Environ Microbiol 67(6):2489–2498. https://doi.org/10.1128/AEM.67.6.2489-2498.2001

Brunner P, Simmons CT (2012) HydroGeoSphere: a fully integrated, physically based hydrological model. Groundwater 50(2):170–176. https://doi.org/10.1111/j.1745-6584.2011.00882.x

Casals-Casas C, Feige JN, Desvergne B (2008) Interference of pollutants with PPARs: endocrine disruption meets metabolism. Int J Obes 32(S6):S53–S61. https://doi.org/10.1038/ijo.2008.207

Cemre Birben N, Bekbolet M (2019) Role of emerging contaminants on solar photocatalytic treatment of organic matter in reverse osmosis concentrate. Catal Today 326:101–107. https://doi.org/10.1016/j.cattod.2018.10.048

Costa EMF, Spritzer PM, Hohl A, Bachega TASS (2014) Effects of endocrine disruptors in the development of the female reproductive tract. Arq Bras Endocrinol Metabol 58(2):153–161. https://doi.org/10.1590/0004-2730000003031

Deschênes L, Lafrance P, Villeneuve J-P, Samson R (1996) Adding sodium dodecyl sulfate and Pseudomonas aeruginosa UG2 biosurfactants inhibits polycyclic aromatic hydrocarbon biodegradation in a weathered creosote-contaminated soil. Appl Microbiol Biotechnol 46(5–6):638–646. https://doi.org/10.1007/s002530050874

Dhangar K, Kumar M (2020) Tricks and tracks in removal of emerging contaminants from the wastewater through hybrid treatment systems: a review. Sci Total Environ 738:140320. https://doi.org/10.1016/j.scitotenv.2020.140320

Diamanti-Kandarakis E, Bourguignon J-P, Giudice LC, Hauser R, Prins GS, Soto AM, Zoeller RT, Gore AC (2009) Endocrine-disrupting chemicals: an endocrine society scientific statement. Endocr Rev 30(4):293–342. https://doi.org/10.1210/er.2009-0002

Dreher KL (2004) Health and environmental impact of nanotechnology: toxicological assessment of manufactured nanoparticles. Toxicol Sci 77(1):3–5. https://doi.org/10.1093/toxsci/kfh041

Ebele AJ, Abou-Elwafa Abdallah M, Harrad S (2017) Pharmaceuticals and personal care products (PPCPs) in the freshwater aquatic environment. Emerg Contam 3(1):1–16. https://doi.org/10.1016/j.emcon.2016.12.004

la Farré M, Pérez S, Kantiani L, Barceló D (2008) Fate and toxicity of emerging pollutants, their metabolites and transformation products in the aquatic environment. TrAC Trends Anal Chem 27(11):991–1007. https://doi.org/10.1016/j.trac.2008.09.010

Fei C, McLaughlin JK, Lipworth L, Olsen J (2009) Maternal levels of perfluorinated chemicals and subfecundity. Hum Reprod 24(5):1200–1205. https://doi.org/10.1093/humrep/den490

Fei C, McLaughlin JK, Lipworth L, Olsen J (2010) Maternal concentrations of perfluorooctanesulfonate (PFOS) and perfluorooctanoate (PFOA) and duration of breastfeeding. Scand J Work Environ Health 36(5):413–421. https://doi.org/10.5271/sjweh.2908

van Ginkel CG (1996) Complete degradation of xenobiotic surfactants by consortia of aerobic microorganisms. Biodegradation 7(2):151–164. https://doi.org/10.1007/BF00114627

Giuliani A, Zuccarini M, Cichelli A, Khan H, Reale M (2020) Critical review on the presence of phthalates in food and evidence of their biological impact. Int J Environ Res Public Health 17(16):5655. https://doi.org/10.3390/ijerph17165655

Gore A (2010) Neuroendocrine targets of endocrine disruptors. Hormones 9(1):16–27. https://doi.org/10.14310/horm.2002.1249

Gunnarsson L, Snape JR, Verbruggen B, Owen SF, Kristiansson E, Margiotta- Casaluci L, Österlund T, Hutchinson K, Leverett D, Marks B, Tyler CR (2019) Pharmacology beyond the patient—the environmental risks of human drugs. Environ Int 129:320–332. https://doi.org/10.1016/j.envint.2019.04.075

Hinkson NC, Whalen MM (2009) Hexabromocyclododecane decreases the lytic function and ATP levels of human natural killer cells. J Appl Toxicol 29(8):656–661. https://doi.org/10.1002/jat.1453

Huber WW, Grasl-kraupp B, Schulte-hermann R (1996) Hepatocarcinogenic potential of Di(2-Ethylhexyl)phthalate in rodents and its implications on human risk. Crit Rev Toxicol 26(4):365–481. https://doi.org/10.3109/10408449609048302

Jahan K, Ordóñez R, Ramachandran R, Balzer S, Stern M (2008) Modeling biodegradation of nonylphenol. Water Air Soil Pollut Focus 8(3–4):395–404. https://doi.org/10.1007/s11267-007-9148-4

Jeong T-Y, Yoon D, Kim S, Kim HY, Kim SD (2018) Mode of action characterization for adverse effect of propranolol in Daphnia magna based on behavior and physiology monitoring and metabolite profiling. Environ Pollut 233:99–108. https://doi.org/10.1016/j.envpol.2017.10.043

Jiménez-Díaz I, Vela-Soria F, Rodríguez-Gómez R, Zafra-Gómez A, Ballesteros O, Navalón A (2015) Analytical methods for the assessment of endocrine disrupting chemical exposure during human fetal and lactation stages: a review. Anal Chim Acta 892:27–48. https://doi.org/10.1016/j.aca.2015.08.008

Joensen UN, Bossi R, Leffers H, Jensen AA, Skakkebæk NE, Jørgensen N (2009) Do perfluoroalkyl compounds impair human semen quality? Environ Health Perspect 117(6):923–927. https://doi.org/10.1289/ehp.0800517

Jovanic BR, Bojovic S, Panic B, Radenkovic B, Despotovic M (2010) The effect of detergent as polluting agent on the photosynthetic activity and chlorophyll content in bean leaves. Health 02(05):395–399. https://doi.org/10.4236/health.2010.25059

Ju X, Jin Y, Sasaki K, Saito N (2008) Perfluorinated surfactants in surface, subsurface water and microlayer from Dalian Coastal Waters in China. Environ Sci Technol 42(10):3538–3542. https://doi.org/10.1021/es703006d

Kabir ER, Rahman MS, Rahman I (2015) A review on endocrine disruptors and their possible impacts on human health. Environ Toxicol Pharmacol 40(1):241–258. https://doi.org/10.1016/j.etap.2015.06.009

Khan HK, Rehman MYA, Malik RN (2020) Fate and toxicity of pharmaceuticals in water environment: an insight on their occurrence in South Asia. J Environ Manag 271(June):111030. https://doi.org/10.1016/j.jenvman.2020.111030

Kim YR, Harden FA, Toms L-ML, Norman RE (2014) Health consequences of exposure to brominated flame retardants: a systematic review. Chemosphere 106:1–19. https://doi.org/10.1016/j.chemosphere.2013.12.064

Krauss M, Hug C, Bloch R, Schulze T, Brack W (2019) Prioritising site-specific micropollutants in surface water from LC-HRMS non-target screening data using a rarity score. Environ Sci Eur 31(1). https://doi.org/10.1186/s12302-019-0231-z

Kristensen DM, Desdoits-Lethimonier C, Mackey AL, Dalgaard MD, De Masi F, Munkbøl CH, Styrishave B, Antignac J-P, Le Bizec B, Platel C, Hay-Schmidt A, Jensen TK, Lesné L, Mazaud-Guittot S, Kristiansen K, Brunak S, Kjaer M, Juul A, Jégou B (2018) Ibuprofen alters human testicular physiology to produce a state of compensated hypogonadism. Proc Natl Acad Sci 115(4):E715–E724. https://doi.org/10.1073/pnas.1715035115

Lapworth DJ, Baran N, Stuart ME, Ward RS (2012) Emerging organic contaminants in groundwater: a review of sources, fate and occurrence. Environ Pollut 163:287–303. https://doi.org/10.1016/j.envpol.2011.12.034

Lei M, Zhang L, Lei J, Zong L, Li J, Wu Z, Wang Z (2015) Overview of emerging contaminants and associated human health effects. Biomed Res Int 2015:1–12. https://doi.org/10.1155/2015/404796

López-Carrillo L, Hernández-Ramírez RU, Calafat AM, Torres-Sánchez L, Galván-Portillo M, Needham LL, Ruiz-Ramos R, Cebrián ME (2010) Exposure to phthalates and breast cancer risk in northern Mexico. Environ Health Perspect 118(4):539–544. https://doi.org/10.1289/ehp.0901091

Majumdar A, Pal A (2020) Recent advancements in visible-light-assisted photocatalytic removal of aqueous pharmaceutical pollutants. Clean Techn Environ Policy 22(1):11–42. https://doi.org/10.1007/s10098-019-01766-1

Marrakchi S, Maibach HI (2006) Sodium lauryl sulfate-induced irritation in the human face: regional and age-related differences. Skin Pharmacol Physiol 19(3):177–180. https://doi.org/10.1159/000093112

Mellanen P, Petänen T, Lehtimäki J, Mäkelä S, Bylund G, Holmbom B, Mannila E, Oikari A, Santti R (1996) Wood-derived Estrogens: studies in vitro with breast cancer cell lines and in vivo in trout. Toxicol Appl Pharmacol 136(2):381–388. https://doi.org/10.1006/taap.1996.0046

Miralles-Marco A, Harrad S (2015) Perfluorooctane sulfonate: a review of human exposure, biomonitoring and the environmental forensics utility of its chirality and isomer distribution. Environ Int 77:148–159. https://doi.org/10.1016/j.envint.2015.02.002

Noguera-Oviedo K, Aga DS (2016) Lessons learned from more than two decades of research on emerging contaminants in the environment. J Hazard Mater 316:242–251. https://doi.org/10.1016/j.jhazmat.2016.04.058

Pal D, Sen S (2023a) Emerging petroleum pollutants and their adverse effects on the environment. In: Impact of petroleum waste on environmental pollution and its sustainable management through circular economy. Springer, Cham, pp 103–137. https://doi.org/10.1007/978-3-031-48220-5_5

Pal D, Sen S (2023b) In-depth coverage of petroleum waste sources, characteristics, environmental impact, and sustainable remediation process. In: Impact of petroleum waste on environmental pollution and its sustainable management through circular economy. Springer, Cham, pp 1–38. https://doi.org/10.1007/978-3-031-48220-5_1

Pelucchi C, Pira E, Piolatto G, Coggiola M, Carta P, La Vecchia C (2006) Occupational silica exposure and lung cancer risk: a review of epidemiological studies 1996–2005. Ann Oncol 17(7):1039–1050. https://doi.org/10.1093/annonc/mdj125

Qu X, Alvarez PJJ, Li Q (2013) Applications of nanotechnology in water and wastewater treatment. Water Res 47(12):3931–3946. https://doi.org/10.1016/j.watres.2012.09.058

Rebello S, Asok AK, Mundayoor S, Jisha MS (2014) Surfactants: toxicity, remediation and green surfactants. Environ Chem Lett 12(2):275–287. https://doi.org/10.1007/s10311-014-0466-2

Rodrigues S, Antunes SC, Nunes B, Correia AT (2019) Histopathological effects in gills and liver of Sparus aurata following acute and chronic exposures to erythromycin and oxytetracycline. Environ Sci Pollut Res 26(15):15481–15495. https://doi.org/10.1007/s11356-019-04954-0

Rodriguez-Narvaez OM, Peralta-Hernandez JM, Goonetilleke A, Bandala ER (2017) Treatment technologies for emerging contaminants in water: a review. Chem Engineer J 323:361–380. https://doi.org/10.1016/j.cej.2017.04.106

Romanelli M, Moraes MC, Villavicencio ALC, Borrely S (2004) Evaluation of toxicity reduction of sodium dodecyl sulfate submitted to electron beam radiation. Radiat Phys Chem *71*(1–2):411–413. https://doi.org/10.1016/j.radphyschem.2004.03.038

Rosety-Rodríguez M, Ordoñez F, Roldan S, Rosety J, Rosety M, Ribelles A, Carrasco C, Rosety I (2009) Acute effects of sodium dodecyl sulphate on the survival and on morpho-histochemical characteristics of the trunk kidney of juvenile turbot Scophthalmus maximus L. Eur J Histochem 46(2):179. https://doi.org/10.4081/1668

Sargis RM, Simmons RA (2019) Environmental neglect: endocrine disruptors as underappreciated but potentially modifiable diabetes risk factors. Diabetologia 62(10):1811–1822. https://doi.org/10.1007/s00125-019-4940-z

Schilling K, Bradford B, Castelli D, Dufour E, Nash JF, Pape W, Schulte S, Tooley I, van den Bosch J, Schellauf F (2010) Human safety review of "nano" titanium dioxide and zinc oxide. Photochem Photobiol Sci 9(4):495–509. https://doi.org/10.1039/b9pp00180h

Shi X, Du Y, Lam PKS, Wu RSS, Zhou B (2008) Developmental toxicity and alteration of gene expression in zebrafish embryos exposed to PFOS. Toxicol Appl Pharmacol 230(1):23–32. https://doi.org/10.1016/j.taap.2008.01.043

Sivaranjanee R, Kumar PS (2021) A review on remedial measures for effective separation of emerging contaminants from wastewater. Environ Technol Innov 23:101741. https://doi.org/10.1016/j.eti.2021.101741

Stuart M, Lapworth D, Crane E, Hart A (2012) Review of risk from potential emerging contaminants in UK groundwater. Sci Total Environ 416:1–21. https://doi.org/10.1016/j.scitotenv.2011.11.072

Thomaidis NS, Asimakopoulos AG, Bletsou AA (2013) Emerging contaminants: a tutorial mini-review. Global NEST J 14(1):72–79. https://doi.org/10.30955/gnj.000823

Tomy GT, Budakowski W, Halldorson T, Whittle DM, Keir MJ, Marvin C, MacInnis G, Alaee M (2004) Biomagnification of α- and γ- hexabromocyclododecane isomers in a Lake Ontario Food Web. Environ Sci Technol 38(8):2298–2303. https://doi.org/10.1021/es034968h

Toolaram AP, Kümmerer K, Schneider M (2014) Environmental risk assessment of anti-cancer drugs and their transformation products: a focus on their genotoxicity characterization-state of knowledge and short comings. Mutat Res Rev Mutat Res 760:18–35. https://doi.org/10.1016/j.mrrev.2014.02.001

Van Boeckel TP, Brower C, Gilbert M, Grenfell BT, Levin SA, Robinson TP, Teillant A, Laxminarayan R (2015) Global trends in antimicrobial use in food animals. Proc Natl Acad Sci 112(18):5649–5654. https://doi.org/10.1073/pnas.1503141112

Villarín MC, Merel S (2020) Paradigm shifts and current challenges in wastewater management. J Hazard Mater 390:122139. https://doi.org/10.1016/j.jhazmat.2020.122139

Waaijers SL, Parsons JR (2016) Biodegradation of brominated and organophosphorus flame retardants. Curr Opin Biotechnol 38:14–23. https://doi.org/10.1016/j.copbio.2015.12.005

Wang Y, Lu J, Mao L, Li J, Yuan Z, Bond PL, Guo J (2019) Antiepileptic drug carbamazepine promotes horizontal transfer of plasmid-borne multi-antibiotic resistance genes within and across bacterial genera. ISME J 13(2):509–522. https://doi.org/10.1038/s41396-018-0275-x

Wilkinson J, Hooda PS, Barker J, Barton S, Swinden J (2017) Occurrence, fate and transformation of emerging contaminants in water: an overarching review of the field. Environ Pollut 231:954–970. https://doi.org/10.1016/j.envpol.2017.08.032

Windels EM, Michiels JE, Fauvart M, Wenseleers T, Van den Bergh B, Michiels J (2019) Bacterial persistence promotes the evolution of antibiotic resistance by increasing survival and mutation rates. ISME J 13(5):1239–1251. https://doi.org/10.1038/s41396-019-0344-9

de Wit CA, Herzke D, Vorkamp K (2010) Brominated flame retardants in the Arctic environment — trends and new candidates. Sci Total Environ 408(15):2885–2918. https://doi.org/10.1016/j.scitotenv.2009.08.037

Yadav D, Rangabhashiyam S, Verma P, Singh P, Devi P, Kumar P, Hussain CM, Gaurav GK, Kumar KS (2021) Environmental and health impacts of contaminants of emerging concerns: recent treatment challenges and approaches. Chemosphere 272:129492. https://doi.org/10.1016/j.chemosphere.2020.129492

Zota AR, Mitro SD, Robinson JF, Hamilton EG, Park J-S, Parry E, Zoeller RT, Woodruff TJ (2018) Polybrominated diphenyl ethers (PBDEs) and hydroxylated PBDE metabolites (OH-PBDEs) in maternal and fetal tissues, and associations with fetal cytochrome P450 gene expression. Environ Int 112:269–278. https://doi.org/10.1016/j.envint.2017.12.030

Chapter 5
Sustainable and Resilient Urban Development Using Environmental Planning Tools

Shreya Proch, Vrinda Sharma, and Mandeep Kour

Abstract Sustainable urban development indicates an approach in which sustainability is bolstered alongside progress, development, and urban transformation. The urban and emerging sub-urban areas have been witnessing the establishment of human settlements on the fringes where the adaptive capacity is minuscule, and statutory and insurance protections are fragmented, thus hampering the competence of the population to generate capital assets. Sustainability with regard to urban development is, hence, essential for the viability of urban and semi-urban cities. Environmental planning is a burgeoning field, in which the extenuation of the deleterious effects of urban developmental projects on human settlements, natural resources, and biodiversity niches such as wildlife sanctuaries, biodiversity hotspots, national parks, community reserves, and conservation reserves is envisaged. Urban resilience or sustainability in urban development can be attained using environmental planning, which is the cooperative tendency of scaling, modulating, and moderating the unbridled use of biodiversity with the involvement of all the stakeholders as government, sharecroppers, farmers, communities, tribal cohorts, landholders, and the indigenous people. This planning can be sought with the help of certain tools known as 'Environmental Planning Tools'. Environmental planning tools can go a long way in fostering a continuum of project planning and evaluation prior to its implementation, owing to which the inimical threats of these urbanized and commercial developmental projects on biodiversity can be diluted if not abated. The present communication is an endeavour to explore the various types of environmental planning tools and their role in ensuring sustainable urban designing and development.

S. Proch (✉) · V. Sharma
Department of Botany, School of Life Sciences, Central University of Jammu, District Samba, Jammu and Kashmir, India

M. Kour
Division of Plant Physiology, Sher-e-Kashmir University of Agricultural Sciences and Technology of Jammu, District Jammu, Jammu and Kashmir, India

S. Dey, S. Bhattacharya (eds.), *Biotechnological Interventions in the Removal of Emerging Pollutants*, Interdisciplinary Biotechnological Advances,
https://doi.org/10.1007/978-981-97-9922-0_5

Keywords Environmental planning tools · Sustainable · Urban development · Biodiversity

5.1 Introduction

Since the epoch of troglodytes, urban areas and cities have remained the boulders of economic development, technological upliftment, and prosperity of human communities (Mumford 1961). As per certain dynamics computed by the Global Health Observatory, which is a public observatory established by the World Health Organization, more than 50% of the world's population has been residing in urban settlements, as up till 2010. It was also projected by extrapolations that by 2030 six of every ten persons would be dwelling in urban settlements and by 2050, it is further forecasted that every six out of ten individuals will dwell in the metropolis and the megapolis by 2030. This proportion will escalate to seven out of 10 by 2050 (WHO 2010).

The burgeoning metropolitan and the Oppidan areas have been central to the establishment of residential dwellings along the peripheral limits of these areas. These outskirts have limited adaptability, flexibility and resilience. These areas also lack cohesive insurance and regulatory safeguards, thus undermining the potential of the residing communities to generate possessions for the individuals as well as the community (Allenby and Fink 2005). In conjunction with this, the population densities in urban areas have been perpetually on the rise, leading to congestion and consequently escalating the pressure on CPR (Common Property Resources). The commercial and remunerative activities in the downtown posh areas release about 70% of Greenhouse gases into the atmosphere (UN Human Settlement Programme 2011).

Climate change has been affecting human health, atmospheric balance, and soil potency in a myriad and interrelated ways. According to the Global Climate Risk Index, Germanwatch, 2021, the human and economic losses to the world in the form of fatalities per 100,000 inhabitants and losses per unit GDP (%) due to climate change in 2019 have been perturbing and alarming as shown in Fig. 5.1. This climatic catastrophe has ultimately resulted in the worsening of the quality of life of people across different regions of the globe as sketched by the respective HDI (Human Development Index) of the countries as shown in Fig. 5.2.

The rampant and indiscriminate cashing in of biodiversity to meet the commercial ends by humans in the prevalent epoch has invoked an immediate overhaul of the extant utilization regimens of biodiversity and its products (Tilman et al. 1996; Tilman and Downing 1994). It has also been collated that the functions performed by the ecosystem and the services tendered by the same as nutrient recycling, biomass production, energy rationalization, biogeochemical recycling, hydrological balance, etc. are very sensitive to the variations that can erupt in biological diversity (Hector et al. 1999). The conservation statistics and metrics reveal that the current dynamics of species are around 5–15 million (Kearns 2010). Out of these species,

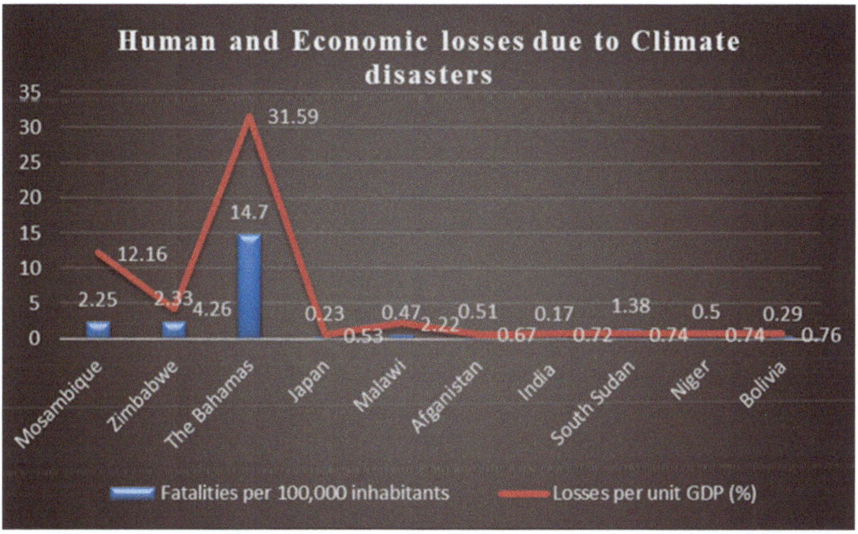

Fig. 5.1 Human and economic losses due to climate disasters in 2019. (Source: Global Climate Risk Index, Germanwatch, 2021)

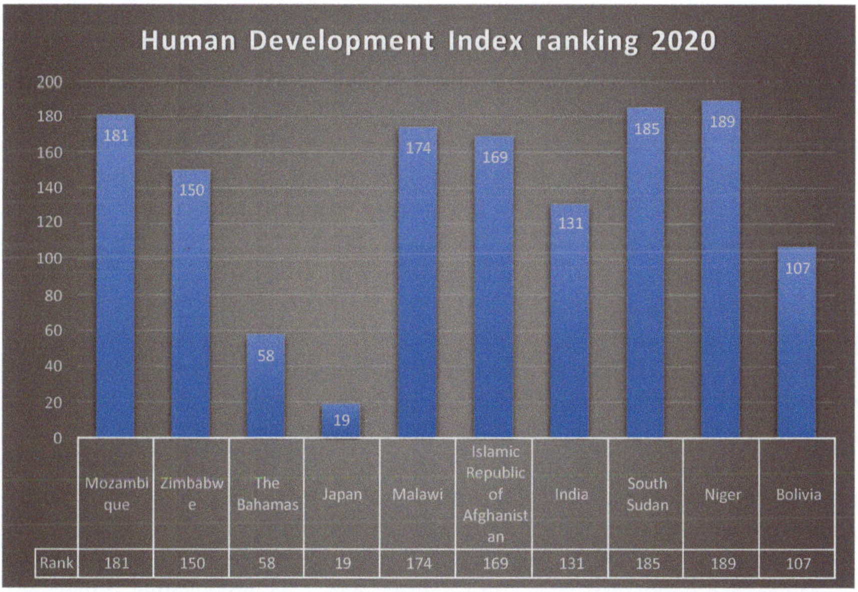

Fig. 5.2 Human Development Index of 10 most affected countries in 201. (Source: Global Climate Risk Index, Germanwatch, 2021)

only 1.8 million species have been registered and documented by the research community (Vié et al. 2009). The rate of addition of new species by discovery every year is around 15,000 (Dirzo and Raven 2003). However, the extinction rates pose a grave threat as they amount to be 100–1000 times greater than background extinction rates (Hambler 2004). It has also been deduced that the pace of speciation has been laggard when collated to the tempo of extinction process and the lag is computed to be around 100 years (Wilson 2002). In this backdrop, the pertinence of urban sustainability using environmental planning gains merit.

Sustainable urban development is a proverbial acclaim which means savouring the services of urban upgradation at a rate such that it complies with the rate at which these services are being replenished in order to certify that the needs of future as well as present generations are not compromised. This concept, not only inculcates the dimensions of biodiversity and resource preservation along with modern development, but it also suggests the significance of a resource sensitive metropolis in vitalizing the life of its residents (Tang and Lee 2016). Urban sustainability upgrades the socio-economic and cultural ethos while preserving environmental limits and upholding fairness within the current generation (Talukdar and Hossain 2011).

Environmental planning can be leveraged for bringing about urban resilience or sustainability, in which there is a balanced synthesis of factors like fiscal advancement, environmental integrity, and an egalitarian society in urban and sub-urban areas (Talukdar and Hossain 2011). It has now been believed that environmental planning has the potential to play a very determining task in bringing forth sustainable urbanization and resource use optimization. This is an emerging field in sustainable bio resources conservation protocols, wherein various environmental planning tools are leveraged to evaluate, screen, scope, and draw mitigations and solutions for the amelioration of the biodiversity utilization projects and endeavours (Sivaramanan 2015).

Planning tools are basically the interrogative or probing tools that can be leveraged by the planning organizations and firms to fulfil the task of planning, designing and drafting. The various components encapsulated in the process of planning are multi-dimensional, diverse, and notable. This includes earmarking and analysis of space related variables, referencing and analysis of process related factors, for instance, resolution in multi-stakeholder domains (Runhaar et al. 2009).

These tools used in the process of environmental planning are basically the means which aid in the task of decision-making pertaining to the environment related aspects and help in analysing and interpreting the knowledge on various environmental issues. The decision which is made in this regard is basically strategic rather than normative (Dale and English 1999). The broad ambit of tools also covers the processes, analytic strategies, methods, and concepts used for environmental amelioration and biodiversity conservation.

The practicality and rationality of tools is very subjective and open to interpretations. The rationality of its usage depends on the premise or the context in which it is being used. The correlation between the tools and the pretext of its requisition is a point of contention in modern day research and must be considered as a very important variable in the fabrication of tools for environmental planning (Dale and English 1999). Another pertinent dimension of the immersive conditions related to the tool usage is the extent and the degree up to which the tools can be brought into operation. It also includes the regulation of the tool usage with the help of legal tools and instruments or some other benchmarked or referenced (Fischer 2003). Different types of environmental planning tools have been in use across the different regions of the globe in order to usher in environmental sustainability in urban planning and development as catalogued in Table 5.1.

Table 5.1 Use of environmental planning tools in bringing viability and sustainability across different areas of the world

S. no.	Case study (place)	Findings	References
1.	Arabian Gulf	EIA used to alleviate the environmental impacts of saltwater desalination plant	Dawoud (2012)
2.	Salzburg province of Austria.	SEA used to develop an integrated waste management plan	Salhofer et al. (2007)
3.	Europe	TIA used in the sustainable transformation of Trans-European Networks Policy	Camagni (2009)
4.	Netherlands	Definite demarcation of environmental targets in the spatiotemporal plans of urban development	Runhaar and Driessen (2007)
5.	Rijnmond, Netherlands	LOGO (Hybrid tool) used in bringing sustainability in the various environmental indicators of the region	De Roo and Visser (2016)
6.	China	CDM proved to be a game-changer in the furtherance of renewable energy deployment.	Wang and Chen (2010)
7.	Iraq	The townified urban communities' wastewater treatment methods' viability was assessed using MCDA.	Omran et al. (2021)
8	Southern Italy (Province Naples)	Maps for priority areas for soil-protection interventions and landscape protection were provided.	Geneletti et al. (2007)
9	Greece (Kalloni, Lesbos)	The methodology developed for evaluating ecosystem services and the involvement of social actors in the administrative process contributed to addressing the deficiencies of the prior conservation planning procedure in Kalloni.	Oikonomou et al. (2011)
10	Vilnius city, Lithuania	Principles of Sustainable transport have been put in place which can lead the cities towards a well-planned and integrated transportation matrix.	Ogryzek et al. (2020)
11.	Halifax, Canada	The townified urban communities' wastewater treatment methods' viability was assessed using MCDA.	Lobaccaro et al. (2019)
12.	Renton, Washington	Establishment of a 'Metropolitan place' using the tool of Transit Transit-oriented development	Talukdar and Hossain (2011)

5.2 Types of Environmental Planning Tools

Tools can be categorized on the basis of their usefulness and adequacy. First set of tools pertains to the planning, strategizing, and overall management of the public sector. This spectrum of tools broadly revolves around participatory ideology. The second set of tools stems from a keen exploration of environmental systems and equations and is inclusive of umpteen number of expert tools and techniques, for instance, LCA (life cycle assessment), SFA (substance flow analysis), and MCA (multi-Criteria analysis). The third group of environmental planning and management tools deals with the amalgamative and cumulative advances made by the corporate sector into the aegis of public sector, thus providing conjugations to the non-conventional or the market regulating mechanisms (Dale and English 1999).

A varied number of tools, for instance, substance-oriented planning tools which include EIA (Environmental Impact Assessment), TIA (Territorial Impact Assessment), and SEA (Strategic Environmental Assessment); process-oriented planning tools like CDM (Clean Development Mechanism), expert methodologies, and Hybrid Planning Tools such as MCDA (Multi-criteria decision analysis), and LOGO (Local area typology and environmental quality), etc. can be used to draw parallels, analyse, and draw solutions to protect the ecosystems niches and fledge the ecosystem biodiversity (Finnveden et al. 2003; Runhaar et al. 2009; Uhde et al. 2015) as shown in Fig. 5.3.

5.2.1 Substance-Oriented Planning Tools

These types of planning tools examine the state of environment using the knowledge of certain parameters, for instance, certain indicators, 'Green decision making', GIS (Geographic Information System), Planning Support Systems (PSSs), which are inclusive of Information Technology-based geoinformation technologies, aimed at envisioning external environmental conditions and their associated effects. The indicators which are employed in these planning tools are pre-cogitated and empirically more reliable (Runhaar et al. 2009; Cerreta and Monno 2010).

5.2.1.1 EIA (Environmental Impact Assessment)

The inception, continuation, and ecesis of this environmental planning tool stemmed out of the perturbation caused due to the alterations taking place in the environment as a consequence of unregulated anthropogenic developmental activities. These alterations were the direct consequences of the economic, geopolitical, social, and commercial changes taking place in the late nineteenth century. The thrust for this formulation of EIA was laid down by the discontent and concern arising out of the

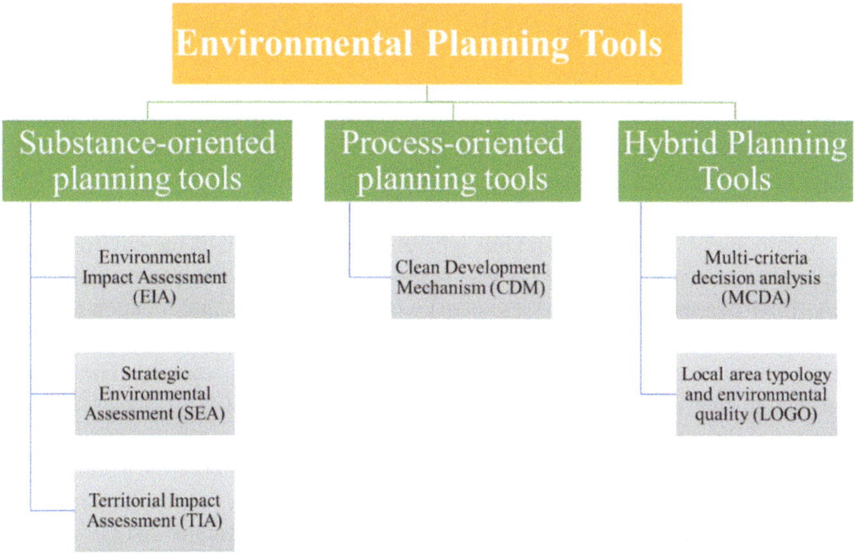

Fig. 5.3 Classification of environmental planning tools

failure of conventional tools and techniques which were in use for environmental decision-making. This was further aided by the participation and representation deficit of the public in biodiversity-specific environmental planning and decision-making. Thus, operationalizing of EIA puts forth a vibrant, viable, reproducible proposition to environmental planning and environment-sensitive decision-making (Weston 2000). The criticism of the rationalist planning and decision-making model, and by extension of mainstream EIA, is a recurring theme in discussions. Also, there is a need to investigate and create models that incorporate novel perspectives on planning and decision-making procedures within their broader social, cultural, political, and economic contexts (Morgan 2012).

The established protocol for EIA (Environmental Impact Assessment) includes the stages of Screening, Scoping, Baseline data collection, Prediction, Assessment, Mitigation, Decision Monitoring, and Auditing. Basically, the process strives to achieve the objective of informing the proponents of the project under consideration about the due impacts of the project on the environmental variables and further suggest the ways to reform and ameliorate the repercussions (Weston 2000).

There are however certain issues that remain inept in the entire protocol of EIA. The prime kingpin of EIA, since its inception, has been mostly on standalone projects (Partidário 1999). The process had been very rigorous, time-consuming, administratively tiring, and monetarily draining in some countries due to which certain flexibilities were introduced in the protocol (Wood 1995). In conjunction to this, Heuvelhof and Nauta (1996) have also pointed out that although the functionality of EIA in decisions pertaining to environmental issues is very significant and conclusive, yet in practice they are employed at a very downstream stage of the entire process due to which the proficiency of decision-making becomes feeble (Hildén et al. 2004).

Stalwart global organizations as the World Bank and UNEP have also noted the implications of EIA as in 1989 World Bank notified that EIA ought to be carried out by the borrower country for major projects. UNEP (United Nations Environment Program) has also put forth certain directives to its member countries relating to the proper framing and implementation of EIA (UNEP 1988).

A case study of the Arabian Gulf wherein the subsidence of the environmental impacts of seawater desalination was registered using this environmental tool of EIA (Environmental Impact Assessment) which was project- and location-specific. Its detrimental environmental impacts include the chemical discharge into the seawater, intensive energy requirements, and the air pollutants released. Thus, a risk assessment matrix composed of the variables, i.e., probability of occurrence and severity of the effect was prepared (Dawoud 2012).

The policy direction defined by individuals with a special curiosity in the continuous mishandling of natural resources is what accounts for the EIA system's flaws rather than oversights or shortcomings in judgment (Morgan 2012).

5.2.1.2 SEA (Strategic Environmental Assessments)

Strategic environmental assessments may be defined as the accredited, extensive, vivid process in which the environmental repercussions of a policy framework, design of a project, program are examined and probed in its totality. It is followed by penning down of a formal report in which all the findings are compiled which are further used in framing people centric and environment-friendly policies which are conveniently accountable (Therivel et al. 2013). Strategic environmental assessments (SEAs) thus render a concerted platform wherein an inclusive involvement of all the collaborators and associators involved in the environment-sensitive decision-making process can undertake relevant strides (Sadler 1995). Strategic Environmental Assessment was initially conceived as a spin-off of the EIA (Environmental Impact Assessment) with added dimension of decision-making by the higher authorities involved in it, with the other technicalities of EIA remaining more or less the same (Lee and Walsh 1992). An important aspect of the employment of SEA as an environmental planning tool is its scale of operation and the integration of various faculties that it offers.

EIA reels under the issue of delayed implementation and planning and this problem can be obliterated by introgression of SEA in the environmental planning and decision-making regimens. EIA in certain cases has failed to provide an environmental sustainability assurance (ESA) and these shortcomings and infirmities of EIA led to the introduction of SEA (Sadler 1999). But the evidence and testimony of the efficacy of Strategic Environmental Assessment are quite frugal and meagre (Runhaar and Driessen 2007).

An SEA case study has been given by Salhofer et al. (2007) in the Salzburg province of Austria. In this study, for minimizing the impacts of a fragmented waste collection regimen amounting to 15 different types of wastes and their expensive separate disposal, the process of SEA was carried out. This process also included the aspects of public discussions and roping in the technical expertise. It culminated in the development of a regional waste management plan which was sensitive to the needs of people, environment and the domestic economy as well.

5.2.1.3 TIA (Territorial Impact Assessment)

In the prevalent epoch, owing to the enormous territorial overhaul that has taken place due to the digital revolution and transnational globalism, there has been an accelerating quantum of focus on geographic contouring and the associated issues therein (Delaney 2009). In this environmental planning tool, the sphere of territorial cooperation is existent. The consolidation of territories endows a global awareness to the environmental planning process, thus making the entire process very inclusive and reverberant (Johnson 2009).

Although TIA is the most complicated, its policy review process has the most promising potential for evaluating initiatives, plans, and policies. Strategies and procedures of territorial development strongly influence policy evaluation techniques at subnational, national, and transnational levels (Medeiros 2020).

The pertinence of this environmental planning tool can be deduced from the fact that the third world developing countries have been trying to subside the inequality in opportunities for their respective principals for quite some time. Thus, they need to enter into an amalgamative territorial collaboration to counteract the hegemonic presence of the Western order (MacKinnon and Cumbers 2018). The typology shows that the Europeanization of planning has an impact on projects but also on governance structures, territory, and domestic policies. Several distinctions were made on types of impacts of EU policies on territory, i.e., direct and indirect impacts. There is an additional distinction between impacts on territory and territorial governance (Zonneveld and Waterhout 2009).

An instance of the usage of territorial impact assessment (TIA) in the European Transport Policy shows how using the environmental planning tool of TIA, a comprehensive, sound, and executable model was prepared. This model called Interactive Simulation Package (SIP) was used with Trans-European Networks Policy of the European Union (EU). It led to the exploration of the objective of territorial cohesion, territorial integrity, and quality (Camagni 2009). In practical terms, TIA methodologies arose from the European Territorial Observatory Network (ESPON) Program's vigorous search for clear causal relationships between public policy investments and territorial development processes, not just from the environmental aspect of development that can be evaluated using SEA and EIA tools (Medeiros 2020).

5.2.2 *Process-Oriented Planning Tools*

Process-oriented planning tools may be defined as the tools which encapsulate the concept of Socialization and interaction (Valentin and Spangenberg 2000). The planning in this kind of tool is facilitated by common goals, coalescing perspectives and innovation-based objectives (Amler et al. 1999). The indicators pimped in this sort of planning tool are not pre-decided and are brought out as a consequence of the negotiations and deliberations (Runhaar and Driessen 2007).

5.2.2.1 CDM (Clean Development Mechanism)

Clean Development Mechanism is one of the three flexible market-based mechanisms enshrined in the Kyoto Protocol that enable developing countries to fulfil their carbon-offsetting commitments. It is inclusive of all those energy efficiency amelioration endeavours which result in Greenhouse Gas emission reductions. Every GHG emission reduction, whether onsite or offsite, qualifies for CDM. The Clean Development Mechanism (CDM) was founded in 1997 and has been in active operation since 2006. After the Clean Development Mechanism (CDM) was ratified as part of the Kyoto Protocol in 1997, this initiative has been referred to by several names, including 'the Kyoto Surprise', 'the win-win mechanism', 'a bridge between North and South', and 'the front-runner of the Kyoto Regime' (Matsuo 2003). The Clean Development Mechanism (CDM) permits projects aimed at reducing emissions (or removing emissions) in Non-Annex I countries (developing nations) to acquire certified emission reduction credits (CERs), each representing one ton of CO2. These CERs are tradable and marketable, providing Annex I countries (industrialized nations) with a cost-effective means to fulfil a portion of their emission reduction targets as outlined in the Kyoto Protocol (Wang and Chen 2010).

In order to mitigate global warming, a carbon market is quickly being recognized as an essential tool for stabilizing greenhouse gas concentrations in the atmosphere. Apart from functioning as a means of cost containment for affluent countries aiming to fulfil their emission reduction obligations under the Kyoto Protocol, the Clean Development Mechanism (CDM) may also be of great significance in obtaining private sector financing for low-carbon technology projects in developing nations. This environmental planning tool involves certain stages that ultimately culminate into assisting the developing countries accomplish the coveted goal of sustainable development. It begins with the identification of the project that is to be considered followed by Government Endorsement, Project Development, validation, registration, monitoring, verification, certification and finally the issuance of CERs (Matsuo 2003).

5.2.3 *Hybrid Planning Tools*

Hybrid planning tools are those environmental planning tools which involve the cumulative essence of both the substance-oriented planning tools and process-oriented planning tools conducive to accomplish the prospects of spatial development, temporal development, and pliable collateral modelling with due regard to the environmental standards (Liew and Sundaram 2009). The prime objective in the employability of these environmental planning tools in the environmental planning process and biodiversity preservation is the expansion of region-specific and area-centred environmental aspirations which are discrete and customized to the auspices of the region or geographical cohort under consideration. Recently, the Netherlands has been known for the development of a hybrid planning tool which properly delineates the environmental ambitions in the spatial plans for specific areas. These are also supplemented with carefully laid plans to achieve these ambitions (Runhaar et al. 2009).

5.2.3.1 MCDA (Multi-criteria Decision Analysis)

Multi-criteria decision analysis (MCDA) is comprehended to be the environmental planning tool that has been fabricated to decipher and simplify complicated and intriguing issues with respect to planning the projects with a sustainable demeanour (Mendoza et al. 1999). MCDA stands as a comprehensive evaluation of sustainability, functioning as an operational assessment and decision support method proficient in addressing complex challenges marked by substantial uncertainty, conflicting objectives, diverse data types, and information sources, multiple interests and perspectives, and the nuanced consideration of dynamic biophysical and socio-economic systems. MCDA has been recruited to narrow down to the criteria and specifications for the furtherance of viable living, resource rationalization and sustainable development (Mendoza and Prabhu 2000). An additional benefit of using this tool is that it also factors in the non-conventional value of the ecosystem services which is bereft of the existent market considerations (Wolfslehner et al. 2005).

5.2.3.2 LOGO (Local Area Typology and Environmental Quality)

LOGO tool for environmental planning was contemplated by the Department of Environment for the Rijnmond area for fine tuning and amending the quality of various environmental indicators of the region. The magnitude of substance in this tool can be deduced from the fact that this tool is used to frame and delineate a caricature for the furtherance of region-sensitive intentions for enhancing environmental quality. These ambitions are in coherence with the spatial and temporal planning process (De Roo and Visser 2016).

5.3 Sustainable Urban Designing Using Environmental Planning Tools

Different types of urban planning are known by different names throughout the world, such as 'smart city' or 'sustainable city', and in some places, they are also called 'liveable city' or 'future city'. Even though the goal of all urban planning is sustainability, inclusive growth, and urban resilience, the higher urban population density renders the planning unsustainable in terms of things like severe traffic jams, air pollution, a decline in green space, and a reduction in amenities (Mallick 2021).

5.3.1 Green Infrastructure and Sustainable Architecture

Urbanism and nature coalesce together to create a healthy and enriching place to occupy, a human settlement that is derived from ecological balance and self-reliance on the community. It can bestow a number of environmental benefits and create a link between the two (Mersal 2016). For example, InSMART concept is an innovative method of sustainable energy planning in urban development. Its primary objective is to bring together scientific and technical knowledge and cities, at the local level, in order to develop novel planning strategies which will pave the way towards implementing plans into action for creating new and sustainable cities. This project's main novelty is accumulation of tools and analyses that can interact with one another and yield distinct, self-consistent results which cannot be obtained in any other way. Ad hoc models and participative techniques are used in an integrated approach to carry out urban planning and energy planning (Gargiulo et al. 2017).

The regions which bear valuable and substantial reserves of natural deep pockets have the innate potential of providing water to a broad spectrum of people using the crafting of management systems which are premised on engineered and blended ecosystem models. But parallelly, it requires an exhaustive, strategic and calibrated stewardship of ecosystem services, relating to the safeguarding and recovery of the ecosystem associated (Green et al. 2015; Harrison et al. 2016).

5.3.2 Low Impact Development Strategies

Implementation of low impact urban design and development is a complex matter and calls for contributions from a variety of fields. As no individual has the necessary skill set to handle every issue, a team approach is essential (Van Roon et al. 2005). For instance, against pluvial flooding, low impact development (LID) techniques are employed as adaptation measures to assist counterbalancing and the increasing flows brought on by climate change, which strengthens the drainage system's carrying capacity for a storm with a five-year return period. Furthermore, when the budget is tight, it makes more sense to allocate LID funds to the lower sub-catchments. These techniques can be used in any community that uses LIDs as a means of adaptation to pluvial floods (Chen et al. 2017). Planning and design stages of low-impact development procedures need to take future urbanization and climate change scenarios into account. According to studies, the combination of increased urbanization and climate change has resulted in up to 20% more runoff generation. Well-established metrics should be identified and quantified for the contribution of low impact development system, to enhance urban resilience and attain fulfilment of UN SDG (Batalini de Macedo et al. 2022).

5.3.3 Sustainable Transportation Planning

The burgeoning attention on sustainability springs up from the effect that humans have caused on the environment from a broader perspective being the pollution which has occupied every sphere of the environment. The new age and the prevalent superior technological outcomes, which are confer energy efficiency and are focussed at ameliorating the negative impact per car and per kilometre driven, do not have the potency to actually reduce the problems associated with the usage of cars, and thus lag behind in bringing about sustainability in transportation (Steg 2007).

Sustainability planning proactively manages challenges rather than crying for it later on. The idea of sustainable transportation is viewed worldwide. The primary goal is to achieve effective communication that reduces the negative environmental effects of automobiles while still being economically beneficial.

'The goal of sustainable transportation is to ensure that environment, social and economic considerations are factored into decisions affecting transportation activity' (MOST 1999). Although transportation wasn't specifically discussed during the conference, the Stockholm Declaration did outline principles that would later form fundamental aspects of the concept of 'sustainable transportation'. To address the increasing number of oil spills, the International Convention for the Prevention of Pollution from Ships (MARPOL) was established in 1973. This was the first convention to specifically address the environmental effects of transportation (Gudmundsson et al. 2016).

In 1992, the notion of sustainable transportation started to emerge during the inaugural Rio Conference. President Clinton and Vice President Gore unveiled a 'Climate Change Action Plan', which outlined a distinct set of measures aimed at addressing transportation in 1993. Promoting the usage of public transportation—such as buses, trains, subways, and other vehicles—can help lower pollution and traffic jams. Reliance on private automobiles is reduced with the investment in public transportation networks that are accessible and efficient. In addition to lowering carbon emissions, promoting cycling, walking, and other non-motorized transportation also enhance public health and eases traffic congestion. Active transportation is promoted by the design of communities with bike lanes, pedestrian-friendly infrastructure, and secure crossings. Transportation systems can lessen its negative environmental consequences, such as air and water pollution and urban heat island effects, by including green infrastructure features like green spaces, permeable pavements, and urban trees. By enforcing laws and policies that encourage ecologically friendly transportation choices—such as fuel taxes, emission standards, congestion pricing, and zoning regulations—it is possible to influence people's behaviour in favour of greener alternatives (Hickman et al. 2013).

5.3.4 Urban Green Spaces

Urban green areas have the potential to play a major role in sustainable development. Urban green spaces have the potential to function as an integrated approach for ensuring the long-term environmental sustainability. They improve air quality and enhance the quality of life, while also boosting property value due to their aesthetic appeal and amenities, and reducing energy costs linked to cooling buildings. Enough vegetation makes a city aesthetically pleasant and appealing to investors as well as locals. Public parks and gardens, among other urban green spaces, offer opportunities for fun and relaxation. Ideally, these spaces facilitate physical relaxation and promote therapeutic mental healing. Above all, the ecological viewpoint views urban green areas as a means of facilitating the reduction of human activity's effect by re-releasing oxygen and absorbing pollutants (Hough 1994).

5.3.5 Resilience-Oriented Urban Planning

Resilience is a concept that is gaining traction as a heuristic to direct conversations about cities in the political and scientific domains. SGD 11 focuses attention on the significance of bolstering the concept of urban resilience by The UN Sustainable Development Goals (SDGs) emphasize the significance of increasing urban resilience SDG 9 emphasizes the creation of resilient infrastructure to assist sustainable development initiatives, while SDG 11 emphasizes the need for cities to adopt resilience-focused plans in line with the Sendai Framework for Disaster Risk Reduction 2015–2030 (UNSDG 2015). The tendency of incorporating resilience in the long term and short-term Resilience thinking gives a new outlook to planning, which enables it to face and handle the uncertainties and complexities involved in comprehending and evaluating urban systems. Green and blue infrastructures, solid waste management, decentralization of infrastructures, etc. are considered important in resilience-oriented infrastructure planning (Rauland and Newman 2011).

5.3.6 Data Science in Cities

The accelerating urban sprawl in the metropolis and its fringes has spin off effects in all the dimensions of urban living. Sectors such as transportation, energy, ecology and environment, are the ones which are being affected by this ballooning urban population in the form of deteriorating quality of life and resource asymmetry. The urban systems are constantly under the monitoring and vigil of performance trackers which record data on a real time basis using distributed sensor resources across the cities. The data generated through these digital sensors is also augmented with the human generated data, pertaining to the human activities in the cities, produced through the social media platforms and electronic gadgets.

This humongous amount of generated data suffers from ambiguities, complexities, and interoperability barriers because of the water tight data parcels in which urban data is collected. This leads to employment of a separate planning approach for each urban meta space and thus leads to fragmentation of efforts, decline in the economies of scale and a subsequent loss of efficiency in city planning endeavours. This calls for the use of environmental planning tools in the efficient collection and collation of information and data so as to bolster the on-going urban planning process using this data science. These tools are based on the premise of allowing the sharing of interoperable data and the amalgamation of data from diverse sources (Psyllidis et al. 2015).

5.3.7 Transit-Oriented Development (TOD)

Transit-oriented development (TOD) is a land use tool which is leveraged for essentially directing the course of development in a synchronized and potentially sustainable manner by designing the residential, commercial and mixed-use hybrid areas in such a manner that the access to public transport in all these complexes is enhanced. It basically involves a development in which there is a combination of nodes and places, i.e., transit points and neighbourhoods respectively. It ultimately extends a behavioural nudge to the stakeholders involved whereby they get discouraged to use their personal vehicles and shift to transit ridership because of a very well-connected street network with active transport. It subsequently results in the attenuation of traffic obstructions, sensible economic development, amelioration of the air quality parameters, and an enhancement in the environmental and economic sustainability (Kamruzzaman et al. 2014).

5.4 Conclusion

The rate of obliteration of species as has been witnessed in the past decade has been very alarming with species like Ivory-billed woodpecker, Western Black Rhinoceros, Smooth Handfish, Bramble Cay Melomys, Lake Lanao freshwater fish, Spix's Macaw, and Splendid poison frog, which formed extremely significant links of the biodiversity and had been declared extinct in the recent time frame. The integration of environmental planning tools into the realm of Sustainable and Resilient Urban Development offers a promising vision for the future of our cities. As global challenges such as climate change, rapid urbanization, and resource scarcity intensify, these tools provide a sophisticated and data-driven approach to address complex urban issues. In this backdrop, environmental planning tools can play a determining role in facilitating a seamless process of project planning before execution, as a consequence of which, the deleterious impacts of these commercial developmental projects can be attenuated. As technology continues to advance, the future may see even more sophisticated environmental planning tools, incorporating real-time data, artificial intelligence, and innovative solutions to enhance the precision and effectiveness of urban development strategies. Ultimately, the vision for Sustainable and Resilient Urban Development using Environmental Planning Tools is a future where cities become dynamic, adaptive, and harmonious ecosystems that thrive in the face of global challenges while providing a high quality of life for their inhabitants.

References

Allenby B, Fink J (2005) Toward inherently secure and resilient societies. Science 309(5737):1034–1036

Amler B, Betke D, Eger H, Ehrich C, Kohler A (1999) Land use planning – methods. In: Strategies and tools. Deutsche Gesellschaftfür Technische Zusammenarbeit (GTZ) GmbH, Eschborn, Germany

Batalini de Macedo M, Nobrega Gomes Junior M, Pereira de Oliveira TR, Giacomoni MH, Imani M, Zhang K et al (2022) Low impact development practices in the context of United Nations sustainable development goals: a new concept, lessons learned and challenges. Crit Rev Environ Sci Technol 52(14):2538–2581

Camagni R (2009) Territorial impact assessment for European regions: a methodological proposal and an application to EU transport policy. Eval Program Plann 32(4):342–350

Cerreta M, Monno V (eds) (2010) Making strategies in spatial planning: knowledge and values, vol 9. Springer Science & Business Media, Dordrecht

Chen PY, Tung CP, Li YH (2017) Low impact development planning and adaptation decision-making under climate change for a community against pluvial flooding. Water 9(10):756

Dale VH, English MR (1999) Tools to aid environmental decision making. Springer Science & Business Media, New York

Dawoud MA (2012) Environmental impacts of seawater desalination: Arabian Gulf case study. Int J Environ Sustain 1(3):22–37

De Roo G, Visser J (2016) Smart methods for environmental externalities: urban planning, environmental health and hygiene in The Netherlands. Routledge, London

Delaney D (2009) Territory and territoriality. In: Kitchin R, Thrift N (eds) International encyclopedia of human geography. Elsevier, Amsterdam 11:196–208
Dirzo R, Raven PH (2003) Global state of biodiversity and loss. Annu Rev Environ Resour 28(1):137–167
Finnveden G, Nilsson M, Johansson J, Persson Å, Moberg Å, Carlsson T (2003) Strategic environmental assessment methodologies—applications within the energy sector. Environ Impact Assess Rev 23(1):91–123
Fischer TB (2003) Strategic environmental assessment in post-modern times. Environ Impact Assess Rev 23(2):155–170
Gargiulo M, Chiodi A, De Miglio R, Simoes S, Long G, Pollard M et al (2017) An integrated planning framework for the development of sustainable and resilient cities–the case of the InSMART project. Procedia Eng 198:444–453
Geneletti D, Bagli S, Napolitano P, Pistocchi A (2007) Spatial decision support for strategic environmental assessment of land use plans. A case study in southern Italy. Environ Impact Assess Rev 27(5):408–423
Green PA, Vörösmarty CJ, Harrison I, Farrell T, Sáenz L, Fekete BM (2015) Freshwater ecosystem services supporting humans: pivoting from water crisis to water solutions. Glob Environ Chang 34:108–118
Gudmundsson H, Hall RP, Marsden G, Zietsman J, Gudmundsson H, Hall RP et al (2016) Transportation and sustainability. In: Sustainable transportation: indicators, frameworks, and performance management. Springer, Heidelberg, pp 81–109
Hambler C (2004) Conservation. Cambridge University Press, Cambridge, UK
Harrison IJ, Green PA, Farrell TA, Juffe-Bignoli D, Sáenz L, Vörösmarty CJ (2016) Protected areas and freshwater provisioning: a global assessment of freshwater provision, threats and management strategies to support human water security. Aquat Conserv Mar Freshwat Ecosyst 26:103–120
Hector A, Schmid B, Beierkuhnlein C, Caldeira MC, Diemer M, Dimitrakopoulos PG et al (1999) Plant diversity and productivity experiments in European grasslands. Science 286(5442):1123–1127
Heuvelhof EF, Nauta CM (1996) (M)erkenning, onderzoek naar de doorwerking van m.e.r. (Impact of Environmental Impact Assessment). Evaluatiecommissie Wet Milieubeheer, The Hague
Hickman R, Hall P, Banister D (2013) Planning more for sustainable mobility. J Transp Geogr 33:210–219
Hildén M, Furman E, Kaljonen M (2004) Views on planning and expectations of SEA: the case of transport planning. Environ Impact Assess Rev 24(5):519–536
Hough M (1994) City form and natural processes, London: Croom Helm. In: Haughton G, Hunter C (eds) Sustainable cities. JKP, London
Johnson CM (2009) Cross-border regions and territorial restructuring in Central Europe: room for more transboundary space. Eur Urban Reg Stud 16(2):177–191
Kamruzzaman M, Baker D, Washington S, Turrell G (2014) Advance transit oriented development typology: case study in Brisbane, Australia. J Transp Geogr 34:54–70
Kearns C (2010) Conservation of biodiversity. Nat Educ Knowledge 3(10):7
Lee N, Walsh F (1992) Strategic environmental assessment: an overview. Project Appraisal 7(3):126–136
Liew A, Sundaram D (2009) Flexible modelling and support of interrelated decisions. Decis Support Syst 46(4):786–802
Lobaccaro G, Croce S, Lindkvist C, Probst MM, Scognamiglio A, Dahlberg J et al (2019) A cross-country perspective on solar energy in urban planning: lessons learned from international case studies. Renew Sust Energ Rev 108:209–237
MacKinnon D, Cumbers A (2018) An introduction to economic geography: globalisation, uneven development and place. Routledge, Milton
Mallick SK (2021) Prediction-adaptation-resilience (PAR) approach- a new pathway towards future resilience and sustainable development of urban landscape. Geogr Sustain 2(2):127–133

Matsuo N (2003) CDM in the Kyoto negotiations. Mitig Adapt Strateg Glob Chang 8(3):191–200
Medeiros E (2020) Introduction: a handbook on territorial impact assessment (TIA). Territorial impact assessment:1–6
Mendoza GA, Prabhu R (2000) Development of a methodology for selecting criteria and indicators of sustainable forest management: a case study on participatory assessment. Environ Manag 26:659–673
Mendoza GA, Macoun P, Prabhu R, Sukadri D, Purnomo H, Hartanto H (1999) Guidelines for applying multi-criteria analysis to the assessment of criteria and indicators, vol 9. CIFOR, Bogor, Indonesia
Mersal A (2016) Sustainable urban futures: environmental planning for sustainable urban development. Procedia Environ Sci 34:49–61
Morgan RK (2012) Environmental impact assessment: the state of the art. Impact Assess Proj Apprais 30(1):5–14
Moving on Sustainable Transportation (MOST) (1999) Transport Canada, www.tc.gc.ca/envaffairs/most
Mumford L (1961) The city in history: its origins, its transformations, and its prospects, vol 67. Houghton Mifflin Harcourt, New York
Ogryzek M, Adamska-Kmieć D, Klimach A (2020) Sustainable transport: an efficient transportation network—case study. Sustain For 12(19):8274
Oikonomou V, Dimitrakopoulos PG, Troumbis AY (2011) Incorporating ecosystem function concept in environmental planning and decision making by means of multi-criteria evaluation: the case-study of Kalloni, Lesbos, Greece. Environ Manag 47:77–92
Omran II, Al-Saati NH, Al-Saati HH, Hashim KS, Al-Saati ZN (2021) Sustainability assessment of wastewater treatment techniques in urban areas of Iraq using multi-criteria decision analysis (MCDA). Water Pract Technol 16(2):648–660
Partidário MR (1999) Strategic environmental assessment-principles and potential. Handbook on environmental impact assessment, vol 1. Blackwell Science, Oxford
Psyllidis A, Bozzon A, Bocconi S, Titos Bolivar C (2015) A platform for urban analytics and semantic data integration in city planning. Communications in Computer and Information Science. Springer, Berlin Heidelberg 527:21–36
Rauland V, Newman P (2011) Decarbonising Australian cities: a new model for creating low carbon, resilient cities. In: Sustaining our future-understanding and living with uncertainty. MODSIM2011: 19th International Congress on Modelling and Simulation Perth, pp 3073–3079
Runhaar H, Driessen PP (2007) What makes strategic environmental assessment successful environmental assessment? The role of context in the contribution of SEA to decision-making. Impact Assess Proj Apprais 25(1):2–14
Runhaar H, Driessen PP, Soer L (2009) Sustainable urban development and the challenge of policy integration: an assessment of planning tools for integrating spatial and environmental planning in The Netherlands. Environ Plan B Plan Design 36(3):417–431
Sadler B (1995) Ecology, economics, and the assessment of sustainability, themes, and approach. In: Balancing the scale: integrating environmental and economic assessment. Foundation for International Training, Toronto, pp 1–36
Sadler B (1999) A framework for environmental sustainability assessment and assurance. In: Handbook of environmental impact assessment, vol 1. Edward Elgar Publishing Limited, Cheltenham, pp 12–32
Salhofer S, Wassermann G, Binner E (2007) Strategic environmental assessment as an approach to assess waste management systems. Experiences from an Austrian case study. Environ Model Softw 22(5):610–618
Sivaramanan S (2015) Conservation and sustainable use of biodiversity through Inclusive Environmental planning. https://doi.org/10.13140/RG.2.1.4574.2564
Steg L (2007) Sustainable transportation: a psychological perspective. IATSS Res 31(2):58–66
Talukdar MH, Hossain Z (2011) Planning tools that can be used to facilitate urban sustainability. Manag Res Pract 3(4)

Tang HT, Lee YM (2016) The making of sustainable urban development: a synthesis framework. Sustain For 8(5):492

Therivel R, Wilson E, Heaney D, Thompson S (2013) Strategic environmental assessment. In: Routledge eBooks, London. https://doi.org/10.4324/9781315070339

Tilman D, Downing JA (1994) Biodiversity and stability in grasslands. Nature 367(6461):363–365

Tilman D, Wedin D, Knops J (1996) Productivity and sustainability influenced by biodiversity in grassland ecosystems. Nature 379(6567):718–720

Uhde B, Andreas Hahn W, Griess VC, Knoke T (2015) Hybrid MCDA methods to integrate multiple ecosystem services in forest management planning: a critical review. Environ Manag 56:373–388

Un-Habitat (2011) Cities and climate change: global report on human settlements 2011. Routledge, London

United Nations Environment Program, Environmental Impact Assessment (1988) Basic procedures for developing countries. UNEP Regional Office for Asia and the Pacific, Bangkok

UNSDG (2015) Transforming our world: the 2030 agenda for sustainable development, http://www.un.org/sustainabledevelopment/sustainable-development-goals/

Valentin A, Spangenberg JH (2000) A guide to community sustainability indicators. Environ Impact Assess Rev 20(3):381–392

Van Roon M, Dixon J, Van Roon H (2005) Reformulating planning tools to promote low impact urban design and development. In: Proceedings of the New Zealand water and waste association 4th south pacific conference on stormwater and aquatic resource protection

Vié JC, Hilton-Taylor C, Stuart SN (eds) (2009) Wildlife in a changing world: an analysis of the 2008 IUCN red list of threatened species. IUCN, Gland, Barcelona

Wang Q, Chen Y (2010) Barriers and opportunities of using the clean development mechanism to advance renewable energy development in China. Renew Sust Energ Rev 14(7):1989–1998

Weston J (2000) EIA, decision-making theory and screening and scoping in UK practice. J Environ Plan Manag 43(2):185–203

Wilson EO (2002) Speciation and biodiversity. American Institute of Biological Sciences. Maximillan publishers

Wolfslehner B, Vacik H, Lexer MJ (2005) Application of the analytic network process in multi-criteria analysis of sustainable forest management. For Ecol Manag 207(1–2):157–170

Wood C (1995) Environmental impact assessment: a comparative review. Longman, Harlow

World Health Organization (WHO) (2010) Unmasking and overcoming health inequities in urban settings. World Health Organization, Kobe

Zonneveld W, Waterhout B (2009, August) EU territorial impact assessment: under what conditions. In: Territorial cohesion of Europe and integrative planning, 49th Congress of the European Regional Science Association, pp 25–29

Chapter 6
Microalgae-Based Bioremediation: A Sustainable Approach for the Removal of Emergent Pollutants

Charu Deepika, Gabriele Frascaroli, Anusha Gowri, and Alla Silkina

Abstract Emergent pollutants pose significant threats to global ecosystems and human health, dispersing across water, air, and soil matrices and causing diverse adverse effects like carcinogenesis and antimicrobial resistance. Addressing their removal is crucial for sustainability, yet traditional methods are limited by high energy consumption and harmful by-products production. Microalgae biotechnology emerges as a cost-effective and energy-efficient alternative, especially in wastewater treatment. Microalgae, with their diversity and adaptability, offer unique capabilities for pollutant sequestration and metabolism. By maximizing the performance of microalgal species for improved bioremediation, a viable solution emerges for addressing pollution caused by contaminants in wastewater. This chapter highlights microalgae's numerous applications in environmental remediation, emphasizing the need for continued research, innovation, and integration with existing technologies to advance this field and mitigate the impact of emergent pollutants effectively. The future success of microalgal bioremediation hinges on collaborative efforts across disciplines and industries to address pressing environmental challenges.

Keywords Microalgae · Bioremediation · Emergent pollutants · Sustainable · Pollution mitigation · Sustainable approaches

---

C. Deepika (✉)
Institute of Molecular Bioscience, The University of Queensland, Brisbane, QLD, Australia

G. Frascaroli
Department of Civil Engineering and Environmental Management, School of Computing, Engineering and Built Environment, Glasgow Caledonian University, Glasgow, Scotland

A. Gowri
SRM Institute of Science and Technology, Chennai, India

A. Silkina
Algal Research Group, Bioscience Department, College of Science, Swansea University, Swansea, UK

S. Dey, S. Bhattacharya (eds.), *Biotechnological Interventions in the Removal of Emerging Pollutants*, Interdisciplinary Biotechnological Advances,
https://doi.org/10.1007/978-981-97-9922-0_6

6.1 Introduction

In a global landscape marked by rapid industrialization and urbanization, rise of emergent pollutants (EPs), spanning from pesticides and pharmaceuticals to heavy metals and synthetic chemicals, presents a substantial threat to environmental equilibrium (Fig. 6.1) (Banik 2021; Tuckett 2021). The increasing occurrence of EPs, characterized by their unique and often intricate chemical compositions, poses a challenge to conventional remediation methods, necessitating an exploration of innovative and sustainable approaches (Yadav et al. 2021). There is a crescent need to explore innovative and sustainable approaches to address the challenges posed by the escalating prevalence of EPs and related environmental concern (Kaşdoğan 2020). Additionally, the adverse effects of EPs on ecosystems and human health, along with their pathways in the environment, particularly through wastewater, underscore the urgency of addressing this environmental challenge.

In response to this pressing environmental crisis, bioremediation has emerged as a pivotal and sustainable strategy to mitigate the adverse impacts of EPs on ecosystems (Saravanan et al. 2023; Agrawal et al. 2020). Bioremediation, defined as the utilization of biological organisms to detoxify or remove contaminants from a

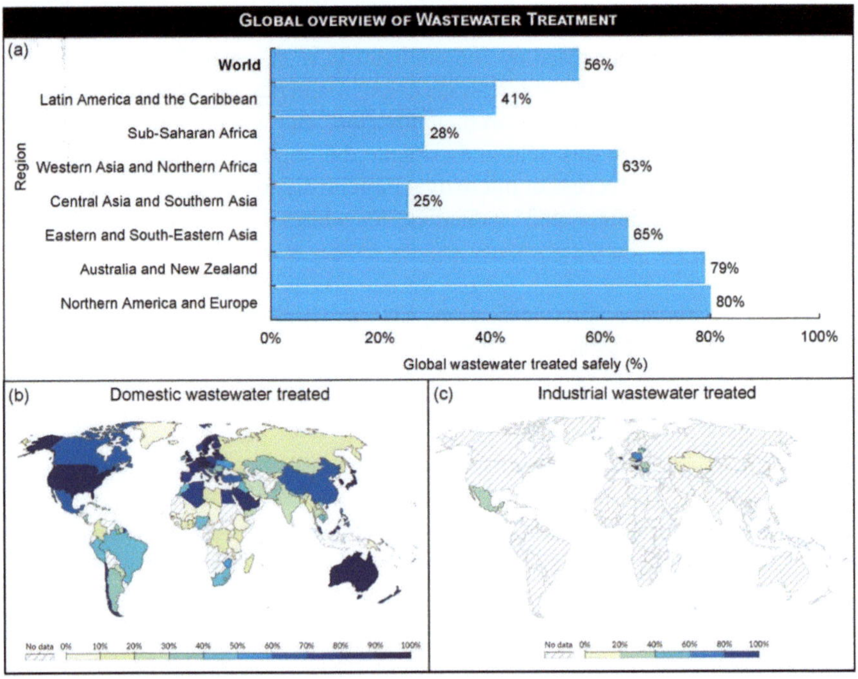

Fig. 6.1 Global overview of wastewater treatment. (**a**) Percentage of global wastewater treated safely in 2020 (WHO and UNICEF) (Fleck 2023). (**b**) Proportion of domestic wastewater safely treated in 2022 (WHO) (Data, O.W.I 2023a). (**c**) Proportion of industrial wastewater treated in 2015 (UN) (Data, O.W.I 2023b)

polluted environment, garners attention for its intrinsic eco-friendly nature and its distinctive ability to harness the metabolic capabilities of living organisms for environmental restoration (Zahra et al. 2023).

Microalgae-based technology, in particular, emerges as frontrunners in this biotechnological revolution due to their evolutionary proved adeptness at removing a diverse range of pollutants originating from industrial, agricultural, and domestic sectors (Klepacz-Smolka et al. 2023). Functioning as versatile cell factories, microalgae harness the abundant energy resource of the sun to assimilate CO_2 and nutrients, sourced from wastewater streams, into a varied array of cellular biocompounds (Choudhary et al. 2022). This process facilitates the production of renewable fuels, food, water, and bioproducts, rendering microalgae both cost-effective and sustainable (Deepika et al. 2022). Within the vast diversity of microalgae species, reported at ~40,000, the high-efficiency production of specific target strains necessitates the optimization of species-specific production parameters (Khan et al. 2018). These parameters include nutrient supply (Radzun et al. 2015; Wolf et al. 2015), light intensity (Sivakaminathan et al. 2018), temperature (Ranglová et al. 2019), mixing (Mendoza et al. 2013), pH (Aditya et al. 2023), and CO_2 concentrations (Aditya et al. 2023). Microalgae exhibit effective remediation abilities against various pollutants, including nutrients such as phosphorus, nitrogen, ammonium, and heavy metals like lead and cadmium. Additionally, they demonstrate efficacy in addressing persistent contaminants such as pesticides (methyl parathion, DDT), pharmaceuticals (carbamazepine, diclofenac, ibuprofen), and dyes (basic fuchsin, methyl red) (Zahra et al. 2023; Goswami et al. 2022).

The concept of circular economy is inherently intertwined with microalgae-based bioremediation, as the coupling of bioproduct production and wastewater treatment using microalgae accelerates the circular economy (Mondal et al. 2022). Microalgae not only play a crucial role in EP removal but also exhibit the potential of incorporating the production of valuable bioproducts from the biomass generated during the remediation process, exemplifying a circular economy model where waste is transformed into valuable resources, emphasizing the sustainability and efficiency of microalgae-based bioremediation (Najar-Almanzor et al. 2023). They harness toxic substances or pollutants for their cellular growth and metabolism (as nutrients), employing various uptake mechanisms such as biosorption, bioaccumulation, and biodegradation (Chan et al. 2022). Their photosynthetic and CO_2 biofixation abilities contribute to reducing atmospheric CO_2 levels, aligning with the principles of circular economy by utilizing carbon resources in an environmentally responsible manner. Furthermore, microalgae photosynthesis generates oxygen, aiding in the reduction of biochemical oxygen demand (BOD) and chemical oxygen demand (COD) levels in wastewater (Zahra et al. 2023).

This chapter aims to provide an in-depth exploration of the comprehensive landscape of microalgae-based bioremediation, scrutinizing its principles, mechanisms and applications, while highlighting its role in accelerating the circular economy. Primary objectives include elucidating the diverse roles microalgae play in pollutant removal, assessing the factors influencing their efficiency, and exploring innovative strategies and technologies. The scope encompasses a synthesis of existing knowledge, recent advancements, and critical insights into the potential of microalgae-based bioremediation as a sustainable and effective tool for addressing the challenges posed by EPs in diverse environmental contexts. Through this exploration, the chapter endeavours to contribute to a deeper understanding of the intricate interplay between microalgae, bioremediation, and environmental sustainability within the framework of a circular economy.

6.2 EPs: A Comprehensive Overview

Infiltrating various environmental domains and the atmosphere, EPs are a group of unregulated chemical substances that bring potential health risks to both living organisms and the environment (Geissen et al. 2015). These pollutants are extensively classified into conventional and unconventional groups. Conventional pollutants include pharmaceuticals, personal care products, endocrine disruptors, pesticides, heavy metals, and synthetic organic chemicals (Geissen et al. 2015). Unconventional pollutants encompass steroids, surfactants, fire retardants, rare earth elements, industrial additives, hormones, nanomaterials, organometallic compounds, perfluorinated compounds, microbeads, toxins from algae and cyanobacteria, and disinfectant by-products.

According to the classification established by NORMAN, a prominent organization facilitating the exchange of information on emerging environmental substances, EPs are categorized into pharmaceutical pollutants, personal care products, microplastics, persistent organic pollutants, endocrine disruption chemicals, and artificial sweeteners. The NORMAN network classification was subsequently reorganized into heavy metals, pharmaceuticals, pesticides, dyes, and microplastics (Dulio et al. 2018). Over 700 EPs, including metabolites and by-products, have been individually characterized and found ubiquitously in all environmental matrices worldwide, such as groundwater, surface water, and wastewater effluents (Iftehimul et al. 2023). Approximately 10–50% of the reported EPs are believed to incrementally impact human, animal, and environmental health, based on their physicochemical properties and ecotoxicological impacts (Yadav et al. 2021; Gaohua et al. 2021; Gao et al. 2022). Pharmaceuticals, with their molecular complexity, present unique challenges compared to other EPs, due to their resilience to conventional wastewater treatment and the possibility of causing cancer, disrupting endocrine systems, and fostering antimicrobial resistance (Zahra et al. 2023). Heavy metals, including lead (Pb), mercury (Hg), arsenic (As), barium (Ba), and uranium (U), stand out as significant EPs due to their distinct physicochemical properties and toxicity levels (leading to nutrient imbalance impacting water and soil) (Khatun et al. 2022). Pesticides, such as fluroxypyr, imidacloprid, naphthalene, and glyphosate, are analysed based on various molecular characteristics and toxicity levels (Zhang et al. 2011). Synthetic dyes, including pyrene, azo dye, crystal violet, malachite green, and others, are categorized similarly to other EPs (Sarkar and Dey 2021). Microplastics, composed of numerous monomers and polymer chains of carbons and hydrogens, pose a significant threat to ecosystems globally (Dusaucy et al. 2021; Padervand et al. 2020).

EPs originate from various sources, including healthcare facilities, industrial sites, agricultural activities, and households (Ahmad et al. 2022). Disposal in soils, animal excretion, direct discharge into sewage, and inefficient water treatment in wastewater plants are among the pathways through which EPs are released into the environment (Ahmad et al. 2022). Various EPs exhibit distinct pathways of entering the natural environment, primarily contingent on emission sources and the chemical characteristics of the compounds.

Pharmaceuticals, including antibiotics, hormones, and veterinary drugs, enter the environment stemming from diverse sources, including improper disposal of medicines, discharge from pharmaceutical manufacturing processes, wastewater effluents, and the spreading of wastewater sludge. These pathways contribute to the widespread presence of pharmaceutical compounds in environmental matrices, highlighting the need for comprehensive strategies to mitigate their impact on ecosystems and human health (Iftehimul et al. 2023). Antibiotics, being persistent in nature, are a significant concern, contributing to the crisis of antimicrobial resistance (Ahamad et al. 2020). The use of hormones and endocrine disruptor drugs in various industries leads to their release in wastewater bodies, posing a potential threat to surface waters (Kasonga et al. 2021). These substances exhibit the capability to disrupt endocrine systems in both aquatic organisms and, consequently impact the ecosystems they inhabit. Heavy metals (such as copper, zinc, cobalt, lead, mercury, arsenic, chromium, and cadmium) enter aquatic environments through waste discharge from industrial operations. Anthropogenic sources, including urban and industrial development, as well as waste from leather, glass manufacturing, pharmaceutical industries, and domestic activities, contribute to heavy metal pollution in aquatic environments (Goswami et al. 2022; Heinemann and Hiscox 2022). Pesticides, employed in agricultural, golf course, and forest settings, infiltrate diverse strata of aquatic ecosystems, demonstrating persistence, bioaccumulation, and biomagnification phenomena (Zaller and Zaller 2020). Agricultural lands, acting as non-point sources, play a significant role in the contamination of water bodies with pesticides. Runoff and leaching processes from these areas lead to the introduction of pesticides into surface waters, thereby contributing substantially to the issue of aquatic pollution. Dyes from textile, silk factories, and printing industries enter aquatic ecosystems through wastewater discharge. Dyes, with complex chemical structures, are challenging to remove by traditional treatment methods (Liu 2020).

Microplastics, particles with the size less than 5 mm, result from increased plastic production and improper waste management. Various factors, including waste dumping, plastic manufacturing, sewage, and fishing activities, contribute to microplastic pollution in aquatic environments (Padervand et al. 2020). Exposure to EPs results in environmental degradation and hazardous impacts, including genotoxicity, carcinogenicity, endocrine disruption, and immunological toxicity across a wide range of living organisms (Sharma et al. 2023). EPs accumulate in living organisms or their tissues and spread through the food web, employing processes such as bioaccumulation and biomagnification. Humans are exposed to EPs through tropic transfer in food webs, leading to dysfunction in various organ systems, including respiratory, cardiovascular, reproductive, immune, and nervous systems, depending on the level and duration of exposure. Bioaccumulation involves the build-up of harmful substances within an organism, while biomagnification amplifies the concentration of a pollutant across a food chain. EPs are transported from non-living surroundings to living organisms, accumulating across trophic levels in food webs, contaminating food chains and posing implications for human health.

EPs contribute significantly to the global water pollution crisis, with the discharge of 2 mn tonnes of sewage, industrial effluents, and agricultural waste daily. This massive volume, equivalent to the entire weight of the world's human population, significantly impacts the annual production of wastewater, estimated at around 1500 km^3 (Vasilachi et al. 2021). Insufficient sanitation practices globally worsen the issue, affecting billion people who lack access to proper sanitation facilities (Zvobgo and Do 2020). This has repercussions beyond human health, endangering aquatic ecosystems, with over 50% of native freshwater fish species in some regions and nearly one-third of the world's amphibians facing extinction risks. The Asian region, home to over 70% of the population without adequate sanitation, faces significant challenges. China, experiencing rapid economic growth, confronts water scarcity issues, exacerbated by the untreated release of 33% of industrial wastewater and 70% of household sewage (Factsanddetails 2022). This lack of proper treatment facilities contributes to water supply contamination in 90% of Chinese urban centres, with a severe environmental and economic toll (Factsanddetails 2022). In the European aquatic environment, over 700 EPs have been identified, including their metabolites and transformation products. These pollutants, falling into more than 20 classes, pose threats due to potential impacts on aquatic life and human health (Arman et al. 2021). The lack of routine monitoring programs and understanding of their behaviour necessitates urgent action. The European Union (EU) adopts a proactive approach, creating a watch list of EPs under the Water Framework Directive (Colzani et al. 2024; Council, E.P 2002). This directive designates 33 substances or compound classes as priority pollutants, requiring effective management strategies (Decision 2455/2001/EC). This introduced a conceptual framework for monitoring programs, fate and risk assessment tools, and policies related to EPs, aiming to support sustainable water resource management practices.

Based on the comprehensive review of EPs and their diverse pathways of entry into the environment, coupled with their adverse effects on ecosystems and human health, it is imperative to adopt proactive measures for their management and mitigation. The integration of robust monitoring programs, advanced treatment technologies, and stringent regulatory frameworks is essential for safeguarding water resources and ensuring the sustainability of aquatic ecosystems in the face of escalating pollution challenges.

6.3 Microalgae as Bioremediation Agents

Every day, various EPs from industries, agriculture, and the general population find their way into sewage systems and subsequently reach wastewater treatment plants, if not directly discharged into the environment. Conventional wastewater treatment plants, not originally designed to eliminate EPs of recent prevalence, lead to the continuous release of these contaminants into the aquatic environment and soils (through sludges spreading), where they exert their toxic effects (Deblonde et al. 2011).

Microalgae-based wastewater treatment techniques emerge as promising solutions due to their cost-effectiveness, versatile capabilities, and the production of valuable co-products (Fig. 6.2a, b). Microalgae thrive in wastewater due to its composition closely resembling the culture media used for their cultivation. Various waste streams, including olive wastewater, mine drainage, and dairy wastes, are under examination for potential microalgae-based treatment systems (Redalje et al. 2018). Optimizing standard wastewater treatment plants by integrating microalgae-based systems in a circular economy framework presents an attractive solution to minimize environmental impact and produce secondary ecological services.

This section delves into the multifaceted role of microalgae in bioremediation, elucidating their mechanisms of pollutant removal, exploring the criteria for selecting microalgae species, highlighting the distinct advantages associated with microalgae-based bioremediation, and comparing these systems with currently available technologies. The ultimate goal is to assess the viability of employing microalgae for treating EPs, discerning both the constraints and prospects of this sector for the future.

6.3.1 Mechanisms of EP Removal Within Microalgae-Based Systems

The pathways for eliminating EPs in microalgae-based systems encompass biotic biosorption, bioaccumulation, biodegradation, and abiotic processes like photolysis, hydrolysis, and volatilization, as outlined in Fig. 6.2c. The microalgae species and characteristics and the physicochemical properties of the EPs mainly determine the choice of mechanism. Understanding these mechanisms is vital for unravelling the efficacy of microalgae in bioremediating EPs.

6.3.1.1 Biosorption

Bioadsorption, characterized as a passive process, entails the interaction of antibiotics with algal cell walls and extracellular polymeric substances (EPSs). This interaction is influenced by factors such as the hydrophobicity and functional groups of

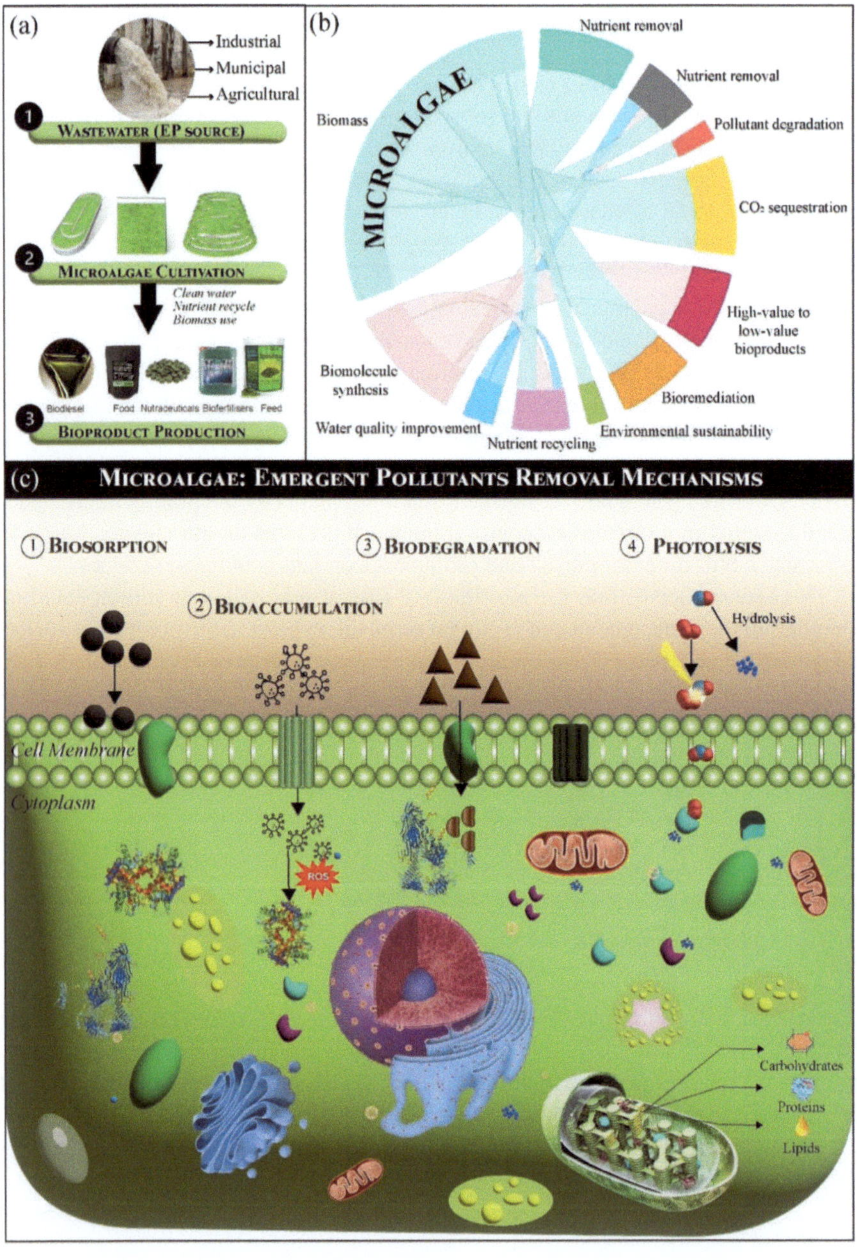

Fig. 6.2 Microalgae-based bioremediation. (**a**) Illustration of the three primary stages: EP source selection, microalgae cultivation, and bioproduct production. (**b**) Chord diagram highlighting the intricate connections between key entities in microalgae bioremediation. (**c**) Five mechanisms employed by microalgae for EP removal, encompassing (1) biosorption, (2) bioaccumulation, (3) biodegradation, and (4) photolysis

the EPs, whose ionization is pH-dependent. In this phenomenon, a non-metabolic interaction occurs between the positively charged EPs and the negatively charged microalgal cell wall or EPS. Consequently, heavy metals like cadmium(II), mercury(II), and lead(II) (Bayramoğlu et al. 2006), as well as lipophilic compounds such as PAHs, are easily absorbed (Ke et al. 2010). Biosorption effectiveness relies heavily on the species-specific compositions of cell walls, EPS in the adopted strain, media composition, and pH. Additionally, it is crucial to note that bioadsorbed pollutants can remain problematic if the biomass is not appropriately disposed of, thereby restricting the potential for reusing exhausted biomass.

6.3.1.2 Bioaccumulation

Bioaccumulation, or biouptake, involves microalgae actively taking up and storing EPs within their cytoplasm and various organelles following biosorption. Nutrients, heavy metals, and lipophilic EPs are readily bioaccumulated in this process along with certain pharmaceuticals (Bai and Acharya 2017). The differentiation between biosorption and bioaccumulation hinges on the fact that in the latter mechanism, EPs are internalized and stored within the cells, making their containment within the algal biomass less problematic regarding desorption potential. However, similar to biosorption, in this scenario, the compounds can persist in their active, unmetabolized form. Therefore, it is crucial to consider the appropriate disposal and utilization of the exhausted biomass.

6.3.1.3 Biodegradation

Biodegradation represents an energy-driven pathway where microalgae enzymatically degrade EPs, converting them into simpler forms through oxidation, reduction, and hydrolysis reactions. The detoxification process involves three phases. In the initial stage, the enzyme Cytochrome P450 alters lipophilic EPs into hydrophilic counterparts by introducing hydroxyl groups. Subsequently, in the second phase, enzymes conjugate these hydrophilic compounds with glutathione, facilitating further degradation. The third phase employs a variety of enzymes to convert the compounds into simpler, detoxified forms. Biodegradation has been observed in the case of numerous EPs, such as phenolic compounds (Pinto et al. 2002). While this process occurs more slowly than the previous ones, it ensures a thorough degradation of the EPs, reducing concerns about the reusability of the biomass. However, it is noteworthy that biological bioremediation systems not always achieve complete degradation.

6.3.1.4 Photolysis, Hydrolysis, and Volatilization

Abiotic degradation mechanisms, including photolysis, hydrolysis, and volatilization, play a significant role in the dissipation of antibiotics. Photolysis entails the breakdown of antibiotics when exposed to light, while the environmental pH influences hydrolysis. These processes heavily rely on the chemical characteristics of the compounds and the aqueous phase, and there is potential for their interaction, especially in cases of pH-dependent photolysis observed in certain antibiotics (Torniainen et al. 1996). Volatilization involves the dissipation of certain EPs, such as nutrients, through air stripping, which is highly relevant in microalgae-based technologies. This phenomenon has been noted for volatile compounds like 4-octylphenol, galaxolide, and tributyl phosphate (Matamoros et al. 2016). It is crucial to note that algae can significantly influence abiotic factors by introducing sheltering effects from light or altering the composition of the media. This influence occurs when EPs adhere to the cell walls, potentially providing protection, or when algae enhance the dissolved oxygen in the aqueous phase, thereby facilitating the oxidation of EPs.

The mechanisms involved in removing EPs are numerous and intricate, encompassing both biotic and abiotic factors. Comprehending these mechanisms through extensive large-scale studies that tackle the challenges posed by dynamic conditions is vital. Such understanding is essential for optimizing removal performance by strategically manipulating all factors involved in the removal process.

6.3.2 Microalgae Strains Employed in Bioremediation

Different microalgae strains, such as *Scenedesmus*, *Chlorella*, *Chlamydomonas*, and *Spirulina* sp., have undergone thorough examination for their bioremediation potential, as outlined in Table 6.1. These species are frequently chosen due to their robustness in wastewater treatment or well-established metabolic pathways and applications. The composition and pH of the contaminated media greatly influence the choice of strains. Extremophilic algae thriving in extreme pH values, salinity, or temperatures can be employed for specific WW treatment applications. Another crucial aspect to consider is the system for treating wastewater containing EPs, which ideally should closely resemble real environmental conditions.

Emphasizing precision in the terminology used to describe EP removal mechanisms is crucial, specifically focusing on distinguishing between abiotic and biotic factors and how they can be manipulated to improve removal. With the vast diversity of microalgae, there remains a necessity for comparative studies between well-known and lesser-known species to assess their potential in EP treatment. To advance the bioremediation through microalgae, there is a call for scaling up experiments to real-world conditions, employing larger systems that operate with authentic wastewater compositions and incorporate mixtures of EPs at concentrations representative of aquatic environments. Furthermore, comprehensive cost–benefit analyses must be undertaken throughout the entire process, encompassing treatment

Table 6.1 Utilization of microalgae strains in the bioremediation of EP

Microalgae strain	EP category	EP compound	Time (h)	Removal (%)	Removal mechanism	Experimental conditions	Ref
Anabaena sp.	Pesticide	Methyl parathion	792	95	Biodegradation	Medium, Flasks	Barton et al. (2004)
Chlamydomonas reinhardtii	Nutrients	Nitrogen	240	83	Biouptake	WW, 9 L PBR	Kong et al. (2010)
		Phosphorous	240	15	Biouptake		
	Herbicide	Fluroxypyr	120	57	Bioaccumulation, biodegradation, abiotic removal	Medium, Flasks	Zhang et al. (2011)
	Microplastics	Polyethylene terephthalate	672	35	Biodegradation	Solution, Tubes	Kim et al. (2020)
Chlorella fusca	Aromatic organic compound	Bisphenol A	168	90	Biodegradation, bioaccumulation	Medium, Test tubes	Hirooka et al. (2005)
Chlorella pyrenoidosa	Pharmaceutical	Carbamazepine	240	10–30	Biosorption	Urine or Anaerobic treated wastewater, Flasks	de Wilt et al. (2016)
		Diclofenac	240	40–60	Photolysis, biosorption, biodegradation		
		Ibuprofen	240	100	Photolysis, biosorption		
		Metoprolol	240	70–100	Biodegradation, biosorption		
		Paracetamol	240	100	Biodegradation, biosorption		
		Trimethoprim	240	40–60	Photolysis, biosorption		

(continued)

Table 6.1 (continued)

Microalgae strain	EP category	EP compound	Time (h)	Removal (%)	Removal mechanism	Experimental conditions	Ref
Chlorella pyrenoidosa	Antibiotic	Cefalexin	48	95	Bioremoval, Self-degradation	AS, Algal reactor	Guo and Chen (2015)
		Cefixime	48	100	Bioremoval, Self-degradation		
		Cefradine	48	90	Bioremoval, Self-degradation		
		Ceftazidime	48	90	Bioremoval, Self-degradation		
Chlorella sp.	Dye	Pyrene	720	79	Biodegradation	Medium, N.S.	Aldaby and Mawad (2019)
Chlorella sp.	Nutrients	Ammonia	312	94	Biouptake	Centrate wastewater with 1 g L^{-1} crude glycerol, 150 L PBR	Min et al. (2011)
		COD	312	77	Biouptake		
		Nitrogen	312	82	Biouptake		
		Phosphate	312	88	Biouptake		
Chlorella vulgaris	Dye	Basic Fuchsin	172	91	Biosorption, biodegradation	Medium, Flasks	El-Sheekh et al. (2009)
	Herbicide	Atrazine	72	83–90	Biosorption, bioaccumulation	Medium, 1.5 Bottles	González-Barreiro et al. (2006)
		Terbutryn	72	85–93	Biosorption, bioaccumulation		
	Heavy metal	Cr	180	87	Biosorption	Medium, 1 L Column	Xie et al. (2014)
Chlorella zofingiensis	Antibiotic	Levofloxacin	720	97	Biodegradation, bioaccumulation	Medium with glucose, Flasks	Peng et al. (2022)
Chroococcus minutus	Dye	Amido Black 10B	624	49	Biodegradation	Medium, Flasks	Parikh and Madamwar (2005)

Cosmarium sp.	Dye	Malachite green	360	92	Biodegradation	Dye solutions, Flasks	Daneshvar et al. (2007)
Elkatohrix viridis	Dye	Basic Fuchsin	173	93	Biosorption, biodegradation	Medium, Flasks	El-Sheekh et al. (2009)
Gloeocapsa pleurocapsoides	Dye	Acid Red 97	624	78	Biodegradation	Medium, Flasks	Parikh and Madamwar (2005)
		FF Sky Blue	624	77	Biodegradation		
		FF Sky Blue	624	90	Biodegradation		
Gloeothece sp.	Dye	Acid Red 97	624	40	Biodegradation	Medium, Flasks	Parikh and Madamwar (2005)
Nannochloropsis oculata	PAH	Acenaphthylene	336	95	Biosorption, bioaccumulation, biodegradation	Medium, 3 L PBR	Marques et al. (2021)
		Benzo(a)pyrene	336	95	Biosorption, bioaccumulation, biodegradation		
		Benzo(b)fluoranthene	336	95	Biosorption, bioaccumulation, biodegradation		
		Naphthalene	336	96	Biosorption, bioaccumulation, biodegradation		

(continued)

Table 6.1 (continued)

Microalgae strain	EP category	EP compound	Time (h)	Removal (%)	Removal mechanism	Experimental conditions	Ref
Nannochloropsis sp.	Antibiotic	Ciprofloxacin	168	100		Medium, Flasks	Bai and Acharya (2017)
		Sulfamethoxazole	168	11	Photolysis, biosorption, bioaccumulation		
		Trimethoprim	168	11	Algae-mediated photolysis		
	Pharmaceutical	Triclosan	168	100	Photolysis, biosorption, bioaccumulation		
		Carbamazepine	168	13	Biosorption		
Navicula sp.	Pharmaceutical	Ibuprofen	840	20–60	Bioaccumulation, Biodegradation	Medium, Flasks	Ding et al. (2017)
Neochloris oleoabundans	Nutrients	Nitrogen	144	95	Biouptake	AD, 50 L PBR	Levine et al. (2011)
Nostoc linckia	Dye	Basic Fuchsin	170	84	Biosorption, biodegradation	Medium, Flasks	El-Sheekh et al. (2009)
		Methyl red	168	82	Biosorption, biodegradation		
Oscillatoria rubescens	Dye	Basic Fuchsin	171	95	Biosorption, biodegradation	Medium, Flasks	El-Sheekh et al. (2009)
		Methyl red	169	81	Biosorption, biodegradation		

Oscillatoria splendida	Dye	FF Sky Blue	624	58	Biodegradation	Medium, Flasks	Parikh and Madamwar (2005)
Oscillatoria sp.		Pyrene	720	95	Biodegradation	Medium, N.S.	Aldaby and Mawad (2019)
Phaeodactylum tricornutum		Acid Red 97	624	88	Biosorption, bioaccumulation, biodegradation	Seawater with medium, N.S.	Deng et al. (2013)
Phormidium ceylanicum	Heavy metal	Hg	144	68	Biodegradation	Medium, Flasks	Parikh and Madamwar (2005)
Phormidium valderianum	Dye	Acid Red	336	90	Biosorption	Dye solutions, Reactors	Shah et al. (2001)
		Acid Red 119	336	90	Biosorption		
Phormidium valderianum	Dye	Direct Black 155	336	90	Biosorption	Dye solutions, Reactors	Shah et al. (2001)
Phormidium autumnale	Dye	Indigo	336	91	Biodegradation	Textile wastewater, 6 L Bioreactor	Dellamatrice et al. (2017)
Scenedesmus abundans	Microplastics	Poly(methyl methacrylate)	144	98	Biosorption	Medium, Flasks	Cheng and Wang (2022)
Scenedesmus abundans	Nutrients	COD	384	80	Biouptake	Sterilized wastewater, 5.5 L PBR	SundarRajan et al. (2020)
		Nitrogen	384	91	Biouptake		
		Phosphate	384	86	Biouptake		
Scenedesmus incrassatulus	Heavy metal	Cd	312	24	N.S.	Medium, Flasks	Pena-Castro et al. (2004)
		Cr	312	53	N.S.		
		Cu	312	32	N.S.		

(continued)

Table 6.1 (continued)

Microalgae strain	EP category	EP compound	Time (h)	Removal (%)	Removal mechanism	Experimental conditions	Ref
Scenedesmus quadricauda	Fungicide	Dimethomorph	96	44	Bioaccumulation, biodegradation	Medium, Flasks	Dosnon-Olette et al. (2010)
		Pyrimethanil	96	23	Bioaccumulation, biodegradation		
	Herbicide	Isoproturon	96	54	Bioaccumulation, biodegradation		
Selenastrum capricornutum	Polycyclic aromatic hydrocarbon (PAH)	Fluoranthene	96	100	Biosorption, bioaccumulation, biodegradation	Medium, Flasks	Chan et al. (2006)
		Phenanthrene	96	96	Biosorption, bioaccumulation, biodegradation		
		Pyrene	96	100	Biosorption, bioaccumulation, biodegradation		
Spirulina maxima	Heavy metal	Cu	240	82	Biosorption	Wastewater effluent, Flasks	Chan et al. (2014)
		Zn	240	0	Biosorption		
Spirulina platensis	Nutrients	Ammonia	120	100	Biouptake	Sago starch wastewater, 60 L HRAPs	Phang et al. (2000)
		Phosphate	120	99	Biouptake		
Synechococcus elongatus	Herbicide	Atrazine	72	70–80	Biosorption, bioaccumulation	Medium, 1.5 Bottles	González-Barreiro et al. (2006)
		Terbutryn	72	>80	Biosorption, bioaccumulation		
Volvox aureus	Dye	Basic Fuchsin	174	85	Biosorption, biodegradation	Medium, Flasks	El-Sheekh et al. (2009)

Consortium of 15 isolates	Nutrients	Nitrate	240	99	Biouptake	Carpet industry wastewater + Municipal wastewater, Flasks	Chinnasamy et al. (2010)
		Phosphate	240	99	Biouptake		
Consortium: Chlorella sorokiniana predominant	Antibiotic	Sulfamethoxazole	168	54	Biodegradation, bioadsorption, photolysis	Municipal wastewater, 24 L PBR	da Silva Rodrigues et al. (2020)
Consortium: Chlorella sp. *predominant*	Antibiotic	Chlortetracycline	144	100	Bioremoval, photolysis	Swine wastewater, 12 L PBR	Michelon et al. (2022)
		Doxycycline	264	91	Bioremoval, photolysis		
		Tetracycline	264	100	Bioremoval, photolysis		
Consortium: Chlorella sp., *Scenedesmus obliquus, Stichococcus* sp.	Anionic surfactant	Anionic surface active substances	672	73	N.S.	Wastewater+Medi um, 100 L Tank	Safonova et al. (2004)
	Nutrients	BOD	672	97	N.S.		
		COD	672	51	N.S.		
	Heavy metal	Cu	672	62	N.S.		
		Fe	672	64	N.S.		
		Mn	672	70	N.S.		
		Ni	672	62	N.S.		
	Aromatic organic compound	Phenols	672	85	N.S.		
	Heavy metal	Zn	672	90	N.S.		

(continued)

Table 6.1 (continued)

Microalgae strain	EP category	EP compound	Time (h)	Removal (%)	Removal mechanism	Experimental conditions	Ref
Consortium: Chlorella sp., *Scenedesmus* sp. *and Stigeoclonium* sp. *Predominant*	Nutrients	Ammonia	192	100	Biouptake	Microalgal digestate, 30 L PBR	Arias et al. (2018)
		Ammonium	192	58	Biouptake		
		COD	192	70	Biouptake		
		Phosphate	192	100	Biouptake		
Consortium: Chlorella vulgaris and Coenochloris pyrenoidosa	Aromatic organic compound	4-Chlorophenol	120	100	Biodegradation	Wastewater with aromatic compounds, Flasks	Lima et al. (2004)
Consortium: Chlorella vulgaris and Spirulina maxima	Heavy metal	Cu	240	77	Biosorption	Wastewater effluent, Flasks	Chan et al. (2014)
		Zn	240	3	Biosorption		
Consortium: Chlorella vulgaris, Phormidium sp., *Scenedesmus acuminatus, Pseudanabaena acicularis, Scenedesmus acutus*	Pharmaceutical	Ibuprofen	96	26–94	N.S.	Synthetic wastewater, 3.75 L anoxic bioreactor and 11.25 L HRAP	López-Serna et al. (2019)
		Naproxen	96	0–52	N.S.		
		Propylparaben	96	73–100	N.S.		
		Salicylic acid	96	34–97	N.S.		
		Triclosan	96	68–100	N.S.		
Consortium: Scenedesmus obliquus and Scenedesmus quadricauda	Fungicide	Dimethomorph	96	77	Bioaccumulation, biodegradation	Medium, Flasks	Dosnon-Olette et al. (2010)
		Pyrimethanil	96	15	Bioaccumulation, biodegradation		
	Herbicide	Isoproturon	96	58	Bioaccumulation, biodegradation		
	Heavy metal	Cu	240	81	Biosorption		Chan et al. (2014)
Consortium: Synechocystis sp., *Chlorella* sp., *Scenedesmus* sp.		Zn	240	28	Biosorption	Wastewater effluent, Flasks	

N.S. Not specified, Bioremoval: a clear removal mediated by algae without distinction of the specific mechanisms, *AS* Activated sludges, Self-degradation: includes abiotic hydrolysis, photolysis and biotic biosorption Consortium: might include the presence of bacteria, *AD* Anaerobic digestate expenses and the environmental benefits of carbon capture and biomass reusability. This holistic approach can ascertain whether microalgae-based solutions can deliver cost-effective, environment-friendly, and innovative alternatives to conventional treatment methods.

6.3.3 Comparison with Conventional Remediation Methods and Advantages of Microalgae-Based Systems

If traditional wastewater treatment plants fall short in eliminating EPs, advanced chemical and physical treatments offer heightened efficiency. Standout methods for effectively treating organic pollutants and ensuring their destruction are advanced oxidation processes. These processes, which include ozonation, Fenton, electrolysis, photolysis, and photocatalysis, facilitate the oxidation of contaminants, yielding simpler compounds suitable for subsequent bioremediation. Another highly effective technique is adsorption, widely recognized for its proficiency in eliminating and treating organic pollutants in wastewater. This effectiveness is attributed to the inherent affinities of adsorbents with targeted compounds, allowing for their accumulation or adsorption from the liquid phase into a solid phase. Membrane technology represents another advanced approach, employing synthetic membranes to separate chemical components efficiently. The advantage of this technique lies in physically segregating pollutants within specialized membranes without introducing any additional chemicals or compounds into the wastewater stream. It should be noted that the effectiveness of these methods varies, potentially producing secondary metabolites that could be more toxic than the original compounds, often linked to endocrine-disrupting effects. Despite their capability to break down EPs, concerns arise due to the persistence of toxicity, posing risks to water streams (Phoon et al. 2020). Furthermore, adopting these techniques may involve significant costs and is constrained by its limited potential to remove pollutants without offering many additional environmental benefits.

Sustainability concerns extend beyond treating EPs in wastewater, encompassing economic feasibility, environmental impact, and energy consumption. A significant portion of electricity in developed nations, around 0.6–3%, is dedicated to wastewater treatment, with subsequent carbon emissions (Mohsenpour et al. 2021). The use of advanced chemical-physical treatments could exacerbate these issues. Currently, there is a demand for eco-friendly technologies that eliminate EPs, reduce the environmental impact of wastewater treatment plants and produce additional environmental benefits. In the context of circular and bio-based economies, algae-based technology emerges as a promising addition to traditional wastewater treatment plants due to the environmental advantages these organisms provide. Microalgae efficiently not only remove EPs but also, as photosynthetic organisms, capture atmospheric CO_2 in their biomass. They generate valuable compounds like proteins, pigments, fatty acids, and carbohydrates, applicable in fertilizer, feeds, biofuels, or biomaterial production. The Sabana initiative, backed by EU funding between 2016 and 2021, illustrates this idea by implementing an extensive biorefinery centred around microalgae. This biorefinery extracted nutrients from wastewater centrate and pig manure, facilitating the production of biostimulants, biopesticides, feed additives, biofertilizers, and aquafeed.

This technology proved effective in removing nutrients and pollutants from wastewater and demonstrated substantial agricultural advantages. Field trials revealed a 10% increase in crop yield and a 12% reduction in fertilizer demand (Horizon 2020).

Despite the growing microalgae sector, its presence in wastewater treatment remains limited due to regulatory constraints. Future research should account for the intricate nature of wastewater and the potential of biological means for EP removal linked to other environmental benefits. Essential considerations include reusing reclaimed water from wastewater after microalgae treatment and proposing guidelines for safe and efficient microalgae use in wastewater treatment, aligning with environmental goals. Addressing these goals will facilitate the development of a regulatory framework for the use of microalgae-based technology as a bioremediation source for EPs removal.

6.4 Real-World Success Stories: Microalgae Bioremediation

The real-world success stories of microalgae bioremediation span diverse geographical locations, showcasing their bioremediation capabilities and the global applicability of this environmentally friendly approach. Beginning in Belgium, Ghent, the utilization of microalgal bacterial flocs in industrial wastewater treatment was demonstrated through sequencing batch reactors, showcasing the effectiveness of microalgae in tackling intricate industrial pollution challenges. This approach not only presented a promising avenue for sustainable solutions but also underscored the significance of balancing inorganic carbon to organic carbon ratios for optimal performance, along with evaluating their efficiency in removing calcium from calcium-rich wastewater in diverse (agro-)industrial wastewater types (Van Den Hende et al. 2014). Moving to Spain, Valladolid, *Oocystis* sp. demonstrated its efficacy in treating fish processing wastewater (Riaño et al. 2011). In Pune, India, the bioremediation of domestic wastewater using *C. vulgaris* and *S. quadricauda* showcased the applicability of this technology in addressing local water quality challenges (Kshirsagar 2013). Shifting to South Korea, Wonju specifically, *S. obliquus*, isolated from municipal wastewater effluent, treated piggery wastewater, underlining its versatility across wastewater types (Ji et al. 2013). Further, in Kuala Lumpur, Malaysia, the use of *C. vulgaris* proved effective in bioremediating textile wastewater (Lim et al. 2010). Highlighting the environmental threat of heavily contaminated distillery effluents (with sugars, dark-coloured compounds like melanoidins, ethanol, volatile organic compounds, suspended solids, etc.), a consortium of cyanobacterial strains, including *Oscillatoria* sp., *Lyngbya* sp., and

Synechocystis sp., demonstrated significant melanoidin decolourization (Singh and Patel 2012). Studies in Tamil Nadu, India, identified *Nostoc muscorum* as a dominant cyanobacteria genus in distillery effluent, showcasing its potential for removing various chemicals and nutrients (Selvam et al. 2011). Together, these studies emphasize the global applicability of microalgae and cyanobacteria in diverse industrial contexts, highlighting their adaptability across various wastewater types and their potential to address local water quality challenges. Moreover, they demonstrate the potential for regional solutions to environmental issues. Further research is needed to address effluents that have not been widely studied. Additionally, scaling up successful microalgae bioremediation projects from different locations could provide valuable insights into scalability and practical implementation on a larger scale.

6.5 Challenges and Limitations

The latest progress in leveraging microalgal-based bioremediation for EPs offers a sustainable and promising shift from traditional physicochemical approaches employed in wastewater treatment. Overcoming different challenges is crucial to positioning microalgae as a competitive, cost-effective, and sustainable solution for EP bioremediation in wastewater.

6.5.1 Nutrient Availability and Algal Growth

Algal nutrient requirements for efficient biomass production and growth are most probably in higher demand as they consume excess nutrients than their proximate growth demands and stock it internally in algal cells. The key essential macro nutrients needed by most algae include nitrogen, phosphorous, sulphur and iron. Nitrogen is the building block for protein and nucleic acid formation, whereas phosphate is part of the backbone of RNA and DNA as well as phospholipids. Nitrogen deficiency in algal culture results in accumulation of lipids, triglycerides, and reduced protein content. This results in enhanced lipid/protein ratio. Phosphate limitations reduce chlorophyll content and overall protein thereby increasing/decreasing absolute carbohydrate content in microalgae cells, resulting in the accumulation of astaxanthin, which eventually hinders algal growth (Procházková et al. 2014). Trace metal deficiencies can cause algal growth limitation, whereas metal concentrations above the threshold leave may damage the cell wall, deplete antioxidants, inhibit growth, and impair photosynthesis. Iron deficiency may result in the substitution of ferredoxin with flavodoxin and is toxic since the catalytic role of ferredoxin is superior to flavodoxin. Non-essential metals such as Pb, Cr and Cd can inhibit various metabolic pathways and hinder algal growth. Whereas essential elements like Zn and Cu over trace quantities can delay nutrient uptake and inhibit carbon fixation

(Liu et al. 2021). The essential nutrient for efficient algal growth is often overlooked, as most of the nutrients are phycosequestrated when present in their natural environment. Correlation between algal biomass cultivation and nutrient load has been inferred exclusively with respect to coastal regions and eutrophication of lakes, but poorly studied with overall biomass productivity in bioreactors or in aquaculture. If terrestrial agriculture is considered as a comparison model, then a major challenge in improving efficiency in biomass cultivation, befalls providing adequate nutrients for scaling up algal cultivation. In the currently operating terrestrial agriculture sector, micro- and macronutrient supplements, or fertilizer, account for large expenditure. Microalgal strains are resilient to the harsh environmental conditions found in municipal and industrial waste. The transition from a linear bioeconomy based on resource diversification to a low-carbon, integrated biorefinery model that can handle various waste streams has also led to the restructuring of conventional wastewater treatment procedures. Major bottlenecks for algal bioremediation over conventional wastewater treatment are lack of data on proper nutrient assimilation and enhanced microalgal population. Phycoremediation with superior nutrient content generates more biomass concentrations with higher value-added products which results in lower overall production cost (Saravanan et al. 2023). Nitrogen and phosphorus supply concentrations for scaled up algal cultures are reported to range from 25 to 5000 mg L^{-1} and 0.98 to 179 mg L^{-1}, respectively, with the average N:P ratio in algal cells stated to be 16:1 (Figler et al. 2021). As previously mentioned, ammonia volatilization in open ponds will cause the nitrogen concentration available for biomass cultivation and the total available N/P ratio of the wastewater to differ from the existent nitrogen contents and N/P ratio of the wastewater, so these variations should be investigated (Chaudry 2021). Most of the experimental and modelling studies still make use of Monod formulations, which neglect internal nutrient storage in algal cells. Furthermore, Monod formulations are employed by most of the researchers due to the simplicity, but under nutrient-limiting conditions, they underpredict the actual algal growth rate. N and P uptake rates in certain algal species wide fluctuations in nutrient uptake help them to withstand environmental variations more efficiently (Procházková et al. 2014). Internal storage and nutrient availability in bulk plays a vital role in P uptake and are independent of immediate growth requirements. The rate limitation on the growth of algal strains can be best analysed by Liebig's law of the minimum and Droop's formula for nutrient uptake, which clearly suggests that algal growth rate at any instant is only limited by utmost negative impact factor on growth and non-multiplicative rule (Shriwastav and Bose 2015). Nutrient stress conditions with respect to different algal species and their effects on its metabolism are discussed in Table 6.2.

In summary, the diverse responses of microalgae strains to various nutrient stress conditions underscore the intricate interplay between environmental factors and algal metabolisms, providing valuable insights for optimizing cultivation strategies and harnessing the potential of microalgae in biotechnological applications.

Table 6.2 Influence of nutrient stress on the metabolism of different microalgae species

Microalgae strains	Nutrient stress conditions	Effect	Ref
Emiliania huxleyi	Exposure to lower concentrations of CO_2	Results in an increase in 22:6 (n−3) PUFA	Riebesell et al. (2000)
Dunaliella salina	Exposure to increased concentrations of CO_2 with changes in composition	Increased fatty acid accumulation	Muradyan et al. (2004)
Chlorella pyrenoidosa	Addition of excess ammonia as a stress condition	Increased amino acid content	Safafar et al. (2016)
Spirulina platensis	Exposure to a deficit of nitrogen	Reduced carbon fixation capacity	Zaparoli et al. (2020)
Dunaliella sp. and *Haematococcus pluvialis*	Depletion of nitrogen condition	Increased amounts of astaxanthin, carotenoids, and acyl esters (up to 13% w/w) accumulation	Zhekisheva et al. (2002)
Scenedesmus sp.	Reduction of total phosphorus from 2.0 mg L^{-1} to 0.1 mg L^{-1}	Increased lipid content (23–53%)	Van der Hoek et al. (2018)
Chlamydomonas reinhardtii	Limitation in phosphorous content	Lead to a reduction in phosphatidylglycerol (PG)	Sugimoto et al. (2008)
Selenastrum minutum	Starvation conditions for phosphorus	Reduced respiration rate	Theodorou et al. (1991)
Chlorella vulgaris	Exposure to higher concentrations of iron	Showed an increase in lipid content	Liu et al. (2008)
Chlorella pyrenoidosa	Exposure to over trace amounts of copper	Inhibited both cell division and photosynthesis	Chen et al. (2016)
Chlorella vulgaris	Salt stress due to NaCl	Lead to an increased lipid content	Gauthier et al. (2020)
Anabaena sp.	Exposure to over trace amounts of zinc	50% inhibition in biomass growth	Tang et al. (2013)
Dyctiosphaerium chlorelloides	Exposure to higher concentrations of chromium (Cr(III))	50% inhibition in biomass growth	D'ors et al. (2010)
Nostoc minutum	Exposure to very high concentrations of arsenic (As(V))	Stimulated biomass growth	Ferrari et al. (2013)
Isochrysis galbana	Exposure to minimal concentrations of aluminium (Al)	50% inhibition in biomass growth	Trenfield et al. (2015)
Pseudokirchneriella subcapitata	Exposure to higher concentrations of copper (Cu(II))	Lead to a loss of membrane integrity	Machado et al. (2015)
Coelastrella sp.	Exposure to minimal concentrations of zinc (Zn(II))	Maximized NH3-N and total phosphorus (TP) removal efficiency	Li et al. (2020)

(continued)

Table 6.2 (continued)

Microalgae strains	Nutrient stress conditions	Effect	Ref
Neochloris oleoabundans and *Nannochloropsis* sp.	Nitrogen deprivation as a stress condition	Increased in cellular lipid accumulation	Martin et al. (2014)
Tetraselmis subcordiformis	Exposure to low concentrations of nitrogen (N)	Induced carbohydrate biosynthesis	Ho et al. (2013)
Acutodesmus dimorphus	Nitrogen starvation as a stress condition	A two-fold increase in protein synthesis	Chokshi et al. (2017)
Chlorella vulgaris	Nutrient deprivation for phosphorus (P)	Increased biomass productivity	Goiris et al. (2015)
Chlorella sorokiniana and *Scenedesmus acuminatus*	Exposure to high concentrations of copper (Cu)	Increased biomass productivity	Hamed et al. (2017)

6.5.2 Toxicity and Bioaccumulation

Aquatic biota function as significant sinks for primary and secondary pollutants, but they are also highly susceptible to pollution (Upadhyay et al. 2020). The bioavailability, toxicity, and bioaccumulation of heavy metals in aquatic systems are influenced by various environmental factors. Metal uptake involves the adsorption of metal ions to cell surfaces, followed by chemical interactions with essential metal functional groups within the cell wall. Microalgal growth inhibition studies, focusing on acute toxicity with algae in wastewater, utilized endpoints like cell membrane integrity and antioxidant enzyme activity. The mechanism of metal ion entry into algal cells includes processes like reverse osmosis, disruption of protein structure, molecular mimicry, or endocytosis (Monteiro et al. 2012). Metal accumulation rates in algal cells are influenced by environmental parameters, with species-specific variations. Algal tolerance to heavy metal accumulation relies on defence responses against oxidative damage caused by reactive oxygen species (ROS). Heavy metal toxicity, especially with cadmium (Cd), is intricate due to pleiotropic effects altering metabolic processes and inducing ultrastructural changes. Cd is absorbed by marine cells as complexes with organic matter and is integrated into both organic and inorganic substances. Cd inhibits phosphorus uptake, and this adverse effect intensifies within the pH range of 5.5 to 8.5. Zinc, another hazardous metal, is rapidly absorbed by algae and integrated into polysaccharide and nucleic acid fractions. Microalgae play a crucial role in bioaccumulating heavy metals due to their large, exposed surface area (Pena-Castro et al. 2004; Chan et al. 2014). They possess significance in adverse environmental conditions, heavy metal tolerance, large-scale biomass production, autotrophic/heterotrophic growth with high photosynthetic efficiency, and genetic manipulation capacity. In the natural environment, algal cell surfaces endure continuous exposure to toxic pollutants, demonstrating the ability to absorb, adsorb, and phytosequester, thereby reducing elements from polluted biota.

6.5.3 Scale-Up Challenges

Efficient algal cultivation places a primary emphasis on extracting valuable by-products like lipids and biomass. Heterotrophic cultivation emerges as the most effective method for enhancing by-product yield, generating substantial interest from researchers and bioenergy companies keen on commercializing algal technology, particularly through the heterotrophic technique. However, the downside of heterotrophic cultivation lies in the potential for cross-contamination with other microorganisms, posing significant challenges in the setup and operation of large-scale installations. The primary utility of this cultivation mode lies in the production of primary or secondary metabolites with high market value, driven by the costliness of pure organic carbon sources. While current commercial microalgal culture systems typically operate at around 1000 L in open pond reactors, the risk of cross-contamination underscores the suitability of closed photobioreactors for effective commercial scaling up.

Operational challenges further complicate the picture, with contamination in open reactors impacting algal biomass concentration, especially with EPs like heavy metals (Cd, Pb, Cu, Zn, Hg), phenols, and pesticides (DDT, Atrazine), as well as organic acids (formic acid). Harvesting microalgal biomass and subsequent processing pose challenges with associated operational costs, while economic feasibility concerns arise when considering the potential use of microalgal biomass for value-added bioproducts post-bioremediation. Scaling up microalgae culture from laboratory to large-scale for commercial production is a complex and comprehensive process, presenting numerous obstacles that demand innovative solutions. Additionally, there are research gaps related to the limited integration of microalgae with other treatment methods, highlighting the need to explore new microalgal species and evaluate their real-time applications, configurations, and designs for enhanced functionality in diverse wastewater treatment scenarios. Addressing these challenges is crucial to unlock the full potential of microalgal-based bioremediation in mitigating the impact of ECs in wastewater. Microalgae cultivation not only holds promise as a sustainable and economically viable technology but also presents comprehensive applications in bioremediation and renewable energy/bioproduct production (Koutra et al. 2023). Achieving these applications requires a focused approach, concentrating on contamination management, resource optimization, efficient light capture, genetic engineering, and a thorough examination of environmental and economic factors. Intensive research and development efforts, coupled with adherence to species-specific standard protocols, will further ensure a brighter and more sustainable microalgae-based industrial growth.

6.6 Strategies for Enhancing Microalgae-Based Bioremediation

Microalgae-based bioremediation holds immense potential, and several strategies can be employed to enhance its effectiveness in removing EPs from wastewater. The continuous improvement of microalgae's capability to absorb and degrade pollutants is crucial for the development of sustainable and efficient bioremediation techniques.

6.6.1 Significance of Strain Selection

The significance of strain selection in microalgae-based bioremediation presents the foundation for the effectiveness and efficiency of the entire remediation process. When considering the diversity of microalgae species available, each possessing unique characteristics and capabilities, strategic strain selection becomes essential. This comprehensive selection process encompasses various factors such as pollutant tolerance, growth rate, nutrient uptake efficiency, biomass productivity, and adaptability to environmental conditions. In the aspect of biotechnological application, the mass cultivation ability of the selected strain is very important (Silkina et al. 2019). The multifaceted aspects of strain selection and its profound implications for enhancing microalgae-based bioremediation efforts are in focus for this section.

One of the primary considerations in strain selection is the inherent ability of microalgae to tolerate and metabolize pollutants present in contaminated environments. As the microalgae are ones of the first organisms that colonized this planet, certain microalgae species have evolved mechanisms to thrive in polluted waters with a high content of nutrients, exhibiting enhanced resistance to toxic substances such as heavy metals, hydrocarbons, pesticides, and industrial effluents (Faluweki et al. 2023). By selecting strains with robust pollutant tolerance, e.g., *Desmodesmus* and *Chlorella,* the remediation process can effectively target specific contaminants while minimizing the risk of adverse effects on the microalgae population (Hu et al. 2021; Hejna et al. 2022).

Moreover, the growth characteristics of microalgae play a pivotal role in determining their suitability for bioremediation applications. Strains with rapid growth rates and high biomass productivity are favoured, as they facilitate the efficient utilization of nutrients and energy sources for pollutant uptake and metabolism, such as green microalgae and cyanobacteria (Agrawal et al. 2020; Forján et al. 2015; Bhatt et al. 2022). Fast-growing microalgae species can accelerate the remediation process, leading to more rapid reductions in pollutant concentrations and shorter remediation durations.

Nutrient uptake efficiency is another critical factor that influences the effectiveness of microalgae-based bioremediation and their mass cultivation. Microalgae strains capable of efficiently assimilating nutrients such as nitrogen, phosphorus, and micronutrients from the surrounding environment can help to compete with other organisms (bacteria and protozoa) and sustain robust growth even in nutrient-deficient/rich conditions. By optimizing nutrient uptake processes, selected strains can maintain high metabolic activity levels, thereby enhancing pollutant removal rates and overall remediation performance.

Furthermore, the adaptability of microalgae to diverse environmental conditions is essential for ensuring sustained remediation efficacy over extended periods. Strains that demonstrate resilience to fluctuations in temperature, pH, salinity, and light availability are better equipped to thrive in dynamic environmental settings commonly encountered in polluted ecosystems. This adaptability confers a competitive advantage to selected strains, allowing them to persist and proliferate under challenging conditions while actively participating in pollutant degradation processes.

6.6.2 Genetic Modification for Improved Pollutant Uptake

Genetic modification represents a cutting-edge strategy for enhancing microalgae-based bioremediation by equipping microalgae with advanced capabilities to tackle pollutants more effectively. This approach involves the targeted manipulation of microalgae genomes to introduce specific genes or modify existing ones, thereby augmenting their pollutant uptake, metabolism, and remediation potential. Genetic modification holds immense promise to change the field of bioremediation and address environmental challenges with unprecedented precision and efficiency.

As an example, a study by Qi et al. (2018) employed UV mutagenesis to induce mutations in *Scenedesmus obliquus*, a type of microalgae. Additionally, they applied low pH conditions as a selective pressure to identify mutants with increased CO_2. The mutant strain demonstrated higher biomass productivities with 255 higher lipid content than natural strain, and light conversion efficiencies when exposed to elevated CO_2 concentrations compared to the parent strains. This indicates that the mutant strain could utilize CO_2 more efficiently for biomass production. This research serves as compelling evidence, affirming the feasibility of employing genetically transformed algae as a proven approach to remediate various types of contaminants.

In the context of microalgae-based bioremediation, genetic engineering enables tailoring and modifying microalgae strains to thrive in polluted environments and actively participate in the removal and degradation of contaminants. One of the primary objectives of genetic modification is to enhance microalgae's ability to uptake, sequester, and metabolize pollutants, thereby accelerating the remediation process and mitigating environmental damage.

One of the most promising applications of genetic modification in microalgae-based bioremediation is the introduction of genes encoding enzymes or transporters that facilitate the breakdown and detoxification of specific pollutants, such as heavy metals. One of the aspects of metal recovery was done by Cheng et al. (2019) using the method of the exogenous DNA by constructing transgenic algal strains (Hassanien et al. 2023). Another example is that microalgae can be engineered to express enzymes capable of metabolizing hydrocarbons, pesticides, pharmaceuticals, and other recalcitrant pollutants commonly found in contaminated ecosystems (Kumar and Bharadvaja 2019). By genetic modification, microalgae could produce specialized enzymes, enabling them to transform harmful pollutants into harmless by products or assimilate them into biomass for subsequent removal.

In addition to improving EP uptake and metabolism, genetic modification can enhance microalgae's tolerance to adverse environmental conditions commonly encountered in polluted ecosystems. By introducing genes associated with stress tolerance, such as heat shock proteins, antioxidants, and osmolyte synthesis enzymes, scientists can bolster microalgae's resilience to factors such as temperature fluctuations, pH changes, salinity variations, and exposure to toxic compounds. This increased resilience enables genetically modified microalgae to thrive in harsh environmental conditions and maintain robust remediation performance over extended periods.

However, it is essential to recognize the potential risks and ethical considerations associated with genetic modification, including unintended ecological consequences, gene flow, and regulatory concerns, when the process could be potentially applied in the field (Sayler and Ripp 2000). Therefore, comprehensive risk assessments, environmental monitoring, and regulatory frameworks must be established to ensure the safe and responsible deployment of genetically modified microalgae in bioremediation applications.

6.6.3 Synergistic Approaches with Other Biotechnological Interventions

Synergistic approaches involving microalgae-based bioremediation and other biotechnological interventions hold significant promise for addressing environmental pollution and restoring ecosystem health. By combining the unique capabilities of microalgae with complementary techniques and microorganisms, synergistic strategies can enhance pollutant removal efficiency, broaden the range of contaminants targeted, and promote more sustainable remediation outcomes.

One of the most promising synergistic approaches involves coupling microalgae-based bioremediation with microbial consortia or co-cultures, where microalgae and bacteria work synergistically to degrade pollutants and enhance remediation efficiency (Fernandes et al. 2022). Certain bacteria possess specialized metabolic pathways capable of breaking down complex organic pollutants, such as hydrocarbons, pesticides, and pharmaceuticals, into simpler compounds that can be readily assimilated by microalgae or further degraded through microbial activity. By co-cultivating microalgae with pollutant-degrading bacteria, synergistic interactions can be leveraged to accelerate pollutant removal rates, improve overall remediation performance, and enhance ecosystem resilience.

Moreover, microbial-assisted bioremediation strategies, such as bioaugmentation and biostimulation, can be integrated with microalgae-based bioremediation to augment pollutant degradation capabilities and promote ecological restoration. Bioaugmentation involves the deliberate introduction of specialized pollutant-degrading microorganisms into contaminated environments, where they can synergistically interact with microalgae to enhance pollutant removal and detoxification. Similarly, biostimulation techniques involve the addition of nutrients, electron acceptors, or other growth-promoting factors to stimulate the growth and activity of indigenous microorganisms, potentially isolated from the treatment site, thereby enhancing their capacity to metabolize pollutants and support microalgae growth (Silkina et al. 2017).

Additionally, the integration of emerging biotechnological innovations, such as nanotechnology, omics technologies, and synthetic biology, holds immense potential for advancing synergistic approaches in microalgae-based bioremediation. Nanomaterials, for example, can be used to enhance pollutant adsorption, promote microbial attachment, and facilitate the targeted delivery of nutrients and growth factors to microalgae-bacteria consortia. Omics technologies, including genomics, transcriptomics, proteomics, and metabolomics, provide valuable insights into the molecular mechanisms underlying synergistic interactions and can inform the design of more effective bioremediation strategies. Similarly, synthetic biology approaches enable the rational design and engineering of microalgae and microbial consortia with tailored traits optimized for pollutant uptake, metabolism, and environmental resilience.

6.6.4 Optimization of Operational Parameters

Optimization of operational parameters is a critical aspect of enhancing microalgae-based bioremediation, as it directly impacts the efficiency, effectiveness, and sustainability of the remediation process. By carefully managing key parameters such as light intensity, temperature, pH, nutrient availability, mixing, and CO_2 supply, it is possible to create optimal conditions for microalgae growth, metabolism, and pollutant uptake. In this discussion, we explore the significance of operational parameter optimization and its role in maximizing the remediation potential of microalgae-based bioremediation systems (Silkina et al. 2015).

Light intensity is a crucial factor influencing microalgae growth and photosynthetic activity. By optimizing light intensity levels within bioremediation systems, sufficient light conditions for algal photosynthesis need to be provided. Natural light access, light-emitting diodes (LEDs), and solar collectors are among the technologies employed to regulate and modulate light intensity, enabling precise control over microalgae growth rates and biomass productivity.

Temperature plays an equally important role in determining microalgae growth rates, metabolic activity, and nutrient uptake kinetics. By maintaining optimal temperature conditions within bioremediation systems, enhanced pollutant removal efficiency could be demonstrated by algal culture. Temperature control mechanisms such as heating and cooling systems, thermal insulation, and thermal stratification can be implemented to stabilize temperature gradients and mitigate fluctuations caused by ambient conditions.

pH levels influence various physiological processes in microalgae, including nutrient uptake, enzyme activity, and metabolic pathways. By regulating pH levels within bioremediation systems, the optimum conditions for microalgae growth and pollutant uptake will be demonstrated. pH adjustment techniques on a large scale are usually controlled by CO_2 injection, as algal culture growth provides the buffering capacity. CO_2 supply is critical for sustaining microalgae photosynthesis, biomass production, and waste compound uptake.

Nutrient availability is a key determinant of microalgae biomass productivity and pollutant uptake capacity. By optimizing nutrient concentrations and ratios within growth media, robust microalgae growth and metabolism while minimizing nutrient limitations or excesses could be sustained. Nutrient supplementation strategies such as batch feeding, continuous dosing, and recycling enable efficient utilization of resources and support long-term sustainability of bioremediation operations.

Mixing and agitation are essential for maintaining uniform nutrient distribution, preventing sedimentation, and promoting gas exchange within bioremediation systems. By optimizing mixing parameters such as mixing system and flow patterns, nutrient diffusion rates could be enhanced, minimizing mass transfer limitations and improving overall process efficiency.

In conclusion, optimization of operational parameters is essential for maximizing the remediation potential of microalgae-based bioremediation systems. By carefully managing key parameters such as light intensity, temperature, pH, nutrient availability, mixing, and CO_2 supply, the optimal conditions for microalgae growth, metabolism, and pollutant uptake can be created. Through strategic parameter optimization, microalgae-based bioremediation small and large systems can achieve higher efficiency, effectiveness, and sustainability, contributing to the restoration and preservation of environmental ecosystems for future generations.

6.7 Environmental and Economic Sustainability of Microalgae-Based Bioremediation

In the future, microalgae-based bioremediation can offer a promising approach for the removal of EPs from wastewater, providing not only environmental benefits but also economic advantages. Evaluating the sustainability of this technology requires a comprehensive assessment of its environmental impact and economic feasibility.

6.7.1 Assessment of Environmental Impact

The assessment of environmental impact is integral to understanding the efficacy and sustainability of microalgae-based bioremediation for the removal of emergent pollutants (Scherer et al. 2017). This process involves evaluating the direct and indirect effects of employing microalgae as a bioremediation tool on surrounding ecosystems, including the chemical composition, toxicity, and persistence of targeted pollutants. Monitoring interactions between microalgae and the ecosystem entails studying changes in water quality, biodiversity, and nutrient cycling within the remediation site to predict long-term ecological consequences and ensure ecosystem health (Zhang et al. 2020). Additionally, assessing the energy and resource requirements of microalgae-based bioremediation is crucial for determining its sustainability, optimizing processes to minimize environmental footprint, and maximizing efficiency. A comprehensive evaluation of environmental impact provides valuable insights into the effectiveness, feasibility, and sustainability of microalgae-based bioremediation, enabling the development of environmentally sound strategies that contribute to the preservation and restoration of ecosystems.

6.7.2 Cost–Benefit Analysis of Microalgae-Based Bioremediation Downgrading Heating Energy

The co-cultivation of microalgae with heterotrophic microorganisms like fungi, bacteria, and yeast involves a symbiotic cooperation, facilitating the exchange of nutrients and metabolites. This collaboration enhances the efficiency of wastewater treatment, leading to improved bioremediation and the growth of algal biomass. Economic assessment of dried algal biomass grown in a high-volume V-shaped pond is estimated USD 0.482 kg^{-1} on diary effluent (Bai and Acharya 2017). Anaerobically digested household effluent and centrate, the projected expenses of wastewater treatment integrated with microalgae are USD 1.69 m^{-3} and USD 0.53 m^{-3}, respectively. For anaerobically digested domestic effluent, the wastewater treatment process was reported to produce 4.3 kg biomass m^{-3} as well as 1.34 kg biomass m^{-3} for centrate (Chaudry 2021). However, the expected cost of treating sewage is USD 0.18 m^{-3}, producing 0.24 kg biomass m^{-3} of effluent. Overall credit generated by biomass over USD 0.73/kg would be used to offset the cost of treating sewage. Similarly, the estimated costs of treating agro-industrial wastewater, AD piggery effluent and AD abattoir effluent with algae are USD 0.24 m^{-3}, USD 0.83 m^{-3}, and USD 0.78 m^{-3}, respectively. Any credit produced by biomass above, USD 0.41 kg^{-1}, USD 0.51 kg^{-1} and USD 0.92 kg^{-1} will be used to offset the wastewater treatment costs for these industries. When compared to either no treatment at all or expensive conventional wastewater treatment (USD 0.22 m^{-3}), phycoremediation shows great potential (Cashman et al. 2014). Phycoremediation process should be analysed and designed based on individual requirements, considering the economic constrains, availability of land, geographic location and use of the biomass value-added products to estimate the total economic sustainability of the process. In developing countries, the phycoremediation process should be optimized to reduce the overall cost along with improved biomass yield. Wet biomass should be consumed animal/aqua culture feed, biofertilizer, biodigestion of algal biomass for value-added products and energy recovery. In contrast, for developed countries, the chief objective of phycoremediation should be a downstream process and should focus on maximum biomass recovery. Moreover, closed photobioreactors and/or CO_2 supplies should be contemplated to boost production and reduce the amount of space needed. A superstructure optimization study would, in any event, provide a thorough examination of the procedure. Furthermore, GHG emission credit for the process needs to be investigated as microalgae can potentially also sequester CO_2.

6.7.3 Comparison with Conventional Remediation Methods

Comparing microalgae-based bioremediation with conventional remediation methods is vital for understanding its competitiveness and advantages. Conventional methods, such as chemical treatments or activated sludge processes, may have established infrastructure but could be energy-intensive and generate sludge. Microalgae-based bioremediation, in contrast, can offer a more sustainable and energy-efficient alternative.

The comparison should consider treatment efficiency, resource recovery, and the adaptability of each method to different types of contaminants. Additionally, the scalability and ease of integration into existing wastewater treatment plants or industrial processes are critical factors in determining the overall feasibility of microalgae-based bioremediation.

In conclusion, a comprehensive evaluation encompassing environmental impact, cost–benefit analysis, and a comparative assessment with conventional methods is essential to gauge the true sustainability of microalgae-based bioremediation. This multidimensional analysis will guide decision-makers, researchers, and industry professionals in advancing and implementing this innovative technology for efficient and sustainable wastewater treatment.

6.8 Future Perspectives

The future of microalgae-based bioremediation holds exciting possibilities, with emerging trends, integration with circular economy concepts, and potential industrial applications shaping the trajectory of research and practical implementation. As technology continues to evolve, several noteworthy trends are surfacing in the field. Genetic engineering and synthetic biology are advancing the creation of tailor-made microalgae strains with heightened pollutant uptake capabilities. Omics technologies, encompassing genomics, transcriptomics, and metabolomics, provide profound insights into molecular mechanisms governing pollutant degradation and cellular responses. Exploring microalgal consortia and their synergistic interactions is gaining prominence for a more versatile approach to combat a broader spectrum of contaminants.

A significant trend involves the integration of artificial intelligence and machine learning for optimizing bioremediation processes. These technologies empower predictive modelling, real-time monitoring, and adaptive control systems, elevating the efficiency and reliability of microalgae-based treatment systems. The alignment of microalgae-based bioremediation with circular economy principles marks a promising avenue for sustainable resource management. Closing the loop by utilizing recovered biomass for applications like biofuel production or biofertilizers echoes circular economy ideals, driving innovative resource recovery approaches within the microalgae bioremediation process.

Circular economy strategies extend to exploring synergies between microalgae-based bioremediation and other waste streams or industrial processes, fostering interconnected systems where outputs from one process become valuable inputs for another. Recognition of the potential industrial applications of microalgae-based bioremediation is growing, with the technology's scalability and adaptability positioning it as a viable option for large-scale industrial wastewater treatment. Tailored microalgae-based solutions are increasingly relevant for industries grappling with diverse pollutant profiles such as pharmaceuticals, heavy metals, or organic pollutants.

Policy implications are critical for shaping the future of microalgae-based bioremediation. Governments and regulatory bodies play a pivotal role in incentivizing the adoption of sustainable technologies. Future research directions should prioritize the development of clear regulatory frameworks, guidelines, and incentives to seamlessly integrate microalgae-based bioremediation into existing wastewater management practices.

In conclusion, the future of microalgae-based bioremediation hinges on refining technologies, embracing interdisciplinary approaches, and aligning with circular economy principles. The seamless integration of these innovations into industrial practices, coupled with supportive policies, holds the key to realizing the full potential of microalgae-based bioremediation as a sustainable and efficient solution for environmental pollution mitigation.

References

Aditya L et al (2023) Role of culture solution pH in balancing CO2 input and light intensity for maximising microalgae growth rate. Chemosphere 343:140255

Agrawal K et al (2020) Bioremediation: an effective technology toward a sustainable environment via the remediation of emerging environmental pollutants. In: Emerging technologies in environmental bioremediation. Elsevier, pp 165–196

Ahamad A et al (2020) Types of water pollutants: conventional and emerging. In: Sensors in water pollutants monitoring: role of material. Springer, pp 21–41

Ahmad HA et al (2022) The environmental distribution and removal of emerging pollutants, highlighting the importance of using microbes as a potential degrader: a review. Sci Total Environ 809:151926

Aldaby E, Mawad A (2019) Pyrene biodegradation capability of two different microalgal strains. Global NEST J 21(3):290–295

Arias DM et al (2018) Integrating microalgae tertiary treatment into activated sludge systems for energy and nutrients recovery from wastewater. Bioresour Technol 247:513–519

Arman NZ et al (2021) A review on emerging pollutants in the water environment: existences, health effects and treatment processes. Water 13(22):3258

Bai X, Acharya K (2017) Algae-mediated removal of selected pharmaceutical and personal care products (PPCPs) from Lake Mead water. Sci Total Environ 581:734–740

Banik B (2021) Emerging contaminants in agriculture and their impacts on environment. In: International Conference on Emotions and Multidisciplinary Approaches- ICEMA

Barton J et al (2004) Reductive transformation of methyl parathion by the cyanobacterium Anabaena sp. strain PCC7120. Appl Microbiol Biotechnol 65:330–335

Bayramoğlu G et al (2006) Biosorption of mercury (II), cadmium (II) and lead (II) ions from aqueous system by microalgae Chlamydomonas reinhardtii immobilized in alginate beads. Int J Miner Process 81(1):35–43

Bhatt P et al (2022) Microalgae-based removal of pollutants from wastewaters: occurrence, toxicity and circular economy. Chemosphere 306:135576

Cashman S, Gaglione A, Mosley J, Weiss L, Ashbolt N, Hawkins T, Cashdollar J, Xue X, Ma C, Arden S (2014) Environmental and cost life cycle assessment of disinfection options for municipal wastewater treatment. U.S. Environmental Protection Agency, Washington, DC

Chan SMN et al (2006) Removal and biodegradation of polycyclic aromatic hydrocarbons by Selenastrum capricornutum. Environ Toxicol Chem: Int J 25(7):1772–1779

Chan A, Salsali H, McBean E (2014) Heavy metal removal (copper and zinc) in secondary effluent from wastewater treatment plants by microalgae. ACS Sustain Chem Eng 2(2):130–137

Chan SS et al (2022) Recent advances biodegradation and biosorption of organic compounds from wastewater: microalgae-bacteria consortium-a review. Bioresour Technol 344:126159

Chaudry S (2021) Integrating microalgae cultivation with wastewater treatment: a peek into economics. Appl Biochem Biotechnol 193(10):3395–3406

Chen Z et al (2016) Toxicity of Cu (II) to the green alga Chlorella vulgaris: a perspective of photosynthesis and oxidant stress. Environ Sci Pollut Res 23:17910–17918

Cheng SY, Show PL, Lau BF, Chang JS, Ling TC (2019) New prospects for modified algae in heavy metal adsorption. Trends Biotechnol 37(11):1255–1268

Cheng Y-R, Wang H-Y (2022) Highly effective removal of microplastics by microalgae Scenedesmus abundans. Chem Eng J 435:135079

Chinnasamy S et al (2010) Microalgae cultivation in a wastewater dominated by carpet mill effluents for biofuel applications. Bioresour Technol 101(9):3097–3105

Chokshi K et al (2017) Nitrogen starvation-induced cellular crosstalk of ROS-scavenging antioxidants and phytohormone enhanced the biofuel potential of green microalga Acutodesmus dimorphus. Biotechnol Biofuels 10:1–12

Choudhary S, Tripathi S, Poluri KM (2022) Microalgal-based bioenergy: strategies, prospects, and sustainability. Energy Fuel 36(24):14584–14612

Colzani L et al (2024) Determination of pollutants, antibiotics, and drugs in surface water in Italy as required by the third EU Water Framework Directive Watch List: method development, validation, and assessment. Environ Sci Pollut Res:1–13

Council EP (2002) Directive 2000/60/EC of the European Parliament and of the Council of 23 October 2000 establishing a framework for community action in the field of water policy. Environmental Research Quarterly. pp 66–106

D'ors A et al (2010) Toxic effects and specific chromium acquired resistance in selected strains of Dyctiosphaerium chlorelloides. Chemosphere 81(2):282–287

Daneshvar N et al (2007) Biological decolorization of dye solution containing Malachite Green by microalgae Cosmarium sp. Bioresour Technol 98(6):1176–1182
Data OWI (2023a) Data from multiple sources compiled by the UN – processed by Our World in Data. "6.3.1 - Proportion of safely treated domestic wastewater flows (%) - EN_WWT_WWDS" [dataset]. Data from multiple sources compiled by the UN [original data]. SDG Indicators Database, United Nations, Department of Economic and Social Affairs
Data OWI (2023b) Data from multiple sources compiled by the UN – processed by Our World in Data. "6.3.1 - Proportion of wastewater treated, by activity and location (%) - EN_WWT_TREATR - Industries" [dataset]. Data from multiple sources compiled by the UN [original data]. SDG Indicators Database, United Nations, Department of Economic and Social Affairs
Deblonde T, Cossu-Leguille C, Hartemann P (2011) Emerging pollutants in wastewater: a review of the literature. Int J Hyg Environ Health 214(6):442–448
Deepika C et al (2022) Sustainable production of pigments from Cyanobacteria. In: Cyanobacteria in biotechnology: applications and quantitative perspectives. Springer, pp 171–251
Dellamatrice PM et al (2017) Degradation of textile dyes by cyanobacteria. Braz J Microbiol 48:25–31
Deng G et al (2013) Studies of biouptake and transformation of mercury by a typical unicellular diatom Phaeodactylum tricornutum. Chin Sci Bull 58:256–265
Ding T et al (2017) Toxicity, degradation and metabolic fate of ibuprofen on freshwater diatom Navicula sp. J Hazard Mater 330:127–134
Dosnon-Olette R et al (2010) Fungicides and herbicide removal in Scenedesmus cell suspensions. Chemosphere 79(2):117–123
Dulio V et al (2018) Emerging pollutants in the EU: 10 years of NORMAN in support of environmental policies and regulations. Environ Sci Eur 30(1):1–13
Dusaucy J et al (2021) Microplastic pollution of worldwide lakes. Environ Pollut 284:117075
El-Sheekh MM, Gharieb M, Abou-El-Souod G (2009) Biodegradation of dyes by some green algae and cyanobacteria. Int Biodeterior Biodegradation 63(6):699–704
Factsanddetails (2022) Water pollution in China. 02/02/2024; Available from: https://factsanddetails.com/china/cat10/sub66/item391.html
Faluweki MK et al (2023) Active Spaghetti: collective organization in cyanobacteria. Phys Rev Lett 131:158303
Fernandes F et al (2022) Microalgae cultivation on nutrient rich digestate: the importance of strain and digestate tailoring under pH control. Appl Sci 12(11):5429
Ferrari SG et al (2013) Arsenic tolerance of cyanobacterial strains with potential use in biotechnology. Rev Argent Microbiol 45(3):174–179
Figler A et al (2021) Effects of nutrient content and nitrogen to phosphorous ratio on the growth, nutrient removal and desalination properties of the green alga Coelastrum morus on a laboratory scale. Energies 14(8):2112
Fleck A (2023) Only 56% of global wastewater was treated safely in 2020 [02/02/2024]; Available from: https://www.statista.com/chart/29547/domestic-wastewater-flow-safely- treated-by-region/
Forján E et al (2015) Microalgae: fast-growth sustainable green factories. Crit Rev Environ Sci Technol 45(16):1705–1755
Gao F et al (2022) Promoting effect of plant hormone gibberellin on co-metabolism of sulfamethoxazole by microalgae Chlorella pyrenoidosa. Bioresour Technol 351:126900
Gaohua L, Miao X, Dou L (2021) Crosstalk of physiological pH and chemical pKa under the umbrella of physiologically based pharmacokinetic modeling of drug absorption, distribution, metabolism, excretion, and toxicity. Expert Opin Drug Metab Toxicol 17(9):1103–1124
Gauthier M, Senhorinho G, Scott J (2020) Microalgae under environmental stress as a source of antioxidants. Algal Res 52:102104
Geissen V et al (2015) Emerging pollutants in the environment: a challenge for water resource management. Int Soil Water Conserv Res 3(1):57–65
Goiris K et al (2015) Impact of nutrient stress on antioxidant production in three species of microalgae. Algal Res 7:51–57

González-Barreiro O et al (2006) Removal of triazine herbicides from freshwater systems using photosynthetic microorganisms. Environ Pollut 144(1):266–271
Goswami R et al (2022) Bioremediation of heavy metals from wastewater: a current perspective on microalgae-based future. Lett Appl Microbiol 75(4):701–717
Guo R, Chen J (2015) Application of alga-activated sludge combined system (AASCS) as a novel treatment to remove cephalosporins. Chem Eng J 260:550–556
Hamed SM et al (2017) Sensitivity of two green microalgae to copper stress: growth, oxidative and antioxidants analyses. Ecotoxicol Environ Saf 144:19–25
Hassanien A et al (2023) Genetic engineering to enhance microalgal-based produced water treatment with emphasis on CRISPR/Cas9: a review. Front Bioeng Biotechnol 10:1104914
Heinemann JA, Hiscox TC (2022) Rethinking the drivers of biotechnologies: a paradigm for holistic climate change solutions. Curr Opin Environ Sustain 59:101222
Hejna M, Kapuścińska D, Aksmann A (2022) Pharmaceuticals in the aquatic environment: a review on eco-toxicology and the remediation potential of algae. Int J Environ Res Public Health 19(13):7717
Hirooka T et al (2005) Biodegradation of bisphenol A and disappearance of its estrogenic activity by the green alga Chlorella fusca var. vacuolata. Environ Toxicol Chem: Int J 24(8):1896–1901
Ho S-H et al (2013) Bioethanol production using carbohydrate-rich microalgae biomass as feedstock. Bioresour Technol 135:191–198
Horizon E (2020) SABANA - Sustainable Algae Biorefinery for Agriculture aNd Aquaculture
Hu N et al (2021) Removal of atrazine in catalytic degradation solutions by microalgae Chlorella sp. and evaluation of toxicity of degradation products via algal growth and photosynthetic activity. Ecotoxicol Environ Saf 207:111546
Iftehimul M et al (2023) Algae-based bioremediation of emerging pollutants. In: Management and mitigation of emerging pollutants. Springer, pp 143–199
Ji M-K et al (2013) Removal of nitrogen and phosphorus from piggery wastewater effluent using the green microalga Scenedesmus obliquus. J Environ Eng 139(9):1198–1205
Kaşdoğan D (2020) Designing sustainability in blues: the limits of technospatial growth imaginaries. Sustain Sci 15(1):145–160
Kasonga TK et al (2021) Endocrine-disruptive chemicals as contaminants of emerging concern in wastewater and surface water: a review. J Environ Manag 277:111485
Ke L et al (2010) Effects of metals on biosorption and biodegradation of mixed polycyclic aromatic hydrocarbons by a freshwater green alga Selenastrum capricornutum. Bioresour Technol 101(18):6950–6961
Khan MI, Shin JH, Kim JD (2018) The promising future of microalgae: current status, challenges, and optimization of a sustainable and renewable industry for biofuels, feed, and other products. Microb Cell Factories 17(1):1–21
Khatun J, Intekhab A, Dhak D (2022) Effect of uncontrolled fertilization and heavy metal toxicity associated with arsenic (As), lead (Pb) and cadmium (Cd), and possible remediation. Toxicology 477:153274
Kim JW et al (2020) Functional expression of polyethylene terephthalate-degrading enzyme (PETase) in green microalgae. Microb Cell Factories 19(1):1–9
Klepacz-Smolka A et al (2023) Microalgae are not an umbrella solution for power industry waste abatement but could play a role in their valorization. Crit Rev Biotechnol:1–29
Kong Q-X et al (2010) Culture of microalgae Chlamydomonas reinhardtii in wastewater for biomass feedstock production. Appl Biochem Biotechnol 160:9–18
Koutra E et al (2023) Biorefinery and bioremediation potential of microalgae. In: Valorization of microalgal biomass and wastewater treatment. Elsevier, pp 197–217
Kshirsagar AD (2013) Bioremediation of wastewater by using microalgae: an experimental study. Int J Life Sci Biotechnol Pharma Res 2(3):339–346
Kumar L, Bharadvaja N (2019) Enzymatic bioremediation: a smart tool to fight environmental pollutants. In: Smart bioremediation technologies. Elsevier, pp 99–118

Levine RB, Costanza-Robinson MS, Spatafora GA (2011) Neochloris oleoabundans grown on anaerobically digested dairy manure for concomitant nutrient removal and biodiesel feedstock production. Biomass Bioenergy 35(1):40–49

Li X et al (2020) Nutrient removal from swine wastewater with growing microalgae at various zinc concentrations. Algal Res 46:101804

Lim S-L, Chu W-L, Phang S-M (2010) Use of Chlorella vulgaris for bioremediation of textile wastewater. Bioresour Technol 101(19):7314–7322

Lima SA et al (2004) Biodegradation of p-chlorophenol by a microalgae consortium. Water Res 38(1):97–102

Liu Q (2020) Pollution and treatment of dye waste-water. In: IOP Conference Series: Earth and Environmental Science. IOP Publishing

Liu Z-Y, Wang G-C, Zhou B-C (2008) Effect of iron on growth and lipid accumulation in Chlorella vulgaris. Bioresour Technol 99(11):4717–4722

Liu T et al (2021) Effects from fe, p, ca, mg, zn and cu in steel slag on growth and metabolite accumulation of microalgae: a review. Appl Sci 11(14):6589

López-Serna R et al (2019) Removal of contaminants of emerging concern from urban wastewater in novel algal-bacterial photobioreactors. Sci Total Environ 662:32–40

Machado MD, Lopes AR, Soares EV (2015) Responses of the alga Pseudokirchneriella subcapitata to long-term exposure to metal stress. J Hazard Mater 296:82–92

Tuckett R (2021) Greenhouse gases and the emerging climate emergency. In: Climate change. Elsevier, pp 19–45

Yadav D et al (2021) Environmental and health impacts of contaminants of emerging concerns: recent treatment challenges and approaches. Chemosphere 272:129492

Saravanan A et al (2023) Strategies for microbial bioremediation of environmental pollutants from industrial wastewater: a sustainable approach. Chemosphere 313:137323

Zahra SA et al (2023) Microalgae-based bioremediation of pharmaceuticals wastewater. In: The treatment of pharmaceutical wastewater. Elsevier, pp 277–309

Radzun KA et al (2015) Automated nutrient screening system enables high-throughput optimisation of microalgae production conditions. Biotechnol Biofuels 8(1):1–17

Wolf J et al (2015) High-throughput screen for high performance microalgae strain selection and integrated media design. Algal Res 11:313–325

Sivakaminathan S et al (2018) High-throughput optimisation of light-driven microalgae biotechnologies. Sci Rep 8(1):11687

Ranglová K et al (2019) Rapid screening test to estimate temperature optima for microalgae growth using photosynthesis activity measurements. Folia Microbiol 64(5):615–625

Mendoza J et al (2013) Fluid-dynamic characterization of real-scale raceway reactors for microalgae production. Biomass Bioenergy 54:267–275

Mondal S et al (2022) Redefining the role of microalgae in industrial wastewater remediation. Energy Nexus 6:100088

Najar-Almanzor CE et al (2023) Microalgae-assisted green bioremediation of food-processing wastewater: a sustainable approach toward a circular economy concept. J Environ Manag 345:118774

Zhang S et al (2011) Bioaccumulation and degradation of pesticide fluroxypyr are associated with toxic tolerance in green alga Chlamydomonas reinhardtii. Ecotoxicology 20:337–347

Sarkar P, Dey A (2021) Phycoremediation–an emerging technique for dye abatement: an overview. Process Saf Environ Prot 147:214–225

Padervand M et al (2020) Removal of microplastics from the environment. A review. Environ Chem Lett 18:807–828

Zaller JG, Zaller JG (2020) Pesticide impacts on the environment and humans. In: Daily poison: pesticides-an underestimated danger. Springer, pp 127–221

Sharma AK et al (2023) Mapping the impact of environmental pollutants on human health and environment: a systematic review and meta-analysis. J Geochem Explor:107325

Vasilachi IC et al (2021) Occurrence and fate of emerging pollutants in water environment and options for their removal. Water 13(2):181

Zvobgo L, Do P (2020) COVID-19 and the call for 'Safe Hands': challenges facing the under-resourced municipalities that lack potable water access-a case study of Chitungwiza municipality, Zimbabwe. Water Res X 9:100074

Redalje DG et al (2018) Algae as ideal waste removers: biochemical pathways. In: Biotreatment of agricultural wastewater. CRC Press, pp 91–110

Pinto G et al (2002) Biodegradation of phenols by microalgae. Biotechnol Lett 24:2047–2051

Torniainen K, Tammilehto S, Ulvi V (1996) The effect of pH, buffer type and drug concentration on the photodegradation of ciprofloxacin. Int J Pharm 132(1–2):53–61

Matamoros V et al (2016) Assessment of the mechanisms involved in the removal of emerging contaminants by microalgae from wastewater: a laboratory scale study. J Hazard Mater 301:197–205

de Wilt A et al (2016) Micropollutant removal in an algal treatment system fed with source separated wastewater streams. J Hazard Mater 304:84–92

Min M et al (2011) Cultivating Chlorella sp. in a pilot-scale photobioreactor using centrate wastewater for microalgae biomass production and wastewater nutrient removal. Appl Biochem Biotechnol 165:123–137

Xie Y et al (2014) Kinetic simulating of Cr (VI) removal by the waste Chlorella vulgaris biomass. J Taiwan Inst Chem Eng 45(4):1773–1782

Peng J et al (2022) Removal of levofloxacin by an oleaginous microalgae Chromochloris zofingiensis in the heterotrophic mode of cultivation: removal performance and mechanism. J Hazard Mater 425:128036

Parikh A, Madamwar D (2005) Textile dye decolorization using cyanobacteria. Biotechnol Lett 27:323–326

Marques IM et al (2021) A photobioreactor using Nannochloropsis oculata marine microalgae for removal of polycyclic aromatic hydrocarbons and sorption of metals in produced water. Chemosphere 281:130775

Shah V, Garg N, Madamwar D (2001) An integrated process of textile dye removal and hydrogen evolution using cyanobacterium, Phormidium valderianum. Springer

SundarRajan P et al (2020) An insight into carbon balance of product streams from hydrothermal liquefaction of Scenedesmus abundans biomass. Renew Energy 151:79–87

Pena-Castro J et al (2004) Heavy metals removal by the microalga Scenedesmus incrassatulus in continuous cultures. Bioresour Technol 94(2):219–222

Phang S et al (2000) Spirulina cultivation in digested sago starch factory wastewater. J Appl Phycol 12:395–400

da Silva Rodrigues DA et al (2020) Biodegradation of sulfamethoxazole by microalgae-bacteria consortium in wastewater treatment plant effluents. Sci Total Environ 749:141441

Michelon W et al (2022) Removal of veterinary antibiotics in swine wastewater using microalgae-based process. Environ Res 207:112192

Safonova E et al (2004) Biotreatment of industrial wastewater by selected algal-bacterial consortia. Eng Life Sci 4(4):347–353

Phoon BL et al (2020) Conventional and emerging technologies for removal of antibiotics from wastewater. J Hazard Mater 400:122961

Mohsenpour SF et al (2021) Integrating micro-algae into wastewater treatment: a review. Sci Total Environ 752:142168

Van Den Hende S et al (2014) Treatment of industrial wastewaters by microalgal bacterial flocs in sequencing batch reactors. Bioresour Technol 161:245–254

Riaño B, Molinuevo B, García-González M (2011) Treatment of fish processing wastewater with microalgae-containing microbiota. Bioresour Technol 102(23):10829–10833

Singh NK, Patel D (2012) Microalgae for bioremediation of distillery effluent. In: Farming for food and water security, pp 83–109

Selvam G, Baskaran R, Mohan P (2011) Microbial diversity and bioremediation of distilleries effluent. J Res Biol 1(3):153–162

Procházková G et al (2014) Effect of nutrient supply status on biomass composition of eukaryotic green microalgae. J Appl Phycol 26:1359–1377

Shriwastav A, Bose P (2015) Algal growth in photo-bioreactors: impact of illumination strategy and nutrient availability. Ecol Eng 77:202–215

Riebesell U et al (2000) The effects of varying CO2 concentration on lipid composition and carbon isotope fractionation in Emiliania huxleyi. Geochim Cosmochim Acta 64(24):4179–4192

Muradyan E et al (2004) Changes in lipid metabolism during adaptation of the Dunaliella salina photosynthetic apparatus to high CO2 concentration. Russ J Plant Physiol 51:53–62

Safafar H et al (2016) Enhancement of protein and pigment content in two Chlorella species cultivated on industrial process water. J Mar Sci Eng 4(4):84

Zaparoli M et al (2020) Cellular stress conditions as a strategy to increase carbohydrate productivity in Spirulina platensis. Bioenergy Res 13(4):1221–1234

Zhekisheva M et al (2002) Accumulation of oleic acid in Haematococcus pluvialis (chlorophyceae) under nitrogen starvation or high light is correlated with that of astaxanthin esters1. J Phycol 38(2):325–331

Van der Hoek JP, Duijff R, Reinstra O (2018) Nitrogen recovery from wastewater: possibilities, competition with other resources, and adaptation pathways. Sustain For 10(12):4605

Sugimoto K et al (2008) Upregulation of PG synthesis on sulfur-starvation for PS I in Chlamydomonas. Biochem Biophys Res Commun 369(2):660–665

Theodorou ME et al (1991) Effects of phosphorus limitation on respiratory metabolism in the green alga Selenastrum minutum. Plant Physiol 95(4):1089–1095

Tang Y et al (2013) Synergistic effects of nano-sized titanium dioxide and zinc on the photosynthetic capacity and survival of Anabaena sp. Int J Mol Sci 14(7):14395–14407

Trenfield MA et al (2015) Aluminium, gallium, and molybdenum toxicity to the tropical marine microalga Isochrysis galbana. Environ Toxicol Chem 34(8):1833–1840

Martin GJ et al (2014) Lipid profile remodeling in response to nitrogen deprivation in the microalgae Chlorella sp.(Trebouxiophyceae) and Nannochloropsis sp.(Eustigmatophyceae). PLoS One 9(8):e103389

Upadhyay AK et al (2020) Bioaccumulation and toxicity of As in the Alga Chlorococcum sp.: prospects for as bioremediation. Bull Environ Contam Toxicol:1–7

Monteiro CM, Castro PM, Malcata FX (2012) Metal uptake by microalgae: underlying mechanisms and practical applications. Biotechnol Prog 28(2):299–311

Silkina A et al (2019) Large-scale waste bio-remediation using microalgae cultivation as a platform. Energies 12(14):2772

Qi F et al (2018) Characterization of a microalgal UV mutant for CO2 biofixation and biomass production. Biomed Res Int 2018:4375170

Sayler GS, Ripp S (2000) Field applications of genetically engineered microorganisms for bioremediation processes. Curr Opin Biotechnol 11(3):286–289

Silkina A et al (2017) Bioremediation efficacy—comparison of nutrient removal from an anaerobic digest waste-based medium by an algal consortium before and after cryopreservation. J Appl Phycol 29:1331–1341

Silkina A et al (2015) Standard operating procedures for analytical methods and data collection in support of pilot-scale cultivation of microalgae. Public Output report WP1A3. 01 of the EnAlgae Project

Scherer MD et al (2017) Environmental study of producing microalgal biomass and bioremediation of cattle manure effluents by microalgae cultivation. Clean Techn Environ Policy 19:1745–1759

Zhang B et al (2020) Microalgal-bacterial consortia: from interspecies interactions to biotechnological applications. Renew Sust Energ Rev 118:109563

Chapter 7
Agro-(Bio)Microplastics: Extraction, Identification, and Quantification Methodologies

Laura Hernández Bautista and Alma Berenice Jasso-Salcedo

Abstract The process of disintegration of biodegradable bioplastics is often studied using the residual materials, which can be collected by the naked eye and examined through changes in crystallinity, functional groups, thermal properties, and molecular weight. However, disintegration products, such as microplastics, nanoplastics, oligomers, monomers, and additives, are often overlooked because of the challenges in identifying and measuring them. This chapter provides an analysis of recent advancements in extraction (high-density saline solutions, modified Fenton at room temperature), identification (microscopy, IR, Raman, DSC), and quantification (DLS, NMR, TOC, MS) methodologies for (bio)microplastics, with a particular focus on those from agricultural settings. These techniques require a library of (bio)microplastics for calibration, thereby providing abundant research possibilities. Strategies currently under experimentation, such as magnetic separation, tracer molecules, biotechnological techniques, and EIS, are also included. The study of agricultural (bio)microplastics, known as agro-(bio)microplastics, is critical for understanding their impact on the environment.

Keywords Bioplastics · Soil extraction · Biotechnological identification · Biomicroplastics quantification · Bioplastics quantification · Agro-biomicroplastics quantification · Agricultural microplastics

---

L. H. Bautista
Biociencias y Agrotecnología, Centro de Investigación en Química Aplicada, Saltillo, Coahuila, México
e-mail: laura.hernandez.ma23@ciqa.edu.mx

A. B. Jasso-Salcedo (✉)
Ingeniería y Química de Materiales, Centro de Investigación en Materiales Avanzados/ Investigadora por México CONAHCYT, Chihuahua, Chih, México
e-mail: alma.jasso@cimav.edu.mx

S. Dey, S. Bhattacharya (eds.), *Biotechnological Interventions in the Removal of Emerging Pollutants*, Interdisciplinary Biotechnological Advances,
https://doi.org/10.1007/978-981-97-9922-0_7

7.1 Overview of Bioplastic Types and Applications in Agriculture

7.1.1 Bioplastic Definition

Bioplastics are defined by EuropeanBioplastics.org as materials made from renewable or nonrenewable biological sources but are not necessarily biodegradable (ability to break down in the environment over time). Figure 7.1 depicts bioplastics that have been applied in the agricultural context, and it distinguishes between bioplastics derived from fossil and biological sources, and whether they are biodegradable or not. Polylactic acid (PLA), poly(hydroxyalkanoate) (PHA), polybutylene adipate terephthalate (PBAT), poly(butylenesuccianate) (PBS) and poly(ecaprolactone) (PCL) are biodegradable, with only PLA and PHA being derived from renewable biological sources. The global production of biodegradable bioplastics accounts for half of the total production in 2022 (1.81 thousand tons). PLA and PHA are expected to experience significant growth by the end of the decade, with PLA being the most widely produced. In terms of PHA, the market prefers short chain biodegradable bioplastics such as poly(hydroxyvalerate) (PHV), poly(3-hydroxybutyrate-co-4-hydroxybutyrate) (P(3HB-co-4HB)) and poly(3-hydroxybutyrate-co-3-hydroxyvalerate) (P(3HB-co-3HV) or PHBV) over the medium chain ones, such as poly(hydroxybutyrate-co-hydrohyhexanoate) (PHBH). Other biobased bioplastics with a growing market are the ones based on proteins, starch, chitin, and cellulose, for instance, the regenerated cellulose with double production of rayon and cellophane films for a variety of applications.

7.1.2 Biodegradable Bioplastics in Agricultural Settings

The utilization of bioplastics in agriculture in place of conventional plastics is a rapidly growing trend. In the ideal situation, these bioplastics are biodegradable, and further research and development at both the research and industrial scales may expand the range of possibilities. The biodegradable bioplastics can largely replace conventional plastics in installation of mulch films and tunnels for crop protection but also in greenhouse materials, shade and protective nets, tubing and accessories for drip irrigation, pipes, clips, tutors and plant clips, string, trimmer lines, seeds protection, nursery trays, plant pots, organic and inorganic fertilizing sacks, phytosanitary dispensers, water and wastewater cleaning coagulants, bins, agrochemical containers, pruning bags, packaging, and postharvest overall (Mansoor et al. 2022; Aguirre-Loredo et al. 2023).

The database for materials at industrial level, such as knowde.com, omnexus.specialchem.com, matweb.com, and marketandmarkets.com, offer comprehensive information on the various industries that manufacture biodegradable bioplastics of different grades. These products have certifications of biodegradability and/or

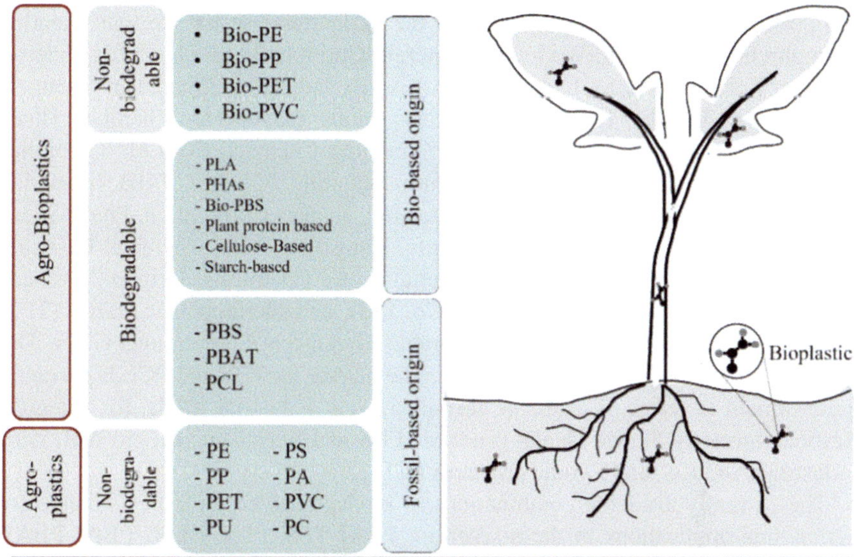

Biodegrabable bioplastics studied in agricultural settings

Separation		Identification		Quantification	
PLA		PLA		PBAT-PLA	PBAT
PBAT		PHAs		PLA	PBS
PHB		PBS		PHB	
PBS		PBAT		PCL	
Opportunities					
PCL	TPS	PCL	TPS	TPS	PHAs
PHAs					

Summary of tecniques for bio-microplastics

Separation/Extraction	Identification *	Quantification	
Saline solutions	Microscopies	DLS	HPLC
Magnetic separation	FTIR	NMR	GC-MS
Fenton modified	Raman	GPC	TG-MS
Enzymatic-oxidative	DSC	TOC	

*New alternatives: Biotechnological (i.e. PLA, PBAT, PBSA), Electrochemical impedance spectroscopy

Fig. 7.1 Agrobioplastics, which may be derived from either fossil or biological sources, are an increasingly important component of the environment. The study of agro-(bio)microplastics, which involves extraction, identification, and quantification techniques, is critical for understanding their impact. This chapter includes a summary of the techniques used as well as future research possibilities. The classification of biopolymers is adapted from Ali et al. (2022)

composting suitable for applications in the agricultural sector. We can mention examples for PLA (NatureWorks, TotalEnergiesCorbion, BASF SE, Cofco, Futerro, Danimer Scientific, Toray Industries Inc., Evonik Industries, Mitsubishi Chemical Corporation, Unitika Ltd., among other minor players as Zhenjian Hisun Biomaterials Co. Ltd., GH Bioplastics, Sumimoto Chemical Co. Ltd., Carbiolice and Floreon), PHAs (Novomer, Danimer Scientific, CJ Bio, PHB Industrial, Metabolix, Biomer, CJ Bio, DaniMer Scientific, Kaneca Corporation, PhaBuilder), PBAT (BASF, Novamont, Willeap, Kingfa, Hangzhou Peijin Chemical Co. Ltd., Zhejiang Biodegradable Advanced Material, Lotte Chemicals, Orinko Advanced Plastics Co. Ltd., Torise Biomaterials Co. Ltd), and thermoplastic starch (TPS) based on starch (Biotec, BioLogiQ, Grabio Greentech Corporation, Green Dot Bioplastics, for biodegradable, Resonac). The market for PBS and PCL is currently quite limited in terms of available formulations at industrial levels, for example, Resonac formerly Showa Denko produces PBS and Eurothane, Elogio AM, Zeus Industrial Products, and Evonic produces PCL.

The currently available combinations of biodegradable bioplastics suitable for agricultural applications in the market are PBAT-TPS, PLA-PBAT, PBAT-PBAT, TPS-PBAT-PLA, and PBS-starch-PLA. Table 7.1 lists the company and formulations for the recommended biodegradable bioplastics in the agricultural sector, primarily for mulching, and highlights three properties (glass transition temperature, melting point temperature, and the density) of the pure and mix bioplastics. The variability on properties aims to provide insight into the key factors that are essential for recovering, identifying, and quantifying these bioplastics in the environment.

The mixtures of TPS-starch sold as Biomax®TPS and Paragon, the TPS-PBAT labeled as Bioplast® and Ecoworks, Starch-PBAT sold as Biosafe® and Ecoflex®, and PLA-based formulations labeled as Biomelle, Ecovio®, and Ingeo® correspond to preparations intended for mulching (Menossi et al. 2021; Campanale et al. 2024). The future trend in the mulch film market indicates that bioplastics being utilized in descending order of PLA, PHA, and PBS-based formulations. On the other hand, the mixture of PLA-PBAT sold as Ecovio® T2206 is useful for filaments and twines in crop production. The Mater-Bi NF803 is a PBAT-TPS-PCL mix, Mater-Bi NF01U is PBAT, and Bioflex F1130 is PLA, and all three have been tested for pipes and drippers (Hiskakis et al. 2011).

7.1.3 Biodegradable Bioplastics Origin and Concentration in Agricultural Settings

In academic environments and among certain entrepreneurs, the commercially available bioplastics are blended with cellulose, starch fibers, bamboo, and other biobased or nonbiodegradable additives to achieve the mechanical properties required for the processing of plastics and shape them for specific uses. The commercial mixtures often fail to disclaim the full content and the most persistent bioplastics accumulate in the soil over time. For instance, the time required for PLA to

Table 7.1 Density and thermal properties of biodegradable bioplastics recommended for applications in agricultural sector

Biopolymer	Density (g/cm³)	Glass transition temperature (T_g, °C)	Melting point (T_m, °C)	Marketed bioplastic mix (tag or grade, supplier)	Application in agriculture
PBS	1.26 Bionolle 1001 MD	−45 to −10	90 to 120	PBS (Bionolle1001MD, Resonac) PBS, starch, and PLA (Bionelle Starcla 10S, Resonac)	Gardening applications, mulching film, trash bags, plant pots, filaments, yarns, nets, bottles, gloves, containers, laminated paper, trays, combs, foamed tubes, pesticides, and sustained-release fertilizer
PCL	1.14	−66	58 Capa 6500 58 to 60	PCL (CAPSUL™ PCL, Sulzer Chemtech) PCL (LACTEL® 25:75 DL-PLC (B6015-1), Evonik) PCL (Capa6500, Perstorp)	Horticulture, agricultural films, postharvest gas control, packaging films
PHA	1 to 1.3 Mirel P1003 1.4 Biocycle 1000 1.2 to 1.25 Biomer P226	5 Nodax PHA −4 3HP 25 to 15	170 to 180 Nodax PHA 125 Biocycle1000 170–175	PHA (Nodax™ and DaniMer 12291 and 12263, DaniMer Scientific) PHA (PHActA1000P, CJ Bio) Poly(3-hydroxypropionate) 3HP (Rinnovo™ 3HP H1000 XP, Novomer) PHA (MirelP1003 and P1004, Metabolix) PHB (Biocycle1000, PHB Industrial) PHB (BiomerP226, Biomer) PHB, PBAT (MirelP5001, Metabolix)	Landscaping weed barrier film, compost bags, hay wraps, mulch films, pesticide coating in agriculture, packaging, trays, biocarriers in the slow release of fertilizers

(continued)

Table 7.1 (continued)

Biopolymer	Density (g/cm^3)	Glass transition temperature (T_g, °C)	Melting point (T_m, °C)	Marketed bioplastic mix (tag or grade, supplier)	Application in agriculture
PLA	1.21 to 1.3 PLLA (1.24 to 1.3), PDLA (1.24) and PDLLA (1.25 to 1.27) Ecovio T2206 1.24 to 1.26 Compostable 7003 1.27	60 to 70	160 to 170 Ecovio T2206 with two 110 to 120 and 140 to 155	PLA (Ingeo) PLA (Mater-Bi EF04P, Novamont) PLA (Compostable 7003, CerePlast) PLA (INZEA F09E and F10BC50S, Nurel) PLA and PBAT (Ecovio T2206 is mix of Ecoflex with 60% PLA, BASF) PLA and bamboo fiber (Transmare BIO-35LA, TransmareGroup) PLA and starch derivatives (Cereplast Compostables7003, Trellis Bioplastics)	Agricultural mulch, package, drip irrigation pipes and conduits, greenhouse tunnels
PBAT	1.25 Ecoflex 1.25 to 1.27 Ecovio M2351 1.37 to 1.4	28	100 to 120 Ecoflex 110 to 120 Ecovio M2351 with two 110 to 120 and 140 to 155 MaterBiNF803N 110	PBAT (Mater-Bi EF04, Novamont) PBAT (Ecoflex, BASF) PBAT and PLA (Ecovio M2351, BASF) PBAT and PLA (Ecovio M2351 with 10% biobased, Organix Solutions) PBAT and PLA (Mater-Bi EF04 with 25% biobased BioAgri, BioBag Americas Inc.) PBAT (Bio360 mulch film, Dubois Agrinovation) PBAT and TPS (BioBlendBC 27130, 27132, BioLogiQ) PBAT, TPS and PCL (MaterBiNF803N, Novamont)	Mulch films, netting, slow release system (fertilizers, fungicides) through mulch.

Starch based	TPS 1.45 to 1.5 Bioplast500 1.28 NuPlastiQ 1.4 BioBlendBC 1.3 Bionelle Starcla 10S 1.3 TerratekBD 1.57	Starch 52-60 TPS -75 to 10 BioBlendBC 81-100	Starch 256 TPS 104 BioBlendBC 130	TPS (Bioplast500 with 57% biobased, Biotec) TPS (NuPlastiQ CG1000, BioLogiQ) TPS, biodegradable polyester and natural plasticizers (CardiaCompostableB-F, CardiaBioplastics) TPS and PBAT (BioBlendBC 27152 BioLogiQ) TPS and PLA (BioBlendBC 27255, BioLogiQ) TPS, PBAT, PLA (Bio-BlendBC 27240, BioLogiQ) TPS (GB090, Grabio) TPS (TerratekBD5100, Green Dot Bioplastics) TPS (Cornopole, JapanCorn- Starch) Starch, PBS, PLA (Bionelle Starcla 25S and 45S, Resonac) Starch, chitosan and biodegradable copolyester (SARAH, Heppe Biomaterial)	Mulch films, greenhouse and tunnel films, heavy duty bags, gardening, plant pots and organic waste containers

(continued)

Table 7.1 (continued)

Biopolymer	Density (g/cm³)	Glass transition temperature (T_g, °C)	Melting point (T_m, °C)	Marketed bioplastic mix (tag or grade, supplier)	Application in agriculture
Cellulose based	1.45 ARBOFORML 1.3	>100	200 to 255 ARBOFORM® L 190	Paper mulch (WeedGuardPlus) Cellulose acetate (ARBO-FORML V100natur, Tecnaro) Cellulose fiber (MimGreen, Mimcord SA) Cellulose fiber (Zelfo HG, ZelfoTechnology) Woodfiber and PLA (Fibrolon F8530, FKuR Kunststoff)	Mulch films, packaging and to be used in combination with the biodegradable bioplastics

These data are adapted from (Ma et al. 2019; Fojt et al. 2020; Astner et al. 2023; Campanale et al. 2024), knowde.com, omnexus.specialchem.com, matweb.com databases, and the corresponding product webpage. For comparison, the density of organic fraction of soil is 0.11 g/cm³, density of mineral fraction of soil is 1.8 g/cm³, and density of stones is 2.4–2.8 g/cm³ (Périé and Ouimet 2008). The bulk density of every soil can be calculated from available models based on the soil organic matter content lose 90% of its mass is approximately 300 days, and this duration differ based on the discrepancies in conditions between laboratory and field trials (Van Roijen and Miller 2022). Currently, these bioplastic residues in agricultural environments are overlooked. This emphasizes the importance of examining both existing and potential scenarios where bioplastics are used and discarded, as well as gathering data on the presence of their end-products in the environment.

Current research tends focus on three type of studies: (1) bioplastics submitted to biodegradation or accelerated aging involving UV radiation, chemical agents, mechanical and/or thermal processes, (2) bioplastics intentionally introduced in burial tests in the field or in the laboratory (Fig. 7.1, and studies of bioplastics spiked in plants), and (3) data collection of bioplastics in existing or abandoned agricultural installations and its products.

The study of concentrations, shapes, size, chemistry, kinetics, and stability of residues is mainly done on standards of pure bioplastics of nonagricultural grade (Krawczyk-Walach et al. 2021; Fojt et al. 2022; Wohlleben et al. 2023). Only a small number of studies have used marketed mulch formulations (Nelson et al. 2020; Serrano-Ruíz et al. 2020; Griffin-LaHue et al. 2022; Meng et al. 2023) for this purpose. The subjects of the investigation are soil, sediments, fauna, and rarely agricultural products, wastewater, and groundwater. In type 1 and 2 studies, bioplastics are typically downsized (pieces of 2 cm^2, 5 mm^2, or micrometric dimensions) and subsequently subjected to test conditions. The process of cleavage of hydrolyzable bonds leads to the formation of soluble and insoluble fractions. The insoluble fraction is bound to exudates, salts, organic matter, soil, and microorganisms. The soluble fraction is often overlooked but is comprised of various plasticizers, compatibilizers, accelerators of degradation, biocides, dyes, antistatic, and fillers, which are in the best scenario within the allowable concentration limits set by regulatory standards, nonetheless accumulable (Cui et al. 2020; Li et al. 2023). The insoluble fraction is identified with the term microplastic (MP) according to the ISO 24187:2023. MP is defined as "any solid plastic particle insoluble in water with dimension between 1 μm and 1000 μm" and the "dimension is related to the longest length of the particle." The standard also defines a large microplastic as MP between 1 mm and 5 mm.

The next paragraph highlights the recent findings of MP of type 3 studies, and the content of Sect. 7.2 includes future research possibilities of the methodologies employed in all three types of studies, with a greater focus on types 1 and 2, which are more prevalent.

7.1.3.1 Concentration on Soil and Sediments of Agricultural Sites

The results suggest that residual bioplastics can accumulate, be detected, and quantified in the agricultural soil within a short time frame (half a decade). Such is the case of studies in China, the United States, and the Netherlands.

A long-term investigation in the northern United States found that the MPs profile of agricultural soils with biodegradable mulches applied annually from 2015 to

2018 varies according to the bioplastic composition. For instance, bioplastics mulches containing on PLA had the highest proportion of MPs (>40% of size 2–6 mm), while those based on PBAT produces residual bioplastics larger than 6 mm (>30% of size 6–12 mm, >20% of size 12–25 mm, and >30% of size 25–75 mm) after 2 years of abandoning the experimental site (Griffin-LaHue et al. 2022). The study also revealed that black mulch generates a larger MPs fraction than clear mulch from the same brand. Although the report fail to quantify the MPs <1 mm, it was possible to demonstrate that disintegration process continues over the years, with an expected time of up to 72 months for 90% of the mass to be reduced.

An independent study found bioplastics in agricultural soils in the south of the Netherlands in quantities ranging from 1 to 53 pieces per gram, with maximum dimensions of 500 μm (Lwanga et al. 2023). Bioplastics were also detected in sediments and airborne dust from the same sites. Three types of bioplastics were identified: poly(ethylene furanoate) (PEF), multilayer polyethylene films with starch, and a mixture of bio-based and conventional materials. Notably, the presence of PEF in the sampling site coincides with the recent announcement of the mass production of this biobased bioplastic in the north of the Netherlands. The presence of PEF was particularly noteworthy, as it is set to become widely used in the agricultural sector and replace food contact polymers. Furthermore, the authors confirmed that organic farming brings higher levels of MP than conventional farming using plastic mulch, which suggest an external input from the compost and organic fertilizers (Lwanga et al. 2023).

Apart of agricultural fresh soils, sediments are also contaminated with bioplastics. For instance, the sediment samples collected from China contained an average of 230 nanograms of PLA per gram of sediment, and the concentration varied depending on the depth of the sample, ranging from 2.5 to 21 cm. At a depth of 2.5 cm, the concentration of PLA was higher than that of PET, while at a depth of 21 cm, the concentration of PLA decreased significantly. The authors of this study suggest that the presence of PLA in sediments coincided with the introduction of this bioplastic in the market in 1990 (Wang et al. 2022). Additionally, residual lactic acid was detected in sediments at a concentration of 0.6 micrograms per gram, along with two other residual bioplastics, PCL and a polymer of polyether group that may correspond with cellophane. A more recent example is the result on soil treated with biosolids in the 1990s with traces of PLA, chitin and cellulose modified (Adhikari et al. 2024). Another example of PLA residues discovered in soil samples is the presence of high levels of lactic acid, lactide, and oligomers, which are common byproducts of PLA biodegradation, in three out of four commercially available soil samples in Poland (Krawczyk-Walach et al. 2021). The results suggest a history of contamination in the sediments by PLA, which has been occurring for several decades, and the contamination has recently increased due to agricultural applications.

7.1.3.2 Concentration on Agricultural Products

The available literature lacks information regarding the identification of bioplastics in agricultural products, particularly in fruit and vegetable tables sourced from mulched or nonmulched sites. Only three independent studies from Germany, Italy, and Turkey have focused on the identification and quantification of conventional MPs in soil-level products, such as lettuce, cabbage, carrots, broccoli, onion, potatoes, cucumbers, tomatoes, asparagus, strawberries, and fruits like apples and pears (Conti et al. 2020; Aydın et al. 2023). These studies have shown contamination with MPs of nonbiodegradable plastics in variable amounts, mainly attributed to the processing of these products. For instance, fresh strawberries contained 226 pieces per kilogram, asparagus had 1150 pieces per kilogram, pear and apple contained 4500 pieces per kilogram, tomato and cucumber had >50,000–200,000 pieces per gram, and lettuce, carrot, and cabbage had >50,000 pieces per kilogram.

Two of the three studies failed to identify the chemistry of the observed MPs, emphasizing the need for nondestructive protocols to quantify and identify all plastic residues, especially bioplastics, in agricultural products. Additionally, the quantification of bioplastics in agricultural fruits and vegetables should consider the bioplastic residues that are directly related to growth in biomulched sites and distinguish them from the contributions of postharvest handling and processing activities in supermarkets and local markets.

7.2 Analysis of Biodegradable Bioplastics in Agricultural Settings

The evidence of poor reliability in microplastics counting in an interlaboratory analysis (Pérez-Guevara et al. 2022) confirms that the sample processing is key to increase the evidence of bioplastic residues in agricultural soil, agricultural products, and fauna linked. The description of microplastics of bioplastics used in the agricultural sector is an active field of research from who collect information of the extraction, identification, and quantification methodologies. The scientific databases such as Google Scholar and SciFinder are a great source to gather recent advances on the area. The terms searched were "bioplastics," "soil AND sediment," "vegetables," "bioplastics AND agriculture," "identification AND microplastics," specific bioplastic name AND the matrix of interest, for instance "PBAT AND soil."

The first critical discussion on the topic of bioplastics in agricultural settings and strategies for microplastic detection took place in 2020 (Fojt et al. 2020), and it was briefly revisited in 2022 (Censi et al. 2022). The methods used to characterize bioplastics produced through microorganisms and those utilized to track their degradation in laboratory settings (Baidurah 2022) can provide valuable insights into the current techniques available for studying bioplastics. Also the reviews of studies types 1 and 2 on bioplastics (biomicroplastics from accelerated aging and from controlled burial tests), including those on their impacts on agricultural ecosystems, provide an actualized guide for studies to look for methodologies for plants, worms, and microorganisms (Serrano-Ruíz et al. 2020; Chah et al. 2022; Campanale et al. 2024), but rarely include the separation and identification of bioplastic residues. Furthermore, those studies are hardly carried out in field conditions and leave out protocols with agrobioplastic formulations. Therefore, it is crucial to revise recent advances and emerging trends in the extraction, identification, and quantification of biomicroplastics.

7.2.1 *Techniques of Extraction from Agricultural Settings*

The process of extracting MPs has been extensively revised for conventional plastics (Bellasi et al. 2021; Picó and Barceló 2021; Astner et al. 2023; Pham et al. 2023), and the discussion on the greenness methodologies is also included. These reviews assess the environmental sustainability of sampling methods, reagents, waste and energy consumption of the techniques of separation, identification, and quantification. However, there is currently a lack of analysis oriented to bioplastics in these discussions.

The next section will compile protocols aimed for bioplastics used in agricultural applications; however, the environmental compatibility of the reagents utilized in these protocols should be thoroughly scrutinized and reevaluated by other researchers.

Bioplastics are often subjected to recovery operations that involve a combination of cleaning, extraction, purification, separation and enrichment protocols to remove them from complex matrices such as soil, soil macrobiota, and vegetative tissue. To generalize these processes, this section refers to them collectively as bioplastic recovery.

7.2.1.1 Separation from Other Bioplastics and Conventional Agroplastics

Most bioplastics possess densities bigger than that of water, thus rendering water as an unsuitable medium for separation of various bioplastics. Instead, researchers have employed saline solutions with varying densities, ranging from 1.5 to 4.2 g/cm^3. However, certain salts, such as NaI (1.9 g/cm^3) and $ZnCl_2$ (1.7 g/cm^3), are expensive and pose environmental hazards, making them less desirable options. In a specific instance, researchers evaluated various salts to separate PBAT M-VERAB5026 MPs from polyamides (PA) and polyethylene terephthalate (PET) in soil. Among the salts tested, CsCl (1.3 g/cm^3) proved effective using isopycnic centrifugation to break up any agglomerates that may form (Jakobs et al. 2023). For the recovery of PLA, PBS, and PBAT from soil containing agroplastics such as PP, LDPE, PS, and PVC, it is recommended to use NaBr (3.21 g/cm^3) for different particle sizes ranging from 200 to 1000 μm (Li et al. 2021). This salt has the advantage of being recyclable, with an average recovery of 95% from the bioplastic material. The authors recorded minor surface changes in bioplastics, in contrast with agroplastics, but did not affect the average particle size of recovered microplastics. Ideally, a recovery method bioplastic-oriented, and a parallel separation method designed for agroplastics, is recommended when dealing with soil and other complex samples. To achieve this strategy, it would be necessary to double the amount of sample and processing time, resulting in a higher bioplastic recovery. An example of this approach was to isolate PE (Dow 310E), PLA (Ingeo2003D), and PBAT (EcoflexFC1200) from soil and earthworms gut. The complex samples were first washed with distilled water and then with a solution of NaH_2PO_4 (1.28 g/cm^3) to recover both bioplastics. In a separate process to recover PE, the sample was treated with 70% ethanol, followed by a washing with distilled water (Meng et al. 2023). The literature has explored the use of lipophilic media in the study of microplastics in general (Picó and Barceló 2021; Pham et al. 2023). However, these media have proven to be ineffective in recovering PLA and cellulose acetate microplastics. The recovery rates for PLA were 17% in sandy soils and 36% in organic-rich soils, while polypropylene was recovered at a rate of 23% in sandy soils. It should be noted that no recovery was observed in organic-rich soils for PLA and polypropylene (PP) (Bottone 2019).

Alternative methods for separating microplastics from sandy soils include magnetic separation, which has shown promise for separating PLA from PP, PET, and polycarbonate (PC). Studies suggest that the concentration of $MnCl_2$ (up to 2.9 g/cm^3) can affect the levitation height of microplastics, with a 3 M concentration being the most effective for PLA (Ren et al. 2022). In proof-of-concept experiments, a two-pole-facing neodymium block magnet was positioned at a distance of 45 mm and facilitated the microplastics recovery.

7.2.1.2 Recovery from Complex Matrices

The physicochemical stability of bioplastics should be ensured during their recovery from soil and other complex samples. While most common protocols, such as 30% w/w H_2O_2 oxidation and enzymatic digestion (Pfohl et al. 2021; Möller et al. 2022), are efficient in removing organic matter, they may produce unwanted reactions that lead to inaccurate estimates of particle size and number in the sample. Temperature plays a critical role in the recovery process for bioplastics. Specifically, the melting points of PBS and PCL are 90 and 58 °C, respectively (refer to Table 7.1). A rise in temperature, particularly due to an exothermic reaction, can impair the thermal stability and shape profile of these bioplastics. Therefore, an alternative protocol based on Fenton oxidation at low temperatures may be more suitable for bioplastics. Surfactants such as SDS, Triton, and Tween 20, are utilized in protocols that involve bioplastics to effectively separate microplastics by reducing their aggregation.

For instance, an enzymatic-oxidative protocol was used to assess the integrity of PLA microplastics alongside with conventional agroplastics, such as polyethylene terephthalate (PET), polyethylene (PE), and polyvinyl chloride (PVC). The microplastics of Ingeo7001D PLA subjected to the protocol underwent particle destruction, resulting in the creation of new microplastics, altered PLA molar mass, and enthalpy of crystallization. The 8 days long protocol (costing 2 USD per 250 g soil) utilized proteases to break down insoluble proteins but facilitated the hydrolysis of PLA (Möller et al. 2022). Thus, the process could potentially be shortened by eliminating the 12-hour protease step. The advantage of the enzymatic-oxidative protocol lies in the reduction of 80% of the lightweight fraction of soil, including the mineral and organic components, and indicates that Fenton oxidation in an ice bath at 40 °C does not cause morphological changes in PLA microplastics. Additionally, the recovery protocol did not interfere with further analyses using techniques such as micro-FTIR, ATR-FTIR, DSC, and GPC. The potential of these protocols should be evaluated in soil that has been mulched with bioplastics, in order to assess their effectiveness and offer a valuable opportunity for research.

Three biodegradable bioplastics marketed by BASF SE (PLA based, PBAT and PBS based), along with six conventional microplastics, were submitted separately to two different oxidation protocols to estimate their physicochemical stability. The alkaline protocol based on a mixture of 50 g/L H_2O_2 and 40 g/L NH_3 at 70°C was contrasted with a Fenton oxidation protocol (250 g/L H_2O_2 and 2.5 g/L $FeSO_4$) at ambient temperature. The authors reported that PLA and PBS suffered degradation and resulted with an increasing number of PLA microplastics using the alkaline protocol. The three bioplastics showed evidence of surface damage, and PBS appeared to be more susceptible to this effect (Pfohl et al. 2021). Although the Fenton protocol can result in residual ferrous particles in microplastics, it is the friendliest protocol for PLA, PBAT, and PBS. Precautions are recommended particularly when the study's main objective is to assess PLA size profile. Also, as suggested by the authors, the choice of protocol should also take into account its impact on the stability of additives, the acceleration of the additive leaching from the bioplastic, and the potential for unexpected interactions between bioplastics and plastics in the sample.

Thus, as suggested by Pfohl et al. (2021), the optimal combination of biopolymer and sample preparation protocol is dependent upon the research questions being addressed. For example, if the aim is to analyze the size profile, the study should focus on physical stability of the microplastic and nanoplastic fractions. On the other hand, if the study aims to conduct a quantitative analysis, it is crucial to prioritize chemical stability.

7.2.2 *Qualitative Analysis of Biodegradable Bioplastics*

The subsequent step after retrieving bioplastic microplastics from the soil, which might exhibit various sizes, shapes, color, and chemical properties, is to conduct identification work. The size range, advantages, disadvantages and cost of the techniques for the identification of microplastics are presented elsewhere (Schwaferts et al. 2019; Primpke et al. 2020); for instance, microscopy is nonlimited because AFM, SEM, and TEM can visualize 1 nm particles; however, the based IR spectroscopy is limited to 10 micrometers and Raman spectroscopy to 1 micrometer. Here, we present the relevant results on biomicroplastics as guide for future research.

It is widely acknowledged that employing a single technique may not suffice in providing a comprehensive understanding of the samples, which is why multiple techniques are commonly recommended to achieve more comprehensive information.

7.2.2.1 Microscopies for Size, Shape, and Color Analysis

Several variations of microscopies are available for biomicroplastics (stereomicroscope, optical microscope, fluorescence microscopy, scanning electron microscopy, and atomic force microscopy) with limitations on the particle size, sample size analysis, time and cost (Schwaferts et al. 2019; Primpke et al. 2020). For the purpose of examining the dimensions, shape, and colors of bioplastics and their microplastics, staining techniques are commonly utilized in conjunction with microscopic visualization. Additionally, other promising approaches include the employment of UV light and the heating of samples in bulk or as individual microplastics.

For the first case, independent studies utilized fluorescence microscopy and fluorophores to identify PHAs and PLA in complex samples. BODIPY 493/503 was found to provide a longer staining time for PHA and PLA particles compared to Nile red and Nile blue A (Cao et al. 2022). The cost of these dyes can vary, but among the most expensive options are BODIPY 493/503, Lipid green, and Nile red. One advantage of using Nile red in environmental samples is that the fluorescence of the organic components of soil is too low for the threshold limit, and inorganic components such as sand show no fluorescence.

Regarding the use of heating for microscopic visualization, one example is PLA, where the fluorescence of microplastics can be achieved without the use of dyes by simply heating the sample. In this case, a photon emission was observed when PLA was subjected to heat treatment, such as a 12-hour treatment at 140 °C. The authors report that better observations are possible if Nile red is used in combination with heat treatment and an excitation of a 440 nm laser (Monteleone et al. 2021). The need for further research on various types of bioplastics is evident, as their T_m values are similar, leading to potential misidentification. This is the case of PBS and PBAT. It is imperative to determine the limitations of this method due to the potential impact of overheating on the morphology and size of the bioplastics.

A comparable method that employs optical microscopy is the hot needle test. It is an effective technique for identifying microplastics that have dimensions larger than 500 micrometers (Beckingham et al. 2023). While this technique has been underutilized for identifying bioplastics in environmental samples, it holds great potential when combined with a chemical analysis such as FTIR or Raman. This approach enables a distinct differentiation between bioplastic fibers and naturally occurring fibers in composted bioplastics and bioplastics that have been subjected to field tests. It is particularly challenging to achieve this differentiation using microscopes alone.

The advancement of in-situ methods for the identification of microplastics in terms of shape and size is of utmost importance. Here we present two examples. First the application of portable 365 nm UV flashlights in combination with optical microscopes and cameras has already been proven to be effective in identifying microplastics in cave sediment (Giardino et al. 2023). The potential usefulness of this approach for distinguishing biomicroplastics has been validated through empirical observations in our laboratory.

Alternative techniques for identifying microplastics by size are UV-vis and X-ray microtomography (Pfohl et al. 2022). The study demonstrated the potential of light-scattering techniques, using UV-vis spectroscopy, in identifying microplastics and water-soluble organics in environmental samples. The research focused on characterizing in-vitro aged plastics and showed that the particles in question exhibit a prominent UV absorption band at 500 nm, while the soluble fraction emits a strong signal at 265 nm. The findings of this study contribute to the advancement of low-cost methods for characterizing microplastics that may find their way into identifying biomicroplastics.

7.2.2.2 Chemical Analysis

The chemical identification of biodegradable bioplastics can be divided into destructive (thermal and chromatographic methods) and nondestructive (microscopies and spectroscopies) techniques.

7.2.2.2.1 IR Spectroscopies

FTIR is a widely employed technique for determining the chemistry of microplastics. A variant of this approach involves coupling the FTIR technique with microscopy, as the micro-FTIR or the direct laser infrared (LDIR) method, which enables simultaneous chemical analysis and particle imaging. All materials possess a unique fingerprint region, which can be identified by comparing the results with a library. An open library is offered in https://openanalysis.org/openspecy/, but also nonfree Agilent Microplastics Starter 2.0 library are useful. This methodology implies that the library contains the spectra of agrobioplastics, aged agrobioplastics, typical compounds of bioplastics degradation, additives of bioplastics, and contaminants derived from the soil matrix.

Particularly, bioplastics exhibit a robust signal in the fingerprint region of 1700–1800 cm^{-1}, which facilitates their distinguishing characteristics. For instance, the signal at 1715 cm^{-1} corresponds to carbonyl band of PBAT, PHA, and PBS, while the signal at 1750 cm^{-1} corresponds to PLA and PCL (Sintim et al. 2020). On the other hand, spectra of PBAT and PET can hardly be differentiated since both contain terephthalate.

The presence of microplastics in complex samples, such as the human placenta, earthworm gut, and soil, has been examined utilizing LDIR. One study utilized LDIR to identify 17 types of microplastics, including PBS (Zhu et al. 2023) and other to identify PLA and PBAT (Meng et al. 2023), demonstrating the technique effectiveness.

Overall, depending on the objectives of the study on bioplastics and the complexity of the sample, it is common to utilize infrared spectroscopy in conjunction with other techniques.

7.2.2.2.2 Raman Spectroscopy

The high specificity of Raman spectroscopy has drawbacks as long-time analysis and interference with organic and silicates components of soil and bioplastic pigments. The use of 785 nm wavelength laser may reduce fluorescence interference compared to 532 nm (Wohlleben et al. 2023). Furthermore, machine learning is been used to improve chemical identification in soil and complex samples (Dong et al. 2022). An in-depth examination of the advantages and drawbacks of aged microplastics was conducted by the Dong research group, who had previously published a Raman spectra database on the subject (Dong et al. 2022). Raman spectroscopy can be employed in conjunction with microscopic techniques to visualize microplastics within the range of 10 to 20 micrometers. This method is commonly referred to as micro-Raman spectroscopy.

Examples of analysis of biomicroplastics by Raman distinguish PLA, PBAT, and PBS from PE, PP, and PS. PLA has specific peaks around 1770, 394 and 409 cm^{-1}. The band at 1617 cm^{-1} is unique of PBAT, while bands at 960 and 830 cm^{-1} correspond to PBS (Li et al. 2024a, b). A library of Raman spectra is necessary in order to facilitate the straightforward identification of biomicroplastics.

7.2.2.2.3 Electrochemical Impedance Spectroscopy

The utility of bioplastics dielectric properties for chemically identifying microplastics in complex samples is a promising prospect. The relative permittivity and dissipation factor of bioplastics and agroplastics are particular for each material, indicating the potential of electrochemical impedance spectroscopy (EIS) (Hegde 2017). The proof of concept, which was tested on conventional agroplastics (PS and

PE) and their mixtures, returned outstanding results (Ching et al. 2022; Pedrero et al. 2024). It is believed that this method can differentiate between microorganisms and biological components in complex samples, and can be employed in static or flow-through devices without the need for sample preparation, providing a quick and cost-effective analysis. This proof of concept demonstrates great potential for bioplastics in this context.

7.2.2.2.4 Thermoanalytical Methods

Differential scanning calorimetry (DSC) is a method that determines the melting temperature (T_m) and glass transition temperature (T_g) characteristic of each bioplastic (see Table 7.1). By comparing the results of this analysis with a reference library, it is possible to identify the bioplastic present in the sample. In order to address the need for a standardized methodology for microplastics of bioplastics, it is crucial to consider the heating rate (10 to 20 °C per minute), as it affects the exothermic or endothermic changes that occur.

An example of DSC used to chemically identify bioplastics in a mixture is the formulation of PLA-PHB 70-30 (from Ningbo Hesu Fibers Co., Ltd), which shows two peaks corresponding to PHB at approximately 175 °C and PLA at approximately 155 °C (Bao et al. 2022). While a standard of PBH CAS 29435-48-1 reports a T_m of 172 °C, experimental measurements show values of 169.7 °C and 169.1 °C for PHB biocomposted (Capitain et al. 2020).

The T_m values for PCL, PBS, PBAT, and LDPE are similar, but T_g is useful for identifying PCL. However, distinguishing between LDPE, PBAT, and PBS is challenging, and a secondary technique such as IR or Raman spectroscopy is required. Similarly, PP, PLA, and PHB have similar T_m values, but distinctive T_g values (see Table 7.1). Therefore, a combination of T_g and T_m is unique for bioplastics, making this technique a valuable tool for identifying and characterizing various types of bioplastics.

7.2.2.2.5 Biotechnological Technologies

Biomonitoring is a method that evaluates the pollution of agricultural soils with biomicroplastics. The microbial community present in the soil is specific to the contaminant, and thus serves as a means of identifying microplastics in the environment. This is because the microbial communities have preferences for certain substrates, and tracking these preferences can help identify the substrate source. The microorganisms that are capable of biodegrading bioplastics, whether alone or in combination, serve as a library for the qualitative analysis of residual bioplastics in the environment, particularly when tested in the field or in the laboratory using agricultural soil. In addition to degrading microorganisms, microbial predators and symbiotic microorganisms also colonize the plastisphere, and as a group, serve as bioplastic biomonitors. Some examples of studies that can be used to hypothesize the use of biomonitors in soil are provided below.

Yi Huang research demonstrated that the microbial communities in the microenvironments of microplastics differ among soil, agrobioplastics such as PE, and bioplastics such as PBAT-PLA. The microenvironment or plastisphere contains monomers and oligomers as products of disintegration, and provides a carbon source for the microbial community. As the plastisphere of bioplastics undergoes changes over time, it ultimately affects the initial community. In the case of PBAT-PLA, the microbial community was found to be highly abundant compared to that of soil and PE, with *Proteobacteria* and *Actinobacteria* being predominant (Li et al. 2023). Another study observed that the plastisphere of PLA microplastics was populated by *Paraburkholderia* because of its lactate preferences and *Rhizobium*, whereas the PBAT plastisphere was rich in nitrogen-fixing bacteria such as *Bradyrhizobium*, *Cupriavidus*, *Rhodopseudomonas*, and *Agrobacterium* (Zhang et al. 2024).

It is crucial to acknowledge that the plastisphere is subject to fast change due to bioplastic decomposition, which in turn affects the microbial community, and ultimately, the microbial profile. For example, in another study, the authors identified early colonizers (*Cladosporium* spp.) and long-term colonizers (*Tetracladium* spp. and nitrogen-fixing bacteria) in the bioplastic poly(butylene succinate-co-adipate) (PBSA) cultivated in agricultural soils (Purahong et al. 2021).

Drug resistance genes could be another biomarker. Since there is evidence of the accumulation of drugs on microplastics, tracing the resistant microbial community inhabiting the plastisphere of biomicroplastics is a fascinating research topic. Indeed, the abundance of antibiotic resistance genes, most of which are assigned to the genus *Bradyrhizobium*, is higher in PBAT-PLA plastisphere than in soil and other agroplastic such as PE (Li et al. 2024a, b). In this sense, bioinformatics, when combined with machine learning, serves the purpose of rapidly acquiring information from databases such as Metacyc, enviPath, and PlasticDB. This allows the identification of microorganisms, proteins, and genes that have the potential to become biomarkers for biomicroplastics in complex systems.

7.2.3 Quantitative Analysis of Biodegradable Bioplastics in Agricultural Settings

7.2.3.1 Light Scattering (DLS)

Dynamic light scattering (DLS) is a valuable tool for assessing the size profile and molecular weight distributions of suspensions. The accuracy of DLS measurements depends on the choice of solvent, suspension concentration, and measurement temperature. There are several modes of light scattering available in the market, such as static light scattering, dynamic light scattering, and multiangle dynamic light scattering and Fraunhofer light scattering, which are designed to increase sensitivity in detecting the hydrodynamic size of particles ranging from 0.3 nm to 10 micrometers. The Zetasizer equipment is commonly used in nanoparticle laboratories for this purpose. Researchers have employed DLS to study the nanoplastics and oligomers surging from the aging and mechanical disintegration of bioplastics. This technique is suitable for bulk quantification, with a minimum volume of 20 μL, and can be used in combination with microscopes for a more detailed analysis or replacement.

In the studies conducted by Tamayo-Belda, the technique of dynamic light scattering (DLS) has been utilized for the purpose of sizing analysis. In the study of PHB, PCL, and PLA nanoparticles, particle sizes ranging from <1 nm to 200 nm, 30 to 150 nm, and 100–1000 nm, respectively, were investigated (Tamayo-Belda et al. 2022). However, it is essential to note that the nanoplastics were free of soil contamination in these bioplastics tested. Nonetheless, it is encouraging for detecting smaller fractions of bioplastics in complex matrices such as plants and soil fauna.

DLS has the potential to efficiently replace time-consuming microscopy techniques, such as SEM (Griffin-LaHue et al. 2022) and optical microscopy, for the quantification of biomicroplastics.

7.2.3.2 Total Organic Carbon (TOC)

The determination of water-soluble nanoplastics, oligomers, and monomers released from the disintegration of bioplastics using TOC is a highly effective method. The application of TOC for the quantification of water-soluble nanoplastics, oligomers, and monomers released from the disintegration of bioplastics is effective. This technique, which is commonly utilized in the characterization of food, wastewater, and soils, is classified as destructive due to the utilization of strong oxidizing reagents and high-temperature catalytic oxidation. The technique calls for TOC equipment, filters with the desired pore size, and an appropriate oxidation method for the specific sample. The quantification is carried out on filtrates, and as a result, the pore size of membranes, syringe filters, and/or ultrafiltration tubes can be adjusted according to the research question. For instance, the quantification of bioplastics using TOC involves studying PCL nanoplastics smaller than 100 nm and PCL oligomers smaller than 50 KDa. The authors utilized the ratio of carbon content per repetitive unit of PCL, which was 0.632 mg of TOC/mg PCL, to successfully transform the results from the mass of total organic carbon to the mass of PCL (Tamayo-Belda et al. 2022).

An enhancement of the protocol involves the use of carbon-free filters, in which the desired fraction of microplastics is preserved, and the TOC is measured on the entire filter. This principle was tested on the quantification of PBAT-PLA MPs smaller than 0.22 micrometers and greater than 0.22 micrometers resulting from mechanical and UV irradiation disintegration (Sun et al. 2024). The discussed examples were mostly carried out under sterile conditions, and no additional carbon sources were present. Consequently, at this point, the TOC technique can be applied to clean recovered samples of bioplastics. It is important to apply TOC to soil samples, as these samples may be affected by organic matter. To accurately measure the size profiles of bioplastics using TOC, it is recommended to use appropriate controls.

7.2.3.3 Nuclear Magnetic Resonance Spectroscopy (NMR)

The use of proton nuclear magnetic resonance spectroscopy (^{1}H-NMR) is a highly effective method for determining the molecular structure and chemical composition of compounds in a sample, making it an ideal tool for the chemical identification of biomicroplastics, oligomers or even monomers in soil. Additionally, the use of an internal quantification standard allows for the precise quantification of protons in biopolymers in samples extracted from soil and other complex matrices. In a study conducted by the Sander group, the extraction and quantification of PBS (Sigma Aldrich), PBAT standard (BASF), and a commercial biodegradable mulch film made up of PBAT and PLA was examined (Nelson et al. 2022). The authors utilized 1,4-dimethoxybenzene as an internal standard for the quantification of protons in the aromatic ring compounds present in the PBAT polymer chain. The extraction process was successful in extracting spiked PBAT-PLA biomulch films, PBAT films, PBAT microplastics (<300 μm), and soluble forms of PBAT from various types of soil and mineral particles. However, the authors did find an inconvenience in the UV-aged standard PBAT, as it presented insoluble fractions in the extraction solvent, which may be due to the lack of UV stabilizers commonly present in mulch films. The use of agro-formulations containing UV stabilizers may help to reduce the appearance of insoluble forms in bioplastics and improve the recovery of residual biomicroplastics using Sander's methodology. Further research is needed to confirm the effectiveness of ^{1}H-NMR for the extraction and quantification of spiked biomulches in soil over extended periods. Nonetheless, the authors have introduced a standardized, single-step procedure for quantifying bioplastics in soil samples, without the need for prior removal of organic or inorganic materials. The starch and cellulose components found in the study did not interfere with the quantification process, as they are insoluble in chloroform. Here, we briefly present Sanders' methodology: the bioplastic sample was first freeze-dried, followed by micro-Soxhlet extraction and reflux using chloroform and methanol, and then analyzed using NMR spectroscopy which took seconds. The use of this quantification approach for microplastics in agricultural environments, specifically in polyester soluble in chloroform, is envisioned by the authors. Examples of such polyesters include PLA, PBSA, PCL, PHA, and PBS. The authors were able to measure residual PBS bioplastics in soil samples, including those that had been composted for over 400 days. Alternative internal standards, such as 1, 2, 4, 5-tetrachlorobenzene, were utilized for the quantification of PLA, PBS, and PBAT in various formulations. The objective of this research was to determine if it is possible to differentiate between these bioplastics in commercial products. According to a study conducted in 2021, it was found that ^{1}H-NMR could be used to measure the contents of PLA, PBS, and PBAT in commercial supermarket bags without any interference (Chen et al. 2021). ^{1}H-NMR technique was utilized to analyze the water-soluble organic components of eight unused biomulches. This approach significantly expanded the library of available biodegradable bioplastic-based commercial mulches.

The potential applications of the extraction and quantification protocol are vast, particularly for soil, sediment, and compost samples that may be contaminated with multiple types of bioplastics of agricultural importance.

7.2.3.4 Chromatographic Techniques (GCP, LC)

The use of chromatographic techniques, such as liquid chromatography (HPLC) and gel permeation chromatography (GPC), is essential in the characterization of oligomers and monomers, including additives, of biopolymers. In addition, these techniques are also utilized in the quantification of microplastics through the use of chromatographies. Two distinct approaches exist for this purpose. The first approach serves to support the measurement of bioplastic residuals in environmental samples by separating fractions of biopolymers based on their molecular weight and subsequently measuring them using mass spectroscopy, such as LC/MS or GPC-GC-MS. The advantage of this approach is that it allows for the determination of the molecular weight distribution of the sample prior to its MS analysis. The disadvantage of LC-MS technique is that the incomplete dissolution of the bioplastic on the solvent, or mobile phase, affects its correct quantification (Fojt et al. 2022).

The second approach involves indirectly estimating the quantity of bioplastic in the environment per area from the losses on the molecular weight of the bigger pieces residuals of bioplastics recovered using GPC. However, it is important to corroborate the results using auxiliary techniques for quantification, such as TOC for bulk quantification and/or NMR for punctual measurement of the biopolymer in the sample. This method has the potential to examine the dynamics of secondary biomicroplastics formation as done for PBAT, PBS and PLA (Pfohl et al. 2021; Möller et al. 2022).

7.2.3.5 Mass Spectroscopy (TG-MS and GC-MS)

Mass spectrometry is widely used for determining the quantity of microplastics and tracer molecules through their degradation products. However, access to this technique may not always be feasible for all researchers. Prior to analysis, samples are subjected to pyrolysis using either thermogravimetry (TG-MS) or pyrolysis coupled with gas chromatography (py-CG-MS). Examples of biomicroplastics quantification in agricultural settings are PLA, PBAT, TPS, and PHB (Serrano-Ruíz et al. 2020; Krawczyk-Walach et al. 2021; Fojt et al. 2022), and an example of tracer molecules quantification is the benzotriazole-type UV stabilizers distinctive of biomulched films (Li et al. 2023). Like many techniques, the effectiveness of this quantification method depends on the availability of a library of biomicroplastics for reference.

7.2.4 *Directory (Table 7.2)*

Table 7.2 For international contacts regarding bioplastics and biomicroplastics extraction, identification, and quantification techniques, the reader might consider reaching out to the following research institutions.

Scientist (email and ORCID)	Expertise
Esperanza Huerta Lwanga ehuerta@ecosur.mx https://orcid.org/0000-0002-4552-5703	Methods for extracting samples of agricultural soil, sediments, animal tissue and water. Identification of bioplastic by LDIR.
Michael Sander (and Taylor Nelson) michael.sander@env.ethz.ch https://orcid.org/0000-0003-3383-2041	Extraction of PBAT and PBS bioplastic from soil. Quantification by ^{1}H-NMR.
Markus Flury flury@wsu.edu https://orcid.org/0000-0002-3344-3962	Long-term studies of bioplastic mulches in field. Microplastics of bioplastic extracted from soil. Identification FTIR, LDIR, and data analysis.
Shruti Venkata Chari shrutiv@geologia.unam.mx https://orcid.org/0000-0001-5284-4107	Water and air extraction methods. Identification by micro-FTIR.
Jakub Fojt and Jiří Kučerík kucerik@fch.vut.cz https://orcid.org/0000-0001-9083-4866	Bioplastics as PBAT and PBS extraction and quantification in soil.
Jiaoxia Sun 990201300012@cqjtu.edu.cn	Bioplastics quantification by membrane filtration-total organic carbon method.
Yi Huang yhuang@pku.edu.cn	Bioplastics recovery from soil. Identification by analysis of bioplastisphere.
Miguel Tamayo-Belda https://orcid.org/0000-0001-5323-8467	Microplastic and nanoplastics from bioplastics PHB, PCL, and PLA in aquatic fauna. Size by DLS, SEM, and TOC.

References

Adhikari K et al (2024) Accumulation of microplastics in soil after long-term application of biosolids and atmospheric deposition. Sci Total Environ 912:168883. https://doi.org/10.1016/j.scitotenv.2023.168883

Aguirre-Loredo RY et al (2023) Agriculture applications of biodegradable polymers. In: Biodegradable polymers: concepts and applications, pp 300–319. https://doi.org/10.1201/9781003230533-21

Ali SS et al (2022) Biowastes for biodegradable bioplastics production and end-of-life scenarios in circular bioeconomy and biorefinery concept. Bioresour Technol:127869. https://doi.org/10.1016/j.biortech.2022.127869

Astner AF et al (2023) Formation, behavior, properties and impact of micro- and nanoplastics on agricultural soil ecosystems (A Review). NanoImpact 31(July):100474. https://doi.org/10.1016/j.impact.2023.100474

Aydın RB et al (2023) Occurrence of microplastics in most consumed fruits and vegetables from Turkey and public risk assessment for consumers. Life 13(8). https://doi.org/10.3390/life13081686

Baidurah S (2022) Methods of analyses for biodegradable polymers: a review. Polymers 14(22). https://doi.org/10.3390/polym14224928

Bao Q et al (2022) Accelerated degradation of Poly(lactide acid)/Poly(hydroxybutyrate) (PLA/PHB) Yarns/fabrics by UV and O2 exposure in South China Seawater. Polymers 14(6):1216. https://doi.org/10.3390/polym14061216

Beckingham B et al (2023) Hot or not: systematic review and laboratory evaluation of the hot needle test for microplastic identification. Microplastics Nanoplastics 3(1):1–13. https://doi.org/10.1186/S43591-023-00056-4

Bellasi A et al (2021) The extraction of microplastics from sediments: an overview of existing methods and the proposal of a new and green alternative. Chemosphere 278:130357. https://doi.org/10.1016/j.chemosphere.2021.130357

Bottone A (2019) Analyzing microplastics in soils: evaluating canola oil extractions. Umea University

Campanale C et al (2024) A critical review of biodegradable plastic mulch films in agriculture: definitions, scientific background and potential impacts. TrAC - Trends Anal Chem 170:117391. https://doi.org/10.1016/j.trac.2023.117391

Cao JS et al (2022) Rapid quantification of intracellular polyhydroxyalkanoates via fluorescence techniques: a critical review. Bioresour Technol 350:126906. https://doi.org/10.1016/j.biortech.2022.126906

Capitain C et al (2020) Differential scanning calorimetry for quantification of polymer biodegradability in compost. Int Biodeterior Biodegradation 149:104914. https://doi.org/10.1016/j.ibiod.2020.104914

Censi V et al (2022) Bioplastics: a new analytical challenge. Front Chem 10(September):1–8. https://doi.org/10.3389/fchem.2022.971792

Chah CN et al (2022) A systematic review on bioplastic-soil interaction: exploring the effects of residual bioplastics on the soil geoenvironment. Sci Total Environ 851:158311. https://doi.org/10.1016/j.scitotenv.2022.158311

Chen X et al (2021) Determination of polylactic acid and other components in biodegradable plastics using quantitative 1H-NMR method. China Plastics 35(8):181

Ching CTS et al (2022) Real-time, economical identification of microplastics using impedance-based interdigital array microelectrodes and k- nearest neighbor model. Biotechnol Bioprocess Eng 8:1–8. https://doi.org/10.1007/s12257-022-0262-y

Conti GO et al (2020) Micro- and nano-plastics in edible fruit and vegetables. The first diet risks assessment for the general population. Environ Res 187:109677. https://doi.org/10.1016/j.envres.2020.109677

Cui H et al (2020) Development of microwave-assisted extraction and dispersive liquid–liquid microextraction followed by gas chromatography–mass spectrometry for the determination of organic additives in biodegradable mulch films. Microchem J 160:105722. https://doi.org/10.1016/j.microc.2020.105722

Dong M et al (2022) Automated analysis of microplastics based on vibrational spectroscopy: are we measuring the same metrics? Anal Bioanal Chem 414:3359–3372. https://doi.org/10.1007/s00216-022-03951-6

Fojt J et al (2020) A critical review of the overlooked challenge of determining micro-bioplastics in soil. Sci Total Environ 745. https://doi.org/10.1016/j.scitotenv.2020.140975

Fojt J et al (2022) A simple method for quantification of polyhydroxybutyrate and polylactic acid micro-bioplastics in soils by evolved gas analysis. Molecules 27(6). https://doi.org/10.3390/molecules27061898

Giardino M et al (2023) Automated method for routine microplastic detection and quantification. Sci Total Environ 859:160036. https://doi.org/10.1016/j.scitotenv.2022.160036

Griffin-LaHue D et al (2022) In-field degradation of soil-biodegradable plastic mulch films in a Mediterranean climate. Sci Total Environ 806. https://doi.org/10.1016/j.scitotenv.2021.150238

Hegde V (2017) Dielectric study of biodegradable and/or bio-based polymeric materials. Universite Grenoble Alpes

Hiskakis M, Babou E, Briassoulis D (2011) Experimental processing of biodegradable drip irrigation systems-possibilities and limitations. J Polym Environ 19(4):887–907. https://doi.org/10.1007/s10924-011-0341-1

Jakobs A et al (2023) A novel approach to extract, purify, and fractionate microplastics from environmental matrices by isopycnic ultracentrifugation. Sci Total Environ 857:159610. https://doi.org/10.1016/j.scitotenv.2022.159610

Krawczyk-Walach M et al (2021) Characterization of potential pollutants from Poly(lactic acid) after the degradation process in soil under simulated environmental conditions. Appl Chem 1(2):156–172. https://doi.org/10.3390/appliedchem1020012

Li C et al (2021) A commonly available and easily assembled device for extraction of bio/non-degradable microplastics from soil by flotation in NaBr solution. Sci Total Environ 759:143482. https://doi.org/10.1016/j.scitotenv.2020.143482

Li K et al (2023) The plastisphere of biodegradable and conventional microplastics from residues exhibit distinct microbial structure, network and function in plastic-mulching farmland. J Hazard Mater 442:130011. https://doi.org/10.1016/j.jhazmat.2024.133428

Li F et al (2024a) Identification and visualization of environmental microplastics by Raman imaging based on hyperspectral unmixing coupled machine learning. J Hazard Mater 465:133336. https://doi.org/10.1016/j.jhazmat.2023.133336

Li K et al (2024b) Potential environmental risks of field bio/non-degradable microplastic from mulching residues in farmland: evidence from metagenomic analysis of plastisphere. J Hazard Mater 465:133428. https://doi.org/10.1016/j.jhazmet.2024.133428

Lwanga EH et al (2023) Microplastic appraisal of soil, water, ditch sediment and airborne dust: the case of agricultural systems. Environ Pollut 316(P1):120513. https://doi.org/10.1016/j.envpol.2022.120513

Ma X et al (2019) Multifunctional flexible composite aerogels constructed through in-situ growth of metal-organic framework nanoparticles on bacterial cellulose. Chem Eng J 356:227–235. https://doi.org/10.1016/j.cej.2018.09.034

Mansoor Z et al (2022) Polymers use as mulch films in agriculture—a review of history, problems and current trends. Polymers 14(23). https://doi.org/10.3390/polym14235062

Meng K et al (2023) Fragmentation and depolymerization of microplastics in the earthworm gut: a potential for microplastic bioremediation? J Hazard Mater 447:130765. https://doi.org/10.1016/j.jhazmat.2023.130765

Menossi M et al (2021) Current and emerging biodegradable mulch films based on polysaccharide bio-composites. A review. Agron Sustain Dev 41(4). https://doi.org/10.1007/s13593-021-00685-0

Möller JN et al (2022) Tackling the challenge of extracting microplastics from soils: a protocol to purify soil samples for spectroscopic analysis. Environ Toxicol Chem 41(4):844–857. https://doi.org/10.1002/etc.5024

Monteleone A et al (2021) New application for the identification and differentiation of microplastics based on fluorescence lifetime imaging microscopy (FLIM). J Environ Chem Eng 9(1):104769. https://doi.org/10.1016/j.jece.2020.104769

Nelson TF et al (2020) Quantification of synthetic polyesters from biodegradable mulch films in soils. Environ Sci Technol 54(1):266–275. https://doi.org/10.1021/acs.est.9b05863

Nelson TF et al (2022) Biodegradation of poly(butylene succinate) in soil laboratory incubations assessed by stable carbon isotope labelling. Nat Commun 13(1):1–16. https://doi.org/10.1038/s41467-022-33064-8

Pedrero D et al (2024) Efficient removal of nanoplastics from water using mesoporous metal organic frameworks. Sep Purif Technol 333. https://doi.org/10.1016/j.seppur.2023.125816

Pérez-Guevara F et al (2022) Coverage of microplastic data underreporting and progress toward standardization. Sci Total Environ 829. https://doi.org/10.1016/j.scitotenv.2022.154727

Périé C, Ouimet R (2008) Organic carbon, organic matter and bulk density relationships in boreal forest soils. Can J Soil Sci 88(3):315–325. https://doi.org/10.4141/CJSS06008

Pfohl P et al (2021) Microplastic extraction protocols can impact the polymer structure. Microplastics Nanoplastics 1(1):1–13. https://doi.org/10.1186/s43591-021-00009-9

Pfohl P et al (2022) Environmental degradation of microplastics: how to measure fragmentation rates to secondary micro- and nanoplastic fragments and dissociation into dissolved organics. Environ Sci Technol 56(16):11323–11334. https://doi.org/10.1021/acs.est.2c01228

Pham DT et al (2023) Greenness of analytical methods for isolation and identification of microplastics in environmental and food samples - A critical review. Trends Anal Chem 167:117263. https://doi.org/10.1016/j.trac.2023.117263

Picó Y, Barceló D (2021) Analysis of microplastics and nanoplastics: how green are the methodologies used? Current Opin Green Sustain Chem 31:100503. https://doi.org/10.1016/i.cogsc.2021.100503

Primpke S et al (2020) Critical assessment of analytical methods for the harmonized and cost-efficient analysis of microplastics. Appl Spectrosc:1012–1047. https://doi.org/10.1177/0003702820921465

Purahong W et al (2021) Back to the future: decomposability of a biobased and biodegradable plastic in field soil environments and its microbiome under ambient and future climates. Environ Sci Technol 55(18):12337–12351. https://doi.org/10.1021/acs.est.1c02695

Ren X, Breadmore MC, Maya F (2022) Biphasic magnetic levitation to detect organic pollutants on microplastics. Anal Chem 94(25):9033–9039. https://doi.org/10.1021/acs.analchem.2c01094

Schwaferts C et al (2019) Methods for the analysis of submicrometer- and nanoplastic particles in the environment. TrAC - Trends Anal Chem 112:52–65. https://doi.org/10.1016/j.trac.2018.12.014

Serrano-Ruíz H et al (2020) Compounds released from unused biodegradable mulch materials after contact with water. Polym Degrad Stab 178:109202. https://doi.org/10.1016/j.polymdegradstab.2020.109202

Sintim HY et al (2020) In situ degradation of biodegradable plastic mulch films in compost and agricultural soils. Sci Total Environ 727:138668. https://doi.org/10.1016/j.scitotenv.2020.138668

Sun J et al (2024) Characterization of the degradation products of biodegradable and traditional plastics on UV irradiation and mechanical abrasion. Sci Total Environ 909:168618. https://doi.org/10.1016/j.scitotenv.2023.168618

Tamayo-Belda M et al (2022) Identification and toxicity towards aquatic primary producers of the smallest fractions released from hydrolytic degradation of polycaprolactone microplastics. Chemosphere 303:134966. https://doi.org/10.1016/j.chemosphere.2022.134966

Van Roijen EC, Miller SA (2022) A review of bioplastics at end-of-life: linking experimental biodegradation studies and life cycle impact assessments. Resour Conserv Recycl 181:106236. https://doi.org/10.1016/j.resconrec.2022.106236

Wang L et al (2022) An in situ depolymerization and liquid chromatography-tandem mass spectrometry method for quantifying polylactic acid microplastics in environmental samples. Environ Sci Technol 56(18):13029–13035. https://doi.org/10.1021/acs.est.2c02221

Wohlleben W et al (2023) Fragmentation and mineralization of a compostable aromatic-aliphatic polyester during industrial composting. Environ Sci Technol Lett 10(8):698–704. https://doi.org/10.1021/acs.estlett.3c00394

Zhang Y et al (2024) Stronger deterministic processes shape the plastisphere microbiota of biodegradable microplastics compared to non-biodegradable microplastics in farmland soil. Appl Soil Ecol 196:105312. https://doi.org/10.1016/j.apsoil.2024.105312

Zhu L et al (2023) Identification of microplastics in human placenta using laser direct infrared spectroscopy. Sci Total Environ 856:159060. https://doi.org/10.1016/j.scitotenv.2022.159060

Chapter 8
Coupling Bioremediation Strategies for Eradication of Soil Organic Pollutants

Aditi Roy, Priya Dubey, Poornima Vajpayee, and Suchi Srivastava

Abstract This study explores the synergistic application of diverse bioremediation strategies to effectively eliminate soil organic pollutants. The generation of persistent and toxic wastes has led to severe consequences in environment. While conventional methods have been employed, their limitations, particularly associated costs, necessitate exploring alternative approaches. Bioremediation techniques have exhibited promising results, but the current paradigm advocates integration of various strategies to enhance degradation rates and reduce overall remediation time. The interdisciplinary nature of this issue calls for amalgamation of knowledge from different bioremediation techniques to achieve more efficient outcomes. This research aims to provide valuable insights into coupling different bioremediation strategies, emphasizing their combined potential for eradicating soil organic pollutants. The study advocates integrated methods to yield synergistic effects, fostering enhanced performance in remediation of contaminated sites. The ultimate goal is to contribute to the development and maintenance of metabolic robustness in degradative microbes within the soil environment.

Keywords Degradation · Bioremediation · Organic pollutants · Microbes · Soil

A. Roy (✉)
Department of Botany (Environmental Science), University of Lucknow, Lucknow, UP, India

Plant Ecology and Environment Technologies, CSIR- National Botanical Research Institute, Lucknow, UP, India

P. Dubey
Department of Biosciences, Faculty of Science, Integral University, Lucknow, UP, India

P. Vajpayee
Department of Botany (Environmental Science), University of Lucknow, Lucknow, UP, India

S. Srivastava
Plant Ecology and Environment Technologies, CSIR- National Botanical Research Institute, Lucknow, UP, India

S. Dey, S. Bhattacharya (eds.), *Biotechnological Interventions in the Removal of Emerging Pollutants*, Interdisciplinary Biotechnological Advances,
https://doi.org/10.1007/978-981-97-9922-0_8

8.1 Introduction

Rapid urbanization and industrial development have led to the excessive consumption and production that have led to the emergence of soil pollution. Among different types of soil contaminants, organic pollutants possess to be highly toxic because of its hydrophobic, lipophilic and high persistence in nature. Due to these factors, the organic pollutants tend to affect our surround and may enter into the food chain increasing the risk to cause more harm. Organic pollutants like polyaromatic hydrocarbons (PAHs), phenols, polychlorinated biphenyls (PCBs), toluene, trinitrotoluene, benzene, different pesticides, and herbicides are persistent in nature and have been reported to cause mutagenicity and carcinogenicity in animals and human through entering into food chain. Conventional remediation techniques majorly focus on ex-situ removal of contaminants through physical or chemical process. However, physical and chemical remediation process like electro reclamation, leaching, landfill, thermal treatment, excavation have higher cost and health risks associated with them. This have led to a paradigm shift towards in-situ remediation technologies where decomposition of xenobiotic compounds via chemical and biological pathways is contained within subsurface environment. The flawed options related to physical and chemical bioremediation techniques have levelled up bioremediation over them.

Soil organic pollutants are dispersed into the earth by air or water, resulting in a slew of major environmental and health issues throughout the globe. Also, humans, agricultural usage, and incorrect waste disposal procedures can all contribute to environmental issues. These pollutants can build up and stay in the environment for a long time. The persistence of these compounds in the environment poses a long-term hazard to human and wildlife health and safety (Roy et al. 2023). Organochlorine pesticides, atrazine, dichlorodiphenyl trichloro ethane, 2,4-dichlorophenoxyacetic acid, and hexachlorocyclohexane are among the most often used herbicides/pesticides in both forestry and agriculture for insect pest control and disease (Roy et al. 2023). All of these chemical compounds have been linked to a wide range of human health issues, including cancer, reproductive issues, and birth defects (Gilden et al. 2010). Pesticides account for nine of the twenty-one persistent organic pollutants. Organochlorine, organophosphate, organometallic, pyrethroids, and carbamates are examples of organic insecticides (Gilden et al. 2010). These contaminants can change the characteristics of soil even if only a small amount is present by affecting the microbial population and metabolic activity. Regrettably, traditional solutions for removing these pollutants from the soil system are prohibitively expensive, thus discouraging enterprises from properly managing their effluent discharge. Numerous chemical and physical engineering procedures, like soil washing, pumping, extraction, air stripping, aeration, and treating polluted water, addition of oxidants including H_2O_2 or $KMnO_4$, or incineration, have been extensively used in the past to remove organic pollutants in situ (Doty 2008). Nonetheless, these applications have the potential to remove the organic pollutants but are not sustainable for long term treatment. Therefore, researchers have stepped up their efforts in this area to create more cost-effective clean-up methods. Phytoremediation and bioremediation are two sophisticated environment-friendly decontamination processes. As organic-degrading organisms, such as bacteria and plants, have been discovered in a variety of situations, a growing number of research have sought to promote biological activity and improve degradation efficiency. This study focuses on a number of significant results in the field of organochloride bioremediation and phytoremediation.

Due to the discovery of a variety of microbial species and their enzymes capable of decomposing organic pollutants, microbial bioremediation is becoming a more feasible and successful strategy for eliminating environmental toxins. Microorganisms may convert or mineralize practically every naturally existing organic molecule, driving the carbon cycle through breakdown of organic materials. Synthetic anthropogenic compounds with non-biotic functional groups or structures, on the other hand, can be durable since microorganisms lack the metabolic machinery to degrade them. Bioremediation technologies were deployed at many polluted sites and have proved the capacity to degrade organic pollutants due to the identification of numerous bacterial isolates that have the ability to breakdown the organic pollutants. However, because environmental factors such as temperature, pH, soil characteristics, and humidity are difficult to manage for optimal biodegradation outcomes, their applicability is restricted. The implementation of microbial bioremediation continues to be hampered by challenges with this method, such as even dispersion of bacteria in the site, competition with local microbes, and monitoring. Plants and their accompanying microorganisms are used to breakdown, accumulate, or immobilize pollutants in soil and water, which is a promising strategy. Metals, insecticides, gasoline, explosives, polyaromatic hydrocarbons (PAH), and other organic chemicals have all been cleaned up with it (Meagher 2000). When opposed to traditional physical or chemical techniques, phytoremediation offers several clear advantages. To begin with, it is quite affordable to implement and administer (Meagher 2000). Second, it is less invasive and cleans up the environment while providing an incredibly attractive environment, carbon sequestration, soil stabilization and wildlife habitats. Finally, phytoremediation plant components can be recycled into woodchips, pulp, or bioenergy resources. Phytoremediation has other advantages. Phytoremediation is expected to gain public acceptance, and the plants' effectiveness can be easily evaluated.

Hazardous organic pollutants endanger the health of humans, animals, and the ecosystem. These contaminants might be dangerous if they are not controlled. Many researchers have increased their efforts to identify more cost-effective and environment-friendly alternatives to the use of hazardous chemicals and treatments to eliminate current hazardous contaminants. Environmental biotechnology, such as phytoremediation and bioremediation, is a promising subject that eliminates hazardous organic wastes using natural resources including such plants and microorganisms. Because of its cost effectiveness and ecologically benign technique, this technology is an appealing alternative to other traditional clean-up procedures. Current biological methods for the removal of organic pollutants, particularly chlorinated hydrocarbons, are discussed in this study, with an emphasis on their limitations and recent initiatives to address them. Environmental biotechnology, such as phytoremediation and bioremediation, is a potential decontamination method that uses natural resources to restore a hazardous environment to a safe state. Without stringent environmental management, hazardous sites are expected to continue to grow. Environmental biotechnology has been used in recent years to develop more eco-friendly and cost-effective ways to lessen the rising number of hazardous sites. Complete bioremediation or phytoremediation, on the other hand, is difficult to achieve since it is impacted by natural factors and difficult to manage for ideal biodegradation conditions. To increase the application of environmental technology, it must be customized to the local native environment, taking into account site-specific features as well as the type of the pollutants. In addition, much study is required to find biological activity in natural resources such as plants and bacteria. For the effective use of environmental biotechnology, it is also necessary to investigate the function of specific genes involved in the metabolism, transport, or absorption of certain contaminants.

8.2 Bioremediation: A Conventional Approach

Over the last few decades, the extensive use of synthetic organic pollutants and their pervasiveness in soil has become a major environmental concern due to their hydrophobic and recalcitrant nature (Meghraj et al. 2011). The adsorption, desorption, and transformation of these organic pollutants are dependent on the physical, chemical, and biological processes of soil. Also, the characteristic of soil, environmental factors, and pollutant properties govern the fate and behaviour of these organic contaminants. The presence of these harmful contaminants in the environment poses a long-term risk to the health and safety of people and wildlife (Haripriyan et al. 2022). Numerous techniques have been developed to address these dangerous pollutants from soil. However, the removal of organic pollutants from soil matrix using traditional engineering techniques is not a cost-effective and environment-friendly approach. Bioremediation is a widely accepted biotechnological technique that has been used by researchers to carry out biologically mediated changes in pollutants through detoxification and mineralization. Bioremediation is the best biological method to remediate soils with loads of organic pollutants using living organisms (micro-organisms, plants, and algae) and their effective enzymes. These living organisms detoxify the hydrocarbons into greener products by transforming them into carbon dioxide, water, and nontoxic intermediates (Meghraj et al. 2011). The metabolic capacity of microorganisms to detoxify or change the pollutant molecule depends on both accessibility and bioavailability in the bioremediation process. The primary advantage of bioremediation is its cost-effectiveness and permanent remediation that could cause the contaminant to completely mineralize. Additionally, it does not affect the existing microbiota and is ecologically safer than other physical and chemical remediation approaches. Hydrocarbon bioremediation is very effective but it can take a few weeks or even months to complete the heterogeneous and complex processes by microbes. Also, several factors are responsible for the working of microorganisms including the physical state and chemical composition of hydrocarbons, pH, optimum temperature, consortia makeup, climate, and indigenous microbial population.

The process of bioremediation on the basis of their treatment site is divided into two: In-situ and ex-situ bioremediation (Fig. 8.1). In-situ bioremediation involves the treatment of hydrocarbons or pollutants on-site under natural conditions. While the ex-situ method uses soil excavation to treat pollutants away from the contaminated site. In-situ methods are preferred more for ecological restoration of soil as it is less expensive and low maintenance is required. In-situ processes for the detoxification of organic pollutants are achieved through intrinsic and engineered bioremediation (Meghraj et al. 2011). Intrinsic or passive remediation does not require human intervention and engineered in-situ remediation may require microbes from outside to enhance the physicochemical condition of the site for the effective degradation of pollutants (Raimondo et al. 2020). Engineered remediation technique includes bioattenuation, biostimulation, bioaugmentation, biosparging, bioventing, and bioslurping. Bioattenuation is a natural process of remediating hydrocarbons by converting them into immobilized/less hazardous forms. This process doesn't require human intervention for monitoring. The immobilization is mostly attributed to biodegradation by microorganisms and, to a lesser extent, interactions with natural chemicals from the geological media (Meghraj et al. 2011). Bioattenuation is generally used to treat the harmful components of fuels. The process of altering the natural environment by introducing the limiting nutrients (nitrogen and phosphorus) and electron acceptors to increase microbial turnover is done through the process of biostimulation. Availability and supply of nutrients such as carbon, nitrogen, phosphorus, and potassium have a significant impact on the concentrations of microbes.

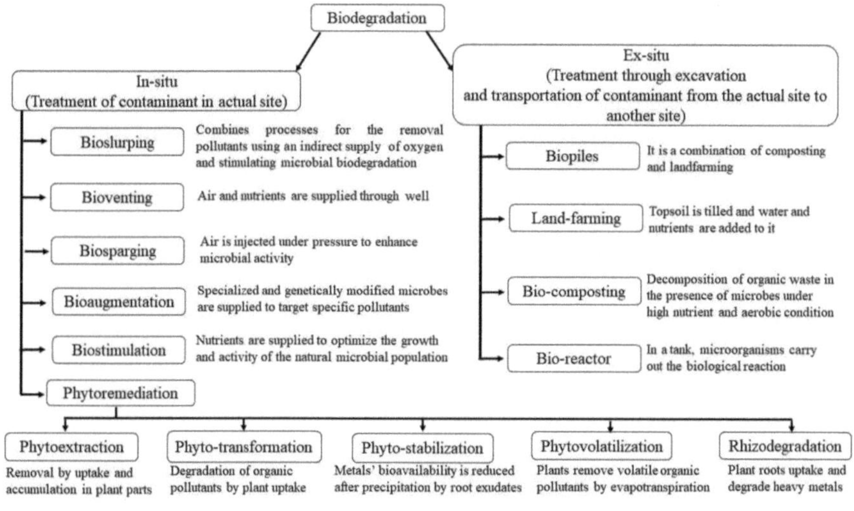

Fig. 8.1 Types of bioremediation techniques

Water-soluble, slow-release, and oleophilic fertilizers are added to promote microbial degradation by changing hydrocarbons into greener forms (Haripriyan et al. 2022).

The process of adding target-specific microorganisms to speed up the depletion of contaminants and promote the functioning of existing microbial populations is known as bioaugmentation. Both exogenous and cultured microorganisms are considered for the bioaugmentation process. The Biosparging technique involves the artificial injection of air beneath the soil surface to enhance the microbial activity for removing pollutants. Air is injected into a saturated zone which may aid in the upward migration of volatile organic molecules to the unsaturated zone to promote the rate of degradation (Höhener and Ponsin 2014). Similar to Biosparging, bioventing is also done to improve the activity of indigenous microbes by supplying oxygen to the unsaturated (vadose) zone. In bioventing, modifications are performed by adding moisture and nutrients to boost bioremediation (Höhener and Ponsin 2014). In order to deal with two different contamination media, bioslurping combines the bioventing and vacuum-enhanced free-product recovery methods. Bioslurping is mostly used for remediating volatile and semi-volatile organic compounds (Höhener and Ponsin 2014; Sharma 2020).

Ex-situ bioremediation strategies include the removal of pollutants from contaminated sites by transferring them to another location for restoration (Sharma 2020). The depth and degree of pollution, cost of treatment, and location of the contaminated site are frequently taken into consideration while performing ex-situ method. Ex-situ bioremediation technology selection is governed by performance standards as well. Ex-situ bioremediation is divided into solid and slurry phase bioremediation (Smith et al. 2015). In solid phase, the contaminated soil is removed from its natural environment and piled up. Ventilation pipes are positioned all over the piles to supply oxygen for microbial respiration. Compared to slurry-phase operations, solid-phase systems required a significant amount of area and took longer to clean up the contaminants. Biopiles, windrows, land farming and, composting are solid-phase treatment techniques (Tomei and Daugulis 2013). The slurry phase bioremediation technique involves bioreactors that blend contaminated soil with water, nutrients, and oxygen, creating an optimal environment for microorganisms to break down soil pollutants. Throughout this process, stones and rubble are separated from the contaminated soil. The amount of additional water used is determined by soil physicochemical characteristics, biodegradation rate, and contaminant concentration. Once the procedure is completed, the soil is extracted and dried using centrifuges, vacuum filters, and pressure filters. Subsequently, the soil is disposed of, and the resulting fluids are treated beforehand (Smith et al. 2015; Sharma 2020; Tomei and Daugulis 2013).

8.3 Factors Affecting Bioremediation

The completion of bioremediation is influenced by a variety of factors. Various environmental factors can influence the effectiveness of microbial absorption and metabolism of hazardous hydrocarbons. To combat pollutants, microorganisms must have access to a variety of compounds that will provide them with energy and nutrients to build more new cells.

The effectiveness of bioremediation is determined by a number of elements, including:

8.3.1 Microbial Adaptability and Nutrient Availability (Acclimatization)

Nutrients like carbon (C), carbon (N), phosphorus (P), potassium (K), iron (Fe), and calcium (Ca) are essential for microbial development and activity; hence, their availability plays significant role in determining rates at which various pollutants degrade. Furthermore, the relative availability of nutrients has a significant regulatory influence on pollutant degradation. For example, site with higher level of organic carbon have more microbial activity which may lead in rapid depletion of bioavailable nutrients such as N, P, K, and Fe (Dhaliwal et al. 2021). Surplus availability of N, P, and K, on the other hand, has been observed to have a deleterious influence on aromatic hydrocarbon decomposition.

The breakdown of petroleum wastes may be influenced by microbial adaptability. Microbial populations that have been exposed to become more efficient in degrading aromatic organic chemicals. In adapted conditions, *Alcaligenes xylosoxidans* Y234 decomposes benzene and toluene more effectively than non-adapted conditions (Pathak et al. 2017). The non-acclimatized microorganisms, on the other hand, had the quickest biodegradation rates in a single component and multicomponent contaminated systems.

8.3.2 Impact of the Environment Conditions

The bioremediation of petrochemical wastes can be influenced by a variety of environmental factors such as temperature, pH, moisture, oxygen availability, salinity, biological factors and toxic compounds. The following are some other environmental elements that play a role:

8.3.2.1 Temperature and pH

Temperature is the most significant physical element in affecting survival of microbes and composition of hydrocarbons (Roy et al. 2023). Under in-situ and ex-situ conditions, temperature has an essential role in hydrocarbon breakdown and microbial metabolism (Margesin and Schinner 2001). In cold climates like the Arctic, oil breakdown by natural processes is especially slow, placing more pressure on microbes to clean up the released petroleum. Most oleophilic bacteria are metabolically inactive due to the sub-zero temperature of the water in this area because their transport channels freeze and cause the entire cytoplasm of the cell to freeze (Macaulay 2015). However, studies show that raising the temperature enhances the solubility of hydrocarbons in the medium, making petrochemical hydrocarbons more comprehensible to microorganisms. The rate of degradation of petrochemical wastes is typically high in the temperature range of 30–40 °C in soil and 15–20 °C in aqueous or marine ecosystems (Al-Hawash et al. 2018).

The microbial decomposition of petrochemical wastes in soil or aqueous media was substantially affected by pH variations. Several investigations show that petroleum hydrocarbons mineralize well at pH levels close to neutral. A little change in pH can have a big influence on the overall biological decomposition processes in an aquatic environment. In very acidic conditions, several fungi and acidophilic microorganisms have been observed to flourish and exhibit biodegradation capacity (Margesin and Schinner 2001). Biodegradation can occur at any pH; however, in most terrestrial and aquatic environments, a pH range of 6.5 to 8.5 is typically ideal for biodegradation (Gangola et al. 2022).

8.3.2.2 Availability of Moisture, Oxygen, and Salinity

Soil moisture serves as a conduit for the movement of soil nutrients and the removal of bacterial metabolic waste products from soil particles (water film). It affects soil toxicity levels and types, soluble materials, aeration, soil pH, osmotic pressure, diffusion processes, and gas transfer, among other things. The presence of hydrocarbon contaminants reduces the soil's porosity and ability to retain water (Sakshi et al. 2019). Due to the direct proportionality between bacterial and soil water activities, which means that as moisture content decreases, bacterial activities also do, and conversely, when soil moisture levels are high, oxygen transport is constrained, this state ultimately inhibits microbial activity. As a result, hydrocarbon breakdown in terrestrial habitats is constrained by the availability of water for microbial growth and metabolism. Because of this, the availability of sufficient moisture in the range of 50–75%, 30–90%, and 50–80% is necessary for hydrocarbon biodegradation (Lee et al. 2018). On the other side, excessive moisture levels are bad for microbial growth and metabolism. This is due to decreased oxygen diffusion in the soil and hampered aerobic hydrocarbon breakdown, which prevents the development of anaerobic soil conditions. In hydrocarbon-contaminated areas, adequate moisture (water availability) is necessary to encourage microbial activity for hydrocarbon breakdown (Kebede et al. 2021).

The amount of oxygen available determines whether a particular environmental condition is anaerobic or aerobic. Petrochemical hydrocarbons have been observed to decompose mostly in aerobic and, in rare circumstances, anaerobic conditions. Anaerobic biodegradation occurs primarily in submerged and aquifers marine sediments, with a low rate of degradation and a focus on halogenated aromatics. During the aromatic ring oxidation process, mono-oxygenase and di-oxygenase enzymes are required for the breakdown of petrochemicals. It is fundamentally necessary to use substituted electron acceptors such as ferrous iron, nitrate, and sulphate to promote aromatic compound oxidation under anaerobic environments. However, in anaerobic circumstances, the reduction of electron acceptors such as ferric iron, nitrate, and sulphate results in the liberation of substantial amounts of phosphorus and ferrous iron, further polluting the natural habitat. Furthermore, anaerobic decomposition of petrochemical hydrocarbons has resulted in the emission of greenhouse gases (CH_4, NO_2, etc.) as well as a rise in pH (Mishra et al. 2021). As a result, oxygen availability is critical in aromatic chemical bioremediation.

Salt deposition and aggregation, whether natural or anthropogenic, reduces microbial hydrocarbon-clastic activity and increases stress for halophile and halotolerant species in their environment (Su et al. 2021). High salt concentrations can raise osmotic pressure, obstructing microbial metabolic activities, solubility, and transportation networks for critical ions, as well as depriving bacteria of oxygen (Su et al. 2021). This might have occurred as a result of the use of a significant number of soluble fertilizers (nitrates or ammonium salts) as well as the discharge of salty solution into the soil. When petroleum compounds pollute hypersaline habitats, halophilic bacteria have the ability to live and breakdown the contaminants. Genes encoding hydrocarbon-clastic enzymes are found in those prospective bacteria, and these enzymes are responsible for their varied hydrocarbon catabolism (Das et al. 2022). The addition of salts like NaCl and NH_4Cl to nutrient-poor soil has been shown to improve microbial hydrocarbon breakdown (Gouthami et al. 2023). Hydrocarbon biodegradation in soil increased with rising cation concentrations due to the presence of such salts. The higher the cation concentrations, the less likely the soil is to support a microbial consortium and so less likely to degrade hydrocarbon pollutants (Das et al. 2022). The type of probable hydrocarbon-degrading bacteria in habitats of hydrocarbon-contaminated soil will therefore determine the optimal salt amount.

8.3.3 Biological Factors and Toxic Compounds

The biotic processes that affect the breakdown of organic pollutants include competition among bacteria for scarce carbon sources, antagonistic interactions between microbes, and predation of microbes by protozoa and bacteriophages. The amount of 'catalyst' and the concentration of contaminants both typically affect how well contamination is degraded. This refers to both the number of organisms that can break down the contaminant and the quantity of enzymes produced by each individual cell (Al-Hawash et al. 2018). Cells produce certain enzymes that can speed up or slow down the rate of contaminant degradation. Additionally, specific enzymes involved in contaminant metabolism must be taken into account, as well as their 'affinity' for the contamination and availability of the contaminant. Toxic effects on microorganisms may occur when hazardous properties of some contaminants are present in high concentrations, slowing the decontamination process. Depending on the toxicants, their quantities, and the microorganisms exposed, toxicity levels and processes change. Various organic and inorganic substances can be harmful to the intended living species.

8.3.3.1 Availability of Degrading Microbes

The concentration of organic pollutants affects biodegradation rates as it influences their uptake, transformation, and mineralization. Excessively high concentrations are recognized for their toxicity, adversely impacting the growth rate and biomass output of degraders, thereby necessitating an extended treatment duration. This is due to the heavy and undispersed nature of high concentration pollutants. Microbes for degrading organic pollutants are abundant and prevalent in contaminated soil because of easy adaptability. Contaminants may act as source of carbon and energy required for their metabolism and growth. Additionally, it is known that soil organic content concentrations higher than 5% can interfere with the C:N:P ratio, oxygen availability, and microbial breakdown. Likewise, exceptionally low concentrations impede biodegradation by suppressing the microbial metabolic genes responsible for producing degradation enzymes. This leads to insufficient carbon supply or availability, hindering microbial growth (Kebede et al. 2021). These microbes are metabolically active and may breakdown either in an aerobic or anaerobic environment. It is important to note, however, that not only the types and nature of hydrocarbon pollutants, but also the surrounding environment influences their distribution and abundance (Lv et al. 2020). As a result, the presence of an ideal concentration (not too high or too low) for efficient mineralization from polluted surroundings is essential for better degradation results.

8.3.3.2 Microbial Competition and Cooperation

Within microbial communities of a specific environment, microbial cooperation and competition are critical forces for survival and stability. In contaminated habitats, microbial populations coexist synergistically and/or antagonistically. However, interspecific or intraspecific competition is a limiting factor for biodegradation effectiveness. Interspecific competition, for example, can occur when hydrocarbon-degrading fungi and hydrocarbon-metabolizing bacteria compete for the use of hydrocarbon pollutants as carbon sources and other limiting resources (N and P) for their development and metabolism. Exogenous bacteria (introduced inocula) have been employed to break-down hydrocarbon pollutants in studies (Díaz-Ramírez et al. 2013), although they are not always efficient. This is due to the fact that they are unable to evade competition from native bacteria, predators, and abiotic factors. It has been found that a synergistic relationship between *Cycloclasticus* and *Alcanivorax borkumensis* improves polyaromatic hydrocarbon degradation, while an antagonistic relationship between *Thalassolituus* and other oleophobic bacteria slows biodegradation (Louvado et al. 2015). The researchers also discovered that cohabitation or cooperation of indigenous and foreign bacteria obstructs each other's metabolic activity, limiting the efficacy of hydrocarbon biodegradation (Kumari and Das 2023). Thus, the interdependence of microbial populations (synergistic effect) is critical for bioremediation to be successful thereby providing the complete views about the modulation of their ability in hydrocarbon biodegradation.

8.3.4 Effects of Pollutants

Pollutant impacts on bioremediation may be separated into two categories. To begin with, the presence of metals and pesticides can have direct effects on organisms. Second, co-existing contaminants have an impact on pollutant bioremediation processes. Depending on the species, development stages, concentrations and characteristics of pollutants, and soil conditions, co-occurrence of heavy metals and pesticides can have a beneficial or detrimental influence on biological growth (microorganisms and plants) (Zhang et al. 2020). Heavy metals have an impact on biological function via a variety of metabolic pathways. As soon as cations are taken up by cells, they can interact with a variety of organic ligands such as cellular membranes, nucleic acids, enzymes, and proteins of microorganisms. In some circumstances, binding is advantageous. Low metal concentrations, such as Ca, Cu, Mg, Mn, and Zn, can increase biological activity (Zhang et al. 2020). Metals, on the other hand, can decrease enzyme or protein function by masquerading/disguising as catalytic active groups or altering protein structures. Toxic metal cations can substitute enzymes' physiologically important cations, and rendering them inactive. Microorganisms are subjected to oxidative pressure by metals. The strong binding of metal ions to sulfhydryl groups of enzymes, which are required for general microbial metabolism, is the most prevalent source of metal poisoning. Many species may flourish in metal-contaminated regions despite the risk for harm due to resistance/tolerance mechanisms. Heavy metals, on the other hand, may have an impact on macronutrient composition (N, P, K, Na, Ca and Mg). In the absence of excessive heavy metals, lowered copper (Cu) concentrations improved the macronutrient content of *Vigna radiata* L.; however, excessive heavy metals blocked macronutrient adsorption (Karmous et al. 2022). The deleterious effects of cadmium (Cd) on maize development are mitigated by the presence of Atrazine, according to data on leaf necrosis, mycorrhizal root colonization, and biomass (Karmous et al. 2022). Pesticides can change the biochemical and physiological behaviour of soil organisms by blocking several enzymes and disrupting the organisms' growth promoting attributes (Shahid and Khan 2022). The production of protein, carbohydrates, and chlorophylls by non-target plants can be inhibited by pesticides. High levels of organic pollutants in contaminated soils may limit the capacity of contaminated soils to give water and nutrients to plants, resulting in lower biomass output (Roy et al. 2023). Notably, while certain bacteria may utilize low concentrations of pesticides as a source of nutrition or energy to reproduce and increase development during bioremediation, others can be poisonous to them (Pathak et al. 2017). Microbial populations can develop tolerance/resistance to pesticides over time and revert to normal.

8.4 Integration of Bioremediation Strategies

Bioremediation strategies individually have been used to decontaminate organic pollutants from soil. Although being economical and environment-friendly process these also have few constraints of its own. Besides the nature and concentration of nutrients, various physical and environmental parameters, chemical compositions also affect the working efficiency of a particular biodegradation process. Moisture, soil texture, pH, organic matter content directly affects the degradation of organic pollutants at variable rates (Raimondo et al. 2020). Thus, any promising design for remediation requires application of different biological strategies. The choice of implementing a certain bioremediation technique depends on degradation ability of biotic factors and extent of contamination present at the site which is to be treated. Properties of organic pollutants like water repulsion, chemical stability, volatilization and low degradation rate have elevated their undesirable consequences in nature. Bioremediation techniques like phytoremediation, mycoremediation, biostimulation, bioaugmentation, and natural attenuation have been operated to remediate soil pollution individually. Recently, integration of these processes has been implemented on fields which had shown better remediation rates. Coupled remediation strategies can help in infusing loopholes of one another. This can also help in escalating removal rates of contaminants than any individual bioremediation process. Integration of bioremediation strategies may provide a new insight in performing better results. The supporting method may act as protective agents against the effect of pH, temperature, solvent, substrate and product inhibition which can help in improved degradation (Pimmata et al. 2013).

8.4.1 Bioaugmentation with Biostimulation

In recent years, studies have been now deviated from inducing single remediation technique to incorporating more than one process to yield better results. Techniques like bioaugmentation and biostimulation have been earlier reported to remediate soil contaminated with organic pollutants specifically with PAH, DDT, Carbofuran, etc. However, there are some cases where these techniques collapse under non-sterile condition or for full scale-field remediation. This is majorly because the microbes which are either stimulated through different nutrient formulation or are augmented in the site fail to survive in highly polluted soil or could not degrade the intermediate products in degradation process. Amalgamation of competent microbial strains individually or through consortia with nutritional amendments hold the promise of fostering degradation rates. Incorporation of bioaugmentation with biostimulation have provided evidence in providing high nutritional content and improving soil quality.

In a study by Ammeri et al. (2022), effective degradation results have been observed on simultaneous application of biostimulation and bioaugmentation in a field contaminated with pentachlorophenol (PCP). This contaminant has been used as pesticide and anti-microbial compound specially to control wood diseases. The study showed PCP contaminated soil when treated with municipal waste compost in organic amendments combined with bacterial consortia reduced PCP concentration to 40 mg kg^{-1} from 82 mg kg^{-1} initially. Also, a significant reduction of 71% was observed when contaminated soil was treated with forest compost along with microbial consortium dropping from 82 mg kg^{-1} to 28.9 mg kg^{-1} (Ammeri et al. 2022). Garg et al. (2016) investigated the effect of biostimulation, bioaugmentation and combine for bioremediation of HCH contaminated soils. According to their results, the combined approach of biostimulation and bioaugmentation provided effective reduction in α- and β-HCH levels. The study reported maximum reduction of total HCH with combined biostimulation and bioaugmentation (58%), followed by biostimulation alone (52%) and finally with bioaugmentation (39%). Full scale field trails through application of microbial consortium consisting of a group of HCH-degrading *Sphingomonads* (*Sphingobium* sp. UM1, *Sphingobium lucknowense* F2, *Sphingobiuum chinhatense* IP26, *Sphingobium* sp. HDIPO4, and *Sphingobium indicum* B90A) along with biostimulation by ammonium phosphate and molasses, was proved to be effective in achieving better results. Another study by Raimondo et al. (2020) reported reduction in half-life of lindane in a condition of integrated bioaugmentation and biostimulation.

Reports by Sayara et al. (2011) showed degradation of total PAHs (fluorene, phenanthrene, anthracene, fluoranthene, pyrene, benzo(a)anthracene, and chrysene) contaminated soil with 1 g of total PAHs kg^{-1} in dry soil through integrated bioaugmentation and biostimulation process. The study suggested incorporation of white rot fungi *Tramestes versicolor* for bioaugmentation with compost of organic fraction of municipal solid waste as bio-stimulant substrate. The biostimulation enhanced the indigenous microbial activity leading to 89% of total PAHs degradation at the end of 30 days incubation period. This result was compared to the control sites (without any biostimulators) which was able to achieve only 29.5% of degradation.

8.4.2 Microbe-Assisted Phytoremediation

Phytoremediation have been extensively exercised for organic pollutant removal from soil. However, it has its own restraints as presence of surplus amount of organic pollutant condenses plant growth and development. This eventually affects phytoremediation effectiveness. However, this loophole can be eliminated if pollutant degrading and plant growth promoting microbial strains are provided in combination with required plants. Hyper-accumulator plant species may provide habitat and nutrition to rhizospheric microbes which in turn can augment plant stress tolerance and enhance soil fertility. Both endophytic and exophytic microbes have been found to assist in phytoremediation of contaminated site. In addition to this, plant roots provide microaerophilic conditions and secretes exudates for enhanced degradation. Studies have been done to express the role of microbes particularly bacteria in association with plant roots for reducing organic pollutants. Degradation potential factors like volatilization, transformation and rhizodegradation mark them to be more susceptible for bioremediation. The most extensively used microbes for micro-assisted phytoremediation belongs to the category of plant growth promoting rhizobacteria (PGPR) which are known for their natural attenuation capability towards organic pollutants (Girolkar et al. 2021). Bacterial species, like *Serratia, Bacillus, Pseudomonas, Burkholderia, Enterobacter, Erwinia, Klebsiella, Beijerinckia, Flavobacterium, Gluconacetobacter, Sphingomonads, Achromobacter, Alcaligenes, Commamonas, Corynebacterium*, and *Rhodococcus,* are commonly used for microbe-assisted phytoremediation. Effective application of microbe-assisted phytoremediation largely depends on the survival rate of pollutant-degrading microbes either applied individually or in consortium (Table 8.1). Prolonged application of microbes in regular intervals consecutively increases degradative potential.

8.4.3 Genetically Engineered Bioremediation

Currently, a number of scientific organizations from across the world are working to remove pollution created by humans by altering the genetic makeup of microorganisms. Genetically modified organisms provide more opportunities for the breakdown of a wide spectrum of contaminants. Contemporary advancements in genetic engineering and omics technology have enabled us to redesign and modify plants, microbes and enzymes for better remediation. Recently, several microbes and enzymes have been developed for biodegradation of persistent organic pollutants, polychlorinated biphenyls, and polycyclic aromatic hydrocarbons. In particular to organic pollutants, alkane-degrading microbes improved with particular genes like alkB, alkB1, alkB2, alkM, aromatic hydrocarbons: xylE and PAHs: ndoB and nidA have been examined for remediation purposes (Wolejko et al. 2016). Genetically modified bacteria with VHB (bacterial haemoglobin) have been tested for remediation of aromatic compounds under anaerobic conditions (Ojuederie and Babalola

Table 8.1 Integration of microbes and plants for bioremediation of organic pollutants

Microbial Species	Plant Species	Target Pollutant	Reference
Actinomycete amycolata sp.	Poplar root	1,4-dioxane	Kelley et al. (2001)
Acinetobacter sp.	*Lolium multiflorum* (Ryegrass)	Phenanthrene and pyrene	Wu et al. (2011)
Azospirillum brasilense	*Tall fescue grass*	Naphthalene	Huang et al. (2004)
Bacillus pumilus	*Lolium perenne*	Chlorpyrifos	Ahmad et al. (2012)
Burkholderi sp.	*Morus* sp.	Biphenyl	Leigh et al. (2002)
Burkholderia fungorum	*Populus deltoides* × *Populus nigra* (*hybrid popular*)	Dibenzothiophene, phenanthrene, naphthalene, and fluorene	Andreolli et al. (2013)
Burkholderia xenovorans LB400	Tall fescue (*Festuca arundinacea*)	Polychlorinated biphenyls (PCBs)	Secher et al. (2013)
B. cepacia strain FX2	Corn and Wheat	Toluene	Wang et al. (2010)
B. cepacia VM1468	Poplar	BTEX, TCE	Taghavi et al. (2011)
Enterobacter sp. *strain PDN3*	Poplar (hybrid)	Trichloroethylene (TCE)	Kang et al. (2012)
Enterobacter ludwigii strains	Italian ryegrass (*Lolium multiflorum var. Taurus*), bird's foot trefoil (*L. corniculatus var. Leo*) and alfalfa (*M. sativa var. Harpe*)	Hydrocarbon	Yousaf et al. (2011)
Enterobacter sp. *12J1*	Wheat (*Triticum* sp.), maize (*Z. mays*) and *Allium macrostemon*	Pyrene	Sheng et al. (2008)
Funneliformis Caledonium, F. mosseae, Rhizophagus intraradices,	Fescue (*Festuca arundinacea*), Alfaalfa (*Medicago sativa*), White clover (*Trifolium repens*), Ryegrass (*Lolium perenne*), Wheat (*Triticum aestivum*)	PAH	Lu and Lu (2015)
Funneliformis Caledonium	Alfaalfa (*Medicago sativa*) *and* Ryegrass (*Lolium perenne*)	PCB	Lu et al. (2014)

(continued)

Table 8.1 (continued)

Microbial Species	Plant Species	Target Pollutant	Reference
Herbaspirillum sp.	*Triticum aestivum*	2,3,4,6-tetrachlorophenol	Mannisto et al. (2001)
Micrococcus species Bacillus species Kurthia species	*Populus deltoides*	Polycyclic aromatic hydrocarbon	Bisht et al. (2015)
Mesorhizobium sp.	*Astragalus sinicus*	3,3,4,4-TCB	Teng et al. (2016)
Pseudomonas putida	Grasses	Naphthalene	Kuiper et al. (2002)
Pseudomonas putida	*Salix purpurea*	Pyrene	Khan et al. (2014)
P. putida W619 TCE	Poplar	TCE	Weyens et al. (2010)
Pseudomonas sp. *strain ITRI53*	Italian ryegrass (*L. multiflorum var. Taurus*)	Hydrocarbon	Andria et al. (2009)
Pseudomonas fluorescens and *Chrysobacterium indologenes*	*Zinnia anguistifolia*	Mefenoxam	Pai et al. (2001)
Pseudomonas sp. SB	*Festuca arundinacea* L.	Oily-sludge contaminated soil	Liu et al. (2013)
Pseudomonas putida VM1441(pNAH7)	*Pisum sativum*	Naphthalene	Kumar et al. (2022)
Rhizophagus irregularis	Wheat	Polycyclic aromatic hydrocarbon	Ingrid et al. (2016)
Rhodococcus ITRH43	Italian ryegrass	Total petroleum hydrocarbons	
Rhizobium meliloti	Alfalfa	Polyaromatic hydrocarbons	Teng et al. (2011)
Streptomyces sp.	Zea mays	Lindane	Alvarez et al. (2015)

2017). However, environmental concerns and regulatory constraints limit their large-scale in-situ as well as ex-situ applications (Megharaj et al. 2011). Apart from microbes, plants engineered with specific genes have provided better alternative for phytoremediation. Kang et al. (2012) have reported better metabolizing ability of transgenic tobacco for trichloroethylene with respect to control. Despite the fact that multiple in vitro bioremediation experiments have been conducted, it has always been difficult to study the fate, behaviour, and removal rate of toxins in the natural ecosystem due to the participation of numerous environmental factors, such as biosafety.

8.4.4 *Rhizoremediation*

The rhizospheric zone consists of most active layer of 1–2 mm around roots, where plants provide positive niche to rhizospheric microbes for degradation (Fig. 8.2). Initially, studies were focused on leguminous plants and grass species because of their extensive branched root system and ability to uphold maximum microbial communities. Microbial growth in rhizospheric region is highly supported by plant root exudates like sugars, amino acids, organic acids, mucilage secretion via root cells in terms of nutrition and triggering a shift in microbial community of the contaminated soils. Rhizodegradation of organic pollutant is generally governed by enzymes secreted by microbial communities and extracellular enzymes secreted by plant exudates like laccases and peroxidases. As a property, organic pollutants are hydrophobic in nature and remain insoluble in water as well as in soil which makes it less bioavailable for microbes. Plant root exudates helps in augmenting solubility of these pollutants and provide proper nutrients and energy sustenance to microbes required for degradation.

Rhizodegradation has been evident for degradation of organic pollutants like nitro-aromatic compounds, petroleum hydrocarbons (PHCs), pyrene, phenanthrene and polycyclic aromatic hydrocarbons (PAHs). Initially, Corgie et al. (2004) found maximum degradation of PHCs near 3 mm zone of root surface depicting utmost microbes' concentration. Kuiper et al. (2002) reported the reduction of PAHs through *Pseudomonas putida* strain PCL1444 by using root exudates. Later, Joner

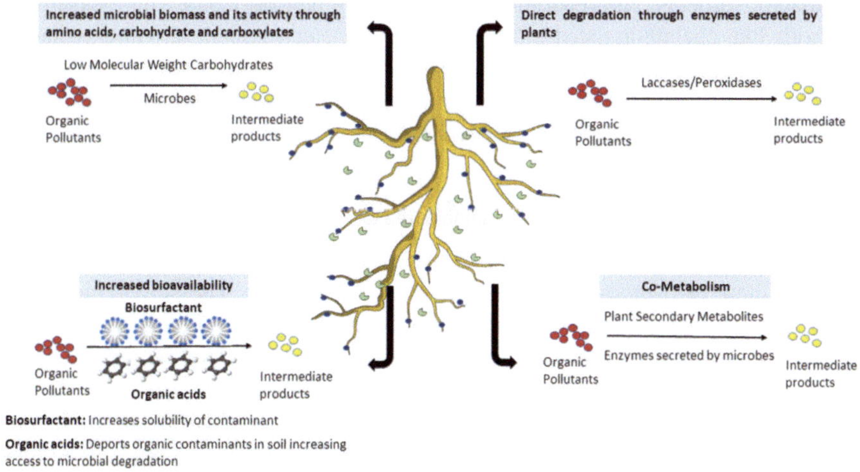

Fig. 8.2 Mechanisms involved in rhizoremediation of organic pollutants

and Leyval (2003) reported maximum degradation of phenanthrene with *Trifolium* L. and *Lolium perenne* in association with indigenous microbial strains. Apart from this, *Burkholderia* sp. along with flavones such as kuwanon C, morusinol, and morusin in roots of the mulberry plant have been reported to degrade PCBs concentration effectively (Gonzalez-Chavez et al. 2004). In some cases, for effective organic pollutant degradation, technique of plant breeding may prove to enhance rhizodegradation by providing stimulation for growth and activity of indigenous microbes (Compant et al. 2019).

8.5 Future Potential and Challenges

Organic pollutants have always been a constant threat to human and environment which demands immediate remediation. With the past studies, techniques of bioremediation have evolved to be better suited for successful removal of pollutants from nature apart from any other physical or chemical remediation methods. Key properties like cost-effectiveness and environment friendly have made it applicable in various instances. Use of plants and indigenous microbial species for exclusion of organic pollutants from a contaminated site have provided positive results. Although plants and microbes can mineralize organic pollutants through secretion of various enzymes, they have their own restrains. Presence of pollutant in higher level may inhibit the growth of plants as well as microbes in contaminated soil. In some cases, end products derived after degradation of organic pollutants shows xenobiotic characteristics and is hardly non-degradable. Thus, there has been an endless search for enhancing bioremediation results specially in ex-situ cases. Researchers have tried to recuperate bioremediation process through the application of various physical and chemical stimulators like nutrients (nitrogen and phosphorus), organic substrate (methane), sorbents, biopreparations, enzymes and nanoparticles. There have been methods like chelate-assisted bioremediation and electrokinetic phytoremediation for supplementing bioremediation; however, they can be considered responsible for generating other secondary environmental contaminants. Keeping in mind the relentless over piling of contaminants, these needs to be degraded rapidly. Opting for single bioremediation technique can help in removing pollutant from soil but at slower pace. Plants and microbes both have their limiting factors which may bound them from demonstrating suitable results. However, mutualistic behaviour of both can assist in gaining required results. Coupling of various bioremediation technique may facilitate in overcoming the loophole of one another. The slow remediating speed to acceptable levels and cumulative effects of organic pollutants to environment have emerged as limiting factors for bioremediation. However, this can be handled through incorporating various bioremediation techniques with natural microflora and phytotechnologies. However, regular conduction of studies is required for testing the efficiency of these methods on a larger field scale. If serious efforts are made, the merger of technologies may provide an efficient, economic, and sustainable green and clean remediation technology for the abatement of organic pollutants form soil.

References

Ahmad F, Iqbal S, Anwar S, Afzal M, Islam E, Mustafa T, Khan QM (2012) Enhanced remediation of chlorpyrifos from soil using ryegrass (*Lollium multiflorum*) and chlorpyrifos-degrading bacterium *Bacillus pumilus* C2A1. J Hazard Mater 237:110–115

Al-Hawash AB, Dragh MA, Li S, Alhujaily A, Abbood HA, Zhang X, Ma F (2018) Principles of microbial degradation of petroleum hydrocarbons in the environment. Egypt J Aquatic Res 44(2):71–76

Alvarez A, Benimeli CS, Sáez JM, Giuliano A, Amoroso MJ (2015) Lindane removal using *Streptomyces* strains and maize plants: a biological system for reducing pesticides in soils. Plant Soil 395:401–413

Ammeri RW, Simeone GDR, Hidri Y, Abassi MS, Mehri I, Costa S et al (2022) Combined bioaugmentation and biostimulation techniques in bioremediation of pentachlorophenol contaminated forest soil. Chemosphere 290:133359

Andreolli M, Lampis S, Poli M, Gullner G, Biró B, Vallini G (2013) Endophytic *Burkholderia fungorum* DBT1 can improve phytoremediation efficiency of polycyclic aromatic hydrocarbons. Chemosphere 92(6):688–694

Andria V, Reichenauer TG, Sessitsch A (2009) Expression of alkane monooxygenase (alkB) genes by plant-associated bacteria in the rhizosphere and endosphere of Italian ryegrass (*Lolium multiflorum* L.) grown in diesel contaminated soil. Environ Pollut 157(12):3347–3350

Bisht S, Pandey P, Bhargava B, Sharma S, Kumar V, Sharma KD (2015) Bioremediation of polyaromatic hydrocarbons (PAHs) using rhizosphere technology. Braz J Microbiol 46:7–21

Compant S, Samad A, Faist H, Sessitsch A (2019) A review on the plant microbiome: ecology, functions, and emerging trends in microbial application. J Adv Res 19:29–37

Corgié SC, Beguiristain T, Leyval C (2004) Spatial distribution of bacterial communities and phenanthrene degradation in the rhizosphere of *Lolium perenne* L. Appl Environ Microbiol 70(6):3552–3557

Das N, Bhuyan B, Pandey P (2022) Correlation of soil microbiome with crude oil contamination drives detection of hydrocarbon degrading genes which are independent to quantity and type of contaminants. Environ Res 215:114185

Dhaliwal SS, Sharma S, Sharma V, Shukla AK, Walia SS, Alhomrani M et al (2021) Long-term integrated nutrient management in the maize–wheat cropping system in alluvial soils of North-Western India: influence on soil organic carbon, microbial activity and nutrient status. Agronomy 11(11):2258

Díaz-Ramírez I, Escalante-Espinosa E, Schroeder RA, Fócil-Monterrubio R, Ramírez-Saad H (2013) Hydrocarbon biodegradation potential of native and exogenous microbial inocula in Mexican tropical soils. In: Biodegradation of hazardous and special products. IntechOpen, p 155178

Doty SL (2008) Enhancing phytoremediation through the use of transgenics and endophytes. New Phytol 179(2):318–333

Gangola S, Bhatt P, Kumar AJ, Bhandari G, Joshi S, Punetha A et al (2022) Biotechnological tools to elucidate the mechanism of pesticide degradation in the environment. Chemosphere 296:133916

Garg N, Lata P, Jit S, Sangwan N, Singh AK, Dwivedi V et al (2016) Laboratory and field scale bioremediation of hexachlorocyclohexane (HCH) contaminated soils by means of bioaugmentation and biostimulation. Biodegradation 27:179–193

Gilden RC, Huffling K, Sattler B (2010) Pesticides and health risks. J Obstet Gynecol Neonatal Nurs 39(1):103–110

Girolkar S, Thawale P, Juwarkar A (2021) Bacteria-assisted phytoremediation of heavy metals and organic pollutants: challenges and future prospects. In: Bioremediation for environmental sustainability. Elsevier, pp 247–267

Gonzalez-Chavez MC, Carrillo-Gonzalez R, Wright SF, Nichols KA (2004) The role of glomalin, a protein produced by arbuscular mycorrhizal fungi, in sequestering potentially toxic elements. Environ Pollut 130(3):317–323

Gouthami K, Mallikarjunaswamy AMM, Bhargava RN, Ferreira LFR, Rahdar A, Saratale GD et al (2023) Microbial biodegradation and biotransformation of petroleum hydrocarbons: progress, prospects, and challenges. In: Genomics approach to bioremediation: principles, tools, and emerging technologies. Wiley, pp 229–247

Haripriyan U, Gopinath KP, Arun J, Govarthanan M (2022) Bioremediation of organic pollutants: a mini review on current and critical strategies for wastewater treatment. Arch Microbiol 204(5):1–9

Höhener P, Ponsin V (2014) In situ vadose zone bioremediation. Curr Opin Biotechnol 27:1–7

Huang H, Bowler BF, Oldenburg TB, Larter SR (2004) The effect of biodegradation on polycyclic aromatic hydrocarbons in reservoired oils from the Liaohe basin. NE China Organic Geochem 35(11–12):1619–1634

Ingrid L, Sahraoui ALH, Frédéric L, Yolande D, Joël F (2016) Arbuscular mycorrhizal wheat inoculation promotes alkane and polycyclic aromatic hydrocarbon biodegradation: microcosm experiment on aged-contaminated soil. Environ Pollut 213:549–560

Joner EJ, Leyval C (2003) Rhizosphere gradients of polycyclic aromatic hydrocarbon (PAH) dissipation in two industrial soils and the impact of arbuscular mycorrhiza. Environ Sci Technol 37(11):2371–2375

Kang JW, Khan Z, Doty SL (2012) Biodegradation of trichloroethylene by an endophyte of hybrid poplar. Appl Environ Microbiol 78(9):3504–3507

Karmous I, Tlahig S, Loumerem M, Lachiheb B, Bouhamda T, Mabrouk M et al (2022) Assessment of the risks of copper-and zinc oxide-based nanoparticles used in *Vigna radiata* L. culture on food quality, human nutrition and health. Environ Geochem Health:1–17

Kebede G, Tafese T, Abda EM, Kamaraj M, Assefa F (2021) Factors influencing the bacterial bioremediation of hydrocarbon contaminants in the soil: mechanisms and impacts. J Chem 2021:1–17

Kelley SL, Aitchison EW, Deshpande M, Schnoor JL, Alvarez PJ (2001) Biodegradation of 1, 4-dioxane in planted and unplanted soil: effect of bioaugmentation with *Amycolata* sp. CB1190. Water Res 35(16):3791–3800

Khan MI, Cheema SA, Shen C, Hassan I, Chen Y (2014) Phytotoxicity assessment of phenanthrene and pyrene in soil using two barley genotypes. Toxicol Environ Chem 96(1):94–105

Kuiper I, Kravchenko LV, Bloemberg GV, Lugtenberg BJ (2002) *Pseudomonas putida* strain PCL1444, selected for efficient root colonization and naphthalene degradation, effectively utilizes root exudate components. Mol Plant-Microbe Interact 15(7):734–741

Kumar S, Thakur N, Singh AK, Gudade BA, Ghimire D, Das S (2022) Microbes-assisted phytoremediation of contaminated environment: global status, progress, challenges, and future prospects. In: Phytoremediation technology for the removal of heavy metals and other contaminants from soil and water. Elsevier, pp 555–570

Kumari S, Das S (2023) Bacterial enzymatic degradation of recalcitrant organic pollutants: catabolic pathways and genetic regulations. Environ Sci Pollut Res:1–30

Lee SH, Ji W, Kang DM, Kim MS (2018) Effect of soil water content on heavy mineral oil biodegradation in soil. J Soils Sediments 18:983–991

Leigh MB, Fletcher JS, Fu X, Schmitz FJ (2002) Root turnover: an important source of microbial substrates in rhizosphere remediation of recalcitrant contaminants. Environ Sci Technol 36(7):1579–1583

Liu W, Sun J, Ding L, Luo Y, Chen M, Tang C (2013) Rhizobacteria (*Pseudomonas* sp. SB) assist phytoremediation of oily-sludge-contaminated soil by tall fescue (Testuca arundinacea L.). Plant Soil 371:533–542

Louvado A, Gomes NCM, Simões MM, Almeida A, Cleary DF, Cunha A (2015) Polycyclic aromatic hydrocarbons in deep sea sediments: microbe–pollutant interactions in a remote environment. Sci Total Environ 526:312–328

Lu YF, Lu M (2015) Remediation of PAH-contaminated soil by the combination of tall fescue, arbuscular mycorrhizal fungus and epigenic earthworms. J Hazard Mater 285:535–541

Lu Y, Huang Q, Meng G, Wu L, Jingjing Z (2014) Label-free selective SERS detection of PCB-77 based on DNA aptamer modified SiO 2@ Au core/shell nanoparticles. Analyst 139(12):3083–3087

Lv M, Luan X, Liao C, Wang D, Liu D, Zhang G et al (2020) Human impacts on polycyclic aromatic hydrocarbon distribution in Chinese intertidal zones. Nature Sustainability 3(10):878–884

Macaulay BM (2015) Understanding the behaviour of oil-degrading micro-organisms to enhance the microbial remediation of spilled petroleum. Appl Ecol Environ Res 13(1):247–262

Männistö MK, Tiirola MA, Puhakka JA (2001) Degradation of 2, 3, 4, 6-tetrachlorophenol at low temperature and low dioxygen concentrations by phylogenetically different groundwater and bioreactor bacteria. Biodegradation 12:291 301

Margesin R, Schinner F (2001) Biodegradation and bioremediation of hydrocarbons in extreme environments. Appl Microbiol Biotechnol 56(5):650–663

Meagher RB (2000) Phytoremediation of toxic elemental and organic pollutants. Curr Opin Plant Biol 3(2):153–162

Megharaj M, Ramakrishnan B, Venkateswarlu K, Sethunathan N, Naidu R (2011) Bioremediation approaches for organic pollutants: a critical perspective. Environ Int 37(8):1362–1375

Mishra M, Singh SK, Kumar A (2021) Environmental factors affecting the bioremediation potential of microbes. In: Microbe mediated remediation of environmental contaminants. Woodhead Publishing, pp 47–58

Ojuederie OB, Babalola OO (2017) Microbial and plant-assisted bioremediation of heavy metal polluted environments: a review. Int J Environ Res Public Health 14(12):1504

Pai SG, Riley MB, Camper ND (2001) Microbial degradation of mefenoxam in rhizosphere of *Zinnia angustifolia*. Chemosphere 44(4):577–582

Pathak B, Gupta S, Fulekar MH (2017) Biodegradation of benzene under anaerobic condition. In: Biodegradation and bioconversion of hydrocarbons. Springer, pp 163–204

Pimmata P, Reungsang A, Plangklang P (2013) Comparative bioremediation of carbofuran contaminated soil by natural attenuation, bioaugmentation and biostimulation. Int Biodeterior Biodegradation 85:196–204

Raimondo EE, Aparicio JD, Bigliardo AL, Fuentes MS, Benimeli CS (2020) Enhanced bioremediation of lindane-contaminated soils through microbial bioaugmentation assisted by biostimulation with sugarcane filter cake. Ecotoxicol Environ Saf 190:110143

Roy A, Vajpayee P, Srivastava S, Srivastava PK (2023) Revelation of bioremediation approaches for hexachlorocyclohexane degradation in soil. World J Microbiol Biotechnol 39(9):243

Sakshi, Singh SK, Haritash AK (2019) Polycyclic aromatic hydrocarbons: soil pollution and remediation. Int J Environ Sci Technol 16:6489–6512

Sayara T, Borràs E, Caminal G, Sarrà M, Sánchez A (2011) Bioremediation of PAHs-contaminated soil through composting: influence of bioaugmentation and biostimulation on contaminant biodegradation. Int Biodeterior Biodegradation 65(6):859–865

Secher C, Lollier M, Jézéquel K, Cornu JY, Amalric L, Lebeau T (2013) Decontamination of a polychlorinated biphenyls-contaminated soil by phytoremediation-assisted bioaugmentation. Biodegradation 24:549–562

Shahid M, Khan MS (2022) Ecotoxicological implications of residual pesticides to beneficial soil bacteria: a review. Pestic Biochem Physiol 105272

Sharma I (2020) Bioremediation techniques for polluted environment: concept, advantages, limitations, and prospects. In: Trace metals in the environment-new approaches and recent advances. IntechOpen

Sheng X, Chen X, He L (2008) Characteristics of an endophytic pyrene-degrading bacterium of *Enterobacter* sp. 12J1 from *Allium macrostemon* Bunge. Int Biodeterior Biodegradation 62(2):88–95

Smith E, Thavamani P, Ramadass K, Naidu R, Srivastava P, Megharaj M (2015) Remediation trials for hydrocarbon-contaminated soils in arid environments: evaluation of bioslurry and biopiling techniques. Int Biodeterior Biodegradation 101:56–65

Su Z, Wang S, Yang S, Yin Y, Cao Y, Li G, Ma T (2021) Genetic and comparative genome analysis of *Exiguobacterium aurantiacum* SW-20, a petroleum-degrading bacteria with salt tolerance and heavy metal-tolerance isolated from produced water of changing oilfield, China. Microorganisms 10(1):66

Taghavi S, Weyens N, Vangronsveld J, van der Lelie D (2011) Improved phytoremediation of organic contaminants through engineering of bacterial endophytes of trees. In: Endophytes of forest trees: biology and applications. Springer, pp 205–216

Teng Y, Shen Y, Luo Y, Sun X, Sun M, Fu D et al (2011) Influence of *Rhizobium meliloti* on phytoremediation of polycyclic aromatic hydrocarbons by alfalfa in an aged contaminated soil. J Hazard Mater 186(2–3):1271–1276

Teng Y, Li X, Chen T, Zhang M, Wang X, Li Z, Luo Y (2016) Isolation of the PCB-degrading bacteria *Mesorhizobium* sp. ZY1 and its combined remediation with *Astragalus sinicus* L. for contaminated soil. Int J Phytoremed 18(2):141–149

Tomei MC, Daugulis AJ (2013) Ex situ bioremediation of contaminated soils: an overview of conventional and innovative technologies. Crit Rev Environ Sci Technol 43(20):2107–2139

Wang Y, Li H, Zhao W, He X, Chen J, Geng X, Xiao M (2010) Induction of toluene degradation and growth promotion in corn and wheat by horizontal gene transfer within endophytic bacteria. Soil Biol Biochem 42(7):1051–1057

Weyens N, Croes S, Dupae J, Newman L, van der Lelie D, Carleer R, Vangronsveld J (2010) Endophytic bacteria improve phytoremediation of Ni and TCE co-contamination. Environ Pollut 158(7):2422–2427

Wołejko E, Wydro U, Łoboda T (2016) The ways to increase efficiency of soil bioremediation. Ecol Chem Eng S 23(1):155–174

Wu FY, Yu XZ, Wu SC, Lin XG, Wong MH (2011) Phenanthrene and pyrene uptake by arbuscular mycorrhizal maize and their dissipation in soil. J Hazard Mater 187(1–3):341–347

Yousaf S, Afzal M, Reichenauer TG, Brady CL, Sessitsch A (2011) Hydrocarbon degradation, plant colonization and gene expression of alkane degradation genes by endophytic *Enterobacter ludwigii* strains. Environ Pollut 159(10):2675–2683

Zhang H, Yuan X, Xiong T, Wang H, Jiang L (2020) Bioremediation of co-contaminated soil with heavy metals and pesticides: influence factors, mechanisms and evaluation methods. Chem Eng J 398:125657

Chapter 9
Microbial Degradation of Polyaromatic Hydrocarbons in Marine and Coastal Ecosystems

Dhruba Bhattacharya and Abhrajyoti Ghosh

Abstract The potential detrimental effects on human health and pervasive distribution of polycyclic aromatic hydrocarbons (PAHs) have drawn increased attention to their biodegradation mechanisms. A variety of sources, including atmospheric fallout, combustion of fossil fuels, industrial runoff, petroleum spills, automobile exhaust, and urban discharge, can release PAHs into aquatic environments. Due to the poor water solubility of PAHs, sediment acts as the ultimate sink for PAH deposition which makes the coastal and marine ecosystems most vulnerable to PAH accumulation. Microbial communities majorly participate in controlling the fate of PAHs, thereby influencing the biogeochemical cycling of carbon in water and sediments. Microbes of coastal and marine ecosystems are deemed to have intrinsic property of PAH degradation owing to their rich diversity and unique adaptations. This chapter summarizes the recent developments in microbial degradation of PAHs in both marine and coastal ecosystems.

Keywords Polycyclic aromatic hydrocarbons (PAHs) · PAH contamination · Marine pollution · Microbial degradation · Bioremediation

9.1 Introduction

PAHs, which are prevalent environmental contaminants, exhibit moderate to poor solubility in water, leading to their absorption onto particles and subsequent accumulation in marine as well as coastal sediments (Ghandourah 2022). The conventional expansion of coastal industries and urban areas has led to the establishment of extensive metropolises, ports, and industrial and fishing facilities, contributing to the degradation of adjacent marine ecosystems. PAHs pose a significant

D. Bhattacharya · A. Ghosh (✉)
Department of Biological Sciences, Bose Institute, Kolkata, WB, India
e-mail: abhrajyoti.ghosh@jcbose.ac.in

S. Dey, S. Bhattacharya (eds.), *Biotechnological Interventions in the Removal of Emerging Pollutants*, Interdisciplinary Biotechnological Advances,
https://doi.org/10.1007/978-981-97-9922-0_9

environmental threat to marine life due to their adverse health impacts on organisms, including their ability to disrupt endocrine systems (Gearing et al. 1980). PAHs infiltrate marine ecosystems via various pathways, including combustion of organic matter (resulting in pyrolytic origin), gradual geothermal alteration of organic matter (such as petroleum hydrocarbons), and the breakdown of biogenic materials through diagenesis. Human-induced actions contribute to the presence of numerous PAHs in aquatic environments, particularly with elevated concentrations detected in estuaries, coastal regions, and zones characterized by intense vessel traffic and oil processing. Nevertheless, these hydrocarbons can be made accessible through processes like resuspension, bioturbation, erosion, and dredging (McGenity 2014). This accessibility presents a dual impact—it boosts the potential for biodegradation while also heightening the risk to the biota. Presently, over 100 distinct PAHs, either in singular or combined forms, have been documented. High-molecular-weight (HMW) PAHs consist of four or more aromatic rings, while low-molecular-weight (LMW) PAHs contain three or fewer aromatic rings. Anthracene, benzo[a]pyrene, benz[a]anthracene, dibenz[a,h]anthracene, fluoranthene, benzo[k]fluoranthene, benzo[b]fluoranthene, naphthalene, indeno[1,2,3-cd]pyrene, fluorene, phenanthrene, and pyrene are among the 16 significant PAHs identified by the United States Environmental Protection Agency (USEPA) considering their potential toxicity upon exposure to living organisms, as detailed in Table 9.1. PAHs, such as chrysene and benzo[a]pyrene, consist of four- and five-ring structures and are recognized for their insolubility in water. These compounds have the ability to adhere to particulate surfaces, facilitating their transfer from terrestrial and atmospheric sources to freshwater environments, as well as from the surface of oceans to deeper layers (Vagge et al. 2018). Marine and coastal ecosystems are among the most endangered and extensively utilized natural systems worldwide (Barbier et al. 2011). Over recent decades, there has been rapid growth and heightened utilization of coastal regions, leading to significant socio-economic and ecological changes. This trend is anticipated to persist into the future. In contrast to inland regions, coastal areas experience more frequent exchange between land and water, enhancing the migration capability of PAHs in the environment. The intrusion of seawater into coastal areas reduces the microbial content in soil and sediment, leading to a decreased natural degradation rate of PAHs compared to inland areas. Consequently, depending solely on natural degradation or traditional methods is ineffective. The majority of PAHs released into coastal and marine environments undergo various processes, including chemical oxidation, volatilization, adsorption, and bioaccumulation onto sediment particles. Microbial (algae, fungi, and bacteria) facilitated degradation of PAHs involves the biotransformation of compounds into simpler metabolites such as H_2O, CO_2 (under aerobic conditions), or CH_4 (under anaerobic conditions). The study of PAH degradation using algae, fungi, and bacteria is extensively pursued due to their rapid adaptation to environmental conditions. The degradation rate of PAHs varies depending on numerous factors including microbial population, chemical structure, cellular transport properties, availability of nutrients, chemical partitioning in growth the medium, scale of acclimation, pH, oxygen levels, and temperature (Haritash and Kaushik 2009). In the past, PAHs were

Table 9.1 Physical properties of the 16 PAHs listed as priority pollutants by USEPA

Polycyclic aromatic hydrocarbons	Molecular formula	Molecular weight	Boiling point (°C)	Melting point (°C)	Appearance	Solubility in water at 25 °C (mg L^{-1})	IARC group
Naphthalene	$C_{10}H_8$	128.18	217.9	80.2	White crystalline flakes	31	2B
Acenaphthylene	$C_{12}H_8$	152.19	280	89.4	Colorless crystalline solid	3.93	Not evaluated
Acenaphthene	$C_{12}H_{10}$	154.21	277.5	93	White needles	3.9	3
Fluorene	$C_{13}H_{10}$	166.22	294	114.76	White leaflets	1.69	3
Phenanthrene	$C_{14}H_{10}$	178.23	338.4	99	Colorless monoclinic crystals	1.15	3
Anthracene	$C_{14}H_{10}$	178.23	341.3	216	White to yellow solid	1.29	3
Fluoranthene	$C_{16}H_{10}$	202.25	384	110.2	Light yellow fine crystals	0.20–0.26	3
Pyrene	$C_{16}H_{10}$	202.25	394	150.62	Colorless solid	0.135	3
Benz(a)anthracene	$C_{18}H_{12}$	228.3	437.6	155–157	Colorless leaflets	9.4×10^{-3}	2B
Chrysene	$C_{18}H_{12}$	228.3	448	255	Crystalline solid	2.0×10^{-3}	2B
Benzo(b)fluoranthene	$C_{20}H_{12}$	252.3	481	168.4	Needles or yellow fluffy powder	Insoluble	2B
Benzo(k)fluoranthene	$C_{20}H_{12}$	252.3	480	217	Pale-yellow needles	8.0×10^{-4}	2B
Benzo(a)pyrene	$C_{20}H_{12}$	252.3	496	179	Pale-yellow crystals	1.62×10^{-3}	1
Dibenz(a,h)anthracene	$C_{22}H_{14}$	278.3	524	269	White crystals or pale-yellow solid	2.49×10^{-3}	2B
Benzo(g,h,i)perylene	$C_{22}H_{12}$	276.3	550	278.3	White crystalline solid	2.6×10^{-4}	2A
Indeno(1,2,3-cd)pyrene	$C_{22}H_{12}$	276.3	536	164	Yellow crystals	1.9×10^{-4}	3

thought to resist biodegradation without oxygen as an electron acceptor. However, recent findings have shown that they can indeed biodegrade under different anaerobic conditions. While there is extensive research on the anaerobic microbial degradation of aliphatic and monoaromatic hydrocarbons, studies on the anaerobic biodegradation of polyaromatic hydrocarbons are still limited (Fuchedzhieva et al. 2008). The enduring presence of PAHs in the environment has been linked to their limited susceptibility to nucleophilic attack, attributed to the dense clouds of pi-electrons surrounding their ring structures (Covino et al. 2010). Additionally, physical characteristics such as low aqueous solubility and high solid/water distribution ratios hinder microbial degradation of these compounds, contributing to their accumulation in soils and sediments.

9.2 Factors Influencing Source, Distribution, and Fate of PAHs

Environmental pollution resulting from anthropogenic activities has infiltrated various ecosystems, including marine and coastal environments. Hypersaline ecosystems are frequently affected by organic pollutants including PAHs which often originating from municipal and industrial effluents. In particular, the petroleum industry serves as a significant source of contaminants in wastewater, with aromatic hydrocarbons such as PAHs being prominent pollutants. The contamination and subsequent biodegradation of these pollutants in extreme environments, such as those with high salinity, have garnered increased attention. Anthropogenic PAHs enter the ocean through various pathways, including river runoff, emissions from industries and households, air deposition, and oil spills. While natural sources of PAHs are typically less prominent than anthropogenic ones (Chen et al. 2012), it is crucial to examine PAH concentrations and distribution within a country's marine and coastal environments. This assessment is necessary to understand the potential consequences on related ecosystems, given the environmental persistence of PAHs and their potential ecotoxicological effects (Soclo et al. 2000). The presence of PAH chemicals detected in marine sediments determines their source and bioavailability (Wang et al. 2014). PAHs can be categorized as either petrogenic or pyrogenic. In coastal marine habitats, oil spills, urban runoff, and atmospheric deposition are the primary sources of PAHs (Fathallah et al. 2012). The levels of organic matter content and particle size play a significant role in determining PAH levels in sediments, influencing the sorption of these pollutants onto the solid phase (Wang et al. 2001; Xia and Ball 1999). El-Naggar et al. (2021) suggested that a sediment organic carbon concentration exceeding $1/10^{th}\%$ is sufficient to significantly enhance PAH sorption onto marine sediments. Additionally, the type of organic materials present in the sediment may also influence PAH concentrations.

PAHs possess physicochemical properties that render them highly mobile in the environment, allowing for long-range transport and contributing to global pollution. These compounds are widely distributed in both abiotic and biotic components

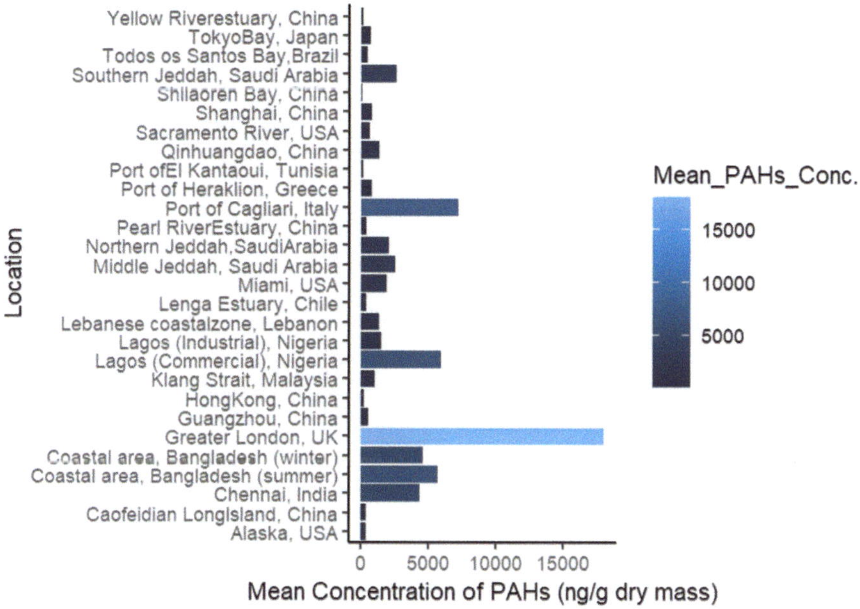

Fig. 9.1 Global distribution of PAHs in recent years. (Adapted from Dai et al. 2022)

worldwide. The distribution of PAHs in the environment is predominantly governed by their solubility and hydrophobicity, which designate sediments as their primary reservoir. The global distribution of PAHs in marine and coastal ecosystems in recent years is illustrated in Fig. 9.1.

The fate of PAHs in marine sediments is determined by two primary physicochemical properties: vapor pressure and solubility (Landrum et al. 1994). UV radiation, temperature, sediment dynamics, and organic carbon content are among the environmental and geochemical factors along with the processes such as resuspension, grain size, and bioturbation which dictate the fate of PAHs in near-bottom water and sediments (Johnson et al. 2001). Among these factors, salinity and temperature are particularly crucial as they influence the adsorption, dissolution, and transport of PAHs in coastal and marine ecosystems. Moreover, the seasonal fluctuations of climatic conditions for instance wind direction along with anthropogenic factors causing seasonal variabilities such as shipping activity, hot weather and intensity of sea water also can influence the composition and content of PAHs in coastal sediments (Meador 2003). Polycyclic aromatic hydrocarbons, are known for their mutagenic and carcinogenic properties, prompting concerns about their impact on both animals and humans (Samanta et al. 2002).

Over recent years, there has been a notable increase in the detection of PAHs in various environmental mediums, including the atmosphere, soil, seawater, and organisms inhabiting polar regions. Remarkably, the concentrations of PAHs found in these remote polar areas are comparable to those observed in urbanized regions, suggesting widespread distribution (Moran et al. 2023; Ren et al. 2021). This dispersion raises significant environmental and health concerns, as PAHs can accumulate in organisms and potentially harm ecosystems, including humans who consume contaminated seafood. To address this issue, regulatory measures, monitoring programs, and remediation efforts are underway, yet continued research and proactive strategies are essential to mitigate the persistent threat of PAH pollution.

9.3 Microbial Biodegradation of PAHs

Conventional methods for addressing the hazards posed by PAHs typically entail the alteration, removal, or isolation of the pollutants. These techniques often involve costly processes such as excavating contaminated soil for incineration or containment. However, such methods may simply transfer the pollutant from one phase to another without fully resolving the issue. In contrast, biodegradation offers a more sustainable approach by utilizing natural biological processes to transform PAH compounds into nonhazardous or less hazardous forms (Ward et al. 2003). This method requires less input of energy, time, and chemicals compared to conventional techniques, making it an attractive option for remediation.

While various processes like adsorption, volatilization, photolysis, and chemical degradation contribute to the degradation of PAHs, microbial degradation stands out as the primary mechanism (Yuan et al. 2001). Microorganisms play a crucial role in bioremediation due to their catabolic activity, although the changes in microbial communities remain unpredictable, leading to the microbial community being often referred to as a "black box" (Dua et al. 2002).

9.3.1 Microorganisms Capable of Degrading PAHs

PAH-degrading microorganisms encompass micro-algae, bacteria, and fungi. They facilitate the breakdown of organic compounds through biotransformation into less complex metabolites and mineralization into inorganic minerals, H_2O, CO_2 under aerobic conditions, or CH_4 under anaerobic conditions (Haritash and Kaushik 2009). The effectiveness and rate of bioremediation depend on several factors including the types and quantity of microorganisms present, environmental conditions, and the chemical structure of the PAH compound being degraded. In aerobic pathways, microorganisms primarily utilize a variety of peripheral enzymes to degrade aromatic contaminants, converting them into major intermediates. On the other hand, anaerobic degradation of aromatic compounds is conducted by various types of bacteria; including manganese-reducing, nitrate-reducing, sulfate-reducing, and iron-reducing bacteria as well as phototrophic bacteria along with methanogenic groups (Bosma et al. 2001). It's worth noting that the metabolism of organic compounds through respiration in aerobic conditions leads to a more efficient utilization of potential chemical energy compared to anaerobic conversion (Reineke

2001). While both aerobic and anaerobic biodegradation play significant roles in removing aromatic pollutants from the environment, aerobic mechanisms are generally preferred due to their faster process.

9.3.2 Bacterial Degradation of PAHs

Bacteria possess the ability to utilize PAHs as their primary energy and carbon source (Reineke 2001). Within this group, both gram-positive and gram-negative bacteria demonstrate the capacity to metabolize specific PAHs. Johnsen et al. (2005) observed that bacteria cultivated in suspended, shake-flask cultures containing PAHs in crystalline form at concentrations surpassing their aqueous solubility as the sole source of carbon and energy exhibit distinctive growth patterns. These patterns are delineated into three phases: the exponential phase, the pseudo-linear phase where the growth rate is mainly constrained by the concentration of PAHs, and the pseudo-stationary phase. However, it's important to note that real heterogeneous media deviate from these idealized conditions and may not exhibit the same characteristics.

The majority of PAH-degrading bacteria utilize dioxygenases enzyme to oxidize PAHs. Dioxygenase is typically composed of multiple components, including ferredoxin, reductase, and terminal oxygenase subunits (Mallick et al. 2011). However, few bacteria, such as *Mycobacterium* sp., possess the capability to oxidize the aromatic rings of PAHs via the cytochrome P450 monooxygenase enzyme, resulting in the formation of trans-dihydrodiols instead of cis-dihydrodiols (Pagnout et al. 2007; Cao et al. 2009). In aerobic aromatic catabolism, oxygen serves as both the final electron acceptor and a cosubstrate for hydroxylation and oxygenolytic ring cleavage of the aromatic ring. Conversely, anaerobic aromatic catabolism employs a distinct strategy, primarily relying on reductive reactions to target the aromatic ring. Primarily, bacteria prefer aerobic conditions for the degradation of PAHs through oxygenase-mediated metabolism, facilitated by either monooxygenase or dioxygenase enzymes. Typically, the initial step in aerobic bacterial PAH degradation involves the hydroxylation of an aromatic ring by a dioxygenase, resulting in the formation of a cis-dihydrodiol. This dihydrodiol is subsequently rearomatized to a diol intermediate through the action of a dehydrogenase. These diol intermediates can then undergo cleavage by intradiol or extradiol ring-cleaving dioxygenases, following either an ortho-cleavage or meta-cleavage pathway. This process leads to the formation of intermediates such as catechols, which are ultimately converted to tricarboxylic acid (TCA) cycle intermediates (Eaton and Chapman 1992; Ghosal et al. 2016). Multiple studies have demonstrated the effectiveness of various bacterial species in utilizing PAHs as sole sources of carbon. These include but are not limited to genus *Aeromonas, Acinetobacter, Arthrobacter, Alcanivorax, Burkholderia, Bacillus, Corynebacterium, Cellulomonas, Gordonia, Enterobacter, Dietzia, Haemophilus, Microbulbifer, Marinobacter, Mycobacterium, Micrococcus, Pseudomonas, Paenibacillus, Sphingomonas, Xanthomonas,* and *Rhodococcus*. Recent studies have provided in-depth insights into the regional distribution of hydrocarbon-degrading bacteria. Coastal anoxic sediments have exhibited microbial signatures associated with the Desulfococcus group representing members of delta-proteobacteria belonging to family Desulfobacteraceae as significant contributors to the initial stages of anaerobic short-chain and long-chain alkane degradation in marine seeps (Kleindienst et al. 2014; Acosta-González and Marques 2016). In coastal seawater and sediments near refineries or harbors of the temperate regions, gamma-Proteobacteria, particularly from genera like *Thalassolituus*, *Alcanivorax*, *Cycloclasticus*, *Oleibacter*, *Pseudomonas*, *Colwellia*, *Marinobacter*, *Oleiphilus*, and *Shewanella,* were commonly found (Cravo-Laureau and Duran 2014; Yakimov et al. 2005). In deep-sea sediments, research has revealed the presence of bacteria from genera like *Acinetobacter*, *Pseudomonas*, *Sphingobium*, *Achromobacter* (Tomasino et al. 2021), as well as *Erythrobacter*, *Parvibaculum*, *Marinobacter*, *Alcanivorax*, *Colwellia*, and *Halomonas* (Perez Calderon et al. 2019). Neethu et al. (2019) discovered two previously unrecognized taxa of hydrocarbon-degrading bacteria: Methylococcales in seawater and *Acinetobacter* in sediments. A subset of 100 bacterial genera known as hydrocarbon degraders have been identified by Wanapaisan et al. (2018) across various environments, particularly in marine settings. These genera are predominantly from the γ and α-Proteobacteria classes and demonstrated a specialized capability for degrading various hydrocarbons. Research on the degradation of polycyclic aromatic hydrocarbons (PAHs) by pure and indigenous mixed cultures isolated from coastal sediments was conducted by Tadros and Hughes (1997). They identified two bacterial species *Pseudomonas* sp. and *Ochrobactrum* sp. which are capable of degrading PAHs in these sediments. Furthermore, they observed a correlation between the concentration of bacteria and the degradation of the PAHs. For instance, the indigenous mixed microorganisms achieved degradation rates of 98.7% for naphthalene and 90.4% for fluorene when used as the sole growth substrate after 8 days of incubation. Hinga (2003) investigated subtidal marine sediments following a fuel oil spill. The study revealed a correlation between the sediment's total organic carbon (TOC) and degradation rate of low-molecular-weight (LMW) polycyclic aromatic hydrocarbons (PAHs). In their research, Shao et al. (2010) investigated the deep-sea subsurface environment, particularly the subsurface sediments along the Mid-Atlantic Ridge. They found that sediment located 2.7 meters beneath the seabed surface at a water depth of 3962 meters, contained a total concentration of 445 ng/g dry mass of PAHs. An uncultured *Gordonia* sp. was identified as a significant high-molecular-weight (HMW) polycyclic aromatic hydrocarbon (PAH) degrader through the detection of a *nidA* dioxygenase gene in subcultures containing pyrene (Gallego et al. 2014). The connection between Actinobacteria and pyrene degradation in coastal areas suggests a contrast with open sea or sediment settings, where hydrocarbon degrading Gamma-proteobacteria, notably *Cycloclasticus* (Gallego et al. 2014; Vila and Grifoll 2009), are primarily responsible for degrading high-molecular-weight (HMW) polycyclic aromatic hydrocarbons (PAHs) (Head et al. 2006; Rodríguez-Blanco et al. 2010). *Planococcus*, belonging to the phylum Firmicutes, exhibit the capability to degrade monoaromatics, alkanes of varying lengths, and polycyclic aromatic hydrocarbons (PAHs) (Li et al. 2006; Al-Awadhi et al. 2012).

9.3.2.1 Bacterial Mechanism for PAH Biodegradation

Naphthalene, anthracene, and phenanthrene are among the most frequently encountered low-molecular-weight (LMW) polycyclic aromatic hydrocarbons (PAHs). Among these, naphthalene is commonly utilized as the sole carbon and energy source in numerous studies. While *Pseudomonas* sp. is a prominent naphthalene degrader among bacteria, other species such as *Mycobacterium*, *Rhodococcus*, and *Bacillus* have also been documented for this capability (Mohapatra and Phale 2021; Huang et al. 2016; Imam et al. 2022). The degradation of naphthalene, including its biochemical pathway and genetic regulatory mechanisms, has been thoroughly explored in various studies. Huang et al. (2016) found that a *Pseudomonas* strain achieved a maximum naphthalene degradation of 77% from an initial concentration of 800 mg L^{-1} within 48 hours. In contrast, the *Bordetella avium* strain exhibited a maximum biodegradation efficiency of 95% from an initial concentration of 7 mM naphthalene ne (Abo-state et al. 2018). Typically, the genetic makeup of naphthalene-degrading bacterial genes consists of three components. One component governs the upper pathway, facilitating the conversion of naphthalene to salicylate and the subsequent opening of the two aromatic rings. The second component includes the regulatory gene responsible for synthesizing salicylate 1-hydroxylase. The third component regulates the lower pathway, which is involved in transforming salicylate into pyruvate, a process that eliminates aromatic rings altogether (Imam et al. 2022). Similar to naphthalene, phenanthrene degradation involves both an upper and a lower pathway. Initially, a multicomponent dioxygenase catalyzes the ring hydroxylation, producing cis-3,4-dihydroxy-3,4-dihydrophenanthrene. Subsequently, this compound undergoes dihydrodiol dehydrogenation, resulting in the formation of 3,4-dihydroxy-phenanthrene (Phale et al. 2019). After this stage, the meta-cleavage of the obtained 3,4-dihydroxy-phenanthrene forms 1-hydroxy-2-naphthoic acid (Xu et al. 2019) which represents the most common upper pathway for phenanthrene degradation. High-molecular-weight (HMW) polycyclic aromatic hydrocarbons (PAHs) pose greater challenges for degradation due to their complex fused aromatic structure, which renders them thermodynamically stable, hydrophobic, and less bioavailable (Subashchandrabose et al. 2019). Unlike low-molecular-weight (LMW) PAHs, only a limited number of microbes, primarily gram-positive species, have been documented for their ability to degrade HMW PAHs. Pyrene, among HMW PAHs, is widely studied for microbial degradation. Pyrene mineralization involves dioxygenase activity, which occurs at either the C-1 and C-2 positions or the C-4 and C-5 positions of the aromatic ring, producing pyrene-1,2 diol and pyrene cis-4,5-dihydrodiol, respectively (Wu et al. 2019). Following the ring aromatization of dihydrodiols and cleavage by dioxygenase, phenanthrene-4,5-dicarboxylate is produced, which eventually undergoes decarboxylation to form phenanthrene-4-carboxylate. An intermediate, cis-3,4-dihydroxyphenanthrene-4-carboxylate, is generated via dioxygenation and subsequently rearomatizes to yield 3,4-dihydroxyphenanthrene. This compound is further metabolized to generate 1-hydroxy-2-naphthoate (Singh et al. 2013). Additionally, intradiol ring cleavage via dioxygenation results in the formation of o-phthalate, which enters the Krebs cycle for further metabolic breakdown. Addressing several aspects of high-molecular-weight (HMW) PAH degradation requires further investigation. This includes delving into the underlying genetic mechanisms, assessing the impact of low-molecular-weight (LMW) PAHs or other hydrocarbons, and conducting microbial community analyses during HMW PAH degradation. Bridging these gaps in understanding is crucial for the development of consortia that are more efficient and effective in PAH degradation.

9.3.3 Fungal Degradation of PAHs

A wide range of fungi, including both ligninolytic and nonligninolytic types, possess the ability to oxidize PAHs. However, fungi do not utilize PAHs as their sole source of carbon and energy. Instead, they transform PAHs cometabolically, as described by Cerniglia (1993). PAH-degrading fungi belong to various classes including Ascomycetes, Zygomycetes, Blastomycetes, Coelomycetes, and Hyphomycetes. In recent years, there has been significant research focus on microbial degradation by ligninolytic fungi. This attention is due to their remarkable capability to degrade lignin, which is an amorphous and complex biopolymer possessing an aromatic structure similar to that of certain environmental pollutants. Ligninolytic fungi achieve this through the production of extracellular enzymes such as peroxidase and laccase with very low substrate specificity (Haritash and Kaushik 2009). Consequently, they are well-suited for the degradation of aromatic compounds like PAHs. Márquez-Rocha (2000) observed that the white rot fungus *Pleurotus ostreatus* metabolized four soil adsorbed polycyclic aromatic hydrocarbons (PAHs) viz. anthracene, pyrene, and phenanthrene, which were mineralized after 21 days. However, benz[a]pyrene was oxidized but not mineralized. They also noted that biodegradation was enhanced with the addition of the surfactant Tween 40. In their study, Chen et al. (2010) utilized fungal biomass from white-rot fungi to assess the efficacy of biosorption in removing five PAHs: naphthalene, fluorene, acenaphthene, phenanthrene, and pyrene. Their findings underscored the importance of biosorption as a significant process by which white-rot fungi influence the fate of PAHs in the environment. As per Reineke (2001), various nonligninolytic fungi have been identified to convert polycyclic aromatic hydrocarbons (PAHs) into metabolites resembling those produced by mammalian enzymes. Nonligninolytic fungi, including *Penicillium janthinellum* and *Cunninghamella elegans* possess the ability to metabolize various PAHs into polar metabolites (Potin et al. 2004). According to Reineke (2001), *Cunningamella elegans* successfully degraded naphthalene, acenaphthene, anthracene, phenanthrene, benzo[a]pyrene, benzo[a]anthracene, fluoranthene, and pyrene. Only a limited number of fungi demonstrate the capability to fully catabolize PAHs into CO_2. Nonligninolytic fungi metabolize PAHs through reactions catalyzed by epoxide hydrolase and cytochrome P450 monooxygenase such as nonheme iron oxygenases and methane monooxygenases (Daccò et al. 2020) leading to the formation of trans-dihydrodiols. Additionally, other metabolites such as quinones, phenols, and their conjugates are also generated (Cerniglia and Sutherland 2001). The marine-derived fungi *Aspergillus sclerotiorum* CBMAI 849 exhibited the most effective performance in depleting pyrene (99.7%) and benzo[a]pyrene (76.6%) after 8 and 16 days, respectively. Furthermore, *Mucor racemosus* CBMAI 847 also achieved significant depletion of benzo[a]pyrene (>50.0%). (Passarini et al. 2011). In another research, the fungus *Aspergillus* sp. BAP14, isolated from marine sediments in China, exhibited degradation capabilities. It degraded 30% and 60% of benzo[a]pyrene (initial concentration: 0.010 mg/mL) after 3 and 10 days of incubation, respectively (Wu et al. 2009). Previous studies have similarly evaluated the tolerance of estuarine sediment-isolated filamentous fungi against phenanthrene and pyrene, showed comparable results (Souza et al. 2017). It is essential to consider salinity as a factor in the application of ligninolytic fungi for detoxifying marine sites contaminated with PAHs, as studied by Valentín et al. (2006, 2007). They evaluated the impact of high salinity typical of coastal areas on fungal growth, enzyme production, and ligninolytic activity. The study focused on four PAHs: phenanthrene, fluoranthene, pyrene, and chrysene. Results indicated a significant inhibitory effect on PAH degradation, varying from strong inhibition for certain fungi to negligible inhibition for others. Fungal remediation of PAH contaminants holds greater promise compared to bacteria. Additionally, the mycelial habitat provides advantages such as rapid substrate colonization and hyphal extensions, which facilitate better access to pollutants.

9.3.4 Biodegradation of PAHs by Microalgae and Cyanobacteria

It is well-established that Cyanobacteria and eukaryotic algae possess the ability to metabolize PAHs. Cyanobacteria, being the largest and most extensively distributed group of photosynthetic prokaryotes on Earth (Reuter and Müller 1993), specifically in marine and coastal ecosystems makes them important biocomponent to consider. Cellular biodegradation and/or bioaccumulation are the primary effective mechanisms involved in eliminating PAHs by algae and cyanobacterial species (Ke et al. 2010). Depending on the algal species, PAH bioremediation involves both dioxygenase and monooxygenase enzymatic pathways, leading to the production of hydroxylated and dihydroxylated metabolites. Several studies have explored the differences among cyanobacteria and microalgae species regarding their ability to degrade PAHs. Cerniglia et al. (1980) investigated the naphthalene metabolism in nine cyanobacteria, four green microalgae, one red microalgae, and two diatoms. They observed that all these organisms could oxidize naphthalene under photoautotrophic conditions. Moreover, Hong et al. (2008) found that *Skeletonema costatum* exhibited a higher tolerance to phenanthrene and fluoranthene compared to *Nitzschia* sp. Additionally, they noted that the toxic effect of fluoranthene on both *S. costatum* and *Nitzschia* sp. was greater than that of phenanthrene. Narro et al. (1992) reported that the marine cyanobacterium *Oscillatoria* sp. strain JCM primarily oxidized naphthalene to 1-naphthol. Other studies have examined the interaction of PAHs with metals. Ke et al. (2010) explored the impact of certain metals on the biosorption and biodegradation of a mixture of five polycyclic aromatic hydrocarbons by *Selenastrum capricornutum*. They observed that for low-molecular-weight PAHs, both the dosage of metals and the duration of exposure had a notable positive effect on their removal. However, for high-molecular-weight PAHs, the presence of metals did not influence the efficiency of removal. Natural algal communities, which are diverse and interact with each other, appear to be more sensitive to PAH contamination or sporadic oil discharges compared to algal populations grown in culture. PAHs typically occur as mixtures of compounds in natural water, and they can induce various effects (additive, synergistic, or antagonistic) on natural phytoplankton. Changes in algal composition have been observed in both marine (Kottuparambil and Agustí 2019) and freshwater ecosystems (Rimet et al. 2004).

9.4 Concluding Remarks and Research Needs

In recent years, research focusing on microbial degradation of PAHs has made significant advancements. The majority of studies in this field are motivated by concerns regarding the teratogenicity, carcinogenicity, immunotoxicity, and ecotoxicity of PAHs. Biological methods for transforming PAHs are highly regarded for being eco-friendly, demanding minimal energy input, and capable of achieving complete degradation in a shorter duration. Overall, PAHs with lower molecular weights are typically degraded more readily than those with higher molecular weights. Among different environmental compartments, soil has been the most extensively studied, followed by sediment. Water, especially freshwater, has received less attention compared to soil and sediment, while seawater has been studied less than freshwater. Modern molecular techniques such as genomics, proteomics, and metabolomics are utilized to unravel the biochemistry of PAH-degrading organisms. Understanding the catabolic potential of microbes, as well as the genes and structure–function correlations within microbial communities involved in PAH degradation, holds promise for developing microbial consortia or genetically engineered microorganisms in an ecologically sustainable and biologically safe manner. Future research aimed at utilizing techniques like in silico studies, data analytics-driven experimental design, and genome editing has the potential to expedite the development of consortia comprising both bacterial and fungal species with diverse biocatalytic functions for PAH removal. The increasing exploitation of aging oil wells through enhanced oil recovery, the prevalence of heavier crudes in the global refining portfolio, and the adoption of higher-severity processes in the petrochemicals sector are expected to contribute to a rise in PAH volumes. Consequently, there will likely be a growing demand for field-proven commercial bioremediation products to address these environmental challenges. Moreover, generating additional sequence information is crucial to enhance our understanding of the overall distribution and PAH degradation capacities of indigenous microbes in marine environments, particularly concerning PAH contamination. This comprehensive understanding will enable the development of bioremediation strategies by promoting the activity of degrading microorganisms that are present in low abundance at sites affected by PAH contamination. Ultimately, this approach will support the complete degradation of all chemical components of PAHs and their derivatives.

References

Abo-State MAM, Riad BY, Bakr AA, Aziz MA (2018) Biodegradation of naphthalene by *Bordetella avium* isolated from petroleum refinery wastewater in Egypt and its pathway. J Radiat Res Appl Sci 11:1–9

Acosta-González A, Marqués S (2016) Bacterial diversity in oil-polluted marine coastal sediments. Curr Opin Biotechnol 38:24–32

Al-Awadhi H, Dashti N, Kansour M, Sorkhoh N, Radwan S (2012) Hydrocarbon-utilizing bacteria associated with biofouling materials from offshore waters of the Arabian Gulf. Int Biodeterior Biodegrad 69:10–16

Barbier EB, Hacker SD, Kennedy C, Koch EW, Stier AC, Silliman BR (2011) The value of estuarine and coastal ecosystem services. Ecol Monogr 81(2):169–193

Bosma T, Harms H, Zehnder A (2001) Biodegradation of xenobiotics in environment and technosphere. In: Beek B (ed) Biodegradation and persistence. Springer, Berlin

Cao B, Nagarajan K, Loh KC (2009) Biodegradation of aromatic compounds: current status and opportunities for biomolecular approaches. Appl Microbiol Biotechnol 85:207–460

Cerniglia CE (1993) Biodegradation of polycyclic aromatic hydrocarbons. Curr Opin Biotechnol 4:331–338

Cerniglia CE, Sutherland JB (2001) Bioremediation of polycyclic aromatic hydrocarbons by ligninolytic and non-ligninolytic fungi. In: de Gadd GM (ed) Fungi in bioremediation. Cambridge University Press, Cambridge

Cerniglia CE, Gibson DT, Van Baalen C (1980) Oxidation of naphthalene by cyanobacteria and microalgae. J Gen Microbiol 116:495–500

Chen B, Wang Y, Hu D (2010) Biosorption and biodegradation of polycyclic aromatic hydrocarbons in aqueous solutions by a consortium of white-rot fungi. J Hazard Mater 179:845–851

Chen HY, Teng YG, Wang JS (2012) Source apportionment of polycyclic aromatic hydrocarbons (PAHs) in surface sediments of the Rizhao coastal area (China) using diagnostic ratios and factor analysis with nonnegative constraints. Sci Total Environ 414:293–300

Covino S, Čvančarová M, Muzikář M, Svobodová K, D'annibale A, Petruccioli M, Federici F, Křesinová Z, Cajthaml T (2010) An efficient PAH-degrading *Lentinus (Panus) tigrinus* strain: effect of inoculum formulation and pollutant bioavailability in solid matrices. J Hazard Mater 183(1–3):669–676

Cravo-Laureau C, Duran R (2014) Marine coastal sediments microbial hydrocarbon degradation processes: contribution of experimental ecology in the omics' era. Front Microbiol 5:39

Daccò C, Girometta C, Asemoloye MD, Carpani G, Picco AM, Tosi S (2020) Key fungal degradation patterns, enzymes and their applications for the removal of aliphatic hydrocarbons in polluted soils: a review. Int Biodeterior Biodegrad 147:104866

Dai C, Han Y, Duan Y, Lai X, Fu R, Liu S, Leong KH, Tu Y, Zhou L (2022) Review on the contamination and remediation of polycyclic aromatic hydrocarbons (PAHs) in coastal soil and sediments. Environ Res J 205:112423

Dua M, Singh A, Sethunathan N, Johri A (2002) Biotechnology and bioremediation: successes and limitations. Appl Microbiol Biotechnol 59:143–152

Eaton RW, Chapman P (1992) Bacterial metabolism of naphthalene: construction and use of recombinant bacteria to study ring cleavage of 1, 2-dihydroxynaphthalene and subsequent reactions. J Bacteriol 174(23):7542–7554

El-Naggar M, Hanafy S, Younis AM, Ghandour MA, El-Sayed AAY (2021) Seasonal and temporal influence on polycyclic aromatic hydrocarbons in the Red Sea coastal water. Egypt Sustain 13(21):11906

Fathallah S, Medhioub MN, Kraiem MM (2012) Photo-induced toxicity of four polycyclic aromatic hydrocarbons (PAHs) to embryos and larvae of the carpet shell clam *Ruditapes decussatus*. Bull Environ Contam Toxicol 88:1001–1008

Fuchedzhieva N, Karakashev D, Angelidaki I (2008) Anaerobic biodegradation of fluoranthene under methanogenic conditions in presence of surface-active compounds. J Hazard Mater 153(1–2):123–127

Gallego S, Vila J, Tauler M, Nieto JM, Breugelmans P, Springael D, Grifoll M (2014) Community structure and PAH ring-hydroxylating dioxygenase genes of a marine pyrene-degrading microbial consortium. Biodegradation 25:543–556

Gearing PJ, Gearing JN, Pruell RJ, Wade TL, Quinn JG (1980) Partitioning of No. 2 fuel oil in controlled estuarine ecosystems. Sediments and suspended particulate matter. Environ Sci Tech 14(9):1129–1136

Ghandourah M (2022) An insightful overview of the distribution pattern of polycyclic aromatic hydrocarbon in the marine sediments of the Red Sea. Open Chem 20:777–784

Ghosal D, Ghosh S, Dutta TK, Ahn Y (2016) Current state of knowledge in microbial degradation of polycyclic aromatic hydrocarbons (PAHs): a review. Front Microbiol 7:1369

Haritash AK, Kaushik CP (2009) Biodegradation aspects of polycyclic aromatic hydrocarbons (PAHs): a review. J Hazard Mater 169(1–3):1–15

Head IM, Jones DM, Röling WF (2006) Marine microorganisms make a meal of oil. Nat Rev Microbiol 4(3):173–182

Hinga KR (2003) Degradation rates of low molecular weight PAH correlate with sediment TOC in marine subtidal sediments. Marine Pollut Bull 46:466–474

Hong YW, Yuan D-X, Lin QM, Yang TL (2008) Accumulation and biodegradation of phenanthrene and fluoranthene by the algae enriched from a mangrove aquatic ecosystem. Marine Pollut Bull 56:1400–1405

Huang H, Wu K, Khan A, Jiang Y, Ling Z, Liu P, Chen Y, Tao X, Li X (2016) A novel *Pseudomonas gessardii* strain LZ-E simultaneously degrades naphthalene and reduces hexavalent chromium. Bioresour Technol 207:370–378

Imam A, Suman SK, Kanaujia PK, Ray A (2022) Biological machinery for polycyclic aromatic hydrocarbons degradation: a review. Bioresour Technol 343:126121

Johnsen AR, Wick LY, Harms H (2005) Principles of microbial PAH-degradation in soil. Environ Pollut 133(1):71–84

Johnson MD, Keinath TM, Weber WJ (2001) A distributed reactivity model for sorption by soils and sediments. 14. Characterization and modelling of phenanthrene desorption rates. Environ Sci Technol 35(8):1688–1695

Ke L, Luo L, Wang P, Luan T, Tam NFY (2010) Effects of metals on biosorption and biodegradation of mixed polycyclic aromatic hydrocarbons by a freshwater green alga *Selenastrum capricornutum*. Bioresour Technol 101:6961–6972

Kleindienst S, Herbst FA, Stagars M, Von Netzer F, Von Bergen M, Seifert J, Peplies J, Amann R, Musat F, Lueders T, Knittel K (2014) Diverse sulfate-reducing bacteria of the *Desulfosarcina/Desulfococcus* clade are the key alkane degraders at marine seeps. ISME J 8(10):2029–2044

Kottuparambil S, Park J (2019) Anthracene phytotoxicity in the freshwater flagellate alga *Euglena agilis* Carter. Sci Rep 9:15323

Landrum PF, Dupuis WS, Kukkonen J (1994) Toxicokinetics and toxicity of sediment-associated pyrene and phenanthrene in *Diporeia spp.*: examination of equilibrium-partitioning theory and residue-based effects for assessing hazard. Environ Toxicol Chem 13(11):1769–1780

Li H, Liu YH, Luo N, Zhang XY, Luan TG, Ji MH, Wang ZY, Pei CW, Chen MJ, Jia QL (2006) Biodegradation of benzene and its derivatives by a psychrotolerant and moderately haloalkaliphilic *Planococcus sp.* strain ZD22. Res Microbiol 157(7):629–663

Mallick S, Chakraborty J, Dutta TK (2011) Role of oxygenases in guiding diverse metabolic path ways in the bacterial degradation of low-molecular-weight polycyclic aromatic hydrocarbons: a review. Crit Rev Microbiol 37(1):64–90

Márquez-Rocha FJ, Hernández-Rodríguez VZ, Vázquez-Duhalt R (2000) Biodegradation of soil-adsorbed polycyclic aromatic hydrocarbons by the white rot fungus *Pleurotus ostreatus*. Biotechnol Lett 22:469–472

McGenity TJ (2014) Hydrocarbon biodegradation in intertidal wetland sediments. Curr Opin Biotechnol 27:46–54

Meador JP (2003) Bioaccumulation of PAHs in marine invertebrates. In: PAHs: an ecotoxicological perspective. Wiley, pp 147–171

Mohapatra B, Phale PS (2021) Microbial degradation of naphthalene and substituted naphthalenes: metabolic diversity and genomic insight for bioremediation. Front Bioeng Biotechnol 9:144

Moran IL, Tidwell L, Barton M, Kile M, Miller P, Rohlman D, Seguinot-Medina S, Ungwiluk B, Waghiyi V, Anderson K (2023) Diffusive fluxes of persistent organic pollutants between Arctic atmosphere, surface waters and sediments. Sci Total Environ 892:164566

Narro ML, Cerniglia CE, Van Baalen C, Gibson DT (1992) Evidence for an NIH shift in oxidation of naphthalene by the marine cyanobacterium *Oscillatoria* sp. strain JCM. Appl Environ Microbiol 58:1360–1363

Neethu CS, Saravanakumar C, Purvaja R, Robin RS, Ramesh R (2019) Oil-spill triggered shift in indigenous microbial structure and functional dynamics in different marine environmental matrices. Sci Rep 9(1):1354

Pagnout C, Frache G, Poupin P, Maunit B, Muller JF, Férard JF (2007) Isolation and characterization of a gene cluster involved in PAH degradation in *Mycobacterium sp.* strain SNP11: expression in *Mycobacterium smegmatis* mc2155. Res J Microbiol 158(2):175–186

Passarini MRZ, Rodrigues MVN, da Silva M, Sette LD (2011) Marine derived filamentous fungi and their potential application for polycyclic aromatic hydrocarbon bioremediation. Mar Pollut Bull 62(2):364–370

Perez Calderon LJ, Gontikaki E, Potts LD et al (2019) Pressure and temperature effects on deep-sea hydrocarbon-degrading microbial communities in subarctic sediments. Microbiol Open 8:e768

Phale PS, Sharma A, Gautam K (2019) Microbial degradation of xenobiotics like aromatic pollutants from the terrestrial environments. In: Pharmaceuticals and personal care products: waste management and treatment technology. Elsevier

Potin O, Veignie E, Rafin C (2004) Biodegradation of polycyclic aromatic hydrocarbons (PAHs) by *Cladosporium sphaerospermum* isolated from an aged PAH contaminated soil. FEMS Microbiol Ecol 51:71–78

Reineke W (2001) Aerobic and anaerobic biodegradation potentials of microorganisms. In: Beek B (ed) Biodegradation and persistance. Springer, Berlin

Ren K, Wei Y, Li J, Han C, Deng Y, Su G (2021) Polycyclic aromatic hydrocarbons (PAHs) and their derivatives (oxygenated PAHs, azaarenes, and sulfur/oxygen-containing heterocyclic PAHs) in surface soils from a typical city, south China. Chemosphere 283:131190

Reuter W, Müller C (1993) New trends in photobiology: adaptation of the photosynthetic apparatus of cyanobacteria to light and CO_2. J Photochem Photobiol B Biol 21:3–27

Rimet F, Ector L, Dohet A, Cauchie HM (2004) Impacts of fluoranthene on diatom assemblages and frustule morphology in indoor microcosms. Vie Milieu 54:145–156

Rodríguez-Blanco A, Vetion G, Escande ML, Delille D, Ghiglione JF (2010) *Gallaecimonas pentaromativorans gen. nov., sp. nov.,* a bacterium carrying 16S rRNA gene heterogeneity and able to degrade high-molecular-mass polycyclic aromatic hydrocarbons. Int J Syst Evol Microbiol 60(3):504–509

Samanta SK, Singh OV, Jain RK (2002) Polycyclic aromatic hydrocarbons: environmental pollution and bioremediation. Trends Biotechnol 20(6):243–248

Shao Z, Cui Z, Dong C, Lai Q, Chen L (2010) Analysis of a PAH-degrading bacterial population in subsurface sediments on the Mid-Atlantic Ridge. Deep-Sea Res I Oceanogr Res Pap 57:724–730

Singh SN, Kumari B, Kumar S, Mishra S, Kumar D (2013) Bacterial degradation of pyrene in minimal salt medium mediated by catechol dioxygenases: enzyme purification and molecular size determination. Bioresour Technol 133:293–300

Soclo HH, Garrigues PH, Ewald M (2000) Origin of polycyclic aromatic hydrocarbons (PAHs) in coastal marine sediments: case studies in Cotonou (Benin) and Aquitaine (France) areas. Mar Pollut Bull 40(5):387–396

Souza HMDL, Barreto LR, Mota AJD, Oliveira LAD, Barroso HDS, Zanotto SP (2017) Tolerance to polycyclic aromatic hydrocarbons (PAHs) by filamentous fungi isolated from contaminated sediment in the Amazon region. Acta Sci Biol Sci 39(4):481–488

Subashchandrabose SR, Venkateswarlu K, Naidu R, Megharaj M (2019) Biodegradation of high-molecular weight PAHs by *Rhodococcus wratislaviensis* strain 9: overexpression of amidohydrolase induced by pyrene and BaP. Sci Total Environ 651:813–821

Tadros M, Hughes J (1997) Degradation of polycyclic aromatic hydrocarbons (PAHs) by indigenous mixed and pure cultures isolated from coastal sediments. Appl Biochem Biotechnol 63–65:865–870

Tomasino MP, Aparício M, Ribeiro I, Santos F, Caetano M, Almeida CMR, de Fátima CM, Mucha AP (2021) Diversity and hydrocarbon-degrading potential of deep-sea microbial community from the Mid-Atlantic Ridge, south of the Azores (North Atlantic Ocean). Microorganisms 9(11):238

Vagge G, Cutroneo L, Castellano M, Canepa G, Bertolotto RM, Capello M (2018) The effects of dredging and environmental conditions on concentrations of polycyclic aromatic hydrocarbons in the water column. Mar Pollut Bull 135:704–713

Valentín L, Feijoo G, Moreira MT, Lema JM (2006) Biodegradation of polycyclic aromatic hydrocarbons in forest and salt marsh soils by white-rot fungi. Int Biodeterior Biodegrad 58:15–21

Valentín L, Lu-Chau TA, López C, Feijoo G, Moreira MT, Lema JM (2007) Biodegradation of dibenzothiophene, fluoranthene, pyrene and chrysene in a soil slurry reactor by the white-rot fungus *Bjerkandera sp.* BOS55. Process Biochem 42:641–648

Vila J, Grifoll M (2009) Actions of *Mycobacterium sp.* strain AP1 on the saturated-and aromatic-hydrocarbon fractions of fuel oil in a marine medium. Appl Environ Microbiol 75(19):6232–6239

Wanapaisan P, Laothamteep N, Vejarano F, Chakraborty J, Shintani M, Muangchinda C, Morita T, Suzuki-Minakuchi C, Inoue K, Nojiri H, Pinyakong O (2018) Synergistic degradation of pyrene by five culturable bacteria in a mangrove sediment-derived bacterial consortium. J Hazard Mater 342:561–570

Wang XC, Zhang YX, Chen RF (2001) Distribution and partitioning of polycyclic aromatic hydrocarbons (PAHs) in different size fractions in sediments from Boston Harbor, United States. Mar Pollut Bull 42(11):1139–1149

Wang Z, Liu Z, Xu K, Mayer LM, Zhang Z, Kolker AS, Wu W (2014) Concentrations and sources of polycyclic aromatic hydrocarbons in surface coastal sediments of the northern Gulf of Mexico. Geochem Trans 15(1):1–12

Ward O, Singh A, Van Hamme J (2003) Accelerated biodegradation of petroleum hydrocarbon waste. J Ind Microbiol Biotechnol 30(5):260–270

Wu YR, He TT, Lun JS, Maskaoui K, Huang TW, Hu Z (2009) Removal of benzo [a] pyrene by a fungus *Aspergillus sp.* BAP14. World J Microbiol Biotechnol 25:1395–1401

Wu F, Guo C, Liu S, Liang X, Lu G, Dang Z (2019) Pyrene degradation by *Mycobacterium gilvum*: metabolites and proteins involved. Water Air Soil Pollut 230:67

Xia G, Ball WP (1999) Adsorption-partitioning uptake of nine low-polarity organic chemicals on a natural sorbent. Environ Sci Technol 33(2):262–269

Xu X, Liu W, Wang W, Tian S, Jiang P, Qi Q, Li F, Li H, Wang Q, Li H (2019) Potential biodegradation of phenanthrene by isolated halotolerant bacterial strains from petroleum oil polluted soil in Yellow River Delta. Sci Total Environ 664:1030–1038

Yakimov MM, Denaro R, Genovese M, Cappello S, D'Auria G, Chernikova TN, Timmis KN, Golyshin PN, Giluliano L (2005) Natural microbial diversity in superficial sediments of Milazzo Harbor (Sicily) and community successions during microcosm enrichment with various hydrocarbons. Environ Microbiol 7(9):1426–1441

Yuan SY, Chang JS, Yen JH, Chang BV (2001) Biodegradation of phenanthrene in river sediment. Chemosphere 43(3):273–278

Chapter 10
Contribution of Cyanobacteria and Microalgae for the Removal of Emergent Pollutants from Wastewater: A Review

Madhulika Gupta

Abstract The advent of emergent pollutants such as pesticides, pharmaceuticals, personal care and industrial chemical products has recently drawn attention of many due to their omni-present nature, stability and resistance to degradation and creates a health hazard for humans and ecosystem. Cyanobacteria and microalgae, due to their ubiquitous nature, being eco-friendly and equipped with bioremediation mechanisms, can be employed in a cost-effective manner in the removal of emergent pollutants from the various sources of wastewater apart from the conventional techniques.

Keywords Emergent pollutants · Bioremediation · Microalgae · Wastewater

10.1 Definition, Characteristics and Categories of Emergent Pollutants

Occurrence of novel pollutants reported to exist without much supervision in ng/l to μg/l level in various ecosystems gained attention of many in recent years as they pose health hazards to human and environment. Such pollutants are termed as emergent pollutants, also known as micropollutants. These unusual chemical compounds with their ubiquitous nature have been detected from various sources such as soil, air, aquatic bodies like groundwater, surface water, wastewater and drinking water. Unwarranted and disproportionate use of pharmaceuticals, personal care products, pesticides, industrial chemicals, hormones, etc. in various fields of daily life has slowly increased the concentration of micropollutants which with long exposure and time resist degradation and became stable and toxic to man and environment.

M. Gupta (✉)
Department of Microbiology, Lady Brabourne College, Kolkata, West Bengal, India
e-mail: madhulika.gupta@ladybrabourne.com

S. Dey, S. Bhattacharya (eds.), *Biotechnological Interventions in the Removal of Emerging Pollutants*, Interdisciplinary Biotechnological Advances,
https://doi.org/10.1007/978-981-97-9922-0_10

10.1.1 Categories of Emergent Pollutants Classification of Emerging Pollutants

Emerging pollutants are classified and categorized as follows (Abdulrazaq et al. 2020; Dey et al. 2019; Yahaya et al. 2020; Samal et al. 2022):

10.1.1.1 Pharmaceuticals

In order to prevent humans and animals from various infections and to alleviate the symptoms of diseases, different bioactive compounds are manufactured in pharmaceutical industries. Pharmaceuticals can be classified into antibiotics analgesics and anti-inflammatories, cardiovascular drugs, antidepressants, antivirals, anticoagulants, sedatives, etc. based on their chemical structure, mode of action and the disease for which it is prescribed. The major sponsor of emerging pollutants being the pharmaceuticals comprising the analgesics and antiinflammatory drugs such as paracetamol, ibuprofen, diclofenac, NSAIDs (Non-steroidal anti-inflammatory drugs). The commonly used antibiotics are penicillins, tetracyclines lincomycins cephalosporins, macrolides, sulfonamides and quinolones. Anti-depressant like SSRIs (Selective serotonin reuptake inhibitors), SNRIs (Serotonin-norepinephrine reuptake inhibitors), SMSs (Serotonin modulator and stimulators), SARIs (Serotonin antagonist and reuptake inhibitors), NRIs (Norepinephrine reuptake inhibitors), NDRIs (Norepinephrine-dopamine reuptake inhibitors), TCAs (Tricyclic antidepressants), TeCAs (Tetracyclic antidepressants) etc. are used for the treatment of depression, persistent pain and obsession (Samal et al. 2022)

10.1.1.2 Personal Care Products

In order to clean, beautify and maintain personal hygiene, the self-care products, such as creams, lotions, soaps, perfumes, shampoos, moisturizers, hair colours, diaper creams, toothpastes, deodorants, sunscreens, mouthwashes, antiperspirants, antiseptics, fragrances, cosmetics, etc., used by humans can be categorised under the term personal care products. The emerging contaminants of personal care products are as follows: alkylphenol polyethoxylates (APEOs), antimicrobials, bisphenols, cyclosiloxanes, ethanolamines, fragrances, glycol ethers, insect repellents, parabens, phthalates, triclosan, plastic microbeads and UV filters (Khalid and Abdollahi 2021; Juliano and Magrini 2017)

10.1.1.3 Disinfection By-products

In order to safeguard public health, treatment of surface and ground water which is subjected to contamination due to various urban and industrial practices is an absolute necessity to be used as potable drinking water. This is achieved through addition of coagulants and flocculants, sedimentation, floatation, filtration and finally disinfection before storage and distribution. The process of disinfection involves application of chemicals such as chlorine, ozone, UV radiation, chlorine dioxide, chloramines, hydrogen peroxide, etc. to kill water-borne pathogens. Amongst these chemicals, the commonly used disinfectant for drinking water is chlorine as it is widely available, economical and has an oxidizing property and a long-lasting effect to further reduce and inhibit the growth of pathogens in the drinking water supply. This remnant chorine when reacts with natural organic matter and inorganic ions leads to the development of disinfection by-products such as trihalomethanes (includes chloroform, bromoform, bromodichloromethane, etc.), haloacetic acids (includes mono, di, trichloroacetic acid, etc.) and haloacetonitriles (includes dibromoacetonitrile, dichloroacetonitrile) (Roque et al. 2023; Kozlowska 2020)

10.1.1.4 Endocrine-Disrupting Compounds

Pituitary, hypothalamus, thyroid, parathyroid, pancreas, adrenal, thymus, ovary and testis constitute the endocrine system that secretes hormones which plays an important role in monitoring the equilibrium and functioning of the organ system in a body. Some synthetic and natural chemicals acting as emergent pollutants which upset the proper functioning of the hormones by imitating the hormones, blocking the receptors, obstruct the pathways for their synthesis are enlisted as endocrine-disrupting compounds (EDCs). The following chemicals can be categorised under EDCs: zearalenone mycoestrogens, polyaromatic hydrocarbons, bisphenol, polyfluoroalkylated substances triclosan, plasticizers, flame retardants, pesticides, Diclofenac, Diethyl phthalate, polychlorinated biphenyls, alkylphenols, caffeine, etc. (Lazofsky and Buckley 2022; Tijani et al. 2013)

10.1.1.5 Steroids and Hormones

Extensive use of steroids and hormones in pharmaceuticals, hospitals, industries, personal care products, livestock and husbandry has led to the contamination of water bodies significantly. The commonly found steroids and hormones in wastewater which act as EDCs are estrogens [Estrone(E1), Estradiol(E2), Estriol (E3), 17-α-ethinylestradiol (EE2)], Glucocorticoids (Prednisone, Prednisolone, Cortisone, Cortisol, Dexamethasone, 6α-methylprednisolone), Androgens (Androsterone, Epiandrosterone, Androstenedione), Progestogens (Norethindrone, Progesterone) and Growth hormones [Zeranol, Trenbolone Acetate (TBA), Melengestrol Acetate (MGA)] (Yazdan et al. 2022)

10.1.1.6 Perfluorinated Compounds (PFCs)

Man-made synthetic class of chemical compounds with fluorine attached carbon chains, with thermal stability, insistent, toxic, shows high rates of accumulation in biota and a property to repel water and oil are termed as perfluorinated compounds. These compounds have diverse utilities in paints, food packaging, non-stick cookware, waterproof clothing, lubricants and fabric stain protector. Perfluorooctanoic acid (PFOA) and perfluorooctane sulfonate (PFOS) are two of the best known PFCs. PFCs have been detected in surface as well as ground waters apart from wastewaters (Mishra et al. 2023)

10.1.1.7 Surfactants

Compounds which have hydrophilic and hydrophobic groups regarded as amphipathic compounds that effectively reduces surface tension, stable and has the capacity to form micelles are termed as surfactants. They are extensively used in households and different industries and are an important part of daily life. They can be classified as cationic (Quaternary ammonium compound-QAC, Cationic Benzalkonium chloride-BAC, Cetylpyridinium bromide-CPB, Cetylpyridinium chloride-CPC Hexadecyltrimethylammonium bromide-HDTMA), anionic (Linear alkylbenzene sulphonic acid-LAS, Anionic Sodium dodecyl sulphate-SDS, Alkyl sulphate-AS, Sodium lauryl sulphate-SLS, Alkylethoxysulphate-AES), non-ionic(Alkylphenol ethoxylate-APE, Alcohol ethoxylate-AE Fatty acid ethoxylate-FAE) and amphoteric types (Amine oxide) based on their chemical nature. They are found in wastewater treatment plants, and pollute surface and groundwater (Arora et al. 2022; Ivanković and Hrenović 2010)

10.1.1.8 Flame Retardants

In order to diminish the peril of fire, chemicals are used in inflammable and ignitable substances to keep life safe and prevent destruction and loss. Such chemical compounds are called flame retardants. These can be classified into four types: phosphorus, nitrogen, halogenated organic containing and inorganic compounds. Amongst these, the most frequently used flame retardants are organophosphate flame retardants (used in furniture, textiles, building materials, electronics, processing chemicals, plasticizers in floor polishes, coatings, engineering thermoplastics and epoxy resins and brominated flame retardants (polybrominated diphenyl ethers (PBDEs) are now removed from the market due to its high toxicity, stability and build-up in biota (Yadav et al. 2023; Yogui and Sericano 2009; Yang et al. 2019; Greaves and Letcher 2014).

10.1.1.9 Plasticizers and Alternative Plasticizers

To make plastics flexible and durable by increasing space between chains of polymer, a number of additives are used, the most common being plasticizers, Plasticizers are synthetic chemicals of which phthalates are the most common liquid plasticizers with diverse applications in homes and consumer merchandises such as PVC (polyvinyl chloride) products, food wraps, kitchen plasticware, nail polish etc. The commonly used phthalates are di(e-ethylhexyl) phthalate (DEHP) and Butyl benzyl phthalate (BBzP). As phthalates are not chemically linked, they easily leak and pollute the environment. Due to being extremely hazardous to humans and ecosystems, they are under strict regulatory controls and their place has been taken by alternative plasticizers such as adipates, benzoates, phosphate esters, citrates, sebacates, terephthalates, trimellitates and many more. Most widely used alternative plasticizers DEHT (Bis(2- ethylhexylterephthalate), DPHP (Di(2-propylheptyl) phthalate), DINCH (1,2- cyclohexanedicarboxylic acid diisonyl ester), Bisphenol-S, Bisphenol-Z (alternative of Bisphenol -A), DDT (dichlorodiphenyl trichloroethane) replaced by organophosphates and cholorpyrifos, etc. were found to be hazardous as per DHEP (Heinlaan et al. 2023; Kimber and Dearman 2010; Qadeer et al. 2022)

10.1.1.10 Pesticides

In order to augment the quality and yield of food products from plants as well as to ward off diseases, weeds, pests, fungi, insects and vectors chemical compounds called pesticides are used. Pesticides are classified as organochlorines (Lindane, Endosulfan, Methoxychlor, DDT, Aldrin, etc.), Organophosphorus (Parathion, Malathion, Phorate, Dimethoate, etc.) carbamates (Carbofuran, Carbendazim, Carbaryl) pyrethrin, and pyrethroids (Cyhalothrin). Large-scale use of pesticides has affected the health of man and ecosystem effectively (Rajmohan et al. 2020)

10.1.1.11 Gasoline Additives

Gasoline additives such as benzene, 1,3-butadiene, and methyl tert-butyl ether (MTBE) are exercised in petrol to increase the burning effectiveness of fuel and reduce carbon monoxide emissions. Due to its stable structure and resists degradation, they are easily found in surface and groundwaters (Lei et al. 2015)

10.1.1.12 Nanomaterials

Nanomaterials are defined according to the International Organization for Standardization (ISO) and the European Union (EU) as a 'material with any external dimension in the nanoscale or having an internal structure or surface structure in the nanoscale'. Nanomaterials can be classified on the basis of materials it is formed of: carbon-based Nanomaterials [carbon nanotubes (CNTs), carbon nanofbers (CNFs), carbon black, fullerenes (C60), graphene (Gr), and nanodiamonds], Inorganic-based Nanomaterials[metal Nanoparticles (e.g. Au and Ag), metal oxide Nanoparticles (e.g. TiO_2 and ZnO), and semiconductors (e.g. Si and ceramics)], Organic-based Nanomaterials (e.g., liposomes, polymeric nanoparticles, micelles, and dendrimers), Composite-based Nanomaterials (multiphase Nanoparticles and nanostructured materials). The surface of nanoparticles with its size, shape, solubility and endurance has made its availability and interaction with humans and ecosystems achievable, which is the cause for its detrimental effects (El-Kalliny et al. 2023; Rauscher et al. 2019; Jeevanandam et al. 2018; Kokkinos et al. 2020)

10.1.2 Effect of Emergent Pollutants on Humans and the Environment

10.1.2.1 Pharmaceuticals

Entry of pharmaceutical contaminants in the environment depends on a number of sources such as discharges from hospitals, households, industries, wastewater treatment plants, agricultural fields, aquaculture services, excretory products of humans and animals, etc. After discharge pharmaceuticals follow different transport pathways which includes leaching, sorption, desorption, degradation, etc. The following consequences can occur once these contaminants reach the aquatic bodies, i.e., compounds can be broken down and converted into water-soluble parental form retaining its bioactivity; secondly, they can be mineralised into carbon dioxide and water; and finally majority of the being lipid soluble becomes resistant to degradation. Exposure of NSAIDs due to long-term use has caused a number of health-related issues in invertebrates and vertebrates, such as modification in antioxidative enzymes, disruption in mitochondrial functions, DNA damage, endocrine disorders. Analgesics cause hypertension, heart related problems, lethargy, feeling of dejection, etc. Extensive use of antibiotics has led to antibiotic resistance in microbial populations which further caused health hazards such as abnormal functioning of protein, enzymes, affecting liver, immune disorders and well as altering the food chain. Veterinary antibiotics used in large scale in poultry, livestock and aquatic animals to treat various diseases and increase feed productivity often are poorly absorbed by the gut and released to the environment through faecal matter in their active forms enter humans through the food chain (Obaid et al. 2023; Kayode-Afolayan et al. 2022; Samal et al. 2022; Patel et al. 2019; Lei et al. 2015)

10.1.2.2 Personal Care Products

Day-to-day activities of consumers such as cleaning, bathing, washing and showering liberate enormous quantities of personal care products into the environment without much modifications, contaminate the waterbodies and cause toxic effects on plants, animals and human beings. These compounds include such as bisphenols (causes alteration in birth weight, anti-androgenic effects, oxidation damage of DNA), triclosan (causes bacterial resistance, foetal deformities and abnormalities, behavioural problems in children, thyroid dysfunction), parabens (act as endocrine disruptor, reduces sperm count and motility, early menarche in girls and boys, breast cancer) and fragrances (cause contact allergy, facial dermatitis). Microplastics from microbeads have been reported to occur in marine organisms such as fishes, bivalves, seabirds, etc. UV filters both organic and inorganic have been extensively associated with aquatic organisms thereby causing immense imbalance in the food chain (Khalid and Abdollahi 2021; Juliano and Magrini 2017).

10.1.2.3 Disinfection By-products

Disinfection by-products were found to fluctuate in drinking waters, highest in surface water whereas lowest in groundwaters. Threat to human health is caused by disinfection by-products (MTBE and THM) through inhalation, contact and ingestion of polluted waters. These chemicals cause respiratory problems such as asthma, reproductive problems, ailments of the liver, kidney, CNS and the immune system, and carcinogenic and mutagenic effects as well (Roque et al. 2023; Arman et al. 2021; Lei et al. 2015)

10.1.2.4 Endocrine-Disrupting Compounds

Endocrine-disrupting compounds acting as emergent pollutants upset the proper functioning of the hormones by imitating the hormones, blocking the receptors and obstructing the pathways for their synthesis, and as a result, pregnancy loss, birth defects, developmental irregularities, testicular and thyroid malfunctioning, growth defects, neurodevelopmental issues have been reported in humans. Aquatic organisms like fishes suffer a shift in gender to feminisation and sexual abnormalities in alligators have been detected (Tijani et al. 2013)

10.1.2.5 Steroids and Hormones

Natural oestrogens released through animal and human waste when used in bulk for organic farming pose a serious health hazard to humans and ecosystems. The steroids and hormones act as endocrine-disrupting compounds which cause hormonal imbalance, reproductive, metabolic and psychological disorders in humans and affect sex ratios, the immune system, egg-laying capacity and growth inhibition in aquatic organisms (Arman et al. 2021; Yazdan et al. 2022; Mishra et al. 2023).

10.1.2.6 Perfluorinated Compounds

Perfluorinated compounds [the most common are PFCAs and perfluorocarbon sulfonic acids (PFSAs), perfluoro octane sulfonate (PFOS) and perfluorooctanoic acid (PFOA)] release from various industrial units without proper treatment pollutes larger distances from the area of focus. These compounds have been recovered from surface and groundwaters through soil leakage (Arman et al. 2021; Mishra et al. 2023; https://www.epa.gov/sites/default/files/2018-11/documents/cecs-pfcs-factsheet.pdf)

10.1.2.7 Surfactants

Disproportionate use of surfactants by industries and households and insufficient treatment and release into waterbodies pose a menace to the aquatic bodies. The toxic effects of various classes of surfactants (SDS, QAC, LAS, SLS, AES, etc.) are due to their binding with macromolecules and changing their biological interactions and causing leakage of compounds by increasing the permeability of the membranes, leading to the death of various algae, bacteria, invertebrates and vertebrates inhabiting the waterbodies. High levels of surfactants cause changes in behaviour, gill damage and loss of sight in fishes. These compounds break the chlorophyll-containing machinery in aquatic plants and affect their growth (Arora et al. 2022; Ivanković and Hrenović 2010)

10.1.2.8 Flame Retardants

In newborn males, presence of polybrominated diphenyl ethers (PBDEs) in mothers' milk caused low birth weight and undescended testicles were reported. Neurological disorders, hormone impairment, foetal abnormalities were observed in animals and humans with Organophosphate Flame-Retardant (OPRF) interactions (Yang et al. 2019; Arman et al. 2021)

10.1.2.9 Plasticizers and Alternative Plasticizers

As phthalate plasticizers are considered to be extremely harmful to both biotic and abiotic ecosystem they were eventually replaced by alternative plasticizers with a hope for lesser cosequences on public health and ecosystem but these substitutes too caused a number of ailments such as Bisphenol-S, F, Z acts as endocrine disruptor, organophosphates and cholorpyrifos causes neurotoxicity, methyl tert-butyl ether caused toxicity in aquatic organisms, hydrofluorocarbons served as greenhouse gas, 1-bromopropane caused Carcinogenicity, etc. (Qadeer et al. 2022)

10.1.2.10 Pesticides

Pesticides upon entry through sewage treatment plants, agricultural fields and industries into aquatic bodies cause a decrease in the dissolved oxygen level as the aquatic plants are destroyed by such pesticides and in turn affects the productivity of fishes to a large extent. Exposure to pesticides such as DDT, Pyrethroid insecticides, Atrazine, organophosphorus, organochloride pesticides and causes respiratory distress, asthma, destruction of bronchial mucosa, endocrinal disorders, DNA damage, affects kidneys, risk of bladder and colon cancer, diabetes, heart-related problems and nerve disorders have been observed (Rajmohan et al. 2020)

10.1.2.11 Gasoline Additives

Gasoline additives such as MTBE pollute waterbodies and are extremely dangerous due to their resistance to degradation and long half-life in groundwater and consequently affect human life through contaminated drinking water. It shows a number of symptoms such as upper respiratory communicable disease, dizziness, anxiety, insomnia, skin rash, palpitations, malaise, allergy, wheezing, asthma, etc. (Lei et al. 2015)

10.1.2.12 Nanomaterials

While adsorbing various inorganic and organic molecules on their surfaces manufactured nanomaterials undergo long-term relocation, alteration processes, and intricate chemical reactions in the environment leading to the formation of new compounds. Animal studies have shown that these new compounds have a carcinogenic effect, pass through blood–brain barrier and placenta thus affects reproductive development, toxicity caused by deposition in liver kidneys and spleen and onset of ailments such as cardiovascular diseases, genetic alterations, neurodegenerative diseases etc. have been reported in humans as well (Lei et al. 2015; El-Kalliny et al. 2023; Yang et al. 2020)

10.2 Role and Mechanisms of Cyanobacteria and Microalgae in the Removal of Emergent Pollutants

Today, emergent pollutants such as pharmaceuticals, personal care products, endocrine-disrupting compounds, steroids, hormones, disinfection by-products, plasticizers, flame retardants, gasoline additives, nanomaterials, etc. being dispersed in aquatic bodies without much regulatory guidelines have created a great anxiety and alarm in the health of aquatic plants, animals and among humans and ecosystem due to their toxicity, resistance to biodegradation, bioaccumulation and high polarity. A variety of conventional techniques such as activated sludge, solvent extraction, chemical precipitation, constructed wetlands, anaerobic bed reactors, etc. have been employed for the removal of emergent pollutants but were not of much success as these waterbodies contain manifold and numerous pollutants. From various literature, it has been confirmed that cyanobacteria and microalgae such as Chlorella, Chlamydomonas, Scenedesmus, Coelastrum and many more due to their fast-growing capacity, ability to withstand harsh environmental conditions, adaptability and metabolic flexibility play an important role in the remediation of heavy metals and nutrients such as carbon, nitrogen and phosphorus from industrial and domestic wastewaters (Nguyen et al. 2021).

The mechanisms by which emergent pollutants can be removed from various aquatic bodies are Bioadsorption, Biodegradation, Photodegradation and Bioaccumulation.

Bioadsorption is a physical process by which removal of materials present in solution occurs by microalgae and cyanobacteria by the use of their cell wall and EPS. The cell wall of microalgae and cyanobacteria being negatively charged containing functional groups such as carboxyl, amine and phosphoryl so it interacts with emergent pollutants carrying cationic groups resulting in biosorption by the process of electrostatic interaction. EPS is a biopolymer consisting of polysaccharides, proteins and lipids which has an adsorption capability. It has been reported that an increase in antibiotic adsorption was observed with an increased EPS concentration (Nguyen et al. 2021; Viancelli et al. 2020; Obaid et al. 2023). It has been observed that the following microalgae such as Chlorella vulgaris, Scenedesmus obliquus, Chlamydomas efficiently removed the following emergent pollutants such as Ofloxacin, Biperiden, Memantine, Levofloxacin Tramadol, Diclofenac, Salicylic acid, Ciprofloxacin, Tetracycline, Paracetamol, Estone, Ibuprofen etc (Gojkovic et al. 2019; Santos et al. 2017; Xie et al. 2020; Maryjoseph and Ketheesan 2020). Fast adsorption of 7-ACA (amino cephelosporanic acid) and its removal from wastewater by lipid accumulating microalgae was reported (Guo et al. 2016; Obaid et al. 2023). Even the dead cells of freshwater microalgae such as Scenedesmus obliquus and Chlorella pyrenoidosa by electrostatic interaction could adsorb progesterone and norgestrel (Peng et al. 2014; Nguyen et al. 2021)

10.2.1 Biodegradation

The most efficacious method by which cyanobacteria and microalgae remove emergent pollutants of aquatic bodies is biodegradation. The microorganisms could breakdown complex parental compounds into simpler ones involving a number of enzymatic interactions (such as carboxylation, hydrogenation, oxidation, hydroxylation, ring cleavage, demethylation etc. carried out by cytochrome P-450, associated with EPS and glutathione-S-transferase providing protection from oxidative stress) with emergent pollutants (Ding et al. 2017; Nguyen et al. 2021; Viancelli et al. 2020; Obaid et al. 2023). No toxic compounds were liberated from degradation of diclofenac by Chlorella sorokiniana, Chlorella vulgaris, and Scenedesmus obliquus for zebrafish embryos (Escapa et al. 2018)

Cyanobacteria such as Anabaena, Phormidium, Chlorococcus forming consortia with microalgae efficiently eliminated emergent pollutants from water bodies (Yadav et al. 2021). It was further shown by a study that biodegradation of emergent pollutants by microalgae could be affected the deficiency of carbon, nitrogen and phosphorus in wastewaters, an ideal ratio for effective biodegradation of C:N:P was 100:18:2 (Maryjoseph and Ketheesan 2020; Sakarika et al. 2020)

10.2.2 Photodegradation

Sunlight is an essential component to carry out photosynthesis by cyanobacteria and microalgae. This sunlight has been used in the degradation and elimination of emergent pollutants by ultraviolet photodegradation in a direct or indirect manner. In direct photodegradation upon absorption by emergent pollutants molecular energy of these compounds enhances that causes rupturing of bonds. In indirect photodegradation, the emergent pollutants dispersed and suspended in wastewaters upon absorption of solar energy creates reactive oxygen species which eventually degrade the target compounds. It was reported that there was an enhanced tetracycline degradation by the process of photodegradation. The method of photodegradation was dependent on pollutant structure, their chemical properties, environmental factors such as seasons, depth of water column and latitude. A number of emergent pollutants such as sulfamethoxazole, caffeine, triclosan, atrazine, diclofenac, simazine etc. are degraded by this process. (Nguyen et al. 2021; Hom-Diaz et al. 2015; Gruchlik et al. 2018; Mathon et al. 2019; Norvill et al. 2017)

10.2.3 Bioaccumulation

In the process of bioaccumulation, emergent pollutants which have low molecular weight, non- polar and lipid-soluble gain entry into the microalgal cells due to their hydrophobic nature of cell membrane by passive diffusion without expenditure of energy. It has been reported that emergent pollutants like sulfamethoxazole, carbamazepine, trimethoprim, and florfenicol entered and accumulated in such algal cells. On exposure to perfluorinated alkyl acid compounds with respect to Cyanobacterium Anabaena CPB4337, these compounds changed the permeability of cell membrane leading not only to its accumulation within the cell by passive diffusion but modified its sensitivities towards other herbicides. For polar compounds, transporter proteins aid in passive diffusion and entry and accumulation into the cells. An increase in the accumulation and subsequent biodegradation within the cell of levofloxacin was reported in case of Chlorella vulgaris (Abdelfattah et al. 2022; Song et al. 2019; Rodea-Palomares et al. 2015; Xiong et al. 2017)

Conventional treatment vs cyanobacteria and microalgal-based treatment of wastewater contaminated with emergent pollutants

Emergent pollutants being chemically stable, resists degradation, present in low concentrations and can pass in humans through food chain causing a number of health ailments and damaging effects on the environments needs to be regulated and eliminated. For this purpose, a number of conventional wastewater treatment techniques such as physical processes involving activated carbon adsorption, coagulation, micro or ultrafiltration, nanofiltration, reverse osmosis, UV treatment etc. chemical processes such as ozonation, AOP (advanced oxidation process, Fenton and Photo-Fenton, Photocatalysis (TiO_2), chemical precipitation, disinfection etc. biological methods such as activated sludge, constructed wetlands, membrane bioreactor etc. are employed. Most of these techniques have a number of drawbacks such as high cost in operation and maintenance, cannot be applied on a large scale, accumulation of large amount of sludge, low recovery rate, formation of oxidative end products, high energy expenditure, accumulation of toxic intermediates etc. (Zhou et al. 2022; Rizzo et al. 2019; Xiong et al. 2018a, b). Cyanobacteria and microalgae possess carbon concentrating mechanisms , acting as carbon sink, carbohydrates, proteins and lipids production due to photosynthesis can be effectively used as raw materials for bioenergy production, plays an important role in detoxification, eco-friendly, cost-effective and can be incorporated with sewage treatment plants and effluent treatment plants have gained attention in the removal of emergent pollutants from waterbodies (Yadav et al. 2021; Zhou et al. 2022; Liu et al. 2021). The removal of emergent pollutants by microalgae-based systems are in the form of suspended cells, immobilized cells and consortia. For suspended cells high-rate algal ponds (HRAPs) were efficient in removing antibiotics and diclofenac. Closed PBRs (Photobioreactor) efficiently removed endocrine disrupting compounds and personal care products employing a number of microalgal strains Chlorella vulgaris, Chlorella sorokiniana, Scenedesmus obliquus. Further, immobilised microalgae in alginate beads efficiently removed bisphenol AF, bisphenol F and 2, 4 dicholorophenol compared to that of suspended cells. Consortia of microalgae and bacteria were able to eliminate emergent pollutants such as 17β-estradiol, salicylic acid, anti- inflammatory drugs, etc. (Nguyen et al. 2021;Villar-Navarro et al. 2018; Escapa et al. 2016; Sole and Matamoros 2016; Hom-Diaz et al. 2017; Lopez-Serna et al. 2019)

10.3 Effect of Emergent Pollutants on Cyanobacteria and Microalgae

Presence of a variety of emergent pollutants in wastewaters have increased the toxicity of such waters to a great extent. It is known that in microalgal based treatment systems often acts as a source of production of microalgal biomass. It has been reported that low concentration of tetracycline did not affect the microalgal biomass synthesis but was inhibited at a higher concentration of tetracycline. Similarly reverse effect was seen in Synechocystis where growth was retarded at a low concentration of tetracycline compared to its higher concentration. Further presence of diclofenac showed increased microalgal biomass as it acted as a carbon source leading to their growth and hence higher concentration of biomass (Nguyen et al. 2021; Xiong et al. 2018a, b; Pomati et al. 2004).

Moreover, it has been shown that presence of a certain emergent pollutant may increase the elimination efficiency of other pollutants. For example, sulfamethazine when coupled with sulfamethoxazole increased the percentage of its removal. Presence of sodium acetate as an electron donor enhanced the removal efficiency of ciprofloxacin by Chlamydomonas mexicana (Maryjoseph and Ketheesan 2020; Xiong et al. 2019, 2017; Leng et al. 2020)

Oxidative damage to biomolecules, cell function impairment and, finally, cell death by excess production of reactive oxygen species can occur due to high concentration of emergent pollutants being accumulated in microalgal cells. Detachment of any unbound emergent pollutants can return to the waterbodies and cause detrimental effects to the environment.

Compared to bio-adsorption and bioaccumulation, biodegradation causes less toxic effect on microalgal cells. Further, microalgal cells that were previously acclimatised to an emergent pollutant before being exposed to it showed improved metabolic functions and cellular activities. For example, Chlorella pyrenoidosa, when exposed priorly to the antibiotic cefradine, showed marked removal capabilities (Maryjoseph and Ketheesan 2020; Rath 2012; Sutherland and Ralph 2019; Chen et al. 2015; Norvill et al. 2016; Tiwari et al. 2017). Thus, microalgal growth and biomass production depend on factors such as type and concentration of emergent pollutants in wastewaters, availability of nutrients that will support its growth, microorganisms present in wastewaters and culture conditions such as physical factors, such as pH, temperature and light intensity (Nguyen et al. 2021; Gatamaneni et al. 2018; Li et al. 2016; Metsoviti et al. 2019; Miazek et al. 2015)

10.4 Future Prospects

The removal of emergent pollutants by microalgae-based treatment system from wastewater bodies is quite satisfactory but accompanied with a number of deficits which needs modifications. First, since most of the studies are carried out under laboratory conditions in a controlled environment, large-scale application of microalgae and cyanobacteria in natural waterbodies is required where there is a constant change in environmental factors such as temperature, pH, oxygen concentration, light intensity and penetration, microalgal interaction with other microorganisms present in water, etc. Second, wastewater bodies contain diverse types of emergent pollutants of different degrees of degradability, concentrations and stability. As from the literature study microalgae and cyanobacteria could biodegrade, sorp, bioaccumulate and photodegrade only particular classes of emergent pollutants of limited range. So further studies are required to find more novel and genome modified microalgae and cyanobacteria with better efficiencies in the removal of emergent pollutants. Third the effect of different emergent pollutants in natural waterbodies are unstable and undergoing change. So, before treatment with microalgae and cyanobacteria these emergent pollutants in wastewaters should be subjected to preceding treatment so that overall performance of microalgae-based treatments could be enhanced. Fourth, due to bioaccumulation of emergent pollutants within the microalgal cells often causes toxicities leading to impaired metabolic activities, cellular dysfunction and death. So, this reduces the chances of using these toxic microalgal cells as animal feed, biofertilizer, in the manufacture of biofuel, and procurement of value-added products. Research is required in this aspect to find alternative methods and techniques by which the toxicity in microalgae and cyanobacteria could be reduced and the process becomes sustainable. Fifth, wastewaters contain diverse range of emergent pollutants which are hazardous to man and ecosystem and as a result the colour of such waters are often dark and murky. This type of colouration of wastewaters does not allow or inhibit the growth of microalgae as in such waters the nutrient balance is often disrupted. Therefore, prior treatment of such wastewaters is necessary along with a check in the nutrient balance that would support the growth of microalgae and cyanobacteria. Sixth, the combination of the treatment of wastewaters will differ considerably with the method of culture of different types of cyanobacteria and microalgae as their requirements will vary with geographical locations and temperature range. So appropriate methods and techniques should be applied which will be best suited for that location and its temperature range. Seventh, a major problem confronted is the effective separation of microalgae from the treated wastewater as well as the quality of such water has been reported to be inferior. Extensive study and research are required to assess the quality of such wastewaters, the by-products formed in it and also evaluation of their toxic effects and whether it will be possible for re-use. Finally, large-scale supervision and surveillance is indispensable in order to run the microalgae-based treatment systems for which new technologies and devices need to be operated.

10.5 Conclusion

Excessive use of pharmaceuticals, personal care products, pesticides, industrial chemicals, hormones, etc. in various fields of daily life has slowly increased the concentration of emergent pollutants, which with long exposure and time have become stable, resist degradation and are toxic to man and environment. In contrast to different conventional methods of wastewater treatments containing such pollutants, microalgae-based treatment emerged as a boon in the field of bioremediation due to its their fast-growing capacity, ability to withstand harsh environmental conditions, adaptability, metabolic flexibility eco-friendly, cost effective and can be incorporated with sewage treatment plants and effluent treatment plants have gained attention in the removal of emergent pollutants from waterbodies. Further, their biomass could be used in the field of biofuel and extraction of value-added products. In spite of all the above advantages microalgae-based treatment of wastewaters still faces some limitations and challenges, which in future need extensive research work to derive maximum benefits, maintain sustainability and betterment of overall performances with the introduction of novel techniques as well as modification of existing treatment processes.

References

Abdelfattah A, Ali SS, Ramadan H, El-Aswar EI, Eltawab R, Ho SH, Elsamahy T, Li S, El-Sheekh MM, Schagerl M, Kornaros M, Sun J (2022) Microalgae-based wastewater treatment: Mechanisms, challenges, recent advances, and future prospects. Environ Sci Ecotechnol 13:100205. https://doi.org/10.1016/j.ese.2022.100205

Abdulrazaq Y, Abdulsalam AA, Rotimi L, Abdulbasit AA, Clifford O, Abdulsalam A, Racheal ON, Joy AA, Victor FO, Johanne ZM, Bilal M, Umar SM (2020) Classification, potential routes and risk of emerging pollutants/contaminant. In: Emerging contaminants. IntechOpen. https://doi.org/10.5772/intechopen.94447

Arman NZ, Salmiati S, Aris A, Salim MR, Nazifa TH, Muhamad MS, Marpongahtun M (2021) A review on emerging pollutants in the water environment: existences, health effects and treatment processes. Water 13:3258

Arora J, Ranjan A, Chauhan A, Biswas R, Rajput VD, Sushkova S et al (2022) Surfactant pollution, an emerging threat to ecosystem: approaches for effective bacterial degradation. J Appl Microbiol 133:1229–1244. Available from: https://doi.org/10.1111/jam.15631

Chen J, Zheng F, Guo R (2015) Algal feedback and removal efficiency in a sequencing batch reactor algae process (SBAR) to treat the antibiotic cefradine. PLoS One 10(7):e0133273

Dey S, Bano F, Malik A (2019) Pharmaceuticals and personal care product (PPCP) contamination—a global discharge inventory. In: Pharmaceuticals and personal care products: waste management and treatment technology. Butterworth-Heinemann, pp 1–26

Ding T, Yang M, Zhang J, Yang B, Lin K, Li J, Gan J (2017) Toxicity, degradation and metabolic fate of ibuprofen on freshwater diatom Navicula sp. J Hazard Mater 330:127–134. https://doi.org/10.1016/j.jhazmat.2017.02.004

El-Kalliny AS, Mahmoud S, Wahed A, El-Zahhar AA, Hamza IA, Gad-Allah TA (2023) Nanomaterials: a review of emerging contaminants with potential health or environmental impact. Discover Nano 18:68. https://doi.org/10.1186/s11671-023-03787-8

Escapa C, Coimbra RN, Paniagua S, Garcıa AI, Otero M (2016) Comparative assessment of diclofenac removal from water by different microalgae strains. Algal Res 18:127–134. https://doi.org/10.1016/j.algal.2016.06.008

Escapa C, Torres T, Neuparth T (2018) Zebrafish embryo bioassays for a comprehensive evaluation of microalgae efficiency in the removal of diclofenac from water. Sci Total Environ 640:1024–1033. https://doi.org/10.1016/j.scitotenv.2018.05.353

Gatamaneni BL, Orsat V, Lefsrud M (2018) Factors affecting growth of various microalgal species. Environ Eng Sci 35(10):1037–1048. https://doi.org/10.1089/ees.2017.0521

Gojkovic Z, Lindberg RH, Tysklind M, Funk C (2019) Northern green algae have the capacity to remove active pharmaceutical ingredients. Ecotoxicol Environ Saf 170:644–656

Greaves AK, Letcher RJ (2014) Body compartment distribution in female Great Lakes herring gulls and in ovo transfer of sixteen bioaccumulative organophosphate flame retardants. Environ Sci Technol 48:7942–7950

Gruchlik Y, Linge K, Joll C (2018) Removal of organic micropollutants in waste stabilisation ponds: a review. J Environ Manag 206:202–214. https://doi.org/10.1016/j.jenvman.2017.10.020

Guo WQ, Zheng HS, Li S, Du JS, Feng XC, Yin RL, Chang JS (2016) Removal of cephalosporin antibiotics 7-ACA from wastewater during the cultivation of lipid- accumulating microalgae. Bioresour Technol 221:284–290. https://doi.org/10.1016/j.biortech.2016.09.036

Heinlaan M, Vija H, Blinova I (2023) Novel plasticizers are emerging contaminants. Proceedings 92:61. https://doi.org/10.3390/proceedings2023092061

Hom-Diaz A, Llorca M, Rodrıguez-Mozaz S, Vicent T, Barcelo D, Blanquez P (2015) Microalgae cultivation on wastewater digestate: b-estradiol and 17a-ethynylestradiol degradation and transformation products identification. J Environ Manag 155:106–113. https://doi.org/10.1016/j.jenvman.2015.03.003

Hom-Diaz A, Jaen-Gil A, Bello-Laserna I, Rodrıguez-Mozaz S, Vicent T, Barcelo D, Blanquez P (2017) Performance of a microalgal photobioreactor treating toilet wastewater: pharmaceutically active compound removal and biomass harvesting. Sci Total Environ 592:1–11. https://doi.org/10.1016/j.scitotenv.2017.02.224

Ivanković T, Hrenović J (2010) Surfactants in the environment. Arh Hig Rada Toksikol 61:95–110. https://doi.org/10.2478/10004-1254-61-2010-1943

Jeevanandam J et al (2018) Review on nanoparticles and nanostructured materials: history, sources, toxicity and regulations. Beilstein J Nanotechnol 9(1):1050–1074

Juliano C, Magrini GA (2017) Cosmetic ingredients as emerging pollutants of environmental and health concern. a mini-review. Cosmetics 4:11. https://doi.org/10.3390/cosmetics4020011

Kayode-Afolayan SD, Ahuekwe EF, Nwinyi OC (2022) Impacts of pharmaceutical effluents on aquatic ecosystems. Sci Afr e01288. https://doi.org/10.1016/j.sciaf.2022.e01288

Khalid M, Abdollahi M (2021) Environmental distribution of personal care products and their effects on human health. Iran J Pharm Res Winter 20(1):216–253. https://doi.org/10.22037/ijpr.2021.114891.15088

Kimber I, Dearman RJ (2010) An assessment of the ability of phthalates to influence immune and allergic responses. Toxicology 271(3):73–82

Kokkinos P, Mantzavinos D, Venieri D (2020) Current trends in the application of nanomaterials for the removal of emerging micropollutants and pathogens from water. Molecules 25(9):2016

Kozlowska KA (2020) Methods used for the removal of disinfection by-products from water. In: Prasad MNV (ed) Disinfection by-products in drinking water: detection and treatment, 1st edn. Butterworth-Heinemann, Oxford, pp 1–22

Lazofsky A, Buckley B (2022) Recent trends in multiclass analysis of emerging Endocrine Disrupting Contaminants (EDCs) in drinking water. Molecules 27:8835. https://doi.org/10.3390/molecules27248835

Lei M, Zhang L, Lei J, Zong L, Li J, Wu Z, Wang Z (2015) Overview of emerging contaminants and associated human health effects Hindawi Publishing Corporation. Biomed Res Int 2015(404796):12. https://doi.org/10.1155/2015/404796

Leng L, Wei L, Xiong Q, Xu S, Li W, Lv S, Lu Q, WanL Wen Z, Zhou W (2020) Use of microalgae based technology for the removal of antibiotics from wastewater: a review. Chemosphere 238:124680

Li J, Zheng X, Liu K, Sun S, Li X (2016) Effect of tetracycline on the growth and nutrient removal capacity of Chlamydomonas reinhardtii in simulated effluent from wastewater treatment plants. Bioresour Technol 218:1163–1169. https://doi.org/10.1016/j.biortech.2016.07.080

Liu R, Li S, Tu Y, Hao X (2021) Capabilities and mechanisms of microalgae on removing micropollutants from wastewater: a review. J Environ Manage 285:112149. ISSN 0301 4797. https://doi.org/10.1016/j.jenvman.2021.112149

Lopez-Serna R, Posadas E, Garcıa-Encina PA, Munoz R (2019) Removal of contaminants of emerging concern from urban wastewater in novel algal-bacterial photobioreactors. Sci Total Environ 662:32–40. https://doi.org/10.1016/j.scitotenv.2019.01.206

Maryjoseph S, Ketheesan B (2020) Microalgae based wastewater treatment for the removal of emerging contaminants: a review of challenges and opportunities. Case Stud Chem Environ Eng 2(2020):100046

Mathon B, Coquery M, Miege C, Vandycke A, Choubert JM (2019) Influence of water depth and season on the photodegradation of micropollutants in a free-water surface constructed wetland receiving treated wastewater. Chemosphere 235:260–270. https://doi.org/10.1016/j.chemosphere.2019.06.140

Metsoviti MN, Papapolymerou G, Karapanagiotidis IT, Katsoulas (2019) Comparison of growth rate and nutrient content of five microalgae species cultivated in greenhouses. Plan Theory 8(8):279. https://doi.org/10.3390/plants8080279

Miazek K, Iwanek W, Remacle C, Richel A, Goffin D (2015) Effect of metals, metalloids and metallic nanoparticles on microalgae growth and industrial product biosynthesis: a review. Int J Mol Sci 16(10):23929–23969. https://doi.org/10.3390/ijms161023929

Mishra RKA, Mentha SSB, MisraY B, Dwivedi N (2023) Emerging pollutants of severe environmental concern in water and wastewater: a comprehensive review on current developments and future research. Water-Energy Nexus 6(2023):74–95

Nguyen HT, Yoon Y, Ngo HH, Jang A (2021) The application of microalgae in removing organic micropollutants in wastewater. Crit Rev Environ Sci Technol 51(12):1187–1220. https://doi.org/10.1080/10643389.2020.1753633

Norvill ZN, Shilton A, Guieysse B (2016) Emerging contaminant degradation and removal in algal wastewater treatment ponds: identifying the research gaps. J Hazard Mater 313:291–309

Norvill ZN, Toledo-Cervantes A, Blanco S, Shilton A, Guieysse B, Munoz R (2017) Photodegradation and sorption govern tetracycline removal during wastewater treatment in algal ponds. Bioresour Technol 232:35–43. https://doi.org/10.1016/j.biortech.2017.02.011

Obaid ZH, Salman JM, Kadhim NF (2023) Review on toxicity and removal of pharmaceutical pollutants using immobilised microalgae. Ecol Eng Environ Technol 24(6):44–60. https://doi.org/10.12912/27197050/166013

Patel M, Kumar R, Kishor K, Todd Mlsna T, Pittman CU Jr, Mohan D (2019) Pharmaceuticals of emerging concern in aquatic systems: chemistry, occurrence, effects, and removal methods. Chem Rev 119:3510–3673. https://doi.org/10.1021/acs.chemrev.8b00299

Peng FQ, Ying GG, Yang B, Liu S, Lai HJ, Liu YS, Chen ZF, Zhou GJ (2014) Biotransformation of progesterone and norgestrel by two freshwater microalgae (Scenedesmus obliquus and Chlorella pyrenoidosa): transformation kinetics and products identification. Chemosphere 95:581–588. https://doi.org/10.1016/j.chemosphere.2013.10.013

Pomati F, Netting AG, Calamari D, Neilan BA (2004) Effects of erythromycin, tetracycline and ibuprofen on the growth of Synechocystis sp. and Lemna minor. Aquat Toxicol 67(4):387–396. https://doi.org/10.1016/j.aquatox.2004.02.001

Qadeer A, Kirsten KL, Ajmal Z, Jiang X, Zhao X (2022) Alternative plasticizers as emerging global environmental and health threat: another regrettable substitution? Environ Sci Technol 56(3):1482–1488. https://doi.org/10.1021/acs.est.1c08365

Rajmohan KS, Chandrasekaran R, Varjani S (2020) A review on occurrence of pesticides in environment and current technologies for their remediation and management. Indian J Microbiol 60(2):125–138. https://doi.org/10.1007/s12088-019-00841-x

Rath B (2012) Microalgal bioremediation: current practices and perspectives. J Biochem Technol 3(3):299–304

Rauscher H et al (2019) Identifcation of nanomaterials through measurements. Publications Office of the European Union

Rizzo L, Malato S, Antakyali D, Beretsou VG, Dolić MB, Gernjak W, Heath E, Ivancev-Tumbas I, Karaolia P, Lado Ribeiro AR, Mascolo G, McArdell CS, Schaar H, Silva AMT, Fatta-Kassinos D (2019) Consolidated vs new advanced treatment methods for the removal of contaminants of emerging concern from urban wastewater. Sci Total Environ 655:986–1008. https://doi.org/10.1016/j.scitotenv.2018.11.265

Rodea-Palomares I, Makowski M, Gonzalo S, Gonzalez-Pleiter M, Leganes F, Fernandez-Pinas F (2015) Effect of PFOA/PFOS pre-exposure on the ~ toxicity of the herbicides 2, 4-D, Atrazine, Diuron and Paraquat to a model aquatic photosynthetic microorganism. Chemosphere 139:65e72

Roque MI, Gomes J, Reva I, Valente AJM, Simões NE, Morais PV, Durães L, Martins RC (2023) An opinion on the removal of disinfection byproducts from drinking water. Water 15:1724. https://doi.org/10.3390/w15091724

Sakarika M, Koutra E, Tsafrakidou P, Terpou A, Kornaros M (2020) Microalgae-based remediation of wastewaters. In: Microalgae cultivation for biofuels production. Academic Press, pp 317–335

Samal K, Hibzur MS, Ali M (2022) Pharmaceutical wastewater as Emerging Contaminants (EC): treatment technologies, impact on environment and human health. Energy Nexus 6:100076. Elsevier Ltd

Santos CE, de Coimbra RN, Bermejo SP, Perez AIG, Cabero MO (2017) Comparative assessment of pharmaceutical removal from wastewater by the microalgae Chlorella sorokiniana, Chlorella vulgaris and Scenedesmus obliquus. In: Biological wastewater treatment and resource recovery. IntechOpen, p 99

Sole A, Matamoros V (2016) Removal of endocrine disrupting compounds from wastewater by microalgae co-immobilized in alginate beads. Chemosphere 164:516–523. https://doi.org/10.1016/j.chemosphere.2016.08.047

Song C, Wei Y, Qiu Y, Qi Y, LiY KY (2019) Biodegradability and mechanism of florfenicol via Chlorella sp. UTEX1602 and L38: experimental study. Bioresour Technol 272:529e534

Sutherland DL, Ralph PJ (2019) Microalgal bioremediation of emerging contaminants – opportunities and challenges. Water Res 114921

Tijani JO et al (2013) A review of pharmaceuticals and endocrine-disrupting compounds: sources, effects, removal, and detections. Water Air Soil Pollut 224:1770. https://doi.org/10.1007/s.11270-013-1770-3

Tiwari B, Sellamuthu B, Ouarda Y, Drogui P, Tyagi RD, Buelna G (2017) Review on fate and mechanism of removal of pharmaceutical pollutants from wastewater using biological approach. Bioresour Technol 224:1–12

Viancelli A, Michelon W, Rogovski P, Cadamuro RD, de Souza EB, Fongaro G, Treichel H (2020) A review on alternative bioprocesses for removal of emerging contaminants. Bioprocess Biosyst Eng 43:2117–2129. https://doi.org/10.1007/s00449-020-02410-9

Villar-Navarro E, Baena-Nogueras RM, Paniw M, Perales JA, Lara-Martın PA (2018) Removal of pharmaceuticals in urban wastewater: high-rate algae pond (HRAP) based technologies as an alternative to activated sludge-based processes. Water Res 139:19–29. https://doi.org/10.1016/j.watres.2018.03.072

Xie P, Chen C, Zhang C, Su G, Ren NQ, Ho SH (2020) Revealing the role of adsorption in ciprofloxacin and sulfadiazine elimination routes in microalgae. Water Res 115475

Xiong JQ, Kurade MB, Jeon BH (2017) Biodegradation of levofloxacin by an acclimated freshwater microalga, Chlorella vulgaris. Chem Eng J 313(2017):1251–1257

Xiong JQ, Kurade MB, Jeon BH (2018a) Can microalgae remove pharmaceutical contaminants from water? Trends Biotechnol 36(1):30–44. https://doi.org/10.1016/j.tibtech.2017.09.003

Xiong Y, Hozic D, Goncalves AL, Simões M, Hong PY (2018b) Increasing tetracycline concentrations on the performance and communities of mixed microalgae-bacteria photo- bioreactors. Algal Res 29:249–256. https://doi.org/10.1016/j.algal.2017.11.033

Xiong JQ, Kim SJ, Kurade MB, Govindwar S, Abou-Shanab RA, Kim JR, Roh HS, Khan MA, Jeon BH (2019) Combined effects of sulfamethazine and sulfamethoxazole on a freshwater microalga, Scenedesmus obliquus: toxicity, biodegradation, and metabolic fate. J Hazard Mater 370(2019):138–146

Yadav P, Gupta RK, Singh RP, Yadav PK, Jalaluddin PAK, Pandey KD (2021) Chapter 9 - Role of cyanobacteria in green remediation. In: Mishra VK, Kumar A (eds) Sustainable environmental clean-up. Elsevier, pp 187–210. ISBN 9780128238288. https://doi.org/10.1016/B978-0-12-823828-8.00009-8

Yadav IC, Devi NL, Syed JH (2023) Editorial: legacy and emerging flame retardants as environmental pollutants: occurrence, toxicity mechanisms, and control. Front Environ Sci 11:1257582. https://doi.org/10.3389/fenvs.2023.1257582

Yahaya A, Sale FJ, Salehdeen MU (2020) Analytical methods for determination of regulated and unregulated disinfection by-products in drinking water: a review. CaJoST 1:25–36

Yang J, Zhao Y, Li M, Du M, Li X, Li Y (2019) A review of a class of emerging contaminants: the classification, distribution, intensity of consumption, synthesis routes, environmental effects and expectation of pollution abatement to Organophosphate Flame Retardants (OPFRs). Int J Mol Sci 20:2874. https://doi.org/10.3390/ijms20122874

Yang W et al (2020) Nanoparticle toxicology. Annu Rev Pharmacol Toxicol 61:269–289

Yazdan MMS, Kumar R, Leung SW (2022) The environmental and health impacts of steroids and hormones in wastewater effluent, as well as existing removal technologies: a review. Ecologies 2022(3):206–224. https://doi.org/10.3390/ecologies3020016

Yogui GT, Sericano JL (2009) Polybrominated diphenyl ether flame retardants in the US marine environment: a review. Environ Int 35(3):655–666. https://doi.org/10.1016/j.envint.2008.11.001

Zhou JL, Yang L, Huang KX, Chen DZ, Gao F (2022) Mechanisms and application of microalgae on removing emerging contaminants from wastewater: a review. Bioresour Technol 364:128049. https://doi.org/10.1016/j.biortech.2022.128049. ISSN 0960-8524. https://www.epa.gov/sites/default/files/2018-11/documents/cecs-pfcs-factsheet.pdf

Chapter 11
Revitalizing Ecosystems: A Sustainable Approach to Mitigating Arsenic Toxicity

Neelanjan Dutta, Amit Ghosh, Subhendu Chakrabarty, and Amlan Das

Abstract Arsenic contamination in the environment poses a significant threat to ecosystems and human health. Bioremediation is an environment-friendly and cost-effective promising approach for treating arsenic contamination. This approach begins with a comprehensive site assessment to understand the extent of contamination and ecological factors. Arsenic is a toxic element that can be found in soil, water, and sediments due to both natural processes and human activities such as mining, industrial operations, and the use of arsenic-based pesticides. Bioremediation involves the use of living organisms, mainly microorganisms, plants, or fungi, to transform or remove contaminants from the environment. The effectiveness depends on various factors, including the type and concentration of arsenic, the specific microorganisms or plants used, environmental conditions, and the scale of contamination. Community involvement and education are essential for building support and fostering transparency. Robust monitoring and adaptive management ensure the effectiveness of remediation efforts, and a long-term stewardship plan prevents recontamination. Regulatory compliance, research, and innovation are key components, all aimed at achieving effective remediation while preserving ecosystems and safeguarding communities for the future. By incorporating these sustainable practices, bioremediation efforts for arsenic contamination can effectively restore the

N. Dutta
Department of Civil Engineering, National Institute of Technology Sikkim, Ravangla, Sikkim, India

A. Ghosh
Department of Biological Sciences, School of Life Sciences and Biotechnology, Adamas University, Adamas Knowledge City, Kolkata, West Bengal, India

S. Chakrabarty
Department of Microbiology, M.U.C. Women's College, Burdwan, West Bengal, India

A. Das (✉)
Department of Biochemistry, Royal School of Biosciences, The Assam Royal Global University, Guwahati, Assam, India
e-mail: adas5@rgu.ac

S. Dey, S. Bhattacharya (eds.), *Biotechnological Interventions in the Removal of Emerging Pollutants*, Interdisciplinary Biotechnological Advances,
https://doi.org/10.1007/978-981-97-9922-0_11

environment, safeguard human health, and promote the long-term well-being of ecosystems. Additionally, sustainable bioremediation can serve as a model for addressing other types of environmental contamination.

Keywords Arsenic toxicity · Bioremediation · Sustainable approach

11.1 Introduction

Arsenic, a naturally occurring metalloid in the Earth's crust, is found in soil, rocks, and minerals (Genchi et al. 2022; Raju 2022). Anthropogenic activities have significantly augmented arsenic release into water and soil, causing widespread contamination and posing threats to ecosystems, biodiversity, and human health (Ghosh et al. 2022). One of the most significant environmental concerns related to arsenic is its contamination of water sources. In many regions, groundwater is a vital source of drinking water and irrigation. Arsenic can leach into groundwater from geological formations, especially in areas with high concentrations of arsenic in the bedrock (Shaji et al. 2021). The contamination of groundwater with arsenic poses a severe threat to human health, as people relying on such water sources may be exposed to elevated levels of arsenic through drinking and food preparation. Surface water bodies are also at risk of arsenic contamination. Runoff from arsenic-laden soil and industrial discharges can introduce this metalloid into rivers, lakes, and streams. Aquatic ecosystems, including fish and other organisms, can accumulate arsenic, leading to adverse effects on the health and diversity of aquatic life. The contamination of water bodies by arsenic disrupts the delicate balance of ecosystems, affecting the entire food web. Agricultural practices, such as the use of arsenic-containing pesticides and fertilizers, contribute to soil contamination (Loukola-Ruskeeniemi et al. 2022). Industries, particularly mining and smelting, release arsenic into the soil, creating localized contamination hotspots. Over time, arsenic in the soil can impact plant growth and alter the composition of microbial communities, influencing nutrient cycling and overall soil health. The ecological impact of arsenic contamination extends to various components of ecosystems. In aquatic environments, arsenic can lead to changes in the composition of communities, with some species becoming more tolerant to arsenic than others (Zhang et al. 2022). This alteration in species composition can have cascading effects throughout the ecosystem, impacting predator–prey relationships and overall biodiversity. In soils, the presence of arsenic can affect microbial communities responsible for nutrient cycling (Li et al. 2021). Changes in soil microbial activity may influence the availability of essential nutrients for plant growth, further exacerbating the impact on vegetation. Certain plant species can accumulate arsenic, leading to changes in plant communities and potentially affecting the herbivores and other organism's dependent on them. Human health is intricately linked to arsenic contamination in water and soil. Chronic exposure to elevated levels of arsenic, primarily through drinking water, has been associated with a range of health issues. Skin lesions, various cancers (such as skin, bladder, and lung cancers), cardiovascular diseases, and developmental effects are among the health risks associated with arsenic exposure (Dutta and Gupta 2022). In areas where rice is a staple food and arsenic levels in the soil are high, the metalloid can accumulate in rice grains, posing additional risks to human health through dietary intake. Addressing arsenic contamination requires a multifaceted approach. Remediation strategies such as phytoremediation, which involves using plants to absorb and accumulate arsenic, and microbial remediation, utilizing bacteria and fungi to transform or immobilize arsenic, show promise in reducing environmental concentrations of this metalloid (Nedjimi 2021; Sabreena et al. 2022). Sustainable agricultural practices, including the judicious use of arsenic-containing fertilizers and pesticides, can help prevent further soil contamination. Effective water treatment technologies, such as filtration and adsorption, are essential for ensuring safe drinking water. Monitoring and regulatory measures are crucial to controlling industrial discharges and preventing further contamination (Dutta and Gupta 2022).

The review of research on revitalizing ecosystems and mitigating arsenic toxicity reveals a landscape rich with innovative and sustainable approaches. One notable avenue of exploration involves the utilization of arsenic hyperaccumulator plants, either through natural selection or genetic modification. These plants exhibit a unique capacity to absorb and accumulate elevated levels of arsenic, offering a natural and environment-friendly method for detoxifying contaminated soils. Concurrently, microbial remediation strategies employing engineered bacteria or fungi show promise in transforming and immobilizing arsenic in the soil, preventing its absorption by plants and subsequent entry into the food chain. Nanotechnology emerges as a transformative force in arsenic mitigation, with nanomaterials exhibiting remarkable efficiency in adsorbing arsenic from water sources. The development of advanced nanomaterials presents a sustainable solution for water purification, crucial for protecting aquatic ecosystems and the communities dependent on them. Integrating traditional knowledge from local communities into scientific arsenic mitigation strategies adds depth and cultural relevance to the research. This collaborative approach identifies indigenous plants and practices historically effective in combating arsenic contamination. Furthermore, smart agriculture practices, including precision farming and controlled-environment cultivation, play a pivotal role in minimizing arsenic uptake by crops while optimizing resource utilization. In essence, the multifaceted strategies reviewed not only address arsenic toxicity but also contribute to the broader goal of ecosystem revitalization, emphasizing the importance of harmonizing human activities with the environment.

11.2 Chemical and Biological Properties of Arsenic

Arsenic is a chemical element with the symbol As, and atomic number 33. It is a metalloid, which means it has properties of both metals and nonmetals. Arsenic is commonly found in the Earth's crust and is distributed widely in the environment. Here are some key aspects of the chemical and biological properties of arsenic:

11.2.1 Chemical Properties

Arsenic has several allotropes, including yellow, black, and gray arsenic. Yellow arsenic is the most stable form at room temperature. Arsenic can exhibit various oxidation states, including −3, 0, +3, and +5 (Gomez-Caminero et al. 2001). The most common oxidation states are −3 and +3 (Dutta et al. 2021). Arsenic reacts with oxygen, sulfur, and metals. It can form arsenides with metals, and it readily combines with sulfur to form arsenic sulfides. Arsenic compounds can be soluble or insoluble in water, depending on their chemical form. Inorganic arsenic compounds are often more soluble than organic arsenic compounds. Arsenic compounds can act as weak acids or bases, depending on the specific compound and conditions.

11.2.2 Biological Properties

Arsenic is highly toxic to most organisms, including humans. The bioavailability of arsenic depends on its chemical form. Inorganic arsenic, especially arsenite (As^{3+}), is more toxic and bioavailable than organic forms such as arsenobetaine and arsenocholine (Liao et al. 2022) (Fig. 11.1). In humans, arsenic can undergo various metabolic transformations, including methylation reactions. Methylation can lead to the formation of more complex organic arsenic compounds. Arsenic can be absorbed through inhalation, ingestion, and dermal contact. Once absorbed, it can distribute throughout the body, with a tendency to accumulate in certain tissues such as the liver, kidneys, and lungs. The primary route of arsenic elimination from the body is through urine (Yi et al. 2020). The metabolism and excretion of arsenic can vary among individuals.

11.3 Arsenic Toxicity and Effect on Environment

Long-term exposure to elevated levels of arsenic, especially through contaminated water or food, can lead to various health issues. These include skin lesions, cancers (lung, bladder, skin), cardiovascular diseases, and neurological effects (Kumar et al. 2021; Naujokas et al. 2013). Short-term exposure to high levels of arsenic can result in acute poisoning, leading to symptoms such as vomiting, abdominal pain, diarrhea, and in severe cases, death (Tournel et al. 2011; Saha et al. 1999). Consuming arsenic-contaminated water, food, or other beverages is a common route of exposure. Breathing in arsenic-containing dust or fumes, especially in occupational settings, can contribute to exposure (Hughes et al. 2011). Arsenic can be absorbed through the skin, though this is generally a less significant route of exposure. Arsenic undergoes metabolism in the body, with some compounds being methylated to less toxic forms. However, certain methylated forms may still pose health risks. The

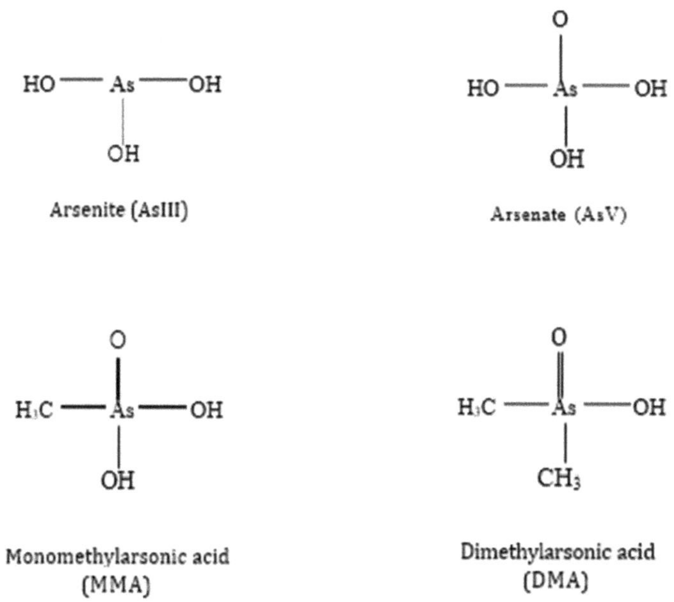

Fig. 11.1 Chemical structures of different arsenic species

susceptibility to arsenic toxicity varies among individuals. Factors such as nutritional status, genetic makeup, and pre-existing health conditions can influence an individual's response to arsenic exposure. Arsenic is naturally present in the Earth's crust and can be released into the environment through geological processes. Human activities, including mining, smelting, agriculture (use of arsenic-containing pesticides), and industrial processes, contribute to environmental arsenic contamination (Tchounwou et al. 2012). Arsenic contamination of groundwater is a significant environmental concern in many regions. This contamination often results from the dissolution of arsenic-bearing minerals in geological formations, leading to elevated levels in well water. Arsenic can accumulate in soil through natural processes or human activities. This can affect plants, leading to the potential transfer of arsenic through the food chain.

Arsenic toxicity can harm aquatic ecosystems, affecting fish and other aquatic organisms (Byeon et al. 2021). The bioaccumulation of arsenic in organisms can lead to ecological imbalances. Some plants can accumulate arsenic in the soil, and this can result in the contamination of crops. Consuming crops grown in arsenic-contaminated soil contributes to human exposure. Arsenic compounds can persist in the environment for extended periods, contributing to long-term exposure risks. Many countries have established regulations to limit arsenic levels in drinking water, soil, and food to protect human health and the environment.

11.4 Arsenic Removal Technologies

Arsenic contamination in water sources is a significant global health concern due to its severe health effects, including cancer and other serious illnesses. Several technologies have been developed for the removal of arsenic from water, each with its advantages and limitations. Among the most widely used technologies are chemical coagulation, electrocoagulation, adsorption, ion exchange, and precipitation. Chemical coagulation is a common method used for arsenic removal, particularly in municipal water treatment plants (Sancha 2006). In this process, coagulants such as ferric chloride or aluminum sulfate are added to water, which react with arsenic and other contaminants to form flocs. These flocs can then be removed through sedimentation or filtration, effectively reducing the arsenic concentration in the water. Chemical coagulation is relatively simple and cost-effective, making it a popular choice for large-scale water treatment applications (Koul et al. 2022). Electrocoagulation (EC) is a variation of chemical coagulation that uses an electric current to destabilize and coagulate contaminants in water. In EC, sacrificial anodes made of materials like iron or aluminum are used to generate coagulating agents in situ, which react with arsenic and other contaminants to form flocs (Dutta et al. 2021). The flocs can then be removed through sedimentation or filtration. EC is effective in removing arsenic from water and has the added advantage of being able to treat water with high levels of organic contaminants. Adsorption is another widely used method for arsenic removal, which involves the use of adsorbent materials to adsorb arsenic from water (Dutta and Gupta 2022). Commonly used adsorbents include activated alumina, and iron oxide (Neisan et al. 2023). These materials have a high affinity for arsenic and can effectively remove it from water through surface interactions. Adsorption is a simple and effective method for arsenic removal and can be used in both small-scale and large-scale water treatment systems (Dutta and Gupta 2023). Ion exchange is a process where arsenic ions in water are exchanged with ions bound to a resin. This process effectively removes arsenic from the water, but the resin eventually becomes saturated with arsenic and needs to be regenerated or replaced. Ion exchange is an effective method for arsenic removal, particularly in areas with low to moderate levels of arsenic contamination (Pillai and Dharaskar 2020). Precipitation methods involve the addition of chemicals to water to convert soluble arsenic into insoluble forms, which can then be removed through sedimentation or filtration. Commonly used chemicals for precipitation include lime, ferric chloride, and alum. Precipitation methods can be effective in removing arsenic from water, but they require careful control of pH and other water chemistry parameters to ensure optimal performance.

11.5 Bioremediation of Arsenic

Bioremediation of arsenic is a sustainable and environment-friendly approach aimed at mitigating arsenic contamination in soil and water (Bertin et al. 2022). This process relies on the natural abilities of living organisms, such as plants and microorganisms, to either accumulate, transform, or immobilize arsenic, thereby reducing its environmental impact. Bioremediation of arsenic encompasses various approaches that leverage the metabolic activities of living organisms to mitigate arsenic contamination. Here are several types of bioremediation strategies for arsenic:

11.5.1 Arsenite Oxidation

Bioremediation of arsenic is a significant environmental process that harnesses the metabolic capabilities of microorganisms to transform toxic inorganic arsenic species, such as arsenite (As(III)), into less harmful forms, primarily arsenate (As(V)) (Preetha et al. 2023). The microbial oxidation of arsenite is not only a detoxification mechanism but also influences the mobility and bioavailability of arsenic in the environment. This microbial activity is crucial in mitigating arsenic contamination in various environments, including soil and water (Ali et al. 2022). Arsenite oxidation is a key step in the biogeochemical cycling of arsenic, which is less likely to leach into water sources. Microorganisms, including bacteria and archaea, play a pivotal role in arsenite oxidation (Yin et al. 2022). Certain bacteria, such as members of the genera Agrobacterium, Pseudomonas, and Bacillus, have demonstrated the ability to oxidize arsenite as part of their metabolic processes (Herrera et al. 2021). The oxidation reactions are often facilitated by enzymes, such as arsenite oxidase, which catalyze the conversion of arsenite to arsenate. The process of arsenite oxidation can occur under aerobic conditions, where oxygen is readily available, or under anaerobic conditions, where alternative electron acceptors like nitrate or iron are used (Sun et al. 2010). Bioremediation strategies leveraging arsenite oxidation often involve the stimulation of indigenous microbial communities or the introduction of engineered microorganisms with enhanced arsenite oxidation capabilities. Additionally, the use of permeable reactive barriers containing arsenite-oxidizing bacteria has been explored to treat arsenic-contaminated groundwater (Zouboulis and Katsoyiannis 2005). Challenges associated with arsenite oxidation in bioremediation include the potential release of arsenic into the environment during intermediate steps of the process and the need for sustainable approaches that can be applied across diverse environmental conditions. Ongoing research focuses on optimizing conditions for microbial arsenite oxidation, understanding the genetic and molecular basis of microbial arsenite metabolism, and exploring ways to enhance the efficiency of bioremediation technologies.

11.5.2 Arsenate Reduction

Bioremediation of arsenic through arsenate reduction is a crucial process that utilizes the metabolic activities of microorganisms to convert arsenate (As(V)), a less toxic form of arsenic, into arsenite (As(III)), which is often more easily immobilized or precipitated in the environment. This microbial-driven reduction is a significant component of the biogeochemical cycling of arsenic and is particularly relevant in environments where arsenate predominates, such as oxygenated surface waters. Microorganisms, including bacteria and archaea, play a key role in arsenate reduction (Drewniak and Sklodowska 2013). Certain bacteria, such as Shewanella and Geobacter species, are known for their ability to respire using arsenate as a terminal electron acceptor (Darma et al. 2022). This respiratory process results in the reduction of arsenate to arsenite, which can be subsequently sequestered or undergo further transformations. Arsenate reduction can occur under anaerobic conditions, where oxygen is limited or absent, and alternative electron acceptors, such as arsenate itself, are utilized by microorganisms. The reduction of arsenate is often facilitated by enzymes, such as arsenate reductase, which catalyze the conversion of arsenate into arsenite (Chrysostomou et al. 2015). Bioremediation strategies employing arsenate reduction may involve stimulating indigenous microbial communities through nutrient amendments or introducing engineered microorganisms with enhanced arsenate reduction capabilities. Additionally, the use of permeable reactive barriers containing microbes capable of arsenate reduction has been explored to treat arsenic-contaminated groundwater (Kanel et al. 2023). While arsenate reduction can contribute to arsenic immobilization and detoxification, challenges include the potential release of arsenic back into the environment and the need for a comprehensive understanding of the environmental conditions influencing microbial activity. Ongoing research focuses on optimizing conditions for microbial arsenate reduction, unraveling the genetic and molecular mechanisms of microbial arsenate metabolism, and developing sustainable and efficient bioremediation technologies.

11.5.3 Phytoremediation

Phytoremediation, a form of bioremediation, is an environment-friendly and sustainable approach used to mitigate arsenic contamination by employing plants to extract, sequester, or transform arsenic from the soil and water (Bhat et al. 2022). This method takes advantage of the unique abilities of certain plant species to absorb and accumulate arsenic in their tissues. The process primarily involves hyperaccumulating plants, which can uptake large amounts of arsenic without displaying significant toxic effects. Several plant species, including ferns, willows, sunflowers, and certain grasses, have demonstrated the ability to hyperaccumulate arsenic (Wang et al. 2002). The plants absorb arsenic through their roots from the soil or water, and the metalloid is then transported and stored in various plant tissues, such as roots, stems, and leaves. This accumulation reduces the concentration of arsenic in the surrounding environment. Phytoremediation offers several advantages, including its cost-effectiveness, minimal environmental impact, and applicability to large-scale sites (Babu et al. 2021). It can be particularly beneficial in areas with moderate arsenic contamination levels. Additionally, the harvested plants can be managed and disposed of properly, minimizing the risk of releasing arsenic back into the environment. However, challenges associated with phytoremediation include the slow rate of arsenic removal, the need for a long-term commitment, and the potential for arsenic to accumulate in edible parts of certain crops, posing risks to human health (Srivastava et al. 2021). Therefore, careful selection of plant species and monitoring of the remediation process are essential. Ongoing research in phytoremediation focuses on identifying and engineering plant species with enhanced arsenic uptake and tolerance, optimizing growth conditions, and developing strategies to increase the efficiency of arsenic removal. Additionally, efforts are directed toward understanding the molecular and genetic mechanisms underlying arsenic accumulation in plants.

11.5.4 Biosorption

Biosorption is a process that involves the use of biological materials, such as microorganisms, algae, or plant residues, to remove heavy metals and other contaminants from water or soil Biosorption is a sustainable approach used to mitigate arsenic toxicity in water. One of the key advantages of biosorption is its cost-effectiveness compared to conventional methods such as chemical precipitation or membrane filtration. Additionally, biosorption is often more environment-friendly, as it does not require the use of harsh chemicals and can be easily integrated into existing water treatment systems. Biosorption of arsenic involves the use of living or nonliving biomass to remove arsenic from water or other solutions (Podder and Majumder 2016). Various biomass sources, including microorganisms, plants, and industrial waste materials, can adsorb arsenic through physical, chemical, and biological mechanisms. The biomass typically possesses functional groups on its surface, such as hydroxyl, carboxyl, and amino groups, that interact with arsenic ions (Xanthopoulou and Katsoyiannis 2023). These materials have a high affinity for arsenic ions, effectively binding them and removing them from the water. The effectiveness of biosorption is influenced by factors such as biomass type, pH, temperature, and the presence of competing ions (Pham et al. 2022). Microbial biosorption often includes the active uptake and accumulation of arsenic by bacteria and fungi, either on cell walls or within cellular structures. Plant-based biosorption, known as phytoaccumulation, involves certain plants accumulating arsenic in their tissues. This process is explored for phytoremediation in arsenic-contaminated areas.

11.5.5 Methylation of Inorganic as Species

Methylation of inorganic arsenic species refers to the process by which methyl groups (–CH_3) are added to inorganic arsenic compounds (Flora 2015). Methylation of arsenic is often observed as a potential mechanism for reducing its toxicity, as the organic forms are typically less toxic and more easily excreted from the body compared to inorganic arsenic. However, it's important to note that some organic arsenic compounds can still pose health risks, and the overall impact of methylation on arsenic toxicity is complex and not fully understood. In terms of sustainability, methylation as a remediation approach has both advantages and challenges.

It can occur naturally and can be stimulated through bioremediation techniques, which can be more environment-friendly compared to conventional chemical treatments. On the other hand, the effectiveness of methylation as a remediation strategy may vary depending on environmental conditions, and there are concerns about the potential for the formation of more toxic organic arsenic species. In the context of arsenic, methylation typically occurs in living organisms, such as bacteria, fungi, and certain plants (Mlangeni 2023). This process is significant because it can lead to the formation of organic arsenic compounds, such as monomethyl arsenic acid (MMA) and dimethyl arsenic acid (DMA). These organic arsenic species are generally considered less toxic and more easily excreted from the body than their inorganic counterparts. The methylation of inorganic arsenic is primarily carried out by arsenic-methylating enzymes, particularly arsenic methyltransferases. These enzymes facilitate the transfer of methyl groups from methyl donors (such as S-adenosylmethionine) to inorganic arsenic species. The methylation process typically involves multiple steps, and various intermediates are formed before the final organic arsenic compounds are produced (Thomas et al. 2007). Methylated arsenic species are more volatile than iAs and can undergo volatilization processes, potentially leading to the atmospheric transport of arsenic. This can contribute to the global distribution of arsenic and its exposure to remote populations. It's worth noting that while methylation can reduce the toxicity of arsenic, some of the methylated arsenic species may still pose health risks, and the overall toxicity of arsenic is dependent on the specific chemical forms present (Fig. 11.2).

Fig. 11.2 An overview of the occurrence of arsenic compounds in the environment

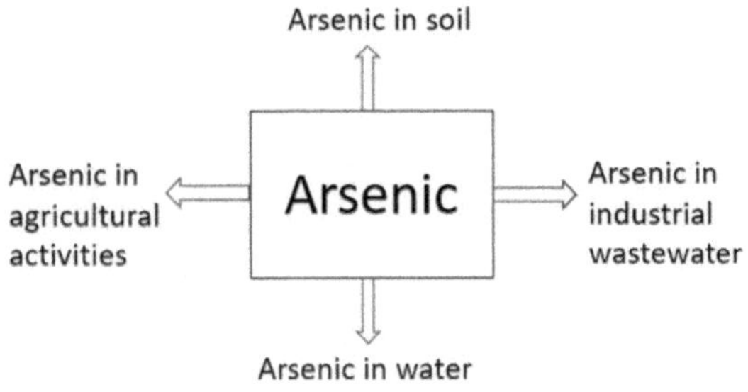

11.6 Conclusions and Future Perspectives

Revitalizing ecosystems offers a sustainable and multifaceted approach to mitigating arsenic toxicity, providing long-term benefits for the environment and human health. Arsenic contamination, primarily from natural geological sources and human activities like mining and agriculture, poses a significant threat globally. Addressing this issue requires a comprehensive strategy that surpasses traditional remediation methods, embracing ecological restoration and conservation practices. Ecosystem revitalization's key advantage lies in its use of natural processes to reduce arsenic levels in soil and water. Bioaccumulating plants, such as ferns and sunflowers, can extract arsenic from the soil, reducing environmental impact while minimizing costly and energy-intensive remediation technologies. Furthermore, ecosystem revitalization considers the interconnectedness of ecosystems and human communities, recognizing the importance of preserving biodiversity. This not only supports ecosystem health but also provides additional benefits like natural pest control and pollination services. By restoring diverse ecosystems, we can create resilient landscapes better equipped to withstand environmental stressors, including arsenic contamination. Adopting sustainable methods for ecosystem revitalization emerges as a promising approach to counteract arsenic toxicity. Strategies such as phytoremediation, microbial bioremediation, and biosorption offer eco-friendly solutions. These approaches leverage the natural processes of living organisms, such as plants and microorganisms, to sequester, transform, or immobilize arsenic, reducing its impact on ecosystems. However, applying these strategies requires careful consideration of factors such as plant and microbial species selection, site-specific conditions, and potential long-term ecological consequences. Future efforts should focus on refining and expanding sustainable approaches, including researching genetic and molecular mechanisms governing arsenic tolerance and uptake in plants and microorganisms. Vigilant monitoring of revitalization efforts' enduring ecological effects is crucial to ensuring the chosen mitigation strategies' viability. Collaborative initiatives involving researchers, environmentalists, and policymakers are essential for formulating comprehensive strategies adaptable to diverse ecosystems and arsenic-contaminated sites globally. Community engagement and awareness also play a pivotal role in executing sustainable arsenic mitigation measures. Educating local communities about these approaches' advantages and potential risks fosters a sense of ownership and ensures sustainability. Integrating these strategies into broader environmental management frameworks, alongside developing innovative technologies, will contribute to a comprehensive and sustainable approach to mitigating arsenic toxicity in ecosystems. Revitalizing ecosystems to counteract arsenic toxicity is not just a scientific challenge; it is a multidimensional undertaking requiring the integration of ecological, technological, and social perspectives. Embracing sustainability as a guiding principle ensures that arsenic mitigation efforts address immediate environmental threats and contribute to the long-term health and resilience of ecosystems.

References

Ali S, Tyagi A, Mushtaq M, Al-Mahmoudi H, Bae H (2022) Harnessing plant microbiome for mitigating arsenic toxicity in sustainable agriculture. Environ Pollut 300:118940

Babu SOF, Hossain MB, Rahman MS, Rahman M, Ahmed AS, Hasan MM et al (2021) Phytoremediation of toxic metals: a sustainable green solution for clean environment. Appl Sci 11(21):10348

Bertin PN, Crognale S, Plewniak F, Battaglia-Brunet F, Rossetti S, Mench M (2022) Water and soil contaminated by arsenic: the use of microorganisms and plants in bioremediation. Environ Sci Pollut Res 29:9462–9489

Bhat SA, Bashir O, Haq SAU, Amin T, Rafiq A, Ali M et al (2022) Phytoremediation of heavy metals in soil and water: an eco-friendly, sustainable and multidisciplinary approach. Chemosphere 303:134788

Byeon E, Kang HM, Yoon C, Lee JS (2021) Toxicity mechanisms of arsenic compounds in aquatic organisms. Aquat Toxicol 237:105901

Chrysostomou C, Quandt EM, Marshall NM, Stone E, Georgiou G (2015) An alternate pathway of arsenate resistance in E. coli mediated by the glutathione S-transferase GstB. ACS Chem Biol 10(3):875–882

Darma A, Yang J, Zandi P, Liu J, Możdżeń K, Xia X et al (2022) Significance of Shewanella species for the phytoavailability and toxicity of arsenic—a review. Biology 11(3):472

Drewniak L, Sklodowska A (2013) Arsenic-transforming microbes and their role in biomining processes. Environ Sci Pollut Res 20:7728–7739

Dutta N, Gupta A (2022) Development of arsenic removal unit with electrocoagulation and activated alumina sorption: Field trial at rural West Bengal, India. J Water Process Eng 49:103013

Dutta N, Gupta A (2023) Characterization and use of waste plastic char for removal of arsenic and COD from aqueous solution. Int J Environ Sci Technol 20(7):7735–7748

Dutta N, Haldar A, Gupta A (2021) Electrocoagulation for arsenic removal: field trials in rural West Bengal. Arch Environ Contam Toxicol 80(1):248–258

Flora SJ (2015) Arsenic: chemistry, occurrence, and exposure. In: Handbook of arsenic toxicology. Academic, pp 1–49

Genchi G, Lauria G, Catalano A, Carocci A, Sinicropi MS (2022) Arsenic: a review on a great health issue worldwide. Appl Sci 12(12):6184

Ghosh D, Ghosh A, Bhadury P (2022) Arsenic through aquatic trophic levels: effects, transformations, and biomagnification—a concise review. Geosci Lett 9(1):1–17

Gomez-Caminero A, Howe PD, Hughes M, Kenyon E, Lewis DR, Moore M et al (2001) Arsenic and arsenic compounds. World Health Organization

Herrera C, Moraga R, Bustamante B, Vilo C, Aguayo P, Valenzuela C et al (2021) Characterization of arsenite-oxidizing bacteria isolated from arsenic-rich sediments, Atacama desert, Chile. Microorganisms 9(3):483

Hughes MF, Beck BD, Chen Y, Lewis AS, Thomas DJ (2011) Arsenic exposure and toxicology: a historical perspective. Toxicol Sci 123(2):305–332

Kanel SR, Das TK, Varma RS, Kurwadkar S, Chakraborty S, Joshi TP et al (2023) Arsenic contamination in groundwater: geochemical basis of treatment technologies. ACS Environ Au 3(3):135–152

Koul B, Bhat N, Abubakar M, Mishra M, Arukha AP, Yadav D (2022) Application of natural coagulants in water treatment: a sustainable alternative to chemicals. Water 14(22):3751

Kumar A, Ali M, Kumar R, Kumar M, Sagar P, Pandey RK et al (2021) Arsenic exposure in Indo Gangetic plains of Bihar causing increased cancer risk. Sci Rep 11(1):2376

Li Y, Zhang M, Xu R, Lin H, Sun X, Xu F et al (2021) Arsenic and antimony co-contamination influences on soil microbial community composition and functions: relevance to arsenic resistance and carbon, nitrogen, and sulfur cycling. Environ Int 153:106522

Liao ZH, Chuang HC, Huang HT, Wang PH, Chen BY, Lee PT et al (2022) Bioaccumulation of arsenic and immunotoxic effect in white shrimp (Penaeus vannamei) exposed to trivalent arsenic. Fish Shellfish Immunol 122:376–385

Loukola-Ruskeeniemi K, Müller I, Reichel S, Jones C, Battaglia-Brunet F, Elert M et al (2022) Risk management for arsenic in agricultural soil–water systems: lessons learned from case studies in Europe. J Hazard Mater 424:127677

Mlangeni AT (2023) Methylation of arsenic in rice: mechanisms, factors, and mitigation strategies. Toxicol Rep 11:295–306

Naujokas MF, Anderson B, Ahsan H, Aposhian HV, Graziano JH, Thompson C, Suk WA (2013) The broad scope of health effects from chronic arsenic exposure: update on a worldwide public health problem. Environ Health Perspect 121(3):295–302

Nedjimi B (2021) Phytoremediation: a sustainable environmental technology for heavy metals decontamination. SN Appl Sci 3(3):286

Neisan RS, Saady NMC, Bazan C, Zendehboudi S, Al-nayili A, Abbassi B, Chatterjee P (2023) Arsenic removal by adsorbents from water for small communities' decentralized systems: performance, characterization, and effective parameters. Clean Technol 5(1):352–402

Pham VHT, Kim J, Chang S, Chung W (2022) Bacterial biosorbents, an efficient heavy metals green clean-up strategy: prospects, challenges, and opportunities. Microorganisms 10(3):610

Pillai P, Dharaskar S (2020) Arsenic removal using nanoparticles from groundwater: a review. In: Handbook of solid waste management: sustainability through circular economy, pp 1–15

Podder MS, Majumder CB (2016) Study of the kinetics of arsenic removal from wastewater using Bacillus arsenicus biofilms supported on a neem leaves/$MnFe_2O_4$ composite. Ecol Eng 88:195–216

Preetha JSY, Arun M, Vidya N, Kowsalya K, Halka J, Ondrasek G (2023) Biotechnology advances in bioremediation of arsenic: a review. Molecules 28(3):1474

Raju NJ (2022) Arsenic in the geo-environment: a review of sources, geochemical processes, toxicity and removal technologies. Environ Res 203:111782

Sabreena, Hassan S, Bhat SA, Kumar V, Ganai BA, Ameen F (2022) Phytoremediation of heavy metals: an indispensable contrivance in green remediation technology. Plan Theory 11(9):1255

Saha JC, Dikshit AK, Bandyopadhyay M, Saha KC (1999) A review of arsenic poisoning and its effects on human health. Crit Rev Environ Sci Technol 29(3):281–313

Sancha AM (2006) Review of coagulation technology for removal of arsenic: case of Chile. J Health Popul Nutr 24(3):267

Shaji E, Santosh M, Sarath KV, Prakash P, Deepchand V, Divya BV (2021) Arsenic contamination of groundwater: a global synopsis with focus on the Indian peninsula. Geosci Front 12(3):101079

Srivastava S, Shukla A, Rajput VD, Kumar K, Minkina T, Mandzhieva S et al (2021) Arsenic remediation through sustainable phytoremediation approaches. Fortschr Mineral 11(9):936

Sun W, Sierra-Alvarez R, Field JA (2010) The role of denitrification on arsenite oxidation and arsenic mobility in an anoxic sediment column model with activated alumina. Biotechnol Bioeng 107(5):786–794

Tchounwou PB, Yedjou CG, Patlolla AK, Sutton DJ (2012) Heavy metal toxicity and the environment. In: Molecular, clinical and environmental toxicology: volume 3: environmental toxicology, pp 133–164

Thomas DJ, Li J, Waters SB, Xing W, Adair BM, Drobna Z et al (2007) Arsenic (+3 oxidation state) methyltransferase and the methylation of arsenicals. Exp Biol Med 232(1):3–13

Tournel G, Houssaye C, Humbert L, Dhorne C, Gnemmi V, Bécart-Robert A et al (2011) Acute arsenic poisoning: clinical, toxicological, histopathological, and forensic features. J Forensic Sci 56:S275–S279

Wang J, Zhao FJ, Meharg AA, Raab A, Feldmann J, McGrath SP (2002) Mechanisms of arsenic hyperaccumulation in Pteris vittata. Uptake kinetics, interactions with phosphate, and arsenic speciation. Plant Physiol 130(3):1552–1561

Xanthopoulou M, Katsoyiannis IA (2023) Enhanced adsorption of chromate and arsenate ions from contaminated water with emphasis on polyethylenimine modified materials: a review. Separations 10(8):441

Yi Y, Gao S, Xia J, Li C, Zhao Y, Zhang Y et al (2020) Study of the accumulation and distribution of arsenic species and association with arsenic toxicity in rats after 30 days of oral realgar administration. J Ethnopharmacol 247:111576

Yin S, Zhang X, Yin H, Zhang X (2022) Current knowledge on molecular mechanisms of microorganism-mediated bioremediation for arsenic contamination: a review. Microbiol Res 258:126990

Zhang W, Miao AJ, Wang NX, Li C, Sha J, Jia J et al (2022) Arsenic bioaccumulation and biotransformation in aquatic organisms. Environ Int 163:107221

Zouboulis AI, Katsoyiannis IA (2005) Recent advances in the bioremediation of arsenic-contaminated groundwaters. Environ Int 31(2):213–219

Chapter 12
Microplastic Pollution: Unveiling the Sources, Pathogenesis, and Effective Remediations

Amit Ghosh, Bhupali Kalita, Jyotirmay Das, Partha Pratim Das, Nabajyoti Das, Subhendu Chakrabarty, Amlan Das, and Neelanjan Dutta

Abstract Microplastics, small particles of plastic, fewer than 5 mm in diameter, have become a universal environmental concern because of their perseverance and impending adverse effects on ecosystems and human health. The creation of effective, scalable, and eco-friendly removal techniques for microplastics is necessary. The challenge of microplastic pollution has prompted the development of several remedial approaches. Using specialized tools like trawl nets, filtering systems, and vacuum pumps, mechanical approaches involve the collecting of microplastics from diverse environmental compartments, such as oceans, rivers, and soils. Contrarily, chemical treatments aim to convert microplastics into inert byproducts by oxidation, hydrolysis, or other chemical processes. Biological remediation techniques take advantage of the ability of bacteria, fungi, and algae to break down or ingest microplastics. Additionally, cutting-edge technologies like nanotechnology and research are being done on artificial intelligence to increase the efficiency of eliminating microplastics. However, each remediation method comes with its challenges

---

Amit Ghosh, Bhupali Kalita, Jyotirmay Das, Partha Pratim Das and Nabajyoti Das contributed equally with all other contributors.

A. Ghosh
Department of Biological Sciences, School of Life Sciences and Biotechnology, Adamas University, Adamas Knowledge City, Kolkata, West Bengal, India

B. Kalita · J. Das · P. P. Das · N. Das · A. Das (✉)
Department of Biochemistry, Royal School of Biosciences, The Assam Royal Global University, Guwahati, Assam, India
e-mail: adas5@rgu.ac

S. Chakrabarty
Department of Microbiology, M.U.C. Women's College, Burdwan, West Bengal, India

N. Dutta
Department of Civil Engineering, National Institute of Technology Sikkim, Ravangla, Sikkim, India

S. Dey, S. Bhattacharya (eds.), *Biotechnological Interventions in the Removal of Emerging Pollutants*, Interdisciplinary Biotechnological Advances,
https://doi.org/10.1007/978-981-97-9922-0_12

and potential environmental consequences. Mechanical methods may inadvertently harm nontarget species, while chemical treatments could produce harmful byproducts or require substantial energy inputs. Biological remediation approaches hold promise but require further research to optimize their efficiency and minimize unintended ecological impacts.

Keywords Microplastics · Environmental concern · Remediation techniques · Mechanical approaches · Chemical treatments · Biological remediation · Bacteria · Fungi · Algae · Nanotechnology · Artificial intelligence · Environmental consequences

12.1 Introduction

Microplastics or nanoplastics are introduced in the environment as a result of the breakdown of bigger plastic objects and other commercial goods including cosmetics, fertilizers, and medicines. Particles less than 1 μm are considered nanoparticles; they typically measure less than 5 mm (Tirkey and Upadhyay 2021). Commonly used polymers for MP formation include polyethylene, nylon polypropylene, polyester, polystyrene acrylic, polyoxyethylene, polyvinyl alcohol, polyvinyl chloride, poly methyl acrylate, polyethylene terephthalate, and polyurethane (Fig. 12.1). The repercussions of microplastics on the environment, oceans, and soil are significant, causing harm to aquatic life and presenting emerging threats globally. Depending on the source they come from, MPs, encompass various types:

(i) Microbeads: Microscopic polyethylene beads commonly used in health and cosmetic products as exfoliants; classified as microplastics and typically sized between 0.1 μm and 1 mm.
(ii) Scrubbers: MPs made from industrial cleaning agents.
(iii) Nurdles: MPs produced from plastic raw materials.
(iv) Plastic powders: MPs derived from powders used in industrial molding processes.
(v) Fibers: MPs created from clothing and cigarette filters.
(vi) Foam-based MPs: Produced from packaging foams, styrofoam cups, and tire wear and tear; the breakdown of bigger plastics produces secondary microplastics.

In 2004, marine biologist Professor Richard Thompson coined the term "microplastics"(Thompson et al. 2004). Subsequent research has extensively explored their presence, accumulation, and effects across ecosystems. Reports indicate over 260,000 tons of plastic waste in ocean basins (Eriksen et al. 2014). Microplastics have been identified in diverse human samples, from breast milk to marine ecosystems (Malafaia and Barcelo 2023). Despite existing policies, the

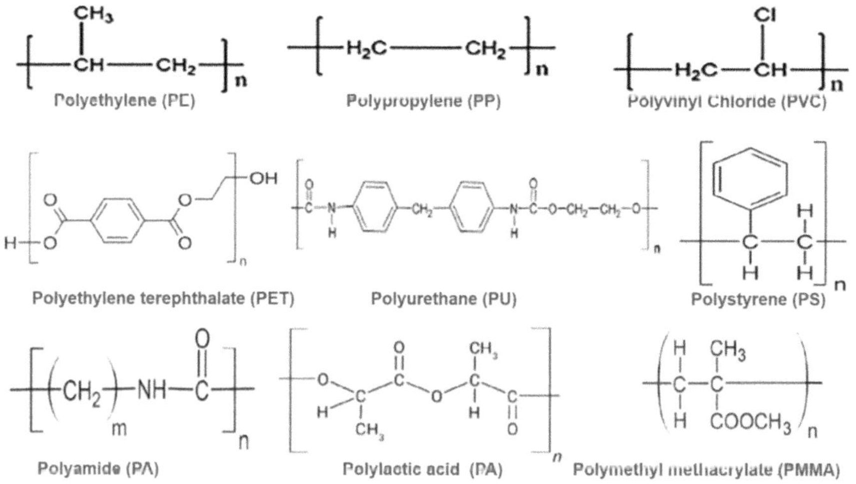

Fig. 12.1 Chemical structure illustration of various types of microplastics. The illustration depicts the chemical structure of various types of microplastics commonly found in the environment. *(1) Polyethylene (PE):* A thermoplastic polymer composed of ethylene monomers, characterized by its high strength and flexibility, commonly used in packaging materials and bottles. *(2) Polypropylene (PP):* A thermoplastic polymer consisting of propylene monomers, known for its resistance to heat and chemicals, often used in food containers and textiles. *(3) Polyvinyl chloride (PVC):* A synthetic polymer containing vinyl chloride monomers, valued for its versatility and durability, widely employed in construction materials and piping. *(4) Polystyrene (PS):* A thermoplastic polymer made from styrene monomers, recognized for its lightweight and insulation properties, commonly utilized in disposable food containers and packaging. *(5) Polyethylene terephthalate (PET):* A polyester polymer composed of terephthalic acid and ethylene glycol units, prized for its clarity and barrier properties, extensively used in beverage bottles and textiles. *(6) Polyurethane (PU):* A polymer formed through the reaction of diisocyanates with polyols, known for its versatility and wide range of applications, including foam insulation, coatings, and adhesives. *(7) Polyamide (PA):* A synthetic polymer composed of repeating amide linkages, noted for its strength, toughness, and resistance to abrasion, commonly used in textiles, automotive parts, and engineering plastics. *(8) Polylactic acid (PA):* Thermoplastic polymer that degrades naturally and is made from renewable materials like sugarcane or corn starch. It is highly composable and biocompatible and is frequently utilized in 3D printing, medical implants, and packaging. *(9) Polymethyl methacrylate (PMPA):* Clear thermoplastic polymer that is often used in acrylic glass, lenses, and medical implants. It is made from methyl methacrylate monomers and is highly valued for its optical clarity and weather resilience. These microplastic polymers persist in the environment due to their resistance to degradation, posing risks to ecosystems and organisms through ingestion and bioaccumulation

threat to human health, wildlife, and oceans persists. Figure 12.1 illustrates common types of microplastics, while Table 12.1 details toxicity based on size. Recent studies reveal organ and tissue toxicity, causing lung damage, hepatotoxicity, immunotoxicity, and reproductive issues (Abarghouei et al. 2021; Amato-Lourenco et al. 2021; Chen et al. 2022; Cheng et al. 2022; Dubey et al. 2022; Huang et al. 2022b). MPs trigger the innate immune response via the endocytosis or pinocytosis process (Liu et al. 2021). Mitochondria become the primary site of ROS production due to MP accumulation, as it interferes with the electron transport chain and oxidative

Table 12.1 Toxicity of MPs as a function of size

Types of microplastics	Size range	Organelle/organ system affected	Medium of study (in vitro/ex vivo/in vivo)	Mechanism of toxicity	References
Polystyrene	1 μ–100 nm	Chloroplasts	Freshwater Cyanobacterium Microcystis Aeruginosa	Inhibition of photosynthesis	Wu et al. (2021)
Polystyrene	50 nm	Brain	*Caenorhabditis elegans* (*C. elegans*)	Oxidative damage and dopamine reduction	Shang et al. (2021)
Polystyrene	200 μm	Intestine	Marine medaka (*Oryzias melastigma*)	Disruption of intestinal microbiota and metabolic homeostasis	Zhang et al. (2021b)
Polystyrene	0.1, 1.0, 5.0, and 20.0 μm	Guts and gills	*L. vannamei*	Oxidative stress	Zhou et al. (2023)
Fragmented MPs	17.23 μm} 3.43 and 34.43 μm} 13.09 μm	Intestine and reproductive system	*Daphnia magna*	Interference in the digestive tract	An et al. (2021)
Polystyrene	5 μm and 20 μm	Liver, kidney, and gut	Mice	Oxidative stress	Deng et al. (2017)
Polyethylene	1–5 μm; 10–29 μm; 38–45 μm	Intestine	Sea urchin *Paracentrotus lividus*	Presence of 1–5 μm MPs in coelomic fluid	Pyl et al. (2022)
Aged PS-MPs		Reproductive system	*Caenorhabditis elegans* (*C. elegans*)	DNA damage-induced cell Apoptosis	
Polystyrene	1–12 μm	Photosynthetic apparatus	*Scenedesmus obliquus,* a unicellular phytoplankton	Oxidative stress	Natarajan et al. (2022)
Polystyrene	0.25 and 8 μm	Liver, intestine	*Carassius auratus* (goldfish)	Oxidative stress	Abarghouei et al. (2021)
Polystyrene	70 nm, 1 μm and 10 μm	Skin, tail, and head	*Eisenia fetida* earthworm	Toxic responses leading to death	Xiao et al. (2022)

High-density polyethylene (HDPE) fragments	14.12 μm, 80.32 μm, 120.97 μm	Intestine and neurodevelopment	Zebrafish larvae	Damage to hair cells and mechanosensory receptors and morphological variations in the gastrointestinal tract	Kim et al. (2022)

The table presents the toxicity of microplastics (MPs) as a function of size. Various sizes of microplastics are categorized, ranging from nano-scale (<100 nm) to macro-scale (>5 mm), with corresponding descriptions of their potential toxicity in organelle/organ systems. Toxicity assessments include the mechanism of toxicity and medium of study phosphorylation. MP exposure can disrupt the mitochondrial membrane potential, structure, function, and also division (Liu et al. 2022).

Destroying plastics in addition to microplastics is critical to deciding how the ecosystem will develop. Nonetheless, there is still much to learn about the breakdown of plastic and the creation of microplastics (Andrady 2017). Presently, numerous studies and advancements are being conducted regarding the breakdown of microplastics through the involvement of biotic elements like fungi, bacteria, and algae, as well as diverse advanced oxidation processes (AOPs) like photocatalytic degradation, photodegradation, and electrochemical degradation. These processes aid in the structural disintegration of microplastics through depolymerization into oligomers, dimers, and monomers. Enzymatic activities, temperature, and specificity are the primary causes of plastic polymer breakdown during biodegradation (Ganesh et al. 2019). Currently, one of the most researched and popular approaches for microplastic degradation is the use of AOPs (Du et al. 2021). Controlling plastic pollution is essential to preserving the sustainability of the ecosystem. The formation of microplastics may be inhibited by traditional methods such as flocculation, coagulation, sedimentation, incineration, etc. because of its persistent character (Zhang et al. 2020a). The method of smoldering trash at high temperatures to turn them into ashes is called incineration. Dioxins are among the most extremely poisonous substances; thus, it is difficult for it to break down as a result of their emission. It disrupts normal metabolism and initiates reproductive and endocrine processes once it enters the body of a person or animal through the food chain. The cytotoxic and neurotoxic properties of microplastic make it fatal (Yang et al. 2012). There is always a chance that MPs might escape if final effluents are dumped straight into local waterways, lakes, rivers, and the ocean. The chemical recycling strategy, which converts wastes into valuable products that may be used as fuel, is currently the most dependable and promising method for the recycling and use of MPs (Miao et al. 2020). The primary mechanisms of MP degradation that will be covered in this paper are mainly mechanical and biological remediation.

12.2 MP and NP Abundances from Oceans to Humans

12.2.1 Accumulation of Microplastics in Ocean Bodies

Approximately 71% of the Earth's surface is covered by marine bodies, with terrestrial sources being the primary contributors to plastic pollution in marine ecosystems (Wong et al. 2020). Plastics in oceans are categorized into mesoplastics (<2.5 cm), microplastics (<5 mm), macroplastics (<1 m), and mega-plastics (>1 m) (Barnes et al. 2009). Larger plastics break down into secondary microplastics through processes like photodegradation and wave abrasion (Zhang et al. 2021a). Improper disposal of plastics in freshwater contributes to their entry into marine ecosystems, and natural disasters further exacerbate oceanic plastic accumulation (Menendez et al. 2021). Salinity and oxygen concentration influence microplastic formation, with high salinity and maximum oxygen promoting their development (Zhang et al. 2021a).

12.2.2 Transfer of MPs/NPs into Human Samples

Microplastics (MPs) and nanoparticles (NPs) are pervasive in the marine ecosystem, posing a threat to human health through the food chain. Notably, Autrup et al. (1975) found MPs and NPs in edible fishes, Daniel et al. (2020) in shrimps, and Cho et al. (2019) in bivalves. Consumption of ocean food introduces these particles into the human body, with MPs detected in various sources like drinking water, soft drinks, and more (Kutralam-Muniasamy et al. 2023). The contamination extends human samples such as feces, meconium, placenta, lung tissue samples, sputum, bronchoalveolar lavage fluid, breast milk, saliva, blood, face, liver, kidney, colon, and facial samples (Malafaia and Barcelo 2023). This highlights the concerning presence of MPs and NPs in both marine organisms and human samples, emphasizing the need for further research and environmental management.

12.2.3 Routes of Entry for Microplastics (MPs) and Nanoplastics (NPs) into the Human Body

Earlier research suggests that MPs and NPs can enter living organisms, as well as humans, through ingestion, inhalation, or skin contact (Hua and Wang 2022; Khan and Jia 2023). Animals and humans are primarily exposed to MPs/NPs through tainted food and water, especially aquatic and terrestrial animals like birds, which often mistake plastics for food (Perkins 2015). Inhalation is another significant entry route for MPs, confirmed by deposition in lung tissues and sputum samples (Huang et al. 2022a; Jenner et al. 2022). The fate of inhaled MPs depends on lifestyle and particle properties, potentially traveling through the bloodstream to various organs (Rist et al. 2018; Yee et al. 2021). Exposure to MPs is increased by cosmetics, body cleansers, medications, and occupational settings (Hua and Wang 2022; Khan and Jia 2023; Yee et al. 2021) and usage of common products like toothpaste, sunscreens, facemasks, and washes pose additional risks (Hua and Wang 2022; Khan and Jia 2023). In summary, there are several methods for MPs and NPs to enter the human body with ingestion, inhalation, and skin contact being key routes. Thus, microplastics (MPs) pose a significant threat to human health, inducing various toxic effects with reproductive and developmental toxicity, oxidative stress, immunological response, cytotoxicity, and multiorgan damage.

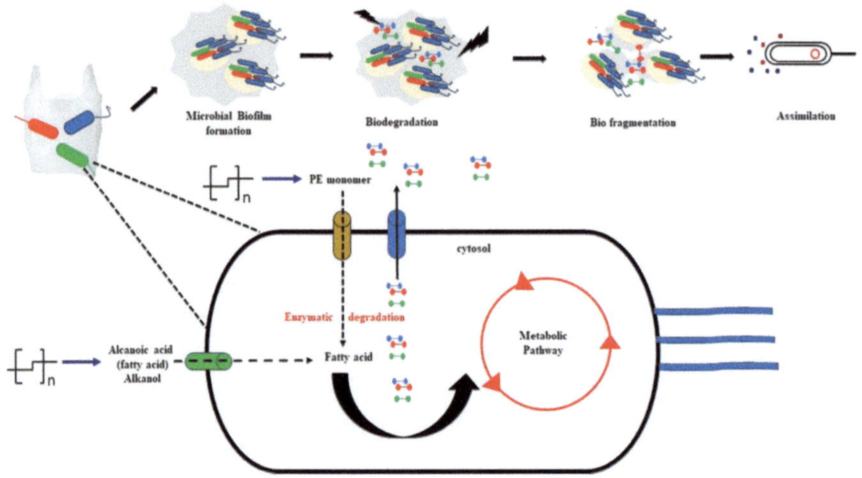

Fig. 12.2 The overall process by which microbes break down MPs. The schematic diagram illustrates the general mechanism of microbial degradation of microplastics. Microorganisms, including bacteria, fungi, and algae, colonize the outside of the microplastic particles. These microbes generate enzymes outside of cells, such as lipases, proteases, and esterases, which catalyze the breakdown of the plastic polymers into smaller fragments. The process involves several stages: *(1) Microbial biofilm formation:* attachment of microorganisms to the microplastic surface, facilitated by biofilm formation. *(2) Biodegradation:* secretion of enzymes by the microorganisms, leading to the dilapidation of polymer chains. *(3) Bio fragmentation:* breakage of the polymer chains into monomers and oligomers, which are smaller molecules. *(4) Assimilation:* the bacteria' use of the broken-down products as sources of carbon and energy for growth and metabolism. This microbial degradation process plays a crucial part in the bioremediation of microplastic pollution in aquatic and terrestrial environments

12.3 Steps and Enzymes of Microorganisms Involve in Microplastic Degradation

Microplastics can be broken down chemically, physically, or biologically. A variety of enzymes are involved in the biological breakdown process (Fig. 12.2) (Bacha et al. 2021; Padervand et al. 2020). The fundamental procedure consists of phases like the collapse of polymers into smaller particles and the further breakdown of the smaller polymers into oligomers, dimers, and monomers. Microbes assist in the subsequent phases of mineralization that come after this breakdown (Fig. 12.2).

When mineralization is fully achieved, carbon dioxide is released along with the creation of a number of intermediary molecules that serve as a source of energy to support the growth of microorganisms. There are several extracellular enzymes that contribute to the breakdown of microplastics, such as lipases, esterases, lignin peroxides, laccases, and manganese peroxides.

These enzymes make microplastics more hydrophilic and transform them into alcohol or carbonyl residues (Taniguchi et al. 2019). By accelerating the chain cleavage events, hydrolase enzymes including lipases, esterase, and cutinase operate on plastic surfaces and break down microplastics. These enzymes act on the polymer's surface, causing fractures to develop, but they are unable to permeate within the polymer. The produced monomers eventually enter several metabolic pathways after being absorbed into the cytoplasm of bacteria.

Very little is currently known about the role of intracellular enzymes in the abasement of microplastics, despite the fact that substantial research has already been done using extracellular enzymes in microplastics biodegradation. Additionally, the pathways involved in the uptake of monomers remain unclear. Breakdown enzymes for macromolecules, such as laccase, amylases, and lipase, can specifically target the polymer chains of microplastics (MPs) and convert them into their monomer, which is subsequently used as a source of carbon in the microorganism energy cycle (Fig. 12.2) (Othman et al. 2021). The capacity to disassemble polymers into monomers is known to exist in a number of microbial enzymes. Enzymes that are associated with PS breakdown include phenylacetaldehyde dehydrogenase, styrene monooxygenase, styrene-oxide isomerase, and serine hydrolase. The last monomer in the tricarboxylic acid (TCA) cycle is acetyl-CoA (Amobonye et al. 2021). Microorganisms' exo-enzymes break down complicated polymers into simpler molecules like dimers and monomers. These molecules are tiny enough to pass through the semipermeable outer membranes of microorganisms and be absorbed by the organisms in the ecosystem as a source of energy and carbon (Fig. 12.2) (Sheel and Pant 2018).

Records indicate that the breakdown of MPs is microbially specific. Regarding this, it was exposed that *Rhodococcus* sp., *Sphingobacterium* sp., *Vibrio* sp., *Xanthomonas* sp., and *Pseudomonas* sp. are strong for the degradation of PS and PP, while *Bacillus cereus*, *Phanerochaete chrysosporium*, and *Trarnetes versicolor* were accountable for breaking down the PE through the action of laccase and manganese peroxidase enzymes (Johnston et al. 2022).

12.4 Mechanical Remediation of Microplastics

The primary goal of microplastic treatment plans is to get the particles out of aquatic environments, where they frequently wind up. Conventional and creative procedures are the two primary strategies for removing microplastics. Adsorption, fast sand filtration, membrane bioreactor, and coagulation are examples of conventional tactics. Innovative techniques for getting rid of microplastics include magnetic separation, electrocoagulation, photocatalytic degradation, and electrochemical oxidation. Every one of these methods has advantages and disadvantages, and the efficacy of eliminating microplastics depends on a number of variables, size, and concentration including pH, water flow velocity, etc.

12.4.1 Coagulation

Coagulation is a commonly employed technique for wastewater treatment. It makes use of a variety of chemical substances, or coagulants, to cause the suspended and dissolved particles to become unstable so that sedimentation can remove them (Shirasaki et al. 2016). There are changes in the removal routes for microplastics among coagulants, for example, those based on iron and aluminum. Nevertheless, conventional techniques for eliminating microplastics, such as adsorption, sweep flocculation, and charge neutralization, are still applicable in explaining how they are removed (Zhou et al. 2021). Despite the fact that coagulation is one of the most used techniques for wastewater treatment, it has several working problems, including producing a large quantity of sludge that further contaminates the environment (Padmaja et al. 2020). This is dangerous since there's a chance the sludge produced by coagulation contains more poisonous substances than the original pollutants, which might lead to requiring expensive follow-up care and removal. Moreover, adding chemicals to enhance coagulation efficiency may raise the total expense of the removal procedure (Bahrodin et al. 2021). One of coagulation's main limitations is the difficulty of treating many contaminants at once. Because different Coagulants need to be added and to effectively treat various types of contaminants, substantial tuning of the reaction parameters is required, the variety in wastewater composition also adds to process costs (Natarajan et al. 2018). These elements may cause the process's overall operating costs to rise to an unaffordable level.

12.4.2 Membrane Bioreactor

Applying additional microbes and nitrifying bacteria, the technique of membrane bioreactors is a trustworthy method for treating wastewater from industry and cities, which often encompasses varying quantities of different toxins (Dvořák et al. 2013). This method was really employed recently by a wastewater treatment plant to remove microplastics. High effluent quality and strong removal efficiency with a high rejection potency toward target contaminants are the obvious benefits of employing membrane technology. Nonetheless, several problems still restrict the efficacy of its elimination, such as membrane fouling, aeration restrictions, and the requirement to feed microorganisms nutrient-rich materials (Al-Asheh and Aidan 2021). These problems might raise operating costs.

12.4.3 Sand Filtration

Sand filtering works quickly to eradicate various contaminants, including viruses and suspended clay particle solids (Nakazawa et al. 2021). Sand filtration technique as a feasible means of eliminating microplastics from wastewater. Because of its high flow rate, poor susceptibility to markers of water quality, and modest land area need, it has been noted that quick sand filtering is a workable method for removing microplastics. However, in the lack of costly flocculating agents, this method's effectiveness is limited, and constant maintenance is required, increasing the overall cost of the filtering procedure (Enyoh et al. 2022).

12.4.4 Adsorption

It has been shown that the adsorption strategy is more effective than other methods for removing microplastics from wastewater. Adsorbents that may be used include chitin and graphene oxide.

Moreover, some materials, including stacked double hydroxides, shown notable adsorption effectiveness, reaching as high as 100% for MPs as well as nanoplastics (Tiwari et al. 2020). The nonselective characteristics of the adsorption pathway, however, restrict the method's overall efficacy. Thus, improving the adsorbent materials' selectivity for microplastics should be the main attention of future investigation efforts in order to improve removal efficiency.

12.4.5 Photocatalytic Degradation

Photodegradation has been acknowledged as a very efficacious and promising technique for the remediation of hazardous organic contaminants most importantly microplastics, in wastewater. During this process, a semiconductor material absorbs visible or ultraviolet light, releasing free radicals that degrade the microplastics. Reactive oxygen species like superoxide and singlet oxygen are examples of these free radicals. The light energy, that is, greater than the photocatalytic semiconductor material's bandgap energy, is absorbed. By forcing an electron to go from the band known as the valence band to the conduction band, it creates positive holes in the valence band. In the end, this process produces hydroxyl with superoxide radicals, which reduce the microplastics. Recently, green synthesis produced iron-zinc oxide nanocomposite has become more and more popular as a semiconductor material (Lam et al. 2021) used for the breakdown of polyethylene using photocatalysis. Despite its efficacy, the photocatalytic approach necessitates vigilant monitoring to avoid any negative impacts on aquatic ecosystems and proper discarding of the leftover sludge formed (Lam et al. 2021).

12.4.6 Electrochemical Oxidation

The two techniques of electrochemical oxidation—indirect cathode oxidation and anodic oxidation—are an endurable and economical way to clean wastewater (Du et al. 2021). Without the need of additional chemical agents, it has been proven that this method may efficiently break down a variety of organic pollutants, such as microplastics, dyes, antibiotics, and antipyretics, into straightforward and nontoxic byproducts like CO_2 and water vapor (Du et al. 2021). Additionally, strong oxidants produced by electrochemical oxidation, such as ozone, H_2O_2, and hydroxyl radicals, effectively disintegration organic contaminants without causing of sludge waste to develop. Several parameters affect the efficacy of electrochemical oxidation: the anode's superficial area and composition, the strength of the current, the kind and quantity of electrolyte utilized, and the degradation process (Kiendrebeogo et al. 2021). Consequently, scientists are now paying close attention to this therapy method.

12.4.7 Electrocoagulation

Combining the benefits of electrochemistry with coagulation, The removal of microplastics from wastewater may be achieved economically, sustainably, and with remarkable effectiveness by the electrocoagulation process. When metallic electrodes are subjected to an electric current, the cations they create form flocs through the process of electrocoagulation. Consequently, Coagulation causes "micro-coagulants" to form and the stability of suspended particles to be lost (Shen et al. 2020). Because electrocoagulation eliminates the prerequisite for chemical coagulant chemicals, it is therefore more effective than traditional coagulation and saves both operating time and expense. Additionally, electrocoagulation provides water with a reduced total dissolved solids content, minimizes the quantity of sludge discarded, and may be effectively used with a variety of wastewater quality. Researchers are becoming interested in this economic and ecological strategy as a substitute for traditional coagulation techniques.

12.4.8 Magnetic Separation

Magnetic separation has proven to be an efficient method for removing microplastics from wastewater due to its long-lasting magnetic effect and high material removal capability (Zhang et al. 2021b). Recent work by Grbic et al. (2019) has used this technology to eliminate microplastics from samples of silt, freshwater, and marine water. This removal technique makes use of a variety of materials termed magnetic seeds, such as iron nanoparticles and magnetic carbon nanotubes. Three processes can regulate magnetic separation: complexation, hydrogen bond formation, and electrostatic interaction (Tang et al. 2021). Microplastics'selectivity and removal effectiveness are, nonetheless, negatively impacted by the existence of additional pollutants, and the size and shape of the particles also affect the dispersion process.

12.4.9 Photodegradation

UV radiation can interrupt the molecular assembly of plastic polymers, oxidize polymers, and produce various byproducts such as gasses through cross-linking processes. The way that microplastics are destroyed by UV radiation varies based on their wavelength and the kind of chemical links that the plastic polymers contain. Since plastics engage their electrons to reach increased reactivity, oxidative decomposition—a prevalent method of MP deterioration in the natural environment—occurs when they are exposed to light radiation. It has been discovered that when MPs were exposed to xenon light, their surface area and hydroxyl content rose, compromising their structural integrity (Luo et al. 2020). Microplastics exposed to UV radiation have been reported to undergo notable morphological and mechanical characteristic changes. Scanning electron microscopy (SEM) was used to validate it, demonstrating the loss of plastic characteristics (Ainali et al. 2021).

12.5 Biological Remediation of Microplastics

It is the process via which biotic elements like bacteria, fungi, and algae intervene to cause the mineralization of plastics. It includes the growth of biofilms on the exterior of plastics, which leads to the depolymerization of plastics by certain enzymes, destroying the polymers' structural integrity. In addition to carbon dioxide, water, and byproducts of depolymerization, the products may contain oligomers, dimers, and monomers. It is possible for biotransformation to take place, which would lead to the making of environmentally hazardous persevering organic pollutants (POPs). The final products of biodegraded polymers include carbon dioxide and water in aerobic settings, and carbon dioxide and methane in anaerobic situations (Giacomucci et al. 2020). MPs are consumed by a number of marine creatures, including zooplankton, top predators, and decomposers despite the fact that it hurts them. Since the majority of studies are carried out ex situ, it is still uncertain how dangerous they would be in the wild; therefore, more research is required to properly validate the findings. The oxygenic photoautotrophic organisms known as algae use plastics as a carbon source to break down the polymers. Years of testing have shown that microalgae, in particular, offer promising potential as microorganisms for MP bioremediation (Roccuzzo et al. 2021). With extremely little labor and based on current research, only a few numbers of algae species strains are known to be capable of degrading plastic particles and microplastics. To break down the waste plastic, algae create extracellular polysaccharides and lignin on their surface. Certain algal sources, such as *Scenedesmus dimorphus* and *Anabena spinosa*, have been seen to colonize plastic surfaces and break down the polymer. Using diatoms, *Navicula pupula*, the same effects were discovered. According to a study, polyethylene terephthalate (PET) degrades more quickly than polyethylene (PE), which suggests that the type of microplastic polymer and when examining deterioration, the possible candidate species might be taken into account. It is possible to get engineered PETase from algae by employing the biotechnological method (Zurier and Goddard 2021). It was found that *Bacillus* sp. and *Paenibacillus* sp. successfully decreased the size of polypropylene (PP) during a 60-day period. Similar to this, under laboratory conditions, *Bacillus cereus* and *Bacillus gottheilii* decreased the size of several MPs components, including polyethylene (PE), polyethylene terephthalate (PET), polypropylene (PP), and polystyrene (PS), in less than 40 days. This indicates a decrease in MPs weight and quantity. Additionally important to the breakdown of plastics was the fungus *Aspergillus flavus* (Zhang et al. 2020b). Because it is economical and environmentally benign, biodegradation has become more popular globally. There are a few disadvantages, though, like the need for a lot of time, the challenge of implementing it on location, species-specific gaps, etc. Therefore, in order to minimize microplastic pollution without endangering the ecosystem, these concerns need to be considered for the degradation of MPs in a sustainable manner.

12.5.1 Role of Algae

The biological breakdown of polymeric material can be facilitated by the employment of microalgae and their toxins and enzymes (Manzi et al. 2022). The primary benefit is that, unlike the bacterial system, they do not require a rich carbon supply for development and can thrive in a range of environments as well as those that are home to the majority of microplastics. Microalgae have been seen to settle on plastic surfaces in wastewater streams. This adherence triggers the development of enzymes called exopolysaccharide and ligninolytic, which break down plastic. These polymers primarily act as a supply of carbon, boost the number of proteins and carbohydrates in cells, and quicken development. Only recently has the disintegration or surface degeneration of low-density polyethylene sheets due to algal colonization been identified using scanning electron microscopy (Sanniyas et al. 2021).

Algal biodegradation mostly happens through a variety of mechanisms, including fouling, hydrolysis, penetration, and corrosion (Chia et al. 2020). Without the need for pretreatment or prooxidative chemicals, it was also discovered that *Phormidium lucidum* and *Oscillatoria subbrevis* could colonize and breakdown the surface of low-density polyethylene. *Chlorella fusca var. vacuolate*, *Chlamydomonas mexicana*, *Stephanodiscus hantzschii*, and *Chlorella vulgaris* were among the bacteria and algae that broke down bisphenol A, an additive with estrogenic activity that is frequently found in polymers (Ji et al. 2014).

Most of the time, the growth of biofilms on the outer surface of polymers is linked to the breakdown into microplastics. The capability to form biofilms on the microplastic polymers was also demonstrated by some of cyanobacterial strains, including those belonging to the genus *Microcystis, Rivularia, Pleurocapsa, Synechococcus, Prochlorothrix, Leptolyngbya Calothrix*, and *Scytonema*. Diatoms are found in the biofilms that support photosynthesis in addition to cyanobacterial species (Amaral-Zettler et al. 2020). Advances in biotechnology have allowed for the creation of many genetically modified microalgal cell factories that are capable of generating and secreting the enzymes required for the breakdown of plastic. Polyethylene terephthalate hydrolase, which can degrade terephthalic acid and polyethylene terephthalate films, was produced as a consequence of genetically modifying the green microalgae *Chlamydomonas reinhardtii* (Kim et al. 2020). Similar alteration was also successfully performed in *P. tricornutum*, which demonstrated catalytic activity against polyethylene terephthalate and the copolymer polyethylene terephthalate glycol, as well as the production of polyethylene terephthalate hydrolase.

In conclusion, microalgae may be helpful microplastic degraders as they are simple to grow and may use plastic monomers as a carbon source by producing degrading enzymes. Using synthetic biology to genetically modify algal strains to increase their capacity for degradation has produced a viable and eco-friendly method for the biological breakdown of polyethylene terephthalate using microalgae.

12.5.2 Role of Fungi

The majority of the broad variety of organisms that make up fungi are saprotrophs, sometimes known as obligatory or opportunistic parasites. Because of their exceptional adaptability, they may flourish in various aquatic and terrestrial environments with a wide array of climatic conditions. They produce a wide variety of extracellular enzymes and natural biosurfactants, such as hydrophobins, that can break down complex polymers into simpler monomers and act as carbon and electrons for microorganisms. This ability to endure various toxic chemicals and metals also helps in the breakdown and mineralization of complex pollutants. *Zalerion maritimum*, *Aspergillus niger*, *Cladosporium*, and *Penicillium simplicissimum* are the main genus linked to the degradation of various polymer types, including polyethylene, polypropylene, and polyethylene terephthalate (Paço et al. 2019). These bacteria use microplastics as their only carbon source after being broken down by extracellular enzymes. They lessen their hydrophobicity and encourage the formation of various chemical linkages with ester, carboxyl, and carbonyl functional groups. Fungal strains including *Aspergillus fumigatus*, *Aspergillus tubingensis*, *Cladosporium pseudocladosporioides*, *Fusarium solani*, *Penicillium chrysogenum*, and strains of *Pestalotiopsis microspora* displayed comparable degradation of polyurethane (Magnin et al. 2020).

Serine hydrolase is usually a key player in the breakdown of polyurethane. Two fungus strains, *Aspergillus tubingensis* VRKPT1 and *Aspergillus flavus* VRKPT2, were found to degrade high-density polyethylene from maritime coastal locations at rates of 6.02 ± 0.2 and $8.51 \pm 0.1\%$, respectively. In recent work, Kunlere et al. (2019) demonstrated that *Aspergillus flavus* and *Mucor circinelloides*, which were isolated from municipal garbage, degraded low-density polyethylene.

Aspergillus niger is known to degrade microplastics like polyethylene more hurriedly when they are pretreated with chemicals like sodium hydroxide and nitric acid. Degradation in low-density polyethylene mediated by *Aspergillus niger* and *Penicillium pinophilum* needed physical pretreatment procedures, such as thermo-oxidization at 80 °C for 15 days. After incubation over 30 months, the degradation showed up as 0.57 and 0.37%. *Aspergillus* and *Lysinibacillus* species demonstrated 29.5 and 15.8%, respectively, of the biodegradation of UV-irradiated and non-UV-irradiated polymer films, respectively (Esmaeili et al. 2013).

Fungi produce a variety of internal and extracellular enzymes that may catalyze a broad range of reactions and degrade petroleum-based polymers. Fungal adaptability and detoxification processes are significantly influenced by intracellular enzymes (Schwartz et al. 2018). The cytochrome P450 family epoxidases and transferases are linked to conjugation and oxidation processes and aid in the metabolism of aromatic, aliphatic, and alicyclic compounds. The cytochrome P450 family of enzymes uses cofactors such as heme, NADPH + H+, and FAD to aid in the development of the spore wall and maintenance of the integrity of the hyphal wall.

On the other hand, hydrolases are among the extracellular enzymes that degrade complicated polymers and increase the solubility of contaminants, hence reducing bioaccumulation. Class II peroxidases, which oxidize a variety of substrates, include laccases, dye-decolorizing peroxidases, manganese peroxidase, and lignin peroxidase. These enzymes can be effective agents for cleaning the environment. According to Straub et al. (2017), lignin-degrading fungi create laccase, which catalyzes the oxidation of both aromatic and nonaromatic substrates, including polymethylmethacrylate and polyhydroxybutyrate, which are nonphenolic or chlorophenolic substances. These enzymes' capacity to withstand high temperatures may encourage the use of them in large-scale reactors for high-kinetic processes involving the breakdown of polypropylene.

Overall, a large range of fungal strains may break down plastics into more ecologically friendly substances because they produce a large number of extracellular and intracellular enzymes, such as hydrolases and oxidases, as well as naturally occurring biosurfactants like hydrophobins.

12.5.3 Role of Bacteria

The decomposition of microplastics by microbes has been the subject of several studies. Bacteria that break down microplastics have been found in a range of environments, such as contaminated sediments, wastewater, sludge, compost, municipal landfills, and harsh climates like Antarctic soils, mangroves, and marine sediments (Awasthi et al. 2020). Furthermore, microorganisms that break down microplastic have also been designated from the earthworms'stomach microbiota. It is extensively known that microorganisms that reside in contaminated areas frequently learn how to activate the enzyme machinery that breaks down microplastics.

For the breakdown of microplastics, bacterial consortiums as well as pure cultures can be employed. Pure cultures, on the other hand, have a number of benefits in the degradation process and provide a practical means of researching the metabolic pathways involved. Furthermore, it is easier to evaluate the effects of external conditions on the degradation process, such as temperature, pH, surfactants, and substrate properties. But the primary drawback is a rather sluggish pace of deterioration. So as to speed up the degradation process, To improve the deteriorating bacterial isolates and optimize the environment, more inventive methods are required. Since it has been established that biodegradation by a single bacterium frequently leads in the creation of hazardous end products, which may be efficiently removed in a stable microbial community (Singh and Wahid 2015), the use of a consortium of bacteria is typically chosen.

The primary degradation process is physicochemical degradation, which shortens the polymer chain and modifies the functional groups of microplastics to increase their susceptibility to the action of microbial enzymes. Lipases, esterases, laccases, amidases, cutinases, hydrolases, and carboxylesterases are among the enzymes that are involved in biodegradation employing enzymes (Inderthal et al. 2021). Therefore, a complete understanding of the metabolic pathways and associated enzymes is necessary to carry out an efficient biodegradation process.

To encourage the biodegradation of plastic, physiochemical pretreatment is advised. This includes thermooxidation, chemical oxidizing agents, and UV radiation of microplastics. To get better the biodegradability of polypropylene, these pretreatments include UV irradiation, nitric acid treatment, and blending with polymers such as starch derivatives, cellulosic esters, polyhydroxybutyrate, poly(3-hydroxybutyrate-co-3-hydroxyvalerate), and polycaprolactone. Research has demonstrated that prooxidative and biodegradable compounds such as starch can improve the biodegradability of low-density, high-density, polyvinyl alcohol, and polystyrene by stimulating amylase activity (Zadjelovic et al. 2020).

Previous studied carried out the first investigation on microplastics biodegrading microorganisms, employing a combination of *Pseudomonas chlororaphis*, *Pseudomonas stautzeri*, and *Vibrio* sp. to break down polypropylene. It was also discovered in the same study that adding starch increased the biodegradation capacity. Subsequently, with a consortium of *Bacillus subtilis*, *B. flexus*, *Pseudomonas stutzeri*, and *Rhodococcus rhodochrous*, respectively, Arkatkar et al. and Fontanella et al. observed biodegradation of polypropylene. According to Kowalczyk et al. (2016), these microbial isolates were discovered to form a biofilm by the isolation of *Achromobacter xylosoxidans*.

Two bacterial strains from the genera *Bacillus* and *Rhodococcus* that were apart from the mangroves sediments in a study by Auta et al. (2018) demonstrated, after 40 days of incubation, polypropylene degradation efficiencies of 4.0 and 6.4%, respectively. Additionally, they stated that microplastics might be broken down by *Bacillus* cereus and *Bacillus gottheilii*. The weight loss percentages caused by *B. gottheilii*-induced microplastics were 6.2%, 3.0%, 3.6%, and 5.8% for polyethylene, polyethylene terephthalate, polypropylene, and polystyrene, in that order.

A few additional bacteria linked to the breakdown of polypropylene were *Bacillus, Pseudomonas, Chelatococcus,* and *Lysinibacillus fusiformis.* These bacteria were identified from a variety of settings, including the habitat of mangroves, compost, cow dung, and soil that had been contaminated by plastic waste. Additionally, it has been noted that the gut microbiota of a number of arthropods, including wax moths (*Galleria mellonella*) (Kong et al. 2019), mealworms (*Tenebrio molitor*), and Indian meal moths (*Plodia interpunctella*), carry microorganisms with the ability to biodegrade microplastics. In research published in 2014, Yang and colleagues discovered two bacteria—*Enterococcus* sp. YT1 and *Bacillus* sp. *YP1*—from the intestines of waxworms that could break down polyethylene by reducing its hydrophobicity and causing damage to polypropylene's surface. A follow-up investigation identified a bacterial strain called *Exiguobacterium* sp. that could break down polystyrene and produce biofilm in the mealworms' intestines.

Strains such as *Microbacterium paraoxydans* and *Pseudomonas aeruginosa* were used to efficiently biodegrade low-density polyethylene; after 2 months of incubation, the amount of degradation was around 61.0 and 50.5%, respectively. Comparably, according to reports, *Pseudomonas* sp. AKS2 biofilm may break down low-density polyethylene up to 5 ± 1% during the course of 45 days of incubation without the need for any kinds of pretreatment (Tribedi and Sil 2013).

The dry weight of microplastics was lowered by 14.7% in 60 days by a bacterial consortium made up of Bacillus and Paenibacillus species. Furthermore, it was studied how low-density polyethylene is degraded by *Lumbicus terrestris*, an earthworm species. The gut isolates included members of the genera Firmicutes and Actinobacteria, which were also inspected individualistically and shown to be accomplished of breaking down low-density polyethylene microplastics and releasing volatile substances such as tricosane, docosane, and eicosane. One hundred twenty days of weight loss was improved by a combination of *Enterobacter* and *Pseudomonas* from cow dung by up to 15% (Skariyachan et al. 2021).

Alcanivorax borkumensis is among the marine hydrocarbonoclastic bacteria that has been revealed to be able to break down branched aliphatic compounds, alkanes, isoprenoid hydrocarbons, and alkyl cycloalkanes (Davoodi et al. 2020). The study was conducted on the same isolate that had previously demonstrated the production of biofilms on low-density polyethylene films when pyruvate, hexadecane, yeast extract, and these substances were present.

Additionally, it was mentioned that the presence of alkanes alters the hydrophilicity of cell membranes, creates biosurfactants that interact with plastic surfaces, and causes C=O and COOH/OH functional groups to develop. The biodegradation of polyethylene was also aided by a number of actinomycetes, such as *Streptomyces* and *Rhodococcus ruber*.

According to Matjašič et al. (2021), among the many bacterial species linked to the decomposition of microplastics, 21% belonged to the *Pseudomonas* genus, about 15% to *Bacillus*, and 17% to combinations of these two genera. Additional bacteria linked to the biodegradation of microplastics included *Exiguobacterium* sp., *Ideonella sakaiensis*, *Pseudomonas chlororaphis*, *Pseudomonas putida AJ*, and *Thermomonospora fusca*. *Streptomyces badius*, *Rhodococcus ruber*, *Comamonasacidovorans*, and *Clostridium thermocellum* were also identified. Table 12.2 provides a comprehensive list of the many bacteria and actinomycetes linked to the breakdown of microplastics.

Table 12.2 Microplastic degradation by fungi, bacteria, and actinomycetes

Strains	Type of microplastic degraded	References
Microplastic degradation by fungi		
Aspergillus sp. *Penicillium* sp.	Polypropylene/ butylene-adipate-co-terephthalate	De Oliveira et al. (2020)
Zalerion maritimum	Polyethylene pellets	Paço et al. (2017)
Bjerkandera adusta	Polypropylene and biomass	Butnaru et al. (2016)
Aspergillus flavus VRKPT2	High-density polyethylene	Devi et al. (2015)
Aspergillus sp. *Paecilomyces Lilacinus, Lasiodiplodia theobromae*	Polypropylene	Sheik et al. (2015)
Aspergillus niger, Aureobasidium pullulans, Paecilomyces varioti, Aspergillus terreus, Trichoderma viride, Penicillium funiculosum, Scopulariopsis brevicaulis, Penicillium ochrochloron	Low-density polyethylene	Nowak et al. (2012)
Fusarium solani, Spicaria spp., *Alternaria solani*, and *Aspergillus flavus*	Polyester polyurethane	Ibrahim et al. (2011)
Penicillium pinophilum ATCC 11,797	Low-density polyethylene powder	Volke-Sepúlveda et al. (2002)
Chaetomium globosum, Aspergillus niger, Pullularia pullulans, Penicillium funiculosum, and Gliocladium virens	Low-density polyethylene	Chandra and Rustgi (1997)
Microplastic degradation by bacteria		
Lysinibacillus sp.	Polypropylene, polyethylene	Jeon et al. (2021)
Enterobacter sp. nov. bt DSCE01, *Enterobacter cloacae* nov. bt DSCE02, and *Pseudomonas aeruginosa* nov. bt DSCE-CD03	Low-density polyethylene, polypropylene	Skariyachan et al. (2021)
Bacillus licheniformis, Bacillus thuringiensis, and Bacillus cereus	Polypropylene and poly-L-lactide	Jain et al. (2022)
Bacillus sp. and *Paenibacillus* sp.	Polyethylene	Park and Kim (2019)
Biofilm composed by Pirellulaceae, Phycisphaerales, Cyclobacteriaceae, and Roseococcus	Polyethylene, polypropylene	Miao et al. (2019)
Bacillus simplex and *Bacillus* sp.	Low-density polyethylene	Huerta Lwanga et al. (2018)
Bacillus sp. strain 27	Polypropylene	Auta et al. (2018)

(continued)

Table 12.2 (continued)

Strains	Type of microplastic degraded	References
Bacillus gottheilii	Polyethylene, polyethylene terephthalate, polypropylene, and polystyrene	Auta et al. (2017)
Bacillus cereus	Polyethylene, polyethylene terephthalate, polystyrene	
Bacillus thuringiensis	Polypropylene and poly-L-lactide	Jain et al. (2018)
Bacillus licheniformis	Polypropylene and poly-L-lactide	Jain et al. (2018)
Microbial consortia (including *Aneurinibacillus* sp. and *Brevibacillus* sp.)	Low-density polyethylene, high-density polyethylene and polypropylene	Skariyachan et al. (2018)
Bacillus cereus	Polypropylene	Helen et al. (2017)
Sporosarcina globispora	Polypropylene	Helen et al. (2017)
Agios Onoufrios and Kalathas	Polystyrene films	Syranidou et al. (2017)
Pseudomonas aeruginosa and *Escherichia coli*	Polyurethanes	Uscátegui et al. (2016)
Stenotrophomonas panacihumi PA3-2	Polypropylene	Jeon and Kim (2016)
Nitrosomonas sp., *Nitrobacter* sp., *Burkholderia* sp., and *Pseudomonas* sp.	High-density polyethylene, low-density polyethylene, polypropylene	Muenmee et al. (2016)
Exiguobacterium sp. strain YT2	Polystyrene	Yang et al. (2015)
B. flexus + *P. azotoformans*	UV treated polymers	Aravinthan et al. (2016)
Bacillus sp. YP_1	Polyethylene films	Yang et al. (2014)
Enterobacter asburiae YT_1		
Chelatococcus sp. E1	Low-molecular-weight polyethylene	Jeon and Kim (2013)
Microplastic degradation by Actinomycetes		
Rhodococcus sp. ADL36 and *Pseudomonas* sp. ADL15	Polypropylene	Habib et al. (2020)
Rhodococcus ruber strain C208	Polyethylene	Sivan et al. (2006)

(continued)

Table 12.2 (continued)

Strains	Type of microplastic degraded	References
Bacterial consortia, Arthrobacter viscosus, Bacillus thuringiensis, Micrococcus lylae, Bacillus pumilus, Micrococcus luteus, Bacillus mycoides, Bacillus cereus	Low-density polyethylene	Nowak et al. (2011)
Microbacterium paraoxydans	Polyethylene (pre-treated with nitric acid)	Rajandas et al. (2012)
Rhodococcus	Polypropylene	Auta et al. (2018)
Actinomadura sp. T16-1 (Enzyme production)	Polylactic acid (production of polylactic acid-degrading enzyme)	Sukkhum et al. (2009)
Rhodococcus ruber	Polystyrene	Mor and Sivan (2008)
Rhodococcus rhodochrous ATCC 29,672	Two polypropylene films (statistical copolymer and block copolymer)	Fontanella et al. (2013)

The table provides an overview of microplastic degradation by fungi, bacteria, and actinomycetes. Each microbial group is listed along with their types of microplastic degradation process

In conclusion, microbiota, polluted sediments, wastewater, sludge, compost, municipal landfills, and harsh circumstances have all been exposed to include bacteria that are capable of break down microplastics. Testing for the breakdown of microplastics by bacteria has been done using both microbial consortiums and pure cultures. Particularly, bacterial consortiums exhibit increased effectiveness and community stability.

12.6 Conclusion

With their detrimental effects on the dynamics of nutrients, microbiota, and living things, microplastics constitute a danger to the integrity of our ecosystems and are transported through both biotic and abiotic channels. The main factors contributing to microplastics‘detrimental effects on their surroundings are their concentration, composition, and form. Furthermore, because additives are quickly released into the environment and do not form a chemical link with microplastics` molecular structure, their leaching from microplastics has a greater detrimental impact than microplastics themselves. Because they host and convey certain microorganisms, microplastics are likely to upset the microbiota’s delicate equilibrium. These bacteria can grow into biofilms on microplastic surfaces, where they become the dominant species in the damaged microbiota by using microplastics as a source of carbon. Microbial biodegradation is regarded as a successful strategy for addressing the worldwide problems in the management of microplastics.

Numerous obstacles and restrictions exist when using bacteria to biodegrade microplastics; however, they can be addressed by various genetic modifications. Still, there is very little research on the effectiveness of genetically engineered bacteria in field settings, and the majority have only been tested in laboratory environments. Furthermore, there's a dearth of information on many metabolic pathways and enzymes. Recent developments in culture-independent techniques and metagenomic analysis of uncultivated microbial communities sampled from contaminated sites can help develop novel bioremediation processes. They can also pave the way for the discovery of novel enzymes and metabolic pathways.

References

Abarghouei S, Hedayati A, Raeisi M, Hadavand BS, Rezaei H, Abed-Elmdoust A (2021) Size-dependent effects of microplastic on uptake, immune system, related gene expression and histopathology of goldfish (Carassius auratus). Chemosphere 276:129977. https://doi.org/10.1016/j.chemosphere.2021.129977

Ainali NM, Bikiaris DN, Markopoulos DA (2021) Aging effects on low- and high-density polyethylene, polypropylene and polystyrene under UV irradiation. J Anal Appl Pyrolysis 158:105207

Al-Asheh BM, Aidan A (2021) Membrane bioreactor for wastewater treatment: a review. Case Stud Chem Environ Eng 4:100109. https://doi.org/10.1016/j.cscee.2021.100109

Amaral-Zettler LA, Zettler ER, Mincer TJ (2020) Ecology of the plastisphere. Nat Rev Microbiol 18(3):139–151. https://doi.org/10.1038/s41579-019-0308-0

Amato-Lourenco LF, Carvalho-Oliveira R, Junior GR, Dos Santos GL, Ando RA, Mauad T (2021) Presence of airborne microplastics in human lung tissue. J Hazard Mater 416:126124. https://doi.org/10.1016/j.jhazmat.2021.126124

Amobonye A, Bhagwat Singh P, Pillai S (2021) Plastic biodegradation: frontline microbes and their enzymes. Sci Total Environ 759(2021):143536. https://doi.org/10.1016/j.scitotenv.2020.143536

An D, Na J, Song J, Jung J (2021) Size-dependent chronic toxicity of fragmented polyethylene microplastics to Daphnia magna. Chemosphere 271:129591. https://doi.org/10.1016/j.chemosphere.2021.129591

Andrady AL (2017) The plastic in microplastics: a review. Mar Pollut Bull 119(1):12e22

Aravinthan A, Arkatkar A, Juwarkar AA, Doble M (2016) Synergistic growth of Bacillus and Pseudomonas and its degradation potential on pretreated polypropylene. Prep Biochem Biotechnol 46(2):109–115. https://doi.org/10.1080/10826068.2014.985836

Auta HS, Emenike CU, Fauziah SH (2017) Screening of Bacillus strains isolated from mangrove ecosystems in Peninsular Malaysia for microplastic degradation. Environ Pollut 231:1552–1559. https://doi.org/10.1016/j.envpol.2017.09.043

Auta HS, Emenike CU, Jayanthi B, Fauziah SH (2018) Growth kinetics and biodeterioration of polypropylene microplastics by Bacillus sp. and Rhodococcus sp. isolated from mangrove sediment. Mar Pollut Bull 127:15–21. https://doi.org/10.1016/j.marpolbul.2017.11.036

Autrup H, Thurlow BJ, Warwick GP (1975) The effects of the continuous administration of N,N-dimethyl-4-phenylazoaniline (DAB) on the activities and the inducibilities of some drug-metabolizing enzymes in rat liver. Chem Biol Interact 11:313–328. https://doi.org/10.1016/0009-2797(75)90001-0

Awasthi AK, Tan Q, Li J (2020) Biotechnological potential for microplastic waste. Trends Biotechnol 38(11):1196–1199. https://doi.org/10.1016/j.tibtech.2020.03.002

Bacha AUR, Nabi I, Zhang L (2021) Mechanisms and the engineering approaches for the degradation of microplastics. ACS ES&T Eng 1(11):1481–1501. https://doi.org/10.1021/acsestengg.1c00216

Bahrodin MB, Zaidi NS, Hussein N, Sillanpää M, Prasetyo DD, Syafiuddin A (2021) Recent advances on coagulation- based treatment of wastewater: transition from chemical to natural coagulant. Curr Pollut Rep 7:379–391. https://doi.org/10.1007/s40726-021-00191-7

Barnes DKA, Galgani F, Thompson RC, Barlaz M (2009) Accumulation and fragmentation of plastic debris in global environments. Philos Trans R Soc Lond Ser B Biol Sci 364:1985–1998. https://doi.org/10.1098/rstb.2008.0205

Butnaru E, Darie-Niţă RN, Zaharescu T, Balaeş T, Tănase C, Hitruc G, Vasile C (2016) Gamma irradiation assisted fungal degradation of the polypropylene/biomass composites. Radiat Phys Chem 125:134–144. https://doi.org/10.1016/j.radphyschem.2016.04.003

Chandra R, Rustgi R (1997) Biodegradation of maleated linear low-density polyethylene and starch blends. Polym Degrad Stab 56(2):185–202. https://doi.org/10.1016/S0141-3910(96)00212-1

Chen X, Zhuang J, Chen Q, Xu L, Yue X, Qiao D (2022) Chronic exposure to polyvinyl chloride microplastics induces liver injury and gut microbiota dysbiosis based on the integration of liver transcriptome profiles and full-length 16S rRNA sequencing data. Sci Total Environ 839:155984. https://doi.org/10.1016/j.scitotenv.2022.155984

Cheng W, Li X, Zhou Y, Yu H, Xie Y, Guo H, Wang H, Li Y, Feng Y, Wang Y (2022) Polystyrene microplastics induce hepatotoxicity and disrupt lipid metabolism in the liver organoids. Sci Total Environ 806:150328. https://doi.org/10.1016/j.scitotenv.2021.150328

Chia WY, Tang DYY, Khoo KS, Lup ANK, Chew KW (2020) Nature's fght against plastic pollution: algae for plastic biodegradation and bioplastics production. Environ Sci Ecotechnol 4:100065. https://doi.org/10.1016/j.ese.2020.100065

Cho Y, Shim WJ, Jang M, Han GM, Hong SH (2019) Abundance and characteristics of microplastics in market bivalves from South Korea. Environ Pollut 245:1107–1116. https://doi.org/10.1016/j.envpol.2018.11.091

Daniel DB, Ashraf PM, Thomas SN (2020) Abundance, characteristics and seasonal variation of microplastics in Indian white shrimps (Fenneropenaeus indicus) from coastal waters off Cochin, Kerala, India. Sci Total Environ 737:139839. https://doi.org/10.1016/j.scitotenv.2020.139839

Davoodi SM, Miri S, Taheran M, Brar SK, Galvez-Cloutier R, Martel R (2020) Bioremediation of unconventional oil contaminated ecosystems under natural and assisted conditions: a review. Environ Sci Technol 54(4):2054–2067. https://doi.org/10.1021/acs.est.9b00906

De Oliveira TA, Barbosa R, Mesquita AB, Ferreira JH, de Carvalho LH, Alves TS (2020) Fungal degradation of reprocessed PP/PBAT/ thermoplastic starch blends. J Mater Res Technol 9(2):2338–2349. https://doi.org/10.1016/j.jmrt.2019.12.065

Deng Y, Zhang Y, Lemos B, Ren H (2017) Tissue accumulation of microplastics in mice and biomarker responses suggest widespread health risks of exposure. Sci Rep 7:46687. https://doi.org/10.1038/srep46687

Devi RS, Kannan VR, Nivas D, Kannan K, Chandru S, Antony AR (2015) Biodegradation of HDPE by Aspergillus spp. from marine ecosystem of Gulf of Mannar, India. Mar Pollut Bull 96(1–2):32–40. https://doi.org/10.1016/j.marpolbul.2015.05.050

Du H, Xie Y, Wang J (2021) Microplastic degradation methods and corresponding degradation mechanism: research status and future perspectives. J Hazard Mater 418:126377

Dubey I, Khan S, Kushwaha S (2022) Developmental and reproductive toxic effects of exposure to microplastics: a review of associated signaling pathways. Front Toxicol 4:901798. https://doi.org/10.3389/ftox.2022.901798

Dvořák L, Svojitka J, Wanner J, Wintgens T (2013) Nitrifcation performance in a membrane bioreactor treating industrial wastewater. Water Res 47:4412–4421. https://doi.org/10.1016/j.watres.2013.03.053

Enyoh CE et al (2022) An overview of physical, chemical and biological methods for removal of microplastics. Microplastics Pollut Aqua Media. https://doi.org/10.1007/978-981-16-8440-1_12

Eriksen M, Lebreton LCM, Carson HS, Thiel M, Moore CJ, Borerro JC, Galgani F, Ryan PG, Reisser J (2014) Plastic pollution in the World's oceans: more than 5 trillion plastic pieces

weighing over 250,000 tons afloat at sea. PLoS One 9:e111913. https://doi.org/10.1371/journal.pone.0111913
Esmaeili A, Pourbabaee AA, Alikhani HA, Shabani F, Esmaeili E (2013) Biodegradation of low-density polyethylene (LDPE) by mixed culture of Lysinibacillus xylanilyticus and Aspergillus niger in soil. PLoS One 8(9):e71720. https://doi.org/10.1371/journal.pone.0071720
Fontanella S, Bonhomme S, Brusson J, Pitteri S, Samuel G, Pichon G et al (2013) Comparison of biodegradability of various polypropylene flms containing pro-oxidant additives based on Mn, Mn/ Fe or Co. Polym Degrad Stab 98:875–884. https://doi.org/10.1016/j.polymdegradstab.2013.01.002
Ganesh KA, Anjana K, Hinduja M (2019) Review on plastic wastes in marine environment biodegradation and biotechnological solutions. Mar Pollut Bull 150:110733
Giacomucci L, Raddadi N, Soccio M (2020) Biodegradation of polyvinyl chloride plastic films by enriched anaerobic marine consortia. Mar Environ Res 158:104949
Grbic J, Nguyen B, Guo E, You JB, Sinton D, Rochman CM (2019) Magnetic extraction of microplastics from environmental samples. Environ Sci Technol Lett 6:68–72. https://doi.org/10.1021/acs.estlett.8b00671
Habib S, Iruthayam A, Abd Shukor MY, Alias SA, Smykla J, Yasid NA (2020) Biodeterioration of untreated polypropylene microplastic particles by antarctic bacteria. Polymers 12(11):2616. https://doi.org/10.3390/polym12112616
Helen AS, Uche EC, Hamid FS (2017) Screening for polypropylene degradation potential of bacteria isolated from mangrove eco-systems in Peninsular Malaysia. Int J Biosci Biochem Bioinform 7:245–251
Hua X, Wang D (2022) Cellular uptake, transport, and organelle response after exposure to microplastics and Nanoplastics: current knowledge and perspectives for environmental and health risks. Rev Environ Contam Toxicol 260:12. https://doi.org/10.1007/s44169-022-00013-x
Huang S, Huang X, Bi R, Guo Q, Yu X, Zeng Q, Huang Z, Liu T, Wu H, Chen Y, Xu J, Wu Y, Guo P (2022a) Detection and analysis of microplastics in human sputum. Environ Sci Technol 56:2476–2486. https://doi.org/10.1021/acs.est.1c03859
Huang T, Zhang W, Lin T, Liu S, Sun Z, Liu F, Yuan Y, Xiang X, Kuang H, Yang B, Zhang D (2022b) Maternal exposure to polystyrene nanoplastics during gestation and lactation induces hepatic and testicular toxicity in male mouse offspring. Food Chem Toxicol Int J Publ Br Ind Biol Res Assoc 160:112803. https://doi.org/10.1016/j.fct.2021.112803
Huerta Lwanga E, Thapa B, Yang X, Gertsen H, Salánki T, Geissen V, Garbeva P (2018) Decay of low-density polyethylene by bacteria extracted from earthworm's guts: a potential for soil restoration. Sci Total Environ 624:753–757. https://doi.org/10.1016/j.scitotenv.2017.12.144
Ibrahim IN, Maraqa A, Hameed KM, Saadoun IM, Maswadeh HM (2011) Assessment of potential plastic-degrading fungi in Jordanian habitats. Turk J Biol 35(5):551–557. https://doi.org/10.3906/biy-0901-9
Inderthal H, Tai SL, Harrison ST (2021) Non-hydrolyzable plastics–an interdisciplinary look at plastic bio-oxidation. Trends Biotechnol 39(1):12–23. https://doi.org/10.1016/j.tibtech.2020.05.004
Jain K, Bhunia H, Reddy MS (2018) Degradation of polypropyl-ene-poly-L-lactide blend by bacteria isolated from compost. Biorem J 22(3–4):73–90. https://doi.org/10.1080/10889868.2018.1516620
Jain K, Bhunia H, Reddy MS (2022) Degradation of polypropylene-poly-L-lactide blends by Bacillus isolates: a microcosm and feld evaluation. Biorem J 26(1):64–75. https://doi.org/10.1080/10889868.2021.1886037
Jenner LC, Rotchell JM, Bennett RT, Cowen M, Tentzeris V, Sadofsky LR (2022) Detection of microplastics in human lung tissue using μFTIR spectroscopy. Sci Total Environ 831:154907. https://doi.org/10.1016/j.scitotenv.2022.154907
Jeon HJ, Kim MN (2013) Isolation of a thermophilic bacterium capable of low molecular–weight polyethylene degradation. Biodegradation 24:89–98. https://doi.org/10.1007/s10532-012-956

Jeon HJ, Kim MN (2016) Isolation of mesophilic bacterium for biodegradation of polypropylene. Int Biodeterior Biodegradation 115:244–249. https://doi.org/10.1016/j.ibiod.2016.08.025

Jeon JM, Park SJ, Choi TR, Park JH, Yang YH, Yoon JJ (2021) Biodegradation of polyethylene and polypropylene by Lysini bacillus species JJY0216 isolated from soil grove. Polym Degrad Stab 191:109662. https://doi.org/10.1016/j.polymdegradstab.2021.109662

Ji MK, Kabra AN, Choi J, Hwang JH, Kim JR, Abou-Shanab RA, Oh YK, Jeon BH (2014) Biodegradation of bisphenol A by the freshwater microalgae Chlamydomonas mexicana and Chlorella vulgaris. Ecol Eng 73:260–269. https://doi.org/10.1016/j.ecoleng.2014.09.070

Johnston B, Adamus G, Ekere AI, Kowalczuk M, Tcuenbou-Magaia F, Radecka I (2022) Bioconversion of plastic waste based on mass full carbon backbone polymeric materials to value-added polyhydroxyalkanoates (PHAs). Bioengineering 9(9):432. https://doi.org/10.3390/bioengineering9090432

Khan A, Jia Z (2023) Recent insights into uptake, toxicity, and molecular targets of microplastics and nanoplastics relevant to human health impacts. iScience 26. https://doi.org/10.1016/j.isci.2023.106061

Kiendrebeogo M, Estahbanati MRK, Mostafazadeh AK, Drogui P, Tyagi RD (2021) Treatment of microplastics in water by anodic oxidation: a case study for polystyrene. Environ Pollut 269:116168. https://doi.org/10.1016/j.envpol.2020.116168

Kim JW, Park SB, Tran QG, Cho DH, Choi DY, Lee YJ, Kim HS (2020) Functional expression of polyethylene terephthalatedegrading enzyme (PETase) in green microalgae. Microb Cell Factories 19(1):1–9. https://doi.org/10.1186/s12934-020-01355-8

Kim SA, Kim L, Kim TH, An YJ (2022) Assessing the size-dependent effects of microplastics on zebrafish larvae through fish lateral line system and gut damage. Mar Pollut Bull 185:114279. https://doi.org/10.1016/j.marpolbul.2022.114279

Kong HG, Kim HH, Chung JH, Jun J, Lee S, Kim HM, Ryu CM (2019) The Galleria mellonella hologenome supports microbiota-independent metabolism of long-chain hydrocarbon beeswax. Cell Rep 26(9):2451–2464. https://doi.org/10.1016/j.celrep.2019.02.018

Kowalczyk A, Chyc M, Ryszka P, Latowski D (2016) Achromobacter xylosoxidans as a new microorganism strain colonizing highdensity polyethylene as a key step to its biodegradation. Environ Sci Pollut Res 23(11):11349–11356. https://doi.org/10.1007/s11356-016-6563-y

Kunlere IO, Fagade OE, Nwadike BI (2019) Biodegradation of low-density polyethylene (LDPE) by certain indigenous bacteria and fungi. Int J Environ Stud 76(3):428–440. https://doi.org/10.1080/00207233.2019.1579586

Kutralam-Muniasamy G, Shruti VC, Perez-Guevara F, Roy PD (2023) Microplastic diagnostics in humans: "the 3Ps" Progress, problems, and prospects. Sci Total Environ 856:159164. https://doi.org/10.1016/j.scitotenv.2022.159164

Lam SM, Sin JC, Zeng H, Lin H, Li H, Chai YY, Choong MK, Mohamed AR (2021) Green synthesis of Fe-ZnO nanoparticles with improved sunlight photocatalytic performance for polyethylene flm deterioration and bacterial inactivation. Mater Sci Semicond Process 123:105574. https://doi.org/10.1016/j.mssp.2020.105574

Liu L, Xu K, Zhang B, Ye Y, Zhang Q, Jiang W (2021) Cellular internalization and release of polystyrene microplastics and nanoplastics. Sci Total Environ 779:146523. https://doi.org/10.1016/j.scitotenv.2021.146523

Liu T, Hou B, Wang Z, Yang Y (2022) Polystyrene microplastics induce mitochondrial damage in mouse GC-2 cells. Ecotoxicol Environ Saf 237:113520. https://doi.org/10.1016/j.ecoenv.2022.113520

Luo J, Chen W, Song H, Liu J (2020) Fabrication of hierarchical layer-by-layer membrane as the photocatalytic degradation of foulants and effective mitigation of membrane fouling for wastewater treatment. Sci Total Environ 699:134398. https://doi.org/10.1016/j.scitotenv.2019.1343

Magnin A, Pollet E, Phalip V, Avérous L (2020) Evaluation of biological degradation of polyurethanes. Biotechnol Adv 39:107457. https://doi.org/10.1016/j.biotechadv.2019.107457

Malafaia G, Barcelo D (2023) Microplastics in human samples: recent advances, hotspots, and analytical challenges. TrAC Trends Anal Chem 161:117016. https://doi.org/10.1016/j.trac.2023.117016

Manzi HP, Abou-Shanab RA, Jeon BH, Wang J, Salama ES (2022) Algae: a frontline photosynthetic organism in the microplastic catastrophe. Trends Plant Sci S1360–1385(22):00157–00161. https://doi.org/10.1016/j.tplants.2022.06.005

Matjašič T, Simčič T, Medvešček N, Bajt O, Dreo T, Mori N (2021) Critical evaluation of biodegradation studies on synthetic plastics through a systematic literature review. Sci Total Environ 752:141959. https://doi.org/10.1016/j.scitotenv.2020.141959

Menendez D, Alvarez A, Peon P, Ardura A, Garcia-Vazquez E (2021) From the ocean to jellies forth and back? Microplastics along the commercial life cycle of red algae. Mar Pollut Bull 168:112402. https://doi.org/10.1016/j.marpolbul.2021.112402

Miao L, Wang P, Hou J, Yao Y, Liu Z, Liu S, Li T (2019) Distinct community structure and microbial functions of bioflms colonizing microplastics. Sci Total Environ 650:2395–2402. https://doi.org/10.1016/j.scitotenv.2018.09.378

Miao F, Liu Y, Ga M, Yu X, Xiao P, Wang M, Wang S, Wang X (2020) Degradation of polyvinyl chloride microplastics via an electro-Fenton - like system with a TiO2/graphite cathode. J Hazard Mater 399:123023

Mor R, Sivan A (2008) Biofilm formation and partial biodegradation of polystyrene by the actinomycete Rhodococcus ruber: biodegradation of polystyrene. Biodegradation 19:851–858. https://doi.org/10.1007/s10532-008-9188-0

Muenmee S, Chiemchaisri W, Chiemchaisri C (2016) Enhancement of biodegradation of plastic wastes via methane oxidation in semi-aerobic landfill. Int Biodeterior Biodegradation 113:244–255. https://doi.org/10.1016/j.ibiod.2016.03.016

Nakazawa Y, Abe T, Matsui Y, Shinno K, Kobayashi S, Shirasaki N, Matsushita T (2021) Diferences in removal rates of virgin/ decayed microplastics, viruses, activated carbon, and kaolin/ montmorillonite clay particles by coagulation, focculation, sedimentation, and rapid sand fltration during water treatment. Water Res 203:117550. https://doi.org/10.1016/j.watres.2021.117550

Natarajan R, Fazari FA, Saadi AA (2018) Municipal waste water treatment by natural coagulant assisted electrochemical technique—parametric efects. Environ Technol Innov 10:71–77. https://doi.org/10.1016/j.eti.2018.01.011

Natarajan L, Soupam D, Dey S, Chandrasekaran N, Kundu R, Paul S, Mukherjee A (2022) Toxicity of polystyrene microplastics in freshwater algae Scenedesmus obliquus: effects of particle size and surface charge. Toxicol Rep 9:1953–1961. https://doi.org/10.1016/j.toxrep.2022.10.013

Nowak B, Pająk J, Drozd-Bratkowicz M, Rymarz G (2011) Microorganisms participating in the biodegradation of modifed polyethylene flms in diferent soils under laboratory conditions. Int Biodeterior Biodegradation 65(6):757–767. https://doi.org/10.1016/j.ibiod.2011.04.007

Nowak B, Pajk J, Karcz J (2012) Biodegradation of pre-aged modifed polyethylene flms. In: Kazmiruk V (ed) Scanning electron microscopy. InTech. https://doi.org/10.5772/35128

Othman AR, Hasan HA, Muhamad MH, Ismail NI, Abdullah SRS (2021) Microbial degradation of microplastics by enzymatic processes: a review. Environ Chem Lett 19(4):3057–3073. https://doi.org/10.1007/s10311-021-01197-9

Paço A, Duarte K, da Costa JP, Santos PS, Pereira R, Pereira ME, Freitas AC, Duarte AC, Rocha-Santos TA (2017) Biodegradation of polyethylene microplastics by the marine fungus Zalerion maritimum. Sci Total Environ 586:10–15. https://doi.org/10.1016/j.scitotenv.2017.02.017

Paço A, Jacinto J, da Costa JP, Santos PS, Vitorino R, Duarte AC, Rocha-Santos T (2019) Biotechnological tools for the efective management of plastics in the environment. Crit Rev Environ Sci Technol 49(5):410–441. https://doi.org/10.1080/10643389.2018.1548862

Padervand M, Lichtfouse E, Robert D, Wang C (2020) Removal of microplastics from the environment. A review. Environ Chem Lett 18(3):807–828. https://doi.org/10.1007/s10311-020-00983-1

Padmaja K, Cherukuri J, Reddy MA (2020) A comparative study of the efciency of chemical coagulation and electrocoagulation methods in the treatment of pharmaceutical efuent. J Water Process Eng 34:101153. https://doi.org/10.1016/j.jwpe.2020.101153

Park SY, Kim CG (2019) Biodegradation of micro-polyethylene particles by bacterial colonization of a mixed microbial consortium isolated from a landfll site. Chemosphere 222:527–533. https://doi.org/10.1016/j.chemosphere.2019.01.159

Perkins S (2015) Nearly every seabird may be eating plastic by 2050. Science. https://doi.org/10.1126/science.aad1694

Pyl M, Taylor A, Oberhansli F, Swarzenski P, Hussamy L, Besson M, Danis B, Metian M (2022) Size-dependent transfer of microplastics across the intestinal wall of the echinoid Paracentrotus lividus. Aquat Toxicol 250:106235. https://doi.org/10.1016/j.aquatox.2022.106235

Rajandas H, Parimannan S, Sathasivam K, Ravichandran M, Yin LS (2012) A novel FTIR-ATR spectroscopy basedtechnique for the estimation of low-density polyethylene biodegradation. Polym Test 31:1094–1099. https://doi.org/10.1016/j.polymertesting.2012.07.015

Rist S, Carney Almroth B, Hartmann NB, Karlsson TM (2018) A critical perspective on early communications concerning human health aspects of microplastics. Sci Total Environ 626:720–726. https://doi.org/10.1016/j.scitotenv.2018.01.092

Roccuzzo S, Beckerman AP, Trögl J (2021) New perspectives on the bioremediation of endocrine disrupting compounds from wastewater using algae-, bacteria-and fungi-based technologies. Int J Environ Sci Technol 18(1):89–106

Sanniyas E, Gopal RK, Gunasekar DK, Raj PP (2021) Biodegradation of low-density polyethylene (LDPE) sheet by microalga, Uronema Africanum Borge. Sci Rep 11(1):1–33. https://doi.org/10.1038/s41598-021-96315-6

Schwartz M, Perrot T, Aubert E, Dumarçay S, Favier F, Gérardin P, Gelhaye E (2018) Molecular recognition of wood polyphenols by phase II detoxifcation enzymes of the white rot Trametes versicolor. Sci Rep 8(1):1–11. https://doi.org/10.1038/s41598-018-26601-3

Shang Y, Wang S, Jin Y, Xue W, Zhong Y, Wang H, An J, Li H (2021) Polystyrene nanoparticles induced neurodevelopmental toxicity in Caenorhabditis elegans through regulation of dpy-5 and rol-6. Ecotoxicol Environ Saf 222:112523. https://doi.org/10.1016/j.ecoenv.2021.112523

Sheel A, Pant D (2018) Microbial depolymerization. In: Waste bioremediation. Springer, Singapore, pp 61–103. https://doi.org/10.1007/978-981-10-7413-4_4

Sheik S, Chandrashekar KR, Swaroop K, Somashekarappa HM (2015) Biodegradation of gamma irradiated low density polyethylene and polypropylene by endophytic fungi. Int Biodeterior Biodegrad 105:21–29. https://doi.org/10.1016/j.ibiod.2015.08.006

Shen M, Song B, Zhu Y, Zeng G, Zhang Y, Yang Y, Wen X, Chen M, Yi H (2020) Removal of microplastics via drinking water treatment: current knowledge and future directions. Chemosphere 251:126612. https://doi.org/10.1016/j.chemosphere.2020.126612

Shirasaki N, Matsushita M, Matsui Y, Marubayashi T (2016) Effect of aluminum hydrolyte species on human enterovirus removal from water during the coagulation process. Chem Eng J 284:786–793. https://doi.org/10.1016/j.cej.2015.09.045

Singh L, Wahid ZA (2015) Methods for enhancing bio-hydrogen production from biological process: a review. J Ind Eng Chem 21:70–80. https://doi.org/10.1016/j.jiec.2014.05.035

Sivan A, Szanto M, Pavlov V (2006) Biofilm development of the polyethylene-degrading bacterium Rhodococcus ruber. Appl Microbiol Biotechnol 72:346–352. https://doi.org/10.1007/s00253-005-0259-4

Skariyachan S, Patil AA, Shankar A, Manjunath M, Bachappanavar N, Kiran S (2018) Enhanced polymer degradation of polyethylene and polypropylene by novel thermophilic consortia of Brevibacillus sps. and Aneurinibacillus sp. screened from waste management landflls and sewage treatment plants. Polym Degrad Stab 149:52–68. https://doi.org/10.1016/j.polymdegradstab.2018.01.018

Skariyachan S, Taskeen N, Kishore AP, Krishna BV, Naidu G (2021) Novel consortia of Enterobacter and Pseudomonas formulated from cow dung exhibited enhanced biodegradation of polyethylene and polypropylene. J Environ Manag 284:112030. https://doi.org/10.1016/j.jenvman.2021.112030

Straub S, Hirsch PE, Burkhardt-Holm P (2017) Biodegradable and petroleum-based microplastics do not difer in their ingestion and excretion but in their biological efects in a freshwater invertebrate Gammarus fossarum. Int J Environ Res Public Health 14(7):774. https://doi.org/10.3390/ijerph14070774

Sukkhum S, Tokuyama S, Kitpreechavanich V (2009) Development of fermentation process for PLA-degrading enzyme production by a new thermophilic Actinomadura sp. T16–1. Biotechnol Bioprocess Eng 14(3):302–306. https://doi.org/10.1007/s12257-008-0207-0

Syranidou E, Karkanorachaki K, Amorotti F, Franchini M, Repouskou E, Kaliva M, Vamvakaki M, Kolvenbach B, Fava F, Corvini PFX, Kalogerakis N (2017) Biodegradation of weathered polystyrene flms in seawater microcosms. Sci Rep 7(1):1–12. https://doi.org/10.1038/s41598-017-18366-y

Tang Y, Zhang S, Su Y, Wu D, Zhao Y, Xie B (2021) Removal of microplastics from aqueous solutions by magnetic carbon nanotubes. Chem Eng J 406:126804. https://doi.org/10.1016/j.cej.2020.126804

Taniguchi I, Yoshida S, Hiraga K, Miyamoto K, Kimura Y, Oda K (2019) Biodegradation of PET: current status and application aspects. ACS Catal 9(5):4089–4105. https://doi.org/10.1021/acscatal.8b05171

Thompson RC, Olsen Y, Mitchell RP, Davis A, Rowland SJ, John AWG, McGonigle D, Russell AE (2004) Lost at sea: where is all the plastic? Science 304:838. https://doi.org/10.1126/science.1094559

Tirkey A, Upadhyay LSB (2021) Microplastics: An overview on separation,identification and characterization of microplastics. Mar Pollut Bull 170:112604. https://doi.org/10.1016/j.marpolbul.2021.112604

Tiwari E, Singh N, Khandelwal N, Monikh FA, Darbha GK (2020) Application of Zn/Al layered double hydroxides for the removal of nano-scale plastic debris from aqueous systems. J Hazard Mater 397:122769. https://doi.org/10.1016/j.jhazmat.2020.122769

Tribedi P, Sil AK (2013) Low-density polyethylene degradation by Pseudomonas sp. AKS2 bioflm. Environ Sci Pollut Res 20(6):4146–4153. https://doi.org/10.1007/s11356-012-1378-y

Uscátegui YL, Arévalo FR, Díaz LE, Cobo MI, Valero MF (2016) Microbial degradation, cytotoxicity and antibacterial activity of polyurethanes based on modifed castor oil and polycaprolactone. J Biomater Sci Polym Ed 27(18):1860–1879. https://doi.org/10.1080/09205063.2016.1239948

Volke-Sepúlveda T, Saucedo-Castañeda G, Gutiérrez-Rojas M, Manzur A, Favela-Torres E (2002) Thermally treated low density polyethylene biodegradation by Penicillium pinophilum and Aspergillus niger. J Appl Polym Sci 83(2):305–314. https://doi.org/10.1002/app.2245

Wong JKH, Lee KK, Tang KHD, Yap PS (2020) Microplastics in the freshwater and terrestrial environments: prevalence, fates, impacts and sustainable solutions. Sci Total Environ 719:137512. https://doi.org/10.1016/j.scitotenv.2020.137512

Wu D, Wang T, Wang J, Jiang L, Yin Y, Guo H (2021) Size-dependent toxic effects of polystyrene microplastic exposure on Microcystis aeruginosa growth and microcystin production. Sci Total Environ 761:143265. https://doi.org/10.1016/j.scitotenv.2020.143265

Xiao X, He E, Jiang X, Li X, Yang W, Ruan J, Zhao C, Qiu R, Tang Y (2022) Visualizing and assessing the size- dependent oral uptake, tissue distribution, and detrimental effect of polystyrene microplastics in Eisenia fetida. Environ Pollut 1987:306. https://doi.org/10.1016/j.envpol.2022.119436

Yang N, Zhang H, Chen M, Shao LM, He PJ (2012) Greenhouse gas emissions from MSW incineration in China: impacts of waste characteristics and energy recovery. Waste Manag 32:2552–2560

Yang J, Yang Y, Wu WM, Zhao J, Jiang L (2014) Evidence of polyethylene biodegradation by bacterial strains from the guts of plastic eating waxworms. Environ Sci Technol 48:13776–13784. https://doi.org/10.1021/es504038a

Yang Y, Yang J, Wu WM, Zhao J, Song Y, Gao L, Yang R, Jiang L (2015) Biodegradation and mineralization of polystyrene by plastic-eating mealworms: part 2. Role of gut microorganisms. Environ Sci Technol 49(20):12087–12093. https://doi.org/10.1021/acs.est.5b02663

Yee MSL, Hii LW, Looi CK, Lim WM, Wong SF, Kok YY, Tan BK, Wong CY, Leong CO (2021) Impact of microplastics and nanoplastics on human health. Nanomaterials 11:496. https://doi.org/10.3390/nano11020496

Zadjelovic V, Chhun A, Quareshy M, Silvano E, Hernandez-Fernaud JR, Aguilo-Ferretjans MM, Bosch R, Dorador C, Gibson MI, Christie-Oleza JA (2020) Beyond oil degradation: enzymatic potential of Alcanivorax to degrade natural and synthetic polyesters. Environ Microbiol 22:1356–1369. https://doi.org/10.1111/1462-2920.14947

Zhang F, Zhao Y, Wang D, Yan M, Zhang J, Zhang P, Ding T, Chen L, Chen C (2020a) Current technologies for plastic waste treatment: a review. J Clean Prod 282:124523

Zhang J, Gao D, Li Q, Zhao Y, Li L, Lin H, Bi Q, Zhao Y (2020b) Biodegradation of polyethylene microplastic particles by the fungus Aspergillus flavus from the guts of wax moth Galleria mellonella. Sci Total Environ 704:135931

Zhang N, Li YB, He HR, Zhang JF, Ma GS (2021a) You are what you eat: microplastics in the feces of young men living in Beijing. Sci Total Environ 767:144345. https://doi.org/10.1016/j.scitotenv.2020.144345

Zhang K, Hamidian AH, Tubić A, Zhang Y, Fang JKH, Wu C, Lam PKS (2021b) Understanding plastic degradation and microplastic formation in the environment: a review. Environ Pollut 274:116554. https://doi.org/10.1016/j.envpol.2021.116554

Zhou G, Wang Q, Li J, Li Q, Xu H, Ye Q, Wang Y, Shu S, Zhang J (2021) Removal of polystyrene and polyethylene microplastics using PAC and FeCl3 coagulation: performance and mechanism. Sci Total Environ 752:141837. https://doi.org/10.1016/j.scitotenv.2020.141837

Zhou N, Wang Z, Yang L, Zhou W, Qin Z, Zhang H (2023) Size-dependent toxicological effects of polystyrene microplastics in the shrimp Litopenaeus vannamei using a histomorphology, microbiome, and metabolic approach. Environ Pollut 316:120635. https://doi.org/10.1016/j.envpol.2022.120635

Zurier HS, Goddard JM (2021) Biodegradation of microplastics in food and agriculture. Curr Opin Food Sci 37:37–44

Chapter 13
Heavy Metals (As, Cd, Hg, Pb) Contamination and Bioaccumulation in Estuarine and Marine Ecosystems with Special Reference to Asia

Arkajyoti Shome, Sayan Bhattacharya, Deep Chakraborty, and Satarupa Dey

Abstract Heavy metal contamination in the aquatic environment is ecologically and economically detrimental, leading to loss of biodiversity and having considerable impacts on human health. Occurrence and accumulation of heavy metals are considered as potential threats to the coastal inhabitants depending on marine resources for food, industry and recreation. In this chapter, four widely used heavy metals, arsenic, cadmium, mercury and lead, are discussed in terms of their toxicological and health impacts on marine environment and humans. Several fishes can bioaccumulate these heavy metals in the tissues, and the heavy metals can be transferred through food chain and can be biomagnified thereafter. Rate of bioaccumulation differs among different species of fishes. Heavy metals are mostly found in the sediments; therefore, benthic species have a higher rate of accumulation rather than pelagic species. Humans are directly exposed to these heavy metals primarily through seafoods; the consequences are evident in several catastrophic breakouts happened previously, like Minamata disaster (in case of mercury) and Itai-Itai disease (in case of cadmium) in Japan. Around a quarter of the human population in South and Southeast Asia are directly or indirectly dependent on fisheries and aquaculture industries and are vulnerable to heavy metal contamination. Several case studies from Asia (South China Sea, Persian Gulf, Mekong Delta, Bay of Bengal) are highlighted in the chapter to show the diverse impacts of heavy metals on marine and estuarine environment. International marine laws regarding waste disposal have

A. Shome · S. Bhattacharya (✉)
School of Ecology and Environment Studies, Nalanda University, Rajgir, Bihar, India
e-mail: sbhattacharya@nalandauniv.edu.in

D. Chakraborty
Department of Environmental Science, Amity School of Life Sciences, Amity University, Gwalior, Madhya Pradesh, India

S. Dey
Department of Botany, Shyampur Siddheswari Mahavidyalaya, Howrah, West Bengal, India

S. Dey, S. Bhattacharya (eds.), *Biotechnological Interventions in the Removal of Emerging Pollutants*, Interdisciplinary Biotechnological Advances,
https://doi.org/10.1007/978-981-97-9922-0_13

been strengthening over the decades to control release and spreading of pollutants. Also, various remediation measures, including physiochemical and biochemical methods, are applied in many cases to remediate heavy metal pollutants in marine and estuarine ecosystems. However, this enormous issue of contamination needs interdisciplinary and sustainable remedial measures, especially collaborative efforts of experts from related disciplines.

Keywords Heavy metals · Bioaccumulation · Biomagnification · Aquaculture · Fisheries

13.1 Introduction

Heavy metals are persistent and widely distributed in the environment, having both natural sources (erosion from rock surfaces, volcanoes, hydrothermal vent) and anthropogenic sources (ship breaking, oil spill, run-off, etc.). Initially, they remain suspended as particulate matter in the ocean water and can be settled down subsequently on the ocean floor (Amin et al. 2009; DeForest et al. 2007; Wang et al. 2013).

The term "heavy metal" includes all the metals and metalloids that have gravity equal to or greater than 4000 kg/m^3 (Vardhan et al. 2019). Most of the heavy metals are considered to be toxic; however, as per the guidelines of the US FDA, heavy metals are divided into two broad categories: essential and nonessential. Metals like manganese, iron, nickel and copper are important for marine organisms to carry out their metabolic activities. Whereas, arsenic, cadmium, lead, and mercury exhibit toxicity and are considered as hazardous (nonessential) metals (Kumar et al. 2012; Vardhan et al. 2019).

Both essential and nonessential minerals get absorbed into the bodies of aquatic organisms naturally through water or by sediment-feeding (Bosch et al. 2016). The factors responsible for the uptake of heavy metals into the body are both metal-specific and organism-specific. Physical and geochemical factors are responsible for metal-specific uptake, whereas several biological factors are responsible for organism-specific uptake of heavy metals (Bosch et al. 2016; Dallinger 1994; Dallinger et al. 1987). Apart from environmental factors, species-specific and organism-specific biological factors also influence the accumulation of heavy metals in the body of the organisms (Bosch et al. 2016; Dallinger 1994; Dallinger et al. 1987).

Most heavy metals are carcinogenic, teratogenic, mutagenic, mostly nonbiodegradable and can bioaccumulate in living organisms. After entering the human body, compounds of heavy metals can execute toxicity even in extremely low concentrations. Therefore, heavy metal pollution remediation has been one of the main focuses in pollution control research (Yan et al. 2022).

Once heavy metals are bioaccumulated in living organisms, they can subsequently move through the food chain to the next trophic levels and can be biomagnified thereafter, resulting in the increase of toxic metal concentrations in the higher trophic levels (fishes, birds, marine mammals, etc.) (Emenike et al. 2022). Fishes are known as useful bioindicators of marine pollution because of their ability to absorb heavy metals even in trace concentrations (Cunningham et al. 2019; Wang et al. 2013). Figure 13.1 shows heavy metal pollution in marine ecosystems; bioaccumulation and biomagnification of heavy metals in marine fishes.

Since the 1950s, the occurrence of catastrophic Minamata diseases and Itai-Itai diseases, caused due to mercury and cadmium pollution, has led to increasing concerns among ecotoxicological researchers regarding heavy metal contamination in the marine systems (Fowler 1990; Gregoria et al. 2017).

Heavy metal pollution in fisheries and aquaculture industries is now one of the major concerns across the globe. South and Southeast Asia, where 70% of the human population lives in the coastal areas, of which 50% is dependent on fisheries and seafood industries directly or indirectly, are severely affected by heavy metal pollution (Amin et al. 2009; Wang et al. 2013). Heavy metal contamination can adversely affect the ecological conditions of the water body by changing water and sediment conditions. Heavy metals affect human health through food chain by the processes of bioaccumulation and biomagnification. Additionally, it also affects the economy by having adverse impacts on fisheries by decreasing revenue and profitability.

Marine ecosystems have diverse assortments of life constitute a great pool of utilizable natural resources. Contamination of these resources has detrimental effects on ecosystems, human health, and economic growth. Therefore, heavy metal contamination needs considerable attention, and mitigation/remedial measures are

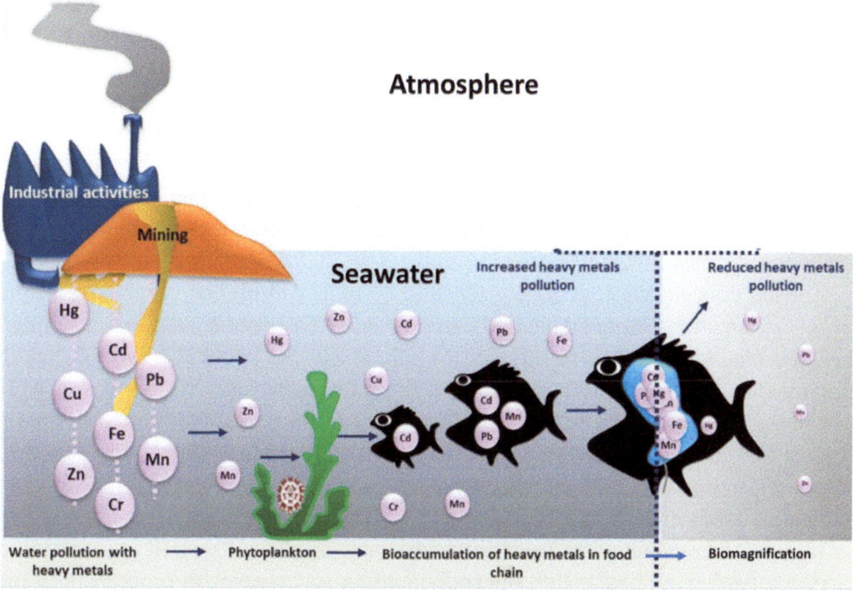

Fig. 13.1 Heavy metal pollution in marine ecosystems; bioaccumulation and biomagnification of heavy metals in marine fishes. (Modified after Mehana et al. 2020)

required to reduce its impacts (Bosch et al. 2016; Rahman et al. 2019; Hassan et al. 2022). This chapter discusses the impacts of the major heavy metal contaminants in marine and estuarine environments, highlighting bioaccumulation and biomagnification in different organisms and humans along with some of the case studies and remediation options with special reference to South and Southeast Asian countries.

13.2 Toxicological and Health Impacts of Heavy Metals in Marine Environment

Heavy metals are generally grouped in two major categories: essential, which includes nickel, manganese, cobalt, and zinc, and nonessential, which includes cadmium, mercury, lead, arsenic, etc. (Kumar et al. 2012). In this chapter, four major nonessential heavy metals— cadmium (Cd), mercury (Hg), lead (Pb), and arsenic (As)—are discussed, which even in lower ionic concentration can exhibit considerable toxicity.

13.2.1 Arsenic

Arsenic (As) is generally considered as a metalloid. Its presence is widely disturbed in nature due to natural resources and anthropogenic activities. Arsenic is generally found in both inorganic and organic form. Till now, more than 30 arsenicals have been found in the marine environment (Kunito et al. 2008). The inorganic form is more stable and can be found in both trivalent (more toxic) and pentavalent states (Bhattacharya et al. 2021; Bosch et al. 2016). Both these forms are toxic and have been classified as Group A carcinogen by US EPA (Castro-González and Méndez-Armenta 2008). In an aquatic environment, arsenic can be found in several other states like methylarsonic acid, methyl arsonous acid, dimethyl arsinic acid, dimethyl arsinous acid, dimethyl arsinoyl acetate, dimethyl arsinoyl ethanol and dimethyl arsinoyl propionate (Kar et al. 2011; Kunito et al. 2008; Wrench et al. 1979).

Arsenic concentration generally tends to be higher in marine organisms; as the inorganic arsenic is more stable and can be readily absorbed by the fishes through the food chain and passes to the next trophic levels. As per WHO (World Health Organization), the permissible limit of arsenic in the muscles of fish is 100 ppb. However, various studies reported that arsenic present in seafoods is in nontoxic organic form (arsenobetaine), and arsenic concentration is found to be maximum in the muscles (Bosch et al. 2016; Kunito et al. 2008; Castro-González and Méndez-Armenta 2008; LeBlanc and Jackson 1973). Most of the research works focused so far on the total arsenic contents in the body muscles of the fishes. It is essential to carry out further studies to find out individual arsenic species in proportion to the total arsenic content in the fish meat. Speciation analysis would be helpful to analyze the concentrations of toxic species of arsenic in fish biomass. Further analyses of the toxic species concentrations in different trophic levels are needed for risk analysis and possible impacts on human population (Mohan and Pittman 2007).

13.2.2 Cadmium

Cadmium (Cd) is considered to be highly toxic to humans due to its carcinogenic nature and long half-life of more than 20 years (Harsha et al. 2019). It has both natural sources like volcanic activities and erosion, and anthropogenic sources due to activities like smelting of metal, burning of fossil fuels, etc. Cadmium can be eventually deposited in the marine environment through different environmental processes (Bosch et al. 2016).

Cadmium was responsible for one of the famous heavy metal poisonings happened in Japan in 1912, known as Itai-Itai disease. Since then, cadmium is well known for its toxic nature. Even in low concentrations, cadmium can readily diffuse through cell membranes, forming stable ligands with the proteins (Bosch et al. 2016; Harsha et al. 2019; Kumar et al. 2012). Cadmium affects cell proliferation, differentiation and apoptosis. Cadmium can induce ROS production, results development of oxidative stress. This mechanism shows the role of cadmium in organ toxicity, carcinogenicity and apoptotic cell death (Rani et al. 2014).

Due to high salinity in seawater, Cd^{2+} quite readily forms complexes ($[CdCl_2^0]$ and $[CdCl^{+1}]$), which reduce the chances of bioaccumulation. However, studies have found that chances of cadmium toxicity among humans are mostly through seafood consumption, although the accumulation of cadmium among the fish species is not homogeneous. Studies have shown accumulation of cadmium is found mostly in benthic organisms like crab and mollusks. It is also hypothesized in earlier studies that uptake of cadmium also can affect the uptake of zinc, which is an essential heavy metal required for metabolic activities. However, not much attention has been given to this aspect (Bosch et al. 2016; Frazier 1979). The United Nations has defined permissible limit of cadmium 0.05 mg kg^{-1} of fresh weight in fishery products, considering its heterogeneous accumulation (Bosch et al. 2016).

13.2.3 Mercury

Natural sources of mercury (Hg) include forest fire and volcanic activities, whereas about 2/3 of the mercury present in the environment comes from anthropogenic activities (paints, fungicides, electrical equipment, mining) (Bosch et al. 2016). Though mercury displays considerable difference in solubility and toxicity depending on different chemical forms, but in general, it has low reactivity, less aqueous solubility, and higher stability as compared to other trace metals (Bosch et al. 2016). Organic mercury is the most stable form in comparison to the inorganic form like mercuric chloride ($HgCl_2$), mercurous chloride (Hg_2Cl_2), mercuric sulfide (HgS) which are insoluble, though they are commonly available in nature. Organic mercury, especially methylated mercury (MeHg), is stable and toxic in nature as it can be readily accumulated in the tissues of humans and fishes.

In the Minamata disaster, which happened in Japan, symptoms of mercury toxicity started in the early 1950s. However, it was later proven that organic mercury caused the disease. It was estimated that around 220 tons of mercury was released with the wastewater in Japan between 1932 and 1958 (Hylander and Goodsite 2006). This is "the heavy metal" that is of most concern for the seafood industry, as seafood is considered to be the prime source of mercury for humans (Bosch et al. 2016).

Due to increase in anthropogenic activities in the coastal areas of the Southeast and South Asian region, the concentration of mercury has increased severalfold as compared to mid-sea zones. Though there are certain regulations to limit the release and disposal of mercury in the sea, especially after the Minamata disaster, still some places have mercury concentration almost 10 times higher than the permissible limit. Future researches need to focus on developing strategies toward efficient detection of methyl mercury levels from total measured mercury in the marine and estuarine environment (Bosch et al. 2016).

13.2.4 Lead

Lead (Pb) is the most widely used metal in the industrial sectors and can be easily detected in all biological systems (Castro-González and Méndez-Armenta 2008). The use of leaded petrol was the main source of anthropogenic lead; tetra-ethyl lead is transformed into aerosols due to the complete combustion of gasoline. These aerosols present in the atmosphere were deposited in the marine environment (Bosch et al. 2016). Regulations adopted to curb the use of leaded petrol resulted in significant reduction of lead concentrations in the marine environment.

The organic form of lead is more toxic in comparison to the inorganic form. Higher number of methyl and ethyl groups linked to lead molecule can increase toxicity. About 50–70% of lead in the marine environment is in organic form. Fishes except for carp accumulate lead in all parts of the body. According to the European Commission, the maximum permissible limit for lead in the fish meat is 0.3 mg kg^{-1} (Bosch et al. 2016). It has been also found that uptake of lead is inversely proportional to the dietary uptake of calcium in marine organisms (Bosch et al. 2016; Castro-González and Méndez-Armenta 2008).

13.2.5 Bioaccumulation of Heavy Metals in Marine and Estuarine Fishes

Bioaccumulation can be broadly defined as, the process by which a chemical pollutant enters into the body of an organism and is not excreted but is deposited in the organism's tissue (Nnaji and Okoye 2007). Marine organisms can readily absorb and accumulate both essential and nonessential metals/minerals. Fishes serve as good bioindicators of marine pollution, especially in conditions related to heavy metal contamination (Dallinger et al. 1987; Nnaji and Okoye 2007; Wang et al. 2013). In marine environments, the availability of heavy metals for organisms such as fish depends both on physicochemical factors and biological factors (Zhang et al. 2022). In fishes, bioaccumulation of heavy metals mostly occurs in visceral tissues (kidney, liver) and to smaller extent to the bones and muscles. In some earlier research works, it was shown that chemical speciation of metals influences the solubility, bioaccumulation potential, and toxicity level of that metal toward the aquatic organisms (Bosch et al. 2016; Dallinger 1994; Fowler 1990; Naser 2013). Additionally, water chemistry also plays an important role in influencing metal speciation, especially hydrogen ion activity, salinity, and hardness (Dallinger et al. 1987).

Among the biological factors that are responsible for influencing bioaccumulation of metals are the feeding behavior and habitat preferences of the fishes and other marine organisms. Physiological factors like the accumulation rate, binding capacity of toxicants and ecological factors such as temperature and feeding behavior play a crucial role in the bioaccumulation of heavy metals in fishes (Dallinger et al. 1987; Kojadinovic et al. 2007).

Bioaccumulation rate differs among fishes in an ecosystem depending on their feeding behavior. Heavy metals tend to accumulate in the sediments; therefore benthic species have a higher rate of accumulation rather than pelagic species (Merola et al. 2021). In some previous studies carried out in the South China Sea and Indian ocean, it was found that the concentrations of nonessential minerals were less in pelagic waters, while the benthic dwellers like crabs, mollusks and shellfish have higher heavy metal contents in their bodies (Kojadinovic et al. 2007; Wang et al. 2013). Age, size, duration of exposure, and homeostatic stability regulate the bioaccumulation factor; therefore metal accumulation can vary within the same species (Velusamy et al. 2014).

It has already been established that marine organisms contain a higher amount of arsenic than terrestrial organisms. The level of arsenic in terrestrial organisms rarely crosses 1 ppm (dry biomass), however, in marine organisms, the amount can reach up to more than 100 ppm (Bosch et al. 2016; Kunito et al. 2008; Neff 1997). In marine animals, 90% of the accumulated arsenic in the body is a nontoxic form known as arsenobetaine (Kunito et al. 2008). The proportion of arsenobetaine found in the body varies with feeding behavior and trophic levels. In the aquatic environment, inorganic arsenic is more prevalent in the free state as compared to organic arsenic. However, as most of the fishes accumulate organic arsenic (arsenobetaine), it reduces the risk for both the fishes and humans. In several studies carried out on different regions, it was observed that in comparison to other heavy metal contaminants, arsenic was found to be mostly accumulated in the fish muscles. FAO (Food and Agriculture Organization) expert committee has recommended that 0.05 mg/kg of organic arsenic intake per day basis is not hazardous (Castro-González and Méndez-Armenta 2008; Kunito et al. 2008).

Cadmium often exceeds the threshold level in the commonly consumed fishes across the globe. After the outbreak of Itai-Itai diseases, Cadmium is one of the major threats to the fishing industry. Cadmium generally accumulates in the gills, muscles, and bones of the fishes, however, the distribution is not homogeneous (Bosch et al. 2016; Shazili et al. 2006). Salinity and temperature affect the bioaccumulation rate, and also, speciation of the metal can be influenced by the salinity in the aquatic environment (Castro-González and Méndez-Armenta 2008).

Mercury, in its elemental form Hg^0, is predominant in the environment. However, it cannot accumulate in fishes in elemental form. The toxic properties of mercury are dependent on its chemical forms. Also, the ability of an organism to accumulate and excrete mercury is dependent on the chemical forms. Organic mercury is highly stable and readily accumulated in the body of the fishes and can move through the food chains to the next trophic levels. Inorganic mercury accumulates in the body of fishes in very low concentration and can be excreted easily (Bosch et al. 2016). The inorganic mercury-like mercuric sulfide (HgS), which is common in nature but are insoluble and nontoxic, can be methylated in the environment through a photochemical process, or can be broken down by microorganisms (bacteria) in the sediments, even in the gills and guts of fishes, to form toxic methyl mercury (MeHg) (Bosch et al. 2016). Methyl mercury has a longer half-life than elemental mercury and is the most widely distributed toxic form of mercury available in the marine environment. It has high affinity for sulfhydryl group of amino acids and can easily move from one trophic level to another by the process of biomagnification. The most consumed fishes are placed in the higher trophic levels, therefore mercury concentrations in their bodies are maximum. As per FAO, the maximum permissible limit for mercury in the fish meat is 0.5 mg/kg (Bosch et al. 2016).

Lead in the aquatic environment is easily absorbed by fishes through gills. It can be accumulated in bones, muscles, kidney, and liver. In previous studies, it was observed that lead can accumulate in the kidney in high amounts (Bosch et al. 2016; Kojadinovic et al. 2007). If the concentration of lead is higher in the marine environment, there are chances of lead getting absorbed and accumulated on the scales of fishes. Dissolved organic matter in the environment enhances the accumulation rate of lead from the aquatic environment (Bosch et al. 2016). Mollusks are excellent bioindicators for lead contamination in the marine environment, as they have higher accumulation capacity for lead (Wang et al. 2013).

Figure 13.2 shows the effects of heavy metal bioaccumulation on marine organisms and human beings.

13.2.6 Impacts on Human Health

Seafood is one of the most common sources of most of the widely distributed heavy metals in the human body. Consumption of fishes and seafoods containing heavy metals of concentration more than the permissible limits will have serious health issues (Harsha et al. 2019; Velusamy et al. 2014; Wang et al. 2013). Consumption of heavy metal contaminated seafoods can adversely affect human health across the world. Malfunctioning of lungs, liver, kidney; nervous disorders; Alzheimer's

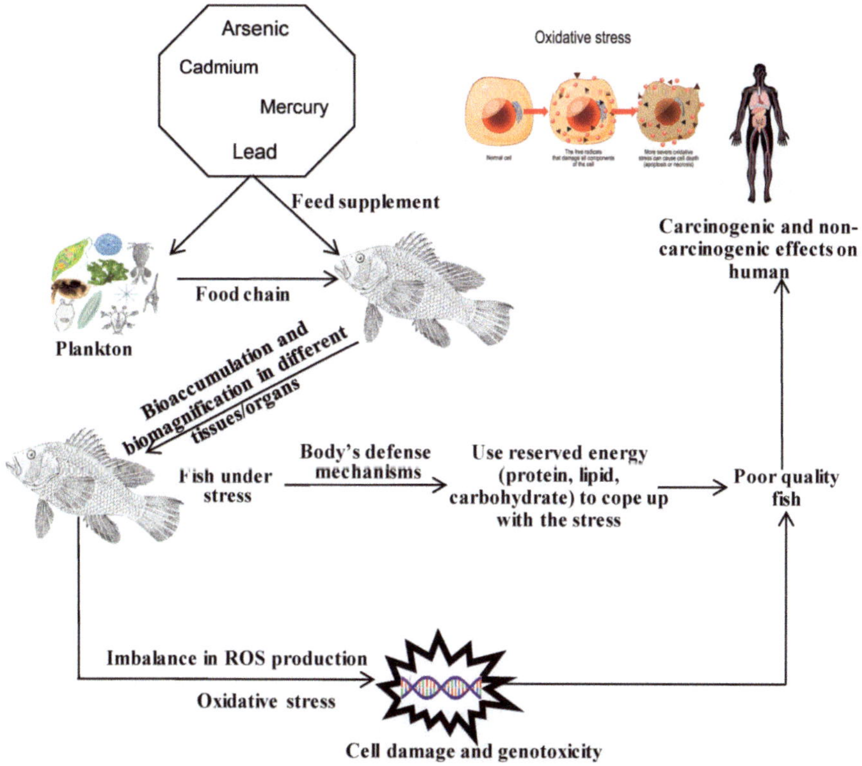

Fig. 13.2 Effects of heavy metal bioaccumulation on marine organisms and human beings; production of ROS and oxidative stress. (Modified after Jamil Emon et al. 2023)

disease is the most common and widely identified adverse effect of heavy metal consumption (Gu et al. 2015; Harsha et al. 2019; Hylander and Goodsite 2006). Most of the arsenic that is accumulated in marine fishes is in the form of nontoxic arsenobetaine, but there are lesser-known inorganic toxic arsenicals found in trace concentration which can pass onto humans through the food chain. Long-term exposure to the toxic arsenicals might be harmful to humans as arsenic in humans can act as a potent carcinogen and endocrine disruptor (Kunito et al. 2008; Sivaperumal et al. 2007).

Cadmium is a nonessential mineral for humans; hence they are not required for any metabolic activity. Cadmium can act as a carcinogen and can disrupt cardiovascular activities, causing skeletal weakness and hypertension (Bosch et al. 2016). Organic cadmium is widely distributed in the fishes and fish products sold around the world and is a potential threat to the local population dependent on marine resources for consumption. After the Itai-Itai outbreak, there has been some bilateral and multilateral treaties regarding proper treatment and check before the discharge of effluent in the water bodies. However, considering the long half-life of organic cadmium (more than 20 years), there might be chances of chronic toxicity in humans feeding on fish products obtained from highly industrialized coastal waters that have been polluted earlier, before the preventive measures were in place.

Methyl mercury is the most common form of mercury consumed by humans through seafoods (Bosch et al. 2016; Hylander and Goodsite 2006; Rahman et al. 2019). Methyl mercury is absorbed by the bloodstream from the intestinal tract and then can be passed on to the target organs, especially to the brain and liver. It even can cross the placental barrier and can harm the neural system of the fetus (Bosch et al. 2016). Symptoms of mercury toxicity include fatigue, headache, and slow body movement. The Minamata disease was the classic and most tragic example of mercury toxicity in humans in which the methyl mercury concentration in the fish was in the range of 1–20 mg/kg (Sivaperumal et al. 2007).

Marine fishes are major sources of lead for humans (Bosch et al. 2016). As per the guidelines issued by the European Commission, the maximum permissible limit for lead in marine fishes is 0.2 mg/kg (Castro-González et al. 2008). The nervous system, circulatory system and excretory system are the most sensitive parts for lead poisoning (Castro-González et al. 2008; Sivaperumal et al. 2007). Due to chronic toxicity of lead, humans can develop other problems like epigastric, constipation, encephalitis, and anorexia. The blood lead level needs to be monitored in case of chronic exposure to lead (Castro-González et al. 2008).

13.3 Global Scenario of Heavy Metal Pollution in Marine and Estuarine Systems

Since the first outbreak of catastrophic Minamata diseases in the 1950s in Japan, the focus on heavy metal contamination in marine fisheries has received much attention (Castro-González and Méndez-Armenta 2008). With the growth in industrialization and population, heavy metal pollution has been identified as a serious threat to the marine ecosystem and a significant challenge for the human society. Since the 1980s, with the growth in analytical techniques, heavy metal pollution has been detected across the globe, especially in the estuarine and marine regions, which hosts the shipping industry, electronic industry, cosmopolitan region, etc. The major areas across the globe which are infamous for heavy pollution includes the Mediterranean region, Southeast Asia, Persian Gulf, Pacific Islands, and Northern Atlantic.

The Arabian Gulf is one of the most anthropogenically disturbed marine regions in the world. Several studies in these regions have found traces of heavy metals, which varied both spatially and temporally. The Arabian Gulf is connected to the Mediterranean Sea through the Suez Canal, which is another region of grave concern in the context of heavy metal pollution. The coast of Egypt is one of the prime locations of heavy metal pollution since the river Nile drains out to the Mediterranean, carrying industrial, agricultural, and domestic wastewater, resulting in elevated levels of heavy metal concentration in the marine ecosystems. The pollution levels could be traced up till the coast near the South Adriatic Sea, near the ports of Apulia and Albania.

In the Northern Atlantic near the Hudson Bay region, heavy metal traces have been found in shell fishes and oysters since 1978; over time the concentration has varied between 7 to 12 times of the permissible limit. The western Coast of the United States in the Pacific also has considerable amount of heavy metal pollution, especially near California. However, there has been temporal variation over the decades in the concentration and speciation of heavy metals due to the change in anthropogenic activities. Heavy metal concentrations have also been found in other parts of the Pacific as well, which include Peter the Great Bay and the Sea of Japan, with exceeding concentrations of heavy metals in algae and mussels.

The coastlines of South and Southeast Asia stretch around 100,000 km and are home to around 700 million people. Around 80% of this population is directly or indirectly related to the seafood industry (Gregoria et al. 2017). The South China Sea, Bay of Bengal, and Pearl River Delta are a few of the major areas which have already been notified as highly contaminated regions by UNEP (United Nations Environment Programme) and other international bodies. Contamination happens mainly due to diverse anthropogenic activities in the ports and shipbreaking industries, release of industrial effluents, deforestation, aquaculture, etc. (Gregoria et al. 2017; Wang et al. 2013). Most of these areas lack proper effluent treatment plants, therefore, discharge the effluents directly into the sea. All these contamination "hotspots" are located in the tropical and subtropical regions which support huge marine biodiversity; approximately 34% of the world's coral reef and 30% of the mangrove area is situated in this region (Gregoria et al. 2017; Wang et al. 2013).

13.3.1 South China Sea and Mekong Delta

The South China Sea covers large areas touching the industrialized and populated zones in the coastal regions of East Asia. Hong Kong and Pearl River estuaries were found to be the most polluted areas and are considered as contamination hotspots. Since the early 1980s, China has experienced rapid economic growth, and consequently, the environmental conditions are compromised (Chan 1995; Cheng et al. 2013; Wang et al. 2013). The tremendous growth in the electronic industry has been the major reason for the generation of heavy metal pollutants. The Pearl River Delta has seen a mammoth expansion in fishing, agricultural and industrial activities, and as a result, huge influx of chemical pollutants is released in that zone for near about four decades. The levels of heavy metals in some of the zones of South China Sea and Pearl River Delta have exceeded to an unimaginable scale in the bodies of micro and macro aquatic organisms (Leung et al. 2014). A substantial percentage of local population is dependent on these aquatic organisms directly or indirectly (Wang et al. 2013). Mangroves are unique and critical habitat; therefore, disturbance in such habitat even in minimal scale can lead to catastrophic changes in the ecosystem. The accumulation of heavy metals has been reported from mangroves across several regions of Hong Kong, Singapore, Vietnam, etc. (Cheng et al. 2013; Cuong et al. 2005; Ray et al. 2006; Wang et al. 2013). In the mangroves of Singapore, the heavy metals Copper and Zinc are found to be in higher concentrations than other heavy metals; therefore, it was advised to not consume the catches from mangroves of Singapore (Cuong et al. 2005).

13.3.2 Arabian and Persian Gulf

The semienclosed Gulf region is one of the stressful regions connecting the western and eastern worlds. Due to the geographical disadvantage, the Gulf zone has comparatively poor water circulation, which adds to the misery of the growing anthropogenic pollution (Cunningham et al. 2019; Naser 2013). The heavy metal pollution in this region is attributed to a number of sources, due to direct or indirect human interventions, baring the few natural ones (Elsagh et al. 2021). The Gulf countries have seen the population explosion in the past few decades due to significant increase in the crude oil business in the post-World War period, which has led to increase in pollution in this region (Cunningham et al. 2019; Naser 2013). Because of water scarcity, the region has saw the aggressive development of desalination plants over the past few decades; simultaneous growth of power plants and subsequent generation of municipal wastes and effluents released in the water. Accidental oil spills and seasonal dust storms do also significantly impact the water quality in the region. This zone has been largely surveyed by many teams to quantify and characterize the pollutants among the aquatic organisms, especially fishes. However, most of the results are not as dangerous compared to the East Asian region. The heavy metals like cadmium, mercury and lead need to be regularly assessed in different zones to control pollution levels and food chain contamination (Cunningham et al. 2019).

13.3.3 Bay of Bengal and Mangrove Ecosystems

The northern part of the Bay of Bengal is another developing region of Southeast Asia. The major source of industrial pollutants in this region is the thermal power plants other than the domestic or municipal wastes (Rahman et al. 2019). Different studies have been carried out in the region, especially in the Indian and Bangladesh regions over the time (Kumar et al. 2012; Mandal and Suzuki 2002; Ray et al. 2006; Rahman et al. 2019). Most of the studies found traces of heavy metals like chromium, lead, mercury, etc.; however, accumulation levels in the body of fishes are not considered as hazardous (Kumar et al. 2012; Rahman et al. 2019).

The Coringa river mangrove ecosystem is one of the significant mangroves in the eastern coast of India, after the Sundarbans. Increase in urbanization and

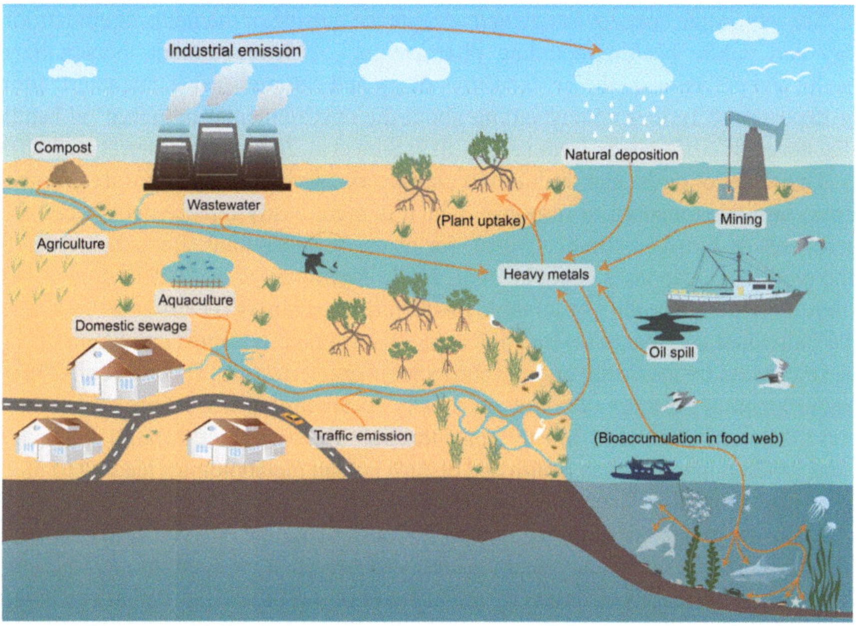

Fig. 13.3 Anthropogenic sources, movement, distribution and bioaccumulation of heavy metals in the coastal environment. (Reprinted with permission from Elsevier, Li et al. 2022, Licence number 5417800774445)

industrialization has led to anthropogenic changes in the landscapes and ecosystems and increase in discharge of effluents in the aquatic systems. Recent studies have led to understanding the biogeochemistry of the region and it was shown that the current pollution levels in the region have not crossed hazardous limits (Ray et al. 2006).

Figure 13.3 demonstrates the overall scenario, showing the anthropogenic sources, movement, distribution and bioaccumulation of heavy metals in the coastal environment.

13.4 Ecosystem Exposome

The term "exposome" was first used to describe the "life-course environmental exposures (including lifestyle factors) of an individual human from the prenatal period onwards" (Wild 2005). A move toward a more thorough characterization of exposure was suggested, with the goal of giving exposure risk factors the same priority as genetic risk factors. The genome was presented as a complement to exposure. Notably, although they correspond to the same idea, the terms "envirome" and "environome" are not as widely recognized and used as the exposome (Toscano and Oehlke 2005).

The exposome concept has gained acceptance outside of the fields of epidemiology and public health, including pharmacology, risk assessment, personalized health and precision medicine, (eco)toxicology and comprehensive characterization of exposure to further understanding about the environmental influence on health (Escher et al. 2017; Niedzwiecki et al. 2019; Sillé et al. 2020; Vermeulen et al. 2020).

The scope of molecular multiomics characterizations to study exposure is sometimes limited compared to response, which is more easily represented by cellular processes or biosignals indicative of, for example, behavioral (Beauchaine 2012) and emotional responses (Behnke et al. 2022; Zaehringer et al. 2020). Many exposures do not elicit readily observable biochemical responses (Peters et al. 2012). At the individual and macro levels, molecular characterizations must complement and be reinforced by structural and psychosocial characterizations (Peters et al. 2021). Environmental exposure of pollutants mostly occurs through ingestion, skin contact and inhalation. Pollutants mostly enter through digestive tract through water and food which impacts human health and disturb human metabolomics leading to morbidity and mortality (Cabral-Pinto et al. 2020).

Exposure to heavy metals are elements present in marine or water ecosystem can have a significant impact on human health because, when ingested through the food chain, they can directly or indirectly affect membrane proteins, particularly enzymes whose functions depend on them, and/or alter pathways connected to things like calcium and redox signaling. Because they are constantly flowing through all the matrices like soil, water, food, and air—heavy metals are essential components of any ecosystem. Estuaries and marine environments that sustain indigenous biomass and have important hydro-climatic and ecological qualities are essential for improving the local economy and ecology. Anthropogenic activities have resulted in ecosystem contamination, with heavy metals such as arsenic, mercury, lead, and cadmium playing a substantial role in their toxicity. Lead and Mercury are heavy metal pollutants constituting frequent exposomes. Lead and mercury exposomes are nonbiodegradable with a non-safe limit of exposure (Ferreira et al. 2022). Arsenic including other heavy metals enters into marine food resources through various activities in coastal area or through river ecosystem which significantly contaminate the sea food. Not only India but also throughout the world coastal community almost fully relies on sea food. The previous studies have also observed that there are numerous instances of food contamination in marine ecosystem in which seafood is acting as a bioindicator (Djedjibegovic et al. 2020; Di Bella et al. 2020).

13.5 Conclusion and Recommendation

Since the 1970s, several multilateral and bilateral treaties are developed involving the Southeast Asian countries to deal with marine pollution issues. The United Nations have also asked the States to join the United Nations Convention on the Law of the Sea (UNCLOS) which is an international framework for the prevention of pollution to harmonize the policies regarding effluent discharge in the international waters.

Majorly, there are three reasons for the marine pollution in South and Southeast Asian regions—(1) high levels of pollutants; (2) wide distribution and movement of pollutants in water; and (3) significant impacts of the pollutants on the local flora and fauna. Due to geographical locations of South and Southeast Asia, the impact on the ecosystem and human health is tremendous compared to other places.

With the increase of heavy metal pollution in the aquatic system, different remedial measures have also developed to facilitate the remediation process at various scales. The major processes include chemical precipitation, electrochemical methods, adsorption, bioremediation, etc. Remediation using various adsorbents has gained much attention in recent years with development of various low-cost absorbents to be used for industries and agricultural sectors. Though the new technologies have several disadvantages, however, some low-cost and environment friendly adsorbents have shown promising results. A detailed review on the absorbents has been published and their efficiency and disadvantages are discussed in recent researches. There are research gaps that need to be addressed in order to develop cost-effective yet efficient and sustainable remedial technologies to remediate heavy metal pollution in marine environments.

Creating public awareness in the coastal areas on detrimental effects of heavy metals in the marine and estuarine environment is the need of the hour. Campaigns need to be carried out by both government and nongovernment sectors. Also, marine pollution laws and their implementation need to be mainstreamed into the more widely understood and implemented conservation strategies. Generating ecological consciousness in the minds of local communities would be helpful to mitigate heavy metal pollution in the marine and estuarine environment.

References

Amin B, Ismail A, Arshad A, Yap CK, Kamarudin MS (2009) Anthropogenic impacts on heavy metal concentrations in the coastal sediments of Dumai, Indonesia. Environ Monit Assess 148(1–4):291–305. https://doi.org/10.1007/s10661-008-0159-z

Beauchaine TP (2012) Physiological markers of emotion and behavior dysregulation in externalizing psychopathology. Monogr Soc Res Child Dev 77:79–86. https://doi.org/10.1111/j.1540-5834.2011.00665.x

Behnke M, Buchwald M, Bykowski A, Kupiński S, Kaczmarek LD (2022) Psychophysiology of positive and negative emotions, dataset of 1157 cases and 8 biosignals. Sci Data 9:1–15. https://doi.org/10.1038/s41597-021-01117-0

Bhattacharya S, Sharma P, Mitra S, Mallick I, Ghosh A (2021) Arsenic uptake and bioaccumulation in plants: a review on remediation and socio-economic perspective in Southeast Asia. Environ Nanotechnol Monit Manage 15:100430. https://doi.org/10.1016/j.enmm.2021.100430

Bosch AC, O'Neill B, Sigge GO, Kerwath SE, Hoffman LC (2016) Heavy metals in marine fish meat and consumer health: a review. J Sci Food Agric 96(1):32–48. https://doi.org/10.1002/jsfa.7360

Cabral-Pinto MMS, Inácio M, Neves O, Almeida AA, Pinto E, Oliveiros B, Eduardo A, da Silva F (2020) Human health risk assessment due to agricultural activities and crop consumption in the surroundings of an industrial area. Expo Health 12:629–640

Castro-González MI, Méndez-Armenta M (2008) Heavy metals: implications associated to fish consumption. Environ Toxicol Pharmacol 26:263–271. https://doi.org/10.1016/j.etap.2008.06.001

Chan KM (1995) Metallothionein: potential biomarker for monitoring heavy metal pollution in fish around Hong Kong. Mar Pollut Bull 31(4–12):411–415. https://doi.org/10.1016/0025-326X(95)00125-7

Cheng Z, Chen KC, Li K-B, Nie XP, Wu SC, Wong CKC, Wong MH (2013) Arsenic contamination in the freshwater fish ponds of Pearl River Delta: bioaccumulation and health risk assessment. Environ Sci Pollut Res 20(7):4484–4495. https://doi.org/10.1007/s11356-012-1382-2

Cunningham PA, Sullivan EE, Everett KH, Kovach SS, Rajan A, Barber MC (2019) Assessment of metal contamination in Arabian / Persian Gulf fish: a review. Mar Pollut Bull 143(April):264–283. https://doi.org/10.1016/j.marpolbul.2019.04.007

Cuong D, Bayen S, Wurl O (2005) Heavy metal contamination in mangrove habitats of Singapore. Mar Pollut Bull 50(12):1727–1732. https://doi.org/10.1016/j.marpolbul.2005.09.007

Dallinger R (1994) Invertebrate organisms as biological indicators of heavy metal pollution. Appl Biochem Biotechnol 48(1):27–31. https://doi.org/10.1007/BF02825356

Dallinger R, Prosi F, Segner H, Back H (1987) Contaminated food and uptake of heavy metals by fish: a review and a proposal for further research. Oecologia 73(1):91–98. https://doi.org/10.1007/BF00376982

DeForest DK, Brix KV, Adams WJ (2007) Assessing metal bioaccumulation in aquatic environments: the inverse relationship between bioaccumulation factors, trophic transfer factors and exposure concentration. Aquat Toxicol 84(2 SPEC. ISS):236–246. https://doi.org/10.1016/j.aquatox.2007.02.022

Di Bella C, Traina A, Giosuè C, Carpintieri D, Lo Dico GM, Bellante A, Del Core M, Falco F, Gherardi S, Uccello MM, Ferrantelli V (2020) Heavy metals and PAHs in meat, milk, and seafood from Augusta area (southern Italy): contamination levels, dietary intake, and human exposure assessment. Front Public Health 8. https://doi.org/10.3389/fpubh.2020.00273

Djedjibegovic J, Marjanovic A, Tahirovic D, Caklovica K, Turalic A, Lugusic A, Omeragic E, Sober M, Caklovica F (2020) Heavy metals in commercial fish and seafood products and risk assessment in adult population in Bosnia and Herzegovina. Sci Rep 10(1). https://doi.org/10.1038/s41598-020-70205-9

Elsagh A, Jalilian H, Aslshabestari MG (2021) Evaluation of heavy metal pollution in coastal sediments of Bandar Abbas, the Persian Gulf, Iran: mercury pollution and environmental geochemical indices. Mar Pollut Bull 167:112314

Emenike, E. C., Iwuozor, K. O., & Anidiobi, S. U. (2022). Heavy metal pollution in aquaculture: sources, impacts and mitigation techniques. Biol Trace Elem Res, 200(10):4476-4492

Escher BI, Hackermüller J, Polte T, Scholz S, Aigner A, Altenburger R, Böhme A, Bopp SK, Brack W, Busch W et al (2017) From the exposome to mechanistic understanding of chemical-induced adverse effects. Environ Int 99:97–106. https://doi.org/10.1016/j.envint.2016.11.029

Ferreira G, Santander A, Chavarría L et al (2022) Functional consequences of lead and mercury exposomes in the heart. Mol Asp Med 87:101048. https://doi.org/10.1016/j.mam.2021.101048

Fowler S (1990) Critical review of selected heavy metal and chlorinated hydrocarbon concentrations in the marine environment. Mar Environ Res 29(1):1–64

Frazier JM (1979) Bioaccumulation of cadmium in marine organisms. Environ Health Perspect 28(February):75–79

Gregoria M, Tiquio JP, Marmier N, Francour P (2017) Ocean & Coastal Management Management frameworks for coastal and marine pollution in the European and South East Asian regions. Ocean Coast Manag 135:65–78. https://doi.org/10.1016/j.ocecoaman.2016.11.003

Gu YG, Lin Q, Wang XH, Du FY, Yu ZL, Huang HH (2015) Heavy metal concentrations in wild fishes captured from the South China Sea and associated health risks. Mar Pollut Bull 96(1–2):508–512. https://doi.org/10.1016/j.marpolbul.2015.04.022

Harsha K, Senthil P, Panda RC (2019) A review on heavy metal pollution, toxicity and remedial measures: current trends and future perspectives. J Mol Liq 290:111197. https://doi.org/10.1016/j.molliq.2019.111197
Hassan S, Sabreena, Khurshid Z, Bhat SA, Kumar V, Ameen F, Ganai BA (2022) Marine bacteria and omic approaches: a novel and potential repository for bioremediation assessment. J Appl Microbiol 133:2299–2313. https://doi.org/10.1111/jam.15711
Hylander LD, Goodsite ME (2006) Environmental costs of mercury pollution. Sci Total Environ 368:352–370. https://doi.org/10.1016/j.scitotenv.2005.11.029
Jamil Emon F, Rohani MF, Sumaiya N, Tuj Jannat MF, Akter Y, Shahjahan M, Abdul Kari Z, Tahiluddin AB, Goh KW (2023) Bioaccumulation and bioremediation of heavy metals in fishes—a review. Toxics 11:510. https://doi.org/10.3390/toxics11060510
Kar S, Maity JP, Jean JS, Liu CC, Liu CW, Bundschuh J, Lu HY (2011) Health risks for human intake of aquacultural fish: arsenic bioaccumulation and contamination. J Environ Sci Health A Tox Hazard Subst Environ Eng 46(11):1266–1273. https://doi.org/10.1080/10934529.2011.598814
Kojadinovic J, Potier M, Le Corre M, Cosson RP, Bustamante P (2007) Bioaccumulation of trace elements in pelagic fish from the Western Indian Ocean. Environ Pollut 146(2):548–566. https://doi.org/10.1016/j.envpol.2006.07.015
Kumar B, Sajwan Ks, Mukherjee DP (2012) Distribution of heavy metals in valuable coastal fishes from North East Coast of India. Turk J Fish Aquat Sci 12:917–924. https://doi.org/10.4194/1303-2712-v12
Kunito T, Kubota R, Fujihara J, Agusa T, Tanabe S (2008) Arsenic in marine mammals, seabirds, and sea turtles. Rev Environ Contam Toxicol 195:31–69. https://doi.org/10.1007/978-0-387-77030-7_2
LeBlanc PJ, Jackson AL (1973) Arsenic in marine fish and invertebrates. Mar Pollut Bull 4(6):88–90. https://doi.org/10.1016/0025-326X(73)90389-5
Leung HM, Leung AOW, Wang HS, Ma KK, Liang Y, Ho KC et al (2014) Assessment of heavy metals/metalloid (As, Pb, Cd, Ni, Zn, Cr, Cu, Mn) concentrations in edible fish species tissue in the Pearl River Delta (PRD), China. Mar Pollut Bull 78(1–2):235–245. https://doi.org/10.1016/j.marpolbul.2013.10.028
Li C, Wang H, Liao X, Xiao R, Liu K, Bai J, Li B, He Q (2022) Heavy metal pollution in coastal wetlands: a systematic review of studies globally over the past three decades. J Hazard Mater 424(Pt A):127312. https://doi.org/10.1016/j.jhazmat.2021.127312
Mandal BK, Suzuki KT (2002) Arsenic round the world: a review. Talanta 58(1):201–235. https://doi.org/10.1016/S0039-9140(02)00268-0
Mehana E-SE, Khafaga AF, Elblehi SS, Abd El-Hack ME, Naiel MAE, Bin-Jumah M, Othman SI, Allam AA (2020) Biomonitoring of heavy metal pollution using acanthocephalans parasite in ecosystem: an updated overview. Animals 10:811. https://doi.org/10.3390/ani10050811
Merola C, Bisegna A, Angelozzi G, Conte A, Abete MC, Stella C et al (2021) Study of heavy metals pollution and vitellogenin levels in brown trout (Salmo trutta trutta) wild fish populations. Appl Sci 11(11):4965
Mohan D, Pittman CU (2007) Arsenic removal from water/wastewater using adsorbents- a critical review. J Hazard Mater 142(1–2):1–53. https://doi.org/10.1016/j.jhazmat.2007.01.006
Naser HA (2013) Assessment and management of heavy metal pollution in the marine environment of the Arabian Gulf: a review. Mar Pollut Bull 72(1):6–13. https://doi.org/10.1016/j.marpolbul.2013.04.030
Neff JM (1997) Ecotoxicology of arsenic in the marine environment. Environ Toxicol Chem 16(5):917–927. https://doi.org/10.1002/etc.5620160511
Niedzwiecki MM, Walker DI, Vermeulen R, Chadeau-Hyam M, Jones DP, Miller GW (2019) The exposome: molecules to populations. Annu Rev Pharmacol Toxicol 59:107–127. https://doi.org/10.1146/annurev-pharmtox-010818-021315
Nnaji J, Okoye FC (2007) Toxicity of heavy metals to fish: an important consideration for succesful aquaculture (21st Annua). Fisheries Society of Nigeria (FISON), New Bussa, Niger State, Nigeria

Peters A, Hoek G, Katsouyanni K (2012) Understanding the link between environmental exposures and health : does the exposome promise too much? J Epidemiol Community Health 66:103–105. https://doi.org/10.1136/jech-2011-200643

Peters A, Nawrot TS, Baccarelli AA (2021) Hallmarks of environmental insults. Cell 184:1455–1468. https://doi.org/10.1016/j.cell.2021.01.043

Rahman SM, Hossain S, Ahmed K, Akther S, Jolly YN, Akhter S et al (2019) Assessment of heavy metals contamination in selected tropical marine fish species in Bangladesh and their impact on human health. Environ Nanotechnol Monit Manage 11(January 2018). https://doi.org/10.1016/j.enmm.2019.100210

Rani A, Kumar A, Lal A, Pant M (2014) Cellular mechanisms of cadmium-induced toxicity: a review. Int J Environ Health Res 24(4):378–399. https://doi.org/10.1080/09603123.2013.835032

Ray AK, Tripathy SC, Patra S, Sarma VV (2006) Assessment of Godavari estuarine mangrove ecosystem through trace metal studies. Environ Int 32(2):219–223. https://doi.org/10.1016/j.envint.2005.08.014

Shazili NAM, Yunus K, Ahmad AS, Abdullah N, Rashid MKA (2006) Heavy metal pollution status in the Malaysian aquatic environment. Aquat Ecosyst Health Manag 9:137–145. https://doi.org/10.1080/14634980600724023

Sillé FCM et al (2020) The exposome—a new approach for risk assessment. ALTEX 37:3–23. https://doi.org/10.14573/altex.2001051

Sivaperumal P, Sankar TV, Viswanathan Nair PG (2007) Heavy metal concentrations in fish, shellfish and fish products from internal markets of India vis-a-vis international standards. Food Chem 102(3):612–620. https://doi.org/10.1016/j.foodchem.2006.05.041

Toscano WA, Oehlke KP (2005) Systems biology: new approaches to old environmental health problems. Int J Environ Res Public Health 2:4–9. https://doi.org/10.3390/ijerph2005010004

Vardhan KH, Kumar PS, Panda RC (2019) A review on heavy metal pollution, toxicity and remedial measures: current trends and future perspectives. J Mol Liq 290:111197

Velusamy A, Satheesh Kumar P, Ram A, Chinnadurai S (2014) Bioaccumulation of heavy metals in commercially important marine fishes from Mumbai Harbor, India. Mar Pollut Bull 81(1):218–224. https://doi.org/10.1016/j.marpolbul.2014.01.049

Vermeulen R, Schymanski EL, Barabási AL, Miller GW (2020) The exposome and health: where chemistry meets biology. Science 367:392–396. https://doi.org/10.1126/science.aay3164

Wang SL, Xu XR, Sun YX, Liu JL, Li HB (2013) Heavy metal pollution in coastal areas of South China: a review. Mar Pollut Bull 76(1–2):7–15. https://doi.org/10.1016/j.marpolbul.2013.08.025

Wrench J, Fowler SW, Yasar Ünlü M (1979) Arsenic metabolism in a marine food chain. Mar Pollut Bull 10(1):18–20. https://doi.org/10.1016/0025-326X(79)90318-7

Wild CP (2005) Complementing the genome with an "exposome": the outstanding challenge of environmental exposure measurement in molecular epidemiology. Cancer Epidemiol Biomarkers Prev. 14(8):1847–50. https://doi.org/10.1158/1055-9965.EPI-05-0456

Yan C, Qu Z, Wang J, Cao L, Han Q (2022) Microalgal bioremediation of heavy metal pollution in water: recent advances, challenges, and prospects. Chemosphere:131870. https://doi.org/10.1016/j.chemosphere.2021.131870

Zaehringer J, Jennen-Steinmetz C, Schmahl C, Ende G, Paret C (2020) Psychophysiological effects of downregulating negative emotions: insights from a meta- analysis of healthy adults. Front Psychol 11. https://doi.org/10.3389/fpsyg.2020.00470

Zhang W, Miao A-J, Wang N-X, Li C, Sha J, Jia J et al (2022) Arsenic bioaccumulation and biotransformation in aquatic organisms. Environ Int 163:107221

Chapter 14
Possible Strategies for Oil Spill Removal from Mangrove Water Emphasizing Bioadsorption Strategies and Its Limitations

Pradipta Kumar Ghosh, Shrayan Bhattacharjee, Arunava Mukherjee, and Banani Mandal

Abstract Oil spills in mangrove ecosystems pose significant threats to biodiversity and ecological integrity, driven primarily by anthropogenic activities such as crude oil extraction and transportation. The unique intertidal habitats of mangroves, crucial for coastal protection, carbon sequestration and habitat provision, are disrupted by the introduction of oil, leading to detrimental consequences. Oil compounds adhere to mangrove roots and sediments, impeding gas exchange and nutrient uptake by vegetation. Toxic components within crude oil, such as polycyclic aromatic hydrocarbons (PAHs), disrupt physiological processes and impair reproductive success of mangrove biota. Sublethal effects persist, compromising population resilience and ecosystem recovery, while oil contamination alters sediment properties, affecting crucial microbial communities. Loss of mangrove-associated fauna disrupts trophic dynamics, cascading throughout the food web and posing risks to human health through contaminated seafood consumption. Mitigating oil spill impacts requires interdisciplinary approaches, prioritizing immediate response efforts for containment and removal, alongside long-term monitoring to assess ecosystem resilience and guide adaptive management strategies. Collaborative research endeavors are crucial to advance understanding of oil spill impacts on mangrove ecosystems, informing evidence-based management practices and policy decisions for sustainable coastal stewardship. Addressing these challenges demands concerted

P. K. Ghosh · S. Bhattacharjee
Ecosystem and Ecology Laboratory, Post Graduate department of Zoology, Ramakrishna Mission Vivekananda Centenary College (Autonomous), Kolkata, West Bengal, India

A. Mukherjee (✉)
Department of Zoology, Ramakrishna Mission Vivekananda Centenary College (Autonomous), Kolkata, India

B. Mandal
Department of Zoology, Jogesh Chandra Chaudhuri College, Kolkata, West Bengal, India

S. Dey, S. Bhattacharya (eds.), *Biotechnological Interventions in the Removal of Emerging Pollutants*, Interdisciplinary Biotechnological Advances,
https://doi.org/10.1007/978-981-97-9922-0_14

efforts to minimize environmental damage, enhance ecosystem resilience, and promote global-scale sustainability.

Keywords Oil spills · mangrove · mitigation · adsorption · technology · interdisciplinary approaches

14.1 Introduction

Sources of oil spill in fragile mangrove ecosystem are definitely anthropogenic in nature. Extraction of crude oil derived oil-leaching and spill over accidents during transportation are the well-known sources of oil pollution in marine environment, including mangroves (Onyena and Sam 2020). The discharge of oil from household and sewages is also not at all ignorable. Being an ecotone zone, mangrove experiences large pollution pressure coning from upper riverine flow and lower marine outputs. Tides in mangroves were considered to be one of the most dynamic regulators for oil pollution (Zhang et al. 2019). Tides mainly deposited oil in mangrove integrated root system and increased its availability in mangrove habitat due to natural relatively depression geomorphology of mangroves.

On a general note, oil spill in natural and its adjacent area are a major problem around the world for ages and even today it is considered as a persistent menace (Asif et al. 2022; Zhen et al. 2020). Marine or coastal oil spill and its ill effects have been studied for a long time (Blumer et al. 1971). Oil spill cleanup is a very time consuming process which also takes toll in manpower and expanses. Hence, oil spill is not only a problem of environmental or ecological arena but also of an economic point of view (Etkin 2000). Oil spill in mangrove area is often regarded as an ecological mayhem, as mangroves around the world are considered as treasure troves of biodiversity and carbon storage (Kumar et al. 2021). Mangroves, acting as a global fish nursery and nutrient disperser, are a mega ecosystem support provider in nutshell. Thus any kind of oil spill in mangroves and its associated coastal or estuarine zone can be a serious problem. Immediate response efforts should prioritize the containment and removal of spilled oil to minimize environmental exposure and empower natural recovery processes (Rönnbäck et al. 2007). Long-term monitoring programs are essential to assess ecosystem resilience and guide adaptive management strategies, ensuring the sustained functionality of mangrove ecosystems in the face of anthropogenic disturbances (Alongi 2014). Collaborative research endeavors with multifaceted approaches are imperative to advanced understanding of oil spill impacts on mangrove ecosystems, informing evidence-based management practices and policy decisions aimed at safeguarding these invaluable coastal habitats (Yáñez-Arancibia et al. 2016) which can promote sustainable coastal stewardship on a global scale. In this chapter, we will majorly focus on the adsorption strategies against oil spill and its limitations. We will also briefly speculate on the available options on the mitigation of oil spills in mangrove zones and its consequent health hazards.

14.2 Mechanical Removal of Oil and Grease from Water Using Biological Modifications or Bio-Derived Components

Mechanical removal of oil suspended on water surface is a very tedious procedure. Though used from ages, different mechanical process of oil separation from water has been time tested before. In modern day several bioinspired or bio-derived technologies have been used to successfully separate water and oil easily. Adhesion of molecules, atoms or ions of gas, liquid, colloid and dissolved solid to a layer or surface is a natural phenomenon, which in term of scientific jargon is known as Adsorption. The latter mentioned surface layer is known as adsorbent and the resultant film of attached material is known as adsorbate. Often confused with absorption process, adsorption is a slightly different process where the fluid is not dissolved by another fluid or solid after contact. Unlike absorption, adsorption is a strict surface restricted phenomenon, where the adsorbent doesn't permit the entry of adsorbate into the surface of the matter. In terms of filtration and removal of oil layer from water samples, adsorption processes are widely used. Several examples and their brief descriptions are stated below (Fig. 14.1).

14.3 Biochar

Biochar (BC) is a solid product from the temperature-specific pyrolysis of organic carbon rich matter under limited aerobic or oxygen environment (Liang et al. 2021). Biochar can be formed from nearly all agricultural, livestock and fishery bio-wastes and by products along with urban kitchen waste, municipal solid biodegradable waste and sewage sludge (Gopinath et al. 2021). Biochar as in general has adsorbent properties, which is very useful for separation of oil layer from water surface effectively. Adsorption is a surface phenomenon by which liquid and gas molecules or their colloid forms attach on the outer surface of the adsorbent material. Recent works regarding biochar from different biotic carbon rich sources have found their reusable adsorption modifications that can adsorb pollutants including spilled oil and grease. Biochar made from cornstalk wastes under 350 degree Centigrade (CSBC350) temperature have been found to be an excellent oil adsorbent due to its profound porosity. Other than cornstalk biochar (CSBC), other alternatives such as corncob biochar (CCBC), *Sophora* sawdust biochar (SSBC) and rice husk biochar (RHBC) have been also used experimentally in studies before (Sun et al. 2022). Further enhancing the CSBC350 with magnetic precursors have amplified their oil adsorption output. In relation with the spills of crude oil, diesel oil and engine oil in the water medium, it has been calculated that it can adsorb oil droplets of 8.77 $g\ g^{-1}$, 4.01 $g\ g^{-1}$ and 4.44 $g\ g^{-1}$, respectively. The sorption mechanisms are largely governed by hydrophobic forces, pore intercepts, pi-pi electron bond and hydrogen-bond interactions with their water repulsion properties. Using Heglig tree

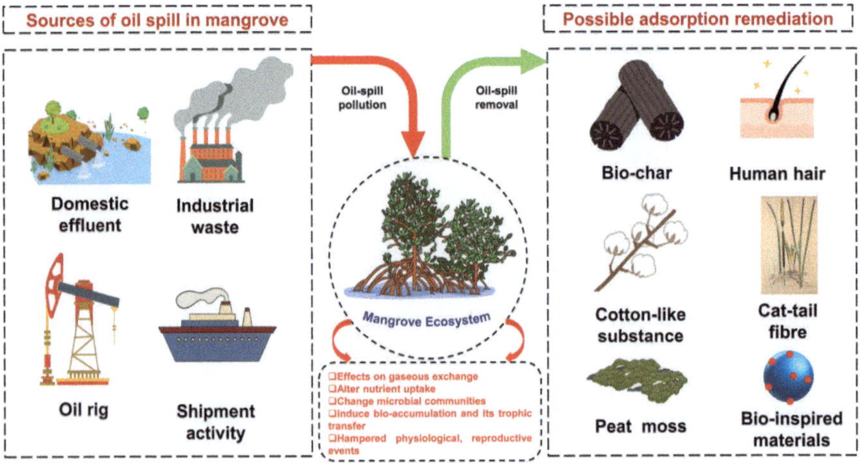

Fig. 14.1 Graphic representation of the sources of oil spill near mangroves and its mitigation options

bark-derived biochar component with cobalt ferrite, it is found that the resultant magnetic BC can remove oil layer successfully from contaminated sea water (Amar et al. 2023). Previous works by Silvani et al. (2017) have also shown that BC sourced from pine wood is much effective than Activated charcoal (AC) regarding the adsorption of toluene under experimental conditions. Even though both materials show high range oil adsorption, but BC found to be superior due to unique textural features and specific sorption abilities (mg sorbed /m^{-2} of surface) through kinetic and equilibrium tests. Using real-time accidental oil spill as an experimental study site (*X-Press Pearl cargo ship accident, 2021, Sri Lanka*), Pallewatta and his group (Pallewatta et al. 2023) demonstrated adsorption properties of wood derived biochar sourced from wood fueled power plant residue on large scale mineral oil spill. All type of wood derived biochar from this study shows positive results with significant oil pulling abilities. Peat-derived biochar is another good competitor for oil spill adsorption, AlAmeri et al. (2019) showed that not only this type of biochar can be effective (32.5 g of crude oil/g) in higher climatic temperature (up to 45 °C), but also it can be reused multiple time to adsorb oil slick. Using commercial hard wood biochar and maple wood biochar (synthesized anoxically in 300–700 °C), it is tested that in both freshwater and in marine water oil spill can be contained as the biochar even attached with oil droplet can be buoyant in nature thus easily gatherable. But it has a limitation also, as it was found oil was less effectively adsorbed in the form of weathered water-in-oil microemulsion ("chocolate mousse") than in water surface spread-out form (Nguyen and Pignatello 2013). Biochar, thus being a natural biodegradable material for oil slick trapping is a better alternative for pollution control in developing and underdeveloped countries. With further modifications, it can be also sustainably reusable to counterattack the oil spill and its resultant economic burden.

14.4 Human Hair

Human hair (Terminal hair or pili on scalp) is a very promising material for the containment of the oil slick on water surfaces. Various studies around the world have shown that effective adsorbent properties of human hair in its dried form can successfully capture and trap oil layer from water surface. Study done by Ifelebuegu et al. (2015) shows that human hair had good adsorption potentials, up to 8000 mg/g in first use. Using artificial sea water as a medium, different oils have been spread out to test the effectiveness of human hair oil traps. It is also interesting to find that wooly hair (African hair) has more affinity with oil slick; hence, it catches much more proportion of oil than other subtypes. The reusability of this hair traps also tested in this study which denotes that, it can be used successfully multiple times, without significant decrease in absorbance. Ukotije-Ikwut et al. (2016) also studied similar works using multiple commercial oil samples. It is also shown that the success rate of oil adsorption decreases with rising temperature Up to a certain degree. Using human hair as a sorbent boom for mitigation of oil contamination in water body is another promising approach, and ex situ study done by Pagnucco and Phillips (2018) used adsorbed booms made from human hair which effectively collect crude mineral oil content from water column (0.84 g of crude oil for every 1 g of sorbent materials).The only problem that arises while using hair-derived boom, that hair tends to be less buoyant while getting heavier due to oil sorption and increasing porosity of the hair.

14.5 Cotton-Like Substance

Cotton waste or cotton-like fibers are another nonconventional natural adsorbent which has been used successfully to combat oil spill pollution. Cotton-containing nonwoven have been used for oil sorption as early as in 1993 (Choi et al. 1993), which shows that, oil sorption capacities of the cotton is slightly greater than those of sorbents made from 100% polypropylene fibers. The cotton sorbents were reusable and can float in ex situ seawater bath for a longer time. Oil sorption by cotton fiber was controlled by adsorptive fiber surface along with capillary actions through its lumen, which is contrary to the polypropylene sorption through capillary bridges between its fibers. Adebajo and Frost (2004) have shown that modification by acetylation of raw cotton has a significant potential for oil spill reduction. Fibers from silk cotton tree (Kapok) have been also used for adsorption strategies for diesel oil spill. Removal of waxy material from the fibers have been shown that the fibers are composed by aliphatic and aromatic hydrocarbons which confers hydrophobic and oleophilic character can reach more than 58.462 ± 6.733 g/g of diesel sorption, which is more effective than its nonmodified counterparts. Low value raw cotton fiber has also been used for heavy oil removal from water (Hussein et al. 2011) which shows a greater number of reduction in oil layer with 22.5 g oil for 1 g of cotton fiber under given limiting time of 15 minutes at 25 °C temperature, after which the rate of adsorption falters. Typical wetland species such as Cat tail (*Typha* sp.) possess cotton like fibrous material on its top most part, which also have adsorption properties. When tested with ex situ oil spill conditions, these fibers show adsorption to engine oil and vegetable oil at a level of 13.4 g/g and 14.6 g/g, respectively. Oil retention capability of cattail fiber hours was over 95% even after 24 hours. The bird down feather (belly feather) like structure on cat tail fibers, were mainly responsible for the oil adsorbance. Though the efficiency of cat tail fibers regarding oil adsorption is inferior to kapok or cotton alternatives (Cui et al. 2014), but it is also notable that cat tails are an inexpensive, semi aquatic weed, which can be harvested with very low cost at any time. Using cattail fibers and 10–20% polypropylene/polyethylene sheath-core composite fibers (ES) complex, oil retention can be possible, which was proved by Dong et al. (2015). Although complex cattail-ES fiber were less effective for oil sorption capacity (11–14 g/g), they offer a total about 90% of oil retention capacity even after 24 h dripping and hold above 80% of initial adsorption capacity even after 10 sorption sessions. Further research regarding synthetic modification of cattail fiber shows that when attached with water repulsing polyimide, more efficient oil separation is facilitated (Wang et al. 2023). The aerogel complex has very light weight, which ensures its buoyancy (29.7 mg/cm^3), superhydrophobicity (water contact angle = 151.3°), and compressibility, along with neutral chemical stability, which are essential for the oil sorption with better amount (25.1–71.0 g/g). It is also recorded that less than 3% decrease in separation efficiency have been found after repeated 10 cycles of sorption, whereas the initial efficiency was calculated at 98%.

14.6 Sorption by Peat Moss

Dry peat moss is a good alternative for oil removal in aquatic medium. Peat moss is a large group of bryophyte moss that inhabits boggy or marshy grounds and has an excellent pollutant storing capabilities. It is also known as sphagnum moss, both of its dead and live cells can hold an enormous quantity of water (4–6.35 kg/kg). Narayanan et al. (2018) have shown that peat moss can be used as an oil remover, by enhancing its sorption limit through aeration, heat treatment and bog formations up to 10%. The overall result of oil sorption is found to be nearly equivalent to 22 times of its own initial weight as it become superhydrophobic and oleophilic in nature. Studies on comparative oil sorption abilities shows that, peat sorbent reach optimum oil saturation within 40 minutes under ex situ conditions with a swelling limit of 11 times its initial weight whereas the sorption capacity was 11.25 g of crude oil per gram of peat sorbent (Ramachandran et al. 2021). Though it was found all type pollutant oils get adsorbed by peat moss in nearly same quantities, peat moss is largely effective for more viscous oils. The study also dealt in the future of oil-soaked peat sorbent, which, despite being a secondary pollutant, can be neutralized and reused as a bio-oil through hydrothermal liquefaction (HTL) process.

14.7 Bioinspired Materials for Oil Spill Adsorption

To mitigate oil spill related contamination in its early period, several bioinspired model has been proposed which have imitate their very structure from nature. By evaluating superhydrophobic self-cleaning lotus leaves (*Nelumbo* sp.) and porous biomaterials, special polyurethane foam have been constructed with simultaneous properties of superhydrophobicity and superoleophilicity, which being buoyant can float in water (Zhang et al. 2016). The outer layer of the shark skin is superhydrophilic and superoleophilic when exposed in air medium but superoleophobic in underwater media (Bixler and Bhushan 2013). Water coats the skin and forms a thin water layer on which oil droplets settle briefly to result in superoleophobicity. But it is also of concern that the shark skin like components are only effective on the water-air surface, which is a major problem for this work (Bhushan 2019). Bioinspired superhydrophobic nanofur surface, which mimics biosurface, is another great alternative for oil absorption, which is fabricated using a hot pulling method, in which fine nano- and microhairs are pulled on the surface of the polymer due to adhesion to a heated sandblasted steel plate (Röhrig et al. 2013). The nanofur component can absorb oil with high capacity, which makes it an ideal material for constructing a porous superoleophobic membrane filter for water-oil separation. The super effective oleophilic nanofur from biodegradable source can also reduce the resultant hazard of decaying synthetic polymer or plastic waste; hence, a biodegradable nanofur is an ecofriendly component, which can be sourced using polylactic acid (PLA) (Mathew et al. 2005). Derived from mostly corn starch or sugarcane waste (bagasse), PLA is a poster boy of bioplastic till this day. PLA nanofur surface is covered in high aspect ratio microhairs, submicron grooves on the PLA microhairs, which is the underlying reason for its superoleophilic nature, thus making it a future candidate for mainstream oil removal strategies.

14.8 Conclusion

Oil spills in mangrove ecosystems worldwide pose significant threats to biodiversity and ecological integrity. Mangroves, characterized by their unique intertidal habitats, provide essential ecosystem services, including coastal protection, carbon sequestration, and habitat provision for numerous species (Alongi 2008). However, the introduction of oil into these sensitive environments disrupts critical biotic and abiotic interactions, leading to detrimental consequences (Giri et al. 2019). The hydrophobic nature of oil compounds facilitates their adherence to mangrove roots and sediments, impeding gas exchange and hindering nutrient uptake by vegetation (Ghazali et al. 2019). Moreover, toxic components within crude oil, such as polycyclic aromatic hydrocarbons (PAHs), exert harmful effects on mangrove biota, disrupting physiological processes and impairing reproductive success (Lin et al. 2020). The ramifications of oil spills extend beyond immediate mortality, as sublethal effects may persist, compromising population resilience and ecosystem recovery (Smith et al. 2018). Additionally, oil contamination alters sediment properties, affecting microbial communities crucial for nutrient cycling and decomposition (Wang et al. 2021). Consequently, the loss of mangrove-associated fauna, including crabs, mollusks, and fish, disrupts trophic dynamics, leading to cascading impacts throughout the food web (Twilley et al. 2001). Furthermore, oil-derived pollutants can bioaccumulate within organisms, posing risks to human health through the consumption of contaminated seafood (Mendelssohn et al. 2020). Effective mitigation strategies for oil spill events in mangrove ecosystems necessitate interdisciplinary approaches integrating ecological, engineering, and socio-economic considerations (Hussain et al. 2019). Among several sorption strategies and alternatives which were elucidated on the previous sections most of them are largely experimented on ex situ conditions. This kind of management often takes real field constraints and challenges in hindsight, which is often impractical for large level of oil spill removal. As global mangrove concentrate around the tropical belt, which is also the poorer counterpart of both global south and north and largely distributed around developing and under developed nations, economic constraints are a constant peril. So the mitigation policy should demand for a more ecofriendly and eco-sustainable alternatives for containment of the oil spill accident aftermaths. Use of inferior and waste agricultural waste for the adsorption of oil is surely a better alternative for these nations. Another problem that can arise from the sorption materials is the secondary pollution from the sorption material itself, which can be mitigated by proper containment policies, ranging from incineration of oil adsorbed biochars to the natural biodegradation of PLA bioplastic sheets. The proper reusability of biomaterials in postsorption scenario with its trapped oil load is of great significance for fuel conservation and eco-protection. Some of the recovery processes are deeply linked with methods like, proper mechanical squeezing, chamber heating, pyrolysis, solvent extraction and vacuum filtration. Controlled temperature up to the boiling point of the adsorbates can also reactivate the solvent-saturated sorbents. Despite alternatives like this, it is largely limited to the low boiling point solvents mostly; hence, more study is required to the removal of high viscous oil from the sorbent itself. On a brighter side, Mangroves being a resilient landscape can also grow back to reclaim the oil polluted stratum and can offer shelter to its regular denizens in both micro and macro levels. Thus it can be concluded with this hope, that our mangrove needs our gentle nudge of bio-strategies to counter its oil spill pollution on its own. Because as famous environmentalist Rachel Louise Carson has quoted once in her famous book, "Silent Spring (1962)" that "In nature, nothing exists alone," which is also true for the man-mangrove bond to fight back pollution.

Acknowledgments All authors cordially acknowledge the Principle of Ramakrishna Mission Vivekananda Centenary College, Rahara, along with the Department of Zoology of the same institution, for supporting the authors during the work. Author SB is greatly thankful to the Zoological Society of Kolkata for giving fellowship.

References

Adebajo MO, Frost RL (2004) Infrared and 13C MAS nuclear magnetic resonance spectroscopic study of acetylation of cotton. Spectrochim Acta A Mol Biomol Spectrosc 60(1-2):449–445. https://doi.org/10.1016/S1386-1425(03)00297-4

AlAmeri K, Giwa A, Yousef L, Alraeesi A, Taher H (2019) Sorption and removal of crude oil spills from seawater using peat-derived biochar: an optimization study. J Environ Manag 250:109465. https://doi.org/10.1016/j.jenvman.2019.109465

Alongi DM (2008) Mangrove forests: resilience, protection from tsunamis, and responses to global climate change. Estuar Coast Shelf Sci 76(1):1–13. https://doi.org/10.1016/j.ecss.2007.08.024

Alongi DM (2014) Carbon cycling and storage in mangrove forests. Annu Rev Mar Sci 6:195–219. https://doi.org/10.1146/annurev-marine-010213-135020

Amar IA, Alzarouq A, Mohammed W, Zhang M, Matroed N (2023) Oil spills removal from seawater surface by magnetic biochar composite derived from Heglig tree bark and cobalt ferrite. World J Eng. (ahead-of-print)

Asif Z, Chen Z, An C, Dong J (2022) Environmental impacts and challenges associated with oil spills on shorelines. J Mar Sci Eng 10(6):762

Bhushan B (2019) Bioinspired oil–water separation approaches for oil spill clean-up and water purification. Phil Trans R Soc A 377(2150):20190120

Bixler GD, Bhushan B (2013) Bioinspired micro/nanostructured surfaces for oil drag reduction in closed channel flow. Soft Matter 9(5):1620–1635

Blumer M, Guillard RRL, Chase T (1971) Hydrocarbons of marine phytoplankton. Mar Biol 8:183–189

Choi HM, Kwon HJ, Moreau JP (1993) Cotton nonwovens as oil spill cleanup sorbents. Text Res J 63(4):211–218. https://doi.org/10.1177/004051759306300401

Cui Y, Xu G, Liu Y (2014) Oil sorption mechanism and capability of cattail fiber assembly. J Ind Text 43(3):330–337

Dong T, Xu G, Wang F (2015) Oil spill cleanup by structured natural sorbents made from cattail fibers. Ind Crop Prod 76:25–33. https://doi.org/10.1016/j.indcrop.2015.06.021

Etkin DS (2000) Worldwide analysis of marine oil spill cleanup cost factors. In: Arctic and marine oilspill program technical seminar, vol 1. Environment Canada, pp 161–174. 1999

Ghazali Z, Aris AZ, Zakaria MP, Juahir H, Yusoff MK (2019) Monitoring of oil pollution in the coastal area of Terengganu, Malaysia. IOP Conf Ser Earth Environ Sci 276(1):012018. https://doi.org/10.1088/1755-1315/276/1/012018

Giri A, Coutriade M, Racaud A, Stefanuto PH, Okuda K, Dane J et al (2019) Compositio nal elucidation of heavy petroleum base oil by GC× GC-EI/PI/CI/FI-TOFMS. J Mass Spectrom 54(2):148–157. https://doi.org/10.1002/jms.4255

Gopinath A, Divyapriya G, Srivastava V, Laiju AR, Nidheesh PV, Kumar MS (2021) Conversion of sewage sludge into biochar: a potential resource in water and wastewater treatment. *Environ Res* 194:110656. https://doi.org/10.1016/j.envres.2021.110656

Hussain AAA, Vincent-Bonnieu S, Bahrim RK, Pilus RM, Rossen WR (2019) The impacts of solubilized and dispersed crude oil on foam in a porous medium. Colloids Surf A Physicochem Eng Asp 579:123671

Hussein F, Janat M, Yakoub A (2011) Assessment of yield and water use efficiency of drip-irrigated cotton (Gossypium hirsutum L.) as affected by deficit irrigation. Turk J Agric For 35(6):611–621

Ifelebuegu AO, Ukpebor JE, Obidiegwu CC, Kwofi BC (2015) Comparative potential of black tea leaves waste to granular activated carbon in adsorption of endocrine disrupting compounds from aqueous solution. Glob J Environ Sci Manag 1(3):205–214. https://doi.org/10.22034/gjesm.2015.01.03.004

Kumar A, Anju T, Archa V, Warrier VP, Kumar S, Goud GS et al (2021) Mangrove forests: distribution, species diversity, roles, threats and conservation strategies. In: Wetlands conservation: current challenges and future strategies, pp 229–271

Liang J, Li Y, Si B, Wang Y, Chen X, Wang X et al (2021) Optimizing biochar application to improve soil physical and hydraulic properties in saline-alkali soils. Sci Total Environ 771:144802. https://doi.org/10.1016/j.scitotenv.2021.144802

Lin Q, Mendelssohn IA, Suidan MT (2020) Interactive effects of oiling and inundation on mangrove seedling root function and soil biogeochemistry. Environ Pollut 264:114729. https://doi.org/10.1016/j.envpol.2020.114729

Mathew AP, Oksman K, Sain M (2005) Mechanical properties of biodegradable composites from poly lactic acid (PLA) and microcrystalline cellulose (MCC). J Appl Polym Sci 97(5):2014–2025

Mendelssohn IA, Lin Q, Suidan MT (2020) The chronic impacts of oil spills on marine organisms. Annu Rev Mar Sci 12:25–48. https://doi.org/10.1146/annurev-marine-010419-011108

Narayanan S, Martin AN, Turrentine FE, Bauer TW, Adams RB, Zaydfudim VM (2018) Mortality after pancreaticoduodenectomy: assessing early and late causes of patient death. *J Surg Res* 231:304–308. https://doi.org/10.1016/j.jss.2018.05.041

Nguyen HN, Pignatello JJ (2013) Laboratory tests of biochars as absorbents for use in recovery or containment of marine crude oil spills. Environ Eng Sci 30(7):374–380. https://doi.org/10.1089/ees.2012.0114

Onyena AP, Sam K (2020) A review of the threat of oil exploitation to mangrove ecosystem: Insights from Niger Delta, Nigeria. Glo Ecol Conserv 22:e00961. https://doi.org/10.1016/j.gecco.2020.e00961

Pagnucco R, Phillips ML (2018) Comparative effectiveness of natural by-products and synthetic sorbents in oil spill booms. J Environ Manag 225:10–16

Pallewatta S, Samarasekara SM, Rajapaksha AU, Vithanage M (2023) Oil spill remediation by biochar derived from bio-energy industries with a pilot-scale approach during the X-Press Pearl maritime disaster. Mar Pollut Bull 189:114813

Ramachandran V, Shriram MK, Mathew ER, Ramkumar K, Prakash DG, Venkatachalam CD (2021) Oil spill remediation and valorization of oil-soaked peat sorbent to biofuel by hydrothermal liquefaction. Biomass Convers Biorefinery 1–13. https://doi.org/10.1007/s13399-021-01851-x

Röhrig M, Schneider M, Etienne G, Oulhadj F, Pfannes F, Kolew A et al (2013) Hot pulling and embossing of hierarchical nano-and micro-structures. J Micromech Microeng 23(10):105014

Rönnbäck P, Troell M, Kautsky N, Primavera JH (2007) Ecosystem goods and services from Swedish coastal habitats: identification, valuation, and implications of ecosystem shifts. Ambio 36(7):534–544. https://doi.org/10.1579/0044-7447(2007)36[534:EGASFS]2.0.CO;2

Silvani L, Vrchotova B, Kastanek P, Demnerova K, Pettiti I, Papini MP (2017) Characterizing biochar as alternative sorbent for oil spill remediation. Sci Rep 7(1):43912

Smith EA, Philipp MA, Gaulke LS (2018) Chronic effects of the Exxon Valdez oil spill on murres: differences between oiled and unoiled areas. Mar Pollut Bull 129(1):448–455. https://doi.org/10.1016/j.marpolbul.2018.02.015

Sun Y, Zhang Y, Lu L, Wu Y, Zhang Y, Kamran MA, Chen B (2022) The application of machine learning methods for prediction of metal immobilization remediation by biochar amendment in soil. Sci Total Environ 829:154668. https://doi.org/10.1016/j.scitotenv.2022.154668

Twilley RR, Chen RH, Hargis T (2001) Carbon sinks in mangroves and their implications to carbon budget of tropical coastal ecosystems. Water Air Soil Pollut 125(1-4):365–381. https://doi.org/10.1023/A:1005253414569

Ukotije-Ikwut PR, Idogun AK, Iriakuma CT, Aseminaso A, Obomanu T (2016) A novel method for adsorption using human hair as a natural oil spill sorbent. Int J Sci Eng Res 7:1754–1765

Wang Z, Yang Z, Deng Y, Hu X (2021) Assessing the impact of oil pollution on the mangrove microbia l community: insights from shotgun metagenomics. Sci Total Environ 754:142206. https://doi.org/10.1016/j.scitotenv.2020.142206

Wang Y, Lv Q, Tian J, Zhang Y, Jiang C, Zhang W (2023) The high genetic yield of Xinjiang cotton is associated with improvements in boll-leaf system photosynthesis. Field Crop Res 304:109176. https://doi.org/10.1016/j.fcr.2022.109176

Yáñez-Arancibia A, Day JW, Currie DJ, Rybczyk JM (2016) Impacts of human activities on mangrove biodiversity: a regional analysis. Environ Manag 57(2):327–349. https://doi.org/10.1007/s00267-015-0607-0

Zhang L, Li L, Dang ZM (2016) Bio inspired durable, superhydrophobic magnetic particles for oil/water separation. J Colloid Interface Sci 463:266–271

Zhang B, Matchinski EJ, Chen B, Ye X, Jing L, Lee K (2019) Marine oil spills—oil pollution, sources and effects. In: World seas: an environmental evaluation. Academic Press, pp 391–406. https://doi.org/10.1016/B978-0-12-805052-1.00018-5

Zhen Z, Li D, Li Y, Chen S, Bu S (2020) Trajectory and weathering of oil spill in Daya bay, the South China sea. Environ Pollut 267:115562

Chapter 15
Removal of Emerging Inorganic Wastewater Pollutants Using Fluidized Bed Bioreactor: A Review

Nikila Bhutia, Priyadarshika Adhikari, and Neelanjan Dutta

Abstract Wastewater treatment is a crucial process for protecting the environment and human health. However, the emergence of new and persistent inorganic pollutants in wastewater has posed a significant challenge to conventional treatment processes. Fluidized bed bioreactors (FBRs) are a promising technology for the removal of these pollutants, due to their ability to provide a large surface area for microbial attachment and biofilm formation. The removal of pollutants is achieved through a combination of mechanisms, including biosorption, bioaccumulation, and biodegradation. However, maintenance of optimal operating conditions (fluidization velocity, pH, temperature), microbial activity, and potential clogging issues need to be carefully addressed to ensure consistent and efficient pollutant removal. Additionally, the choice of microbial species and their adaptability to the specific pollutants in the wastewater are critical factors influencing the success of the bioreactor. The current review explores into the efficiency and applicability of fluidized bed bioreactors in addressing the challenge of eliminating inorganic contaminants, such as heavy metals and metalloids, from wastewater streams. The review highlights the potential of fluidized bed bioreactors as a promising and sustainable solution for the removal of emerging inorganic pollutants from wastewater. By addressing key challenges and optimizing operational parameters, this technology could play a pivotal role in mitigating water pollution concerns on a broader scale.

Keywords Emerging pollutants · Fluidized bed bioreactor · Comparison of FBR · Factor influencing performance · Wastewater treatment · Heavy metal removal

N. Bhutia · P. Adhikari · N. Dutta (✉)
Department of Civil Engineering, NIT Sikkim, Ravangla, South Sikkim, India
e-mail: neel@nitsikkim.ac.in

S. Dey, S. Bhattacharya (eds.), *Biotechnological Interventions in the Removal of Emerging Pollutants*, Interdisciplinary Biotechnological Advances,
https://doi.org/10.1007/978-981-97-9922-0_15

15.1 Introduction

Wastewater treatment is a crucial process for protecting the environment and human health. However, the emergence of new and persistent inorganic pollutants in wastewater has posed a significant challenge to conventional treatment processes. The demand for new products, new technologies, and new production methods has increased as a result of the population's rapid rise as well as technological and industrial advancements. Global civilization has become increasingly industrialized in recent times, which has led to the production of massive amounts of wastewater that contain POPs (permanent organic pollutants) such as chemicals, dyes, plasticizers, and medications (Garcia-Segura et al. 2016) as well as persistent inorganic pollutants. Water resources can be severely harmed by pollutants and bacteria found in wastewater, which can further harm people and other animals who come into touch with contaminated water. Due to the continuous increase in the human population, there would be a rapid increase in the production of wastewater as well, so to address the growing demand for wastewater treatment, it will be necessary to build new plants and upgrade and expand current ones. In densely populated areas, these new facilities and expansions will require additional room; yet, as cities grow, treatment plant space will become more limited. To address this issue, more effective treatment procedures that can handle greater wastewater quantities faster than traditional techniques will be required.

Advanced oxidation processes (AOPs) are one of the potential technologies that have drawn a lot of attention for treating wastewater that is resistant to treatment (Glaze et al. 1987). These technologies are predicated on the production of potent oxidants via a variety of techniques. The most prevalent oxidant is the hydroxyl radical (OH-), which may efficiently break down organic contaminants. Biological treatment is another appealing method that is frequently used to treat wastewater from both homes and businesses. While biological processes are thought to be low-cost and environmentally benign, AOPs are of interest because of their efficacy in breaking down resistant contaminants. For the technology to be applied correctly in any scenario, a reliable contact device is necessary. One such technology that has been shown to have high efficiency in wastewater treatment is the fluidized bed bioreactors (FBRs) (Nelson et al. 2017). A fluidized bed reactor (FBR) has proven to be an effective reactor in the applications of both AOPs (Tisa et al. 2014) and biological processes (Zou et al. 2016). Fluidization is the process of moving a fluid through a bed of solid particles that is static at a surface velocity fast enough to allow the particles to suspend and behave like a fluid. When a low-velocity fluid is added to a static bed, it just goes through the spaces between the solid particles and the bed stays in place. The bed extends with increasing velocity until the buoyancy force equalizes the drag and gravity forces, causing the particles to become suspended. The bed becomes suspended when the pressure drops across it reaches the weight of the particles, or the minimum fluidization velocity (Khan et al. 2014). FBR is a contacting device that functions on the fluidization concept. In many ways, it is comparable to the widely used packed bed reactors; however, the packing material expand when the fluid moves uphill or lower (Burghate and Ingole 2013). The size, density, and viscosity of the particles as well as the fluid's upflow velocity all affect how much the bed expands. Mass transfer, heat transfer, momentum exchange and solid–liquid–gas or solid–liquid–gas multiphase flow systems are all possible in fluidized bed reactors (FBRs). FBR is one of the most significant reactor systems utilized in chemical and biotechnology applications due to its superior features (Si et al. 2011) (Table 15.1).

The removal of emerging inorganic pollutants from wastewater using fluidized bed bioreactors (FBBRs) constitutes a multifaceted research frontier. This study delves into the identification of previously overlooked inorganic pollutants, potentially incorporating nanoparticles and unconventional heavy metals. Furthermore, it explores the optimization of FBR design, considering novel reactor configurations, bed materials, and hydraulic conditions to enhance pollutant removal efficiency. The investigation extends to the realm of microbial consortia and biofilm dynamics within the FBBR, aiming to harness robust biological processes for efficient pollutant transformation. The study also examines synergistic treatment approaches by

Table 15.1 Comparison between different types of reactors

Aspects	Activated sludge process ASP) (Hreiz et al. 2015)	Sequencing batch reactor (SBR) (Al-Rekabi et al. 2007)	Moving bed biofilm reactor (MBBR) (di Biase et al. 2019)	Fluidized aerobic bioreactor (FABR) (Nelson et al. 2017)	Trickling filter (TF) (Daigger and Boltz 2011)
Operational mode	Continuous flow	Batch operation	Continuous flow	Continuous flow	Continuous flow
Process efficiency	Good at removing organic matter	Efficient nutrient removal	Efficient organic matter removal	Efficient organic matter removal	Limited removal efficiency
Aeration	Continuous aeration	Intermittent aeration	Aeration may vary	Continuous aeration	No aeration
Nutrient removal	Yes, including nitrogen and phosphorus	Yes, including nitrogen and phosphorus	Limited nutrient removal	Yes, including nitrogen and phosphorus	Limited nutrient removal
Process control	Quires consistent operation and control	Flexible operation phases	Moderate control required	Requires careful control	Less control required
Sludge production	Moderate	Moderate	Minimum	Minimum	Minimum
Applications	Municipal and industrial wastewater treatment	Municipal wastewater treatment	Municipal and industrial wastewater treatment	Wastewater treatment, organic waste degradation	Domestic and small-scale industrial wastewater treatment
Energy consumption	Moderate	Variable (lower during non-aeration)	Variable	Moderate	Low

integrating FBBR with advanced oxidation processes or electrochemical methods, emphasizing the potential for heightened pollutant removal. Additionally, researchers delve into the development of advanced monitoring and sensor technologies for real-time assessment of FBBR performance. Long-term stability assessments, life cycle analyses, and considerations of economic viability and scalability contribute to a holistic exploration, ultimately aiming to establish guidelines and standards for regulatory compliance in the sustainable application of FBBRs for inorganic pollutant removal.

15.2 Occurrence and Types of Emerging Inorganic Pollutants

Over the past decade, there has been a notable shift in the standards used to define safe water, moving beyond conventional contaminants such as viruses, turbidity, odor, and heavy metals. This transformation is primarily attributed to advancements in analytical technology, which have enabled the identification of chemicals at remarkably low concentrations, often measured in nanograms per liter (ng/l). Consequently, a new class of pollutants, termed "emerging pollutants," has garnered significant attention in the realm of environmental research due to the insights provided by these sophisticated measurements and the consequential issues they bring to light. The occurrence of emerging inorganic pollutants can vary depending on geographical location, industrial activities, and local environmental conditions. Continuous monitoring and research are essential to understand their distribution, behavior, and potential impacts on ecosystems and human health. Additionally, regulatory measures may be developed to manage and mitigate their effects as scientific understanding progresses (Table 15.2).

15.2.1 Heavy Metals

Elements having an atomic density of more than 6 g/cm^3 are referred to as heavy metals, and they are among the most persistent contaminants in sewage systems. The metallic elements of the periodic table are sometimes known as trace elements (Salem et al. 2000). The poisonous heavy metals arsenic, lead, mercury, cadmium, chromium, copper, nickel, silver, and zinc are the most frequently found in wastewater. Significant health and environmental issues are brought on by the discharge of large levels of heavy metals into water bodies, which can also raise the cost of wastewater treatment. Heavy metals also occur in small amounts naturally and may infiltrate into aquatic systems through the leaching of rocks, airborne dust, forest fires, and vegetation (Fernandez-Leborans and Herrero 2000). Their presence and buildup in the environment are caused by either direct or indirect human activity,

Table 15.2 Sources and effects of different types of emerging inorganic pollutants

Name of the emerging inorganic pollutants	Source	Effect	References
Heavy metals	Industrial discharges, urban runoff, agricultural activities, domestic wastewater, atmospheric deposition, mining and ore processing, landfills, accidental spills	Environmental contamination, toxicity to aquatic life, bioaccumulation, disruption of ecosystems, impacts on human health, groundwater contamination, soil degradation, corrosion of infrastructure	Milly et al. (2002), Schwarzenbach et al. (2010)
Metalloids	Natural geological processes, mining and ore processing, agricultural activities, industrial discharges, coal combustion, atmospheric deposition, geothermal activity	Aquatic toxicity, bioaccumulation, human health risks, groundwater contamination, ecosystem disruption, impaired soil quality, reservoir sediment contamination, aesthetic and recreational issues	Rai et al. (2018)
Per- and polyfluoroalkyl substances (PFAS)	Industrial discharges, firefighting foam, landfills, commercial products, textile manufacturing, electronics manufacturing	Environmental persistence, bioaccumulation, human exposure, health concerns, drinking water contamination, concerns for agricultural lands	Tavasoli et al. (2021), Kurwadkar et al. (2022)
Pharmaceutical s and personal care products (PPCPs)	Domestic sewage, healthcare facilities, agricultural runoff, aquaculture operations, industrial discharges, industrial discharges, landfills, household and consumer use, livestock farming	Environmental impact, aquatic toxicity, antibiotic resistance, endocrine disruption, bioaccumulation, drinking water contamination, impacts on microbial communities, human exposure	Agnihotri and Thathola (2019)

(continued)

Table 15.2 (continued)

Name of the emerging inorganic pollutants	Source	Effect	References
Nanoparticles	Consumer products, industrial processes, medical wastewater, urban runoff, electronic waste, nanomaterial production and research, agricultural practices, landfills	Environmental impact, toxicity to aquatic life, bioaccumulation, groundwater contamination, effects on soil microorganisms, human exposure, impact on wastewater treatment processes, regulatory challenges	Spanò et al. (2019), Menicagli et al. (2022)
Flame retardants	High production and use of OPFRs, replacement of brominated FRs, use in consumer products, presence in various products, lack of knowledge on OPFR toxicity, continuous introduction into the environment, lack of studies on OPFR effects	Increased environmental presence, regulatory shift, persistence, fire prevention in various goods, plasticizers, antifoaming agents, antiwear agents, diverse applications in lacquers, ecological risks, limited understanding of toxicity compared to brominated FRs, potential health risks due to exposure	Pantelaki and Voutsa (2019), Ricklund et al. (2009)
Antibiotics resistance genes (ARGs)	Anthropogenic selection pressures, microbial diversity in WWTPs, continuous introduction of genetic material, horizontal gene transfer dynamics, chemical composition of wastewater, integration of ARGs on MGEs, facilitative conditions for ARG evolution	Favorable conditions, concentrated microbes, continuous influx, facilitate HGT, ARG spread, additional selective pressures, clustered ARGs on MGEs, development of MDR, WWTPs as Hotspots, spread of complex ARG structures	Nguyen et al. (2021), Batt et al. (2006)

like quick industrialization, urbanization, and anthropogenic factors. Since heavy metals are poisonous and nonbiodegradable, they persist in wastewater (Ng 2006). Heavy metals can have detrimental effects on plants. For example, cadmium can reduce seed germination and lipid content; chromium can decrease enzyme activity and plant growth; copper and mercury can inhibit photosynthesis; nickel can reduce seed germination; and lead can reduce chlorophyll production and plant growth (Gardea-Torresdey et al. 2005). Reduced growth and development, cancer, organ damage, nervous system impairment, and in severe cases, death is among the effects on animals. Effluents must be sufficiently treated before being released into receiving water bodies to mitigate the harmful impacts of heavy metal toxicity in wastewater.

15.2.2 Metalloids

Water contaminated by metals is turning into a major global hazard. Heavy metals that are found in nature have atomic weights that are five times larger than water and have high densities. Owing to their extensive use in residential, industrial, medicinal, agricultural, and technical applications, they are extensively dispersed throughout the ecosystem and have demonstrated the ability to affect both the atmosphere and human health. The key variables that affect a metal's toxicity are its dose and mode of exposure, as well as an individual's age, gender, and nutritional status. Cadmium, chromium, arsenic, mercury, and lead are very hazardous to humans and have a significant impact on their health. Humans are increasingly vulnerable to major health problems as a result of rising amounts of heavy metal ions in their diet. It was discovered in the past that metals are environmental pollutants that have an impact on both human and ecological health. Besides human exposure has grown substantially due to the rising usage of numerous agricultural, industrial, pharmaceutical residential effluents, and technological applications and according to this the sources of heavy metal ions grows progressively (He et al. 2005). The primary causes of environmental contamination include well-known industries such mining, foundries, and metal-based manufacturers (Shallari et al. 1998). Some heavy metals are regarded as trace elements and are required for biological processes. Examples of these metals are zinc, copper, selenium, chromium, cobalt, iodine, manganese, molybdenum, aluminum, and nickel. They are referred to as vital metals since they are required for a healthy life (Pacyna 2023). Physicochemical mechanisms are unique to each metal, and biological factors such as species traits, trophic interactions, and physiological features also play a significant effect. Nonetheless, there are certain carcinogenic and harmful mechanistic features of heavy metals. Therefore, contamination resulting from metals or heavy metals, such as mercury, lead, and cadmium, poses a serious risk.

15.2.3 Per- and Polyfluoroalkyl Substances (PFAS)

Per- and polyfluoroalkyl substances (PFAS) form a broad category of >4000 partially or fluorinated linear, branched, or cyclic molecules (Sunderland et al. 2019). They are employed as stain and water repellents in the paper goods, adhesives, firefighting foams, cosmetics, and textile sectors. PFAS are utilized in the production of semiconductors, lubricants, coating additives, surfactants, agricultural applications, erosion inhibitors, and aqueous film-forming foams (AFFFs) for firefighting operations (which make up 1–5% w/w). They are also used in the aviation industry as erosion inhibitors (Zhang et al. 2021). PFAS are employed in a variety of home items, including cookware and food packaging, due to their hydrophobic and oleophobic qualities. The majority of PFAS chemicals have not had a comprehensive evaluation, and it is unclear how they may negatively affect ecosystem or human health (Buck et al. 2011). Exposure to per- and polyfluoroalkyl substances (PFAS) has been linked to several adverse effects, including immunosuppression, altered cytokine production, decreased spleen and thymus weights and cellularity, and decreased production of particular antibodies. There are no limitations on the use of PFAS in consumer goods or industrial applications, even though they have been linked to the onset of cancer and many other human illnesses. Furthermore, because PFAS have extended persistence durations, bioaccumulation of these substances is simple. Bai and Son (2021) classify them as emerging environmental pollutants.

15.2.4 Pharmaceuticals and Personal Care Products (PPCPs)

Pharmaceutical and personal care products, or PPCPs, are extensively utilized in a variety of industries, including the medical area, business, aquaculture, livestock production, and daily living. Because of their wide range of applications and inadequate removal by traditional biological wastewater treatment plants, they are becoming increasingly commonplace in the environment. According to their many uses, they can be divided into several classes (Daughton and Ternes 1999). The 2000s have seen a rise in interest in PPCPs. In 1993, more than 1 × 106 tons of PPCPs were produced globally (Daughton and Ternes 1999). A total of 28,000 tons of penicillin were produced annually in 2003, accounting for 60% of the antibiotics used worldwide (Richardson et al. 2005). For the two reasons listed above, the fate of PPCPs in the environment is receiving more and more attention these days. The first is that PPCPs are ubiquitous in the environment due to their extensive use; the second is that advances in analytical technology allow for the detection of PPCPs at the trace level. Numerous investigations have demonstrated that PPCPs were present in surface water at concentrations between ng/l and μg/l (Roberts et al. 2016); in groundwater at concentrations between ng/l and mg/l (Cabeza et al. 2012); in sediment at concentrations up to μg/kg (Richardson et al. 2005); and in soil at μg/kg (Karnjanapiboonwong et al. 2011).

15.2.5 Nanoparticles

Materials with sizes ranging from 1 to 100 nm (1 nm 51,029 m) are typically categorized as nanoparticles (Hochella Jr 2002). However, others advise that the upper limit of the size should be set based on changes in fundamental attributes from the bulk material (Banfield and Zhang 2001). Due to their high specific surface area, nanoparticles in this size regime may have special characteristics such as strong sorption, fast dissolution, high reactivity, and discontinuous characteristics like magnetism and the quantum confinement effect (Gehrke et al. 2015). In addition to being created artificially, nanoparticles can also be found in the environment because the Earth is rich in naturally occurring nanomaterials, such as nanosized mineral particles. Natural nanomaterials occur naturally in Earth's crust by multiple biological and geochemical processes, and it is anticipated that at least 1000 Mt of nanoparticles are cycled annually through the hydrosphere (Hochella Jr et al. 2019). To date, man-made nanoparticles have been implicated in the majority of environmental concerns. Because of their diameters, both artificial and natural nanoparticles can pass through membrane filters with 0.2 μm hole sizes. Filtration using membranes with pore sizes of 0.2 and 0.45 μm is used to characterize dissolved organisms in water functionally. Consequently, the amounts of chemicals detected in filtered water may comprise nanoscale entities that are both dissolved and unaggregated.

15.2.6 Flame Retardants

Chemicals called flame retardants are added to various materials to prevent or reduce the spread of fire. These drugs are useful in avoiding injuries, property losses, and death. Despite these advantages, a lot of flame retardants are poisonous and may be bad for the environment and human health. In this context, polybrominated diphenyl ethers (PBDEs) are developmental neurotoxicants, possible endocrine disruptors (Costa et al. 2008) and bioaccumulate in biota (Wu et al. 2012). Because of their harmful qualities, PBDEs are prohibited in Europe and are subject to EU Water Framework Directive regulations. Because of these bans, PBDEs have been substituted by other chemicals, such as new brominated flame retardants (NBFRs) and organophosphorus flame retardants (OPFRs) (Van der Veen and de Boer 2012). Since these substances have been found in every environmental compartment and many of them have hazardous qualities, they are regarded as emerging contaminants. Industrial and Wastewater Treatment Plant (WWTP) discharges (Ricklund et al. 2009), material releases (Birgul et al. 2012), atmospheric deposition (Melymuk et al. 2011), and runoff are identified as sources of flame retardants in the environment. As a result, it is anticipated that several flame retardants will coexist in the aquatic environment. The proportion of each retardant in water or sediment will vary depending on the factors that affect its manufacturing, use, and discharge as well as its permanence and partition qualities.

15.2.7 Antibiotic Resistance Genes (ARGs)

Antibiotics are frequently used as food additives to speed up an animal's growth or to safeguard human and animal health. Most antibiotics are released into the environment unaltered. Thus, concerns about the potential impact of antibiotic residues in the aquatic environment have grown in recent years (Wright 2007; Kemper 2008). Except for the pristine site in the highlands, before the rivers or streams pass through urban or agricultural regions, it is impossible to discover a place in surface water where antibiotics cannot be detected (Yang and Carlson 2003). According to (Batt

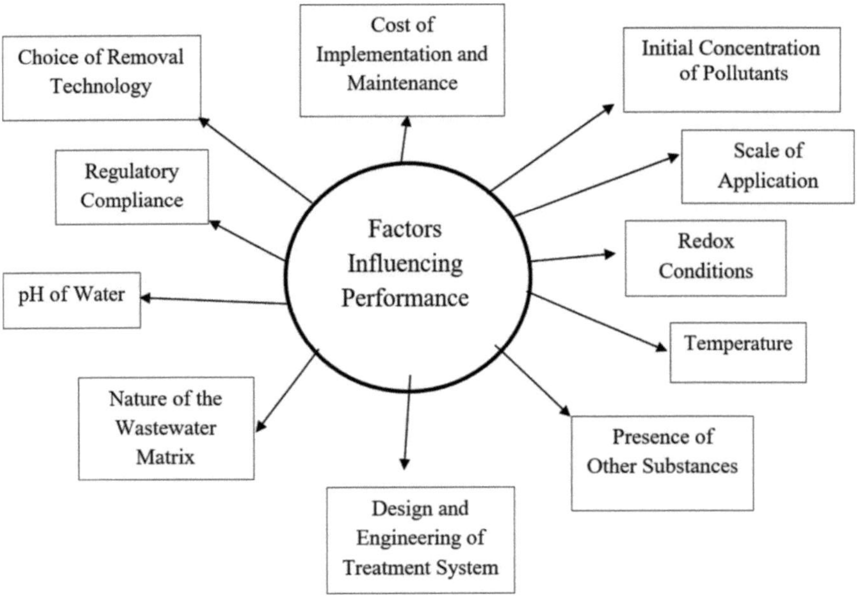

Fig. 15.1 Flow diagram of factors influencing the performance of FBR

et al. 2006), some antibiotics can be discovered in groundwater as deep as more than 10 m. In addition to the chemical pollution that antibiotics themselves create, the usage of antibiotics may hasten the emergence of bacteria and antibiotic resistance genes (ARGs), which may pose health hazards to both humans and animals (Kemper 2008). These microorganisms have the potential to spread from the surroundings to humans through direct or indirect contact (Rodríguez et al. 2006), endangering human life.

15.3 Factors Influencing Performance (Fig. 15.1)

The performance of removal technologies for emerging inorganic pollutants is influenced by various factors. Understanding and carefully considering these factors are essential for developing and implementing effective strategies for the removal of emerging inorganic pollutants from the environment. Treatment technologies should be tailored to the specific conditions of the contaminated site to achieve optimal results. The chemical properties of specific inorganic pollutants play a significant role. Factors such as solubility, reactivity, and speciation can affect the choice of removal technologies (Romero 1959). The pH of the water or soil can impact the speciation and mobility of inorganic pollutants. Some removal technologies are more effective under specific pH conditions, and adjusting the pH may enhance treatment efficiency (Briones et al. 2014). The initial concentration of pollutants in the environmental matrix influences the choice and design of removal technologies (Shieh et al. 1981). Higher concentrations may require more robust and efficient treatment methods. The nature of the environmental matrix, whether it's water, soil, or air, affects the selection of appropriate removal technologies (Grace and Sun 1991). Different matrices may require different treatment approaches. Redox conditions, including the presence of reducing or oxidizing agents, can significantly impact the behavior of certain inorganic pollutants. Redox-sensitive pollutants may require specific treatment strategies (Tschoepe et al. 2020). Temperature can influence the reaction rates and efficiency of removal technologies (Milacic et al. 2022). Some processes may be more effective at certain temperatures, and extreme temperatures may affect the stability of treatment agents. The presence of other substances in the environment, such as organic matter, competing ions, or other pollutants, can interfere with removal processes (Werther 2000). Interactions between different substances may affect treatment efficiency. The choice of removal technology is crucial. Different technologies, such as adsorption, precipitation, ion exchange, membrane filtration, and advanced oxidation processes, have varying effectiveness for different pollutants and under different conditions. For certain removal processes, such as adsorption, the particle size and surface area of the adsorbent material play a crucial role. Larger surface areas may provide more sites for adsorption (Sun and Grace 1990). The design and engineering of the treatment system, including factors like residence time, flow rates, and reactor configuration, influence the overall performance of removal technologies. The cost of implementing and maintaining a removal technology is a practical consideration. Sustainable and cost-effective technologies are often preferred for large-scale applications (Flegkas et al. 2019). Adherence to environmental regulations and standards may influence the selection of removal technologies. Some methods may be more suitable for meeting specific regulatory requirements. The scale at which the removal technology is applied, whether at a laboratory scale, pilot scale, or full scale, can impact its performance and efficiency (Rüdisüli et al. 2012).

15.4 Removal Mechanism for Emerging Inorganic Pollutants

15.4.1 Removal of Heavy Metal

The groundbreaking study conducted by (Nielsen et al. 1997) sought to pioneer an innovative method for the removal of heavy metals from industrial wastewater, recognizing the pressing need for effective solutions in environmental remediation. The envisioned technique was centered around the utilization of a fluidized bed reactor, a cutting-edge technology that held the potential to revolutionize the conventional approaches to heavy metal removal. Specifically, the proposed method involved intricate adsorption and coprecipitation processes within the fluidized bed reactor. In the intricate design, the focus was on the creation of adsorptive granules, selecting either ferric oxyhydroxide or manganese dioxide, that would form a resilient coating on the surface of a carrier material. This coating, crucial for the success of the process, aimed to enhance the interaction between heavy metal contaminants and the reactive materials, ensuring efficient adsorption and coprecipitation. The introduction of two types of reagents, namely ferrous or manganese ions, along with an oxidation agent, was a pivotal step in achieving the desired coating on the carrier surface, as outlined in the comprehensive study by (Nielsen et al. 1997). Building upon this pioneering work, Aktor (1994) undertook a similar methodology in the realm of chromate removal, acknowledging the versatility of the fluidized bed reactor. The objective of Aktor's study was to exploit the reactor's capabilities to catalyze the redox reaction between ferrous and chromate ions. To achieve this, catalyzing iron oxyhydroxides (FeOOH) were strategically fixed as a coating on the carriers within the fluidized bed, showcasing the adaptability of the fluidized bed reactor across various heavy metal removal applications. Further expanding the knowledge base in this field, Seckler et al. (1996) contributed significantly by addressing the removal of phosphate, recognizing its role in controlling eutrophication. Their method, intricately described in their study, involved the precipitation of calcium phosphate within a fluidized bed reactor. This demonstrated the versatility of the fluidized bed reactor not only in heavy metal removal but also in mitigating broader environmental concerns related to nutrient control. The overarching goal of these collective efforts was to engineer a fluidized sand bed that could effectively function as a "heavy metal stripper," a term coined to emphasize the innovative and efficient nature of the proposed method. This ambitious objective required a multifaceted approach, involving the simultaneous injection of a carbonate solution and the upward pumping of wastewater containing metals through the fluidized bed of sand. The fluidized sand bed, a key component of the devised method, provided a vast surface area for heavy metal precipitation. The precipitation process, occurring as a result of the introduced reactive agents, covered the sand surface with a layer of immobilized heavy metals. Consequently, the heavier sand particles laden with metals settled at the bottom of the reactor, while larger coated sand particles, posing challenges to fluidization, gradually drained from the reactor, ensuring the sustained functionality of the system over time. Remarkably, this one-step-through process offered two pivotal benefits that addressed critical practical challenges associated with traditional treatment methods. Firstly, the use of a fluidized bed reactor significantly reduced the physical space required for the treatment system, providing a compact and efficient alternative. Secondly, and perhaps more importantly, the innovative technique eliminated the need for sludge dewatering, a traditionally resource-intensive and environmentally impactful aspect of wastewater treatment. The heavy metals, being coated on the top layer of the sand, obviated the necessity for additional sludge management efforts.

In conclusion, this pioneering approach in heavy metal removal from industrial effluent demonstrated a synergy of innovative design, advanced reactor technology, and a nuanced understanding of the underlying chemical processes. The method showcased not only the effectiveness of the fluidized bed reactor in heavy metal removal but also its potential to revolutionize the broader landscape of wastewater treatment by addressing space constraints and mitigating sludge management challenges.

15.4.2 Removal of Nutrient

In the expansive domain of wastewater treatment, the role of inorganic nutrients emerges as a critical determinant of the environmental balance, encapsulating a diverse array of essential elements and compounds imperative for the sustenance and growth of living organisms. This intricate tapestry includes nitrogen compounds like ammonium, nitrate, and nitrite, phosphorus compounds exemplified by orthophosphate, sulfur compounds embodied in sulfate, and an assortment of trace elements such as iron and zinc. The introduction of these nutrients into wastewater stems from a myriad of sources, ranging from the discharge of domestic sewage and industrial effluents to runoff laden with agricultural byproducts. While these nutrients form the building blocks of life, their heightened concentrations in wastewater pose formidable environmental challenges, notably contributing to the pervasive issue of eutrophication in aquatic ecosystems. Within the expansive realm of wastewater treatment methodologies, a particularly intriguing and innovative approach takes center stage—the utilization of a fluidized bed bioreactor. This specialized system is meticulously designed to address the intricate task of cleaning up wastewater, with a specific emphasis on the removal of inorganic nutrients. The underlying mechanism orchestrating nutrient removal within this system hinges upon the collaborative endeavors of minuscule microorganisms that inhabit carrier particles meticulously distributed within the confines of the bioreactor. These microorganisms, in their microscopic grandeur, weave a dynamic and intricate community, commonly referred to as a biofilm, ensconcing the surface of the carrier particles. The primary mission entrusted to these microorganisms is to tackle the undesirable elements present in the wastewater, honing in on the intricate dance of nitrogen and phosphorus compounds. Within this microbial ballet, some of these microorganisms assume the role of specialized cleaners, executing a biochemical choreography that transforms nitrogen compounds into harmless gasses. Meanwhile, others adopt the persona of efficient vacuum cleaners, adept at absorbing and storing phosphorus, effectively curbing its impact on the ecosystem.

What elevates the allure and complexity of this process is the dynamic nature inherent in the fluidized bed bioreactor. The carrier particles, serving as microorganism habitats, embark on a perpetual journey of movement within the water column, propelled by the inherent upward flow characteristic of the system. This perpetual motion creates a dynamic and responsive environment, ensuring extensive and consistent opportunities for microorganisms to intimately interact with and eliminate nutrients from the wastewater. The carrier particles, doubling as sanctuaries for microorganisms, unveil a remarkable capacity to harbor a dense concentration of these diminutive life forms. As these microorganisms diligently engage in the intricate ballet of nutrient removal and undergo cycles of multiplication, the fluidized bed undergoes a dynamic expansion, ingeniously designed to gracefully accommodate the burgeoning population of these microscopic caretakers. This strategic expansion effectively mitigates potential challenges associated with bed clogging, ensuring the sustained efficacy of the bioreactor over extended periods. Furthermore, the fluidized bed bioreactor is a marvel of intentional engineering, meticulously crafted to thwart the disruptive displacement of microorganisms by the flowing water. This deliberate design feature serves as a guardian of the persistent integrity of the microbial community within the reactor, empowering them to maintain their concerted contribution to the wastewater treatment process. In essence, the nutrient removal mechanism within the fluidized bed bioreactor unfolds as a captivating symphony orchestrated by the coordinated efforts of microorganisms within a biofilm, dynamic mobile habitats, and a fluidization process reminiscent of a precisely choreographed dance. In summation, the fluidized bed bioreactor emerges as a highly effective and innovative instrument for wastewater treatment and the mitigation of harmful nutrients.

This orchestrated and collaborative methodology not only showcases the prowess of the bioreactor in addressing the intricate challenges of nutrient removal but also underscores its potential as a transformative force in the ongoing quest for sustainable and environmentally conscious wastewater treatment practices.

15.4.3 Removal of Industrial Waste

The process of removing industrial wastewater through a Fluidized Bed Reactor-Fenton (FBR-Fenton) system represents a sophisticated and specialized method meticulously designed to tackle the inherent challenges associated with conventional Fenton reactions, specifically the production of sludge. In recent years, the emergence of FBR-Fenton has captured the attention of the scientific and industrial communities alike, positioning itself as a noteworthy approach for effectively treating recalcitrant industrial wastewater that poses unique challenges to conventional treatment methods. At the heart of this transformative process lies the fluidized bed reactor, a pivotal component that operates on the fundamental principle of fluidization. This principle allows the materials within the reactor to dynamically expand through the upward or downward movement of the fluid, creating an environment that is uniquely conducive to enhanced reaction rates and, consequently, efficient treatment processes. Within the intricate framework of the FBR-Fenton system, the materials subjected to fluidization can take the form of inert solids or, intriguingly, active catalysts known as carriers, as underscored by the comprehensive study conducted by (Cai et al. 2021). These carriers play a fundamental and multifaceted role in facilitating the Fenton reaction within the fluidized bed reactor. The primary objective of incorporating carriers is to harness the synergistic benefits derived from both homogeneous Fenton reactions and the observed reduction in sludge production characteristic of heterogeneous Fenton processes. The Fenton reaction, which stands as the linchpin of this mechanism, involves the generation of highly reactive oxidative species, prominently hydroxyl radicals (OH^-), that prove to be instrumental in the degradation of organic contaminants prevalent in industrial wastewater.

To investigate further into the intricacies of the mechanism, the FBR-Fenton process ingeniously employs carriers within the fluidized bed reactor to efficiently promote the Fenton reaction. These carriers serve as active facilitators, enhancing the contact between contaminants and Fenton reagents, thereby fostering the generation of hydroxyl radicals. These radicals, in turn, play a pivotal role in effectively breaking down complex organic molecules into simpler substances, primarily water and carbon dioxide. The nonselective nature inherent in advanced oxidation processes (AOPs), including FBR-Fenton, endows the system with the remarkable capability for the comprehensive degradation of various recalcitrant compounds that persist in industrial wastewater, contributing to its challenging nature. Numerous operational and design parameters intricately shape the efficiency of the FBR-Fenton system, as expounded upon in the study by (Cai et al. 2021). These parameters encompass Fenton reagents dosage, operational pH, water matrix, bed expansion rate, and reactor configuration. The meticulous consideration and control of these factors are imperative, constituting essential elements that ensure the success and optimization of FBR-Fenton for oxidation and iron crystallization processes, thereby underscoring the intricate interplay between the operational nuances and the overarching treatment goals. Despite the promising attributes of FBR-Fenton, the path to widespread implementation on a larger scale is not without its challenges. Issues such as the substantial consumption of Fenton reagents and the imperative need for strict pH control have been identified as potential impediments. These challenges, as elucidated in the referenced study, have spurred active and innovative exploration of steps and strategies aimed at overcoming limitations and enhancing the overall cost-effectiveness of the FBR-Fenton process, reflecting the adaptive and forward-thinking nature of ongoing research endeavors. In conclusion, FBR-Fenton emerges as a beacon of promise within the realm of industrial wastewater treatment, offering a cost-effective and scalable solution to address the complexities posed by challenging industrial wastewater. The mechanism driving its efficacy revolves around the unique capabilities of the fluidized bed reactor in facilitating the Fenton reaction, with carriers assuming a pivotal role in optimizing the efficiency of the process. As research and development in this field continue to unfold, ongoing efforts seek to address existing challenges, further refine the FBR-Fenton process, and solidify its standing as a viable and transformative option for the treatment of industrial wastewater.

15.5 Conclusion

In the conclusive evaluation of wastewater treatment practices, it becomes evident that addressing the issues posed by newly discovered inorganic contaminants is not just a matter of environmental responsibility but a crucial imperative for safeguarding both the environment and public health. The exponential growth of persistent inorganic pollutants (POPs) and permanent organic pollutants (POPs), encompassing heavy metals, fertilizers, and per- and polyfluoroalkyl substances (PFAS), necessitates innovative and adaptive wastewater treatment methods. The rapid expansion of industrial activities and the surging global population further intensify the urgency to improve and expand wastewater treatment facilities worldwide. The escalating challenges posed by emerging contaminants underscore the pressing need for more effective and versatile treatment methods, and in this landscape, fluidized bed bioreactors (FBRs) emerge as a compelling solution. Within the panorama of wastewater treatment, FBRs exhibit clear efficacy, especially in the elimination of newly emerging inorganic contaminants. The study underscores the multifaceted nature of persistent inorganic pollutants and permanent organic pollutants, emphasizing that an optimal response requires treatment methods capable of handling diverse contaminants effectively. The versatility of FBRs becomes apparent as they prove to be dynamic systems capable of adapting to various wastewater streams, effectively tackling industrial waste, heavy metals, and nutrient removal. The complexity of factors influencing FBR performance is highlighted, emphasizing the critical importance of optimal temperature, air velocity, humidity, and airflow rate for achieving effective fluidization. The dynamic nature of FBRs and their unique ability to foster the production of biofilms with the assistance of carrier particles further underscores their effectiveness in addressing a wide array of contaminants. The removal processes elucidated in the study illuminate how FBRs excel in handling different types of wastewater streams, showcasing their adaptability and versatility in the ever-evolving landscape of emerging contaminants. Moreover, the success of mitigation plans hinges on a comprehensive understanding of newly emerging inorganic contaminants and their prevalence. Robust monitoring programs, sustainable behaviors, and the continuous development of cleaner technologies are identified as crucial components in this quest for effective mitigation. The study advocates for a holistic approach, emphasizing that cooperative efforts are indispensable to ensure the creation and implementation of strategies that guarantee a healthy and sustainable environment for future generations. In summary, the adoption of novel technologies like FBRs and the meticulous adjustment of their operating parameters provide a promising route toward a more resilient and sustainable future for our water resources and ecosystems. This becomes especially pertinent in the face of urgent challenges posed by emerging inorganic pollutants and the escalating demand for wastewater treatment globally. The intricate details explored in the study not only underscore the efficacy of FBRs but also emphasize the need for a comprehensive, collaborative, and forward-thinking approach to navigate the complexities of emerging contaminants and ensure the continued well-being of our environment.

Acknowledgments The authors are thankful to NIT Sikkim for all of its help in making this work possible. Sincere gratitude is also given to every researcher in the references section for providing pertinent data that helped with this study project.

Corresponding Author Correspondence to Neelanjan Dutta.

Disclaimer Any opinions, findings, and conclusions expressed in this material are those of the author(s) and do not necessarily reflect the views of the affiliated universities of the authors.

FundingNo funding was received to assist with the preparation of this manuscript.

References

Agnihotri V, Thathola P (2019) Pharmaceutical and personal care products (PPCPs) in waste water and fresh water sources: distribution and related health concerns. Himalayan Ecol 27:63

Aktor H (1994) Continuous high-rate removal of chromate in a fluidised bed without sludge generation. Water Sci Technol 30(7):31

Al-Rekabi WS, Qiang H, Qiang WW (2007) Review on sequencing batch reactors. Pak J Nutr 6(1):11–19

Bai X, Son Y (2021) Perfluoroalkyl substances (PFAS) in surface water and sediments from two urban watersheds in Nevada, USA. Sci Total Environ 751:141622

Banfield JF, Zhang H (2001) Nanoparticles in the environment. Rev Mineral Geochem 44(1):1–58

Batt AL, Snow DD, Aga DS (2006) Occurrence of sulfonamide antimicrobials in private water wells in Washington County, Idaho, USA. Chemosphere 64:1963–1971

Birgul A, Katsoyiannis A, Gioia R, Crosse J, Earnshaw M, Ratola N et al (2012) Atmospheric polybrominated diphenyl ethers (PBDEs) in the United Kingdom. Environ Pollut 169:105–111

Briones RM, de Luna MDG, Su CC, Lu MC (2014) Factors affecting Fenton oxidation of acetaminophen in a fluidized-bed reactor. J Environ Eng 140(1):77–83

Buck RC, Franklin J, Berger U, Conder JM, Cousins IT, De Voogt P et al (2011) Perfluoroalkyl and polyfluoroalkyl substances in the environment: terminology, classification, and origins. Integr Environ Assess Manag 7(4):513–541

Burghate SP, Ingole NW (2013) Fluidized bed biofilm reactor–a novel wastewater treatment reactor. Int J Res Environ Sci Technol 3(4):145–155

Cabeza Y, Candela L, Ronen D, Teijon G (2012) Monitoring the occurrence of emerging contaminants in treated wastewater and groundwater between 2008 and 2010. The Baix Llobregat (Barcelona, Spain). J Hazard Mater 239:32–39

Cai QQ, Lee BCY, Ong SL, Hu JY (2021) Fluidized-bed Fenton technologies for recalcitrant industrial wastewater treatment–recent advances, challenges and perspective. Water Res 190:116692

Costa LG, Giordano G, Tagliaferri S, Caglieri A, Mutti A (2008) Polybrominated diphenyl ether (PBDE) flame retardants: environmental contamination, human body burden and potential adverse health effects. Acta Biomed 79(3):172–183

Daigger GT, Boltz JP (2011) Trickling filter and trickling filter-suspended growth process design and operation: a state-of-the-art review. Water Environ Res 83(5):388–404

Daughton CG, Ternes TA (1999) Pharmaceuticals and personal care products in the environment: agents of subtle change? Environ Health Perspect 107(suppl 6):907–938

di Biase A, Kowalski MS, Devlin TR, Oleszkiewicz JA (2019) Moving bed biofilm reactor technology in municipal wastewater treatment: a review. J Environ Manag 247:849–866

Fernandez-Leborans G, Herrero YO (2000) Toxicity and bioaccumulation of lead and cadmium in marine protozoan communities. Ecotoxicol Environ Saf 47(3):266–276

Flegkas S, Birkelbach F, Winter F, Groenewold H, Werner A (2019) Profitability analysis and capital cost estimation of a thermochemical energy storage system utilizing fluidized bed reactors and the reaction system $MgO/Mg(OH)_2$. Energies 12(24):4788

Garcia-Segura S, Bellotindos LM, Huang YH, Brillas E, Lu MC (2016) Fluidized-bed Fenton process as alternative wastewater treatment technology—a review. J Taiwan Inst Chem Eng 67:211–225

Gardea-Torresdey JL, Peralta-Videa JR, De La Rosa G, Parsons JG (2005) Phytoremediation of heavy metals and study of the metal coordination by X-ray absorption spectroscopy. Coord Chem Rev 249(17–18):1797–1810

Gehrke I, Geiser A, Somborn-Schulz A (2015) Innovations in nanotechnology for water treatment. In: Nanotechnology, science and applications, Dove Press, Macclesfield, United Kingdom pp 1–17

Glaze WH, Kang JW, Chapin DH (1987) The chemistry of water treatment processes involving ozone, hydrogen peroxide and ultraviolet radiation. Ozone: Sci Eng 9(4):335–352

Grace JR, Sun G (1991) Influence of particle size distribution on the performance of fluidized bed reactors. Can J Chem Eng 69(5):1126–1134

He ZL, Yang XE, Stoffella PJ (2005) Trace elements in agroecosystems and impacts on the environment. J Trace Elem Med Biol 19(2–3):125–140

Hochella MF Jr (2002) Nanoscience and technology: the next revolution in the earth sciences. Earth Planet Sci Lett 203(2):593–605

Hochella MF Jr, Mogk DW, Ranville J, Allen IC, Luther GW, Marr LC et al (2019) Natural, incidental, and engineered nanomaterials and their impacts on the earth system. Science 363(6434):eaau8299

Hreiz R, Latifi MA, Roche N (2015) Optimal design and operation of activated sludge processes: state-of-the-art. Chem Eng J 281:900–920

Karnjanapiboonwong A, Suski JG, Shah AA, Cai Q, Morse AN, Anderson TA (2011) Occurrence of PPCPs at a wastewater treatment plant and in soil and groundwater at a land application site. Water Air Soil Pollut 216:257–273

Kemper N (2008) Veterinary antibiotics in the aquatic and terrestrial environment. Ecol Indic 8:1–13

Khan MJH, Hussain MA, Mansourpour Z, Mostoufi N, Ghasem NM, Abdullah EC (2014) CFD simulation of fluidized bed reactors for polyolefin production–a review. J Ind Eng Chem 20(6):3919–3946

Kurwadkar S, Dane J, Kanel SR, Nadagouda MN, Cawdrey RW, Ambade B et al (2022) Per-and polyfluoroalkyl substances in water and wastewater: a critical review of their global occurrence and distribution. Sci Total Environ 809:151003

Melymuk L, Robson M, Diamond ML, Bradley LE, Backus S (2011) Wet deposition loadings of organic contaminants to Lake Ontario: assessing the influence of precipitation from urban and rural sites. Atmos Environ 45(28):5042–5049

Menicagli V, Castiglione MR, Balestri E, Giorgetti L, Bottega S, Sorce C et al (2022) Early evidence of the impacts of microplastic and nanoplastic pollution on the growth and physiology of the seagrass Cymodocea nodosa. Sci Total Environ 838:156514

Milacic E, Manzano MN, Madanikashani S, Heynderickx GJ, van Geem KM, van de Greef AAM et al (2022) Experimental study on the temperature distribution in fluidised beds. Chem Eng Sci 248:117062

Milly PCD et al (2002) Increasing risk of great floods in a changing climate. Nature 415(6871):514–517

Nelson MJ, Nakhla G, Zhu J (2017) Fluidized-bed bioreactor applications for biological wastewater treatment: a review of research and developments. Engineering 3(3):330–342

Ng WJ (2006) Industrial wastewater treatment. World Scientific, Imperial College Press, London

Nguyen AQ, Vu HP, Nguyen LN, Wang Q, Djordjevic SP, Donner E et al (2021) Monitoring antibiotic resistance genes in wastewater treatment: current strategies and future challenges. Sci Total Environ 783:146964

Nielsen PB, Christensen TC, Vendrup M (1997) Continuous removal of heavy metals from FGD wastewater in a fluidised bed without sludge generation. Water Sci Technol 36(2–3):391–397

Pacyna JM (2023) Monitoring and assessment of metal contaminants in the air. In: Toxicology of metals, vol 1. CRC Press, Boca Raton, pp 9–28

Pantelaki I, Voutsa D (2019) Organophosphate flame retardants (OPFRs): a review on analytical methods and occurrence in wastewater and aquatic environment. Sci Total Environ 649:247–263

Rai PK, Lee J, Kailasa SK, Kwon EE, Tsang YF, Ok YS, Kim KH (2018) A critical review of ferrate (VI)-based remediation of soil and groundwater. Environ Res 160:420–448

Richardson BJ, Lam PK, Martin M (2005) Emerging chemicals of concern: pharmaceuticals and personal care products (PPCPs) in Asia, with particular reference to Southern China. Mar Pollut Bull 50(9):913–920

Ricklund N, Kierkegaard A, McLachlan MS, Wahlberg C (2009) Mass balance of decabromodiphenyl ethane and decabromodiphenyl ether in a WWTP. Chemosphere 74(3):389–394

Roberts J, Kumar A, Du J, Hepplewhite C, Ellis DJ, Christy AG, Beavis SG (2016) Pharmaceuticals and personal care products (PPCPs) in Australia's largest inland sewage treatment plant, and its contribution to a major Australian river during high and low flow. Sci Total Environ 541:1625–1637

Rodríguez C, Lang L, Wang A, Altendorf K, García F, Lipski A (2006) Lettuce for human consumption collected in Costa Rica contains complex communities of culturable oxytetracycline- and gentamicin- resistant bacteria. Appl Environ Microbiol 72:5870–5876

Romero JB (1959) Factors affecting fluidized bed quality. University of Washington, United States

Rüdisüli M, Schildhauer TJ, Biollaz SM, van Ommen JR (2012) Scale-up of bubbling fluidized bed reactors—a review. Powder Technol 217:21–38

Salem HM, Eweida EA, Farag A (2000) Heavy metals in drinking water and their environmental impact on human health. In: International conference on the environmental hazards mitigation, Cairo University, Egypt, pp 542–556

Schwarzenbach RP, Egli T, Hofstetter TB, Von Gunten U, Wehrli B (2010) Global water pollution and human health. Annu Rev Environ Resour 35:109–136

Seckler MM, Van Leeuwen MLJ, Bruinsma OSL, Van Rosmalen GM (1996) Phosphate removal in a fluidized bed—II. Process optimization. Water Res 30(7):1589–1596

Shallari S, Schwartz C, Hasko A, Morel JL (1998) Heavy metals in soils and plants of serpentine and industrial sites of Albania. Sci Total Environ 209(2–3):133–142

Shieh WK, Sutton PM, Kos P (1981) Predicting reactor biomass concentration in a fluidized- bed system. J Water Pollut Control Fed 53:1574–1584

Si C, Zhou J, Guo Q (2011) Characterization of pressure fluctuation signals in an acoustic bubbling fluidized bed. J Taiwan Inst Chem Eng 42(6):929–936

Spanò C, Bottega S, Sorce C, Bartoli G, Ruffini Castiglione M (2019) TiO_2 nanoparticles may alleviate cadmium toxicity in co-treatment experiments on the model hydrophyte Azolla filiculoides. Environ Sci Pollut Res 26:29872–29882

Sun G, Grace JR (1990) The effect of particle size distribution on the performance of a catalytic fluidized bed reactor. Chem Eng Sci 45(8):2187–2194

Sunderland EM, Hu XC, Dassuncao C, Tokranov AK, Wagner CC, Allen JG (2019) A review of the pathways of human exposure to poly-and perfluoroalkyl substances (PFASs) and present understanding of health effects. J Expo Sci Environ Epidemiol 29(2):131–147

Tavasoli E, Luek JL, Malley JP, Mouser PJ (2021) Distribution and fate of per-and polyfluoroalkyl substances (PFAS) in wastewater treatment facilities. Environ Sci: Processes Impacts 23(6):903–913

Tisa F, Raman AAA, Daud WMAW (2014) Applicability of fluidized bed reactor in recalcitrant compound degradation through advanced oxidation processes: a review. J Environ Manag 146:260–275

Tschoepe A, Wyrwoll M, Schneider M, Mandel K, Franzreb M (2020) A magnetically induced fluidized-bed reactor for intensification of electrochemical reactions. Chem Eng J 385:123845

Van der Veen I, de Boer J (2012) Phosphorus flame retardants: properties, production, environmental occurrence, toxicity and analysis. Chemosphere 88(10):1119–1153

Werther J (2000) Fluidized-bed reactors. In: Ullmann's encyclopedia of industrial chemistry, Germany

Wright GD (2007) The antibiotic resistome: the nexus of chemical and genetic diversity. Nat Rev Microbiol 5(3):175–186

Wu J, Zhang Y, Luo X, She Y, Yu L, Chen S, Mai B (2012) A review of polybrominated diphenyl ethers and alternative brominated flame retardants in wildlife from China: levels, trends, and bioaccumulation characteristics. J Environ Sci 24(2):183–194

Yang S, Carlson K (2003) Evolution of antibiotic occurrence in a river through pristine, urban and agricultural landscapes. Water Res 37:4645–4656

Zhang W, Cao H, Liang Y (2021) Plant uptake and soil fractionation of five ether-PFAS in plant-soil systems. Sci Total Environ 771:144805

Zou G, Papirio S, Lakaniemi AM, Ahoranta SH, Puhakka JA (2016) High rate autotrophic denitrification in fluidized-bed biofilm reactors. Chem Eng J 284:1287–1294

Chapter 16
Advanced Strategies in Microbial Bioremediation of Dyes from Industrial Wastewater

Rupesh Dutta Banik, Sibashish Baksi, and Pritha Pal

Abstract Industry expansion and the worldwide urbanization trends are producing enormous amounts of wastewater globally. Rivers are emptied of around 80% of the generated industrial effluent, deteriorating the ecosystem. Organic pollution from industrial processes, including dyes, is one of the main reasons why water quality is declining. Several industries release highly colored wastewater-containing dyes during their diverse processes, such as dying, tanning, and subsequent washing stages that pose a threat to environment, plants, animals, and humans as they interfere with many processes, such as photosynthesis, inhibit plant growth, enter the food chain, and have a potential for mutagenicity and carcinogenicity. The majority of traditional physicochemical methods fail to extract dyes from industrial effluents entirely. Technological and microbiological developments have made bioremediation a viable and eco-friendly alternative to traditional treatment methods in this field. This study evaluated potential microbiological mechanisms for developing remediation technologies to prevent environmental pollution caused by wastewaters containing dyes.

Keywords Wastewater · Industrial effluent · Dyes · Carcinogenesis · Physicochemical techniques · Bioremediation

R. D. Banik · P. Pal (✉)
Department of Microbiology, School of Life Sciences, Swami Vivekananda University, Barrackpore, West Bengal, India
e-mail: prithap@svu.ac.in

S. Baksi
Department of Biotechnology, School of Life Sciences, Swami Vivekananda University, Barrackpore, West Bengal, India

S. Dey, S. Bhattacharya (eds.), *Biotechnological Interventions in the Removal of Emerging Pollutants*, Interdisciplinary Biotechnological Advances,
https://doi.org/10.1007/978-981-97-9922-0_16

16.1 Introduction

Rapid industrialization and urbanization around the world have provided the foundation for a link between environmental degradation and citizen health (Kumar et al. 2016). It goes without saying that a clean water source is necessary for the development and upkeep of a variety of human activities. Aside from aquatic life, water resources provide important food and irrigation for agricultural productivity. Nonetheless, the majority of water sources worldwide are contaminated by liquid and solid pollutants produced by industrial processes and human populations (Nahiun et al. 2021).

Mauveine, the first synthetic dye, was developed by "W.H. Perkin" in 1856. Since then, a great variety of synthetic dyes have been produced, and many of them find employment in numerous industries. There are likely over 100,000 different types of dyes and pigments produced annually. The yearly consumption of commercial synthetic dyes is greater than 7×10^5 metric tons, and they are used in a wide range of industries, including those dealing with textiles, leather, paper printing, paint, food, cosmetics, pigments, plastics, and pharmaceuticals (Sudarshan et al. 2023). Many of these industries simply dump their waste into the rivers and lakes nearby without treating it first or only following initial treatment (Nahiun et al. 2021). There may be a connection between the lack of clean water in many developing nations and improper wastewater treatment or discharge of wastewater below a permissible level that is compatible with encompassing waterbodies. Dye, heavy metals, toxicants, and other organic and inorganic dissolved solids found in textile industry effluent degrade water quality and have an impact on the surrounding landscapes. Particularly in densely populated nations like India, the quality of the water has declined to concerning degrees, and dyes have long been thought to be hazardous pollutants (Saxena and Gupta 2020). Every year, wastewater from the textile sector releases approximately 280,000 tons of textile dyes into the environment worldwide (Berradi et al. 2019).

Several environmental issues can arise from the release of untreated wastewater into aquatic bodies without any treatment procedures, including the following: (i) aquatic life depends on dissolved oxygen in the water stream to thrive, but large amounts of organic matter in untreated wastewater will deplete the dissolved oxygen required to meet the wastewater's biochemical oxygen demand (BOD); (ii) untreated wastewater typically has high concentrations of hazardous substances and pathogenic or disease-causing bacteria that can reside in the human digestive system and endanger human health; (iii) a certain quantity of nutrients found in wastewater may also encourage the formation of algal blooms and aquatic plant life, eutrophicating lakes and streams; (iv) odorous gases can be produced in huge numbers as a result of the organic components in wastewater breaking down (Nahiun et al. 2021).

Due to the strong tinctorial value of the dye, water bodies can be seen colored even at concentrations below 1 mg L^{-1}. To make matters worse, this contamination renders the water unfit for human use. Textile dyes may have ecotoxic, mutagenic, and carcinogenic effects if these wastewaters are returned untreated to aquatic habitats. They may also interfere with photosynthesis, bioaccumulate and enter the food chain (Sudarshan et al. 2023).

Therefore, treating wastewater contaminated by harmful textile dyes is currently one of the most urgent problems. Traditional methods have several serious drawbacks, such as the production of hazardous intermediates, the need to transport contaminated water or soil for treatment, the high cost of treatment, and the poor restoration of wildlife and natural habitats. Reducing and neutralizing the contaminants from tannery wastewater through typical biological processes, either by aerobic or anaerobic processes, is feasible and economical when using biodegradation and bioremediation techniques that involve biological systems like microbes (bacteria, fungi, and Actinomycetes, etc.) or their products. Microbes can be utilized to degrade contaminated areas and clean up contaminated ecosystems. Enzymes and metabolic reaction rates are released in substantial volumes, even in highly contaminated environments. For a stressed-out ecosystem, this is quite advantageous (Banik et al. 2023).

16.2 Categorization and Attributes of Dyes

The textile industry uses around 10,000 different dyes and pigments, and about 7×10^5 tons of synthetic dyes are produced annually worldwide (Bhatia et al. 2017). Dyes are used to add color to textiles because they include chromophoric groups in their structure. Chromophores play a major role in the dye molecules' capacity to absorb light, while auxochromes serve as color enhancers (Sudarshan et al. 2023). Something more that sets these dyes apart from each other is their water solubility. Fibers on which the dye is to be imprinted and the chemical and physical makeup of the dyes are the main factors used to categorize dyes. Auxochromes, a type of color enhancer, are created through the removal or donation of electrons, while chromophores are mostly made up of a group of atoms that give a dye its particular hue. Auxochromes like $-NH_3$, $-COOH$, $-OH$, and $-SO_3H$ have functional groups, while chromophores have groups like $-N{=}N-$, $-C = O$, $-NO_2$, and $O = (C_6H_4) = O$ (quinoid assembly. It has been stated that the auxochromes' sulfonate groups provide the dyes their great solubilizing capacity. Several auxochromes are categorized into several types of reactants, which can be either dissolved or undissolved, colored, vat, anionic, or in grain form. They can also be acidic or basic, astringic, or dispersion. Additionally, dyes are classed according to the reactive groups that create covalent bonds with $-OH$ and $-SH$ groups. These dyes are then applied to various fabrics, including cotton, silk, nylon, and wool (Bhatia et al. 2017).

Industrial dyes can be categorized according to their chemical makeup, application techniques, color and provenance (naturally occurring organic compounds versus synthetically produced goods). Chemical structures of dyes further subdivide them into categories such as triphenylmethane, azo, nitro, anthraquinone, indigoid, nitroso, and phthalein. Table 16.1 lists the numerous more categories into which

Table 16.1 Category of various types of dye and classification of dye compounds based on chemical structure (Saxena and Gupta 2020; Bhatia et al. 2017)

Types of dye	Category
Acid dyes	Color acid sodium salts with phenolic group or sulfonic acid
Basic dyes	Basic amino groups protonate in an acidic environment and create salt bonds
Direct dyes	Includes sulfuric acid groups, although these are not the attachment points
Mordant dyes	Has a group capable of retaining metal in coordination complexes or chelate groups
Vat dyes	Vat dyes are similar to sulfur dyes; however, they are utilized in a reduced state following a reduction agent treatment
Reactive dyes	Creates a covalent link with fibers that have amino or hydroxyl groups, such as dyes containing an atom of chlorine
Disperse dyes	Small, water-insoluble dyes with an amino or hydroxyl group to provide limited water solubility at a specific temperature
Solvent dyes	Insoluble in water, soluble in an organic solvent
Chemical structure-based categorization	
Class of dyes	*Chromophore*
Anthraquinone dyes	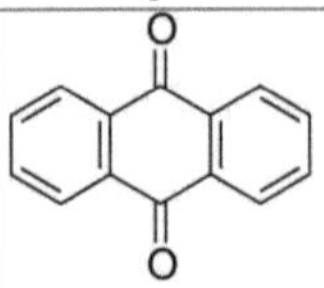
Azo dyes	—N=N—
Nitro dyes	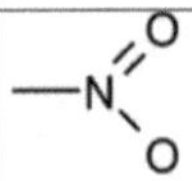
Triphenylmethane dyes	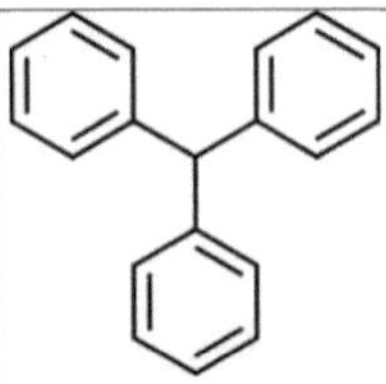
Indigoid dyes	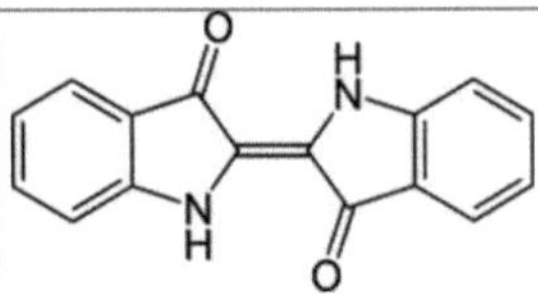
Nitroso dyes	—N=O
Phthalein dyes	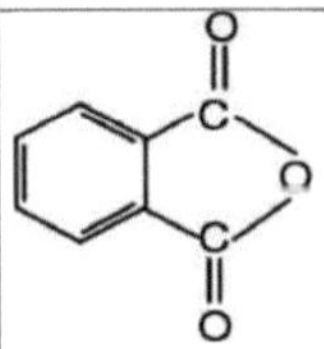

dyes can be divided based on the chromophore. They can be divided into three groups according to the charge of the particles: cationic forms (basic dyes), anionic forms (direct, acidic, and reactive dyes), and non-ionic forms (disperse dyes) (Vikrant et al. 2018) (Ihsanullah et al. 2020). Azo dyes make up to 60–70% of all known dye compounds and are typically composed of one or more azo groups (–N = N–). Anthraquinone dyes are the second most utilized dyes after azo dyes because of their excellent dyeing ability, simplicity of use, and low cost of production. There are thought to be 25 different kinds of chromophore-based dye groups accessible (Sudarshan et al. 2023) (Varjani et al. 2020).

16.3 Toxicity of Dye Industry Wastes

Textile dyes have been linked to a multitude of illnesses, such as dermatitis, bleeding disorders, nausea, mucous membrane and skin ulcerations, nasal septal perforations, and respiratory tract irritations. Inhaling them can also result in breathing problems and, in rare instances, nausea, vomiting, diarrhea, gastritis, and mental disorientation. When textile dye effluent is improperly dumped into water bodies, the water becomes very colored. This lowers light penetration in the aquatic environment, which in turn affects photosynthetic activity. Aquatic life is negatively impacted by decreased dissolved oxygen levels brought on by increasing heterotrophic activity. A specific dye's teratogenic, mutagenic, and carcinogenic properties also have a negative impact on aquatic life (Sudarshan et al. 2023).

16.3.1 Effects of Frequently Used Dyes on Environment and Public Health

Azo dyes find use in the paper printing, paint, leather, food, cosmetics, and pharmaceutical sectors. They are substances that are genotoxic, carcinogenic, and mutagenic. As shown by chromosomal abnormalities in mammalian cells and experimental animals, workers handling azo dyes have been documented to have bladder cancer, splenic sarcoma, and hepatic carcinomas. Metanil yellow, a monoazo dye, is extensively utilized as a colorant in food, drink, soap, spirit lacquer, shoe polish, wood paints, leather dyeing, ink, and colored paper manufacture (Anjaneya et al. 2011). It has been proven to slow down the rate of spermatogenesis, change gene expression, and cause cancer. When taken orally, individuals can develop cyanosis and toxic hemoglobinemia. It's a crucial dye and a very harmful carcinogen.

Methyl orange is an additional azo class dye that is utilized as a pH indicator in laboratories and in the textile industry. Aquatic life may be affected both acutely and chronically by it, and it is said to be toxic and mutagenic (Tan et al. 2016). Triphenylmethane class dye methylene blue is used in the textile industry as well as medicine to treat duck hepatitis B, psoriasis, methemoglobinemia, and West Nile virus infections. Induced tissue necrosis in humans can result in cyanosis, jaundice, elevated blood pressure, cardiac depression, and the formation of Heinz bodies. Triphenylmethane, or crystal violet, has been used historically as a bacteriostatic and dermatological agent in humans. Today, it is utilized as a biological stain in laboratories. In chicken production, it was frequently added to feed to help reduce fungal infections, intestinal parasites, and mold growth. When this material is exposed to certain fish species, the likelihood of developing tumors increases. Under extreme conditions, it is mitotically toxic, carcinogenic, and mutagenic. It can irritate the skin, gastrointestinal tract, and perhaps cause respiratory and renal failure. Preferred by the paper, textile, rubber, and plastic industries, brilliant green is a triphenylmethane dye, contaminants, and mutators that affect aquatic life as well as humans (Sudarshan et al. 2023).

A triphenylmethane dye called malachite green is utilized in aquaculture as an antifungal and antiprotozoal agent. In addition, it finds extensive application in the food, paper, paint, textile, and leather sectors. It has an impact on people's immunological and reproductive systems. According to Srivastava et al. (2004), it has genotoxicity, carcinogenicity, and teratogenic qualities in addition to being extremely cytotoxic to mammalian cells. Rhodamine B, a commercial xanthene class dye, is used extensively as a food colorant, water tracer dye, and in the paper, paint, leather, and textile industries. It is also carcinogenic, neurotoxic, and genotoxic to people and animals. Drinking water tainted with radon B can result in subcutaneous tissue-borne sarcoma. It irritates the skin, eyes, and respiratory system and is dangerous if eaten (Sudarshan et al. 2023).

16.4 Elimination of Dyes from Environment

Usually, textile dyes are made to withstand fading from chemicals, light, and oxidizing agents. Consequently, standard physical-chemical methods of decolorizing dye effluent are not very effective, have limited applicability, and have significant operating costs. But these techniques produce a lot of toxic sludge, which makes disposal an issue. Complex, aromatic compounds that are frequently very stable and resistant to deterioration make up synthetic dyes. This clarifies why it is challenging to treat dye-containing effluents using traditional physicochemical treatment techniques (Sudarshan et al. 2023).

16.4.1 Physicochemical Techniques

A number of different physicochemical treatment methods have been used to remove colors from wastewater. These consist of membrane separation, adsorption, coagulation and flocculation, ion exchange, photodegradation, and oxidation

(Fig. 16.1). However, there are a number of technical and financial challenges with these technologies, including their high cost, production of large amounts of sludge and harmful byproducts, and lack of commercial viability.

Wastewater that contains dye can be effectively decolored using physical techniques based on flocculation and coagulation. Typically, the coagulation process uses ferrous sulfate and ferric chloride to facilitate the dye uptake from textile effluent. Yet, research has also documented the effective use of additional coagulants for textile wastewater cleanup, including magnesium chloride, polyaluminum chloride, chitosan, and aluminum chloride. But coagulation has drawbacks as well, including poor decolorization effectiveness, high expense, and significant sludge production.

Because of its high efficiency, affordability, and convenience of application, the adsorption process is one of the most often used methods in water treatment for the removal of different types of contaminants. Due to its higher decolorization efficacy

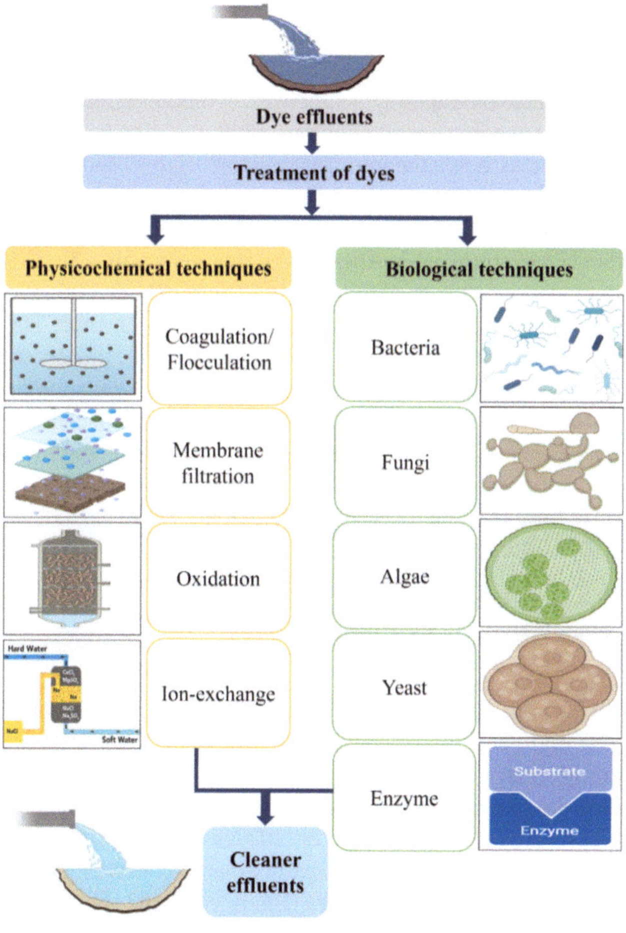

Fig. 16.1 Categorization of frequently employed dye treatment methods

for a variety of dyes from textile wastewater, adsorption techniques have garnered a great deal of attention. For use in water treatment applications, an adsorbent should have high affinity, the capacity to absorb specific dyes, low cost, and exceptional regeneration ability. The fact that adsorption does not result in the disintegration of dye molecules is a significant drawback. The treatment of wastewater discharged from the textile industry has also made use of membrane filtration techniques such as reverse osmosis (RO), nanofiltration (NF), and ultrafiltration (UF). The benefits of membrane-based dye treatment include minimal maintenance requirements, a small environmental footprint, and simplicity of setup and use. The literature has reports on a variety of sophisticated membranes that are used to selectively separate different dye molecules. There are some disadvantages to membrane-based dye treatment, including the inability to process large amounts of dyes, the production of concentrated waste, and membrane clogging.

Additional customary methods typically utilized for the deterioration of textile wastewaters comprise oxidation methods including chemical oxidation and advanced oxidation procedures (AOP). AOP reactions result in the production of hydroxyl radicals, which oxidize color molecules. For the mineralization of organic contaminants, chemical oxidation typically uses an oxidizing agent, such as H_2O_2 or O_3. These oxidizing agents generate copious amounts of hydroxyl radicals, which have the ability to break down the intricate structure of dyes and form smaller, non-chromophoric molecules. Chemical oxidation techniques are appropriate for treating textile effluent containing double-bonded color molecules. Because fewer hydroxyl radicals are produced by these oxidizing agents than by AOP reactions, their rate of degradation is lower. Ozonation is a commonly used procedure in water treatment to break down organic contaminants that are harmful. The benefits of ozonation include the absence of sludge development and the preservation of wastewater volume. However, the two biggest drawbacks of the ozonation process are its high cost and the production of harmful byproducts. Salt concentration, pH, and temperature all have an impact on how stable ozone is. Textile wastewater treatment is another application for ion-exchange resins. Though, this method is less effective and rather costly for removing dyes (Ihsanullah et al. 2020).

16.4.2 Green Strategy to Reduce Risks

Environmental degradation is pervasive, the number of polluted environments is estimated to be high, and the ongoing discovery of these environments in recent years has spurred international efforts to clean up many of these environments, either to reduce the risk of harmful health or environmental effects resulting from pollution or to prepare the area for reuse or redevelopment. Using greener solutions to prevent pollution and other risks is necessary due to concerns about environmental contamination caused by the industries (Banik et al. 2023).

Bioremediation is an effective substitute for traditional physicochemical methods for treating textile effluent. Its benefits include low operating costs and the generation of non-toxic and harmless products. Through the process of bioremediation, organic wastes are either reduced to a tolerable concentration or biologically broken down into innocuous molecules. During the bioremediation process, microorganisms are used (Fig. 16.1). They break down the environmental toxins by eating them. The metabolism and breakdown of potentially dangerous compounds depend on the availability of nutrients and other molecules. The primary agents that facilitate metabolic processes are enzymes. Different enzymes are shown to have better ability to break down a wide range of dyes. The microbial growth that is necessary for the bioremediation process is influenced by the ambient conditions, which are crucial to the process. The ambient parameters can be changed to encourage microbial development and, in turn, increase the microorganisms' degrading efficiency for a successful bioremediation procedure.

16.4.2.1 Bioremediation of Dyes Using Bacterial Strains

A total breakdown of dyes employing many bacterial strains was observed in several experiments. The ability of bacteria to decolorize dye has been shown in both pure cultures and consortiums. In contrast to pure isolates, the bacterial consortia typically demonstrated higher dye removal efficiency. Numerous colors have been degraded by a variety of bacterial species.

When bacteria decolorize azo dyes, they often cleave the azo links in anaerobic environments to produce colorless aromatic amines. However, further degradation is necessary to lessen the harmful effects of these aromatic amines because they are naturally mutagenic and carcinogenic. Through further degradation by bacteria in an aerobic treatment system, the aromatic amines become innocuous chemicals (Ihsanullah et al. 2020).

It has been demonstrated that bacteria and cyanobacteria are the most efficient degraders of synthetic dyes because of their short life cycles, low production of secondary waste, and adaptability to a range of substrates. Additionally, because they are photosynthetic bacteria, these cyanobacteria do not require organic carbon as an energy source. As microorganisms adapt to the dangerous chemical environment, naturally occurring dye-resistant strains arise that subsequently change a variety of extremely deadly compounds into less toxic or non-toxic forms. Under certain environmental conditions, microorganisms have the innate capacity to degrade, detoxify, and mineralize synthetic dyes (Goud et al. 2020). The 1970s saw the first reports of pure bacterial cultures of *Bacillus cereus*, *Aeromonas hydrophila*, and *Bacillus subtilis* that were able to break down azo dyes. These dye-degrading bacteria are now known to exist in a wide variety of settings and sources. The isolated strain proved more effective at decolorizing hues than strains that were sold commercially. Levels of degradation can reach 100% (Table 16.2).

One way to carry out dye breakdown is either with a pure bacterial culture or with mixed cultures, which are commonly referred to as "consortiums." The latter utilizes synergistic metabolic effects, which has been specifically demonstrated to facilitate efficient dye decomposition. The usage of breakdown products produced

Table 16.2 An overview of the microorganisms used for different dye degradation along with their removal efficiency (Sudarshan et al. 2023; Ihsanullah et al. 2020)

Dye	Microorganisms	Removal efficiency
Bacterial spices		
Reactive green 19 A	Micrococcus glutamicus NCIM 2168	100
Reactive violet 5	*Paracoccus* sp.	100
Reactive Red 239	*Bacillus* sp. strain CH12	100
Reactive red 2	*Pseudomonas* sp. SUK1	96
Reactive yellow 145	*Pseudomonas aeruginosa*	100
Reactive red 120	*Shewanella haliotis* RDB_1	100
Malachite green	*Kocuria rosea* MTCC 1532, Enterobacter sp. CV-S1, Micrococcus sp. BD15	100
Malachite green	*Sphingomonas paucimobilis, Synechococcus elongatus, Pseudomonas pulmonicola* YC32, *Pseudomonas* sp. DY1, *Bacillus cereus* KM201428	82.49, 99.5, 82.5, >96, >98
Congo Red	*Enterococcus faecalis* R1107, Bacterial consortium 1. *Pseudomonas stutzeri* L1 2. *Acinetobacter baumannii* L2, *Arthospira maxima*	90.17, 97, 46
Direct blue-6	*Pseudomonas desmolyticum* NCIM 2112	92 (COD–88.95%)
Reactive black 5	*Enterobacter* sp. EC3, *Bacillus* sp. YZU1, *Pseudomonas* sp. RA20	92.56, 95, 98.5
Red HE7B	*Pseudomonas desmolyticum* NCIM 2112	95 (COD–71%)
Reactive red 180	*Citrobacter* sp. CK3	96.2
Acid Orange	*Bacillus megaterium* PMS82	100
Reactive red BLI	*Pseudomonas* sp. SUK1	99.28
Acid blue 113	*Pseudomonas stutzeri* AK6	86.2
Crystal violet	*Pseudomonas putida, Enterobacter* sp. CV–S1	78.5–80, 100
Reactive orange 16	Bacterial consortium DAS 1. *Pseudomonas* sp. SUK1 2. *Pseudomonas* sp. LBC2 3. *Pseudomonas* sp. LBC3	100
Gentian violet	Bacterial consortium 1. *Pseudomonas stutzeri* L1 2. *Acinetobacter baumannii* L2	95
Acid black 1	*Oscillatoria curviceps* BDU92191	84
Indigo dye	*Phormidium autumnale* UTEX1580	100
Scarlet R	Bacterial consortium-GR 1. *Proteus vulgaris* NCIM-2027 2. *Micrococcus glutamicus* NCIM-2168	100
Procion red H-3B	*Pseudomonas stutzeri*	96
Disperse blue-284	*Klebsiella pneumoniae* GM-04	95
Orange II	*Pseudomonas putida* SKG-1	92.8
Direct black 38	*Enterococcus gallinarum*	85
Methyl red	*Staphylococcus saprophyticus* AUCASVE3	97
Metanil yellow	*Bacillus* sp. AK1, *Lysinibacillus* sp. AK2	100, 100
Reactive Black 5	*Enterococcus faecalis* R1107	96.82

(continued)

Table 16.2 (continued)

Dye	Microorganisms	Removal efficiency
Green HE4BD	Bacterial consortium GR 1. *Proteus vulgaris* NCIM-2027 2. *Micrococcus glutamicus* NCIM-2168	100
Basic fuchsin	*Hydrocoleum oligotrichum*	92.44
FF sky blue	*Gloeocapsa pleurocapsoides*	90
Reactive blue	Bacterial consortium PMB11 1. *Bacillus odysseyi* SUK32. *Morganella morganii* SUK53. *Proteus* sp. SUK7	100
Reactive red 198	Immobilized *Acinetobacter baumannii*	96.2
Reactive Black 5	*Aeromonas hydrophila*	76
Remazol red	*Lysinibacillus* sp. RGS	100 (COD–92%)
Direct Black 38	*Enterococcus faecalis* R1107	81.95
Methylene blue	*Bacillus subtilis* MTCC 441	91.68
Reactive red HE8B	*Pseudomonas aeruginosa*	86
Direct black 22	Bacterial consortium: 1. *Pseudomonas aeroginosa*, 2. *Stenotrophomonas maltophila*, 3. *Pseudomonas mirabilis*	91
Indigo BANN 30	*Phormidium* sp. CENA135	100
Reactive Yellow 174	Bacterial consortium: 1. *Sphingomonas paucimobilis*, 2. *Pseudomonas putida*, 3. *Lactobacillus acidophilus*	90 (COD–86%)
Acid red 97	*Phormidium ceylanicum*	89
Miscellaneous Acid red 119	*Phormidium valderianum* BDU 20 041	>90
Fungal species		
Remazol Brown, Remazol Red	*Aspergillus foetidus*	90–95
Red azo dye	*Aspergillus niger*	99.69
RB	*Rhizopus oryzae*	90
AB40	*Thuja orientalis*	97.06 mg/g
RBB	*Rhizophus arrhizus*	86.9
Gryfalan Black RL	*Rhizopus arrhizus*	666.7 mg/g
Gemazol Torquoise Turquoise Blue-G	*Rhizopus arrhizus*	85.9
Sandal-fix Red C4BL	Immobilized purified manganese peroxidase (MnP) produced from *Ganoderma lucidum* IBL-05	87.5
RB19, RB49, AV43	*Funalia trogii*	96.3, 100, 96.2
A mixture of Azure B, CR, Trypan Blue and RBBR	*Thermomucor indicaeseudaticae*	79.28
MG	Laccase of *Cerrena* sp.	91.6
Reactive Green, Reactive Blue 3	*Rhizopus nigricans*	86, 83

(continued)

Table 16.2 (continued)

Dye	Microorganisms	Removal efficiency
RB5, CR, MG	Activated carbon Immobilized by *Aspergillus niger/ Penicillium* sp.	98.2, 84.6, 82.6
CR, RB5, AO7	*Curvularia clavata* NZ2	88–92
AO7, AR8, RB5	*Schizophyllum commune*	12.3, 27, 40.7
RB19	Mycelium pellets of *Penicillium oxalicum*	91
AR97	Laccase from *Peroneutypa scoparia*	75
Cotton Blue	*Penicillium ochrochloron* MTCC 517	93
BG	*Aspergillus* sp.	99.27
RO16	*Irpex lacteus*	95
C.I. Direct Blue 199	*Aspergillus niger*	29.96 mg/g
Acid Red 151/ OII	Aspergillus flavus SA2	97.70, 55.68
AR57	*Neurospora crassa*	98.78
Microalgal species		
Methylene blue	*Chlamydomonas* sp.	99
Malachite green	*Cosmarium* sp.	92.4
Xanthene Rhodamine B	*Coelastrella* sp.	80
Malachite green	*Chlorella* sp., *Chlamydomonas* sp., *Chlorella vulgaris*, Immobilized *Desmodesmus* sp.	99.4, 99.2, >90, 89.1
Methylene blue	*Chlorella* sp.	99
Yeast		
Acid Blue 93	*Candida tropicalis*	100
Basic Violet 3 (BV3)	*Candida krusei*	100
MR	*Galactomyces geotrichum* MTCC 1360	100
RB13	*Candida rugopelliculosa* HXL-2	90
ARB	*Magnusiomyces ingens* LH-F1, *Pichia* sp. TCL	97.37, 90
Navy Blue HER	*Trichosporon beigelii* NCIM-3326	100
Acid Brilliant Scarlet GR	*Candida tropicalis* TLF1	97.2
Acid Scarlet 3R	*Scheffersomyces spartinae* TLHS-SF1	98.14
Fast Blue RR	*Trametes trogii* BAFC 463	>85
BV3	*Candida tropicalis*	85.3
RB5	*Trichosporon akiyoshidainum* HP2023	89
RBBR	*Paraconiothyrium variabile*	93
Reactive Black B	*Phormidium* sp.	60
Enzymes		
Disperse Red 17, Disperse Brown 1	Calcium alginate-pectin entrapped bitter gourd (*Momordica charantia*) peroxidase	98
Methyl orange	Horseradish peroxidase	98
AR18, DR81, Reactive Yellow 15, Disperse Blue 56	Laccase from Paraconiothyrium variabile	97, 3, 2
DR23, Direct Blue 80	Tomato (*Lycopersicon esculentum*) peroxidase	93

(continued)

Table 16.2 (continued)

Dye	Microorganisms	Removal efficiency
Reactive Blue 221	Trichosporon akiyoshidainum	72
DB38, Direct Blue 53	Horseradish peroxidase	83
Direct Yellow	Citrus limon peroxidase	89.47
Malachite green	MnP from Irpex lacteus F17	96
Direct Red 23 (DR23)	Turnip peroxidase	88

by one strain of microorganisms for further decomposition may be the cause of this synergistic interaction between microbes or their enzymes. A pure culture may not exhibit the same range of activities as a bacterial consortium. For example, Saratale et al. (2010) created the bacterium consortium GR, which included *Proteus vulgaris* and *Micrococcus glutamicus*. This consortium showed that six harmful sulfonated reactive dyes could be degraded, including Green HE4BD, Golden yellow HE4R, Orange 3R, Violet 5R, Red ME4BL, and Red M2BN. Furthermore, the consortium led to a faster decolorization of Green HE4BD than any of the individual bacterial strains could. Abou-El-Souod and El-Sheekh (2016) investigated the biodegradation of two azo dyes, basic fuchsin and methyl red, using cyanobacterial strains, *Hydrocoleum oligotrichum* and *Oscillatoria limnetica*. Wang et al. (2022) looked into the dye-degrading capabilities of Enterococcus faecalis R1107 in relation to the decolorization of diazo dye (Congo red and reactive black 5) and triazo dye (direct black 38).

Citrobacter CK3 (isolated from textile mill activated sludge) is reported to have degraded >95% of reactive red 180 in 48 hours, whereas heat-killed cells only produced 29.7% decolorization after 120 hours of incubation, illustrating the benefit of biodegradation over biosorption. Similar results were obtained by Wang et al. (2013) using reactive black 5, an azo dye that was destroyed by living *Bacillus* sp. YZU1 after 120 hours of incubation, but only 3.1% of the dye was decolored by heat-killed cells. Azole reductase was in charge of the significantly greater efficiency in the live cells. In addition, *Pseudomonas desmolyticum* NCIM 2112 was able to completely biodegrade the diazo dye direct blue-6 through the use of the oxidative enzymes lignin peroxidase, laccase, and tyrosinase. These enzymes are highly active in *Micrococcus* sp. BD15 to biodegrade malachite green. Colors could be decolorized by enzymes such as laccase, tyrosinase, veratryl alcohol oxidase, manganese peroxidase, azoreductase, peroxidase, NADH-DCIP reductase, and catalase.

Direct black 38 was shown to be degraded by *Enterococcus gallinarum* produced azoreductase upon the removal of the azo linkage. Benzidine and 4-aminobiphenyl (4-ABP) are the products of this. Compared to the original dye, these secondary metabolites pose less of a risk and cause mutagenesis. Similar to this, the indigo dye was completely broken down by the cyanobacterium *Phormidium autumnale* UTEX 1580 into two secondary metabolites, which were identified as isatin or indole-2,3 dione (m/z 421) and anthranilic acid (m/z 399). The oxidative rupture of the central link is the initial step in the breakdown of indigo; this produces isatin as a metabolite, which is subsequently hydrolyzed, reduced, and decarboxylated to yield anthranilic acid. There was no risk associated with these metabolites (Dellamatrice et al. 2017).

A single strain of *P. valderianum*, a marine cyanobacterium, has been shown to successfully decolorize three textile dyes (Acid red, Acid red 119, and Direct black 155). In certain cases, this strain is also capable of biologically eliminating numerous chemicals. This species decolored over 90% of these dyes effectively through adsorption. It has been demonstrated that *Paracoccus* sp. GSM2 is the best strain for treating textile wastewater that contains reactive azo dyes (Bheemaraddi et al. 2014). In 38 hours, a combination of six distinct azo dyes may also decolorize with >70% efficiency, in addition to reactive violet 5. Shah (2014) found that, in static conditions, an isolated strain of *Bacillus megaterium* PMS82 completely degraded acid orange as a single source of carbon. In just 38 hours, the same bacterial strain was able to decolorize five different azo dyes with varying physical structures by more than 70%. It illustrates the variety of uses for microbial strains in the treatment of textile effluent. Sometimes co-substrates are required, even when in other situations the dye acts as the only or primary source of carbon, nitrogen, or phosphorus. For example, Karim et al. (2018) provided evidence of this by investigating the decolorization of five commercially available textile reactive dyes (Novacron orange FN, Novacron brilliant blue, Bezema yellow, and Bezema red) and their mixture by indigenous bacterial species using native bacterial isolates such as *Neisseria*, *Vibrio*, *Bacillus* sp., and *Aeromonas* sp. The isolates' inability to initiate decolorization was determined by the lack of co-substrates, namely glucose and yeast extract, as per the results. It implies that the microbes did not primarily obtain their energy from the dyes.

The findings of this study imply that the decolorization efficiency of microorganisms may be lowered by an increase in azo linkages in synthetic dyes and an increase in dye concentration. It was found that the type of microorganism used in the experiment, the dye's concentration, and the molecular structure of the dye all had an impact on the pace of degradation.

16.4.2.2 Dye Degradation by Fungal Strains

A variety of textile dyes have been successfully mineralized and broken down by fungal cultures thanks to their ability to adapt their metabolism to changing environmental conditions, their sturdy morphology, and their effective enzymatic machinery, which includes an extracellular ligninolytic enzyme system. These extracellular and intracellular enzymes support metabolic activity, which is essential to their survival, and have the ability to break down several color types found in effluent from textile production. The hyphae of fungus contain extracellular enzymes that cause the dye to degrade by adhering to the surface and then rupturing chemical connections within the dye molecules. Because of their nonspecific lignin-modifying enzyme, white-rot fungi are among the fungi that are frequently employed to remove colors (Ihsanullah et al. 2020). The breakdown of colors by white-rot fungus does, however, come with a number of inherent drawbacks, including the need for constrictive nitrogen conditions, a lengthy growth phase, and inconsistent enzyme output. Degradation in the exclusive presence of fungi is likewise not advised since the system and substrate do not stay stable for very long (20–30 days). The reason for this is because once the bacteria start growing, fungi won't be able to control the system and break down the specific dyes (Jonstrup et al. 2013). Initial growth circumstances, nutrient dose/concentration, temperature, pH, oxygen, dye wastewater properties, pretreatment, and the presence of carbon and nitrogen molecules are some of the factors that influence the degradation of dyes by fungal strains. The impact of nitrogen and carbon sources on *Curvularia clavata* NZ2's degradation of dyes [Congo Red (CR), Acid Orange 7 (AO7), and RB5] was noted that the majority of the colors were broken down by the fungus in both the presence and absence of carbon and nitrogen sources, with the exception of the AO7 dye, which remained colorless in the presence of the nitrogen source. The equilibrium was reached in roughly 40 minutes, and the adsorption capacity decreased as the temperature rose. The presence of surplus hydroxyl ions and competition between negatively charged dye ions for biosorption sites caused the biosorption to be reduced at higher pH values. Similar results were noted for the biosorption of the textile dye Acid Blue 40 (AB40) by *Thuja orientalis* cone biomass. The percentage of decolorization rose as the C/N ratio grew and fell at high dye concentrations. The dye could be degraded by the fungal strain up to an acidic pH of 7, after which the breakdown stopped. Similarly, dye degradation rose as temperature increased, reaching a maximum at 30 °C, and then steadily dropped after 30 °C. Spore production is how fungi spread and procreate. *Aspergillus niger* fungus pellets were used by Li et al. (2019) to promote the growth of fungal pellets while eliminating acidic anionic dyes. Acid dye at a 200 mg/L concentration was eliminated by 98% using the fungal pellets. New fungus spores or pellets were obtained in about 3 days, and the growth cycle of *Aspergillus niger* was studied. In 6 days, the process of creating colored dyes within the fungus pellets was finished at the same time. During the first 3 days, the rate of decolorization was sluggish; but, after that, it quickly rose. Electrostatic contacts and complexation were key components of the biosorption process. Fungi can also create enzymes that aid in the breakdown of pollutants. By releasing the pollutants and adhering to the root, the fungal species offer a large surface area for the adsorption of contaminants through their hyphae and spores (Ihsanullah et al. 2020).

More than 81–399% of the different metals present in the aqueous phase were eliminated by the aquatic strains of *Mucor hiemalis*'s microbiomes, germinated spores, and dead, insoluble cell walls. Biosorption and precipitation of various metal nanoparticles are actively facilitated by the outer cellular surface of sporangiospores. Metal ions were better able to be biosynthesized when positively charged metal cations interacted with the electronegative spots on spore surfaces. But the spore production in the fungus decreased the membrane's pure water permeability, which was used to treat textile effluent using filamentous fungus *Aspergillus carbonarius* M333 (Isik et al. 2019).

16.4.2.3 Bioremediation of Dyes Using Microalgae

Microalgae are eukaryotic, photosynthetic, chlorophyll-bearing microorganisms that inhabit freshwater, marine, and brackish environments. Using algae to transform dangerous pollutants found in soil or water into non-hazardous substances is known as "phyco-remediation." Algae are promising due to their high growth rate in a basic media and affordability, much like bacteria. Three different processes can be used by microalgae to remove dyes: (1) producing algal biomass; (2) converting colored to non-colored molecules causing the dye molecules to mineralize into H_2O and CO_2; and (3) chromophores adhering to the algal biomass (dye 2 Er; Ajaz et al. 2020). The dye degradation tests used enzymes produced by microalgae, namely peroxidase, malate dehydrogenase, catalase, bromo, and chloroperoxidase. In extracts without cells, it was demonstrated that three marine microalgal species exhibited peroxidase activity: *Porphyridium purpureum, Phaeodactylum tricornutum,* and *Dunaliella tertiolecta.* Rhodamine B could be biodegraded by the green algae *Coelastrella* sp. into its most basic, non-toxic forms. It may decolorize 80% of the dye in 20 days, and the peroxidase enzyme catalyzed the biodegradation process by cleaving the aromatic ring structure of rhodamine B. More than 30 distinct azo dye compounds can be biodegraded and decolorized into simpler aromatic amines by *Chlorella pyrenoidosa, C. vulgaris* and *Oscillatoria tenuis* when the azo reductase enzyme is present. The algal azo reductase enzyme breaks down the azo link, producing an aromatic amine as a byproduct in the process. When these metabolites finally mineralize, simple chemical molecules and CO_2 are produced. In terms of photosynthetic efficiency, microalgae surpassed terrestrial plants, producing biofuels like biodiesel with more efficiency from carbon dioxide and solar radiation. After the bioremediation study, algae species used for biodegradation, including *Coelastrella* sp., generate a significant amount of biomass and are suitable for the production of biodiesel (Sudarshan et al. 2023).

16.4.2.4 Yeast-Mediated Dye Degradation

Yeasts offer a number of benefits over filamentous fungi and bacteria when it comes to the breakdown of textile colors. As an illustration, they can withstand adverse conditions and grow quickly. Studies on yeasts' ability to decolorize and break down colors are scarce. A yeast strain can extract a significant number of different dyes from wastewater. The two main processes via which yeast strains degrade textile colors are biosorption and azo bond reductive cleavage. CR, Acid Red B (ARB), RB5, MG, RB, and RBB textile dyes are among those that are typically decolored using yeast. At pH 5, RB5's decolorization efficiency reached 100% after 18 hours. Nonetheless, the efficacy of decolorization declined in both highly acidic and alkaline pH environments. Similarly, temperature had an impact on the decolorization efficiency; after 18 hours, 100% decolorization was seen at 30 °C (Ihsanullah et al. 2020).

16.4.2.5 Enzyme-Mediated Dye Degradation

Bio catalysts in the liquid state, enzymes are favored in situations where precipitation is the primary method of removing difficult-to-remove target contaminants. In recent years, biocatalysts have gained a lot of popularity for a variety of industrial uses. The enzymatic breakdown of textile effluents has several advantages over traditional techniques, including high efficiency, low prices, easy accessibility, green chemistry, substrate specificity, and selectivity. Using enzymatic breakdown processes also has the advantage of being non-toxic, environmentally beneficial, and having the possibility to be reused. Nevertheless, the deactivation of the biocatalyst due to denaturation is one of the main difficulties in enzymatic degradation. Oxygen transfer, pH, temperature, dye structure, presence of redox mediators, and concentration of both the enzyme and the dye are some of the critical operational parameters that affect the efficiency and degradation of enzymatic decolorization of dyes. Immobilization of enzyme is essential for bioremediation of textile dye in a biological reactor. Enzyme mobilization techniques have advanced significantly to improve their suitability for various applications, and this evolution continues. Cross-linking is a widely used immobilization method because it works well with nearly any type of enzyme. Several researches have recently been carried out to investigate the efficacy of different sources of enzymes for the breakdown of dyes. The average removal efficiency of horseradish peroxidase (HRP) immobilization on calcium-alginate for three separate synthetic dye degradation processes was reported to be over 75% in a recent study. In a different investigation, the investigator suggested employing unique ionic liquids to improve laccase-mediated indigo carmine (IC) breakdown properties. In 30 minutes, the degrading efficiency of 82% was reached (Ihsanullah et al. 2020).

16.5 The Breakdown Mechanisms of Dyes by Microorganisms

Microorganisms such as bacteria, algae, and fungus that can adsorb and then destroy the pollutants in wastewater can be used to enable a combination of biological adsorption and enzyme-mediated degradation.

Several techniques, including adsorption, accumulation, and degradation, are used by microbes to remove textile dyes; these techniques are briefly described in Fig. 16.2. According to Kumar et al. (2006), both living and dead cells can be used to extract nonbiodegradable dyes from aqueous solutions. In comparison to living organisms, dead microbial matter has advantages such as resistance to toxicity, ability to regenerate and reuse, and lack of reliance on an ongoing nutrition source. Furthermore, the potential of dead cells to accumulate pollutants is on par with or higher than that of active cells. The adsorption capacity is influenced by the type of biomass, the chemistry of the dye, the dose of biosorbent, the biosorbent's surface quality, and environmental elements like temperature, ionic strength, pH, and the presence of competing molecules. Wastes from industry and agriculture have been treated with biosorbents, which can be made from algae, bacteria, yeasts, and other fungi. Microorganisms work well as an adsorbing agent for waste removal because of their innate capacity to break down waste and their resilience in the face of adversity. A physiologically independent process, biosorption by dead cells mostly takes place in the cell walls of the organism and involves ion exchange, complexation, micro-precipitation, chelation, and physical and chemical adsorption. Although the pH of the water must be carefully regulated because it can affect both the adsorbent surface properties and the degree of ionization of the contaminants, biosorption technologies are highly recommended due to their efficiency, selectivity, and superior removal performance (Sudarshan et al. 2023).

Microorganisms that are adapted to grow in the presence of toxic dye compounds are best suited to perform biodegradation of dyes isolated from environments contaminated with textile dyes, such as soil, wastewater treatment plant sludge, and effluent. As a source of carbon, bacteria generally break down the dye molecules. For example, certain microalgae may break down dye molecules and use this as a source of phosphorus or nitrogen. Production of algal biomass and CO_2 fixation are aided by the use of microalgae in textile dye wastewater bioremediation. Organic molecules are transformed into carbon dioxide and water during the entire mineralization or biodegradation process. Microbial enzymes that break down the intricate chemical structures of harmful dyes into less hazardous, simpler molecules are responsible for decolorizing wastewater containing textile dyes (Vikrant et al. 2018).

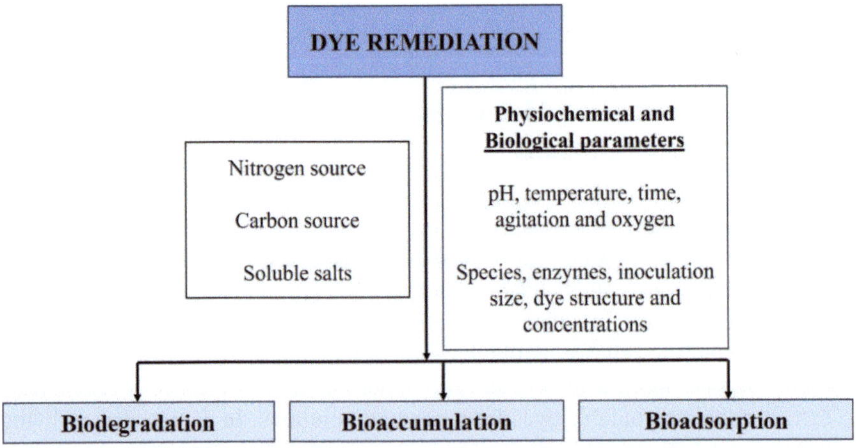

Fig. 16.2 Methods by which textile dyes can be bioremediated by microorganisms (Sudarshan et al. 2023)

16.6 Variables Influencing the Biodegradation of Dyes

There are two categories of factors that affect dye bioremediation. Environmental factors come first, then nutrition. The species used for bioremediation, biomass concentration, pH, temperature, and other physicochemical parameters are all related to the biological decolorization of dyes. Developing a commercial-scale textile dye removal technique can greatly benefit from optimization of the bioremediation process' physicochemical parameters. The principal operational characteristics that restrict the rate of dye bioremediation in industrial wastewater include elevated temperatures, low pH levels, and anaerobic environments.

16.6.1 Impact of pH

The pH of the treatment environment has a significant impact on the biological decolorization process. Algal cell walls are important for adsorption because they act as a place for dye molecules to be attracted to one another electrostatically. The microbial biomass's surface charge is contingent upon functional groups such as carboxyl and hydroxyl groups. These collectives typically function as sites or adsorbing agents. Both the dye binding sites at the cell surface and the ionic form of dye in water are influenced by the pH of the solution. Depending on the kind of dyes and salts employed, the pH can be neutral, acidic, or alkaline. The bacterial strain *Bacillus* sp. SR-2–1/1 at pH 7.0 caused the most decolorization of Congo red. The chosen azo dyes are comparatively less decolorized by the strain SR-2–1/1 at extremely alkaline and acidic pH values. It was discovered that the best pH levels for maximum decolorization were neutral or slightly alkaline, and that the rates of decolorization tend to decline quickly in strong alkaline or acidic pH circumstances. Wastewater from textiles is frequently buffered to improve microbial cells' capacity to remove dye. The activity of the azoreductase enzyme and the survival of the bacterial cells are greatly impacted by pH values that are excessively high or low, which lowers the pace of decolorization. After 108 hours of incubation, *Enterobacter* sp. EC3 demonstrated a high decolorization rate (92.6%) at pH 7.0 for synthetic dyeing effluent containing reactive black 5, while the rate of dye removal was significantly reduced (<30%) at acidic pH 4.0 (Sudarshan et al. 2023; Majhi et al. 2021).

16.6.2 Impact of Temperature

The best temperatures for bacterial growth are commonly stated to be between 30 °C and 40 °C. At these temperatures, the decolorization efficiency is great. *Bacillus* sp. SR-2–1/1 was the bacterial strain that was able to decolorize reactive red 120 and reactive black 5 to the largest extent at 30 °C. Xie et al. (2020) found that when the temperature rose from 30 to 40 °C, the degree of reactive black 5 decolorization increased. Reactive black 5's decolorization effectiveness dropped to 43% after 48 hours above 45 °C, while the maximum decolorization rate of 82% was attained at 40 °C. This could be caused by an abrupt spike in enzyme activity that occurs when the temperature rises to a certain point. However, extremely high temperatures may result in the thermal inactivation of the decolorization-inducing enzymes, which would reduce the pace of decolorization by damaging cell components like the membrane and causing cell viability to be lost, even following a cooling or heat-exchange step, rather high temperatures (between 50 and 60 °C). This suggests that the biomass of living microorganisms used, whether they are living or dead, needs to be active in order to decolorize the wastewater within this temperature range. The maximum dyes removal temperature shifts toward higher values when a microbial cell culture is immobilized on a support material, like sodium alginate beads, due to changes in the microenvironment of the cells. The advantages of immobilizing cells include increased rates of dye breakdown, high biomass loading, and defense against the harmful effects of dye and its byproducts. The dyes of textile effluent could be removed by immobilized *C. vulgaris* in alginate up to 48.9% more effectively than by free suspension cultures of algal cells (34.9%). Majhi et al. (2021) also found that live *C. pyrenoidosa* algae cells that had been immobilized had exceptional thermal stability at 60 °C, with a 97.2% decolorization efficiency. This could be because algal cells that attach to alginate have a higher temperature stability and provide a different monolayer for the adsorption of dye molecules.

16.6.3 The Impact of Initial Dye Concentration and Molecular Makeup

High dye concentrations in aquatic environments have the potential to obstruct light transmission, which could have an effect on photosynthesis and the growth of aquatic microalgae. Higher dye concentrations can be toxic, which can lower the rate, survival, and effectiveness of microbial decolorization. The percentage of dye decolorization by *Kocuria rosea* was significantly reduced when the concentration of methyl orange was increased from 10 to 100 mg L^{-1}. This could be because removing greater amounts of color requires a higher concentration of biomass. The xanthophyta alga *Vaucheria species*, which showed a decrease in decolorization effectiveness when the starting dye concentration was raised, was used to biotreat the triphenylmethane dye malachite green. The oil-producing freshwater microalgae *Chorella sorokiniana* degraded dispersed blue 2BLN at a slower pace as the dye concentration increased, according to Xie et al. (2016). The pace of decolorization is dependent on the dye's molecular structure and initial concentration. The direct relationship between the dye concentration in an aqueous solution and the accessible sites on an adsorbent surface dictates the effect of the initial dye concentration in the bioremediation process. Enzymes produced by microorganisms that break down dyes may have missed low dye concentrations. Conversely, excessive concentrations of dye are poisonous to microorganisms and have an adverse effect on the breakdown of dye by saturating adsorption sites on the adsorbent surface, blocking enzyme active sites, and toxically affecting microorganisms that degrade dye.

For reactive red 180, the maximum decolorization extent was 96.20%. For the other six dyes, azo dyes (Reactive black 5; Reactive yellow 81; Reactive blue 17); and anthraquinone dyes (Reactive blue 19; Reactive blue 49; Reactive blue 74), the bacterial strain *Citrobacter* sp. strain CK3 produced decolorization values ranging from 73.11% to 85.28%. The findings showed that anthraquinone dyes exhibit greater resistance to degradation than azo dyes. In a similar vein, 100 mg L^{-1} of azo dyes, Reactive black 5, Reactive orange 16, Reactive red 120, and Reactive yellow 2, were decolored in 72 hours by *Pseudomonas* sp. RA20 culture to 98.6, 92.6, 23.7, and 49.6%, respectively. This variation in the percentage of dyes that decolorize could be explained by the molecular structures of the dyes; dyes with a lower molecular weight and a simpler structure decolored more quickly than those with a large molecular weight and a more complex structure (Sudarshan et al. 2023).

16.6.4 The Impact of Oxygen

How oxygen affects dye decrease and cell growth is the most crucial thing to take into account. *Bacillus* sp. strain CH12 was shown to degrade azo dyes 90–100% under anoxic and anaerobic conditions. In shaking conditions, the decolorization rate was discovered to be lower (2–18.6%). Generally, bacteria have two ways to break down azo dyes: aerobic and anaerobic mechanisms. Aqueous dyes'anaerobic decolorization was usually impeded by dissolved oxygen, and oxygen was a better terminal electron acceptor than azo groups. Since the anaerobic dye reduction method inhibits dye reduction since oxygen was the most favored electron acceptor during the procedure. In a similar vein, azoreductase needs electrons in order to reduce the azo bonds in reactive black-5 dye, and that the enzyme is inhibited by oxygen from obtaining electrons from NADH/NADPH. Azole dyes degrade anaerobically quickly; however, the metabolites that are produced are colorless aromatic amines. When the "N = N" (azo) link is broken, no more dye molecule disintegration is seen, and as a consequence, it releases colorless aromatic amines that are poisonous, carcinogenic, and mutagenic. However, in aerobic conditions, the suppression of the azo dye degradation process is an irreversible impact that lasts for a short while. The aromatic amines that result from this process are further broken down into harmless compounds by monooxygenase or dioxygenase in a biological treatment system that uses aerobic bacteria. This allows for additional breakdown and mineralization of the dyes. It takes aerobic conditions to fully mineralize the reactive azo dye. Malachite green (1000 mg L^{-1}) may be decolored in less than 24 hours by *Pseudomonas* sp. strain DY1, with a high decolorizing capability (90.3%). Nevertheless, under a static state, only 78.9% of malachite green at

1000 mg L^{-1} in 24 hours is decolorized. The safest way for the full mineralization of dye molecules is therefore the combination of anaerobic treatment and aerobic treatment, which is highly recommended (Pandey et al. 2020).

16.6.5 The Impact of Nutrition

The majority of dyes have low carbon contents. This makes the procedure of removing color using microorganisms without the need for an additional carbon source challenging. Carbon sources like glucose, starch, and yeast increased the pace of bacterial growth and metabolism, resulting in a larger percentage of dye decolorization. Greater than 96% of the reactive red 180 has been decolored by *Citrobater* sp., and the best amount of carbon source for this process is the additional glucose at 4 g L^{-1}, while *Enterobacter* sp. EC3 decolorizes reactive black 5 with 90% effectiveness in 48 hours when glucose is added at a rate of 2 g L^{-1}. The dye‘s molecular makeup and the microbe that causes the decolorization could have an impact on the variation in nutrient requirements. In this investigation, yeast extract was utilized as a carbon source, and *Bacillus endophyticus* LWIS1 was utilized to decolorize 98% of the azo dye Remazol black-B after 10 hours of incubation. It demonstrates that yeast extract augments *Bacillus endophyticus* LWIS1's decolorization activity and functions as a co-substrate. The rate of dye removal and the growth response were both slowed down by changes in the carbon source's concentration. Reducing the amount of carbon sources was insufficient to support bacterial strains' growth and metabolic demands. A high concentration of carbon causes the suppression of dye decolorization and the preferential use of carbon sources for other metabolic processes.

Necessary nitrogenous supplements in the decolorization study include ammonium sulfate, ammonium nitrate, ammonium chloride, and peptone. When it came to *Pseudomonas putida* SKG-1 (MTCC 10510) decolorizing monoazo dye orange II, ammonium sulfate at 0.1% (w/v) worked best, resulting in a 92.8% efficiency (Sudarshan et al. 2023).

16.7 Conclusion

One significant worldwide cause of water contamination is the textile industry's effluents. Being xenobiotics, they are hazardous to both people and aquatic life. In the textile industry, bioremediation can employ yeast, fungi, algae, and bacteria as adsorbents. Bioremediation is a cheap, efficient, and environmentally safe method of treating textile effluents. Microorganisms like bacteria and algae that have been isolated from habitats contaminated with textile dyes can be employed effectively for bioremediation because they have well-developed mechanisms including oxidative and reductive enzymes that assist split the chemical molecules of dye.

Microorganisms may also be able to treat wastewaters contaminated with dyes due to their numerous other benefits, which include shorter life cycles, a large surface area that facilitates excellent biosorption, and the presence of chemical groups on their cell walls that provide a multitude of sites for electrostatic adsorption. While many issues require immediate attention, more study in this field may be able to fix the majority of the present problems. Most bioremediation techniques now in use are lab-based. A significant technological challenge is integrating these technologies for widespread commercial use. A number of aspects need to be taken into account while adopting bioremediation, such as the kind of effluent, the toxicity of metabolites, the associated expenses, and the planned use of the treated water. Precise control over variables like temperature, pH, biomass concentration, agitation, and dye concentration can improve the effectiveness of bioremediation; these factors need to be considered. The best combination will come from carefully weighing the benefits of each process and the possible interactions between combining different technologies; using a range of strains and consortiums isolated from dye-contaminated sites; isolating novel microorganisms, such as halotolerant or thermophilic ones; or adapting already-existing ones to use dye molecules as their carbon and nitrogen sources in a way that yields the desired outcomes. Further studies focused on the identification of a variety of microorganisms living in contaminated areas and their subsequent application in the total breakdown and mineralization of the dangerous textile dyes would therefore undoubtedly constitute a cost-effective, useful, and environmentally beneficial strategy.

16.8 Future Scope

In the past 20 years, bioremediation of dyes using microorganisms has gained popularity. Using bacteria and microalgae as biodegrading agents or biosorbents could be one way to lessen the harmful effects of dye effluent. Bioremediation techniques can be used to remediate wastewater containing textile dyes in an efficient manner. Nevertheless, a few problems prevent bioremediation from being used in practice. To increase their efficacy, bioremediation systems for commercial and industrial uses in the treatment of wastewater containing textile dyes must solve the following issues:

(a) To achieve increased decolorization success, a thorough understanding of the experimental parameters for bioremediation of resistant dyes, favorable conditions for the development of microorganisms, the dye degradation process, and the toxicity of degraded products is necessary.
(b) There is a lack of data regarding the bioremediation of real textile dye effluent, even though most research is based on wastewater containing synthetic dyes made in laboratories. Much research needs to be done to determine whether employing single microorganisms or groups of microorganisms for the bioremediation of real wastewater is feasible.

(c) The possibility of hazardous biomass accumulation during the bioremediation process, as well as the possibility of producing biomass and using created biomass as animal feed, must all be investigated.

Acknowledgments I would like to thank the higher authorities of Swami Vivekananda University for granting me permission to conduct this study.

Conflict of Interest There isn't any conflict of interest with this study.

Author Contribution Statement Acquisition, interpretation of data, visualization, designing, writing, reviewing, and editing of this study are done by Mr. Rupesh Dutta Banik. Formal analysis, revising, and editing of this study are done by Dr. Sibashish Baksi. Conception, supervision, revising, and editing of the article are done by Dr. Pritha Pal.

References

Abou-El-Souod GW, El-Sheekh MM (2016) Biodegradation of basic fuchsin and methyl red. Environ Eng Manag J 15:279–286

Ajaz M, Shakeel S, Rehman A (2020) Microbial use for azo dye degradation—a strategy for dye bioremediation. Int Microbiol 23:49–159. https://doi.org/10.1007/s10123-019-00103-2

Anjaneya O, Souche SY, Santoshkumar M, Karegoudar TB (2011) Decolorization of sulfonated azo dye Metanil Yellow by newly isolated bacterial strains: *Bacillus sp.* strain AK1 and *Lysinibacillus sp.* strain AK2. J Hazard Mater 190(1–3):351–358. https://doi.org/10.1016/j.jhazmat.2011.03.044

Banik RD, Pal P, Baksi S (2023) Green strategy for reclaiming ecosystem: microbial remediation of leather industry wastes. J Surv Fish Sci. https://doi.org/10.53555/sfs.v10i1S.2161

Berradi M, Hsissou R, Khudhair M, Assouag M, Cherkaoui O, El Bachiri A et al (2019) Textile finishing dyes and their impact on aquatic environs. Heliyon 5(11). https://doi.org/10.1016/j.heliyon.2019.e02711

Bhatia D, Sharma NR, Singh J, Kanwar RS (2017) Biological methods for textile dye removal from wastewater: a review. Crit Rev Environ Sci Technol 47(19):1836–1876. https://doi.org/10.1080/10643389.2017.1393263

Bheemaraddi MC, Patil S, Shivannavar CT, Gaddad SM (2014) Isolation and characterization of *Paracoccus sp.* GSM2 capable of degrading textile azo dye reactive violet 5. Sci World J:1–9. https://doi.org/10.1155/2014/410704

Dellamatrice PM, Silva-Stenico ME, Moraes LABD, Fiore MF, Monteiro RTR (2017) Degradation of textile dyes by cyanobacteria. Braz J Microbiol 48:25–31. https://doi.org/10.1016/j.bjm.2016.09.012

Goud BS, Cha HL, Koyyada G, Kim JH (2020) Augmented biodegradation of textile azo dye effluents by plant endophytes: a sustainable, eco-friendly alternative. Curr Microbiol 77:3240–3255. https://doi.org/10.1007/s00284-020-02202-0

Ihsanullah I, Jamal A, Ilyas M, Zubair M, Khan G, Atieh MA (2020) Bioremediation of dyes: current status and prospects. J Water Process Eng 38:101680. https://doi.org/10.1016/j.jwpe.2020.101680

Isik Z, Arikan EB, Bouras HD, Dizge N (2019) Bioactive ultrafiltration membrane manufactured from *Aspergillus carbonarius* M333 filamentous fungi for treatment of real textile wastewater. Bioresour Technol Rep 5:212–219. https://doi.org/10.1016/j.biteb.2019.01.020

Jonstrup M, Kumar N, Guieysse B, Murto M, Mattiasson B (2013) Decolorization of textile dyes by *Bjerkandera sp.* BOL 13 using waste biomass as carbon source. J Chem Technol Biotechnol 88(3):388–394. https://doi.org/10.1002/jctb.3852

Karim ME, Dhar K, Hossain MT (2018) Decolorization of textile reactive dyes by bacterial monoculture and consortium screened from textile dyeing effluent. J Genet Eng Biotechnol 16(2):375–380. https://doi.org/10.1016/j.jgeb.2018.02.005

Kumar KV, Ramamurthi V, Sivanesan S (2006) Biosorption of malachite green, a cationic dye onto *Pithophora sp.*, a fresh water algae. Dyes Pigments 69(1–2):102–107. https://doi.org/10.1016/j.dyepig.2005.02.005

Kumar SS, Shantkriti S, Muruganandham T, Murugesh E, Rane N, Govindwar SP (2016) Bioinformatics aided microbial approach for bioremediation of wastewater containing textile dyes. Eco Inform 31:112–121. https://doi.org/10.1016/j.ecoinf.2015.12.001

Li S, Huang J, Mao J, Zhang L, He C, Chen G et al (2019) In vivo and in vitro efficient textile wastewater remediation by *Aspergillus niger* biosorbent. Nanoscale Adv 1(1):168–176. https://doi.org/10.1039/C8NA00132D

Majhi PK, Kothari R, Pandey A, Tyagi VV (2021) Adsorptive behavior of free and immobilized *Chlorella pyrenoidosa* for decolorization. Biomass Convers Biorefinery 11:3023–3036. https://doi.org/10.1007/s13399-020-00770-6

Nahiun KM, Sarker B, Keya KN, Mahir FI, Shahida S, Khan RA (2021) A review on the methods of industrial waste water treatment. Sci Rev 7(3):20–31. https://doi.org/10.32861/sr.73.20.31

Pandey K, Saha P, Rao KB (2020) A study on the utility of immobilized cells of indigenous bacteria for biodegradation of reactive azo dyes. Prep Biochem Biotechnol 50(4):317–329. https://doi.org/10.1080/10826068.2019.1692219

Saratale RG, Saratale GD, Chang JS, Govindwar SP (2010) Decolorization and biodegradation of reactive dyes and dye wastewater by a developed bacterial consortium. Biodegradation 21:999–1015. https://doi.org/10.1007/s10532-010-9360-1

Saxena A, Gupta S (2020) Bioefficacies of microbes for mitigation of Azo dyes in textile industry effluent: a review. Bioresources 15(4):9858–9881

Shah MP (2014) Bioremedial application of *Bacillus megaterium* PMS82 in microbial degradation of acid orange dye. Int J Environ Bioremediat Biodegrad 2:93–99. https://doi.org/10.12691/ijebb-2-3-1

Srivastava S, Sinha R, Roy D (2004) Toxicological effects of malachite green. Aquat Toxicol 66(3):319–329. https://doi.org/10.1016/j.aquatox.2003.09.008

Sudarshan S, Harikrishnan S, RathiBhuvaneswari G, Alamelu V, Aanand S, Rajasekar A et al (2023) Impact of textile dyes on human health and bioremediation of textile industry effluent using microorganisms: current status and future prospects. J Appl Microbiol 134(2):lxac064. https://doi.org/10.1093/jambio/lxac064

Tan KA, Morad N, Ooi JQ (2016) Phytoremediation of methylene blue and methyl orange using *Eichhornia crassipes*. Int J Environ Sci Develop 7(10):724. https://doi.org/10.18178/ijesd.2016.7.10.869

Varjani S, Rakholiya P, Ng HY, You S, Teixeira JA (2020) Microbial degradation of dyes: an overview. Bioresour Technol 314:123728. https://doi.org/10.1016/j.biortech.2020.123728

Vikrant K, Giri BS, Raza N, Roy K, Kim KH, Rai BN et al (2018) Recent advancements in bioremediation of dye: current status and challenges. Bioresour Technol 253:355–367. https://doi.org/10.1016/j.biortech.2018.01.029

Wang ZW, Liang JS, Liang Y (2013) Decolorization of Reactive Black 5 by a newly isolated bacterium *Bacillus sp.* YZU1. Int Biodeterior Biodegradation 76:41–48. https://doi.org/10.1016/j.ibiod.2012.06.023

Wang R, Li H, Liu Y, Chen J, Peng F, Jiang Z et al (2022) Efficient removal of azo dyes by *Enterococcus faecalis* R1107 and its application in simulated textile effluent treatment. Ecotoxicol Environ Saf 238:113577. https://doi.org/10.1016/j.ecoenv.2022.113577

Xie L, Zhou L, Liu T, Xu X (2016) Degradation of disperse blue 2BLN by oleaginous *C. sorokiniana* XJK. RSC Adv 6(108):106935–106944. https://doi.org/10.1039/C6RA21915B

Xie XH, Zheng XL, Yu CZ, Zhang QY, Wang YQ, Cong JH et al (2020) High-efficient biodegradation of refractory dye by a new bacterial flora DDMY1 under different conditions. Int J Environ Sci Technol 17:1491–1502. https://doi.org/10.1007/s13762-019-02582-2

Chapter 17
Mycoremediation: A Sustainable Approach to Curb Emerging Pollutants

Venkata Kanaka Srivani Maddala, Supriya Chatla, and Lurdhumary Kunduri

Abstract Pollutants like pharmaceuticals, plastics, personal care products (cosmetics, soaps, dyes, detergents, shampoos), pesticides and heavy metals were manufactured to lead a comfortable life, keep us healthy and ensure food and material supply for the growing human population. These pollutants are released into the environment and degrade the terrestrial and aquatic environment. Many methods were developed to mitigate these pollutants that led to negative effects and created havoc in the environment. Today, fungi and its role in bioremediation are cost effective and eco-friendly and have drawn attention all over the world. It shows more resistance to higher level of contamination compared to bacteria and algae. There are many fungal species that degrade these pollutants without effecting the environment. Some of the fungal species like *Trametes versicolar*, *Phanerochaete chrysosporium* and *Asperigillus sydowii* degrade pesticides. *Fomes fomentarius*, *Phyllotopsis nidulans* and *Hypholoma fasciculare* can degrade pharmaceuticals and personal care products. *Asperigillus clavatus*, *Asperigillus flavus*, and *Asperigillus terreus* degrade *polyethylene. Fusarium solani* and *Schizophyllum commune* degrade polychlorinated biphenyls. *Asperigillus flavus* degrades dyes and surfactants. This book chapter explains the current literature of various fungal species that degrade the emerging pollutants, highlights the role of enzymes in pollutant degradation and throws light on mechanism of mycoremediation against emerging pollutants for sustainability.

V. K. S. Maddala (✉)
Department of Chemistry, Vignans Foundation for Science Technology and Research, School of Applied Science and Humanities, Vadlamudi, Guntur, Andhra Pradesh, India

S. Chatla
Department of Pharmaceutics and Biotechnology, Nirmala College of Pharmacy, Atmakuru, Andhra Pradesh, India

L. Kunduri
Department of Pharmaceutical Sciences, Vignans Foundation for Science Technology and Research, Vadlamudi, Guntur, Andhra Pradesh, India

S. Dey, S. Bhattacharya (eds.), *Biotechnological Interventions in the Removal of Emerging Pollutants*, Interdisciplinary Biotechnological Advances,
https://doi.org/10.1007/978-981-97-9922-0_17

Keywords Emerging pollutants and its impacts · Significance of mycoremediation · Mechanism of mycoremediation · Fungal enzymes and its role · Role of fungi in degradation of pollutants

17.1 Introduction

Increase in population, rapid industrialization and urbanization resulted in intense rise in the level of pollutants. Advancements in chemical, agriculture, pharmaceutical and plastic industries have made human life comfortable, and the negative consequences of these created unsustainability. Many of the harmful chemicals that were intentionally or unintentionally released into the terrestrial and aquatic environment degraded its quality. Emerging pollutants consist of synthetic chemicals that have adverse effects on man and environment (Vasilachi et al. 2021). Many compounds from the emerging pollutants were used to satisfy human needs to lead a comfortable life, but they destroy the environmental quality (Pujari and Kapoor 2021; Roy et al. 2022; Lebreton et al. 2017). Current statistics show that approximately 144 million individuals utilise contaminated water and 2.2 billion people suffer due to lack of safe water availability. It has been found in the research that half of our world population would be living in water stress conditions by 2025. Rapid industrialization and urbanization degraded 22 million hectares of land mass. Other issues like serious health problems, loss of biodiversity, pollution worsen the conditions and lead to unsustainability on earth. To curb this situation many methods were developed which are expensive, generate harmful products and insufficient to control them. So bioremediation, which is the process of using the bacteria, algae, fungi, and plants to detoxify pollutants in water and soil, has gained proper attention. Among these, fungi have been recognised to play a vital role in removing the pollutants.

Bioremediation by fungi is called mycoremediation, and it plays a vital role in the degradation of emerging pollutants. It has the ability to degrade pesticides, pharmaceuticals, personal care products, plastics and heavy metal . It's also called fungal treatment of fungal based technology It is the application of fungi and has the capability to degrade these pollutants in various ways. In mycoremediation process undesirable product is identified transformed, sequestered and removed completely without causing harm to the nature. Removal of these pollutants and their consequences is the promising technology today in the scientific world. Fungi degrades various emerging pollutants within a short time without disturbing the natural ecosystems. This chapter throws light on the consequences of various emerging pollutants and also highlights the role of fungal species in the degradation of pesticides, plastics, microplastics, personal care products, and other industrial effluents.

17.2 Emerging Pollutants and Their Impacts on Environment

Emerging pollutants are the significant environment pollutants on earth. These may occur naturally or synthetically and do not harm the environment under normal conditions. But unfortunately, due to excess growth in the industrial, agricultural and domestic sectors, concentration of these pollutants has increased and disturbed our environment, so they should be removed immediately to protect environment.

Polycyclic aromatic hydrocarbons release from incomplete combustion of coal, wood and petroleum. It leads to immunity suppression and different types of cancer. Application of pesticides and herbicides in agricultural activities effects environment and public health. These pollutants lower dissolved oxygen levels and damage aquatic life by causing eutrophication. Many harmful residues increase in fruits and vegetables, making their way into the food chain and thus damaging health.

Application of pesticides causes serious threat to our environment. Pesticides such as organochlorines, organophosphates and carbamates persist in the environment for a long time and cause adversely affect nontarget species (Kumar et al. 2019; Kumari et al. 2014). organochlorines, organophosphates, and carbamates also leads to bioaccumulation and long range transmission. Pesticide exposure develop different types of cancer and dysfunction of endocrine and nervous system (Kim et al. 2017; Shah and Parveen 2021).

Dyes are harmful and cause serious threat to the ecosystem. It causes skin rashes, allergy and irritation in man. Plastic pollutants in water bodies cause adverse effects on environment. For example, plastics will effect marine biodiversity ie ingestion and entanglement of plastic litter (Kurtela and Antolović 2019). Microplastics affect the health of soil fauna and degrade the soil quality and its function. It shows toxic effects on fish and other aquatic organisms.

Plastics and microplastics affect terrestrial and aquatic ecosystems, kill coral reefs and cause the ecosystem to collapse. Personal care products lead to depletion of natural resources, damage marine water quality, carcinogenic, effects metabolism pollution of air, water and soil. The impacts of emerging pollutants were mentioned in Fig. 17.1.

Many environment contaminants released from deodorants, detergents, toothpastes, soaps, shampoos (personal care products). Phthalates and phthallic acid esters are used as additives and plasticizers in personal care products. Phthalates like diethyphthalate, dimethylphthalate, diisobutylphthalate, and di-n-butyl phthalate are present in daily used fragrnaces, baby products, cosmetics, and nail polishes (Koniecki et al. 2011). Many plastic fragments were observed in the stomach regions of sea birds like *phalacrocoraxbougainvilli*, *Pelecanoides garnotii*, *Pelecanoides urinatrix*, *Pelecanus thagus*, *Spheniscus humboldti* and more ingestion capacity detected in *Larus dominicanus* as it feeds upon plastic containers, waster material and fishing nets (Thiel et al. 2018). Microplastics and nanoplastics

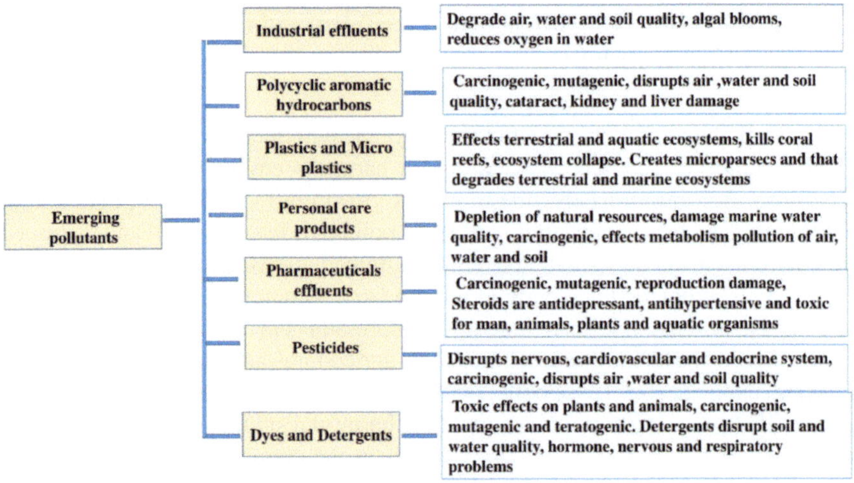

Fig. 17.1 Emerging pollutants and its effects on environment

degrade the soil quality and alter the physical, chemical and biological properties of soil (Pathan et al. 2020; Rillig et al. 2017). Pesticides effects food chain and kills non target species instead of pests (Lozowicka et al. 2016; Rani 2015). PAH are carcinogenic and mutagenic; disrupt air, water and soil quality; cataract, kidney and liver damage.

Pharmaceutical effluents disrupts the reproductive system and causes cancer and muragenesis.Steroids are antidepressant, antihypertensive and toxic for man, animals, plants and aquatic organisms. Dyes and detergents show toxic effects on plants and animals and are carcinogenic, mutagenic and teratogenic. Detergents disrupt soil and water quality and cause hormone, nervous, and respiratory problems. Once discharged, these polycyclic aromatic hydrocarbons, plastics, pesticides, personal care products, pharmaceuticals, and heavy metals persist in the environment for long time and cause serious threat to marine and aquatic environment, and it is difficult for the natural microbiome to remove or neutralise them. Many strategies were developed to tackle this problem and remained ineffective. These are toxic and more attention was not given in recent times (Baybil et al. 2022). Many methods of remediation were developed to combat these issues but they are ineffective and not up to the mark.

17.3 Mycoremediation and its significance

Mycoremediation is an important promising technology in controlling the pollution and conservation of environment. It utilises the fungi to decontaminate the environment and has the capability to break down the substances that are resistant to bacterial Degradation, for example, fragrances and pharmaceuticals. It also degrades toxic substances in waste water and personal care products without causing harm to the environment. It is environmentally sound technology to curb all the ever-increasing problems arising from emerging pollutants. Its also used in microfiltration and industrial effluent filtration and has the ability to convert waste to valuable products.

17.3.1 Process of Mycoremediation

It is the process of using the fungi that take part in degrading the harm ful pollutants into harm less substances and restore the environment. The following steps are involved in this process.

Emerging pollutants present in soil or water bodies were observed.

Before starting the remediation project, the fungal species (mushrooms) are selected and then the feasibiltiy studies are conducted to find out the process and toxicity in that fungi later the fungal strains are inoculated in the contaminated soil and the fungi/mushroom grows uses enzymes and degrade the pollutants, absorb them and metabolize into harmless substances. Figure 17.2 shows how the pollutants are degraded and converted into harmless products in mycoremediation process.

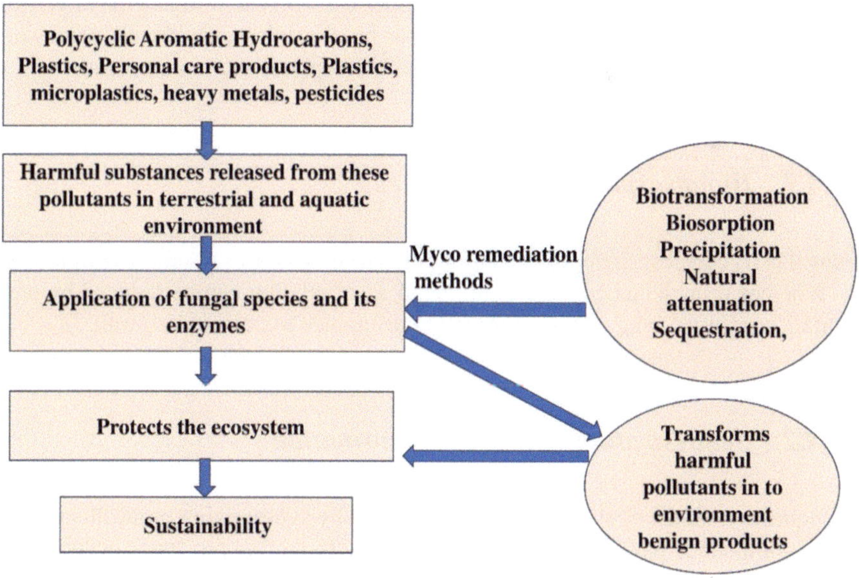

Fig. 17.2 Application of fungi and its enzymes in converting harmful pollutants to harmless products

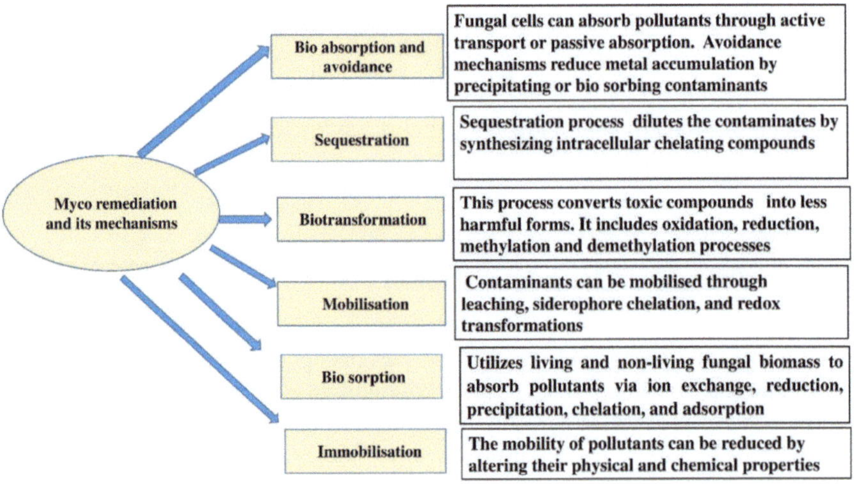

Fig. 17.3 Mycoremediation and its mechanism

17.4 Mechanism of Mycoremediation

Many pathways and mechanisms employed by fungi to reduce emerging pollutants. The detailed mechanism and its methods are shown in Fig. 17.3.

Fungal cells employ several strategies to interact with and mitigate the effects of pollutants. These include mechanisms such as bioabsorption, avoidance, sequestration and biotransformation, mobilization and biosorption, and immobilization.

17.4.1 Bioabsorption and Avoidance

Fungal cells can absorb pollutants through active transport or passive absorption.

Whereas in avoidance mechanism, metal accumulation can be reduced by precipitation or biosorbing contaminants and thus reduce toxicity in cell wall.

17.4.2 Sequestration and Biotransformation

In sequestration, the contaminants can be diluted by synthesizing intracellular chelating compounds with in the fungal cell. Biotransformation processes include oxidation, reduction, methylation, and demethylation and convert toxic compounds in to harmless ones.

17.4.3 Mobilization and Biosorption

Microbes mobilize contaminants through processes such as leaching, siderophore chelation, and redox transformations. Biosorption, on the other hand, utilizes living and nonliving fungal biomass to absorb pollutants via ion exchange, reduction, precipitation, chelation, and adsorption.

17.4.4 Immobilization Process

Pollutant mobility can be reduced by altering their physical and chemical properties, making them less available in biological systems. Fungal cell walls trap the contaminants through interactions with functional groups.

17.5 Role of Fungal Enzymes

Fungal enzymes play an important role in biodegradation. It has the capability to break down complex organic substances in to simple substances. The cultivation and production of enzymes need a deep understanding of various fungal species involved and the environmental conditions that optimise their activity. The following sources are used in the fungus enzyme cultivation: harsh environments, soil, compost, dead decomposing matter and industrial waste Fungi has extra- and intracellular enzymatic machinery with the capability of excreting acids to attack both organic and inorganic pollutants (Deshmukh et al. 2016; Harms et al. 2011).

17.5.1 Intracellular and Extracellular Enzymatic Activity

Whenever pollutants are detected in the environment, then application of fungi detects the pollutants by using sensory mechanisms and producing enzymes to break down the pollutants. Fungi use extra- and intracellular enzymatic machinery and secrete acids that helps in attacking organic and inorganic pollutants. Fungi show enzymatic machinery both inside and outside the cell walls. The detailed process of extra- and intracellular enzymatic machinery is shown in Fig. 17.4.

Types of enzymes in intracellular enzymatic activity are: (i) metabolic enzymes break down the nutrient for energy production, for example, amylases break down starch into glucose and lipases break down lipids into fatty acids; (ii) signal transduction enzymes transmit the signals within the cell and regulate cellular activities, for example, protein kinases; (iii) cell division enzymes facilitate the cell division and growth, for example, DNA polymerases and RNA polymerases; (iv) respiratory

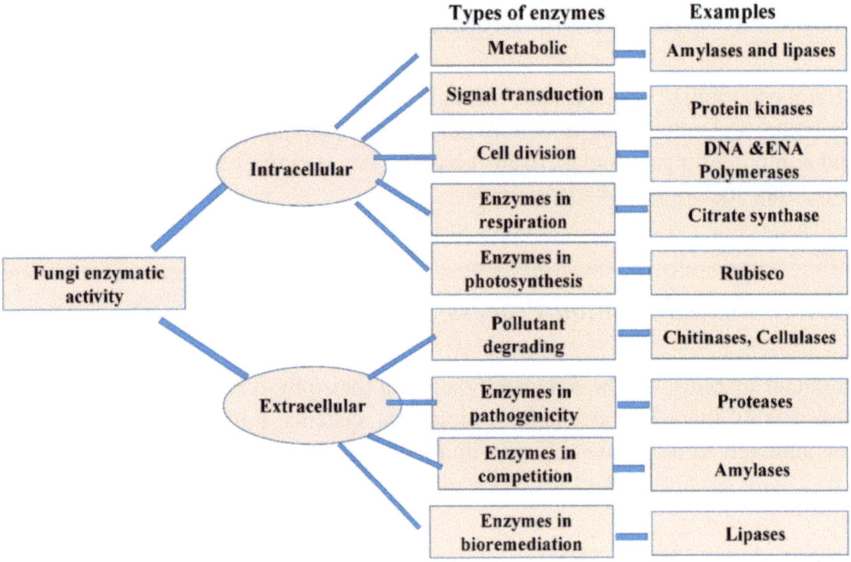

Fig. 17.4 Fungal intracellular and extracellular enzymatic activity

enzymes convert organic compounds to energy, for example, citrate synthase; (v) photosynthetic enzymes convert light into chemical energy, for example, rubisco.

Extracellular enzymatic activity: (i) pollutant-degrading enzymes break down pollutants into harmless substances, for example, chitinases and cellulases; (ii) enzymes like proteases break down proteins (enzymes in pathogenicity); (iii) enzymes involved in competition with other organisms break down nutrients to outcompete other organisms, for example, amylases; (iv) lipase enzymes break down the pollutants in the contaminated site (bioremediation).

17.6 Fungal Species in Degradation of Emerging Pollutants

Fungi have intra- and extracellular enzymatic machinery, and they have the capacity to extract acids and attack a variety of organic and inorganic pollutants in environment (Deshmukh et al. 2016; Harms et al. 2011). Fungi degrades environmental pollutants because it produces enzymes that will break lignin and cellulose. Many species of fungi can degrade pollutants in different environment. Fungal enzymes results in organic material production, hydrolysis of lignin related compounds, decolorization of dyes, treatment of wastes, bleaching of pulp and wood pulping (Yesilada et al. 2018; Goala et al. 2021). *Aspergillus flavus*, *Marasmius cladophyllus*, *Phlebia acerina*, *Bjerkandera adusta degrades dyes* by using laccase, manganese peroxidase and lignin peroxidase enzymes (Moore et al. 2001; Kim et al. 2007; Bhattacharya et al. 2011). Herbicides and pesticides can be degraded by

Botryoshaeria larieina, *Asperigillus glaucus*, *Tremetes pavonia*, *Penicillium verruculosum* with the help of Ligninolytic enzymes by esterification, dioxygenation, dehydrogenation, dechlorination, demethylation mediated degradation mediated degradation (Azubuike et al. 2016; Igiri et al. (2018); Fomina and Gadd 2014; Urík et al. 2007; Abu-Elsaoud et al. 2017; Oladipo et al. 2018). *Aspergillus species*, *Rhizomucor species*, *Fusarium species*, *Emericella species*, *Funneliformis geosporum*, *Pleurotus ostreatus*, *Trichoderma harzianum*, *Trichoderma ghanense* and *Penicillium rubens* play a vital role in degradation of heavy metals and antioxidants using ligninolytic enzymes and prevent the damage caused by oxidative stress. *Phanerochaete chrysosporium*, *Trametes versicolor*, *Pleurotus ostreatus*, *Pleurotus eryngii* and *Cochliobolus lunatus* degrade polycyclic aromatic hydrocarbons with the help of ligninolytic enzymes, cytochrome P-450 monooxygenase, dioxygenase, dehydrogenases, FAD-dependent monooxygenases, glutathione transferase and epoxide hydrolases (Samanta et al. 2002; Abdel-Shafy and Mansour 2016; Bhattacharya et al. 2014; Schamfuß et al. 2013; Young et al. 2015). Pharmaceuticals can be degraded by the enzymes like ligninolytic enzymes and cytochrome 450 produced by *Mucor hiemalis*, *Trametes versicolar*, *Phanerochaete chrysosporium* and *Lentinula edodes* (Dai et al. 2018; Mohamed et al. 2014; Gao et al. 2017; Du et al. 2015; Larsson 2014; Esterhuizen-Londt et al. 2017).

Phthalates are degraded by *Fusarium oxysporium*, *Fusarium subglutinans*, *Penicillium brocae* and *Purpureocillium lilacinum* with the help of cutinase enzyme (Kryczyk-Poprawa et al. 2019; Prieto et al. 2011; García-Galán et al. 2011; Cvancarová et al. 2013; Benjamin et al. 2015). Lipase enzymes produced from *Asperigillus*, *Aeremonium*, *Alternaria*, *Beauveria*, *Candida*, *Eremothecium*, *Fusarium*, *Geotrichum*, *Humicola mucor*, *Penicillium*, *Rhizomucor*, *Rhizopus*, and *Trichoderma* degrade fatty acids (Masse et al. 2001; Takamoto et al. 2001). *Humicola insolens* degrades polyethylene terephthalate by using cutinase enzymes (Ronkvist et al. 2009). *Pleurotus ostreatus*, *Leptospaherulina* sp., *Irpex lacteus*, and *Lentinula edodes* degrade antibiotics using versatile peroxidase, laccase, manganese peroxidase and cytochrome 450 enzymes (Esterhuizen-Londt et al. 2017; Marco-Urrea et al. 2010; Asif et al. 2017; Muszyńska et al. 2019; Yang et al. 2013). Polypropylene is degraded by *Asperigillus species*, *Paecilomyces lilacinus*, and *Lasiodiplodia theobromae* using laccase enzyme (Sheik et al. 2015). Pentachlorophenol, which is used as a biocide and wood preservative, is degraded by the fungus *Phlebia acanthocystis*, yielding methylated and oxidized products such as pentachlorophenol anisole and *p*-tetrachlorohydroquinone (enzymes: Ligninperoxidase and Mn peroxidase) (Xiao and Kondo 2020). *Choloterans* degrades high-density polyethylene (HDPE) using extracellular oxidoreductase enzyme (Di Napoli et al. 2023). *Trichoderma harzianum* degrades polyethylene with peroxidase enzyme (Sowmya et al. 2014). Urethane hydrolases enzyme by *Asperigillus terreus*. *Chaetomium globosum* degrades polyurethane (Boubendir 1993). *Penicillium citrinum* and *Beauveria brongniartii* degrade polyethylene terephthalate by using polyesterases (Almansa et al. 2008). *Pleurotus erynngii* degrade lignin, cellulose, and PAH. It also removes heavy metals such as lead and cadmium through lignocellulosic enzymes (Bempah et al. 2013). *Asperigillus niger*, *Eurotium repens*, *Cladosporium pseudoclado sporioides*, *Penicillium chrysogenum*, *Bipolaris* sp., *A. nidulans*, and *Epicoccumnigrum* use α-1-rhamnohydrolase, β-N-acetyl hexoseaminidase and urease and decompose nonsteroid anti-inflammatory drugs such as diclofenac, ibuprofen, diflunisal, mefenamic, and piroxicam (Gonda et al. 2016). Dyes are harmful and cause serious threat to the ecosystem. Fungi play a vital role in degradation and decolurization of organic dyes with the help of fungl enzymes and processes such as adsorption, absorption, and aggregation. Azodyes are used in textile and carpet coloring. These dyes, when discharged into the water bodies, affect plants, animals, and human beings. *Peroneutypa scoparia* degrades acid red 97 by using enzyme laccase (Pandi et al. 2018). *Phomatropica*, *Dichotomomyces cejpii* degrades congo red and methyl red by using laccase enzyme (Krishnamoorthy et al. 2018). *Asperigillus nigar*, *Penicillium chrysogenum* degrade phthalates like Di-2-ethylhexylphthalate (DEHP). Dibutylphthalate (DBP), benzyl butyl phthalate (BBP) by using phthalase enzyme.

Laccase enzyme from *Asperigillus flavus* degrades dyes and surfactants, Ligninolytic enzyme from *Trametes pavonia*, *Penecillium verruculosum*, *Asperigillus glaucus* degrades pesticide and herbicides (Singh et al. 2015; Young et al. 2015; Prieto et al. 2011; García-Galán et al. 2011; Cvancarová et al. 2013; Benjamin et al. 2015). Ligninolytic enzymes produced by the *Asperigillus species*, *Fusarium species*, *Rhizomucor species*, *Emericella species*, *Funneliformis geosporum*, *Pleurotus ostreatus*, *Trichoderma harzianum*, *Trichoderma ghanense*, and *Penicillium rubens* help in the degradation of heavy metals and antioxidants (Tchounwou et al. 2012; Chibuike and Obiora 2014; Singh et al. 2015; Young et al. 2015). Different types of fungal species degrading emerging pollutants and their enzymes involved in degradation are mentioned in Table 17.1.

Table 17.1 Fungal species involved in degradation of emerging pollutants

S. no.	Name of the fungi	Enzymes secreted	Pollutant degraded	Process of degradation
1	*Aspergillus flavus*, *Marasmius cladophyllus*, *Phlebia acerina*, *Bjerkandera adusta*	Laccase, manganese peroxidase and lignin peroxidase in degradation of dyes	Dyes	Moore et al. (2001); Kim et al. (2007); Bhattacharya et al. (2011)
2	*Botryoshaeria larieina*, *Asperigillus glaucus*, *Tremetes pavonia*, *Penicillium verruculosum*	Ligninolytic enzymes	Herbicides and pesticides	Azubuike et al. (2016); Igiri et al. (2018); Fomina and Gadd (2014); Urík et al. (2007); Abu-Elsaoud et al. (2017); Oladipo et al. (2018)

(continued)

Table 17.1 (continued)

S. no.	Name of the fungi	Enzymes secreted	Pollutant degraded	Process of degradation
3	*Aspergillus species, Rhizomucor species, Fusarium species, Emericella species, Funneliformis geosporum, Pleurotus ostreatus, Trichoderma harzianum, Trichoderma ghanense, Penicillium rubens*	Ligninolytic enzymes	Heavy metals	Tchounwou et al. (2012); Singh et al. (2015); Young et al. (2015)
4	*Phanerochaete chrysosporium, Trametes versicolor, Pleurotus ostreatus, Pleurotus eryngii, Cochliobolus lunatus*	Ligninolytic enzymes, cytochrome P-450 monooxygenase, dioxygenase, dehydrogenases, FAD-dependent monooxygenases, glutathione transferase and epoxide hydrolases	Polycyclic aromatic hydrocarbons	Samanta et al. (2002); Abdel-Shafy and Mansour (2016); Bhattacharya et al. (2014); Schamfuß et al. (2013); Young et al. (2015)
5	*Mucor hiemalis, Trameted versicolor, Phanerochaete chrysosporium, Lentinulaeododes*	Ligninolytic enzymes and cytochrome 450	Pharmaceuticals	Dai et al. (2018); Mohamed et al. (2014); Gao et al. (2017); Du et al. (2015); Larsson (2014); Esterhuizen et al. (2016)
6	*Fusarium oxysporium, Fusarium subglutinans, Penicillium brocae, Purpureocillium lilacinum*	Cutinase	Pthalates	Kryczyk-Poprawa et al. (2019); Prieto et al. (2011); García-Galán et al. (2011); Cvancarová et al. (2013); Benjamin et al. (2015)
7	*Asperigillus acremonium, Alternaria, Beauveria, Candida eremothecium, Fusarium geotrichum, Humicola, Mucor, Ophiostoma, Penicillium, Rhizo mucor, Rhizopus* and *Trichoderma*	Lipases	Lipases can degrade fatty wastes	Masse et al. (2001); Takamoto et al. (2001)
8	*Humicola insolens*	Polyethylene terephthalate	Cutinase	Ronkvist et al. (2009)

(continued)

Table 17.1 (continued)

S. no.	Name of the fungi	Enzymes secreted	Pollutant degraded	Process of degradation
9	*Pleurotus ostreatus, Leptospaherulina sp, Irpex lacteus, Lentinula edodes*	Antibiotics	Versatile peroxidase, la manganese perox cytochrome 450	Esterhuizen-Londt et al. (2017); Marco-Urrea et al. (2010); Asif et al. (2017); Muszyńska et al. (2019); Yang et al. (2013)
10	*Asperigillus species, Paecilomyces, Lilacinus, Lasiodiplodia theobromae spergillus* sp. *Paecil omyces, Lilacinus, Lasiodiplo dia theobromae*	Polypropylene	Enzyme: laccase	Sheik et al. (2015)
11	*Phlebia acanthocystis*	Pentachloropheno l used as biocide	Ligninperoxidase and Mn peroxidase	Xiao and Kondo (2020)
12	*Choloterans*	Degrades high-density polyethylene	Extracellular oxidoreductase	Di Napoli et al. (2023)
13	*Trichoderma harzianum*	Polyethylene	Peroxidases	Sowmya et al. (2014)
14	*Penicillium citrinum, Beauveria brongniartii*	PET (polyethylene terephthalate)	Polyesterases	Almansa et al. (2008); Liebminger et al. (2007)
15	*Asperigillusterreus. Chaetomium globosum*	Urethane hydrolases	PUR (Polyurethane)	Boubendir (1993)
16	*Pleurotus erynngii*	Degrades lignin, cellulose and polycyclic aromatic hydrocarbons (PA H) and curb environmental pollution It also removes heavy metals like lead and cadmium through lignocellulosic enzymes	Lignocellulosic enzymes like lignin peroxidase (LiP), manganese peroxidase (MnP), and laccase	Bempah et al. (2013)
17	*Asperigillus niger, Eurotium repens, Cladosporium pseudocladosporioid es, Penicillium chrysogenum, Bipolaris* sp., *A. nidulans*, and *Epicoccum nigrum*	α-1-rhamnohydrolase, β-N-acetyl hexoseaminidase, urease	Decompose nonsteroid anti-inflammatory drugs like diclofenac, ibuprofen, diflunisal, mefenamic, and piroxicam	Gonda et al. (2016)

(continued)

Table 17.1 (continued)

S. no.	Name of the fungi	Enzymes secreted	Pollutant degraded	Process of degradation
18	*Peroneutypa scoparia*	Acid red 97	Laccase	Pandi et al. (2018)
19	*Phoma tropica* and *Dichotomomyces cejpii* degrade Congo red and methyl red by using laccase enzyme	Congo red and methyl red	Laccase	Krishnamoorthy et al. (2018)
20	*Asperigillus nigar* and *Penicillium chrysogenum* degrade by using phthalates like Di-2-ethylhexylphthalate (DEHP)	Dibutyl phthalate (DBP), benzylbutyl phthalate (BBP)	Phthalase enzyme	
21	*Asperigillus species, Fusarium species, Rhizomucor species, Emericella species, Funneliformi s geosporum, Pleurotus ostreatus, Trichoderma harzianum, Trichoderma ghanense, Penicillium rubens*	Degradation of heavy metals and antioxidants	Ligninolytic enzymes	Tchounwou et al. (2012); Chibuike and Obiora (2014); Singh et al. (2015); Young et al. (2015)

17.7 Future Aspects

Mycoremediation methods are eco-friendly, cost-effective, and can be used with or without plant interactions. To improve mycoremediation, it is encouraged to use new fungal species that can degrade various pollutants in marine and terrestrial environments. There are so many fungal species and their importance was unknown that are used to be evaluated in the future for further benefit and sustainability. Many innovative methods should be developed and implemented to make mycoremediation more economical and feasible. There are some difficulties in mycoremediation so more research is required to determine the variety of mycoremediation for the future benefit. Mycoremediation processes can be successful by selecting right and desired fungal species that has the capability to break down the contaminants. Maintenance of environmental conditions such as pH temperature, moisture, and oxygen availability can curb the negatives on its growth. Mycoremediation can be more effective if the cocultivation and synergistic relationships are maintained. Utilizing advance technical methods like enzyme engineering, next generation sequencing and Micro array technologies can increase the efficiency in mycoremediation. Some of the alternative techniques, such as phytoremediation and natural attenuation, can reduce contamination-induced stress in ecosystems. The efficiency and application of mycoremediation can be improved by using nanoparticles and other delivery systems to enhance uptake of fungal activity. It has more potential and is sustainable, so proper care and actions are needed to remove the barriers for the benefit of society. Additionally, Mycoremediation has a promising future with many chances for innovation and application.

17.8 Conclusion

Mycoremediation is effective and well-designed method for removal of emerging pollutants like peroxy acetyl nitrate, dyes and detergents, plastics and micro plastics, pesticides and heavy metal removal. Here, the fungal species and their role on degradation of pollutants were highlighted. There is no zero pollution on earth, and it's a continuing process, and this mycoremediation is the promising sustainable technology for restoring environment in the future. Many native fungal species in polluted environment give valuable targets for further research. It not only protects the natural resources but also gives assurance of human health and resilience of ecosystem. There are some difficulties, such as incomplete removal of pollutants, delaying processes, and time consumption, which can be overcome by further research and implementing innovative methods. Mycoremediation uses fungi, which are a renewable resource, and it never depends on any synthetic chemicals and helps in enhancing biodiversity, improves soil health, nutrient cycling, and benefits the ecosystems in local areas. Let us hope that the challenges of mycoremediation can be controlled by further advanced research to make more interesting and sustainable technology.

References

Abdel-Shafy HI, Mansour MSM (2016) A review on polycyclic aromatic hydrocarbons: source, environmental impact, effect on human health and remediation. Egypt J Pet 25:107–123

Abu-Elsaoud AM, Nafady NA, Abdel-Azeem AM (2017) Arbuscular mycorrhizal strategy for zinc mycoremediation and diminished translocation to shoots and grains in wheat. PLoS One 12:e0188220

Almansa E, Heumann S, Eberl A, Fischer-Colbrie G, Martinkova L, Marek J, Cavaco-Paulo A, Guebitz GM (2008) Enzymatic surface hydrolysis of PET enhances bonding in PVC coating. Biocatal Biotransformation 26:365–370. https://doi.org/10.1080/10242420802357613

Asif MB, Hai FI, Singh L, Price WE, Nghiem LD (2017) Degradation of pharmaceuticals and personal care products by white-rot fungi—a critical review. Curr Pollut Rep 3:88–103

Azubuike CC, Chikere CB, Okpokwasili GC (2016) Bioremediation techniques – classification based on site of application: principles, advantages, limitations and prospects. World J Microbiol Biotechnol 32:180

Baybil HK, Teshome FT, Li YC (2022) Emerging contaminants in soil and water. Front Environ Sci 10:873499

Bempah CK, Ewusi A, Obiri-Yeboah S, Asabere SB, Mensah F, Boateng J, Voigt HJ (2013) Distibution of arsenic and heavy meatals from mine tailing dams at Obuasi municipality of Ghana. Am J Eng Res 2(05):61–70

Benjamin S, Pradeep S, Sarath Josh M, Kumar S, Masai E (2015) A monograph on the remediation of hazardous phthalates. J Hazard Mater 298:58–72
Bhattacharya S, Das A, Prashanthi K, Lazur J (2011) Mycoremediation of Congo red dye by filamentous fungi. Braz J Microbiol 42:1526
Bhattacharya S, Das A, Palaniswamy M, Angayarkanni J (2014) Mycoremediation of Benzo[a]pyrene by Pleurotus ostreatus in the presence of heavy metals and mediators. 3 Biotech 4:205–211
Boubendir A (1993) Purification and biochemical evaluation of polyurethane degrading enzymes of fungal origin. Diss Abstr Int 53:4632
Chibuike GU, Obiora SC (2014) Heavy metal polluted soils: effect on plants and bioremediation methods. Appl Environ Soil Sci 2014:1–12
Cvancarová M, Moeder M, Filipová A, Reemtsma T, Cajthaml T (2013) Biotransformation of the antibiotic agent flumequine by ligninolytic fungi and residual antibacterial activity of the transformation mixtures. Environ Sci Technol 47:14128–14136
Dai W, Chen X, Wang X, Xu Z, Gao X, Jiang C, Deng R, Han G (2018) The algicidal fungus Trametes versicolor F21a eliminating blue algae via genes encoding degradation enzymes and metabolic pathways revealed by transcriptomic analysis. Front Microbiol 9:826
Deshmukh R, Khardenavis AA, Purohit HJ (2016) Diverse metabolic capacities of fungi for bioremediation. Indian J Microbiol 56:247–264
Di Napoli M, Silvestri B, Castagliuolo G, Carpentieri A, Luciani G, Di Maro A, Sorbo S, Pezzella A, Zanfardino A, Varcamonti M (2023) High density polyethylene (HDPE) biodegradation by the fungus *Cladosporium halotolerans*. FEMS Microbiol Ecol 99:fiac148. https://doi.org/10.1093/femsec/fiac148
Du J, Pu G, Shao C, Cheng S, Cai J, Zhou L et al (2015) Potential of extracellular enzymes from Trametes versicolor F21a in Microcystis spp. degradation. Mater Sci Eng C 48:138–144
Esterhuizen-Londt M, Schwartz K, Pflugmacher S (2016) Using aquatic fungi for pharmaceutical bioremediation: Uptake of acetaminophen by Mucor hiemalis does not result in an enzymatic oxidative stress response. Fungal biology 120(10):1249–1257
Esterhuizen-Londt M, Hendel AL, Pflugmacher S (2017) Mycoremediation of diclofenac using Mucor hiemalis. Toxicol Environ Chem 99:795–808
Fomina M, Gadd GM (2014) Biosorption: current perspectives on concept, definition and application. Bioresour Technol 160:3–14
Gao X, Wang C, Dai W, Ren S, Tao F, He X et al (2017) Proteomic analysis reveals large amounts of decomposition enzymes and major metabolic pathways involved in algicidal process of Trametes versicolor F21a. Sci Rep 7(1):3907
García-Galán MJ, Rodríguez-Rodríguez CE, Vicent T, Caminal G, Díaz-Cruz MS (2011) Biodegradation of sulfamethazine by white rot fungi in solid-state fermentation: mycelial growth kinetics and removal ability. J Hazard Mater 192:1197–1205
Goala M, Yadav KK, Alam J, Adelodun B, Choi KS, Cabral-Pinto MM, Hamid AA, Alhoshan M, Ali FAA, Shukla AK (2021) Phytoremediation of dairy wastewater using Azolla pinnata: application of image processing technique for leaflet growth simulation. J Water Process Eng 42:102152
Gonda S, Kiss-Szikszai A, Szücs Z, Balla B, Vasas G (2016) Efficient biotransformation of non-steroid anti-inflammatory drugs by endophytic and epiphytic fungi from dried leaves of a medicinal plant, Plantago lanceolata L. Int Biodeterior Biodegradation 108:115–121
Harms H, Schlosser D, Wick LY (2011) Untapped potential: exploiting fungi in bioremediation of hazardous chemicals. Nat Rev Microbiol 9:177–192
Igiri BE, Okoduwa SIR, Idoko GO, Akabuogu EP, Adeyi AO, Ejiogu IK (2018) Toxicity and bioremediation of heavy metals contaminated ecosystem from tannery wastewater: a review. J Toxicol 2018:2568038
Kim YH, Seo HS, Min J, Kim YC, Ban YH, Han KY, Park JS, Bae KD, Gu MB, Lee J (2007) Enhanced degradation and toxicity reduction of dihexyl phthalate by Fusarium oxysporum f. sp. pisi cutinase. J Appl Microbiol 102:221–228

Kim K-H, Kabir E, Jahan SA (2017) Exposure to pesticides and the associated human health effects. Sci Total Environ 575:525–535

Koniecki D, Wang R, Moody RP, Zhu J (2011) Phthalates in cosmetic and personal care products: concentrations and possible dermal exposure. Environ Res 111:329–336

Krishnamoorthy R, Arul Jose P, Ranjith M, Anandham R, Suganya K, Prabhakaran J, Thiyageshwari S, Johnson J, Gopal NO, Kumutha K (2018) Decolourisation and degradation of azo dyes by mixed fungal culture consisted of Dichotomomyces cejpii MRCH 1-2 and Phoma tropica MRCH 1-3. J Environ Chem Eng 6(1):588–595

Kryczyk-Poprawa A, Zmudzki P, Maslanka A, Piotrowska J, Opoka W, Muszynska B (2019) Mycoremediation of azole antifungal agents using in vitro cultures of Lentinula edodes. 3 Biotech 9:207

Kumar SS, Ghosh P, Malyan SK, Sharma J, Kumar V (2019) A comprehensive review on enzymatic degradation of the organophosphate pesticide malathion in the environment. J Environ Sci Health C Environ Carcinog Ecotoxicol Rev 37:288–329

Kumari M, Ghosh P, Joshi S, Thakur I (2014) Microcosmic study of endosulfan degradation by paenibacillus sp. ISTP10 and its toxicological evaluation using mammalian cell line. Int Biodeterior Biodegradation 96:33–40

Kurtela A, Antolović N (2019) The problem of plastic waste and microplastics in the seas and oceans: impact on marine organisms. Croat J Fish 77:51–56

Larsson DGJ (2014) Pollution from drug manufacturing: review and perspectives. Philos Trans R Soc B Biol Sci 369:20130571

Lebreton LCM, van der Zwet J, Damsteeg J-W, Slat B, Andrady A, Reisser J (2017) River plastic emissions to the world's oceans. Nat Commun 8:15611

Liebminger S, Eberl A, Sousa F, Heumann S, Fischer-Colbrie G, Cavaco-Paulo A, Guebitz GM (2007) Hydrolysis of PET and bis-(benzoyloxyethyl) terephthalate with a new polyesterase from *Penicillium citrinum*. Biocatal Biotransformation 25:171–177. https://doi.org/10.1080/10242420701379734

Lozowicka B, Jankowska M, Hrynko I, Kaczynski P (2016) Removal of 16 pesticide residues from strawberries by washing with tap and ozone water, ultrasonic cleaning and boiling. Environ Monit Assess 188:1–19

Marco-Urrea E, Pérez-Trujillo M, Blánquez P, Vicent T, Caminal G (2010) Biodegradation of the analgesic naproxen by *Trametes versicolor* and identification of intermediates using HPLC-DAD-MS and NMR. Bioresour Technol 101:2159–2166

Masse L, Kennedy KJ, Chou SP (2001) Effect of temperature and moisture content on the biological stability of municipal solid waste. Compost Sci Util 9:150–161

Mohamed ZA, Hashem M, Alamri SA (2014) Growth inhibition of the cyanobacterium Microcystis aeruginosa and degradation of its microcystin toxins by the fungus Trichoderma citrinoviride. Toxicon 86:51–58

Moore RW, Rudy TA, Lin TM, Ko K, Peterson RE (2001) Abnormalities of sexual development in male rats with in utero and lactational exposure to the antiandrogenic plasticizer di(2-ethylhexyl) phthalate. Environ Health Perspect 109:229–237

Muszyńska B, Dąbrowska M, Starek M, Zmudzki P, Lazur J, Pytko-Polończyk J, Opoka W (2019) Lentinula edodes mycelium as effective agent for piroxicam mycoremediation. Front Microbiol 10:313

Oladipo OG, Awotoye OO, Olayinka A, Bezuidenhout CC, Maboeta MS (2018) Heavy metal tolerance traits of filamentous fungi isolated from gold and gemstone mining sites. Braz J Microbiol 49:29–37

Pandi A, Kuppuswami GM, Ramudu KN, Palanivel S (2018) A sustainable approach for degradation of leather dyes by a new fungal laccase. J Clean Prod 211:590–597

Pathan SI, Arfaioli P, Bardelli T, Ceccherini MT, Paolo Nannipieri P, Pietramellara G (2020) Soil pollution from micro- and nanoplastic debris: a hidden and unknown. Sustainability 12:7255

Prieto A, Möder M, Rodil R, Adrian L, Marco-Urrea E (2011) Degradation of the antibiotics norfloxacin and ciprofloxacin by a white-rot fungus and identification of degradation products. Bioresour Technol 102(23):10987–10995

Pujari M, Kapoor D (2021) Heavy metals in the ecosystem: sources and their effects. In: Heavy metals in the environment. Elsevier, pp 1–7

Rani S (2015) Studies on the degradation and mineralization of organophosphate pesticides employing advanced oxidation processes. Thesis, Sant Longowal Institute of Engineering and Technology, p 255

Rillig MC, Ziersch L, Hempel S (2017) Microplastic transport in soil by earthworms. Sci Rep 7:1362

Ronkvist AM, Xie WC, Lu WH, Gross RA (2009) Cutinase catalyzed hydrolysis of poly(ethylene terephthalate). Macromolecules 42:5128–5138

Roy P, Mohanty AK, Misra M (2022) Microplastics in ecosystems: their implications and mitigation pathways. Environ Sci Adv 1:9

Samanta SK, Singh OV, Jain RK (2002) Polycyclic aromatic hydrocarbons: environmental pollution and bioremediation. Trends Biotechnol 20:243–248

Schamfuß S, Neu TR, Van Der Meer JR, Tecon R, Harms H, Wick LY (2013) Impact of mycelia on the accessibility of fluorene to PAH-degrading bacteria. Environ Sci Technol 47:6908–6915

Shah ZU, Parveen S (2021) Pesticides pollution and risk assessment of river ganga: a review. Heliyon 7:e07726

Sheik S, Chandrashekar KR, Swaroop K, Somashekarappa HM (2015) Biodegradation of gamma irradiated low density polyethylene and polypropylene by endophytic fungi. Int Biodeterior Biodegrad 105:21–29

Singh M, Srivastava PK, Verma PC, Kharwar RN, Singh N, Tripathi RD (2015) Soil fungi for mycoremediation of arsenic pollution in agriculture soils. J Appl Microbiol 119:1278–1290

Sowmya HV, Krishnappa M, Thippeswamy B (2014) Degradation of polyethylene by Trichoderma harzianum—SEM, FTIR, and NMR analyses. Environ Monit Assess 186:6577–6586

Takamoto T, Shirasaka H, Uyama H, Kobayashi S (2001) Lipase-catalyzed hydrolytic degradation of polyurethane in organic solvent. Chem Lett 30:492–493

Tchounwou PB, Yedjou CG, Patlolla AK, Sutton DJ (2012) Heavy metal toxicity and the environment. EXS 101:133–164

Thiel M, Luna-Jorquera G, Alvarez-Varas R, Gallardo C, Hinojosa IA, Luna N, Miranda-Urbina D, Morales N, Ory N, Pacheco AS, Portflitt-Toro M, Zavalaga C (2018) Impacts of marine plastic pollution from continental coasts to subtropical gyres—fish, seabirds, and other vertebrates in the SE Pacific. Front Mar Sci 5:238

Urík M, Cernanský S, Ševc J, Šimonovicová A, Littera P (2007) Biovolatilization of arsenic by different fungal strains. Water Air Soil Pollut 186:337–342

Vasilachi I, Asiminicesei D, Fertu D, Gavrilescu M (2021) Occurrence and fate of emerging pollutants in water environment and options for their removal. Water 13:181

Xiao PF, Kondo R (2020) Biodegradation and biotransformation of pentachlorophenol by wood-decaying white rot fungus *Phlebia acanthocystis* TMIC34875. J Wood Sci 66:2. https://doi.org/10.1186/s10086-020-1849-6

Yang S, Hai FI, Nghiem LD, Roddick F, Price WE (2013) Removal of trace organic contaminants by nitrifying activated sludge and whole-cell and crude enzyme extract of Trametes versicolor. Water Sci Technol 67:1216–1223

Yesilada O, Birhanli E, Geckil H (2018) Bioremediation and decolorization of textile dyes by white rot fungi and laccase enzymes. In: Mycoremediation and environmental sustainability. Springer, Cham, pp 121–153

Young D, Rice J, Martin R, Lindquist E, Lipzen A, Grigoriev I, Hibbett D (2015) Degradation of bunker C fuel oil by white-rot fungi in sawdust cultures suggests potential applications in bioremediation. PLoS One 10:e0130381

Chapter 18
Earthworms: The Environmentally Preferred Bioengineers for Managing Heavy Metals

Urmila, R. K. Gupta, Shweta Sharma, and Rahul Kumar

Abstract Exponentially increasing industries, urbanization, agriculture practices and anthropogenic activities generate huge amount of various types of waste. Heavy metals are one of the substantive components of the waste. Due to feeble waste management practices, these wastes are dumped onto the soil which not only causes the land filling but also contaminates the various components of the environment, especially soil, air and water. This accumulated waste directly or indirectly deteriorates the quality of soil, so it becomes the major concern to manage these wastes through scientific and eco-friendly technique. Nature had already gifted us the environmentally preferred bioengineers, the earthworms, which help us to alienate various soil pollutants. Earthworms have the ability to bioaccumulate various type of soil pollutants in their body. Different species of earthworms have the specificity and capacity to accumulate and regulate various types of soil pollutants in their body. Due to their bioaccumulating ability and wide range of distribution, they become the most preferred creature to remove different types of soil pollutants. Also they are proven to be good biological indicator of soil quality. They not only remove the soil pollutants but also improve the soil physical, chemical and biological properties. Earthworm's high reproduction rate, wide range distribution, adaptation to different environmental conditions and bioaccumulation ability render this creature as bioengineer to alienate soil pollutants.

Keywords Earthworms · Bioaccumulation · Soil pollutants · Bioengineer · Bioremediation · Heavy metals · Vermiremediation

---

Urmila · R. K. Gupta · S. Shweta
Department of Zoology, Chaudhary Charan Singh Haryana Agricultural University, Hisar, Haryana, India

R. Kumar (✉)
Department of Zoology, Chaudhary Charan Singh Haryana Agricultural University, College of Agriculture, Bawal, Rewari, Haryana, India

S. Dey, S. Bhattacharya (eds.), *Biotechnological Interventions in the Removal of Emerging Pollutants*, Interdisciplinary Biotechnological Advances,
https://doi.org/10.1007/978-981-97-9922-0_18

Abbreviations

BAF	Bioaccumulation Factor
BAC	Bioaccumulation Coefficient
GDP	Gross Domestic Product
PCBs	Polychlorinated Biphenyls
DDT	Dichlorodiphenyltrichloroethane
HMs	Heavy Metals
Cd	Cadmium
Pb	Lead
Cu	Copper
As	Arsenic
Hg	Mercury
Cr	Chromium
Ni	Nickel
V	Vanadium
Sb	Antimony
MMA	Monomethylarsonic acid
DMA	Dimethylarsenic acid
BCF	Bioconcentration Factor
Cd-MTs	Cadmium Metallothionein
PAHs	Polycyclic Aromatic Hydrocarbons
Pb-MTs	Lead Metallothionein

18.1 Introduction

The "soil" is the soul of infinite organisms because it promotes and accommodates various fauna and flora in it. The soil is one of the most diverse and dynamic component of natural ecosystem. It lies at the interface between biotic component (life) and abiotic component (air and water) of ecosystem. Unfortunately, many anthropogenic activities produce various types of nonbiodegradable pollutants in the environment. Gradually, these pollutants accumulate in the various components of the ecosystem, especially in the soil, which affects the services provided by the soil to the ecosystem. The pollutants get transferred from soil to various organisms, including human beings, via various routes and deleteriously affect almost all the living creatures (Needelman 2013). Since the industrial revolution, the pollutant load on soil increase at alarming rate. Although we know that the soil is a nonrenewable natural resource, still we use itas sink and dump waste and other contaminants in it. When the element in the soil exceeds its recommended level for living beings then it becomes pollutant. These contaminants and waste not only pollute the soil but also deteriorate soil's physical, Chemical, and biological properties, which further affects soil fertility. Due to accumulation of pollutants at greater extent soil becomes almost infertile and bare which results in decline productivity of various crops. So in order to preserve soil fertility and productivity, there is urgency for controlling soil pollution by blocking various routes of pollutants (Havugimana et al. 2017). These routes include exponential increase in industries, urbanization, agriculture practices and anthropogenic activities, which generate a huge amount of waste every day. This waste is the main carrier of numerous pollutants in the soil stratum. Pollutants mainly include polythene bags, plastic canes, bottles, chemical pesticides, insecticides, fertilizers, construction-related contaminants, trash canes, and heavy metals. Almost all the above mentioned pollutants are non biodegradable type of pollutants and can persist in the soil stratum for a longer duration. Some of the above stated pollutants enters into the food web through various pathways and routs and leads to biomagnifications from one trophic level to other. Due to their biomagnification property, they deleteriously affect almost all the living organisms occupying various trophic levels. Their long persistent nature puts them in the category of one of the major threat to various component of the environment. Also all around the world due to feeble waste management practices these wastes are dumped onto the soil which not only causes the land filling but also affects the creatures inhabiting the soil. This dumped waste deteriorates soil texture, nutrient profile and microfauna of soil. This deteriorated soil quality makes the soil infertile which ultimately impacts the GDP of entire agriculture sectors and agriculture based countries. A large proportion of tillable land deteriorated due to intense mining activities performed on those land areas, exorbitant application of agrochemicals in farms, disposal of septic waste in open lands without treatment. Already there are millions of contaminated sites all over the globe still their number increases day by day. The millions of ton waste and contaminants generated in day to day life aids in the increase in the number of contaminated sites. These contaminated sites act as the major sources of long persisted pollutants such as PCBs, DDT, PAH, and several heavy metals such as arsenic, cadmium, lead, mercury, copper, chromium, zinc, nickel, etc. in the various component of the ecosystem. These contaminated sites serve as the major source which aid in trophic transfer of contaminants. This trophic transfer of pollutants results in substantial reduction in the biodiversity both above and below the ground. Due to the devastating effects of these pollutants on soil and living beings there is a need to either remove or detoxify them. Now these days, these pollutants are dissevered from the soil by utilizing different techniques. Mechanically polluted soil is excavated for removal of pollutants. Mechanical methods are neither eco-friendly nor cost-effective because the risk of leaching to other component of ecosystems is there and also it requires billions of dollars to decontaminant the soil. So, there is a need to approach and implement the cost-effective and eco-friendly techniques. Recently, various bioremediation techniques such as phycoremediation, mycoremediation, phytoremediation and vermiremediation are extensively used to decontaminant the soil. The organisms used in these techniques are mainly bacteria, algae, fungi, plants and earthworms. In this chapter, we mainly focus on heavy metal remediation from soil by using earthworms. The routes which make use of earthworms for removing pollutants from the soil are known as vermiremediation. Vermiremediation is a cost-effective, eco-friendly and convenient technique for combating the soil pollutants (Sinha et al. 2008). Over 600 million years, earthworms are true soldiers of Darwin sir because "they work continuously in the soil and helps in soil management by removing contaminants and restoring soil physical, chemical and biological properties. On the basis of their habitat, earthworms are categorized as follows: (1) epigeic, (2) anecic and (3) endogeic. Almost all the earthworms, epigeic, anecic, and endogeic, have the capability to bioaccumulate and tolerate a wide range of chemical contaminants in their body tissue. Several scientific studies indicate that earthworms have enormous power to bioaccumulates the toxic pollutants and convert these pollutants into less toxic forms. Earthworm inhabits the soil near roadsides, and mining sites have higher concentration of bioaccumulated pollutants, especially the metallic pollutants in various part of the body. Following the study of Ramos et al. (2006), pollutants such as heavy metals (Cd, Pb, Cu and Hg), pesticides, and PAHs were removed from the soil in eco-friendly manner by using different species of earthworms like *E. foetida, E. tetraedra, L. terrestris, L. rubellus* and *A. chlorotica*. These earthworms bioaccumulate the above mentioned pollutants either by absorbing through moist body wall or by ingesting the contaminated soil. After bioaccumulation, they either biodegrade or biotransform these contaminants into less toxic forms. Biodegradation and biotransformation rendering the toxic contaminants harmless in earthworm's body. Simultaneously the quality of soil gets improved by the action of earthworms. They feed on large amount of soil and produce excreta in the form of vermicast, which is the nutritive material loaded with NPK, micronutrients, and microbes, especially the nitrogen fixers. They upturn and mix the soil thoroughly and make burrow, which improves aeration in the soil (Sinha et al. 2008). This chapter has been written by studying all the abovementioned pollutants in the soil and various remediation techniques. The chapter mainly focuses on heavy metal pollutants and use of soil dwelling earthworms in bioremediation of these pollutants. It also provides valuable information to students and the researchers regarding the utilization of bioengineers to alienate the soil pollutants in an eco-friendly manner.

18.2 Heavy Metals

These pollutants are one of the major pollutants found in the soil stratum. Their long persistence and bioaccumulative nature, large toxicity even at part per million (ppm) concentrations and their extensive distribution in the environment from either natural or man-made sources render them in the category of hazardous pollutants. As per Csuros and Csuros (2002), a metal is categorized as heavy metal if "the metals have density greater than 5 g/cm^3 means their specific gravity must be greater than 5." As per Duffus (2002), the word "heavy metals" is describes those metals and metalloids that aid in contaminating the environment and have potential ecotoxicity." Now these days several scientists defined heavy metals as "the metals and metalloids present in earth's crust naturally and have density greater than 5 g cm^{-3} and also their atomic number more than 20. Since time Earth's formation heavy metals are occurring on the soil stratum naturally. Numerous man-made sources are there that aid in heavy metal generation on the Earth's crust. Heavy metals are discharged in the soil stratum either through natural process like weathering of rocks containing metals and volcanic eruption or through man-made processes like mining of mineral resources, multifarious industrial and agricultural practices. The increased rate of mobilization of heavy metals in the soil stratum is due to increase in mining for extraction of mineral resources. This accumulation of HMs in the various components of the ecosystem is becoming one of the major havocs all around the world. Due to their bioaccumulative nature, they require special attention with regard to public health concern. They not only affect the abiotic component of ecosystem but also contaminants the food chains and food webs. Various types of hazardous metals found in our surrounding but As, Ni, Cr, Cu, Hg, Cd, Zn, and Pb come under the category of superlative hazardous heavy metals. A trophic transfer interaction exists between various type of food chains and food webs. This type of interaction aids in the mobilization of hazardous heavy metals into the various component of the environment. So it requires a greater attention to appraise the impact of these hazardous non biodegradable pollutants onto the abiotic component and onto the residing biota of different habitat. As per the numerous ecotoxicological studies on the deleterious impacts of these heavy metals and metalloids on the environment and living beings, it becomes the urgency to take some steps to either eliminate or mitigate these hazardous materials from our environment (Ali et al. 2019; Yadav et al. 2023).

18.3 Environmentally Persistent Heavy Metal and Their Impacts

Heavy metals produced either naturally/geogenically/lithogenically or anthropogenically can persist in the environment for many years. Rapid growth of industries and urban areas all over the globe results in increase anthropogenic share of hazardous materials such as Cu, Cr, Pb, Ni, Hg, Cd and As in our soil. The sources and impact of various HMs are discussed below:

18.3.1 Arsenic

The occurrence of arsenic in our environment can be easily detected at low concentrations as it is a universal element. Arsenic can exist in both inorganic and organic forms. In inorganic forms it mainly exists as trivalent arsenite and pentavalent arsenate. In organic forms it mainly found as methylated metabolites such as MMA, DMA and trimethylarsine oxide. Arsenic follows both natural and anthropogenic routes to enter in the environment and results in environmental pollution at exceeding rates. The natural phenomena such as volcanic outburst and eroded soil also add arsenic in the environment (ATSDR 2000). The major anthropogenic source of are arsenic are the multifarious industries which mainly produce arsenic-containing compounds. These compounds are further utilized for the production of various products makes use in agriculture sector such as insecticides, herbicides, fungicides, etc. Arsenic-containing compounds are also used to manufacture wood preservatives, dye-stuffs, and sheep dips (Tchounwou 1999). Arsenic is not only utilized in the products of agricultural importance but also used in cosmetics products such as eye shadows, lipsticks, and some of the skin care products, such as body lotions, etc. The inorganic forms of arsenic have many harmful impacts as such as skin diseases, cancer, fatigue and nausea. Risk of several types of cancers is associated with chronic exposure to arsenic (Witkowska et al. 2021).

18.3.2 Cadmium

Cadmium is also one among the hazardous HMs. The concentration around 0.1 mg/kg Cd found in soil stratum. Sedimentary rocks are considered to be natural source of cadmium in the soil stratum. On an average 15 mg/kg of cadmium was estimated to be present in these sources (GESAMP 2015). Due to major industrial applications cadmium is most widely and frequently used heavy metal. Cadmium is utilized by these multifarious industries for the manufacture of batteries, alloys, pigments, etc. (Wilson 1988). Due to its wide distribution and high concentration in earth's crust it draw special attention of the scientists. Cadmium is mobilized from the environment to residing biota through the route of trophic transfer. In living being it get bioaccumulated in the body tissues and causes various health problems especially it increases the risk of heart disease in human beings. Even the short-term exposure of cadmium causes serious health problems. Chronic exposure to cadmium adversely affects human beings and causes various type of cancers and insomnia and also affects kidney and bones (Choong et al. 2014).

18.3.3 Chromium

Chromium is also one among those heavy metals which found in the soil stratum naturally. Cr (II), Cr (III), Cr (IV), Cr (V) and Cr (VI) are the oxidation states of chromium (Jacobs and Testa 2005). Chromium can be traced in almost all the environmental matrices. From these environmental matrices it enters into food chains and food webs and get bioaccumulate in plants and living organisms occupying various trophic level. Beside the natural resource, few anthropogenic resources such as industries also release largest amount of Cr in the environment. The major contribution of chromium through industrial processes includes production of pigments containing chromes and ferrochromes, tannery industries, welding related works, etc. The metallurgical, refractory and chemical industries release chromium mainly as Cr (VI) oxidation state in the environment. This hexavalent form of Cr is very septical to human beings as it acts as a major carcinogen in this form (ATSDR 2012; IARC 1990; EPA 1992).

18.3.4 Lead

The bluish-gray color heavy metal called Lead (Pb) found on the earth's crust is also considered as toxic heavy metal. Anthropogenic as well as natural sources are the root causes of these heavy metals in our ecosystem. Anthropogenic activities, especially excessive use of fossil fuels and mineral extraction through mining and factories, are considered to be main contributor of Pb in the soil stratum. Pb is widely applicable in multifarious industries for the production of batteries containing Pb-acids, medical appliances which shield X-rays, soldering of metals, etc. All these earlier described sources and products aid in its concentration increases in various matrices of environment. This elevation in the amount of Pb becomes the major environmental concern. Lead also has bioaccumulative properties, which results in severe health problems (Tchounwou et al. 2012).

18.3.5 Mercury

Hg is a prevailing heavy metal and most hazardous toxicant in found in our environment. In the periodic table, it is placed among the series of transition elements. Due to its peerless existence in nature in organic, inorganic and elemental form it is considered as most toxic pollutant Mercury is supposed to have unique profile of ecotoxicity (Clarkson et al. 2003). Soil stratum Hg exists as mercuric form (inorganic mercury). Soil-dwelling microorganisms act on this mercuric form and convert it into organic form called methylmercury (Dopp et al. 2004). All the abovementioned forms of Hg come in our ecosystem through various industrial routes, especially electrical industries, which utilize mercury for the manufacturing of batteries, switches, and thermostats and release the mercury waste in the environment. Numerous other sources like dentistry (dental amalgams), chemical utilized by wooden industries (e.g., antifungal products), Hg containing preservative used by pharmaceutical industries to preserve their products, nuclear reactors and solvent used for reactive and precious metal extraction also release mercury in the various matrices of the environment (Tchounwou et al. 2003). Prevalent mercury existence in the environment makes it major pollutant and toxicant. Short-term and long-term exposure to different forms of mercury results in long standing health-related problems in different organisms, including human beings. Exposure to elevated level of mercury leads to alterations in the body tissue, especially in soil-dwelling organisms (Sarkar 2005) (Fig. 18.1).

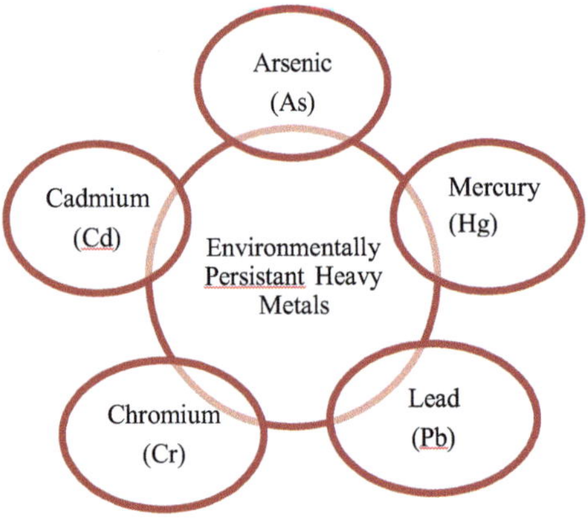

Fig. 18.1 Environmentally persistent heavy metal (HMs)

18.4 Trophic Level Transfer of Heavy Metals

Almost all the heavy metals found on earth's crust have long persistent nature and bioaccumulative properties. They accumulate in almost all the biota of this earth but their extent of bioaccumulates is greater in soil-dwelling organisms such as earthworms. Due to their high bioaccumulation in residing biota they get easily mobilized from one organism to other occupying different trophic level either in food chains or food webs. The extent of bioaccumulation at various trophic levels depends on a number of factors such as the bioavailability of that heavy metal at that trophic level, the half-life of different heavy metals in different species, their accumulation rate in the body, and their elimination rate from the body. The basic route of transfer of hazardous pollutants at the highest trophic level is the abiotic component of ecosystem. Consider the example of soil-dwelling organisms such as earthworms, the entry route of heavy metals into earthworm's tissue is either through reabsorption by body wall or by ingesting contaminated soil. From earthworms it takes the route of trophic transfer and gets accumulated in various organisms. The concentration of a particular heavy metal may vary between various trophic levels in same or different food chains. The retention time of heavy metals at various trophic level is different for different organisms as it depends on the mechanism of regulation and detoxification adopted by that particular organism. The organic form of mercury, i.e., methylmercury, is lipid soluble in nature, which makes it highly accumulated heavy metals in the biota (Ali and Khan 2018).

18.5 Mensuration of Heavy Metals in Biotic Components

The extent of bioaccumulation Toxicants, especially heavy metals, from habitual environment to organism can be quantified by using various mensuration techniques. A few of them are BCF, BAF, and BAC.

18.5.1 Bioconcentration Factor

BCF represents the concentration of HMs in the tissue of biota in relation to the abiotic medium in which they inhabit. BCF can also be represented in the form of ratio between the concentration of HMs in biotic tissue and its habitat. The following equation helps in the calculation and quantification of bioaccumulated heavy metals:

$$\mathrm{BCF} = C_{\text{biotic tissue}} / C_{\text{habitat}}$$

In the abovementioned equation, $C_{\text{biotic tissue}}$ represents the concentration of a particular heavy metal in biotic tissue and C_{habitat} represents its concentration in the abiotic medium.

Few scientists use the term bioaccumulation factor (BAF) instead of bioconcentration factor (BCF) to calculate and quantify different bioaccumulated toxicants and heavy metals in biota. Altogether we can say both the indices are used for the measurement of accumulated heavy metals in organisms with respect to its habitat (Ali et al. 2019). Bioaccumulation factor (BAF) is mainly considered for the quantification of HMs in earthworm's tissue with respect to the substrate. It is represented as

$$\mathrm{BAF} = C_{\text{earthworm's tissue}} / C_{\text{substrate}}$$

Here $C_{\text{earthworm's tissue}}$ represents the concentration of heavy metal in earthworm's tissue and $C_{\text{substrate}}$ represents its concentration in the substrate (Mountouris et al. 2002).

18.5.2 Bioaccumulation Coefficient

BAC is mainly used for the calculation and quantification of heavy metal in plants with respect to soil during phytoremediation technique. It is represented as:

$$\mathrm{BAC} = C_{\text{plant}} / C_{\text{soil}}$$

Here, C_{plant} represents the heavy metal's concentration in the accumulator plants and $C_{\text{substrate}}$ represents the concentration in the soil (Liang et al. 2016).

The abovementioned mensuration equations are compared to unity. If the value of BCF, BAF, or BAC is greater than one, it indicates that the metal concentration is high in the living components (organism tissue, earthworms or plants) as compared to their habitat or substrate. If the value is less than one it means the heavy metal quantity is higher in the habitat (Mountouris et al. 2002). As per to Wang et al.

(2017), the BCF, BAF, and BAC value is in inverse relation to the heavy metals concentration in the various matrices of our environment. For example, the value of BAF may be higher in the body tissue of earthworms inhabiting the less contaminated soil as compared to earthworms inhabiting more contaminated substrate. Few scientific studies reported that the value of bioconcentration factor for most of the hazardous heavy metals in crops decline exponentially with their average concentration in soil (Wang et al. 2017).

18.6 Expulsion of Heavy Metals Via Conventional Techniques

A vast number of conventional tools and techniques followed worldwide for the expulsion of heavy metals from the different matrices of ecosystem. These techniques involved electro-dialysis, ion exchange, adsorption, precipitation, etc. According to Gray (1999), in chemical precipitation, heavy metals are precipitated as suspended particles by adding anions for their precipitation. But this technique is only effective to the higher concentration of heavy metals in the environment otherwise it is not the efficient technique. The ion exchange technique becomes useful even the concentration of heavy metals is available in parts per billion (ppb) in the environment (Hassan et al. 2010). Although they are the most widely used techniques for heavy metals expulsion, they have many limitations. These conventional techniques are more expensive, requirs more energy for processing, are slow and less efficient, and also produce contaminated sludge, which aids in pollution (Gunatilake 2015; Ayangbenro and Babalola 2017; Firdousi 2017; Ahalya et al. 2003).

So, there is a need to find and implement cost-effective, eco-friendly, and efficient techniques for the expulsion of almost all the types of contaminants, especially heavy metals from different matrices of our environment. Bioremediation is one of the eco-friendly and efficient approach. Bioremediation is used as an alternative method for the heavy metal expulsion. The use of various types of natural bioengineers such as microbes, fungus, plants and animals draw a special attention of scientific society to depollute our environment.

18.7 Expulsion of Heavy Metals by Using Biota

18.7.1 Bioremediation

Bioremediation is the technique of utilizing various type of biota for the expulsion of HMs amended in different environment niches. Organisms used in bioremediation process involve microbes, algae, fungi, plants and invertebrates such as earthworms. These living organisms either neutralize or eradicate these heavy metals through various metabolic and physiological actions in their body. The mechanism of bioremediation adopted by most of the living beings is to detoxify, degrade, reduce, immobilization or to transform the more toxic pollutant to less toxic one. The entire mechanism and process of removing pollutants highly depends on what type of pollutant among the following insecticides, pesticides, heavy metals, and xenobiotics will have to be removed (Sharma 2020).

18.8 Various Organisms Used for Bioremediation of Metallic Pollutants

18.8.1 Bioremediation by Using Microbe

Application of microbes to detoxify the soil is widespread, efficient, eco-friendly and cost-effective bioremediation technique. This microbe-mediated technique utilizes fungi, aerobic and anaerobic bacteria to eradicate or immobilize various toxicants. The microbes easily clear the organic form of toxicants but adopt special mechanism for removing inorganic compounds such as metallic pollutants. Bioremediation of HMs by microorganisms depends on their metabolizing capabilities like few bacteria utilize these metals as essential micronutrients. For example, anaerobic bacteria use Fe^{2+} as essential micronutrients. Ahemad (2014) examines the capability of soil microorganisms to exclude or detoxify toxic products. Due to ubiquitous presence of bacteria they become the most extensively used microbes for the exclusion of nonbiodegradable elements especially heavy metals. Most of the bacteria adopt the biosorption technique for removing heavy metals (Mustapha and Halimoon 2015). The Gram-positive and -negative bacteria have ionic functional group in their cell wall which interacts with charged heavy metals, thus helping in their removal. The widely used species of bacteria for bioremediation are *Pseudomonas*, *Bacillus*, *Geobacter*, etc. (Sherbet 1978). Mycoremediation is extensively used bioremediation technique which involves the utilization of fungus for removing ecotoxicants. Pervasive presence and bioaccumulative nature of fungus makes it the useful biota to utilize it for bioremediation. Fungi can bioaccumulate heavy metals in their fruiting bodies and thus help in their clearance from the environmental niches. For example, *Saccharomyces cerevisiae* can bioaccumulate Pb and Cd from contaminated soil (Ogbo and Okhuoya 2011; Damodaran et al. 2011).

18.8.2 Bioremediation by Using Plants (Phytoremediation)

Phytoremediation is another technique extensively applicable for the expulsion of metallic compound from the soil stratum and also from aquatic ecosystem. Plants found everywhere on terrestrial as well as aquatic ecosystem so, they becomes the highly preferred bioengineers for removing both organic and inorganic form of toxic elements (Ali et al. 2013). Plants absorb heavy metals such as Hg and Se and convert both these metals into the volatile state and finally liberate these volatile forms in the surrounding environment. Due to this phytovolatilization of Hg and Se they again enter in the environment (Karami and Shamsuddin 2010). Recently, hyperaccumulator plants draw attention due to their efficient mechanism of degrading and detoxifying toxicants (Ahemad 2014). According to Prasad and Freitas (2003), hyperaccumulators have more tolerance and high capability to bioaccumulate a broad range of heavy metal when compared with the other plants. For example, cadmium has been phytoremedized by using *Arabidopsis halleri* (Dahmani-Muller et al. 2000) and *Solanum nigrum* L. (Wei et al. 2005). Plants called metallophytes have the ability to grow in a metal-rich environment. These metallophyte may be excluders, indicators, and hyperaccumulators, and each of them adopts special mechanism to stand up with higher concentration of concentration (Ali and Khan 2018, Kumar et al. 2020). Although hyperaccumulators are widely used for bioremediation, due to their slow growth, scientists use soil-dwelling organisms, especially earthworms, to eradicate heavy metals from soil.

18.8.3 Bioremediation by Using Earthworms (Vermiremediation)

Due to versatile nature, fast growing rate, high reproduction rate, adaptation to different environmental condition, and availability in different soil profiles, earthworms are considered as the right choice for the clearance of metallic pollutants from soil stratum. Earthworm species are widely distributed in the different horizons of soil, thus helping in removal of toxicants from deeper soil also. The utilization of earthworms for remediation of contaminants from soil is called vermiremediation. Vermiremediation is a cost-efficacious, eco-friendly, and most adroit technique among all the bioremediation techniques as it not only helps in removing toxicants from the soil but simultaneously helps in reclamation of soil by ameliorating the physical, chemical, and biological properties of the soil. Earthworms live and feed on soil which results in bioaccumulation of toxicants in their body tissue. Toxicants such as HMs take the route of entering the body either through moist skin or by ingesting contaminated substrate (Nannoni et al. 2011; Pattnaik and Reddy 2011).

18.8.3.1 Classification of Earthworms

Earthworms are classified in reference to their habitat into three groups: epigeic, anecic, and endogeic. Epigeic earthworms mainly inhabit the soil liter layer and feeds on organic matter there. They are voracious feeders and do not make permanent burrows as they are surface dwellers. *Eisenia fetida* and *Eudrilus euginae* are epigeic earthworms (Sizmur et al. 2011a). Anecic earthworms live deep in the soil. They make permanent burrows of depth about 3 m vertically as well as laterally. *Lumbricus terrestris* is one of the well-studied examples of anecic earthworms (Sizmur et al. 2011a). Endogeic earthworms inhabit the subsoil and make nonpermanent horizontal burrows. *Aporrectodea caliginosa, Aporrectodea rosea*, and *Allolobophora chlorotica* are endogeic earthworms (Sizmur et al. 2011a). Almost all these groups of earthworms are used in vermiremediation.

18.8.3.2 Biotics of Earthworms

Earthworms are soft-bodied, metamerically segmented, long, Cylindrical, and bilaterally symmetrical organisms and belong to phylum annelida. Morphologically, The earthworm's body is dark brown or reddish in color and covered in delicate cuticle. After 8–10 weeks of development, they attain weight approximately 1400–1500 mg (Visvanathan et al. 2005). The lifespan of earthworms varies from species to species and their ecological situations, but on an average, they have lifespan of about 3–7 years (Edward and Lofty 1972). The earthworm's gut harbors millions of decomposer microbes, which is why we can say that theearthworm's gut is a "consortium of useful microbes." The bimolecular profile indicates that in earthworm's body the major contributor biomolecule is protein (65%), then carbohydrates and fats (14%), and ash around 3% (Gerard 1960).

18.8.3.3 Ecology of Earthworms

Usually, earthworms inhabit a wide range of habitats, but they mostly prefer dark and moist environment because they are sensitive to light and their soft skin needs moisture. Earthworms are soil dwelling organisms so they make horizontal, vertical, and lateral burrows by ingesting soil. Although earthworms are widely distributed but their distribution depends on moisture content of soil, pH, and availability of organic waste materials such as humus, cow dung, leaf litters, and various type of organic wastes. Earthworms can't survive in acidic, coarse and high clay soil (Gunathilagraj 1996). The optimal temperature ranges 20–25 °C and moisture content 40–60% are required for reproduction, cocoon laying, cocoon hatching and for proper growth and development of their juveniles (Hand 1988; Edwards and Bohlen 1996).

18.8.3.4 Mechanisms of Vermiremediation

According to the study of Sinha et al. (2008), the major biomolecule in earthworm's bodies is protein (65% on dry weight basis). This protein interacts with heavy metals and aids in the bioaccumulation of these metals in their tissue. Metallothionein protein present in earthworm's body tissues interact with toxic inorganic state of

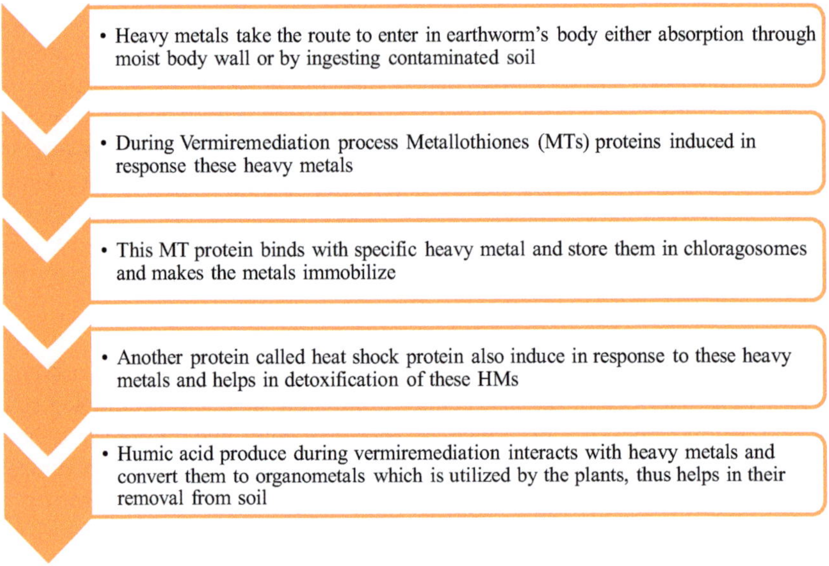

Fig. 18.2 Entry route and vermiremediation mechanism of heavy metals (HMs)

HMs and convert these into least toxic organic state (Alloway 2013; Sahariah et al. 2015). Study of Ireland (1979) indicates that earthworms *L. terrestris* and *D. rubidus* bioaccumulates Cd and Pb in chloragogen cells and by utilizing the mechanism of metallothionein proteins transform them as Cd-MTs and Pb-MTs respectively. The chloragogen cells of earthworms absorb these metals from the intestine and store them in chloragosomes due to this metal remain immobilized. Beside these metallothionein proteins, another protein called heat shock protein also helps in eradicating the heavy metals. Both these metallothionein proteins and heat shock proteins help in detoxification of contaminated soil (Alloway 2013). Utilization of humic acid for the removal of metals and metalloids from the soil is alternate mechanism adopted by the earthworms during the process of vermiremediation. When earthworm feeds on organic matter, it produces humic acid which interacts with various metals and transforms them to organometals, which are utilized by the plants; thus, heavy metal concentration decreases in that soil (Sizmur and Hodson 2009) (Fig. 18.2).

18.8.3.5 Why Earthworms Are Right Choice for Soil Remediation

Use of earthworms for bioremediation and detoxification of toxicants from soil enamors the scientific society. Recently, a number of researchers extensively utilize various species of earthworms for the bioremediation of various contaminants from soil. This soil-dwelling creature of nature has numerous qualities, which makes it the right choice for the remediation of soil. Earthworms have high reproductive rate, short life cycle, high survival rate, and rapid growth; inhabit various soil profile, and can withstand harsh environmental conditions, and their easy acclimatization and diversification aids in their use as organism for soil remediation. As earthworms are the soil-dwelling organisms, they remain in direct physical contact with toxicants. As they are soft-bodied organisms, they absorb heavy metals through their body wall or they feed on soil so by ingestion such that the heavy metals enter their body. They have high bioaccumulative capacity for a wide range of toxicants and also adopt different mechanisms either to detoxify or neutralize those toxicants. Simultaneously with the bioaccumulation of heavy metals they not only help in remediation of soil but also renovating soil texture, mineral content and biota especially the microbes in the soil. Through their burrowing ability they improve soil aeration, water holding capacity, thoroughly mixing the nutrients by upturning the soil. After feeding on soil they excrete it as valuable vermicompost which is highly nutritive material having high NPK, beneficial microbes in it and also devoid of contaminants (Visvanathan et al. 2005). All these characteristics make the use of earthworms as the right choice over all the other organisms used in various bioremediation techniques.

18.9 Background Studies on Vermiremediation of Heavy Metals

As we previously discussed that recently earthworms the most extensively utilizing tool for the remediation of metallic compounds from their habitat “the soil.” They have good bio-accumulative capability for almost all the heavy metals without disturbing the physiology of various organs of their body (Ireland 1983). Several scientific studies reveals that earthworms have enormous power to accommodate heavy metals such as Cd, Cr, Hg, Pb, Ni, Cu and Zn in their body tissue. Due to mobile nature of Cd it gets easily diffuse to and accommodate in the soft, noncalcareous tissues of earthworms body (Table 18.1).

18.10 Sustainability of Vermiremediation Technique

Vermiremediation technology is an eco-friendly, cost-effective, efficient, and self-regulated technique. This technique is easy to operate and maintain as the organism (earthworms) used in this technique require less attention and care. This technique is not generating any type of waste in the environment which makes it as zero-waste technique. This technique is regarded as “value addition technique” by scientists as compared to other bioremediation technique as it not only removes the pollutants and waste but also helps in soil reclamation and produces vermicast which improves soil fertility (Hand 1988). The earthworms used in this technique are easily

Table 18.1 Credit of different earthworms in bioremediation of heavy metals

Earthworms	Earthworms species	Heavy metal bioaccumulation	References
Epigeic	*E. fetida* *E. fetida* *E. fetida* *E. fetida* *Eudrilus eugeniae, E. fetida* and *Perionyx excavatus* *Eisenia fetida Eudrilus eugeniae* *Dendrobaena octaedra*	Cu, Zn, Cd, Pb Zn, Cu Cr, Cu, Cd, Pb Mn, Fe, Cu, Zn, Pb Cu, Zn, Cd, Pb, Mn Cd, Zn, Ni, Pb, Cr Zn, Mn, Cu, Ni, Cd, V, Cr, Pb, Hg, As Cu, Cd, Pb	Zhang et al. (2016) Nirola et al. (2016) Suthar et al. (2014) Sahariah et al. (2015) Pattnaik and Reddy (2011) Aleagha and Ebadi (2011) Ekperusi and Aigbodion, (2015) Holmstrup et al. (2010)
Endogeic	*Amynthas morrisi* *Allolobophora rosea* and *Nicodrilus caliginosus* *Metaphire posthuma* *Pontoscolex corethrurus* *Metaphire posthuma*	Cu, Zn, Cd, Pb Cu, Zn, Cd, Pb, As, Sb Mn, Fe, Cu, Zn, Pb Pb Cd and Pb	Zhang et al. (2016) Nannoni et al. (2011) Sahariah et al. (2015) Jusselme et al. (2012) Panday et al. (2016)
Anecic	*Lumbricus terrestris* *Aporrectodea longa*	Pb, Zn, Cu, As Cu	Sizmur et al. (2011b, c) Ezemonye et al. (2006)

available on various vermiculture farms. It initially requires some cost to buy these earthworms but later they increase their number due to their high reproductive and growth rate and easily acclimatization to new environment (Sinha and Sinha 2007). All these qualities make this creature as economically and environmentally sustainable organism suitable for vermiremediation process. Mechanical and chemical techniques for excavation of heavy metal are very expensive. These techniques require more energy to operate and maintain and also liberate waste material in the environment. Earthworms used in vermiremediation process not only accumulating harmful metals from the soil but also feeds on numerous type of organic waste which otherwise end up as landfills which further requires high cost for disposal. On feeding this organic waste earthworm convert it into valuable vermicompost and also cut off the cost of its disposal which further aids this organism as economical and environmental sustainable one (Kumar and Singh 2022).

18.11 Conclusion

Vermiremediation is a cost-effective, eco-friendly, efficient, economic, and environmentally sustainable technique that not only helps in removal of toxicants from soil but also helps in reclamation of soil. It acts as remedy for the soil as earthworms used in this technique improve different soil properties by their action. This technique also helps to get rid off huge amount of waste in landfills. This action changes waste-filled lands into fertile land.

Acknowledgments All the authors collaborate in writing and development of this book chapter.

Conflicts of Interest The authors declare no conflict of interest.

References

Agency for Toxic Substances and Disease Registry, ATSDR (2000) Toxicological profile for Arsenic TP-92/09. Center for Disease Control, Atlanta, Georgia

Agency for Toxic Substances and Disease Registry, ATSDR (2012) Toxicological profile for Chromium. U.S. Department of Health and Human Services, Public Health Service, Atlanta

Ahalya N, Ramachandra TV, Kanamadi RD (2003) Biosorption of heavy metals. Res J Chem Environ 7(4):71–79

Ahemad M (2014) Remediation of metalliferous soils through the heavy metal resistant plant growth promoting bacteria: paradigms and prospects. Arab J Chem. https://doi.org/10.1016/j.arabjc.2014.11.020

Aleaghal MM, Ebadi AG (2011) Study of heavy metals bioaccumulation in the process of vermicomposting. Afr J Biotechnol 10(36):6997–7001

Ali H, Khan E (2018) What are heavy metals? Long-standing controversy over the scientific use of the term' heavy metals'-proposal of a comprehensive definition. Toxicol Environ Chem 100(1):6–19

Ali H, Khan E, Sajad MA (2013) Phytoremediation of heavy metals-concepts and applications. Chemosphere 91:869–881

Ali H, Khan E, Ilahi I (2019) Environmental chemistry and ecotoxicology of hazardous heavy metals: environmental persistence, toxicity, and bioaccumulation. J Chem 12:1–14

Alloway BJ (2013) Heavy metals in soils. Springer, Dordrecht. ISBN: 978–94–007-4469-1 (Print) 978–94–007-4470-7 (Online), 51–160, available at https://link.springer.com/book/10.1007/978-94-007-4470-7. Accessed 23 Sep 2016

Ayangbenro AS, Babalola OO (2017) A new strategy for heavy metal polluted environments: a review of microbial biosorbents. Int J Environ Res Public Health 14(1):16. Available from:. https://doi.org/10.3390/ijerph14010094

Choong G, Liu Y, Templeton DM (2014) Interplay of calcium and cadmium in mediating cadmium toxicity. Chem Biol Interact 211:54–65

Clarkson TW, Magos L, Myers GJ (2003) The toxicology of mercury-current exposures and clinical manifestations. N Engl J Med 349:1731–1737. PubMed: 14585942

Csuros M, Csuros C (2002) Environmental sampling and analysis for metals. Lewis Publishers, Boca Raton

Dahmani-Muller H, Van-Oort F, Gelie B, Balabane M (2000) Strategies of heavy metal uptake by three plant species growing near a metal smelter. Environ Pollut 109:231–238

Damodaran D, Suresh G, Mohan R (2011) Bioremediation of soil by removing heavy metals using Saccharomyces cerevisiae. In: 2nd international conference on environmental science and technology (ICEST) Singapore. IACSIT Press, Singapore, pp 22–27

Dopp E, Hartmann LM, Florea AM, Rettenmier AW, Hirner AV (2004) Environmental distribution, analysis, and toxicity of organometal (loid) compounds. Crit Rev Toxicol 34:301–333. PubMed: 15239389

Duffus JH (2002) "Heavy metals" a meaningless term? IUPAC technical report. Pure Appl Chem 74(5):793–807

Edward CA, Lofty JR (1972) Biology of earthworms. Chapman and Hall, London, p p283

Edwards CA, Bohlen PJ (1996) Biology and ecology of earthworms, 3rd edn. Chapman and Hall, London

Ekperusi OA, Aigbodion IF (2015) Bioremediation of heavy metals and petroleum hydrocarbons in diesel contaminated soil with the earthworm: *Eudrilus eugeniae*. Springer Plus 4:540

EPA (1992) Environmental criteria and assessment office. United States Environmental Protection Agency. Integrated Risk Information System (IRIS), Cincinnati

Ezemonye LIN, Agbedia CU, Illechie I (2006) The effect of organic matter on the toxicity of Cu to earthworms (*Aporrectodea longa*). Afr J Environ Pollut Publ Health 5(1):59–67

Firdousi SA (2017) Bioaccumulation and bioabsorptions of heavy metals by the mushroom from the soil. J Med Chem Drug Discov 2(3):25–33

Gerard BM (1960) The biology of certain British earthworms in relation to environmental conditions. Ph.D Thesis,. University of London

GESAMP (2015) "Sources, fate and effects of microplastics in the marine environment: a global assessment" (Kershaw PJ ed.) (IMO / FAO / UNESCO-IOC / UNIDO / WMO / IAEA / UN /UNEP / UNDP Joint Group of Experts on the Scientific Aspects of Marine Environmental Protection). Rep. Stud. GESAMP No: 90–96

Gray NF (1999) Water technology: an introduction for environmental scientists and engineers. Hodder Headline Group, London, UK, pp 473–474

Gunathilagraj K (1996) Earthworm: an introduction. Indian Council of Agricultural Research Training Program. Tamil Nadu Agriculture University, Coimbatore

Gunatilake SK (2015) Methods of removing heavy metals from industrial wastewater. J Multidiscip Eng Sci Stud 1(1):12–18

Hand P (1988) Earthworm biotechnology. In: Greenshields R (ed) Resources and application of biotechnology: the new wave. MacMillan Press Ltd, USA

Hassan SH, Awad YM, Kabir MH, Oh SE, Joo JH (2010) Bacterial biosorption of heavy metals. In: Biotechnology cracking new pastures. MD Publications Pvt. Ltd, New Delhi, pp 79–110

Havugimana E, Bhople BS, Kumar A, Byiringiro E, Pierremugabo J, Kumar A (2017) Soil pollution-major sources and types of soil pollutants. Environ Sci Eng 11:53–86

Holmstrup M, Sorensen JG, Overgaard J, Bayley M, Anne-Mette B, Slotsbo S, Fisker KV, Maraldo K, Waagner D, Labouriau R, Asmund G (2010) Body metal concentrations and glycogen reserves in earthworms (*Dendrobaena octaedra*) from contaminated and uncontaminated forest soil. Environ Pollut. In Press

IARC (1990) IARC monographs on the evaluation of carcinogenic risks to humans. IARC Scientific Publications, Lyon, France. IARC 49 Chromium, nickel and welding

Ireland MP (1979) Metal accumulation by the earthworms *Lumbricus rubellus*, *Dendrobaena veneta* and *Eiseniella tetraedra* living in heavy metal polluted sites. Environ Pollut 19:201–206

Ireland MP (1983) Heavy metals uptake in earthworms; earthworm ecology. Chapman and Hall, London

Jacobs JA, Testa SM (2005) Overview of chromium(VI) in the environment: background and history. In: Guertin J, Jacobs JA, Avakian CP (eds) Chromium (VI) handbook. CRC Press, Boca Raton, pp 1–22

Jusselme MD, Poly F, Miambi E, Mora P, Blouin M, Pando A, Rouland-Lefèvre C (2012) Effect of earthworms on plant *Lantana camara* Pb-uptake and on bacterial communities in rootadhering soil. Sci Total Environ 416:200–207

Karami A, Shamsuddin ZH (2010) Phytoremediation of heavy metals with several efficiency enhancer methods. Afr J Biotechnol 9:3689–3698

Kumar R, Sharma P, Gupta RK, Kumar S, Sharma MM, Singh S, Pradhan G (2020) Earthworms for eco-friendly resource efficient agriculture. In: Kumar S, Meena RS, Jhariya MK (eds) Resources use efficiency in agriculture. Springer, Singapore

Kumar R, Singh D (2022) Insecticide-tolerant bacterial population in Eisenia Fetida's gut and Vermicast exposed to Chlorantraniliprole and Fipronil, Appl Biol Res 24(3):273-279. https://doi.org/10.5958/0974-4517.2022.00033.7

Liang SN, Gao Z, Li S, Shen LJ (2016) Investigation of correlativity between heavy metals concentration in indigenous plants and combined pollution soils. Chem Ecol 32(9):872–883

Mountouris A, Voutsas E, Tassios D (2002) Bioconcentration of heavy metals in aquatic environments: the importance of bioavailability. Mar Pollut Bull 44:1136–1141

Mustapha MU, Halimoon N (2015) Microorganisms and biosorption of heavy metals in the environment: a review paper. J Microb Biochem Technol 7(5):253–256

Nannoni F, Protano G, Riccobono F (2011) Uptake and bioaccumulation of heavy elements by two earthworm species from a smelter contaminated area in northern Kosovo. Soil Biol Biochem 43:2359–2367

Needelman BA (2013) What are soils? Nat Educ Knowl 4(3):2

Nirola R, Megharaj M, Saint C, Aryal R, Thavamani P, Venkateswarlu K, Naidu R, Beecham S (2016) Metal bioavailability to *Eisenia fetida* through copper mine dwelling animal and plant litter, a new challenge on contaminated environment remediation. Int Biodeterior Biodegrad 113:208–216

Ogbo EM, Okhuoya JA (2011) Bio-absorption of some heavy metals by *Pleurotus tuber-regium Fr. Singer* (an edible mushroom) from crude oil polluted soils amended with fertilizers and cellulosic wastes. Int J Soil Sci 6(1):34–48

Panday R, Bhatt PS, Bhattarai T, Shakya K, Sreerama L (2016) Aldehyde dehydrogenase expression in *Metaphire posthuma* as a bioindicator to monitor heavy metal pollution in soil. BMC Res Notes 9:491

Pattnaik S, Reddy MV (2011) Heavy metals remediation from urban wastes using three species of earthworm (*Eudrilus eugeniae*, *Eisenia fetida* and *Perionyx excavatus*). J Environ Chem Ecotoxicol 3:345–356

Prasad MNV, Freitas HMDO (2003) Metal hyperaccumulation in plants-biodiversity prospecting for phytoremediation technology. Electron J Biotechnol 6:285–321

Ramos C, Silvia M, Bernal A, Dioselina LD (2006) *Eisenia fetida* increased removal of polycyclic aromatic hydrocarbons (PAHs) from soil. Environ Pollut 141:396–401

Sahariah B, Goswami L, Kim KH, Bhattacharyya P, Bhattacharya SS (2015) Metal remediation and biodegradation potential of earthworm species on municipal solid waste: a parallel analysis between *Metaphire posthuma* and *Eisenia fetida*. Bioresour Technol 180:230–236

Sarkar BA (2005) Mercury in the environment: effects on health and reproduction. Rev Environ Health 20:39–56. PubMed: 15835497

Sharma I (2020) Bioremediation techniques for polluted environment: concept, advantages, limitations, and prospects. In: Trace metals in the environment- new approaches and recent advances, pp 1–21

Sherbet GV (1978) The biophysical characterisation of the cell surface. Academic press, London, p 414

Sinha RK, Sinha R (2007) Environmental biotechnology (role of plants, animals and microbes in environmental management and sustainable development). Aavishkar Publisher, Jaipur, p 315

Sinha RK, Bharambe G, Ryan D (2008) Converting wasteland into wonderland by earthworms-a low-cost nature's technology for soil remediation: a case study of vermiremediation of PAHs contaminated soil. Environmentalist 28:466–475

Sizmur T, Hodson ME (2009) Do earthworms impact metal mobility and availability in soil?–a review. Environ Pollut 157:1981–1989

Sizmur T, Tilston EL, Charnock J, Palumbo-Roe B, Watts MJ, Hodson ME (2011a) Impacts of epigeic, anecic and endogeic earthworms on metal and metalloid mobility and availability. J Environ Monit 13:266–273

Sizmur T, Palumbo-Roe B, Hodson ME (2011b) Impact of earthworms on trace element solubility in contaminated mine soils amended with green waste compost. Environ Pollut 159:1852–1860

Sizmur T, Palumbo-Roe B, Watts MJ, Hodson ME (2011c) Impact of the earthworm *Lumbricus terrestris* (L.) on As, Cu, Pb and Zn mobility and speciation in contaminated soils. Environ Pollut 159:742–748

Suthar S, Sajwan P, Kumar K (2014) Vermiremediation of heavy metals in wastewater sludge from paper and pulp industry using earthworm *Eisenia fetida*. Ecotoxicol Environ Saf 109:177–184

Tchounwou PB, Wilson B, Ishaque A (1999) Important considerations in the development of public health advisories for arsenic and arsenic-containing compounds in drinking water. Rev Environ Health 14(4):211–229. PubMed: 10746734

Tchounwou PB, Ayensu WK, Ninashvilli N, Sutton D (2003) Environmental exposures to mercury and its toxicopathologic implications for public health. Environ Toxicol 18:149–175. PubMed: 12740802

Tchounwou PB, Yedjou CG, Patlolla AK, Sutton DJ (2012) Heavy metals toxicity and the environment. EXS 101:133–164. https://doi.org/10.1007/978-3-7643-8340-4_6

Visvanathan C, Traenklar J, Joseph K, Nagendran R (eds) (2005) Vermicomposting as an eco-tool in sustainable solid waste management; project report of the Asian Regional Research Program on Environmental Technology (ARRPET). Pub. of Asian Institute of Technology (AIT). Anna University, India 36

Wang S, Wu W, Liu F, Liao R, Hu Y (2017) Accumulation of heavy metals in soil-crop systems: a review for wheat and corn. Environ Sci Pollut Res 24(18):15209–15225

Wei SH, Zhou QX, Wang X (2005) Cadmium-hyperaccumulator *Solanum nigrum* L. and its accumulating characteristics. Environ Sci 26:167–171

Wilson DN (1988) Association Cadmium. Cadmium-market trends and influences, London. Cadmium 87 Proceedings of the 6th international cadmium conference 9–16

Witkowska D, Słowik J, Chilicka K (2021) Heavy metals and human health: possible exposure pathways and the competition for protein binding sites. Molecules 26:6060. https://doi.org/10.3390/molecules26196060

Yadav R, Kumar R, Gupta RK, Kaur T, Kour A, Kaur S, Rajput A (2023) Heavy metal toxicity in earthworms and its environmental implications: a review. Environ Adv 12:100374. https://doi.org/10.1016/j.envadv.2023.100374

Zhang C, Mora P, Dai J, Chen X, Giusti-Miller S, Ruiz-Camacho N, Velasquez E, Lavelle P (2016) Earthworm and organic amendment effects on microbial activities and metal availability in a contaminated soil from China. Appl Soil Ecol 104:54–66

Chapter 19
Cellulase Enzymes: Transforming Lignocellulosic Waste into Sustainable Solutions

Kongkana Goswami, Vishal Das, Sangeeta Biswas, and Sosanka P. Sandilya

Abstract Cellulase enzymes serve as crucial tools in tackling the environmental issues caused due to lignocellulosic waste materials. These wastes comprising cellulose, hemicellulose, and lignin have traditionally led to environmental issues due to their improper disposal in urban areas. In this aspect, cellulase provides a sustainable solution to numerous industries, including biofuel production, waste management, paper and pulp, detergent manufacturing, and food and beverage processing. The incorporation of cellulase in these industrial processes not only promotes economic viability but also complies with the goals of global sustainability. Cellulase catalyzes the transformation of cellulosic wastes into fermentable sugars which is an essential step in the production of biofuels like ethanol and butanol; therefore, valorization of these wastes utilizing cellulase has drawn attention globally to promote cleaner energy choices while reducing dependency on fossil fuels and greenhouse gas emissions. In recent years, the use of cellulase enzymes to support environment-friendly practices has undergone an extensive change due to some current advances such as the production of bioengineered cellulases with improved characteristics and immobilized enzymes designed for the effective breakdown of complex lignocellulosic materials. Furthermore, pretreatment techniques have been developed to improve the susceptibility of lignocellulosic wastes to enzymatic hydrolysis, and enzymes can be made more stable and reusable by immobilizing them on solid supports. The fabrication of biofuels has been streamlined by simultaneous fermentation and saccharification processes, and cellulase-based methods

K. Goswami (✉) · V. Das
Department of Medical Laboratory Technology, School of Health Sciences, The Assam Kaziranga University, Jorhat, Assam, India

S. Biswas
Department of Zoology, Royal School of Life Sciences, The Assam Royal Global University. Betkuchi, Guwahati, Assam, India

S. P. Sandilya
Department of Microbiology, School of Health Sciences, The Assam Kaziranga University, Jorhat, Assam, India

S. Dey, S. Bhattacharya (eds.), *Biotechnological Interventions in the Removal of Emerging Pollutants*, Interdisciplinary Biotechnological Advances,
https://doi.org/10.1007/978-981-97-9922-0_19

are being increasingly included in biorefineries, enabling the production of important biochemicals and bioplastics in addition to biofuels. With these developments, there is the possibility of a cleaner environment, less reliance on fossil fuels, and a shift to a more sustainable industrial landscape.

Keywords Cellulase · Environmental issues · Green house emission · Lignocellulose · Waste valorization

19.1 Introduction

The dawn of the twenty-first century is marked not only by the highly increased population but also by a rising need for goods for everyday use, both of which are expected to contribute to the depletion of fossil fuels and other carbon resources. The negative effects of fossil fuel-derived energy sources, which include pollution of the environment, global warming, and ecological imbalances, cast a shadow and pose significant threats to our ecosystem. This confluence of pressing issues necessitates a paradigm change toward alternate routes for the synthesis of chemicals, fuels, and materials—a change that considers sustainability, ecological compatibility, and economic viability from an interdisciplinary perspective (Chukwuma et al. 2020). It highlights the urgent need for innovative platform products made from plentiful biorenewable carbon resources to be produced in an inexpensive, accessible, and high-quality manner. In this case, lignocellulosic biomass emerges to be a viable solution that fulfills these requirements and addresses the global threats of greenhouse gas emissions and environmental pollution by addressing the challenges involved in the production and effective bioconversion of lignocellulosic waste (Goswami et al. 2021).

Lignocellulosic biomass constitutes a huge portion of residential, agricultural, and industrial waste streams, often left unutilized and disposed of in the urban areas, thereby exerting adverse effects on the environment such as global warming, emission of carbon dioxide and other harmful gases and therefore, valorization of these wastes into beneficial products removes pollution on one hand and helps in industrial aspect by acting as raw materials on the other hand (Zhang and Zhang 2013). Though lignocellulosic biomass displays great potential, its rigid nature presents an important challenge that calls for the production of highly selective and promising biocatalytic systems as well as green reaction bioprocesses. A sustainable method of using ligninolytic enzymes to biotransform lignocellulosic biomass into high-value products (Kuhad et al. 2011; Gupta et al. 2013). The scientific community needs to support new approaches in response to the global energy and ecological crises. One such approach is the use of microbial cellulases to effectively utilize lignocellulosic biomass and bioconvert its waste into valuable resources (Goswami et al. 2022; Dar et al. 2019).

19.1.1 Source of Lignocellulose

Plant cell walls are a rich source of lignocellulosic material, a robust composite that comes from a variety of sources, including waste streams and naturally occurring plants. Unlocking the potential of lignocellulose for environment-friendly applications and sustainable economies requires an understanding of its origins.

A. *Natural Plant Sources*

(i) *Agricultural residues:* A substantial portion of lignocellulose stocks is derived from crop residues. Considerable amounts of cellulose, hemicellulose, and lignin can be found in the stalks, leaves, and husks of crops like wheat, rice, and corn (Wang and Tester 2023).
(ii) *Forestry byproducts:* One of the main sources of lignocellulose is wood, branches, and other by-products from forest areas. The cellulose content of hardwood and softwood trees is high, providing the basis for biomass and wood-based industries (Table 19.1).
(iii) *Energy crops:* Significant sources of lignocellulose involve plants grown especially for energy, such as the switch grass and miscanthus. These perennial grasses have a high cellulose content, which makes them promising for the production of bioenergy.

B. *Waste Biomass Sources*

(i) *Municipal solid waste:* Materials rich in lignocellulose can be found in urban waste, such as paper, cardboard, and food waste. By maintaining these materials away from landfills, efficient waste management techniques help promote a circular economy.
(ii) *Textile waste:* Lignocellulose is frequently found in discarded textiles, particularly those made of natural fibers like cotton, linen, and jute. In the fashion and textile sectors, recycling and repurposing these textiles offers sustainable solutions.
(iii) *Agricultural and animal wastes:* Straw, bagasse, and manure are examples of agricultural and animal-related wastes that provide additional sources of lignocellulose. Initiatives for waste-to-energy and agricultural sustainability are supported by the efficient utilization of these wastes (Wang and Tester 2023).

Table 19.1 The chemical composition of softwood, hardwood, and grass (Shen and Sun 2021)

Sl. no.	Cellulose	Hemicellulose	Lignin
Softwood	43–47%	25–35%	16–24%
Hardwood	40%	25–29%	25–31%
Grass	25–29%	26–32%	16–21%

19.1.2 Composition of Lignocellulose

The building block of plant cell walls is lignocellulose, a complex and robust structural material that gives plant cell walls their strength and rigidity. Lignocellulose, which is made up of three essential components (cellulose, hemicellulose, and lignin), is crucial in determining the physical and mechanical features of different plant structures. A significant source of lignocellulose is waste biomass, which is generated from crops, leaves, animal feces, textile fibers, residential garbage, and urban waste materials like paper, wood, and food waste. This plentiful resource is an essential raw material used in the production of a wide variety of industrial goods. The composition of cellulose, hemicellulose, and lignin varies depending on whether the source is hardwood, softwood, or grass (Table 19.1). Agricultural waste can be used in a variety of ways in factories because it usually contains 10–25% lignin, 20–30% hemicelluloses, and 40–50% cellulose, according to studies by Malherbe and Cloete (2002), Kumar et al. (2009), and Iqbal et al. (2011).

(a) *Cellulose:* About 40–50% of plant material is made up of cellulose, the most widely distributed plant polymer in the world. It is made up of linear chains of glucose molecules connected by β-1,4-glycosidic bonds. The number of glucose subunits in a cellulose chain is determined by the degree of polymerization. Because of intra- and intermolecular hydrogen bonds as well as Van der Waal's force, cellulose molecules form intricate microfibril structures and fibers that render them insoluble in organic solvents or water. Effective enzymatic hydrolysis is necessary to transform cellulose into useful products, providing a cost-effective and environmentally safe substitute for physical and chemical processes. Cellulose molecules can be divided into: microfibrils and matrix components.

   (i) *Microfibrils:* The main building block of cellulose molecules is known as microfibrils which are long, narrow fibril structures that contain both crystalline and amorphous regions. This microfibril structure provides tensile strength via hydrogen bonds and is confirmed by electron microscopy and X-ray diffraction crystallography (Rongpipi et al. 2019).
   (ii) *Matrix component:* It is a heterogeneous mixture that varies between species and consists of hemicelluloses, lignin, pectin, protein, and phenolic compounds.

(b) *Hemicellulose:* Hemicelluloses are made up of pentose and hexose units and make 20–30% of lignocellulosic biomass. Together with glucomannan and other polysaccharides, glucuronoxylan is the second most common natural polymer. Hemicelluloses, which are cross-linked with cellulose microfibrils and come in different forms (such as arabinan, galactan, and xylan in grasses and straw, and mannan in hardwood and softwood to help build strong plant cell walls (Brigham et al. 2018).
(c) *Lignin:* About 10–25% of plant biomass is made up of lignin, which is a plant polymer with three monomers. It gives cell walls structural stability,

impermeability, and defense against microbial infections (Anwar et al. 2014). Lignin strengthens plant cell walls by binding cell components and occupying the space between cellulose fibrils. The intricate structure of lignin, consisting of numerous carbon-carbon bonds, presents difficulties for its breakdown into more basic forms. Depending on the source of the material, the long chain of lignin can have one of three structures (I, II, or III) made up of phenyl-propane units connected by ether groups. Structures I and II are seen in grasses and conifers, and structure III is seen in deciduous wood.

19.1.3 Lignocelluloses and Industrial Applications

Plant cell walls contain lignin, cellulose, and hemicellulose together in varying amounts to form lignocellulose, a complex structural material. Because of its special qualities, it is a resource that is abundant and versatile, uncovering use in a wide range of industries. The following are some of the main applications for lignocellulose in various industries:

(a) *Biofuel production:* A major feedstock for the production of biofuels such as bioethanol and biodiesel is lignocellulose. The components of cellulose and hemicellulose can be broken down chemically or enzymatically into sugars, which can subsequently be fermented to produce ethanol (Kuhad et al. 2011; Mood et al. 2013).
(b) *Paper and pulp industry:* The pulp and paper sector is one of the biggest lignocellulose consumers (Behera et al. 2017). Wood fibers rich in cellulose are harvested from trees and employed to make paper and other paper-based products.
(c) *Textile industry:* The textile industry makes extensive use of lignocellulosic fibers, including cotton. These fibers give textiles their strength, resilience, and absorbency (Behera et al. 2017). The cellulose component in these fibers is essential for their mechanical strength.
(d) *Biocomposites and bioplastics*: Studies are underway to develop biocomposites and bioplastics using lignocellulosic materials. It is possible to generate environment-friendly materials with an extensive variety of uses in the packaging, and other industries by combining lignocellulose with biodegradable polymers.
(e) *Medicine and healthcare*: Lignocellulose is a source of therapeutic substances. Due to their prebiotic qualities and possible health advantages, xylooligosaccharides derived from hemicellulose have potential use in the pharmaceutical and healthcare sectors (Kuhad et al. 2011; Gupta et al. 2013).
(f) *Agriculture and animal feed:* Animal feed is made from lignocellulosic leftovers like wheat straw and corn stover. Furthermore, these residues can be applied to crops as crop covers and as soil amendments in agriculture, which supports sustainable farming methods (Kuhad et al. 2011; Gupta et al. 2013).

19.1.4 Complexity in Cellulose Hydrolysis

Plant cell walls are primarily made of cellulose, which has a complex and resistant structure that makes cellulose hydrolysis a complex biochemical process. The structure of crystalline cellulose is made up of long linear chains joined by hydrogen bonds, which form a strong, hard barrier that prevents enzymes from easily accessing and breaking the β-1,4-glycosidic bonds. There are crystalline and amorphous regions in cellulose, and the crystalline portions are very resistant to enzyme breakdown (Chami Khazraji and Robert 2013). The overall resistance to enzymatic hydrolysis is influenced by the degree of polymerization, or the length of cellulose chains, which varies among various plant sources. The formation of hydrogen bonds within individual chains as well as between cellulose chains in the crystalline lattice adds to the strength and stability of cellulose. Because cellulose's molecules are arranged into fibrils, its fibrillar structure makes enzymatic hydrolysis more difficult and necessitates the use of specialized enzymes to break down the material. Endoglucanases and cellobiohydrolases (exoglucanases) are two enzymes that hydrolyze cellulose, but their accessibility is hampered by the complex cellulose matrix. Furthermore, the crystalline portions of cellulose decrease the amount of enzyme adsorption onto the substrate, requiring cellulase enzymes to have cellulose-binding domains (CBD) to aid in binding to the cellulose surface. Research is currently being conducted to improve enzyme systems, pretreatment techniques, and bioprocessing strategies to address these complexities and develop efficient processes for biomass conversion, biofuel production, and other biotechnological applications (Zhang and Zhang 2013).

19.1.5 Cellulase and Microbes in Lignocellulose Hydrolysis

The production of cellulase by microorganisms is vital because it plays a crucial part in the hydrolysis of lignocellulose, a complex polysaccharide present in plant cell walls, into simpler sugars (Rathore 2014; Dar et al. 2019; Legodi et al. 2019). Microorganisms provide a sustainable and economical method of producing cellulase. The fact that lignocellulose serves as a primary carbon source for cellulolytic microorganisms highlights how environment-friendly this process is. There are numerous uses for this enzymatic process in a variety of industries, such as waste treatment, food processing, manufacturing of textiles, and biofuel production. Many different types of microorganisms can be used to produce cellulase, including bacteria and fungi. Fungal species like *Aspergillus, Trichoderma reesei, Trichoderma viride, Penicillium pinophilum, Fusarium solani, Phanerochaete chrysosporium,* etc. are recognized for their ability to secrete cellulase enzymes, which makes them excellent candidates for industrial applications (Mrudula and Murugammal 2011; Ellila et al. 2017; Zhao et al. 2018). Bacteria such as *Bacillus, Acidothermus cellulyticus, Clostridium acetobutylicum, Clostridium thremocellum, Pseudomonas, Novosphingobium,* and *Cellulomonas* sp., grow rapidly, which increases the interest in using microbial cellulase production as a scalable and effective solution for commercial practices (Gautam and Sharma 2012; Kim et al. 2012; Ahmad et al. 2013; Rathore 2014). The development of sustainable solutions like biofuels, bioplastics, and other bio-based materials is made possible by this degradation process (Dar et al. 2019). Using microbes to break down lignocellulose presents a viable path toward solutions that are both economically and environmentally sustainable for a variety of industries (Iqbal et al. 2013). However, since different microorganisms have different cellulolytic abilities, the industrial use of microbial cellulase is currently restricted to a limited number of strains. Progress in research has been demonstrated by the efforts to isolate and screen effective microbial strains that can produce the whole cellulase complex (Goswami et al. 2022). The intricate composition of cellulases, comprising numerous subunits, poses a challenge that scientists are presently attempting to address to increase the number of industrial applications and maximize enzymatic efficiency (Table 19.2).

Table 19.2 List of bacterial strains converting lignocellulosic substances into value-added products

Sl no.	Bacterial strains	Lignocellulosic substances	Value-added products
1.	*Streptomyces, Bacillus* and *Paenibacillus*	Saw dust	Cellulase
2	*Pseudomonas* sp. GO2	Corn stover	Bioflocculant
3	*Ochrobactrum* sp.	Palm oil mill effluent	CMCase and Xylanase
4	*Streptomyces lividans*	Sunflower stalks and rape straw	Triacylglycerol
5	*Bacillus* sp. BS-5	Corn cob	Xylanase, endoglucanase
6	*Actinobacteria*	Olive pomace	Laccase, xylanase
7	*Cellulomonas, Pseudomonas, Bacillus, Clostridium*, and *Fibrobacter*	Peat	NA
8	Anaerobic microbial consortium TC-5	Untreated wheat straw	Methane
9	*Paenibacillus polymyxa* ND25	CMC, avicel, corn starch, rice straw and sugarcane bagasse	Endoglucanase, exoglucanase, and β-glucosidase
10	*Clostridium butyricum* I5-42	Starch free fiber from cassava pulp and xylan	1,3-propanediol
11	*Achromobacter* sp. (PS1)	Rice-straw, wheat-straw and sugarcane-bagasse	Biosurfactant
12	*Caldicellulosiruptor bescii*	Switch grass	Acetate
13	*Enhydrobacter*	Rice husk	Compost
14	*Proteobacteria, Firmicutes, Bacteroidetes Dysgonomona, Sedimentibacter, Comamonas*	Energy grass	Biogas
15	*Paenibacillus* sp.	Rice straw, corn straw, soybean straw, and sugarcane bagasse	–
16	Clostridium thermocellum, *C. stercorarium*, and *Thermoanaerobacter thermohydrosulfurics*	Wheat and Cattail biomass	Ethanol

19.1.6 Cellulase Structure and Activity

Plant cell walls are rich in lignocellulose, a complex structural component that presents a significant bioconversion challenge because of its complex cellulose, hemicellulose, and lignin composition. The intricate structure is a result of the intertwining of hemicellulose and cellulose microfibrils, which are all encased in a lignin matrix and resist easy degradation. However, the challenge lies in providing efficient access to cellulose microfibrils encased within the protective matrix of hemicellulose and lignin. This matrix must be penetrated by cellulase for hydrolysis to proceed properly. In this process, cellulase becomes a key component in the hydrolysis of lignocellulose. Cellulase is an enzymatic resource consisting of enzyme subunits; such as endoglucanases, cellobiohydrolases, and β-glucosidases specialized in breaking down lignocellulose. Targeting the link points within the cellulose chain, the hydrolysis process cleaves the β-1,4-glycosidic linkages in cellulose to release glucose molecules (Dar et al. 2019). The process is started by endoglucanase, which randomly cleaves β-1,4-linkages to release cellulose ends that are both reducing and nonreducing. Cellobiose is then released to produce sugar molecules when these products are hydrolyzed by cellobiohydrolases (Fig. 19.1) (Goswami et al. 2021). CAZy database groups cellulase under glycosyl hydrolases into 133 families. Endoglucanases belong to families GH 5, 6, 7, 8, 9, 12, 45, and 74, cellobiohydrolases in GH 6, 7, 8, 9, and 48, and β-1,4-glucosidase in 1, 3, and 9, according to their functional characteristics. Thereby, cellulase enhances the efficiency of hydrolysis to break down the entire lignocellulosic structure are essential step toward unlocking the full potential of lignocellulose for sustainable bioconversion and bio-based industries' global challenges. In bacterial and fungal cellulases, a catalytic and a noncatalytic linker region, rich in hydroxyl amino acids, connect cellulose-binding domains (CBD) (Quinlan et al. 2010; Sadhu and Maiti 2013). There are 180 distinct forms of CBD that are divided into 13

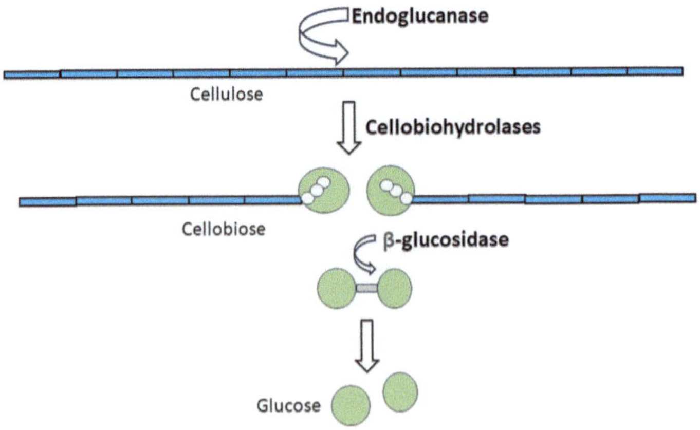

Fig. 19.1 Cellulose hydrolysis mechanism

families according to their amino acid sequences; the arrangement of CBD differs between organisms and cellulase types (Tomme et al. 1995). CBD from families II and III is found in bacterial cellulases, whereas family I, which is found in fungal species, is composed of tiny polypeptides (Carrard et al. 2000). Certain anaerobic bacteria, such as *Fibrobacter succinogens*, use adhesins to bind the cellulosic substrate to the outer membrane of the bacterium (Miron et al. 2001). A linker region rich in threonine and serine connects additional domains that are frequently present in bacterial cellulases, including S-layer homologous, fibronectin type III, Nod-B-like, and unknown regions (Tomme et al. 1995). The hydrolysis process is facilitated by the active targeting of the enzyme toward cellulose by the catalytic and cellulose-binding domains, among others (Reyes-Ortiz et al. 2013).

19.1.7 Mode of Lignocellulose Hydrolysis

Numerous mechanisms are involved in the hydrolysis of cellulose by bacteria, one of which is adhesion, which is aided by structures such as fimbriae, glycocalyx, and cellulose-binding domains (CBDs) (Sadhu and Maiti 2013). These characteristics improve the bacterial attachment to cellulose substrates, which facilitates effective enzymatic degradation. An outline of the role of these elements in cellulose hydrolysis is provided below:

(a) *Attachment:* To establish proximity between the bacterial cell and the substrate, bacteria attach to the cellulose fibers with the help of glycocalyx and fimbriae structures. Bacterial cells have hair-like projections called fimbriae that are frequently made of proteinaceous subunits. An important part of adhesion to the fibers of cellulose is played by bacterial fimbriae. Bacteria can get adjacent to the target material by attaching to the cellulose substrate, which makes it easier for enzymatic reactions to occur later on (Fenno et al. 1989). Bacterial cell walls are also surrounded by a layer called glycocalyx, which is made up of polysaccharides and glycoproteins. As an extracellular matrix, the glycocalyx aids in the adhesion of bacteria to surfaces. Glycocalyx acts as an extra adhesive layer during cellulose hydrolysis, which facilitates the binding of bacterial cells to cellulose fibers (Pell and Schofield 1993). In addition to that, a structural domain called CBD is present in some cellulase enzymes that are made by fungi and bacteria. The CBD binds to cellulose molecules in a specific way that improves cellulase enzyme attachment to the substrate. Efficient enzymatic hydrolysis is promoted by this targeted binding, which raises the local concentration of cellulases on the cellulose surface.
(b) *Enzyme localization:* Due to the presence of these adhesion-promoting structures, the bacteria's cellulase enzymes, which frequently contain CBDs, bind specifically to the cellulose surface. This domain is essential for the effective hydrolysis of cellulose because it enables the enzyme to attach to the intricate fiber structure of cellulose. This facilitates the cleavage of glycosidic bonds by

the enzyme, which increases catalytic activity and improves the conversion of cellulose into sugars that can be fermented and used in various industrial applications.

(c) *Hydrolysis:* The localized amount of cellulases facilitates the enzymatic hydrolysis of cellulose at the point of attachment. Enzymatic activity breaks down cellulose into smaller, soluble sugar units so that the bacteria can continue to metabolize it. The process is started by endoglucanase, which randomly cleaves β-1,4-linkages to release cellulose ends that are both reducing and nonreducing. Cellobiose is then released to produce sugar molecules after hydrolysis of the products by cellobiohydrolases (Fig. 19.1).

19.1.8 Enhancement of Cellulase Activity for Industrial Utilization

Improving the efficiency of cellulose degradation in a variety of industrial processes, including the production of biofuel, textiles, and the processing of food, requires efficient cellulolytic activity. The following are some of the strategies used to increase the enzymatic activity.

(a) *Strain selection and improvement*: To optimize the production of cellulase for industrial applications, strain selection and improvement are essential steps. First, specifically selected microbial strains that naturally produce cellulase are isolated from cellulose-rich environments and the enzyme activity enhancement strategies are implemented after strain selection.

  (i) *Strain improvement* via *mutation*: Improvement of strains is required to boost cellulase production to commercially viable levels. For reliable industrial applications, engineering cellulases to maximize their properties is frequently necessary. Filamentous fungi naturally release diverse cellulases when cultivated on lignocellulosic waste. The classical method of enhancing cellulase expression involves random mutagenesis. To modify enzymes for particular industrial requirements, this method has been used in conjunction with site-specific mutagenesis, or both (Dalby 2007). Over the past two decades, the *Penicillium oxalicum* mutant (JU-A10-T) has been extensively utilized in China for industrial-scale cellulase production (Liu et al. 2013).

  (ii) *Strain improvement* via *genetic engineering*: One particularly potent tool that makes it possible to combine several desirable traits into a single organism is genetic engineering. This method is widely used to modify microorganisms and increase the production of metabolites, which can range from basic proteins to highly targeted therapeutic proteins. The process may become more difficult despite its potential due to the intricacy of the targeted metabolites or the organisms themselves. A thorough understanding of the genetic composition of the organism is a vital requirement for successful genetic engineering. The identification of target sites for genetic alterations is made easier by an understanding of the functions of the entire genome sequence. Effective techniques for introducing directed genetic alterations are necessary for metabolic engineering or targeted strain engineering to improve cellulase production in fungi. The problem has been the usually poor efficiency of gene targeting, which is mostly because of the difficulties relating to glycosylation and the complexities of the genes' introns. Because filamentous fungi can grow on inexpensive substrates, they are a promising source for industrial applications due to their wide range of metabolite production. Apart from secreting proteins, fungi can also carry out posttranslational modifications such as glycosylation and di-sulfidation (Wang and Lu 2021). Although plasmids integrated into the genomes of filamentous fungi are frequently used to transform them for stability, nonhomologous recombination problems have been solved by deactivating parts of the non homologous end-joining (NHEJ) DNA repair pathway (Guangtao et al. 2009; Steiger et al. 2011). Notably, the *Aspergillus* genus and the *T. reesei* strain have become preferred hosts for the expression of a variety of genes in both fungal and nonfungal organisms (Singh et al. 2017).

(iii) *Metabolic engineering*: Metabolic engineering strives to engineer microbes into efficient biocatalysts, enabling the enhanced and cost-effective production of numerous secondary metabolites. These metabolites serve as valuable resources for industrial chemicals, pharmaceuticals, and fuels. The success of metabolic engineering relies on a comprehensive examination of cellular physiology and metabolism through high-throughput genomic experimental techniques. Additionally, it necessitates a deep understanding of cellular processes, often articulated through mathematical models. However, a significant challenge in metabolic engineering lies in the absence of detailed models for specific pathways, such as the protein secretion pathway, limiting the precision of design and optimization strategies (DasGupta et al. 2020). To increase the cellular machinery in responsible for enzyme production, the genes involved in the production of cellulases are over expressed. This could entail using potent promoters or making more copies of the cellulase genes.

(b) *Optimization Conditions*

Culture Conditions To achieve the best possible culture conditions for the highest production of cellulase, temperature, pH, carbon source, nitrogen source, and inoculum percentage need to be optimized. Studies on *Bacillus*, *Pseumononas*, and *Novosphingobium,* etc. have already confirmed the importance of culture condition optimization for enhancing cellulase production by utilizing lignocellulosic waste materials (Dar et al. 2019; Goswami et al. 2021).

Fermentation Process Implementation of fermentation processes, such as fed-batch or continuous culture systems, to maintain optimal conditions throughout cellulase production can also enhance biomass and enzyme yields.

By employing these strategies individually or in combination, researchers and industrial practitioners aim to enhance cellulase activity, enabling more efficient and sustainable processes for the conversion of cellulose-rich biomass into valuable products.

19.1.9 Industrial Application of Microbial Cellulase

Cellulase is essential for breaking down cellulose, the main building block of plant cell walls, into simpler sugars. Because these enzymes can catalyze the hydrolysis process of lignocellulose and unleash its potential for a multitude of applications, they are used in a variety of industries (Fig. 19.2a). The key applications of cellulase in various industries include the following:

(a) *Biofuel production:* A large number of biofuels, especially ethanol, are produced using cellulase enzymes. Cellulase aids in the fermentation process by dissolving cellulose into sugars that can be fermented, which produces ethanol—a sustainable and renewable fuel that can replace conventional fossil fuels (Fig. 19.2b) (Dar et al. 2019; Kuhad et al. 2011).
(b) *Biopolishing of fabrics:* Cellulase enzymes are employed in the textile industry for the biopolishing of fabrics. This procedure gives the outer layer of the fabric a smoother texture by eliminating the appearance of fuzz and protruding fibers. It reduces pilling, improves the look and quality of textiles, and gives the final goods a softer feel (Kuhad et al. 2011).
(c) *Paper and pulp industry:* To enhance pulp quality and brightness of paper, the paper and pulp industry uses cellulase enzymes. They aid in the cellulosic

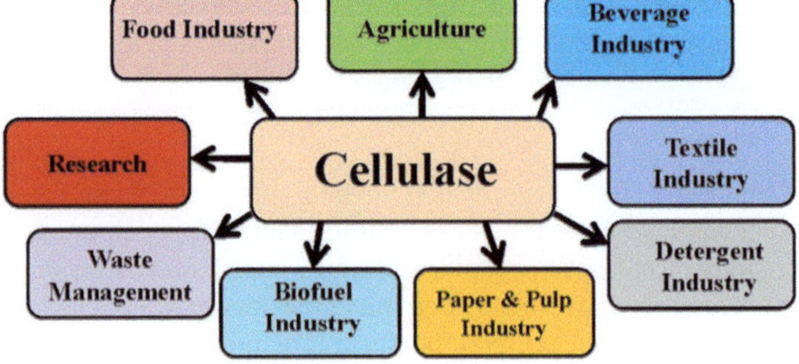

Fig. 19.2 (**a**) Industrial application of cellulase enzyme. (**b**) Biofuel production on lignocellulosic biomass

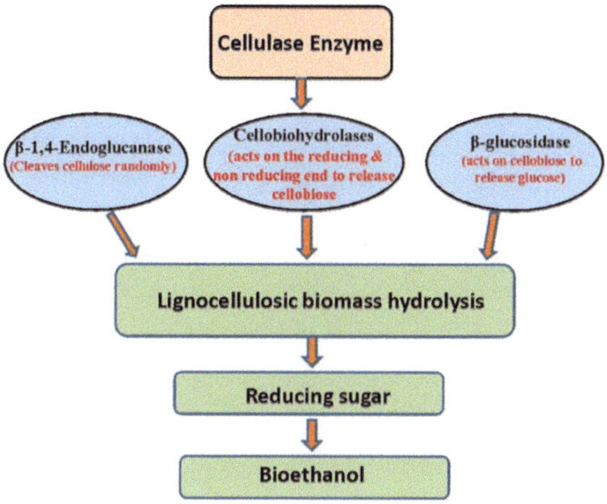

Fig. 19.2 (continued)

fibers' breakdown, improving the refining procedure and lowering the energy needed to produce paper (Saranraj et al. 2012).

(d) *Detergent industry:* Laundry detergents contain cellulase enzymes as agents to help eliminate stains. Stains are broken down by them and thus become easier to remove with water. This enhances the effectiveness of laundry detergents and helps to remove stains effectively (Bhat 2000).

(e) *Food and beverage industry:* Cellulase, along with pectinases and xylanases, is employed in combination in the food industry to expedite the processing of food products with minimal processing time. Utilizing these enzymes in the maceration of different fruit pulps enhances juice production, contributing to more efficient food processing methods. Cellulase enzymes are utilized in the beverage industry to extract fruit juices and hydrolyze the cellulosic wall of the fruits. Cellulase increases the efficiency of the fruit juice extraction process by dissolving the cellulosic structure and facilitating the release of juices (Kashyap et al. 2001).

(f) *The animal feed industry:* To improve the digestibility of plant materials in livestock's digestive systems, animal feed contains cellulase enzymes. This enhances the availability and utilization of nutrients, improving the general health of the animal and increasing feed conversion efficiency (Bharadwaj et al. 2021).

(g) *Waste treatment and bioremediation*: In municipal solid waste treatment plants, cellulose-rich organic waste is broken down with the help of cellulase enzymes. By dissolving cellulose components, they speed up the composting process and reduce the amount of waste production. Cellulase is utilized in bioremediation to break down cellulose-based contaminants in contaminated areas (Lee et al. 2019; Almuharef et al. 2019).

(h) *Agriculture*: In agriculture, various enzyme formulations with different combinations of cellulases, hemicellulases, and pectinases show promise for managing plant diseases and accelerating crop growth (Bhat 2020). To isolate plant or fungal protoplasts and to enable the development of hybrid strains with desired traits, microbial hydrolases are utilized. Cellulases and related enzymes derived from particular fungi can break down plant pathogen cell walls, which helps manage plant diseases.
(i) *Textile effluent treatment:* Cellulase enzymes can help remove color from wastewater when treating textile effluents. By dissolving leftover dyes, this enzymatic treatment minimizes the environmental effect of textile dyeing procedures (Kuhad et al. 2011).
(j) *Biotechnology and research:* In the laboratories conducting biotechnological research, cellulase enzymes play a crucial role. They are employed in several bioprocessing techniques, such as the enzymatic hydrolysis of cellulose to produce bio-based chemicals, textiles, protoplast isolation, and biofuels (Gupta et al. 2013).

Since cellulase enzymes can catalyze the breakdown of lignocellulose in a variety of industries, they are indispensable and contribute to developing more environment-friendly and sustainable processes. Research is still being done to find new uses and improve cellulase efficiency for various industrial applications.

19.2 Future Prospects

Future developments in cellulase and lignocellulosic wastes have the potential to completely transform environment-friendly practices in a variety of sectors. Because of their essential function in breaking down cellulose into useful components, cellulase enzymes have the potential to propel the development of bio-based production. Cellulase has the potential to improve the efficiency with which lignocellulosic biomass is converted into biofuels in the field of bioenergy, thereby lowering dependency on fossil fuels and promoting the use of renewable energy sources. Furthermore, customized cellulase formulations could result in innovations in biomaterials and bioplastics, which would promote the creation of environment-friendly substitutes. Lignocellulosic wastes, which are commonly regarded as agricultural/residential residues or byproducts, are becoming more and more valued resources. In the future, these wastes will be effectively valorized by using cellulase enzymes to create pulp, textiles, and bio-based products. Cellulase and lignocellulosic wastes become essential elements in the global shift to a circular economy, leading to innovation, waste minimization, and environment-friendly solutions in various industrial sectors. It is expected that interdisciplinary approaches, precision engineering advancements, and cooperative efforts among researchers will open up new avenues and help to shape a future in which lignocellulosic wastes and cellulase play crucial roles in a world that is more resource-efficient and sustainable.

19.2.1 Conclusion

This book chapter has explored the capacity of cellulase enzymes to transform lignocellulosic waste into some sustainable solutions. Cellulase enzymes are at the forefront of the sustainable bioeconomy, as demonstrated by the numerous discussions on cellulase applications in waste management, biofuel production, bioplastics synthesis, textile and pulp industries, and beyond. The future looks bright, as discoveries and developments in technology will likely improve the efficiency, specificity, and range of applications of cellulase even more. When cellulase enzymes are present, lignocellulosic waste—which was previously seen as a problem—becomes an invaluable resource. Beyond waste valorization, cellulase has a wide range of applications in addressing global issues such as the need for renewable energy, plastic pollution, and environmental degradation. To fully utilize cellulase enzymes and open the door to innovative, eco-friendly solutions across a range of industries, researchers, technologists, and stakeholders must work together as we navigate the path toward a more sustainable future. This chapter provides evidence of the revolutionary potential of cellulase enzymes, igniting a common goal for a circular economy that supports both environmental stewardship and industrial progress.

References

Ahmad B, Nigar S, Shah SSA, Bashir S, Ali J, Yousaf S, Bangash JA (2013) Isolation and identification of cellulose degrading bacteria from municipal waste and their screening for potential antimicrobial activity. World Appl Sci J 27:1420–1426

Aini FN, Laconi EB, Suhartono MT (2018) The potential of agricultural by-products as a carbon source in the growth of Paenibacillus sp. AIP Conf Proc 2021:030004

Almuharef I, Rahman MS, Qin W (2019) High production of cellulase by a newly isolated strain *Paenibacillus* sp. IM7. Waste Biomass Valor 11:6085–6094

Anwar Z, Gulfraz M, Irshad M (2014) Agro-industrial lignocellulosic biomass a key to unlock the future bio-energy: a brief review. J Radiat Res Appl Sci 7:163–173

Apiwatanapiwat W, Vaithanomsat P, Thanapase W, Ratanakhanokchai K, Kosugi A (2018) Xylan supplement improves 1,3-propanediol fermentation by clostridium butyricum. J Biosci Bioeng 125:662–668. https://doi.org/10.1016/j.jbiosc.2017.12.006

Behera BC, Sethi BK, Mishra RR, Dutta SK, Thatoi HN (2017) Microbial cellulases–diversity & biotechnology with reference to mangrove environment: a review. J Genet Eng Biotechnol 15:197–210

Bhardwaj N, Kumar B, Agrawal K, Verma P (2021) Current perspective on production and applications of microbial cellulases: a review. Bioresour Bioprocess 8:1–34

Bhat MK (2000) Cellulases and related enzymes in biotechnology. Biotechnol Adv 18:355–383

Bhat R, Ahmad A, Jõudu I (2020) Applications of lignin in the agri-food industry. In: Sharma S, Kumar A (eds) Lignin. Springer series on polymer and composite materials. Springer, Cham. https://doi.org/10.1007/978-3-030-40663-9_10

Bohra V, Dafale NA, Purohit HJ (2018) Paenibacillus polymyxa ND25: candidate genome for lignocellulosic biomass utilization. 3 Biotech 8:248. https://doi.org/10.1007/s13205-018-1274-3

Brigham JS, Adney WS, Himmel ME (2018) In Hemicellulases: diversity and applications. In: Handbook on bioethanol, vol 2018. Routledge, pp 119–141

Buraimoh OM, Ilori MO, Amund OO, Michel FC, Grewal SK (2015) Assessment of bacterial degradation of lignocellulosic residues (sawdust) in a tropical estuarine microcosm using improvised floating raft equipment. Int Biodeterior Biodegrad 104:186–193. https://doi.org/10.1016/j.ibiod.2015.06.010

Carrard G, Koivula A, Soderlund H, Beguin P (2000) Cellulose-binding domains promote hydrolysis of different sites on crystalline cellulose. Proc Natl Acad Sci 97:10342–10347

Chami KA, Robert S (2013) Self-assembly and intermolecular forces when cellulose and water interact using molecular modeling. J Nanomater 2013:1–12

Chukwuma OB, Rafatullah M, Tajarudin HA, Ismail N (2020) Lignocellulolytic enzymes in biotechnological and industrial processes: a review. Sustain For 12:7282

Dalby PA (2007) Engineering enzymes for biocatalysis. Recent Pat Biotechnol 1:1–9

Dar MA, Pawar KD, Rajput BP, Rahi P, Pandit RS (2019) Purification of a cellulase from cellulolytic gut bacterium, *Bacillus tequilensis* G9 and its evaluation for valorization of agro-wastes into added value byproducts. Biocatal Agric Biotechnol 20:101219

Dasgupta A, Chowdhury N, De RK (2020) Metabolic pathway engineering: perspectives and applications. Comput Methods Prog Biomed 192:105436

Dulermo T, Coze F, Virolle MJ, Méchin V, Baumberger S, Froissard M (2016) Bioconversion of agricultural lignocellulosic residues into branched-chain fatty acids using Streptomyces lividans. OCL 23:A202. https://doi.org/10.1051/ocl/2015052

Ellila S, Fonseca L, Uchima C (2017) Development of a low-cost cellulase production process using *Trichoderma reesei* for Brazilian biorefineries. Biotechnol Biofuels 10:1–17

Fenno JC, LeBlanc DJ, Fives-Taylor P (1989) Nucleotide sequence analysis of a type I fimbrial gene of *Streptococcus sanguis* FW213. Infec Immun 57:3527–3533

Froese AG, Nguyen TN, Ayele BT, Sparling R (2020) Digestibility of wheat and cattail biomass using a co-culture of thermophilic anaerobes for consolidated bioprocessing. Bioenergy Res 13:325–333. https://doi.org/10.1007/s12155-020-10103-0Gautam R, Sharma J (2012) Optimization, purification of cellulase produced from Bacillus subtilis subsp. inaquosorum under solid state fermentation and its potential applications in denim industry. Int J Sci Res 3:1759–1763

Goswami K, DekaBoruah HP, Saikia R (2021) Production of cellulase by Novosphingobium sp. Cm1 and its potential application in lignocellulosic waste hydrolysis. Prep Biochem Biotechnol 2021:1–12

Goswami K, Deka Boruah HP, Saikia R (2022) Purification and characterization of cellulase enzyme produced by Novosphingobium sp. Cm1 and its waste hydrolysis efficiency and biostoning potential. J Appl Microbiol 132:1–11

Guangtao Z, Hartl L, Schuster A, Polak S, Schmoll M, Wang T, Seidl V, Seiboth B (2009) Gene targeting in a nonhomologous end joining deficient *Hypocrea jecorina*. J Biotechnol 139(2):146–151

Guo H, Hong C, Zheng B, Lu F, Jiang D, Qin W (2017) Bioflocculants' production in a biomass-degrading bacterium using untreated corn stover as carbon source and use of bioflocculants for microalgae harvest. Biotechnol Biofuels 10:306. https://doi.org/10.1186/s13068-017-0987-6

Gupta R, Meheta G, Deswal D, Sharma S, Jain KK, Singh A, Kuhad RC (2013) In: Kuhad RC, Singh A (eds) Biotechnology for environmental management and resource recovery, pp 89–106

Iqbal HMN, Ahmed I, Zia MA, Irfan M (2011) Purification and characterization of the kinetic parameters of cellulase produced from wheat straw by *Trichoderma viride* under SSF and its detergent compatibility. Adv Biosci Biotechnol 2:149–156

Iqbal HMN, Kyazze G, Keshavarz T (2013) Advances in valorization of lignocellulosic materials by bio-technology: an overview. Bio Resour 8:3157–3176

Joy S, Rahman PKSM, Khare SK, Sharma S (2019) Production and characterization of glycolipid biosurfactant from Achromobacter sp. (PS1) isolate using one-factor-at-a-time (OFAT) approach with feasible utilization of ammonia-soaked lignocellulosic pretreated residues. Bioprocess Biosyst Eng 42:1301–1315. https://doi.org/10.1007/s00449-019-02128-3

Kanokratana P, Wongwilaiwalin S, Mhuantong W, Tangphatsornruang S, Eurwilaichitr L, Champreda V (2018) Characterization of cellulolytic microbial consortium enriched on Napier grass using metagenomics approaches. J Biosci Bioeng 125:439–447. https://doi.org/10.1016/j.jbiosc.2017.10.014

Kashyap DR, Vohra PK, Chopra S, Tewari R (2001) Applications of pectinases in the commercial sector: a review. Bioresour Technol 77:215–227

Kim YK, Lee SC, Cho YY, Oh HJ, Ko YH (2012) Isolation of cellulolytic Bacillus subtilis strains from agricultural environments. Int Sch Res Notices 2012:1–9

Kong X, Du J, Ye X, Xi Y, Jin H, Zhang M, Guo D (2018) Enhanced methane production from wheat straw with the assistance of lignocellulolytic microbial consortium TC-5. Bioresour Technol 263:33–39. https://doi.org/10.1016/j.biortech.2018.04.079

Kuhad RC, Gupta R, Singh A (2011) Microbial cellulases and their industrial applications. Enzyme Res 2011:280696

Kumar P, Barrett DM, Delwiche MJ, Stroeve P (2009) Methods for pretreatment of lignocellulosic biomass for efficient hydrolysis and biofuel production. IndEng Chem Res 48:3713–3729

Lee S, Kang M, Bae JH, Sohn JH, Sung BH (2019) Bacterial valorization of lignin: strains, enzymes, conversion pathways, biosensors, and perspectives. Front Bioeng Biotechnol 7:209

Legodi LM, Grange LD, Rensburg VJEL, Ncube I (2019) Isolation of cellulose degrading fungi from decaying Banana Pseudostem and Strelitzia Alba. Enzyme Res 19:1–10

Liu G, Zhang L, Qin Y, Zou G, Li Z, Yan X, Wei X, Chen M, Chen L, Zheng K, Zhang J, Ma L, Li J, Liu R, Xu H, Bao X, Fang X, Wang L, Zhong Y, Lio W, Zheng H, Wang S, Wang C, Xun C, Zhao GP (2013) Long-term strain improvements accumulate mutations in regulatory elements responsible for hyper-production of cellulolytic enzymes. Sci Rep 3:1569

Malherbe S, Cloete TE (2002) Lignocellulose biodegradation: fundamentals and applications. Rev Environ Sci Biotechnol 1:105–114

Medouni-Haroune L, Zaidi F, Medouni-Adrar S, Roussos S, Azzouz S, Desseaux V, Kecha M (2017) Selective isolation and screening of actinobacteria strains producing lignocellulolytic enzymes using olive pomace as substrate. Iran J Biotechnol 15:74–77. https://doi.org/10.15171/ijb.1278

Miron J, Ben-Ghedalia D, Morrison M (2001) Invited review: adhesion mechanisms of rumen cellulolytic bacteria. J Dairy Sci 84:1294–1309

Mood SH, Golfeshan AH, Tabatabaei M, Jouzani GS, Najafi G, Gholami M, Ardjmand M (2013) Lignocellulosic biomass to bioethanol, a comprehensive review with a focus on pretreatment. Renew Sust Energ Rev 27:77–93

Mrudula S, Murugammal R (2011) Production of cellulase by Aspergillus Niger under submerged and solid-state fermentation using coir waste as a substrate. Braz J Microbiol 42:1119–1127

Neoh CH, Lam CY, Ghani SM, Ware I, Sarip SHM, Ibrahim Z (2016) Bioremediation of high-strength agricultural wastewater using Ochrobactrum sp strain SZ1. 3 Biotech 6:143. https://doi.org/10.1007/s13205-016-0455-1

Pell AN, Schofield P (1993) Microbial adhesion and degradation of plant cell walls. In: Hatfield RD, Jung HG, Ralph J, Buxton DR, Mertens DR, Weimer PJ (eds) Forage cell wall structure and digestibility. ASA-CSSA-SSSA, Madison, pp 397–423

Quinlan RJ, Teter S, Xu F (2010) Development of cellulases to improve enzymatic hydrolysis of lignocellulosic biomass. In: Bioalcohol production. Woodhead Publishing, pp 178–201

Rathore P (2014) Cellulase production by liquid state fermentation of *Bacillus* species isolated from woody forest soil. Int J Res Stud Biosci 2:8–13

Reyes-Ortiz V, Heins RA, Cheng G, Kim EY, Vernon BC, Elandt RB, Adams PD, Sale KL, Hadi MZ, Simmons BA, Kent MS (2013) Addition of a carbohydrate-binding module enhances cellulase penetration into cellulose substrates. Biotechnol Biofuels 6:1–13

Rongpipi S, Ye D, Gomez ED, Gomez EW (2019) Progress and opportunities in the characterization of cellulose–an important regulator of cell wall growth and mechanics. Front Plant Sci 9:1894

Sadhu S, Maiti TK (2013) Cellulase production by bacteria: a review. Br Microbiol Res J 3:235–258

Saranraj P, Stella D, Reetha D (2012) Microbial cellulases and its applictions. Int J Biochem Biotech Sci 1:1–12

Shen X, Sun R (2021) Recent advances in lignocellulose prior-fractionation for biomaterials, biochemicals, and bioenergy. Carbohydr Polym 261:117884

Singh A, Patel AK, Adsul M, Mathur A, Singhania RR (2017) Genetic modification: a tool for enhancing cellulase secretion. Biofuel Res J 4:600–610

Solikhah F, Assavalapsakul W, Zulaika E (2018) Peat endogenous lignocellulolytic bacteria for humic waste decomposition. J Phys Conf Ser 1108:030004

Steiger MG, Vitikainen M, Uskonen P, Brunner K, Adam G, Pakula T, Penttilä M, Saloheimo M, Mach RL, Mach-Aigner AR (2011) Transformation system for *Hypocrea jecorina* (*Trichoderma reesei*) that favors homologous integration and employs reusable bidirectionally selectable markers. Appl Environ Microbiol 77:114–121

Straub CT, Khatibi PA, Otten JK, Adams MWW, Kelly RM (2019) Lignocellulose solubilization and conversion by extremely thermophilic Caldicellulosiruptor bescii improves by maintaining metabolic activity. Biotechnol Bioeng 116:1901–1908. https://doi.org/10.1002/bit.26993

Thiyageshwari S, Gayathri P, Krishnamoorthy R, Anandham R, Paul D (2018) Exploration of rice husk compost as an alternate organic manure to enhance the productivity of blackgram in typic haplustalf and typic rhodustalf. Int J Environ Res Public Health 15:358. https://doi.org/10.3390/ijerph15020358

Tomme P, Warren RAG, Gilkes NR (1995) Cellulose hydrolysis by bacteria and fungi. Adv Microb Physiol 37:1–81

Wang T, Lu X (2021) Chapter 8- Overcome saccharification barrier: advances in hydrolysis technology. In: Advances in 2nd generation of bioethanol production. Woodhead Publishing Series in Energy, pp 137–159

Wang K, Tester JW (2023) Sustainable management of unavoidable biomass wastes. Green Energy Resour 1:100005

Xu J, Gao Z, Wu B, He B (2017) Lactose-induced production of a complete lignocellulolytic enzyme system by a novel bacterium Bacillus sp. BS-5 and its application for saccharification of alkali-pretreated corn cob. Cellulose 24:2059–2070. https://doi.org/10.1007/s10570-017-1247-4

Zhang XZ, Zhang YHP (2013) Cellulases: characteristics, sources, production, and applications. Bioprocessing technologies in biorefinery for sustainable production of fuels, chemicals, and polymers. In: Yang S-T, El-Enshasy HA, Thongchul N (eds) , 1st edn. Wiley

Zhao CH, Liu X, Zhan T, He J (2018) Production of cellulase by *Trichoderma reesei* from pretreated straw and furfural residues. RSC Adv 8:36233–36238

Chapter 20
Fungal Oxidoreductases: Efficient Tool for Eco-Friendly and Sustainable Pesticide Bioremediation

Tina Roy, Chandana Paul, Nilasish Pal, and Nirmalendu Das

Abstract Nowadays, it is hard to imagine life without pesticides due to their high efficiency in agricultural practices to get rid of unwanted crop loss and and increase production to feed a population exceeding 8.1 billion. Pesticide formulations have different xenobiotic compounds which are hazardous not only for the present living beings but also for our future generations. The current conventional remediation methods are not only pricey but also sluggish, as well as produce some toxic intermediates. Microbes-assisted bioremediation involves environment-friendly manner of contaminant removal methods, which would be cost-effective and sustainable. Generic diversity, varied metabolic/physiochemical pathways, and adaptation strategies in various ecosystems are inherent to fungal communities and build potential to inhabit on diverse substrate range. Oxidoreductases are the pivotal catalytic weapon of fungal communities that help them to exist in the polluted ecological niche by degrading different xenobiotic compounds such as pesticides. Fungi, specially the filamentous fungi, viz., *Trichoderma*, *Phanerochaete*, *Pleurotus*, *Trametes*, etc., produce a number of oxidoreductases like tyrosinases, laccases, peroxidases, cyrochrome P450 monooxygeneses, etc. These enzymes can efficiently degrade pesticides of different chemical classes like carbamate, pyrethroid, organochlorine, organophosphate, etc. in a sustainable manner. In this chapter, we focus on pesticide mycoremediation assisted by different oxidoreductases, portray the detailed chemical conversion models of representative pesticides of different chemical classes, and

T. Roy
Plant-Microbe Interaction and Molecular Biology Laboratory, Division of Biotechnology, CSIR-Institute of Himalayan Bioresource Technology, Palampur, Himachal Pradesh, India

C. Paul
Department of Microbiology, St. Xavier's College, Kolkata, West Bengal, India

Department of Botany, Barasat Government College, Barasat, Kolkata, West Bengal, India

N. Pal
Department of Chemistry, Seth Anandram Jaipuria College, Kolkata, West Bengal, India

N. Das (✉)
Department of Botany, Barasat Government College, Barasat, Kolkata, West Bengal, India

S. Dey, S. Bhattacharya (eds.), *Biotechnological Interventions in the Removal of Emerging Pollutants*, Interdisciplinary Biotechnological Advances,
https://doi.org/10.1007/978-981-97-9922-0_20

also highlight modern biotechnological intercession such as immobilization, enzyme engineering techniques, omics-based approaches, etc. for removal of these emergent pollutants.

Keywords Pesticides · Oxidoreductase · Laccase · Peroxidase · Tyrosinase · Bioremediation

20.1 Introduction

Anthropogenic and natural sources are the main causes of soil pollution, which is a global issue. Since the usage of chemicals, substances, and chemical agents has become necessary due to urbanization, industrialization, and rising food consumption, toxins have been dispersed throughout the environment and accumulated over time (Fig. 20.1). These chemicals have detrimental impacts on both people and other living things. Specifically, pesticides are dangerous chemicals found in soil and water environments that can enter the human body and lead to a number of illnesses. Alkyl halide pesticides are commonly used; examples include dieldrin, lindane, heptachlor, aldrin, and mirex. Because these pesticides accumulate and are harmful to people, animals, and soil, their use has been curtailed or stopped in the United States, and many other nations. Chlorpyrifos is a typical organophosphorus

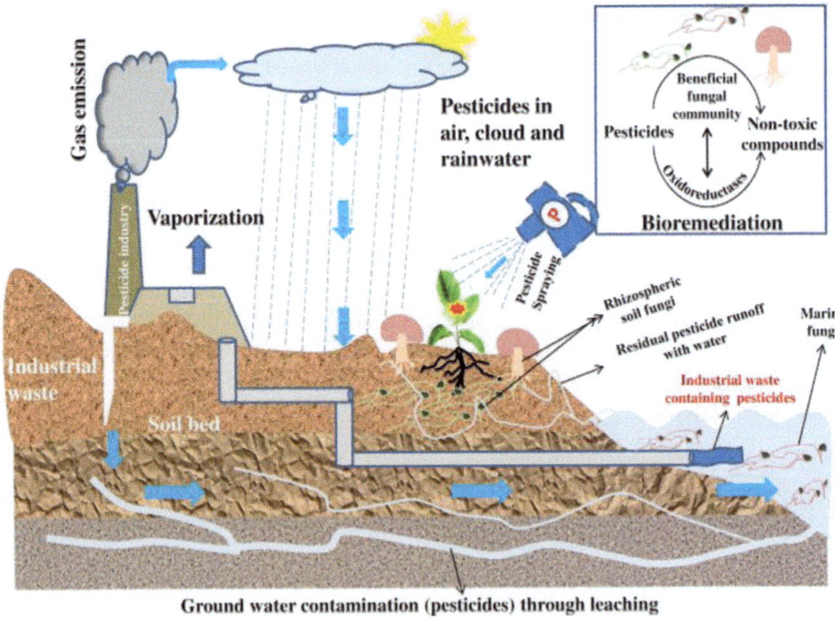

Fig. 20.1 Schematic representation of pesticide contamination and bioremediation through fungus-derived oxidoreductases

pesticide used on maize, peppers and tomatoes to control pests that live in the soil (Nteping et al. 2022). Additionally, chewing insects and sucking pests are managed with malathion organophosphate insecticide, which also deters parasites and mosquitoes. The lungs, skin, and digestive system can all be harmed by the careless use of chemicals. It has an impact on fish liver and blood, vertebrates' immune systems, and invertebrates' central nervous systems. When human blood, lymphocytes, and white blood cells are grown, mutations have been discovered (Parra et al. 2015). When the maximum residue limit (MRL) for the pyrethroid-cyfluthrin insecticide is exceeded, it has detrimental consequences on both human health and the environment. This pesticide leaves poisonous residues on commonly ingested raw vegetables. For this reasons remediation methods that work must be employed to lower the concentration of residual pesticides in the soil. Bioremediation is an inexpensive, environmentally friendly, and highly effective process that can be used in place of more costly and hazardous techniques including chemical and physical procedures. The elimination process in biodegradation can be accomplished through the utilization of microorganism's microbiological activity (Fig. 20.1). Pesticides are broken down by microorganisms into less complex chemicals such as CO_2, water, oxides, or mineral salts that can be used as a source of energy, carbon, or minerals. Because they function as catalysts in various reactions, enzymes are crucial (Raffa and Chiampo 2021).

Due of their extensive potential for bioremediation applications, fungi are the organisms that garnering more and more attention. In the tree of life, fungi are a varied kingdom that are part of the eukaryotic realm. They are chemoheterotrophic organisms that are found in both aquatic and terrestrial settings. As the most advanced decomposers of complex organic matter, fungi are known to break down wood, lignin, cellulose, and other plant-based materials that are frequently used as agricultural waste. Fungi have a broad range of compound classes, including inorganic and organic contaminants, that they can attack and metabolize because of their intra- and extracellular enzymatic machinery and acid excretion capacity. Low substrate specificity is a common characteristic of fungal enzymes. Contrary to popular belief, this is advantageous since nonspecific enzymes catalyze a variety of processes, allowing fungus to utilize a wide range of substances as sources of carbon and energy (Harms et al. 2011). Fungal biodegradation abilities are nearly always associated with oxidative enzymatic activities that are facilitated by a variety of oxidases and peroxidases (Hofrichter 2002). As a result, for fungus to thrive, anoxic environment is usually necessary by releasing enzymes specifically laccases, which utilize oxygen, and peroxidases (manganese peroxidase, lignin peroxidase), which use H_2O_2 as terminal electron acceptor are called oxidoreductases. Other than these numerous oxidation events occur inside of cells and are catalyzed by cytochromes. Fungi can serve as effective bioremediation agents in combating hazardous metal contamination in the environment. Fungi are therefore ideal options for industrial bioremediation projects and biotechnological applications. One of the most well-known processes that degrade extracellular pollutants is oxidation, wherein hydroxyl radicals cleave double bonds in aliphatic or cyclic structures. Following the action of laccases or peroxide-dependent hydroxylations, quinone cycling produces the hydroxyl radicals. These reactions, which are exergonic in nature, help break down complicated pollutant structures into more easily degradable metabolite intermediates. Pesticides, such as pyrethroids, carbamates, organochlorines, endosulfan, DDT (Purnomo et al. 2010), atrazine (Bastos and Magan 2009), dieldrin, aldrin (Xiao et al. 2011), and others have been reported to be degraded by mushrooms. Some fungi have even been proposed as potential degraders of fungicides, including fluxapyroxad (Podbielska et al. 2020), a broad-spectrum pyrazole-carboxamide fungicide that is frequently used to treat fungal diseases in apple trees. The word "pesticide" refers to a broader variety of chemical substances; the breakdown of these is not limited to any particular fungal taxon. Various fungal groups, including filamentous fungi, brown rot fungi, white rot fungi (WRF), and yeasts, are known to degrade pesticides. This chapter aims at highlighting the potential of fungi in bioremediation processes to combat pesticide pollution with special references to their oxidoreductase enzymes.

20.2 Overview of Oxidoreductases

Oxidoreductases play a pivotal role in facilitating electron exchange between donor and acceptor molecules during reactions involving electron transfer, proton/hydrogen extraction, hydride transfer, oxygen insertion, and various other essential steps. Demonstrating remarkable ubiquity in nature, these enzymes exhibit a broad spectrum of catalytic activities with notable specificity, efficiency, and selectivity (Younus 2019). The catalytic mechanism of oxidoreductases involves two half-reactions, one oxidative and one reductive, with a minimum of two substrates, one reducing and one oxidizing, undergoing activation or transformation. These enzymes cooperate a crucial role in facilitating electron exchange between donor and acceptor molecules, engaging in reactions encompassing electron transfer, proton/hydrogen extraction, hydride transfer, oxygen insertion, and other pivotal steps. This comprehensive enzymatic activity underscores their significance in various biological processes across different organisms. The majority of oxidoreductases are nicotinamide cofactor-dependent, displaying a pronounced affinity for NAD or NADP. This enzyme class is further categorized into six major classes: oxidases, dehydrogenases, hydroxylases, oxygenases, peroxidases, and reductases. Each class possesses distinct redox-active centers that underpin their functional diversity (Younus 2019). Oxidoreductases play pivotal roles in both aerobic and anaerobic metabolic processes, showcasing versatility with a broad spectrum of substrates encompassing both organic (e.g., alcohols, amines, and ketones) and inorganic (e.g., small anions like sulfite and metals such as mercury) compounds. These enzymes exhibit diverse redox-active centers crucial for their physiological functions. The inherent liability of these centers is shielded by the polypeptide backbone of oxidoreductases, emphasizing the protective role played by the enzyme structure (Younus 2019). Redox reactions are abundant in living cells and are integral to vital biological processes such as tricarboxylic acid cycle, glycolysis, amino acid metabolism, and oxidative phosphorylation. As such, oxidoreductases are considered housekeeping enzymes crucial for the life and functioning of cells. This group of enzymes distributed across various life forms, including microbes, plants, and animals, are systematically classified under EC 1 in the enzyme commission (EC) number system. This classification is noteworthy, as oxidoreductases constitute approximately one-third of the enzymatic activities documented in the BRaunschweig Enzyme Database (BRENDA) (Sellés Vidal et al. 2018).

20.2.1 Fungal Oxidoreductases

Fungal oxidoreductases have been identified to play a significant role in both pathogenicity and defense against host mechanisms (Yu et al. 2020). Moreover, activity of several oxidoreductases, including, laccase, Cytochrome P450, polyphenol oxidases, peroxidase, and dehydrogenase, has been associated with pesticide degradation.

20.2.1.1 Cytochrome P450 Enzymes

The involvement of fungal Cytochrome P450 monooxygenases (EC 1.14.14.1) extends across numerous vital cellular processes, demonstrating their diverse functional roles. These enzymes facilitate the transformation of hydrophobic intermediates within both primary and secondary metabolic pathways. Furthermore, they contribute to the detoxification of natural and environmental pollutants, enabling fungal adaptation to various environmental conditions. The completion of fungal genome sequencing projects has significantly advanced our understanding, leading to the annotation of numerous novel Cytochrome P450 enzymes. Among these, a substantial proportion represents previously unrecognized families, thereby expanding the known diversity of these enzymes. Cytochrome P450 enzymes (P450s) are heme-thiolate proteins ubiquitously present across diverse life forms, spanning prokaryotes (archaea, bacteria), lower eukaryotes (fungi and insects), and higher eukaryotes (plants and animals, including humans). These enzymes typically facilitate regio- and stereospecific conversions of a broad spectrum of lipophilic compounds, converting them into more hydrophilic derivatives by incorporating an oxygen atom derived from molecular oxygen. The collective involvement of P450s has been established in the metabolism of aliphatic, alicyclic, and aromatic molecules, resulting in various reactions such as hydroxylation, epoxidation, dealkylation, sulfoxydation, deamination, desulfuration, dehalogenation, and N-oxide reduction (Sono et al. 1996).The majority of P450s engage in catalytic reactions with one or more protein components, serving as electron transfer mediators from NADH or NADPH to the P450s. However, some P450s do not necessitate additional protein components, achieving the reductive activation of oxygen independently. The classification of P450 monooxygenase systems is based on the proteins involved in electron transfer, containing heme as a cofactor (Hannemann et al. 2007). CYP51 and CYP61 represent fungal P450 families that are both structurally and functionally conserved, playing crucial roles in membrane ergosterol biosynthesis. While the CYP51 family is ubiquitously found across all kingdoms, the CYP61 proteins are exclusive to fungi, though functional homologues (i.e., CYP710 family) have been identified in higher plants. The initial discovery and functional characterization of the first CYP51 enzyme were conducted in *Saccharomyces cerevisiae*. This enzyme catalyzes the 14α–demethylation of lanosterol in yeast and eburicol in filamentous fungi. This three-step reaction results in the removal of the methyl group as formic acid, producing a $\Delta^{14,15}$ desaturated intermediate in the ergosterol pathway. Notably, specific inhibitors targeting this monooxygenase, such as ketoconazole and fluconazole, hold significant commercial importance as antifungal compounds. These inhibitors selectively target fungal CYP51 enzymes without affecting their mammalian and plant counterparts. Homology modeling has facilitated the construction of three-dimensional models for CYP51 in *Candida albicans* and *Aspergillus fumigatus*. These models have been employed to investigate the binding modes of both natural and antifungal substances. Utilizing the model of the *C. albicans* enzyme, novel fungal-specific inhibitors were designed and synthesized, demonstrating the potential for targeted antifungal drug development (Črešnar and Petrič 2011).

Phenylurea herbicides (PUHs), with aromatic and amine groups in structure, have been widely used for crop weed control in agricultural regions over the last few decades (Dan et al. 2021).

PUHs are frequently detected in surface water (Morawski et al. 2020), groundwater (Dan et al. 2021), soil and sediment (Barbieri et al. 2021) due to their long half-life which can reach several months or even years (Lai et al. 2022). According to previous studies of degradation mechanism of PUHs (Fenoll et al. 2013), the enzymatic transformation processes of PUHs are proposed as follows (Scheme 20.1). The two successive N-demethylations occurs by the action of cytochrome P450 enzymes and then the C-N bond cleaved by hydrolases. Two oxygen added on aromatic ring by aniline dioxygenase and the aromatic ring further cleaved to form open chain dicarboxylic acids by catechol 1,2-dioxygenase. Finally, this dicarboxylic acid transformed to CO_2 and biomass via tricarboxylic acid cycle.

20.2.1.2 Polyphenol Oxidases

Tyrosinase (EC 1.14. 18.1), a member of polyphenol oxidases contains copper that facilitate the o-hydroxylation of monophenols to o-diphenols. The other member simultaneously engage in the oxidation of o-diphenols to quinones, showcasing catecholase (EC1.10.3.2) activity in the presence of oxygen. As fungal tyrosinase has been comprehensively reviewed recently (Halaouili et al. 2006), this discussion refrains from delving into its distribution, a feature seemingly ubiquitous across fungal species. It is pertinent, however, to note the isolation of the latent form of polyphenol oxidase (PPO) in the ascocarp of *Terfezia claveryl*, specifically in the

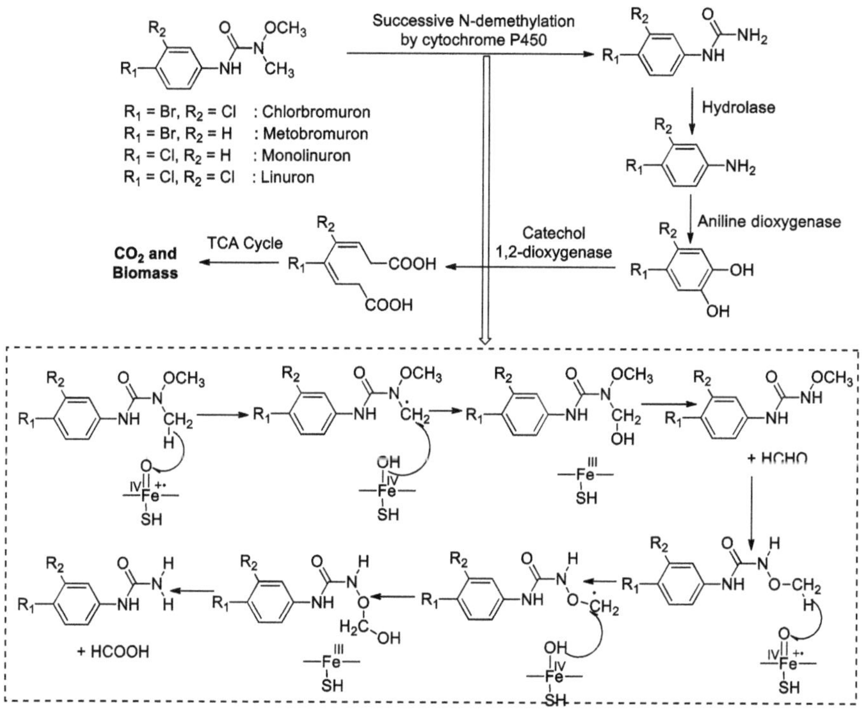

Scheme 20.1 Degradation of phenylurea herbicides by cytochrome P450 enzymes

context of a truffle (Perez-Gilabert et al. 2001). The general behavior of this PPO aligns with that observed in other fungal PPOs. The subcellular localization of fungal polyphenol oxidase (PPO) presents some ambiguity. Conversion of phenol to ortho-quinone via catechol requires the participation of two polyphenol oxidases-tyrosinase and catechol oxidase is shown in Scheme 20.2. Typically, it seems to function as a cytoplasmic enzyme. Notably, a PPO derived from *Pycnoporus*, overexpressed in *Aspergillus niger*, demonstrated the ability to be directed to the extracellular growth medium (Halaouili et al. 2006). Another variant of PPO was identified in the mycelium of *Pycnoporus saguineus*, exhibiting notably high tyrosinase activity and the capacity to convert coumaric acid to caffeic acid in vitro (Halaouili et al. 2005). Unlike other PPOs from Pycnoporus, this particular enzyme manifested as four isoforms with a molecular weight of 45 kDa and lacked N-glycosylation. In certain instances, fungal PPO appears to be associated with the cell wall, seemingly within the extracellular matrix (Rast et al. 2003). The molecular weights of fungal PPOs display considerable diversity, influenced partly by genetic factors, as evidenced by phylogenetic tree clustering (Halaouili et al. 2006), and partly by potential artifacts arising during enzyme isolation, a matter discussed by Halaouili et al. (2006). Fungal PPO, similar to plant PPO, can exist in a latent form that can be activated by SDS, proteolysis, or acid shock. SDS activation in beet root PPO is suggested to be reversible, while trypsin activation is not (Perez-Gilabert

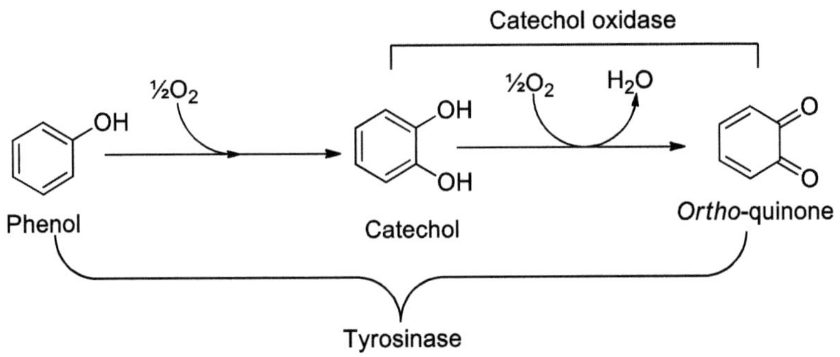

Scheme 20.2 Conversion of phenol to ortho-quinone via catechol requires the participation of two polyphenol oxidases-tyrosinase and catechol oxidase

et al. 2004). Gandia-Herrero et al. (2005) propose a common peptide involvement in activation by both SDS and trypsin, although the evidence is not conclusive. Older research on sugar beet chloroplast PPO indicated that even peptides with a molecular weight of around 10 kDa retained significant PPO activity, highlighting that the entire protein may not be necessary for activity (Mayer et al. 1966). Insights into the activation of latent PPO come from the study of Kanade et al. (2006) on PPO from *Dolichos lablab*, indicating that both acid and SDS alter the environment of a single glutamic residue near the di-copper active site. This unblocks or opens the active site, enhancing enzyme activity. The authors suggest, though unproven, that localized acidification caused by wounding or methyl jasmonate could lead to the conversion of latent to active enzyme. Two tyrosinases from *Agaricus* contain putative glycosylation and phosphorylation sites, although no glycosylation or phosphorylation has been reported for fungal PPO so far (Wichers et al. 2003). Processing by proteolysis is evident in fungal PPOs, where the synthesized form of the enzyme undergoes trimming. However, the exact events during the conversion of latent to active forms of fungal PPOs remain unclear. In vivo, tyrosinases from crustaceans form hexamers, consisting of a single subunit with a molecular weight of 71 kDa (Jaenicke and Decker 2003). This structure recalls older reports of plant and fungal PPO associations into aggregates, yet the hexameric form has not been demonstrated in vivo for any plant or fungal PPO. The structural similarities of this hexameric form with hemocyanin are apparent (Gerdemann et al. 2002). Nonetheless, aggregation into a hexameric form existing in vivo has not been confirmed for any plant or fungal PPO.

Though laccase belongs to polyphenol oxidase group of enzymes but due to their large-scale involvement in pesticide degradation by filamentous fungi, it is separately discussed.

20.2.1.3 Laccases (EC 1.10. 3.2)

Laccases, enzymes widely distributed in plants, fungi, bacteria, and insects, belong to the multicopper oxidase family and facilitate the conversion of aromatic and non-aromatic compounds while reducing molecular oxygen to water utilizing cupper as cofactor. Their involvement spans various processes, including biosynthesis, lignin degradation, morphogenesis, and pigment biosynthesis. This review delves into key aspects of fungal laccases, exploring the existence of gene families, the escalating interest in understanding molecular regulatory mechanisms, and factors influencing laccase production, given their promising biotechnological applications. Furthermore, we conduct a comparative analysis of the structural features of laccases, highlighting both similarities and differences. Emphasis is placed on the general attributes of the active site, revealing laccases' tendency to function as monomeric proteins with highly conserved cupredoxine-type domains. Despite identified inter- and intraspecific differences, the structural distinctions among fungal laccases remain elusive, primarily relying on Crystallography and X-ray diffraction techniques. Initially, the enzymatic activity attributed solely to higher plants and fungi has expanded beyond this narrow scope. Laccases, once considered exclusive to certain organisms, are now recognized as nearly ubiquitous enzymes. They have been identified in plants, various fungal groups (Ascomycetes, Basidiomycetes, and Deuteromycetes), prokaryotes, and arthropods (Giardina et al. 2010). In insects, laccases play a crucial role in sclerotization during epidermal cuticle synthesis. In bacteria, laccases exhibit diverse functions, such as involvement in morphogenesis processes, copper homeostasis, pigment biosynthesis (including melanin and brown spore pigment), and protection of spores against UV light and hydrogen peroxide (Strong and Claus 2011). Fungal laccases contribute to various processes, including sporulation, pigment production, fruit body formation, and plant pathogenesis (Alcalde 2007). Particularly noteworthy are the white rot Basidiomycetes, recognized for their efficient decomposition of lignin, cellulose, and hemicellulose, transforming them into carbon dioxide (Baldrian 2006). Consequently, basidiomycetes have become the focus of extensive study. Over 100 laccases from basidiomycetes and ascomycetes fungi have been purified and characterized, highlighting their significance in diverse biological processes. However, the purification of laccases from plant crude extracts has been relatively underexplored due to its complexity (Strong and Claus 2011). Studies on the characterization of fungal laccases align with the observed gene multiplicity and the production of numerous isoenzymes across different species. Notable instances of fungal laccase gene families include the identification of two distinct laccases (encoded by lcc1 and lcc2 genes) in *Trametes villosa* (Yaver et al. 1996), five (lcc1-5) in *Trametes sanguinea* (Hoshida et al. 2001), four (lcc1-4) in *Rhizoctonia solani* (Wahleithner et al. 1996), three (lcc1-3) in *Trametes* sp. I62 (Mansur et al. 1997), three (lccA-C) in *Trametes* sp. AH28-2 (Xiao et al. 2006), and two (LAC1-2) in *Gaeumannomyces graminis* (Litvintseva and Henson 2002). Additionally, laccase gene families have been elucidated in *Pleurotus* species, with four genes (lac1-4) identified in *Pleurotus sajor-caju* (Soden et al. 2002), two (pel3-4) in *Pleurotus eryngii* (Rodríguez et al.

2008), and six (poxa1a, poxa1b, poxC, poxa2, poxa3a, and poxa3b) in *Pleurotus ostreatus* (Palmieri et al. 2003). These findings underscore the diversity and complexity of laccase gene families within the fungal kingdom. Laccases (EC 1.10.3.2), also known as p-diphenol: dioxide oxidoreductases, belong to the category of blue multicopper oxidases (MCOs). They exhibit the ability to catalyze the oxidation of a wide array of organic aromatic compounds, simultaneously reducing molecular oxygen to water (Ruiz-Duenas and Martínez 2009). While most laccase substrates are phenolic compounds, including ortho and para-diphenols, methoxysubstituted phenols, polyphenols, aromatic amines, benzenethiols, hydroxindols, 1-naphthol, and syringaldazine, the enzyme's activity can be extended to nonphenolic compounds through the use of mediators such as ABTS (2,2'-azino-bis(3-ethylbenzothiazoline-6-sulfonic acid)) and HBT (1-hydroxybenzotriazole) (Solís-Oba et al. 2008). Very recently, we have through review the degradation of pesticides by laccase (Paul et al. 2024). In Scheme 20.3, detoxification of isoproturon herbicide was elucidated by laccase-mediator system according to Zeng et al. (2017).

20.2.1.4 Peroxidases

Peroxidases represent a class of oxidoreductases that employ hydrogen peroxide to facilitate oxidative reactions, feature a heme cofactor within their active sites, or alternatively, possess redox-active cysteine or selenocysteine residues.. Numerous peroxidases have been identified within fungal species, and ongoing research is dedicated to characterizing them at the molecular level. The breakdown of lignin by fungi serves as a crucial pathway in the recycling of carbon fixed through photosynthesis. The oxidative mechanisms utilized in this process hold promise for various biotechnological applications. Lignin peroxidases (LiPs) (EC. 1.11.1.14), manganese peroxidases (MnPs) (EC.1.11.1.13), and similar enzymes found in white rot

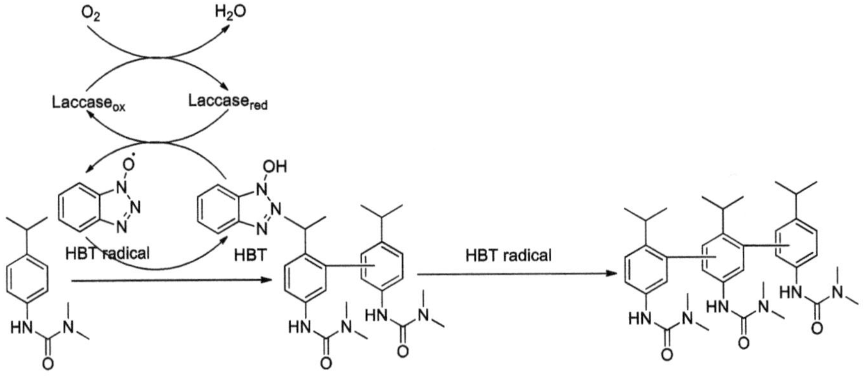

Scheme 20.3 Detoxification of isoproturon herbicide by laccase-mediator system (Zeng et al. 2017)

basidiomycetes are probable participants in the fungal degradation of lignin (Conesa et al. 2002). Lignin peroxidases (LiPs) were initially identified in the extracellular medium of nitrogen-limited Phanerochaete chrysosporium cultures. These enzymes are monomeric hemoproteins with molecular masses of approximately 40 kDa. Structurally, they share similarities with classical peroxidases, such as the horseradish enzyme, wherein their Fe (III) is pentacoordinated to the four heme tetrapyrrole nitrogens and a histidine residue. Similar to classical peroxidases, LiPs undergo oxidation by H_2O_2, leading to the formation of a two-electron-oxidized intermediate. In this intermediate, the iron is in the Fe (IV) state, and a free radical is present on the tetrapyrrole ring or in proximity to an adjacent amino acid (Kirk and Farrell 1987). Following the formation of first intermediate, the subsequent one-electron oxidation of a donor substrate takes place, resulting in the generation of a substrate-free radical and the transformation into second intermediate. In second intermediate, the iron remains in the Fe (IV) state, but the tetrapyrrole no longer bears a radical. Second intermediate then initiates the oxidation of a second donor substrate, producing another substrate-free radical and returning the peroxidase to its resting state. A notable functional distinction between LiPs and classical peroxidases lies in their ability to oxidize aromatic rings. LiPs exhibit the capacity to oxidize aromatic rings that are only moderately activated by electron-donating substituents. In contrast, classical peroxidases exclusively act on strongly activated aromatic substrates. For instance, both LiPs and horseradish peroxidase can oxidize 1,2,4,5-tetramethoxybenzene, as well as numerous phenols and anilines. However, LiPs stand out by their ability to abstract an electron from aromatics with only two or three ether substituents, resembling the predominant nonphenolic structures of lignin. These oxidations yield transient cation radical intermediates, which undergo spontaneous fragmentation. The majority of scissions occur between Ca and Cb of the propyl sidechain, leading to the formation of benzaldehydes. These benzaldehydes are plausible precursors to the benzoic acid residues observed in lignin decayed by white rot fungi (Kersten et al. 1990).

The notion of lignin peroxidases (LiPs) playing a central role in lignin degradation encounters a persistent challenge, as many white rot fungi seem to lack these enzymes. An alternative group of secreted enzymes, manganese peroxidases (MnPs), has gained attention and extensive research as a potential substitute. Unlike LiPs, MnPs are more prevalent and exhibit robust oxidizing properties, undergoing a classical peroxidase cycle. However, they do not directly oxidize nonphenolic lignin-related structures due to the absence of the invariant tryptophan residue necessary for electron transfer to aromatic substrates. In Scheme 20.4, degradation of 2,4-dichlorophenol was shown by synergistic action of both lignin peroxidase and manganese peroxidase in white rot filamentous fungus *P. chrysosporium.* MnPs feature a manganese-binding site composed of several acidic amino acid residues along with one of the heme propionate groups. In the MnP cycle, one-electron transfer to Compound I occurs from bound Mn^{2+}. The resulting product, Mn^{3+}, is released from the active site, and its stability against disproportionation to Mn^{2+} and insoluble Mn^{4+} is ensured when various bidentate chelators are available. Oxalate, an

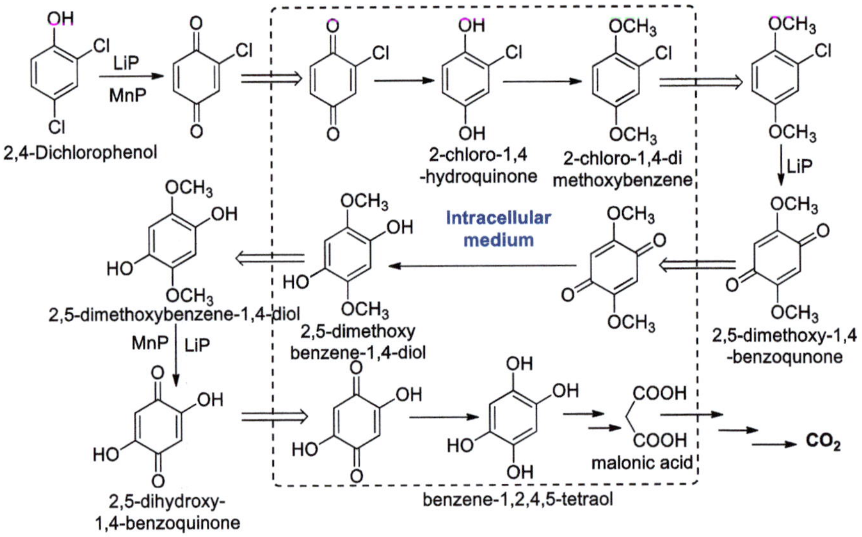

Scheme 20.4 : Degradation pathway of 2,4-dichlorophenol by lignin peroxidase and manganese peroxidase from *P. chrysosporium*

extracellular metabolite produced by many white rot fungi, is believed to be the physiological chelator in this context (Hammel and Cullen 2008).

Other than the abovementioned peroxidases within the Class II heme-peroxidases, notable examples include the nonligninolytic *Arthromyces ramosus* peroxidase (ARP) and *Coprinus cinereus* peroxidase (CIP). These enzymes, sharing catalytic similarities with horseradish peroxidase (HRP), have garnered significant attention for their potential application in the removal of phenolic compounds from wastewater. Studies have demonstrated the efficacy of ARP or CIP in removing various phenolic compounds from aqueous solutions (Ikehata et al. 2004). Another category of peroxidases, less explored but promising in organopollutant degradation, includes heme-thiolate peroxidases (HTPs) and dye decolorization peroxidases (DyPs). Despite possessing considerably divergent protein sequences and unique catalytic properties, these enzymes cannot be classified within the Class II peroxidase family. Instead, they are affiliated with separate (super) families of heme peroxidases (Sugano and Yoshida 2021). Two prominent enzymes from the HTPs family are Caldariomyces fumago chloroperoxidase (CfuCPO) and aromatic peroxygenases (APOs). CfuCPO primarily exhibits activity in oxidizing chloride (Cl^-) into hypochlorous acid at pH < 5, acting as a potent oxidant capable of chlorinating organic compounds. The exact catalytic purpose of this reaction remains unclear. Initially, it was believed that CfuCPO was involved in synthesizing chlorinated metabolites like caldariomycin (Ali et al. 2019).

20.3 Fungi in Pesticide Bioremediation

20.3.1 Fungal Species Used in Bioremediation

Numerous genera of different group such as brown rot fungi, white rot fungi (WRF), filamentous fungi, and yeasts are well known for pesticide degradation. Some examples including *Fusarium, Pleurotus Cladosporium, Penicillium, Aspergillus, Rhizopus, Trametes, Mortiella, Phlebia* (Bastos and Magan 2009; Xiao et al. 2011; Purnomo et al. 2010), etc. are known to degrade pesticides. Due to their nonspecific enzymes, WRF have been extensively examined for their capacity to digest a wide spectrum of pesticides and structurally highly changeable organic contaminants. Because WRF contain a range of extracellular and oxidative ligninolytic enzymes, including laccase (Lac), manganese peroxidase (MnP), versatile peroxidase (VP), and lignin peroxidase (LiP), they are among the most researched options for pesticide bioremediation (Zhuo and Fan 2021). Agricultural wastewater treated with *T. versicolor* has been demonstrated to be free of the herbicides Diuron and Bentazon. The first known fungus to break down organophosphate pesticides (pyrazophos, pirimiphos-methyl, malathion, dimethoate, profenofos, lancer) and use these substances as the only source of phosphorus were two fungi, *Aspergillus flavus* and *A. sydowii*, which were identified from wheat straw (Hasan 1999). At pH 8 and 30 °C, *A. flavus* has also been shown to decompose monocrotophos. It was discovered that a tolerant *A. niger* MRU01 could break down malathion the most effectively, with dimethoate, chlorpyrifos, and parathion following suit (Mohapatra et al. 2021). Zhu et al. (2010) reported that after five days of treatment, *Cunninghamella elegans* ATCC36112 could decompose fenitrothion by up to 81%. According to Liu et al. (2001), a tolerant strain of *A. niger* had a moderate level of chemical specificity since it could not break down dichlorovos or parathion, but it could break down formothion and malathion virtually equally well. *Cunninghamella aelegans, Fusarium solani, Talaromyces atroroseus, A. oryzae, Penicillium* sp., and *Trichoderma atroviride* were found to be effective in degrading dichlorvos (Zhang et al. 2019). The ability to breakdown chlorpyrifos and use 50 mg/l of the compound as the only carbon source was demonstrated by the novel fungal strain *Cladosporium cladosporioides* Hu-01. Aldrin and dieldrin could be broken down by the white-rot fungus *Pleurotus ostreatus* through hydroxylation and epoxidation processes. According to some research, these pesticides undergo esterification, deoxygenation, dehydrogenation, dechlorination, and demethylation reactions when they are transformed by fungi (Purnomo et al. 2017a, b). As a consortium rather than as a solo bioremediatory, many species of *Aspergillus, Penicillium*, and *Fusarium* have been used to remove sulfonylurea herbicides more effectively (Javaid et al. 2016). Another study by Sondhia et al. (2013) revealed two pyrazosulfuron ethyl-degrading fungi, *Penicillium chrysogenum* and *A. niger*, were isolated from rice field soil and were able to break down the chemical in soil under laboratory conditions via a degradation pathway that included hydrolytic cleavage and sulfonylurea bridge. The breakdown of atrazine, which was prompted by a nitrogen shortage, was caused by basidiomycetian strains such as *Pycnoporus sanguineus* MCA16, *Pluteus cubensis* SXS320, *Datronia caperata* MCA5, *Gloelophyllum striatum* MCA7, and *Polyporus tenuiculus* MCA11. In *P. chrysosporium*, the breakdown of the chlorinated insecticide endosulfan can occur through either an oxidative or hydrolytic pathway (Kullman and Matsumura 1996). A cytochrome P450 inhibitor was used to partially restrict oxidative metabolism, causing it to proceed through a hydrolytic pathway instead. It was demonstrated that the P450 system contributes to the deterioration of lindane (Mougin et al. 1996). Fipronil, an insecticide from the phenylpyrazole chemical class with limited aqueous solubility, is likewise broken down by *T. versicolor*. The synthesis of hydroxylated and glycosylated transformation products indicated that cytochrome P450 was responsible for the intracellular metabolism (Wolfand et al. 2016). It has also been demonstrated that this enzyme breaks down the very polar pesticides imidacloprid and acetamiprid (Hu et al. 2022). Acetamiprid, Clothianidin, Imidacloprid, and Thiacloprid are the four neonicotinoids that has also been demonstrated to breakdown using fungal cytochrome P450 isozymes. The list of the pesticide and the degrading organism is given in Table 20.1.

20.3.2 Advantages of Fungi in Pesticide Degradation

Bioremediation employs the biological approach to clean up contaminants using microorganisms. It is simple and needs less labour, cost-effective approach to toxic waste cleaning when compared to other traditional procedures. In a short amount of time, fungi will proliferate within the soil and begin to reproduce on their own. There are no additional costs required to begin the mycoremediation process. Only a small amount of fungus or spores from fungi can help clean the entire polluted surface. Nutrients (both organic and inorganic wastes) are always available and can be easily administered to promote the fast growth of microorganisms. This cleanup procedure is safer than the others because it does not include the digging and disposal process. Furthermore, there is no need for additional steps because this process does not generate secondary waste streams. Because no machinery is needed, it's a simpler and safer way than other methods. Because the outcomes are benign, neither humans nor animals are impacted. This method doesn't disrupt adjacent residents, it is better for the environment and the surrounding habitats. By concentrating on odor reduction and improvement at the performance location, the outcomes are evident. It is natural; the goal of the fungal system is to neutralize toxic compounds caused by pollutants, eventually returning the system to its normal function and homeostasis while bringing pollutants into balance with the environment (Sharma et al. 2020).

Table 20.1 Remediation of different types of pesticide by fungi

Pesticides	Fungal biodegrader species	References
Herbicide		
Diuron/herbicide	*Ganoderma lucidum*, *A. fumigatus*, *A. terreus*, *Mortierella* sp.	Oliveira et al. (2015)
Diuron and bentazon/ herbicides	*Trametes versicolor*	Beltrán-Flores et al. (2021)
Lancer	*A. flavus*, *A. niger* and *A. sydowii*, *A. flavus*, *A. tamari*	Hasan (1999)
Alachlor	*Trichoderma koningii*, *Trichoderma* spp.	Nykiel-Szymanska et al. (2020)
Atrazine	*A. fumigatus*, *A. ustus*, *A. flavipes*, *Rhizopus stolonifer*, *Fusarium moniliforme*, *F. roseum*, *F. oxysporum*, *Penicillium decumbens*, *P. janthinellum*, *P. rugulosum*, *P. luteum*, *T. viride*, *A. fumigatus*, and *A. terreus*,	Oliveira et al. (2015), Singh et al. (2008)
Metsulfuron methyl	*A. niger*, *Penicillium* sp. (DS11F), *Trichoderma* sp.	Vazquez and Bianchinotti (2013)
Glyphosate	*A. terreus*, *A. niger*, *T. viridae*, *Fusarium oxysporum*,	Adelowo et al. (2014)
	A. oryzae A-F02, Penicillium IIR, *P. chrysogenum*, *P. notatum*, and *T. harzianum*	
Metolachlor	*A. flavus*, *A. terricola*, *Trichoderma* spp., *P. oxalicum* MET-F-1	Nykiel-Szymanska et al. (2020)
Sulfosulphuron	*Trichoderma* sp.	Yadav and Choudhury (2014)
Isoproturon	*A. fumigates*, *A. terreus*	Oliveira et al. (2015)
Insecticide		
Dieldrin/organochlorinated	Marine *Penicillium miczynskii* CBMAI 930, *A. flavus*, *P. notatum*	Birolli et al. (2015), Mehrotra et al. (2004)
Methyl parathion/ organophosphate	*Aspergillus sydowii*, *Penicillium decaturense*, *Penicillium raistrickii*, Marine *Penicillium citrinum* (DL4M3), *P. citrinum* (DL9M3) and *Fusarium proliferatum* (DL11A)	Rodrigues et al. (2016)
Chlorpyrifos	*Aspergillus* sp., *A. niger*, *A. terreus*, *A. niger* MRU01, *T. viride*, *Alternaria alternata*, *Cladosporium cladosporioides*, *Fusarium* sp., *C. cladosporioides* Hu-01; *Acremonium* sp., *Fusarium oxysporum*, *P. funiculosum*, *Acremonium strictum*	Hussain et al. (2007), Mohapatra et al. (2021)

(continued)

Table 20.1 (continued)

Pesticides	Fungal biodegrader species	References
Pentachlorophenol	*T. harzianum, T. piluliferum, T. aureoviride*	Sing et al. (2014)
Chlorpyrifos/ organophosphate, cypermethrin/ pyrethroid, dicofol/organochlorinated	*Trametes versicolor*	Hu et al. (2020)
Chlorpyrifos, methyl parathion and profenofos/ organophosphates	*Aspergillus sydowii* CBMAI 935	Soares et al. (2021)
Monocrotophos	*A. orysae, A. niger*, and *A. flavus*	Jain et al. (2014)
Lambda-cyhalothrin/ pyrethroids	Marine *Aspergillus* spp. CBMAI 1829, *Acremonium* spp. CBMAI 1676, *Microsphaeropsis* spp. CBMAI 1675 and *Westerdykella* spp. CBMAI 1679	Birolli et al. (2018)
Dimethoate	*A. niger, A. flavus, A. sydowii, A. niger* MRU01	Hasan (1999), Liu et al. (2001), Mohapatra et al. (2021)
Pirimiphos-methyl	*A. fumigatus, A. flavus, A. niger*, and *A. sydowii*	Hasan (1999)
Chlorfenvinphos	*P. citrinum, A. fumigatus, A. terreus*, and *T. harzianum*	Oliveira et al. (2015)
Endosulfan	*A. sydoni, A. niger, A. tamarii* JAS9, *A. terreus, A. flavus, P. chrysogenum, T. harzianum*	Goswami et al. (2009), Hussaini et al. (2013)
DDT	*T. harzianum, T. viride, A. flavus, A. parasitacus*	Mehrotra et al. (2004)
b-Cypermethrin	*A. niger, A. oryzae* M-4	Zhao et al. (2016)
3-phenoxybenzoic acid	*A. oryzae* M-4	Zhu et al. (2016)
Malathion	*A. flavus, A. sydowii, A. niger* MRU01	Hasan (1999), Mohapatra et al. (2021)
Heptachlor	*A. niger*	Singh and Dwivedi (2004)
b-cyfluthrin	*T. viride* 2211, *A. niger, A. terricola, Phanerochaete chrysoporium*	Saikia and Gopal (2004)
Pyrazophos	*A. flavus, A. niger*, and *A. sydowii*	Hasan (1999)
Fenitrothion	*Cunninghamella elegans* ATCC36112	Zhu et al. (2010)
Lindane	*A. niger*	Hussaini et al. (2013)
a-HCH	*Aspergillus* sp.	Javaid et al. (2016)
Formothion	*A. niger*	Liu et al. (2001)
Aldrin	*A. flavus, P. notatum*	Mehrotra et al. (2004)
Dichlorovos	*A. niger*	Liu et al. (2001)
Profenofos	*A. flavus, A. sydowii*	Hasan (1999)

20.3.3 Challenges and Limitations

The mechanism's transfer from pilot-to-large-scale application is challenging. However, compared to conventional procedures, bioremediation techniques such as reactor designs might be more expensive. Compared to alternative treatment options, a bioremediation therapy takes longer. The bioremediation process may be slowed down by using minimal or no nutrient supplements. Toxic nutrients and supplements for the microorganisms can hinder the bioremediation process. Not all materials break down completely and quickly, especially inorganic pollutants. The right performance criteria for bioremediation remain a source of contention among regulators. Certain biodegradation products survive in the environment and may even be more toxic than the parent substances in certain situations (Azubuike et al. 2016). Researchers looked into the capacity of *Phanerochaete chrysosporium* and *Phanerochaete sordida* to use a land farming method to decompose DDT tainted soil from a discontinued pesticide production plant. In experimental settings, neither of these white-rot fungi could improve the soil despite their rapid growth (Safferman et al. 1995). Other research has demonstrated that white-rot fungi break down DDT, although not significant mineralization (Osano et al. 1999). The breakdown of some pesticides, like DDT, produces the hazardous metabolites DDD and DDE. These findings suggest that further investigation is required before white-rot fungi are routinely employed to clean up pesticide-contaminated areas.

20.3.4 Case Studies on Fungal Bioremediation of Pesticides

Hu et al. (2020) reported three distinct hydrophobic pesticides—chlorpyriphos, dicofol, and cypermethrin were broken down by *Trametes versicolor*. These are a synthetic pyrethroid, an organochlorine, and an organophosphorus, in that order. Over the course of 14 days, their elimination potential at 25 °C was 95% for chlorpyriphos, 88% for dicofol, and 93% for cypermethrin. Every pesticide that was examined had its quick initial adsorption to fungal pellets seen, and then its biodegradation. Depending on the herbicide, hydrolyzation, dichlorination, and oxidation were suggested as significant degradation mechanisms based on the detected metabolites (Hu et al. 2020). The report confirmed that cytochrome P450 was shown to degrade the highly polar pesticides Acetamiprid and Imidacloprid. Beltrán-Flores et al. (2021) demonstrated that *T. versicolor* remove the herbicides Bentazon and Diuron from agricultural effluent. Additionally, *T. versicolor* breaks down Fipronil, an insecticide that is part of the phenylpyrazole chemical class and has a low solubility in water. The synthesis of hydroxylated and glycosylated transformation products indicated that cytochrome P450 was responsible for the intracellular metabolism of this (Wolfand et al. 2016).

It was reported that *P. chrysosporium* uses two cytochrome P450 isozymes to break down four neonicotinoids: acetamiprid, clothianidin, imidacloprid, and thiacloprid *P. chrysosporium* has roughly 150 P450 monooxygenase genes in its genome. According to Kullman and Matsumura (1996), *P. chrysosporium* can break down the chlorinated insecticide endosulfan by an oxidative or hydrolytic mechanism. A cytochrome P450 inhibitor was used to partially limit oxidative metabolism, causing it to proceed through a hydrolytic pathway instead. According to Mougin et al. (1996), the P450 system was also demonstrated to be involved in the degradation of lindane. It was reported that agricultural soils contaminated with various organochlorinated pesticides (OCPs) and polycyclic aromatic hydrocarbons (PAHs) were remedied by using WRF *P. chrysosporium* in combination with the boron silicate mineral tourmaline. While separate treatments with either *P. chrysosporium* or tourmaline exhibited lesser efficiency of 34% and 26%, respectively, the synergistic effects resulted in a 44% elimination of OCPs from the agricultural soils. It has been proposed that tourmaline contributes to soil enzyme activity and the richness of the active soil microorganisms. *P. chrysosporium* was reported to degrades DDT, and when the incubation period was extended, degradation reached 35% in 28 days. It's interesting to note that the fungus used two unique metabolic pathways for DDT metabolism that are not commonly used by other fungi: the minor pathway, which is regulated by the P450 system and creates monohydroxy-DDTs, and the major pathway, which is the unknown mechanism that converts DDT to DDD. It has been reported that *P. ostreatus* was the most effective WRF in the breakdown of OP pesticides. Kulshrestha and Kumari (2011) observed that when growth was examined in nutritional media containing 0.1% sodium nitrate, mixed fungal cultures isolated from several soil types (alluvial, red, and black) used chlorpyrifos as a partial source of carbon/nitrogen. The potential of the mixed culture to breakdown chlorpyrifos by cometabolism was demonstrated by the authors, who were able to obtain degradation as high as 78.2% in broth with normal carbon and nitrogen contents. It has been reported that TCP was building up as a transit metabolite and that there was a rapid loss of chlorpyrifos during the first two days following application—more than 50% loss was seen in all fungal cultures. Chlorpyrifos initially inhibited the activity of laccases and MnP in *P. ostreatus*. However, the fungus quickly adapted and the enzyme activities recovered five days after exposure.

20.4 Regulatory Mechanisms of Pesticide Degradation

20.4.1 Environmental Conditions

Since the emergence of focused attention on the industrial-scale production of laccase by companies such as Novozymes and other biotechnology industry leaders, a surge of interest has captivated research scientists and entrepreneurial ventures. The allure lies in the potential for cost-effective enzyme production, particularly with the prospect of minimal capital investment. This heightened interest has prompted exploratory studies investigating the potential applications of laccase across various fields, albeit mostly at the laboratory scale (Chauhan and Jha 2018; Chen et al. 2019; Mandic et al. 2019). However, it is unfortunate that the majority of these studies merely extrapolate findings from laboratory model studies to predict outcomes in larger-scale applications, potentially overlooking the nuances of uncontrolled environments. As a result, the practical application of most large-scale enzyme production has been limited to domestic uses, distinct from the well-recognized laboratory-scale applications. Consequently, it is imperative to shed light on the factors critical to the successful environmental application of microbial laccases in situ. First, a thorough examination should be dedicated to fundamental environmental factors, such as pH and temperature, which exert interconnected influences on laccase performance. Maintaining consistent enzyme performance, even in immobilized states, becomes challenging due to anticipated fluctuations. Therefore, acquiring a versatile biocatalyst proves immensely valuable in navigating such scenarios. Fortunately, laccases exhibiting polyextremotolerant capabilities have been identified, although their large-scale testing is still pending. Second, the time required for the satisfactory degradation of a diverse range of pollutants in wastewater cocktails leads to variable removal rates for different pollutants. For instance, a substantial difference in biotransformation rates between diclofenac and flufenamic acid was observed, with the former achieving transformation in 3 hours compared to the latter's 24-hour timeframe (Hahn et al. 2018). Third, consideration should be given to the type of substrate or support for the reaction of the free enzyme. This involves not only evaluating the immobilization material but also exploring other materials that could function as cosubstrates with the wastewater. A notable example is the study by Singh et al. (2015), where detectable laccase activity persisted in soil samples despite an extended estrogen transformation of 264 hours. While many researchers may deem it suitable for soil remediation processes, its potential in wastewater and water treatment should be examined, especially if combined with potential humus-forming materials. This approach not only purifies water and wastewater but also contributes to the generation of soil organic matter through laccase-mediated cross-coupling reactions.

20.4.2 Substrate Specificity

Study validates the computationally predicted oxidation of xenobiotic contaminants by employing commercially available pure laccase sourced from *Trametes versicolor*. The selection of contaminants as potential targets for laccase oxidation was based on in silico docking predictions. The measurement of laccase oxidation was conducted by assessing the change in absorbance at specific λ max for each compound. Sinapic acid and tyrosine served as positive and negative controls, respectively. Oxidation was observed in m-chlorophenol, 2,4 dichlorophenol, 2,4,6 tri-chlorophenol, captan, atrazine, and thiodicarb, with the exception of malathion, which exhibited no activity. It can be inferred that predicted substrates showing oxidation shared structural and chemical homology with positive control compounds. In the case of malathion, the lack of structural homology with sinapic acid may explain its inactivity toward laccase, warranting further structural analysis. Consequently, a remediation tool proposing an advanced approach, combining theoretical in silico methods and subsequent experimental validation using pure laccase, is suggested. As the number and diversity of xenobiotics increase, the impracticality of experimentally screening them for bioremediation also escalates. This approach proves useful in reducing the time and cost associated with other screening methods (Dhall et al. 2014).

20.5 Applications and Future Prospects of Pesticide Bioremediation

Mycoremediation can be applied in two ways: either on the spot, known as in-situ mycoremediation, or by digging or removing the contaminant or contaminated soil, sediment, or water matrix to a different area, where ex situ mycoremediation is carried out. According to Azubuike et al. (2016), the most popular in situ bioremediation techniques include bioaugmentation, bioventing, biosparging, and natural attenuation. The application of biopiling, composting, land farming, and bioreactors for solid matrixes and water treatment facilities for liquid media like slurries and water are examples of ex situ bioremediation techniques that are contingent on the matrix. The environmental circumstances, the type and extent of the pollution, the related expenses, and the accessibility of appropriate fungal candidates for cleanup all influence the viability of different remediation techniques.

20.5.1 In Situ Mycoremediation

Since in situ mycoremediation is done on-site, it is inextricably linked to the matrix. The native chemical, physical, and biological systems have already undergone alterations as a result of the pollution input in the multifactor mycoremediation system on the spot. In addition to the pollutant, the addition of a live bioremediation component will impact the site's physicochemical conditions and maybe higher species and microbial populations. Furthermore, the chemical reactions that occur following the application of fungus may be impacted, either positively or negatively, by the microbial communities present at the location.

Because in situ application of fungi causes less disturbance to the contaminated site than other methods, such as soil excavation, it is favorable. This has already been used successfully as a bioremediation technique, treating sites contaminated with hydrocarbons, dyes, heavy metals, and chlorinated solvents with bacteria. As fungal hyphae can develop in air-soil interfaces and penetrate both soil and rock matrices, the dispersal of fungi in soils is not dependent on water-saturated)-soil phases (in contrast to bacteria). Additionally, on-site treatment reduces the amount of shear pressures that fungi are exposed to, which could happen, for instance, during stirring in bioreactors. According to Harms et al. (2011), this permits mycelia to grow and permeate soils without changing. Filamentous fungus are adapted to a variety of environmental situations and have the ability to expand their hyphae networks across hundreds of kilometers of soil. Fungi can also act as an adjuvant to bacterial degradation by allowing bacteria to disperse along fungal mycelium, or they can act as fungal highways, breaking down physical barriers in air-soil interfaces where water transport is restricted (Wick et al. 2007). This is made possible by the ability of fungi to penetrate. This has been shown for bacteria that break down hydrocarbons by means of common soil fungus like *F. oxysporum* or *Rhexocercosporidium* spp. Furthermore, it has been demonstrated that *Pythium ultimum* promotes the chemostatic spread of *Pseudomonas* species that degrade PAHs along their mycelia (Furuno et al. 2012). On land, in situ mycoremediation is a simpler process to implement, whereas the dilution and dispersion of both the fungus and the contaminant make successful fungal applications in aquatic, particularly marine, environments more difficult.

20.5.2 Ex Situ Mycoremediation

Due to its ability to monitor and manage environmental factors, fungal growth, and performance, ex situ mycoremediation generally provides for greater control over the process. Examples of environmental factors that can be altered in bioreactors include pH and oxygen levels, which affect how well some pesticides are degraded by fungi. Because most fungi have a high oxygen need and because it is more expensive and practicable to facilitate this in situ, ex situ mycoremediation is frequently more viable. Moreover, higher temperatures—such as those attained in ex situ facilities—usually lead to quicker and better cleanup. The use of fungi (*T. versicolor*) or consortia of fungi and bacteria in aqueous medium to clean up hospital wastewater is an area of much research and development for integrated biological treatment approaches. It has been demonstrated that the first coagulation-flocculation pretreatment processes in bioreactor settings increase the viability of fungi inside the reactor (Mir Tutusaus et al. 2016). It has been demonstrated that adding redox mediators, including quinones, increases the clearance of medications. This seems to be species-specific because 2,6-dimethoxy-1,4-benzoquinone (DMBQ) and gallic acid were the most effective redox mediators for *T. versicolor* and *G. lucidum*, respectively. Furthermore, genetically modified organisms (GMOs) that are particularly engineered to target one or more contaminants could be beneficial for ex situ mycoremediation.

20.5.3 Genetic Engineering of Fungi

Omics like transcriptomics, proteomics, metabolomics, and genomics support systems biology investigations of microbes in order to analyze genetic control at the level of bioremediation (da Silva et al. 2023). The unique genes implicated in the biodegradation pathways of different persistent pollutants have been resolved by advances in sequencing with high throughput sequencing (HTS) and next-generation sequencing (von Netzer et al. 2018). The use of microorganisms in feed, food, agriculture, medicine, and other fields has advanced thanks to genetic engineering using gene editing technologies (Yadav et al. 2018). Gene editing tools possess significant potential to enhance bioremediation procedures, including the removal of xenobiotics, transformation of harmful substances into less harmful ones, and breakdown of pesticides into simpler parts (Basu et al. 2018). Fungal enzymes can be engineered and modified to enhance their activity and facilitate effective bioremediation. For more effective "green" applications, Wong et al. (2013) created a strong expression platform from *L. edodes* and applied guaiacol as a screening substrate, all in relation to the genomes of fungal detoxifying enzymes like laccases. Parallel to this, Kalyani et al. (2015) produced a recombinant laccase from the yeast *Yarrowia lipolytica* that is extremely effective at hydrolyzing biomass found in wood. In a different investigation, *A. niger* showed strong expression of the *Phanerochaete flavido-alba* laccase gene and a good concentration of active recombinant enzyme (rLac-LPFA) (30 mg/L). Furthermore, if the genes for their potential enzymes, such as laccases and peroxidases, are successfully expressed in plants, fungus may find use in phytoremediation (Sonoki et al. 2012). The expression of *F. oxysporum* cytochrome P450 monooxygenase in *Saccharomyces cerevisae* is another illustration of this. Potential strain *Penicillium oxalicum* SAR-3 has numerous unique genes for azo dye degradation, including stress-responsive genes and ABC transporters and peroxidases, which allowed it to catabolize a variety of azo colors over a broad range. In *P. pastoris*, constitutive expression of the synthetic genes GlLCC1 and POXA 1B from *Ganoderma lucidum* laccase from *P. ostreatus* was obtained (Rivera-Hoyos et al. 2015). The aforementioned research demonstrate that fungal genomics has potential applications in the future for determining pollution levels in different environmental matrices. This might be accomplished by tracking particular marker genes that are involved in the detoxification of contaminants.

20.5.4 Synergistic Approaches with Bacteria

Instead of existing in an axenic environment, microorganisms are a part of intricate microbial communities that exist in their surroundings. Bacteria and fungus frequently coexist in foods, environments, animals, and even human bodies (Getzke et al. 2019). The interactions between fungus and bacteria can range from synergistic (friendly) to antagonistic (competitive), depending on the effect that the fungal-bacterial interactions (FBIs). Because of their well-known capacity to break down a broad variety of organic substances, from simple substrates to complex molecules, fungi and bacteria are excellent choices for breaking down organic contaminants such as hydrocarbons, pesticides, and other xenobiotics (Harms et al. 2011; Semple et al. 2007). Fungi and bacteria can generate synergistic metabolic or cometabolic activities in mixed microbial cocultures, which may result in a faster and more thorough breakdown of organic molecules as well as greater tolerance to environmental fluctuations. Complex organic substances are broken down into smaller molecules or end products (such carbon dioxide and water) by microbes in biodegradation process. It is possible that certain fungal or bacterial species or communities lack the metabolic resources required to completely degrade complex contaminants. Nonetheless, it has been demonstrated that certain mixed multidomain communities can effectively accomplish the total breakdown of these contaminants. In these mixed microbial cultures, every microbe may play a distinct part in the degrading process, whereby certain microorganisms produce intermediates that are used by other members.

Few bacterial species such as *Enterobacter*, *Stenotrophomonas* sp., *Sphingomonas* sp., *Bacillus cereus*, and *Aminobacter* sp. have been utilized to breakdown chlorpyrifos and 2,6-dichlorobenzamide (BAM), a degradation product of the herbicide dichlobenil; bacterial degradation of pesticides is typically only partial (Yang et al. 2006). It has been documented that fungal-bacterial cocultures can be used in both aqueous and porous media to break down pesticides that contain both organochlorine and organophosphate. In aqueous systems, pesticide degradation was mostly due to the degradative activity of both fungal and bacterial cocultures, whereas in porous media, the activity of bacteria linked to the fungal mycelium was responsible for pesticide breakdown. In research on the synergistic fungal-bacterial degradation process, DDT has been the most often used pesticide. Cocultures containing *P. ostreatus–P. aeruginosa* (Purnomo et al. 2017a, b), *F. pinicola–B. subtilis* (Sariwati et al. 2017), *Pleurotus eryngii–R. picketii* (Purnomo et al. 2019a, b), and *G. lingzhi–B. subtilis* (Xu et al. 2007) were found to achieve degradation removal efficiencies in the range of approximately 75–86% of DDT. Purnomo et al. (2019a, b) employed the WRF *P. eryngii* and the biosurfactant-producing bacterium *R. picketii* in a different investigation to degrade DDT. After 7 days of incubation, the coculture degraded DDT to a level of 78%, while the single culture of fungus only showed 43% degradation. A pure bacterial culture produced DDE (1,1-dichloro-2,2-bis(4-chlorophenyl) ethylene), a pure fungal culture produced DDE and DDMU (1-chloro2,2-bis(4-chlorophenyl) ethylene), and a fungal-bacterial coculture produced DDE, DDMU, and DDD (1,1-dichloro-2,2-bis(4-chlorophenyl) ethane). These findings were discovered through analysis of the metabolic products resulting from the degradation of DDT. Building a stable fungal-bacterial community that can successfully form under nonaxenic conditions and is robust enough to treat numerous pollutants at once is essential to the full-scale use of fungal-bacterial bioreactors.

20.5.5 Biotechnological Applications: Nanotechnology in Pesticide Bioremediation

The science of nanotechnology studies artificially created particles that are less than 100 nm in size. At the start of the twenty-first century, nanoparticles and nanotechnology were first used in agriculture. By employing nanoparticles to speed up bioremediation, nanobioremediation focused on cleaning the environment (Bhatt et al. 2021). Nanobioremediation used nanoparticles to speed up bioremediation, hence focusing on cleansing the environment (Bhatt et al. 2021). Utilizing catalyst in the form of nanoparticles to degrade pollutants is the fundamental idea behind nanobioremediation. Due to their minuscule size, nanoparticles can engage in deeper interactions with their surroundings and have a higher surface area to mass ratio than other particle sizes. This increases the rate and effectiveness of bioremediation.

Fungi are used in green nanotechnology to create nanoparticles, and they are essential for eliminating organic pollutants and hazardous substances. Researchers from all over the world have recently become quite interested in the synthesis of metal nanoparticles from fungus (Sunny et al. 2022). Metal nanoparticles made with fungus have a number of benefits, including a greater capacity for metal uptake, easy and inexpensive synthesis, resistance to metals, high scaling, and high stability. Numerous fungal genus, including *Aspergillus*, *Fusarium*, *Verticillium*, and *Penicillium*, have been considered as possible candidates for the manufacture of metallic nanoparticles (Ovais et al. 2018). Research has indicated that a variety of metals, including silica, gold, silver, titanium, platinum, selenium, and palladium, can be used to create nanoparticles. The characteristics of silver nanoparticles (AgNPs) derived from *F. oxysporum* differ from those derived from *Aspergillus fumigates* (Alves and Murray 2022). Research has documented the production of AgNPs with *Trichoderma reesei* and *Coriolus versicolor*. Using the fungus *Cylindrocladium floridanum*, gold nanoparticles (AuNPs) have been synthesized (Narayanan and Sakthivel 2011). *Fusarium oxysporum* was utilized to create silica nanoparticles, while *N. crassa* was utilized to generate platinum nanoparticles. According to Zhang et al. (2019), selenium nanoparticles were produced using *Mariannaea* sp.

Numerous fungal species shown the ability to bind metals, including *Allescheriella* sp., *Botryosphaeria rhodina* sp., and *Stachybotrys* sp. (Benjamin et al. 2019). *Rhizopus oryzae*-derived AgNPs are synthesized for a variety of ecological applications, including pesticide adsorption and wastewater remediation. According to Rasha (2017), certain fungi, such as *Fusarium solani*, are more resistant to certain heavy metals like cadmium, nickel, and lead. They are also better able to synthesize nanoparticles. The bioremediation potential of nanoparticles combined with white-rot fungi (WRF) is enormous (He et al. 2017). Research has shown that harmful pollutants can be removed from wastewater and that WRF-magnetic nanoparticles exhibit good stability as a result of effective immobilization. *Echinodontium taxodii*-Fe3O4 nanoparticles have been shown to breakdown antibiotics such as sulfonamide (Shi et al. 2015). Immobilizing Phanerochaete chrysosporium and using titanium oxide nanoparticles allowed for the cleanup of cadmium and 2,4-dichlorophenol. According to Huang et al. (2017), *P. chrysosporium* and Fe2O3 nanoparticles demonstrated the breakdown of phenols. According to Espinosa-Ortiz et al. (2016), zinc remediation using selenium nanoparticles and *Phanerochaete chrysosporium* was also successful. Thus, it can be said that the use of fungus and nanoparticles together accelerated the cleanup process. Therefore, additional investigation and study should be done in order to build such successful strategies utilizing nanotechnology and ecological microbiology.

20.5.6 Future Research Directions

Additional targeted study is required to maximize mycoremediation techniques that are cost-effective, eco-friendly, and able to be applied with or without plant associations in cocontamination. In terrestrial and aquatic habitats, this will increase decontamination rates and remedial effectiveness. Many important future research avenues could be pursued in the subject of mycoremediation. The efficiency of mycoremediation is limited in the field because it is still in its infancy at the laboratory/greenhouse level. The mycoremediation capacities of each species must thus be assessed in their natural habitat prior to the commercialization of this green technology.

Further investigation into insecticidal microorganisms is necessary in order to: discover new sources of insecticidal microorganisms; expand their range of insecticidal activity; reduce growth times; ensure that they can proliferate both horizontally and vertically; and improve the sustainability of the environment.

20.6 Concluding Remarks

The present review offers a thorough exploration of the potential applications of fungal oxidoreductases as biocatalysts. These enzymes show great promise in mitigating environmental damage by eliminating various toxic chemicals including pesticides, optimizing energy and resource usage, introducing innovative functionalities, and minimizing adverse ecological impacts. The strategic expansion of applications of oxidoreductase is crucial for driving the growth of industrial biocatalysts and promoting sustainable, eco-friendly practices.

Additional targeted research is essential to maximize the efficiency of mycoremediation techniques, ensuring they are cost-effective, eco-friendly, and adaptable for use with or without plant associations in cocontaminated environments. Enhancing decontamination rates and remedial effectiveness in terrestrial and aquatic habitats will be the prevailing approach. Numerous avenues for future research in mycoremediation exist, as its field application is still in its infancy, primarily limited to laboratory and greenhouse settings. Comprehensive assessments of mycoremediation capacities in natural habitats are necessary before commercializing this green technology. Additionally, investigating the ecological stability of these microorganisms in their natural environment is crucial. Future research on microbial pesticide degradation should prioritize cultivating, developing, and utilizing highly efficient fungi for pesticide degradation in natural settings, as well as extracting relevant genes for cloning and expressing genes for degradative enzymes in common species like *E. coli.* Constructing engineered bacteria to enhance degradation capabilities and promoting the field application of bio-pesticides are also key steps. Emphasizing applied research on microbial methods for contaminant degradation in real-world scenarios is vital.

This chapter offers a comprehensive review of the molecular aspects of oxidoreductases, providing insight into the structural basis of fungal bioremediation capabilities and delving into the associated biochemistry and protein homology. Future research could focus on elucidating the biodegradation of pesticides such as chlorpyrifos, cypermethrin, pyrethroids, and organochlorines etc.. Microbial degradation stands out as the predominant process influencing pesticide breakdown in soils, offering a secure, minimally disruptive, and cost-effective approach. Extensive research in this field is essential for advancing the production of catalytic enzymes like various members of oxidoreductases, responsible for breakdown of different pollutants including pesticides.

References

Adelowo FE, Olu-Arotiowa OA, Amuda OS (2014) Biodegradation of glyphosate by fungi species. Adv Biosci Biotechnol 2(1):104–109. https://doi.org/10.4236/abb.2014.51014

Alcalde M (2007) Laccases: biological functions, molecular structure and industrial applications. In: Industrial enzymes: structure, function and applications. Springer Netherlands, Dordrecht, pp 461–476. https://doi.org/10.1007/978-1-4020-5377-4_24

Ali M, Husain Q, Ishqi HM (2019) Chapter-2 fungal peroxidases mediated bioremediation of industrial pollutants. In: Fungal bioremediation: fundamentals and applications, vol 269(11). https://doi.org/10.1007/978-981-13-6524-9_2

Alves MF, Murray PG (2022) Biological synthesis of monodisperse uniform-size silver nanoparticles (AgNPs) by fungal cell-free extracts at elevated temperature and pH. J Fungus 8(5): 439

Azubuike CC, Chikere CB, Okpokwasili GC (2016) Bioremediation techniques- classification based on site of application: principles, advantages, limitations and prospects. World J Microbiol Biotechnol 32:180. https://doi.org/10.1007/s11274-016-2137-x

Baldrian P (2006) Fungal laccases–occurrence and properties. FEMS Microbiol Rev 30(2):215–242. https://doi.org/10.1016/j.femsre.2005.07.001

Barbieri MV, Peris A, Postigo C, Moya-Garc'es A, Monllor-Alcaraz LS, RamblaAlegre M, Eljarrat E, Lopez′ De Alda M (2021) Evaluation of the occurrence and fate of pesticides in a typical Mediterranean delta ecosystem (Ebro River Delta) and risk assessment for aquatic organisms. Environ Pollut 274:115813

Bastos AC, Magan N (2009) *Trametes versicolor*: potential for atrazine bioremediation in calcareous clay soil, under low water availability conditions. Int Biodeterior Biodegrad 63:389–394. https://doi.org/10.1016/j.ibiod.2008.09.010

Basu S, Rabara RC, Negi S, Shukla P (2018) Engineering PGPMOs through gene editing and systems biology: a solution for phytoremediation? Trends Biotechnol 36:499–510. https://doi.org/10.1016/j.tibtech.2018.01.011

Beltrán-Flores E, Sarrà M, Blánquez P (2021) Pesticide bioremediation by *Trametes versicolor*: application in a fixed-bed reactor, sorption contribution and bioregeneration. Sci Total Environ 794:148386. https://doi.org/10.1016/j.scitotenv.2021.148386

Benjamin SR, Lima FD, Florean EOPT, Guedes MIF (2019) Current trends in nanotechnology for bioremediation. Int J Environ Pollut 66:19–40. https://doi.org/10.1504/IJEP.2019.104526

Bhatt P, Pandey SC, Joshi S, Chaudhary P, Pathak VM, Huang Y et al (2021) Nanobioremediation: a sustainable approach for the removal of toxic pollutants from the environment. J Hazard Mater 427:1–18. https://doi.org/10.1016/j.jhazmat.2021.128033

Birolli WG, Yamamoto KY, De Oliveira JR, Nitschke M, Seleghim MH, Porto AL (2015) Biotransformation of dieldrin by the marine fungus *Penicillium miczynskii* CBMAI 930. Biocatalysis Agric Biotechnol 4:39–43. https://doi.org/10.1016/j.bcab.2014.06.002

Birolli W, Vacondio B, Da Silva N, Seleghim M, Porto A (2018) Enantio selective biodegradation of the pyrethroid (±)-lambda-cyhalothrin by marinederived fungi. Chemosphere 197:651–660. https://doi.org/10.1016/j.chemosphere.2018.01.054

Chauhan PS, Jha B (2018) Pilot scale production of extracellular thermo-alkali stable laccase from *Pseudomonas* sp. S2 using agro waste and its application in organophosphorous pesticides degradation. J Chem Technol Biotechnol 93(4):1022–1030. https://doi.org/10.1002/jctb.5542

Chen X, Zhou Q, Liu F, Peng Q, Teng P (2019) Removal of nine pesticide residues from water and soil by biosorption coupled with degradation on biosorbent immobilized laccase. Chemosphere 233:49–56. https://doi.org/10.1016/j.chemosphere.2019.05.073

Conesa A, Punt PJ, vanden Hondel CAMJJ (2002) Fungal peroxidases: molecular aspects and applications. J Biotechnol 93:143–158. https://doi.org/10.1016/S0168-1656(01)00394-7

Črešnar B, Petrič Š (2011) Cytochrome P450 enzymes in the fungal kingdom. Biochimica et Biophysica Acta (BBA)-Proteins and Proteomics 1814(1):29–35. https://doi.org/10.1016/j.bbapap.2010.08.011

Dan Y, Ji M, Tao S, Luo G, Shen Z, Zhang Y, Sang W (2021) Impact of rice straw biochar addition on the sorption and leaching of phenylurea herbicides in saturated sand column. Sci Total Environ 769:144536

da Silva NM, Reis GF, de Fátima Costa F, Grisolia ME, Geraldo MR, Lustosa BPR, de Souza Lima BJF, Weiss VA, de Souza EM, Li R, Song Y (2023) Genome sequencing of Cladophialophora exuberans, a novel candidate for bioremediation of hydrocarbon and heavy metal polluted habitats. Fungal Biol 127(5):1032-1042

Dhall P, Kumar R, Kumar A (2014) Validation of computationally predicted substrates for laccase. Brazilian Arch Biol Tech 57:803–809. https://doi.org/10.1590/S1516-8913201402624

Espinosa-Ortiz EJ, Shakya M, Jain R, Rene ER, van Hullebusch ED, Lens PN (2016) Sorption of zinc onto elemental selenium nanoparticles immobilized in Phanerochaete chrysosporium pellets. Environ Sci Pollut Res 23:21619–21630. https://doi.org/10.1007/s11356-016-7333-6

Fenoll J, Sabater P, Navarro G, P'erez-Lucas G, Navarro S (2013) Photocatalytic transformation of sixteen substituted phenylurea herbicides in aqueous semiconductor suspensions: intermediates and degradation pathways. J Hazard Mater 244–245:370–379

Furuno S, Remer R, Chatzinotas A, Harms H, Wick L (2012) Use of mycelia as paths for the isolation of contaminant-degrading bacteria from soil. Microb Biotechnol 5(1):142–148. https://doi.org/10.1111/j.1751-7915.2011.00284.x

Gandia-Herrero F, Jiménez-Atiénzar M, Garcia-Carmona F, Escribano J (2005) Evidence of a common regulation in the activation of a polyphenol oxidase by trypsin and sodium dodecyl sulfate. Biol Chem 386:601–607. https://doi.org/10.1515/BC.2005.069

Gerdemann C, Eicken C, Galla HJ, Krebs B (2002) Comparative modeling of the latent form of catechol oxidase using a molluskan hemocyanin structure. J Inorg Biochem 89:155–158. https://doi.org/10.1016/S0162-0134(01)00438-0

Getzke F, Thiergart T, Hacquard S (2019) Contribution of bacterial-fungal balance to plant and animal health. Curr Opin Microbiol 49:66–72. https://doi.org/10.1016/j.mib.2019.09.002

Giardina P, Faraco V, Pezzella C, Piscitelli A, Vanhulle S, Sannia G (2010) Laccases: a never-ending story. Cell Mol Life Sci 67:369–385. https://doi.org/10.1007/s00018-009-0169-1

Goswami S, Vig K, Singh DK (2009) Biodegradation of alpha and beta endosulfan by *Aspergillus sydoni*. Chemosphere 75(7):883–888. https://doi.org/10.1016/j.chemosphere.2009.01.035

Hahn V, Meister M, Hussy S, Cordes A, Enderle G, Saningong A, Schauer F (2018) Enhanced laccase-mediated transformation of diclofenac and flufenamic acid in the presence of bisphenol A and testing of an enzymatic membrane reactor. AMB Express 8:1–1. https://doi.org/10.1186/s13568-017-0530-4

Halaouili S, Mi A, Kruus K, Guo L, Hamdi M, Sigoillot JC et al (2005) Characterization of a new tyrosinase from *Pycnoporus* species with high potential for food technological applications. J Appl Microbiol 98:332–343. https://doi.org/10.1111/j.1365-2672.2004.02463.x

Halaouili S, Asther M, Sigoillot JC, Hamdi M, Lomascolo A (2006) Fungal tryrosinases: new prospects in molecular characteristics, bioengineering and biotechnological applications. J Appl Microbiol 100:219–232. https://doi.org/10.1111/j.1365-2672.2005.02791.x

Hammel KE, Cullen D (2008) Role of fungal peroxidases in biological ligninolysis. Curr Opin Plant Biol 11(3):349–355. https://doi.org/10.1016/j.pbi.2008.02.003

Hannemann F, Bichet A, Ewen KM, Bernhardt R (2007) Cytochrome P450 systems biological variations of electron transport chains. Biochim Biophys Acta 1770:330–344. https://doi.org/10.1016/j.bbagen.2006.07.018

Harms H, Schlosser D, Wick LY (2011) Untapped potential: exploiting fungi in bioremediation of hazardous chemicals. Nat Rev Microbiol 9:177–192. https://doi.org/10.1038/nrmicro2519

Hasan HAH (1999) Fungal utilization of organophosphate pesticides and their degradation by *Aspergillus flavus* and *A. sydowii* in soil. Folia Microbiol 44(1):77–84. https://doi.org/10.1007/BF02815089

He K, Chen G, Zeng G, Huang Z, Guo Z, Huang T et al (2017) Applications of white rot fungi in bioremediation with nanoparticles and biosynthesis of metallic nanoparticles. Appl Microbiol Biotechnol 101:4853–4862. https://doi.org/10.1007/s00253-017-8328-z

Hofrichter M (2002) Review: Lignin conversion by manganese peroxidase (MnP). Enzyme Microbial Technol 30:454–466. https://doi.org/10.1016/S0141-0229(01)00528-2

Hoshida H, Nakao M, Kanazawa H, Kubo K, Hakukawa T, Morimasa K, Akada R, Nishizawa Y (2001) Isolation of five laccase gene sequences from the white-rot fungus *Trametes sanguinea* by PCR, and cloning, characterization and expression of the laccase cDNA in yeasts. J Biosci Bioeng 92(4):372–380. https://doi.org/10.1016/S1389-1723(01)80274-7

Hu K, Barbieri MV, López-García E, Postigo C, López De Alda M, Caminal G et al (2022) Fungal degradation of selected medium to highly polar pesticides by *Trametes versicolor*: kinetics, biodegradation pathways, and ecotoxicity of treated waters. Anal Bioanal Chem 414:439–449. https://doi.org/10.1007/s00216-021-03267-x

Hu K, Peris A, Torán J, Eljarrat E, Sarrà M, Blánquez P, Caminal G (2020) Exploring the degradation capability of Trametes versicolor on selected hydrophobic pesticides through setting sights simultaneously on culture broth and biological matrix. Chemosphere 250:126293

Huang Z, Chen G, Zeng G, Guo Z, He K, Hu L et al (2017) Toxicity mechanisms and synergies of silver nanoparticles in 2,4-dichlorophenol degradation by *Phanerochaete chrysosporium*. J Hazard Mater 321:37–46. https://doi.org/10.1016/j.jhazmat.2016.08.075

Hussain S, Arshad M, Saleem M, Khalid A (2007) Biodegradation of alpha- and beta- endosulfan by soil bacteria. Biodegradation 18(6):731–740. https://doi.org/10.1007/s10532-006-9073-6

Hussaini SS, Shaker M, Iqbal MA (2013) Isolation of fungal isolates for degradation of selected pesticides. Bull Environ Pharmacol Life Sci 2(4):50–53. https://doi.org/10.1016/j.biortech.2014.07.079

Ikehata K, Buchanan ID, Smith DW (2004) Recent developments in the production of extracellular fungal peroxidases and laccases for waste treatment. J Environ Eng Sci 3(1):1–9. https://doi.org/10.1139/s03-049

Jaenicke E, Decker H (2003) Tyrosinases from crustaceans form hexamers. Biochem J 371(2):515–523. https://doi.org/10.1042/BJ20030014

Jain R, Garg V, Yadav D (2014) In vitro comparative analysis of monocrotophos degrading potential of Aspergillus flavus, Fusarium pallidoroseum and Macrophomina sp. Biodegradation 25(3):437–446

Javaid MK, Ashiq M, Tahir M (2016) Potential of biological agents in decontamination of agricultural soil. Scientifica 2016:1598325

Kalyani D, Tiwari MK, Li J, Kim SC, Kalia VC, Kang YC (2015) A highly efficient recombinant laccase from the yeast *Yarrowia lipolytica* and its application in the hydrolysis of biomass. PLoS One *10:e0120156*

Kanade SR, Paul B, Rao AA, Gowda LR (2006) The conformational state of polyphenol oxidase from field bean (Dolichos lablab) upon SDS and acid-pH activation. Biochem J 395(3):551–562. https://doi.org/10.1042/BJ20051541

Kersten PJ, Kalyanaraman B, Hammel KE, Reinhammar B, Kirk TK (1990) Comparison of lignin peroxidase, horseradish peroxidase and laccase in the oxidation of methoxybenzenes. Biochem J 268(2):475–480. https://doi.org/10.1042/bj2680475

Kirk TK, Farrell RL (1987) Enzymatic" combustion": the microbial degradation of lignin. Ann Rev Microbiol 41(1):465–501. https://doi.org/10.1146/annurev.mi.41.100187.002341

Kullman SW, Matsumura F (1996) Metabolic pathways utilized by phanerochaete chrysosporium for degradation of the cyclodiene pesticide endosulfan. Appl Environ Microbiol 62:593–600. https://doi.org/10.1128/aem.62.2.593-600.1996

Kulshrestha G, Kumari A (2011) Fungal degradation of chlorpyrifos by Acremonium sp. strain (GFRC-1) isolated from a laboratory enriched red agricultural soil. Biol Fertil Soils 47(2):219–225

Lai F, Tian F, Xu B, Ye W, Gao Y, Chen C, Xing H, Wang B, Xie M, Hu X (2022) A comparative study on the degradation of phenylurea herbicides by UV/ persulfate process: kinetics, mechanisms, energy demand and toxicity evaluation associated with DBPs. Chem Eng J 428:132088

Litvintseva AP, Henson JM (2002) Cloning, characterization, and transcription of three laccase genes from *Gaeumannomyces graminis* var. *tritici,* the take-all fungus. Appl Environ Microbiol 68(3):1305–1311. https://doi.org/10.1128/AEM.68.3.1305-1311.2002

Liu YH, Chung YC, Xiong Y (2001) Purification and characterization of a dimethoate- degrading enzyme of Aspergillus Niger ZHY256, isolated from sewage. Appl Environ Microbiol 67(8):3746–3749

Mandic M, Djokic L, Nikolaivits E, Prodanovic R, O'Connor K, Jeremic S, Topakas E, Nikodinovic-Runic J (2019) Identification and characterization of new laccase biocatalysts from *Pseudomonas* species suitable for degradation of synthetic textile dyes. Catalysts 9(7):629. https://doi.org/10.3390/catal9070629

Mansur M, Suárez T, Fernández-Larrea JB, Brizuela MA, Gonzalez AE (1997) Identification of a laccase gene family in the new lignin-degrading basidiomycete CECT 20197. Appl Environ Microbiol 63(7):2637–2646. https://doi.org/10.1128/AEM.63.7.2637-2644.1997

Mayer AM, Harel E, Ben-Shaul R (1966) Assay of catechol oxidase—a critical comparison of methods. Phytochemistry 5(4):783–789. https://doi.org/10.1016/S0031-9422(00)86276-4

Mehrotra S, Sandhir R, Chandra D (2004) Degradation of xenobiotics and bioremediation. In: Singh DP, Dwivedi SK (eds) Environmental microbiology and biotechnology. New Age International (P) Limited, New Delhi, pp 59–74

Mir Tutusaus JA, Sarrà M, Caminal G (2016) Continuous treatment of non-sterile hospital wastewater by *Trametes versicolor*: how to increase fungal viability by means of operational strategies and pretreatments. J Hazard Mater 318:561–570. https://doi.org/10.1016/j.jhazmat.2016.06.056

Mohapatra D, Rath SK, Mohapatra PK (2021) Evaluating a preparation of malathion- tolerant *Aspergillus Niger* MRU01 for accelerated removal of four organophosphorus insecticides. J Chem Technol Biotechnol 96(6):1603–1610. https://doi.org/10.1002/jctb.6714

Morawski FDM, Winiarski JP, de Campos CEM, Parize AL, Jost CL (2020) Sensitive simultaneous voltammetric determination of the herbicides diuron and isoproturon at a platinum/chitosan bio-based sensing platform. Ecotox Environ Safe 206:111181

Mougin C, Pericaud C, Malosse C, Laugero C, Asther M (1996) Biotransformation of the insecticide lindane by the white rot basidiomycete *Phanerochaete chrysosporium*. Pesticide Sci 47:51–59. https://doi.org/10.1002/(SICI)1096-9063(199605)47:1<51::AID-PS391>3.0.CO;2-V

Narayanan KB, Sakthivel N (2011) Synthesis and characterization of nano-gold composite using *Cylindrocladium floridanum* and its heterogeneous catalysis in the degradation of 4-nitrophenol. J Hazard Mater 189:519–525. https://doi.org/10.1016/j.jhazmat.2011.02.069

Nteping TS, Alakeh MN, Tamungang NEB, Awah TM (2022) Analysis of pesticide residues in some market Garden crops in Santa, North west Cameroon. J Environ Protect 13(5):331–343. https://doi.org/10.4236/jep.2022.135025

Nykiel-Szymanska J, Bernat P, Słaba M (2020) Biotransformation and detoxification of chloroacetanilide herbicides by *Trichoderma* spp. with plant growth-promoting activities. Pestic Biochem Physiol 163:216–226. https://doi.org/10.1016/j.pestbp.2019.12.008

Oliveira BR, Penetra A, Cardoso VV, Benoliel MJ, Crespo MB, Samson RA, Pereira VJ (2015) Biodegradation of pesticides using fungi species found in the aquatic environment. Environ Sci Pollut Res Int 22(15):11781–11791. https://doi.org/10.1007/s11356-015-4362-1

Osano AA, Siboe GM, Ochanda JO, Kokaro JO (1999) Biodegradation of DDT: the role of *Pleurotus* sp., A Lingnicolous fungus. In: Alleman BC, Leeson A (eds) Bioremediation of Nitroaromatic and Haloaromatic compounds. Battelle Press, Columbus, pp 237–251

Ovais M, Khalil AT, Ayaz M, Ahmad I, Nethi SK, Mukherjee S (2018) Biosynthesis of metal nanoparticles via microbial enzymes: a mechanistic approach. Int J Mol Sci 19:4100. https://doi.org/10.3390/ijms19124100

Palmieri G, Cennamo G, Faraco V, Amoresano A, Sannia G, Giardina P (2003) Atypical laccase isoenzymes from copper supplemented *Pleurotus ostreatus* cultures. Enzym Microb Technol 33(2–3):220–230. https://doi.org/10.1016/s0141-0229(03)00184-7

Parra D, Reyes-Lopez FE, Tort L (2015) Mucosal immunity and B cells in teleosts: effect of vaccination and stress. Front Immunol 6:354. https://doi.org/10.3389/fimmu.2015.00354

Paul C, Pal N, Maitra M, Das N (2024) Laccase-assisted bioremediation of pesticides: scope and challenges. Mini Rev Organ Chem 21(6):633–654. https://doi.org/10.2174/1570193X20666221117161033

Perez-Gilabert M, Morte A, Honrubia M, Garcia-Carmona F (2001) The partial purification and characterization and histochemical localization of fully latent desert truffle (Terfezia claveryl, Chatin) polyphenol oxidase. Physiol Plant 49:1922–1927. https://doi.org/10.1034/j.1399-3054.2001.1120521.x

Perez-Gilabert M, Morte A, Garcia-Carmona F (2004) Histochemical and biochemical evidences of the reversibility of tyrosinase activation by SDS. Plant Sci 166:365–370. https://doi.org/10.1016/j.plantsci.2003.09.004

Podbielska M, Książ ek P, Szpyrka, E (2020) Dissipation kinetics and biological degradation by yeast and dietary risk assessment of fluxapyroxad in apples. Sci Rep 10:21212. https://doi.org/10.1038/s41598-020-78177-6

Purnomo AS, Mori T, Kamei I, Nishii T, Kondo R (2010) Application of mushroom waste medium from *Pleurotus ostreatus* for bioremediation of DDT contaminated soil. Int Biodeterior Biodegrad 64:397–402. https://doi.org/10.1016/j.ibiod.2010.04.007

Purnomo AS, Ashari K, Hermansyah FT (2017a) Evaluation of the synergistic effect of mixed cultures of white-rot fungus *Pleurotus ostreatus* and biosurfactant-producing bacteria on DDT biodegradation. J Microbiol Biotechnol 27(7):1306–1315

Purnomo AS, Nawfa R, Martak F, Shimizu K, Kamei I (2017b) Biodegradation of aldrin and dieldrin by the white-rot fungus *Pleurotus ostreatus*. Curr Microbiol 74(3):320–324

Purnomo AS, Maulianawati D, Kamei I (2019a) Ralstonia pickettii enhance the DDT biodegradation by *Pleurotus ostreatus*. Curr Microbiol 76(1):134–139. https://doi.org/10.1007/s00284-018-1611-8

Purnomo AS, Maulianawati D, Kamei I (2019b) *Ralstonia pickettii* enhance the DDT biodegradation by *Pleurotus eryngii*. J Microbiol Biotechnol 28;29(9):1424–1433. https://doi.org/10.4014/jmb.1906.06030

Raffa CM, Chiampo F (2021) Bioremediation of agricultural soils polluted with pesticides: a review. Bioengineering 8(7):92

Rasha FM (2017) Intracellular siderophore detection in an Egyptian, cobalt-treated F. solani isolate using SEM-EDX with reference to its tolerance. Pol J Microbiol 66:235–243. https://doi.org/10.5604/01.3001.0010.7856

Rast DM, Baumgarten D, Mayer C, Hollenstein GO (2003) Cell wall-associated enzymes in fungi. Phytochemistry 64:339–366

Rivera-Hoyos CM, Morales-Álvarez ED, Poveda-Cuevas SA, Reyes-Guzmán EA, Poutou Piñales RA, Reyes-Montaño EA et al (2015) Computational analysis and low-scale constitutive expression of laccases synthetic genes GlLCC1 from *Ganoderma lucidum* and POXA 1B from *Pleurotus ostreatus* in *Pichia pastoris*. PLoS One 10(5):e0125864. https://doi.org/10.1371/journal.pone.0125864

Rodríguez E, Ruiz-Dueñas FJ, Kooistra R, Ram A, Martínez ÁT, Martínez MJ (2008) Isolation of two laccase genes from the white-rot fungus *Pleurotus eryngii* and heterologous expression of the pel3 encoded protein. J Biotechnol 134(1–2):9–19

Rodrigues GN, Alvarenga N, Vacondio B, de Vasconcellos SP, Passarini MR, Seleghim MH, Porto AL (2016) Biotransformation of methyl parathion by marine-derived fungi isolated from ascidian Didemnum ligulum. Biocatal Agric Biotechnol 7:24-30

Ruiz-Dueñas FJ, Martínez ÁT (2009) Microbial degradation of lignin: how a bulky recalcitrant polymer is efficiently recycled in nature and how we can take advantage of this. Microb Biotechnol 2(2):164–177

Safferman SI, Lamar RT, Vonderhaar S, Neogy R, Haught RC, Krishnan ER (1995) Treatability study using *Phanerochaete sordida* for the bioremediation of DDT contaminated soil. Toxicol Environ Chem 50:237–251. https://doi.org/10.1080/02772249509358013

Saikia N, Gopal M (2004) Biodegradation of beta-cyfluthrin by fungi. J Agric Food Chem 52(5):1220–1223

Sariwati A, Purnomo AS, Kamei I (2017) Abilities of co-cultures of brown-rot fungus *Fomitopsis pinicola* and *Bacillus subtilis* on biodegradation of DDT. Curr Microbiol 74(9):1068–1075

Sellés Vidal L, Kelly CL, Mordaka PM, Heap JT, Green J (2018) BRENDA in 2019: a European ELIXIR core data resource. Nucleic Acids Res 47(D1):D542–D549. https://doi.org/10.1093/nar/gky1048

Semple KT et al (2007) Microbial interactions with organic contaminants in soil: definitions, processes and measurement. Environ Pollut 150(1):166–176

Sharma PK, Mayank M, Ojha CS, Shukla SK (2020) A review on groundwater contaminant transport and remediation. ISH J Hydraul Eng 26(1):112–121

Shi C, Zhu N, Cao Y, Wu P (2015) Biosynthesis of gold nanoparticles assisted by the intracellular protein extract of *Pycnoporus sanguineus* and its catalysis in degradation of 4- nitroaniline. Nanoscale Res Lett 10:1–8. https://doi.org/10.1186/s11671-015-0856-9

Sing NN, Zulkharnain A, Roslan HA, Assim Z, Husaini A (2014) Bioremediation of PCP by *Trichoderma* and *Cunninghamella* strains isolated from sawdust. Braz Arch Biol Technol 57(6):811–820

Singh DP, Dwivedi SK (2004) Environmental microbiology and biotechnology. New Age, Kolkata

Singh SB, Lal SP, Pant S, Kulshrestha G (2008) Degradation of atrazine by an acclimatized soil fungal isolate. J Environ Sci Health 43(1):27-33

Singh N, Basu S, Vankelecom IF, Balakrishnan M (2015) Covalently immobilized laccase for decolourization of glucose-glycine Maillard products as colourant of distillery wastewater. Appl Biochem Biotechnol 177:76–89. https://doi.org/10.1007/s12010-015-1814-4

Soares PRS, Birolli WG, Ferreira IM, Porto ALM (2021) Biodegradation pathway of the organophosphate pesticides chlorpyrifos, methyl parathion and profenofos by the marine-derived fungus *Aspergillus sydowii* CBMAI 935 and its potential for methylation reactions of phenolic compounds. Mar Pollut Bull 166:112185. https://doi.org/10.1016/j.marpolbul.2021.112185

Soden BJ, Wetherald RT, Stenchikov GL, Robock A (2002) Global cooling after the eruption of Mount Pinatubo: a test of climate feedback by water vapor. Science 296(5568):727–730

Solís-Oba M, Almendáriz J, Viniegra-González G (2008) Biotechnological treatment for colorless denim and textil wastewater treatment with laccase and ABTS. Rev Int Contam Ambient 24(1):5–11

Sondhia S, Waseem U, Varma RK (2013) Fungal degradation of an acetolactate synthase (ALS) inhibitor pyrazosulfuron-ethyl in soil. Chemosphere 93(9):2140–2147

Sono M, Roach MP, Coulter ED, Dawson JH (1996) Heme-containing oxygenases. Chem Rev 96(7):2841-2888

Sonoki T, Kajita S, Iimura Y (2012) Phytoremediation of bis-phenol A via secretory fungal peroxidases produced by transgenic plants. INTECH Open Access Publisher

Strong PJ, Claus H (2011) Laccase: a review of its past and its future in bioremediation. Crit Rev Environ Sci Technol 41(4):373–434

Sugano Y, Yoshida T (2021) DyP-type peroxidases: recent advances and perspectives. Int J Mol Sci 22(11):5556. https://doi.org/10.3390/ijms22115556

Sunny NE, Kaviya A, Kumar SV (2022) Mechanistic approach on the synthesis of metallic nanoparticles from microbes. In: Agri-waste and microbes for production of sustainable nanomaterials. Elsevier, pp. 577-602

Vazquez M, Bianchinotti MV (2013) Isolation of metsulfuron methyl degrading fungi from agricultural soils in Argentina. Phyton 1(82):113–118

von Netzer F, Granitsiotis MS, Szalay AR, Lueders T (2018) Nextgeneration sequencing of functional marker genes for anaerobic degraders of petroleum hydrocarbons in contaminated environments. In: Anaerobic utilization of hydrocarbons, oils, and lipids. Handbook of hydrocarbon and lipid microbiology

Wahleithner JA, Xu F, Brown KM, Brown SH, Golightly EJ, Halkier T, Kauppinen S, Pederson A, Schneider P (1996) The identification and characterization of four laccases from the plant pathogenic fungus *Rhizoctonia solani*. Curr Genet 29:395–403

Wichers HJ, Recourt K, Hendriks M, Ebbelaar CFM, Biancone G, Hoeberichts FA et al (2003) Cloning, expression and characterisation of two tyrosinase cDNAs from *Agaricus bisporus*. Appl Microbiol Biotechnol 61:336–341

Wick LY, Remer R, Würz B, Reichenbach J, Braun S, Schäfer F et al (2007) Effect of fungal hyphae on the access of bacteria to phenanthrene in soil. Environ Sci Technol 41:500–505. https://doi.org/10.1021/es061407s

Wolfand JM, Lefevre GH, Luthy RG (2016) Metabolization and degradation kinetics of the urban-use pesticide fipronil by white rot fungus *Trametes versicolor*. Environ Sci Proc Impacts 18:1256–1265. https://doi.org/10.1039/C6EM00344C

Wong KS, Cheung MK, Au CH, Kwan HS (2013) A novel *Lentinula edodes* laccase and its comparative enzymology suggest guaiacol-based laccase engineering for bioremediation. PLoS One 8:e66426. https://doi.org/10.1371/journal.pone.0066426

Xiao YZ, Hong YZ, Li JF, Hang J, Tong PG, Fang W, Zhou CZ (2006) Cloning of novel laccase isozyme genes from *Trametes* sp. AH28-2 and analyses of their differential expression. Appl Microbiol Biotechnol 71:493–501

Xiao P, Mori T, Kamei I, Kiyota H, Takagi K, Kondo R (2011) Novel metabolic pathways of organochlorine pesticides dieldrin and aldrin by the white rot fungi of the genus *Phlebia*. Chemosphere 85:218–224. https://doi.org/10.1016/j.chemosphere.2011.06.028

Xu G et al (2007) Mineralization of chlorpyrifos by co-culture of *Serratia* and *Trichosporon* spp. Biotechnol Lett 29(10):1469–1473

Yadav U, Choudhury P (2014) Biodegradation of sulfosulphuron in agricultural soil by *Trichoderma* sp. Lett Appl Microbiol 59(5):479–486

Yadav R, Kumar V, Baweja M, Shukla P (2018) Gene editing and genetic engineering approaches for advanced probiotics: a review. Crit Rev Food Sci Nutr 58:1735–1746. https://doi.org/10.1080/10408398.2016.1274877

Yang C et al (2006) Cloning of mpd gene from a chlorpyrifos-degrading bacterium and use of this strain in bioremediation of contaminated soil. FEMS Microbiol Lett 265(1):118–125

Yaver DS, Xu F, Golightly EJ, Brown KM, Brown SH, Rey MW, Schneider P, Halkier T, Mondorf K, Dalboge H (1996) Purification, characterization, molecular cloning, and expression of two laccase genes from the white rot basidiomycete *Trametes villosa*. Appl Environ Microbiol 62:834–841

Younus H (2019) Oxidoreductases: overview and practical applications. In: Husain Q, Ullah M (eds) Biocatalysis. Springer, Cham. https://doi.org/10.1007/978-3-030-25023-2_3

Yu QK, Han LT, Wu YJ, Liu TB (2020) The role of oxidoreductase-like protein Olp1 in sexual reproduction and virulence of Cryptococcus neoformans. Microorganisms 8(11):1730. https://doi.org/10.3390/microorganisms8111730

Zeng S, Qin X, Xia L (2017) Degradation of the herbicide isoproturon by laccase- mediator systems. Biochem Eng J 119:92–100. https://doi.org/10.1016/j.bej.2016.12.016

Zhang H, Zhou H, Bai J, Li Y, Yan J, Ma Q et al (2019) Biosynthesis of selenium nanoparticles mediated by fungus *Mariannaea* sp. HJ and their characterization. Colloids Surf A Physicochem Eng Asp 571:9–16. https://doi.org/10.1016/j.colsurfa.2019.02.070

Zhao J, Chi Y, Xu Y, Jia D, Yao K (2016) Co-metabolic degradation of b-cypermethrin and 3-phenoxybenzoic acid by co-culture of *Bacillus licheniformis* B-1 and *Aspergillus oryzae* M-4. PLoS One 11(11):e0166796

Zhu YZ, Keum YS, Yang L, Lee H, Park H, Kim JH (2010) Metabolism of a fungicide mepanipyrim by soil fungus *Cunninghamella elegans* ATCC36112. J Agric Food Chem 58(23):12379–11234

Zhu Y, Li J, Yao K, Zhao N, Zhou K, Hu X, Zou L, Han X, Liu A, Liu S (2016) Degradation of 3-phenoxybenzoic acid by a filamentous fungus *Aspergillus oryzae* M-4 strain with self-protection transformation. Appl Microbiol Biotechnol 100(22):9773–9786

Zhuo R, Fan F (2021) A comprehensive insight into the application of white rot fungi and their lignocellulolytic enzymes in the removal of organic pollutants. Sci Total Environ 778:146132. https://doi.org/10.1016/j.scitotenv.2021.146132

Chapter 21
Microbial Fuel Cell as Biotechnological Interventions in Removal of Emergent Pollutants: A Comprehensive Review

Rishi Raj and Sangeeta Tiwari

Abstract Emergent contaminants and their possible negative impacts on ecosystems and human health have become a significant source of worry in recent years. Traditional wastewater treatment methods have faced challenges in effectively removing these contaminants, thereby creating an urgent demand for innovative solutions. Among the emerging technologies, microbial fuel cells (MFCs) have gained prominence as a promising biotechnological approach for addressing the elimination of emergent pollutants from diverse wastewater sources.

This comprehensive chapter reviews profoundly into the utilizations of MFCs in the expulsion of new contaminations, enveloping a wide exhibit of risky substances, including drugs, individual consideration items, endocrine-disturbing synthetic compounds, and weighty metals. Not only does it highlight the practical applications, but it also explores the underlying mechanisms that make MFCs effective in this situation, clarifying recent advancements and developments in the field.

The survey indicated that growing use of MFCs creates challenges, but it also provides insights on MFCs' feasibility and safety to the ecosystem, beginning by emphasizing their potential to reduce the natural impact of emerging contaminants.

Keywords Microbial fuel cells · Emergent pollutants · Biodegradation · Adsorption · Biotechnological interventions · Wastewater treatment · Sustainability · Removal efficiency · Innovative approaches

R. Raj (✉) · S. Tiwari
Amity Institute of Applied Sciences, Amity University, Noida, India
e-mail: rishi.raj8@s.amity.edu

S. Dey, S. Bhattacharya (eds.), *Biotechnological Interventions in the Removal of Emerging Pollutants*, Interdisciplinary Biotechnological Advances,
https://doi.org/10.1007/978-981-97-9922-0_21

21.1 Introduction

Emergent pollutants, often referred to as emerging contaminants or contaminants of emerging concern (CECs), comprise a diverse array of organic and inorganic substances that have recently come to light as possible hazards to both the environment and human well-being (Suresh et al. 2022). This category of pollutants encompasses pharmaceuticals, personal care products, endocrine-disrupting compounds, heavy metals, and various other substances (Sanchez and Egea 2018). The conventional wastewater treatment methods frequently prove insufficient in efficiently eliminating these pollutants from wastewater effluents, prompting the need for the exploration of innovative and sustainable technologies, such as microbial fuel cells (MFCs) (Suresh et al. 2022; Contaminants of Emerging Concern 2023).

21.1.1 Background and Context of Emergent Pollutants

Emergent pollutants, also recognized as emerging contaminants, constitute a diverse assemblage of chemical substances that have recently garnered attention as potential environmental hazards (Vinayak et al. 2021). This gathering of pollutants encompasses a broad spectrum of compounds, incorporating pharmaceuticals, personal care products, endocrine-disrupting chemicals, and heavy metals, among others (Sanchez and Egea 2018). A few variables contribute to the emergence of these pollutants, including advances in detection methods, evolving consumption patterns, and their inclination for chemical persistence. Consequently, emergent pollutants are increasingly being detected in surface waters, groundwater, and aquatic ecosystems, thereby presenting environmental and health risks (Antunes et al. 2021; Yadav et al. 2021).

Addressing the presence of emanant pollutants in the environment requires effective expulsion strategies. A review of the literature reveals various resources that explain the types and specific categories of emerging contaminants, encompassing industrial chemicals, pesticides, pharmaceuticals, and personal care products (Vinayak et al. 2021; Abdulrazaq et al. 2020). The most prevalent classes/categories of emergent pollutants include industrial chemicals, disinfection by-products, pesticides, sweeteners, endocrine-disrupting chemicals, and heavy metals (Manisalidis et al. 2020).

The search results also underscore the environmental and health risks associated with emergent pollutants. For instance, exposure to air pollution can impact soil and water quality by contaminating precipitation, infiltrating water and soil environments, and leading to respiratory and cardiovascular diseases, reproductive disorders, central nervous system dysfunction, and cancer (10 Main Causes of Air Pollution That Make Everyone Worry | AQI India 2023). Animal studies have further demonstrated that certain emergent pollutants may induce testicular, uterine,

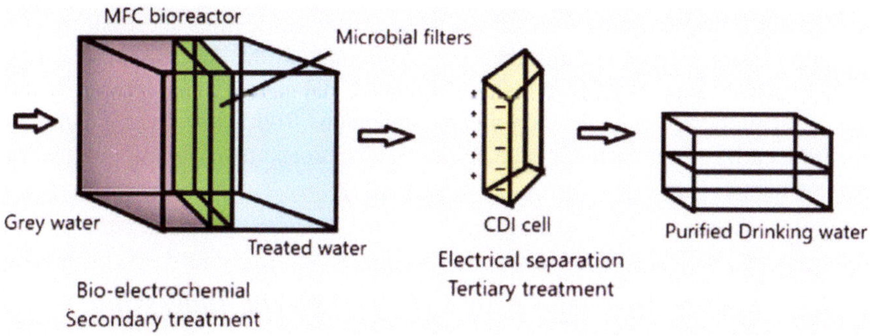

Fig. 21.1 Purified water produced by MFC

and kidney cancer and adversely affect the kidney, immune system, liver, and central nervous system (Lei et al. 2015).

In efforts to mitigate the damages caused by emergent pollutants, various countries have instituted chemical management systems and regulatory frameworks that rely on risk assessment processes utilizing standardized assays for characterizing the toxicological and ecotoxicological hazards of chemicals (Chandel et al. 2022). Nonetheless, the search results indicate that the research and understanding of emergent pollutants remain limited, necessitating further studies to assess the connections between human health outcomes and emerging environmental contaminants (Fig. 21.1).

21.1.2 Significance of Emergent Pollutant Removal

The significance of eliminating emergent pollutants from the environment cannot be overstated due to the numerous critical concerns they pose. Their presence, most importantly, represents a significant gamble to human well-being, as mixtures like drugs and endocrine-disturbing substances can invade drinking water sources, possibly endangering human well-being and upsetting hormonal equilibrium (Evidence rising: the emerging pollutants poisoning our environment 2023; Where Does Air Pollution Come From? - Air (U.S. National Park Service) 2023).

Besides, these contaminations can unleash devastation on environments, prompting a large number of issues. They disturb pecking orders, hurt oceanic life forms, and add to the decrease of biodiversity, accordingly, compromising the fragile equilibrium of amphibian conditions (Pereira et al. 2015). The problem is compounded by the fact that many emergent pollutants are incredibly persistent, accumulating in the environment over time and resulting in long-term ecological consequences (Evidence rising: the emerging pollutants poisoning our environment 2023).

The recognition of these pollutants has required the fixing of regulations connected with water quality principles, in this manner including more strain wastewater treatment offices to address these arising difficulties successfully (Air Pollution Facts, Causes and the Effects of Pollutants in the Air 2023). To handle these issues, it becomes basic to foster imaginative and manageable innovations that can proficiently eliminate developing contaminations from wastewater, guaranteeing the security of our current circumstance and the prosperity of the two people and environments.

21.1.3 Role of Microbial Fuel Cells (MFCs) in Addressing Emergent Pollutants

Microbial fuel cells (MFCs) have emerged as a promising technology for wastewater treatment and energy generation. These bioelectrochemical systems harness the metabolic activities of microorganisms to degrade organic matter and pollutants while simultaneously generating electricity (Evidence rising: the emerging pollutants poisoning our environment 2023).

The unique attributes of MFCs make them well-suited for addressing emergent pollutants:

- Biodegradation capabilities: MFCs support diverse microbial communities capable of biodegrading a wide range of emergent pollutants (Vinayak et al. 2021; Bexfield et al. 2019).
- Synergistic mechanisms: The synergy between pollutant biodegradation and electricity generation provides a sustainable and energy-efficient approach to wastewater treatment (Where Does Air Pollution Come From? - Air (U.S. National Park Service) 2023).
- Resource recovery: MFCs can recover valuable resources, such as electricity and certain metals, from wastewater, enhancing their sustainability (Pereira et al. 2015) (Fig. 21.2).

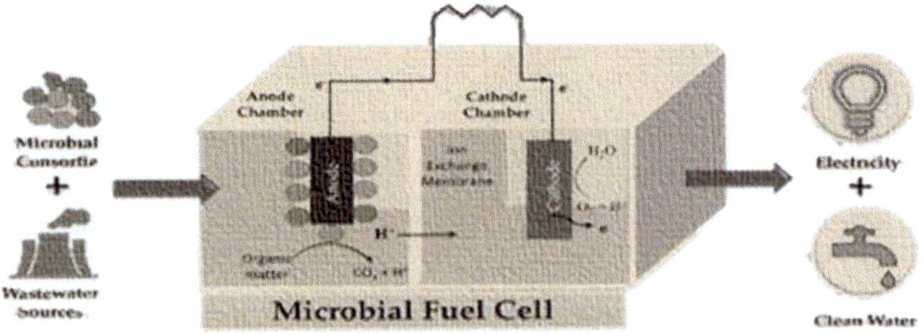

Fig. 21.2 A diagram of microbial fuel cell

21.1.4 Objectives and Organization of the Review

The goals of this survey chapter are complex, expecting to offer a careful assessment of microbial power devices (MFCs) with regards to rising pollutant removal. Most importantly, it tries to give an comprehensive outline of the critical role MFCs play in handling developing pollutants, revealing insight into their importance (Pereira et al. 2015; Emerging Pollutants in Wastewater: An Increasing Threat 2023). Furthermore, it digs into the complex components by which MFCs work with the expulsion of explicit developing pollutants, going from drugs and individual consideration items to endocrine-disturbing chemicals and weighty metals (A Brief History: Pollution Tutorial 2023; N. O. and A. A n.d.). In doing as such, it looks to unwind the science behind their adequacy (Pereira et al. 2015). The chapter likewise directs concentration toward the different variables impact the presentation of MFCs in this imperative undertaking, offering experiences into their pragmatic application (Evidence rising: the emerging pollutants poisoning our environment 2023). Additionally, it highlights ongoing headways and inventive methodologies inside MFC innovation, outlining how this field is developing. Environmental and economic sustainability considerations are likewise tended to, giving an all-encompassing point of view. Finally, the chapter expects to distinguish difficulties and constraints while illustrating future possibilities for MFCs in the evacuation of rising pollutants, proposing important proposals for additional innovative work (Cristina and Silveira Brandão n.d.).

21.2 Microbial Fuel Cells: An Overview

Microbial Fuel Cells (MFCs) represent an innovative bioelectrochemical technology that converts organic matter directly into electricity through the metabolic activities of microorganisms. MFCs certainly stand out in recent years because of their true capacity for sustainable power generation and their applications in wastewater treatment, bioremediation, and presently, as biotechnological mediations for eliminating new emergent pollutants. This segment gives an outline of MFCs, including their essentials, instruments of power generation, and key components (Wang et al. 2022).

21.2.1 Basics of MFCs

Microbial fuel cells (MFCs) represent an innovative biotechnological solution for the removal of emergent pollutants, operating on the fundamental principle of harnessing redox reactions within microbial metabolism to generate electrical power. The intricate components of MFCs are elucidated below:

(a) *Anode*: The anode functions as the epicenter for the oxidation of organic substrates by microorganisms. This pivotal process sees the breakdown of organic matter, releasing electrons into the anode compartment (Zhou et al. 2013).
(b) *Cathode*: Concurrently, the cathode plays a crucial role in facilitating the reduction of electron acceptors, predominantly oxygen or alternative oxidants. By accepting electrons from the anode, this reduction reaction culminates in the completion of the overall electrochemical circuit within MFCs (Malik et al. 2023).
(c) *Proton exchange membrane (PEM)*: In some MFC designs, a PEM separates the anode and cathode compartments. The PEM allows protons (H+) generated at the anode to selectively pass through while blocking the movement of electrons, ensuring that they travel through an external circuit, thus creating an electric current (Bala et al. 2022; Briffa et al. 2020).
(d) *External circuit*: The external circuit serves as the conduit connecting the anode and cathode, facilitating the orchestrated flow of electrons generated at the anode to the cathode. This regulated electron flow through an external load or device translates into the generation of electricity, offering diverse applications in environmental remediation (Abbasi et al. 2016).
(e) *Microbial consortia*: In the intricate realm of MFCs, microorganisms stand as the linchpin, acting as biocatalysts for the oxidation of organic matter. The engagement of diverse microbial communities, particularly exoelectrogenic bacteria, is paramount for the successful electron transfer process within MFC environments (Wu et al. 2020) (Fig. 21.3).

21.2.2 Mechanisms of Electricity Generation

The perplexing process of electricity generation in microbial fuel components (MFCs) unfurls through a progression of essential steps, showcasing the entrancing collaboration between microorganisms, anode materials, and specialized membranes. At the heart of this mechanism, microorganisms inhabiting the anode play a pivotal role. These microorganisms, through a process akin to their breathing known as microbial respiration, adeptly break down pollutants found in wastewater or organic materials. This microbial respiration not only contributes to environmental remediation but also initiates the release of electrons and protons into the MFC system (Badwal et al. 2014).

The liberated electrons embark on a journey, migrating to the anode material, typically composed of conductive substances such as carbon cloth or graphite. This movement of electrons simulates an electric current, a crucial aspect of harnessing the generated electrical power. Concurrently, the protons produced during microbial respiration traverse a specialized membrane towards the cathode. This membrane, often a proton exchange membrane (PEM), selectively allows protons to pass while impeding electron flow, ensuring a controlled and directed movement.

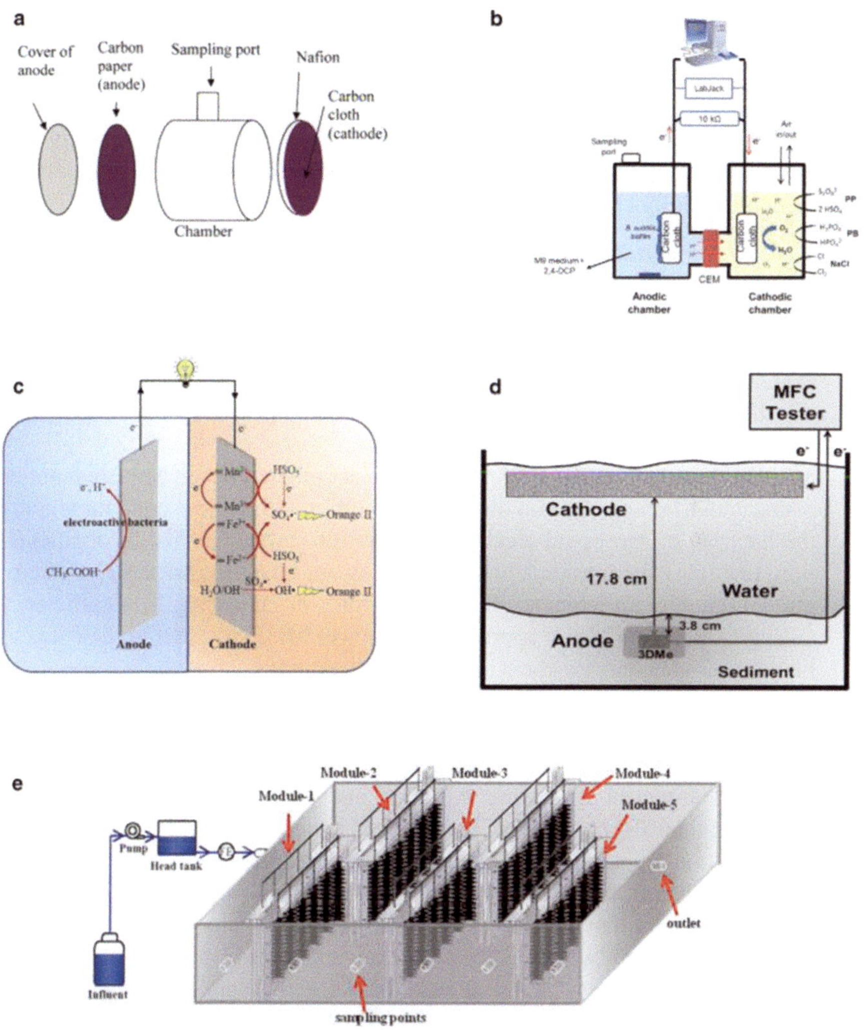

Fig. 21.3 MFC setup

At the cathode, a collaborative interplay unfolds, where electrons, protons, and an electron acceptor (commonly oxygen or an alternative) come together. This orchestrated teamwork results in redox reactions, ultimately producing substances with a lower electron count, such as water. This harmonious collaboration between electrons and protons ensures the sustained flow of electricity.

The continuous movement of electrons through the wire provides a stable electric current, ready to power a myriad of applications, from electronic devices to driving specific chemical reactions (Obileke et al. 2021). Fundamentally, this definite clarification catches the multi-layered components by which MFCs, through the metabolic activities of microorganisms, transform microbial respiration into a sustainable and eco-friendly wellspring of electricity.

21.2.3 Key Components of MFCs

Microbial fuel cells (MFCs) stand at the forefront of biotechnological innovations, leveraging the metabolic prowess of microorganisms to generate electricity. To unlock their full potential, a comprehensive understanding of the key components and meticulous design considerations is imperative.

(a) *Electrode materials*: The selection of electrode materials holds paramount importance in shaping MFC performance. Carbon-based electrodes, such as carbon cloth, graphite, and carbon nanotubes, dominate the landscape due to their commendable attributes—high conductivity and biocompatibility (Obileke et al. 2021). These materials serve as critical interfaces where microbial respiration transmutes into electrical energy, underscoring the significance of their judicious choice.
(b) *Microbial consortia*: The vitality of microbial communities in MFCs cannot be overstated. Efficient electron transfer capabilities within these communities are the linchpin of successful electricity generation. Strategies like bioaugmentation, involving the introduction of specific exoelectrogenic bacteria, or microbial community enrichment, play pivotal roles in enhancing MFC efficiency (Wang et al. 2022). These approaches empower MFCs to excel in their role as both power generators and environmental remediators.
(c) *Proton exchange membrane (PEM)*: In certain MFC configurations, a proton exchange membrane (PEM) takes center stage by segregating the anode and cathode compartments. This separation works with particular proton transport, a basic component impacting particle conductivity and overall durability (Vishwanathan 2021). The selection of PEM material arises as a crucial design consideration, impacting the viability of the MFC framework.
(d) *Reactor configuration*: The architectural blueprint of MFCs encompasses diverse configurations, such as single-chamber or dual-chamber systems. Each configuration comes with its unique set of advantages and disadvantages, necessitating a thoughtful selection aligned with specific applications and performance requirements (Naha et al. 2023). This customized approach guarantees that the MFC configuration harmonizes with its expected purpose, be it energy generation or pollutant elimination.

21.3 Application of MFCs in Removal of Emergent Pollutants

Microbial fuel cells (MFCs) have demonstrated great promise in addressing the removal of emergent pollutants from various wastewater sources. These pollutants encompass a range of organic and inorganic compounds, each posing unique challenges for traditional wastewater treatment methods. Below, we explore the applications of MFCs in the removal of specific categories of emergent pollutants:

21.3.1 Pharmaceuticals

Pharmaceuticals, which include both prescription and over-the-counter medications, are frequently found in wastewater because many people use them. If not appropriately taken out, these substances can harm the climate and individuals' well-being. The incorporation of Microbial Energy components (MFCs) arises as a promising strategy for the proficient evacuation of drugs, offering multi-layered benefits.

Primarily, MFCs cultivate diverse microbial communities, particularly those adapt at pharmaceutical degradation. Inside the anode compartment, microorganisms participate in biodegradation, actually using drug compounds and changing them into less harmful substances. This microbial-driven biodegradation process stands as a pivotal mechanism facilitated by MFCs in pharmaceutical removal (Malik et al. 2023).

Furthermore, MFCs employ a unique electrochemical process to enhance the breakdown of pharmaceuticals. The anodic conditions within MFCs generate a robust electric potential, instigating electrochemical oxidation. This cycle induces responses that result in the fragmentation of pharmaceutical compounds, adding to their powerful expulsion from wastewater.

Crucially, MFCs operate synergistically with other treatment methods, complementing strategies such as the use of specialized materials for pharmaceutical capture and advanced chemical processes. This cooperative methodology guarantees the extensive expulsion of even the most unmanageable pharmaceuticals. MFCs, situated as the underlying stage in wastewater treatment, pave the way for resulting processes, ensuring the thorough elimination of pharmaceutical deposits, and shielding both environmental trustworthiness and human well-being.

21.3.2 Personal Care Products (PCPs)

Personal care products (PCPs), which frequently contain fixings like artificial scents, additives, and antimicrobial specialists, are as often as possible tracked down in wastewater, representing an expected danger to oceanic life and biological systems. The creative utilization of Microbial fuel cells (MFCs) arises as a promising avenue for the evacuation of PCPs, offering a complex methodology.

In tandem with pharmaceutical removal, MFCs demonstrate the capability to facilitate microbial metabolism, enabling the breakdown and elimination of certain PCP components. Within the anode compartment, specific microbial communities exhibit proficiency in dismantling these compounds, effectively diminishing their presence in wastewater (Sen Thapa et al. 2022).

MFCs can also act as special containers that boost the work of microorganisms that break down PCPs. They create a good environment for these helpful microbes, and by providing the right conditions at the anode, MFCs make it easier for the microbes to break down PCPs through a process called biodegradation. This synergistic interplay between MFCs and microbial communities accentuates the efficacy of PCP removal processes.

Even for resilient PCPs resistant to facile degradation, MFCs contribute substantially. They act as platforms to immobilize these compounds onto electrodes or induce chemical transformations, leading to their alteration or complete elimination within the MFC environment. This dual mechanism ensures a comprehensive approach to PCP removal within the MFC setup.

21.3.3 Endocrine-Disrupting Compounds (EDCs)

Endocrine-disrupting compounds (EDCs), exemplified by bisphenol A (BPA) and phthalates, represent a huge danger by impeding hormonal frameworks in both oceanic organisms and people. This subtopic digs into the multi-layered job of Microbial Fuel Cells (MFCs) in contributing to the evacuation of EDCs, utilizing different mechanisms focused on ecological remediation.

- *Microbial Transformation*

A primary mechanism involves the microbial transformation of EDCs within the confines of MFCs. Microorganisms inhabiting the anode compartment of MFCs exhibit the capability to metabolize specific EDCs. Through intricate biodegradation pathways, these microorganisms break down EDCs into less harmful byproducts, minimizing their impact on the environment (Wang et al. 2022).

- *Adsorption and Precipitation*

MFCs facilitate the removal or transformation of EDCs through adsorption and precipitation mechanisms. Certain EDCs possess the propensity to adsorb onto the electrode materials present in MFCs. Additionally, electrochemical reactions induced within the MFC environment can lead to the precipitation of EDCs, rendering them less detrimental to aquatic ecosystems (Selvasembian et al. 2022).

- *Electro-Fenton Processes*

An advanced strategy involves the modification of MFCs to incorporate iron-based electrodes, enabling Electro-Fenton processes. This modification enhances the efficiency of EDC degradation within the MFC setup. Electro-Fenton reactions, driven by iron-based electrodes, offer an effective means of breaking down recalcitrant EDCs, further highlighting the versatility of MFCs in combating environmental contamination (Thulluru et al. 2023).

In summary, the integration of MFCs presents a robust approach to addressing the menace of EDCs, underscoring their potential in environmental bioremediation.

21.3.4 *Heavy Metals*

Heavy metals, like lead, cadmium, and mercury, present persistent and risky contaminations that can possibly accumulate in the climate. Microbial fuel cells (MFCs) offer a multi-layered way to deal with heavy metal evacuation, utilizing a few components. Certain electroactive microorganisms, right off the bat, can mediate the reduction of heavy metal particles to less poisonous structures, successfully immobilizing them on the electrode surface. Furthermore, cathode materials with practical gatherings can adsorb heavy metal particles from wastewater, giving an alternative system for removal. Additionally, microbial cells themselves can go about as biosorbents, sequestering heavy metals through cell surface communications.

21.4 Mechanisms of Emergent Pollutant Removal

Emergent pollutants, including pharmaceuticals, personal care products, endocrine-disrupting compounds, and heavy metals, can be effectively removed from wastewater using various mechanisms within microbial fuel cells (MFCs). Below, we delve into these removal mechanisms:

21.4.1 *Biodegradation*

Biodegradation remains as a principal and perplexing system at the core of Microbial Fuel Cell (MFCs) with regards to the crucial task of eliminating organic emergent pollutants. This fundamental interaction includes the breakdown of complex organic compounds, which might incorporate a wide exhibit of environmental contaminants, into simpler, less harmful substances. This ground-breaking work is done by a different range of microorganisms present inside the MFCs. These microorganisms, frequently alluded to as "biodegraders," have explicit proteins and metabolic pathways that permit them to handle and metabolize new contaminations effectively (Edwards et al. 2022).

One of the critical aspects of biodegradation in MFCs is the presence of microbial consortia inside these frameworks. These consortia are basically powerful networks of microorganisms, each contributing in its unique way to the course of biodegradation. These communities are complicatedly adjusted ecosystems where different species coordinate and synergize to work with the total breakdown of pollutants (Obileke et al. 2021).

In terms of the metabolic pathways involved, biodegraders employ a multitude of strategies to transform emergent pollutants. For instance, pharmaceutical compounds may be metabolized through a series of enzymatic reactions, leading to the formation of less toxic metabolites. This enzymatic transformation is a remarkable example of the microbial capacity to degrade complex organic molecules, making them less harmful to the environment (Roy et al. 2023).

One of the most fascinating aspects of biodegradation in MFCs is the cooperative behavior exhibited by the microbial communities. Different species within these communities work together, often in a highly coordinated manner, to address various stages of the biodegradation process. This synergy enhances the overall efficiency of emergent pollutant removal, demonstrating the power of nature's teamwork.

Moreover, within the realm of MFCs, some microorganisms, known as electroactive bacteria or exoelectrogens, take on a dual role. They take part in the biodegradation cycle as well as add to the exchange of electrons to the anode. This dual usefulness is significant in light of the fact that it considers the generation of electricity during the biodegradation process (Chaturvedi and Verma 2016). The capacity of these microorganisms to deliver electric current and flow while separating pollutants is a promising component for sustainable energy generation and wastewater treatment.

In the quest to optimize MFCs for the effective removal of emergent pollutants, researchers are actively engaged in selecting and enriching microbial communities with enhanced biodegradation capabilities. This optimization process involves the identification of specific biodegraders and fostering their growth within the MFCs. By homing in on these highly effective biodegraders, researchers can target and break down emergent pollutants with precision and efficiency, paving the way for more efficient and environmentally friendly wastewater treatment and energy production systems.

21.4.2 Adsorption Effect of Electrode

Adsorption, a fundamental process in Microbial Fuel Cells (MFCs), involves the physical or chemical adherence of emergent pollutants to the surfaces of solid materials within the system. Several critical aspects of adsorption as a removal mechanism deserve attention. Firstly, the choice of electrode materials is pivotal, as specific materials exhibit high adsorption capacities for particular emergent pollutants (Obileke et al. 2021). Electrodes, especially those modified with activated carbon or graphene, are known to enhance adsorption efficiency, making the selection of electrode materials a crucial factor (Nawaz et al. 2022).

Surface chemistry also plays a significant role in adsorption. The surface properties of electrode materials, such as their area, porosity, and chemical functional groups, influence the capacity and selectivity for different emergent pollutants. The interplay between these surface properties and the emergent pollutant's chemical composition determines the success of the adsorption process (Roy et al. 2023).

Additionally, environmental factors like pH and ionic strength of the wastewater solution can influence the electrostatic interactions between emergent pollutants and the electrode surface, ultimately affecting the efficiency of adsorption (Wang et al. 2022). Manipulating these parameters can help fine-tune the adsorption process to suit the specific characteristics of the pollutants in question.

Notably, adsorption is a reversible process, and researchers have developed MFC systems with the capability to regenerate or desorb adsorbed pollutants from the electrode surfaces. This regeneration feature extends the lifespan of the electrodes and enhances the sustainability of MFCs for pollutant removal, making adsorption a versatile and efficient mechanism in these systems.

21.4.3 Reduction and Precipitation

Reduction and precipitation processes are regularly used to take out destructive inorganic pollutants, like heavy metals, from different sources, with a primary spotlight on wastewater treatment. These processes involve conversion of soluble and toxic metal ions into insoluble and less destructive forms (Palanisamy et al. 2019). Within microbial fuel cells (MFCs), there are essential components of these reduction and precipitation methods:

- *Electrochemical reactions*: The materials of the electrodes in MFCs can facilitate electrochemical reactions, where electrons generated by microbial activities or external electrical input are utilized to transform metal ions into less soluble, often solid, states (Koo and Jung 2021).
- *Redox potential*: The electrical potential on the surface of the electrode is critical in determining the feasibility and speed of reduction reactions. Adjusting this potential can be used to optimize the removal of specific heavy metals.
- *pH control*: Controlling the acidity or alkalinity (pH) in the anode or cathode compartments can influence the solubility and precipitation of heavy metal ions. Adjusting the pH to encourage precipitation can improve the efficiency of pollutant removal.
- *Electrocoagulation*: In certain MFC designs, electrocoagulation processes are applied. These processes involve the co-precipitation of metal ions with iron or aluminum hydroxide particles formed at the electrode's surface.
- *Solid-phase recovery*: Once heavy metals have precipitated, they can be readily separated from the liquid phase, simplifying their retrieval for safe disposal or potential reuse (Koleva et al. 2022) (Fig. 21.4).

In conclusion, microbial fuel cells utilize various mechanisms, including biodegradation, adsorption, reduction, and precipitation, to effectively eliminate emerging pollutants from wastewater.

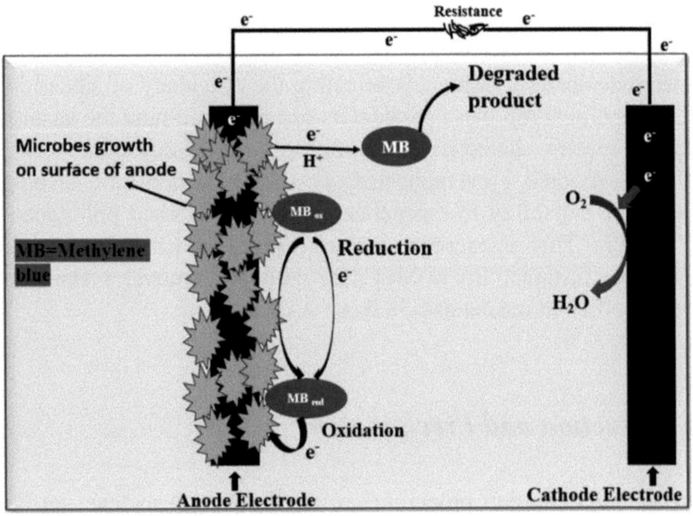

Fig. 21.4 Reduction process

21.5 Factors Influencing MFC Performance

The performance of Microbial fuel cells (MFCs) in removing emergent pollutants and generating electricity is influenced by several key factors. Understanding and optimizing these factors are essential for achieving efficient MFC operation. Here, we explore these factors in detail:

21.5.1 Microbial Consortia

Microbial consortia, complex networks of microorganisms occupying the anode compartment of microbial fuel cells (MFCs), play an essential part in molding the performance and proficiency of these bio electrochemical frameworks. The variety present in the anode biofilm is of fundamental significance. Different microbial networks add to the flexibility of contamination biodegradation, growing the extent of foreign substances that can be successfully treated inside MFCs. This variety, in turn, adds to the strength and stability of MFC activity, making them vigorous instruments for wastewater treatment. Reduction and precipitation processes are regularly used to take out destructive inorganic pollutants, like heavy metals, from different sources, with a primary spotlight on wastewater treatment. These processes involve conversion of soluble and toxic metal ions into insoluble and less destructive forms (Li et al. 2019).

A key subset within these consortia is the presence of exoelectrogenic bacteria, a group capable of transferring electrons to the anode. This electron transfer process is fundamental for the generation of electricity within MFCs. The synergy exhibited by microbial communities further accentuates their significance. Cooperative interactions among different species enhance metabolic pathways, leading to more efficient pollutant degradation and electron transfer. This collaborative behavior results in an overall improvement in the pollutant removal efficiency and electricity generation capacity of MFCs (Koleva et al. 2022).

The development of a well-formed biofilm on the anode surface emerges as another critical aspect. The thickness, composition, and architecture of the biofilm significantly influence mass transport and electron transfer rates, directly impacting the overall performance of MFCs (Koo and Jung 2021). Researchers, recognizing the importance of microbial consortia, have explored bioaugmentation strategies. Bioaugmentation involves the selective introduction of exoelectrogenic bacteria or microorganisms with pollutant-degrading capabilities. This strategic introduction aims to optimize MFC performance, especially concerning specific pollutants or under unique environmental conditions (Fig. 21.5).

21.5.2 Electrode Materials

The performance of microbial fuel cells (MFCs) is intricately linked to the careful selection of electrode materials. This subtopic delves into the multifaceted considerations and impacts associated with electrode material choices in MFCs, providing a detailed analysis for researchers and practitioners.

- *Conductivity*: One of the paramount factors influencing MFC performance is the electrical conductivity of electrode materials. High conductivity is crucial for efficient electron transport from the anode to the cathode. Carbon-based sub-

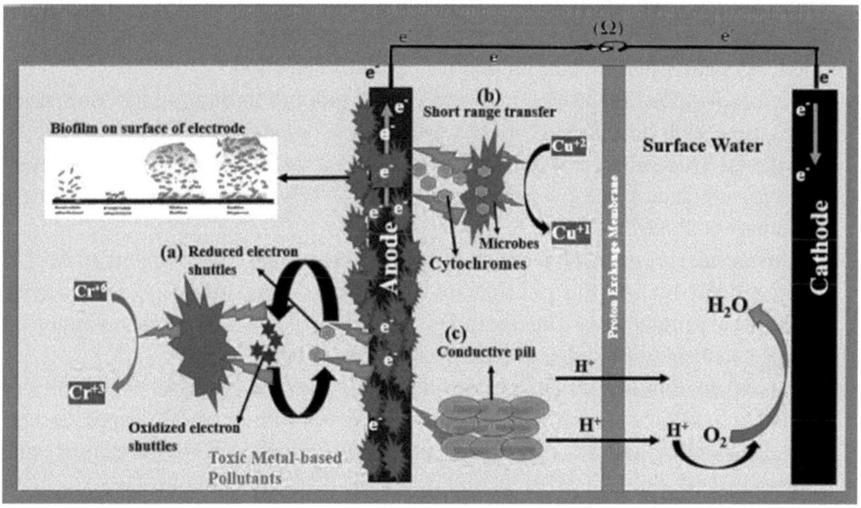

Fig. 21.5 Biofilm on surface of electrode

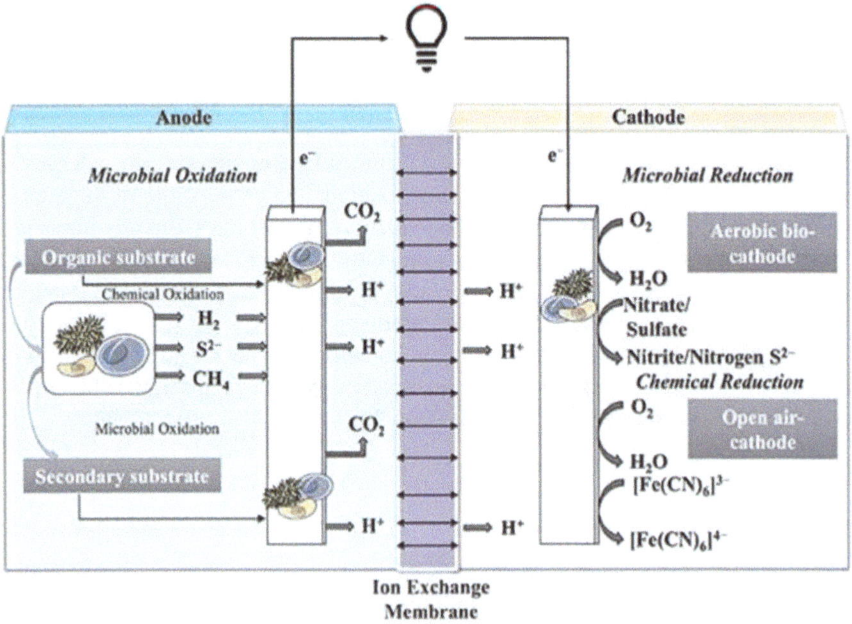

Fig. 21.6 MFC reaction performance

strates, such as carbon cloth and graphite, along with conductive polymers, stand out as common materials in this category (Huang et al. 2021).

- *Surface area*: The surface area of electrodes plays a pivotal role in MFCs. Larger surface areas offer more sites for microbial attachment and pollutant adsorption. This characteristic upgrade electricity generation as well as the evacuation of contaminations. The choice of cathode material altogether impacts this viewpoint, accentuating the requirement for cautious selection.
- *Functionalization*: Electrode materials can be tailored through functionalization to augment their adsorption capacity, selectivity, or catalytic activity for specific pollutants. This strategic modification provides a targeted approach to pollutant removal, enhancing the overall effectiveness of MFCs in wastewater treatment (Agrahari et al. 2022).
- *Corrosion resistance*: Given the challenging operational environments of MFCs, characterized by varying pH and ionic strength, corrosion-resistant electrode materials are imperative. The durability and stability of MFCs hinge on materials that can withstand these harsh conditions (Hamed et al. 2020).
- *Cost and availability*: Practical considerations like the cost and availability of electrode materials are integral to the economic feasibility of MFC applications. The accessibility and affordability of materials contribute to the scalability and widespread adoption of MFC technology (Fig. 21.6 and Table 21.1).

Table 21.1 Findings on electrode materials in microbial fuel cells

Aspect	Description	Common materials	References
Conductivity	Electrical conductivity is crucial for efficient electron transport. Carbon-based substrates and conductive polymers are common materials in this category (Sauvé and Desrosiers 2014).	Carbon cloth, Graphite, Conductive Polymers	(i)
Surface area	Larger surface areas enhance microbial attachment and pollutant adsorption, improving electricity generation and pollutant removal. The choice of electrode material significantly influences this aspect (Naddeo 2020).	Various, depending on material properties	(Huang et al. 2021) (ii)
Functionalization	Tailoring electrode materials through functionalization enhances adsorption capacity, selectivity, and catalytic activity for specific pollutants (Shaffer et al. 2021).	Modified materials with functional groups	(iii)
Corrosion resistance	Corrosion-resistant electrode materials are imperative due to challenging operational environments. Durability and stability of MFCs depend on materials that withstand harsh conditions (Dealing with Emerging Contaminants: Types and Specific 2023).	Corrosion-resistant materials	(Hamed et al. 2020) (iv)
Cost and availability	Practical considerations such as cost and availability are integral to the economic feasibility of MFC applications. Accessibility and affordability contribute to scalability.	Economically viable and readily available	(Agrahari et al. 2022) (v)

21.5.3 Operational Parameters

In the intricate realm of microbial fuel cells (MFCs), achieving optimal functionality demands meticulous control over various operational parameters. This subtopic delves into the nuanced considerations and critical factors that shape the performance of MFCs, shedding light on the intricacies of water treatment and electricity generation.

The nature of the water being treated emerges as a foundational consideration. The composition and concentration of contaminants within the water profoundly impact the efficacy of both pollutant removal and electricity generation processes within the MFC (Li et al. 2019). Changes in water quality necessitate adaptive strategies to ensure consistent and efficient performance.

pH, a measure of water acidity or basicity, stands out as a pivotal parameter influencing MFC dynamics. It intricately affects the functionality of the microbial communities within the MFC and the kinetics of electrochemical reactions (Zhou et al. 2023). Ionic strength, representing the salt content in water, introduces another layer of complexity. Variations in ionic strength alter mass transport and adsorption dynamics, influencing the overall effectiveness of the MFC (Hamed et al. 2020).

Temperature, a seemingly mundane parameter, emerges as a critical determinant of MFC efficiency. The rate of microbial processes and electrochemical reactions is intricately linked to temperature, requiring careful consideration to optimize pollutant removal (Vishwanathan 2021; Li et al. 2019).

The electrical load, representing the external devices connected to the MFC, introduces a dynamic aspect. It directly influences the quantity of electricity generated, necessitating a delicate balance to maximize energy yield (Obileke et al. 2021).

Finally, the hydraulic retention time, depicting the duration water remains in the MFC, emerges as a critical parameter. While prolonged retention times enhance pollutant removal, they may introduce operational trade-offs by potentially slowing down the overall process. Striking the right balance becomes imperative for efficient water treatment and energy generation.

21.5.4 Reactor Configurations

The design and configuration of microbial fuel cell (MFC) reactors wield a substantial influence over their operational efficacy. This subsection scrutinizes the intricate considerations and choices involved in setting up MFC reactors, elucidating the profound impact on performance metrics such as power generation, wastewater treatment efficiency, and pH control.

Initially, we can choose between single-chamber or dual-chamber systems. Dual-chamber MFCs use a special membrane to separate the two parts of the MFC, while single-chamber ones don't have this membrane. The choice affects how efficient the reactor is, how much power it can produce, and how we control the pH (how acidic or basic it is). We can also stack multiple MFC units together in a row or side by side. Stacked ones are used when we need more voltage and power, while when we want to clean up water, we might use them in a row. The way the liquid moves inside the MFC is important too. It affects how well the bad stuff in the water reaches the part where it's cleaned up and how much oxygen gets to the other part. Getting the flow pattern right can make the MFC work better.

The spatial arrangement of MFC units further diversifies the design landscape. Stacking multiple MFCs in series enhances voltage and power output, while aligning them side by side proves beneficial for water treatment applications (Nawaz et al. 2022). The fluid dynamics within the MFC, encompassing the flow pattern of the liquid, emerges as a critical consideration. Achieving an optimal flow pattern ensures efficient transport of contaminants to the treatment zone and facilitates adequate oxygenation, enhancing overall MFC performance.

Tailoring MFC size and shape according to specific needs becomes pivotal. Factors such as electrode spacing and surface area profoundly impact reactor efficiency, requiring a delicate balance between size and performance objectives (Roy et al. 2023).

21.6 Recent Advances and Innovative Approaches

As Microbial Fuel Cells (MFCs) keep on developing as a technology for emergent pollutant expulsion and electricity generation, different recent advances and innovative methodologies have arisen to upgrade their performance and flexibility. Here are some notable developments:

21.6.1 Bioaugmentation Strategies

Bioaugmentation is a valuable strategy in microbial fuel cell (MFC) technology, involving the introduction of specific microorganisms or purposefully designed microbial communities into MFCs to enhance their ability to remove pollutants from wastewater. Recent advancements have delivered a few essential bioaugmentation strategies (Wang et al. 2022). First and foremost, there's the advancement of "designer microbial consortia," where researchers make custom microbial networks with particular pollutant-degrading abilities, fitting them for effective evacuation of specific pollutants. Furthermore, "Synthetic Biology" has made it conceivable to design microorganisms with upgraded pollutant degrading enzymes and metabolic pathways, known as genetically modified organisms (GMOs), which can be introduced into MFCs with work on their performance in pollutant removal and electricity generation (Vishwanathan 2021). At last, the arising field of "microbiome engineering" investigates the control of the microbiome inside MFCs to advance the development and activity of electricity producing exoelectrogenic bacteria and pollutant debasing microorganisms. These innovative bioaugmentation methodologies hold great commitment for making MFCs more adaptable and viable in addressing different water treatment and energy generation challenges.

21.6.2 Nanomaterial Modification

The use of nanomaterials in the change of electrode materials has arisen as a promising way to deal with enhancement of microbial fuel cell (MFC) performance in various aspects. These nanomaterials, including carbon nanotubes, graphene, and conductive polymers, are being incorporated into electrode materials to help their electrical conductivity, working with effective electron transfer (Savla et al. 2020). Moreover, nanomaterials play a part in expanding the surface area and adsorption capacity of the electrodes, which thus upgrades the adsorption of emerging pollutants. They likewise act as catalysts for electrochemical responses, speeding up the degradation of contaminations. Furthermore, nanomaterials can impact the arrangement and strength of biofilms on the electrodes, influencing microbial action and, subsequently, overall MFC performance. In addition, there is developing interest in sustainable nanomaterials like biocompatible nanocellulose, which are being investigated for their similarity with MFC technology, promising headways in eco-friendly MFC designs (Kausar et al. 2023).

21.6.3 Integrated Systems

Integrating microbial fuel cells (MFCs) with other treatment technologies or processes presents a promising avenue for enhancing the overall efficiency of pollutant removal in wastewater treatment (Roy et al. 2023):

- *MFC-bioreactor hybrid systems*: Combining MFCs with traditional bioreactors forms a powerful partnership. MFCs act as a pre-treatment step, preparing wastewater for subsequent biological treatment (Suresh et al. 2022). This collaboration enhances the biodegradation of pollutants, ultimately improving the effectiveness of the entire treatment process.
- *Advanced oxidation processes (AOPs)*: MFCs can be seamlessly integrated with advanced oxidation techniques like electro-Fenton or photocatalysis, enabling the achievement of advanced pollutant degradation and mineralization. This integration is particularly effective for tackling persistent or challenging pollutants in wastewater.
- *Resource recovery*: Integrated systems provide the opportunity to recover valuable resources, such as metals or biogas, from wastewater. This not just upgrades the economic feasibility of MFC technology yet in addition lines up with maintainability objectives by reusing resources, making wastewater treatment more cost-effective and environmentally responsible.

21.6.4 Bioremediation Genes

The incorporation of bioremediation genes or genetic elements into microbial fuel cell (MFC) systems represents a cutting-edge strategy to bolster pollutant removal:

- *Genetically engineered microorganisms*: Microbes can be genetically modified to express specific enzymes or metabolic pathways tailored for the degradation of emerging pollutants (Vinayak et al. 2021), making MFCs more adept at breaking down contaminants.
- *Gene transfer*: Techniques like horizontal gene transfer can be harnessed to introduce pollutant-degrading genes into the microbial community of MFCs, broadening their bioremediation capabilities.
- *Gene expression control*: Advanced genetic engineering allows precise regulation of gene expression in response to pollutant presence, optimizing degradation even under changing conditions (Abbas et al. 2019).

- *Sensor-actuator systems*: MFCs can be equipped with sensor-actuator systems that detect pollutant levels and trigger gene expression in real-time, enabling adaptive and responsive pollutant removal.
- These new progressions and creative strategies feature the capability of MFCs to turn out to be much more incredible tools for eliminating emergent pollutants and producing sustainable energy (Wang et al. 2022).

21.7 Challenges and Limitations in Microbial Fuel Cells

Microbial fuel cells (MFCs) stand at the forefront of sustainable technologies, offering innovative solutions for emergent pollutant removal and sustainable energy generation. However, their implementation is not without challenges and limitations that necessitate careful consideration for their widespread effectiveness. This section delves into some of the key challenges and limitations faced by MFCs:

21.7.1 Low Removal Efficiency

- Challenge 21.7.1.1: *Incomplete Biodegradation*

One significant challenge lies in the potential for incomplete biodegradation of specific emergent pollutants within MFCs. Despite the microbial processes at play, certain pollutants may not undergo thorough degradation, resulting in the formation of partially degraded or recalcitrant compounds. These compounds could persist in the environment, posing potential risks to ecosystems and human health (Obileke et al. 2021).

- Challenge 21.7.1.2: *Variability in Pollutant Removal*

The efficiency of pollutant removal in MFCs is subject to variability influenced by several factors. The type of pollutant, the composition of the microbial community within the MFC, and the specific operating conditions all contribute to the variability in removal rates. Different pollutants may exhibit distinct removal efficiencies, and optimizing conditions for one pollutant may not necessarily translate to effective removal for others. This variability poses a challenge in achieving consistent and high removal rates across a spectrum of emergent pollutants (Nawaz et al. 2022).

As the field of MFCs advances, addressing these challenges becomes paramount for refining the technology's efficacy in emergent pollutant removal and sustainable energy generation (Wang et al. 2022).

21.7.2 Toxicity and Inhibition Effects in Microbial Fuel Cells

- Challenge 21.7.2.1: *Toxicity to Microorganisms*

The intricate microbial consortia within microbial fuel cells (MFCs) face the challenge of toxicity from certain emergent pollutants, notably heavy metals and specific organic compounds. These pollutants exhibit toxic effects, posing a threat to microbial activity within the MFC environment. The consequences extend to a reduction in electricity generation and hindered efficacy in pollutant removal processes (Suresh et al. 2022; Obileke et al. 2021).

- Challenge 21.7.2.2: *Inhibition of Electrochemical Reactions*

Toxicity not only impacts microbial activity but also interferes with essential electrochemical reactions occurring at the anode or cathode interfaces. This hindrance of electrochemical responses is especially unsettling in MFCs intended for heavy metal evacuation applications. The compromised electrochemical processes add to diminished overall MFC execution, featuring the complex connection between microbial activity and electrochemical reactions (Suresh et al. 2022; Bala et al. 2022).

21.7.3 Scaling Up Challenges

- Challenge 21.7.3.1: *Limited Scalability*

While the success of MFC technology is evident at the laboratory scale, transitioning to larger systems for practical applications introduces significant challenges. Achieving scalability without compromising microbial activity and mass transport becomes a complex engineering problem. Maintaining the delicate balance required for optimal MFC performance on a larger scale is a critical consideration in overcoming this challenge (Zhou et al. 2013; Wang et al. 2022).

- Challenge 21.7.3.2: *Space and Infrastructure Requirements*

The implementation of large-scale MFC systems is constrained by substantial space and infrastructure requirements. This challenge becomes particularly relevant in urban or densely populated areas where available space is limited. Addressing the spatial and infrastructural demands is essential for the practical deployment of MFC technology in diverse environments (Malik et al. 2023).

- Challenge 21.7.3.3: *Operational Stability*

Scaling up MFCs introduces increased operational complexity, bringing forth the risk of system instability. Effective management and control systems are imperative to mitigate this risk and ensure operational stability in larger installations. The intricacies of scaling up MFCs necessitate advanced control strategies and robust management practices to uphold their stability (Vishwanathan 2021).

21.7.4 Cost Considerations in Microbial Fuel Cells

Microbial fuel cells (MFCs) present a revolutionary approach to emergent pollutant removal and energy generation; however, they are not without economic challenges. Addressing these challenges is crucial for their widespread adoption and effectiveness.

- Challenge 21.7.4.1: *High Capital Costs*

Constructing MFC systems entails significant initial capital costs, mainly attributed to the procurement of electrode materials, proton exchange membranes (PEMs), and the installation of monitoring/control equipment. These costs pose a barrier to the broader implementation of MFC technology (Roy et al. 2023; Tan et al. 2021).

- Challenge 21.7.4.2: *Maintenance and Operating Costs*

Ongoing maintenance is an essential aspect of MFC operation, involving the regular monitoring of microbial communities, periodic replacement of electrodes, and the resolution of fouling issues. These maintenance requirements contribute to the overall operational expenses of MFC systems, impacting their economic feasibility (Nawaz et al. 2022; Kumar et al. 2019).

- Challenge 21.7.4.3: *Energy Balance*

A persistent challenge in MFC technology is achieving a positive energy balance. In systems with low organic loading rates, MFCs often consume more energy for operation than they can generate. Balancing energy input and output is particularly daunting for smaller-scale applications, requiring innovative solutions for sustainable and efficient performance (Tsekouras et al. 2022).

- Challenge 21.7.4.4: *Economic Viability*

Demonstrating the economic viability of MFC technology remains a significant hurdle, especially when compared to traditional wastewater treatment methods. Establishing a compelling return on investment is crucial for encouraging the widespread adoption of MFCs in various applications (Tan et al. 2021; Kumar et al. 2019) (Table 21.2).

Table 21.2 Challenges and limitations in microbial fuel cells

Challenge	Description	Sources
21.7.1. Low Removal Efficiency	21.7.1.1: Incomplete Biodegradation—Potential for incomplete breakdown of emergent pollutants	(Roy et al. 2023)
	21.7.1.2: Variability in Pollutant Removal—Removal efficiency affected by pollutant type and operating conditions	(Nawaz et al. 2022), (Tan et al. 2021)
21.7.2. Toxicity and Inhibition Effects	21.7.2.1: Toxicity to Microorganisms—Challenge of toxicity from pollutants impacting microbial activity	(Nawaz et al. 2022), (Obileke et al. 2021)
	21.7.2.2: Inhibition of Electrochemical Reactions—Toxicity affecting essential electrochemical processes	(Nawaz et al. 2022), (Obileke et al. 2021)
21.7.3. Scaling Up Challenges	21.7.3.1: Limited Scalability—Transitioning to larger systems poses challenges in maintaining optimal performance	(Obileke et al. 2021), (Sen Thapa et al. 2022)
	21.7.3.2: Space and Infrastructure Requirements—Large-scale implementation constrained by space and infrastructure	(Obileke et al. 2021), (Malik et al. 2023)
	21.7.3.3: Operational Stability—Increased complexity in scaling up introduces the risk of system instability	(https://doi.org/10.1016/B978-0-12-817493-7.00016-3), (Obileke et al. 2021)
21.7.4. Cost Considerations	21.7.4.1: High Capital Costs—Significant initial costs in procuring materials and equipment	https://doi.org/10.1016/j.scitotenv.2018.05.136), (https://doi.org/10.3390/pr9060985)
	21.7.4.2: Maintenance and Operating Costs—Ongoing maintenance contributes to operational expenses	(Nawaz et al. 2022), (https://doi.org/10.1016/j.scitotenv.2018.05.136)
	21.7.4.3: Energy Balance—Challenge in achieving a positive energy balance, especially in low organic loading rates	(https://doi.org/10.3389/fenrg.2022.843768)
	21.7.4.4: Economic Viability—Demonstrating economic viability compared to traditional methods is a significant hurdle	(Kumar et al. 2019)

21.8 Environmental and Economic Sustainability

Microbial fuel cells (MFCs) offer the potential for both environmental and monetary sustainability by tending to emergent pollutant expulsion while at the same time producing renewable power. Here are key aspects of their sustainability:

21.8.1 Energy Recovery

Microbial fuel cells (MFCs) stand out as a sustainable and innovative solution for energy generation. This innovation tackles the power of microorganisms to change organic matter, including emergent pollutants, into electricity. The implications for sustainable energy are significant, offering a renewable source that adds to key environmental and climate objectives.

MFCs play a pivotal role in reducing the carbon footprint by using waste materials as fuel, MFCs effectively add to controlling greenhouse gas emissions. This lines up with globe efforts to battle environmental change, emphasizing the significance of MFCs in the more extensive setting of environmental responsibility (Priya et al. 2022).

A crucial aspect of MFCs is their role in promoting energy independence. By supporting decentralized energy production, MFCs reduce dependence on fossil fuels and centralized power grids. This shift toward decentralized energy aligns with sustainability objectives, providing more resilience in the face of energy challenges (Roy et al. 2023).

Integration into wastewater treatment plants represents a groundbreaking application of MFCs. Beyond energy generation, MFCs have the potential to offset energy consumption in these facilities, making them energy-neutral or even energy-positive. This dual functionality of wastewater treatment and energy generation marks a substantial step forward in achieving sustainable and environmentally responsible energy practices (Wang et al. 2022).

21.8.2 Resource Recycling Through Microbial Fuel Cells

Microbial power modules broaden their effect beyond power generation, offering a comprehensive way to deal with resource recycling from wastewater. This multilayered job incorporates the expulsion of heavy metals from wastewater, introducing a chance for their extraction and subsequent recycling for safe removal. This aligns with the principles of circular economy and resource sustainability, addressing a critical environmental concern.

In addition to metal recovery, MFCs exhibit the capability to generate biogas, such as methane, as a by-product. This biogas can be efficiently harnessed for additional energy production, adding to the overall resource efficiency of MFC systems.

Furthermore, MFCs actively participate in nutrient removal, particularly nitrogen and phosphorus. This opens up avenues for nutrient recovery, providing opportunities for their use as fertilizers. By aligning wastewater treatment with resource conservation and sustainability goals, MFCs contribute to the broader objective of creating environmentally responsible and efficient wastewater management systems (Ezziat et al. 2019).

21.8.3 Life Cycle Assessment

A comprehensive assessment of microbial fuel cell (MFC) systems through a life cycle assessment (LCA) is essential for a holistic understanding of their environmental impact from manufacturing through operation to disposal. This evaluative methodology gives significant insights of knowledge in a few key areas. Firstly, it measures the environmental advantages of MFC execution, including reductions in greenhouse gas emissions, energy savings, and contamination prevention, underlining their true capacity as sustainable innovations. Also, by examining the resource necessities, like materials, water, and energy, LCAs reveal opportunities for resource streamlining and maintainability improvements, adding to more effective and eco-friendly activity, ultimately advancing the broad reception of MFCs in different applications (Fatehbasharzad et al. 2022).

21.8.3.1 Technological Advancements

LCAs can also guide research and development efforts to improve the sustainability of MFCs by identifying areas where environmental impacts are most significant and where innovation can make the most substantial contributions to sustainability (Tsekouras et al. 2022).

21.8.3.2 Comparative Analysis

LCAs allow for comparisons between MFC technology and other wastewater treatment and energy generation methods, helping stakeholders make informed decisions about adopting MFCs based on sustainability criteria (Obileke et al. 2021).

21.9 Future Prospects and Recommendations

Microbial fuel cells (MFCs) offer exciting opportunities for addressing emergent pollutant removal and sustainable energy generation. To unlock their full potential and ensure their successful integration into environmental management, here are some future prospects and recommendations:

21.9.1 Enhanced Biodegradation Pathways

In the pursuit of optimizing microbial fuel cell (MFC) technology for enhanced biodegradation of emergent pollutants, two key recommendations are proposed.

Recommendation 21.9.1.1: *Biodegradation enhancement*: Efforts should be directed towards cutting-edge research aimed at unraveling and augmenting biodegradation pathways within MFCs, specifically tailored for emergent pollutants. This includes a multi-faceted approach, including the distinguishing proof of microbial consortia with intrinsic or engineered capacities for superior biodegradation. Scientific examinations can focus on grasping microbial community dynamics, metabolic pathways, and enzymatic exercises related with the breakdown of designated poisons. This suggestion lines up with work of Rafeeq et al., featuring the significance of microbial consortia in biodegradation processes (Rafeeq et al. 2023). Additionally, insights from Tan et al. emphasize the potential of genetic engineering to enhance microbial capabilities for pollutant removal in bioremediation efforts (Cao et al. 2022).

Recommendation 21.9.1.2: *Bioremediation genes*: Delving deeper into the realm of genetic modifications, researchers are encouraged to explore the intricate landscape of microorganisms within MFCs. The objective is to leverage genetic engineering techniques to induce the efficient expression of pollutant-degrading enzymes and pathways. This tailored approach aims to enhance the overall bioremediation capacity of MFCs. The work of Bala et al. underscores the significance of genetic modification strategies in biodegradation processes, emphasizing the role of engineered genes in pollutant removal (Bala et al. 2022). The controlled articulation of bioremediation genes, especially in response to pollutant presence, remains as a promising road for accomplishing focused on and effective contaminant degradation inside MFCs.

21.9.2 Synergistic Approaches

As microbial fuel cells (MFCs) evolve as innovative tools for pollutant removal and sustainable energy generation, two strategic recommendations are proposed for enhancing their efficacy through synergistic approaches.

Recommendation 21.9.2.1: *Integrated systems*: The integration of MFCs into broader treatment frameworks is paramount for maximizing their potential. Ongoing research should focus on the development and optimization of integrated systems that combine MFCs with diverse treatment technologies. This includes the exploration of advanced oxidation processes (AOPs) and traditional bioreactors, strategically designed to work synergistically with MFCs. Such integrated systems are anticipated to achieve heightened removal efficiencies for emergent pollutants while simultaneously facilitating resource recovery. This recommendation aligns with the findings of Roy et al., emphasizing the versatility of MFCs in wastewater treatment and their capacity for integration with complementary technologies (Roy et al. 2023).

Recommendation 21.9.2.2: *Sensor-actuator systems*: Real-time monitoring of pollutant levels within MFCs is pivotal for responsive and efficient operation. To achieve this, the implementation of sensor-actuator systems is recommended. These systems enable continuous monitoring of pollutant concentrations, allowing for prompt adjustments to operating conditions (Sen Thapa et al. 2022). Additionally, responsive actions, such as initiating gene expression pathways for enhanced pollutant removal, can be triggered. This aligns with the insights from Naik et al., emphasizing the potential of sensor-equipped microbial fuel cells for water quality detection (Naik and Eswari 2021).

21.9.3 Regulatory and Policy Considerations

Recognizing the broader implications of MFC technology, recommendations are made for proactive engagement with regulatory and policy frameworks.

Recommendation 21.9.3.1: *Environmental regulations*: It is imperative to collaborate with regulatory agencies for the development and updating of environmental regulations that explicitly acknowledge and incentivize the utilization of MFCs for emergent pollutant removal and sustainable energy generation. The work of Sivasankar et al. underscores the reliability and renewability of MFCs, emphasizing their role in sustainable power generation (Sivasankar et al. 2019).

Recommendation 21.9.3.2: *Policy support*: Advocacy efforts should be directed toward securing policy support mechanisms to foster the widespread adoption of MFC technology. This involves lobbying for tax incentives, grants, subsidies, and other supportive measures to encourage its application in both wastewater treatment and energy sectors. Zhu et al.'s work highlights the pivotal role of MFCs in addressing environmental pollution and energy shortages, making them deserve of policy backing (Zhu et al. 2022).

21.9.4 Knowledge Gaps and Research Directions

Recommendation 21.9.4.1: *Fundamental research*: Invest in fundamental research to address knowledge gaps in MFC technology, including the optimization of electrode materials, microbial consortia, and reactor configurations for emergent pollutant removal.

Recommendation 21.9.4.2: *Long-term studies*: Conduct long-term studies to assess the performance, stability, and robustness of MFCs in real-world conditions, considering varying influent qualities, temperature fluctuations, and other factors that can impact system performance (Obileke et al. 2021).

Recommendation 21.9.4.3: *Interdisciplinary collaboration*: Encourage interdisciplinary collaboration between microbiologists, engineers, chemists, and environmental scientists to foster innovation and holistic approaches to MFC research and development.

Recommendation 21.9.4.4: *Data sharing and standards*: Promote data sharing and the establishment of standards for assessing MFC performance and sustainability, allowing for more meaningful comparisons and advancements in the field (Vishwanathan 2021).

Recommendation 21.9.4.5: *Education and outreach*: Invest in educational programs and outreach efforts to raise awareness about the potential of MFCs among researchers, practitioners, policymakers, and the public. This can foster greater acceptance and support for MFC technology (Fatehbasharzad et al. 2022) (Table 21.3).

Table 21.3 Recommendations for microbial fuel cells (MFCs) in emergent pollutant removal

Future prospect	Recommendation	Description	References
Enhanced Biodegradation Pathways	21.9.1.1: Biodegradation Enhancement	Direct efforts towards cutting-edge research to unravel and augment biodegradation pathways in MFCs, tailored for emergent pollutants. Focus on microbial consortia and engineered capabilities.	(i), (ii)
	21.9.1.2: Bioremediation Genes	Explore genetic modifications to leverage engineering techniques for efficient expression of pollutant-degrading enzymes and pathways in MFCs. Controlled gene expression for targeted pollutant degradation.	(iii)
Synergistic Approaches	21.9.2.1: Integrated Systems	Integrate MFCs into broader treatment frameworks, focusing on the development and optimization of systems combining MFCs with diverse treatment technologies. Emphasize synergies for heightened pollutant removal and resource recovery.	(iv)
	21.9.2.2: Sensor-Actuator Systems	Implement sensor-actuator systems for real-time monitoring of pollutant levels in MFCs. Enable continuous monitoring, allowing prompt adjustments to operating conditions for efficient pollutant removal.	(v)
Regulatory and Policy Considerations	21.9.3.1: Environmental Regulations	Collaborate with regulatory agencies for the development and updating of environmental regulations, explicitly acknowledging and incentivizing the use of MFCs for emergent pollutant removal and sustainable energy generation.	(vi)
	21.9.3.2: Policy Support	Direct advocacy efforts to secure policy support mechanisms for the widespread adoption of MFC technology. Lobby for tax incentives, grants, subsidies, and other measures to encourage MFC application.	(vii)

(continued)

Table 21.3 (continued)

Future prospect	Recommendation	Description	References
Knowledge Gaps and Research Directions	21.9.4.1: Fundamental Research	Invest in fundamental research to address knowledge gaps in MFC technology, optimizing electrode materials, microbial consortia, and reactor configurations for emergent pollutant removal.	(viii)
	21.9.4.2: Long-Term Studies	Conduct long-term studies assessing the performance, stability, and robustness of MFCs in real-world conditions. Consider varying influent qualities, temperaturefluctuations, and other factors impacting system performance.	(ix)
	21.9.4.3: Interdisciplinary Collaboration	Encourage interdisciplinary collaboration between microbiologists, engineers, chemists, and environmental scientists. Foster innovation and holistic approaches to MFC research and development.	(x)
	21.9.4.4: Data Sharing and Standards	Promote data sharing and establish standards for assessing MFC performance and sustainability. Allow for meaningful comparisons and advancements in the field.	(xi)
	21.9.4.5: Education and Outreach	Invest in educational programs and outreach to raise awareness about the potential of MFCs. Target researchers, practitioners, policymakers, and the public for greater acceptance and support of MFC technology.	(xii)

21.10 Conclusion

Microbial fuel cells offer a promising biotechnological solution for the removal of emergent pollutants from wastewater streams. By harnessing the metabolic exercises of microorganisms, MFCs can biodegrade, adsorb, and lessen many developing foreign substances (Suresh et al. 2022). While huge headway has been made in grasping the components and streamlining MFC execution, a few difficulties remain, including further developing evacuation proficiency, tending to poisonousness concerns, and increasing the innovation. Nevertheless, the environmental and economic sustainability of MFCs, along with their potential for synergistic approaches and integration into existing wastewater treatment systems, make them a valuable tool in the ongoing battle against emergent pollutants (Roy et al. 2023). Proceeded with exploration, advancement, and coordinated effort across multidisciplinary fields are fundamental to understanding the maximum capacity of MFCs in shielding our current circumstance and general well-being (Venkatramanan et al. 2021).

References

10 main causes of air pollution that make everyone worry | AQI India. Accessed: 6 Dec 2023. [Online]. Available: https://www.aqi.in/blog/here-are-the-10-main-causes-of-air-pollution/

A brief history: pollution tutorial. Accessed 6 Dec 2023. [Online]. Available: https://oceanservice.noaa.gov/education/tutorial_pollution/02history.html

Abbas SZ, Rafatullah M, Khan MA, Siddiqui MR (2019) Bioremediation and electricity generation by using open and closed sediment microbial fuel cells. Front Microbiol 10:427179. https://doi.org/10.3389/FMICB.2018.03348/BIBTEX

Abbasi U et al (2016) Anaerobic microbial fuel cell treating combined industrial wastewater: correlation of electricity generation with pollutants. Bioresour Technol 200:1–7. https://doi.org/10.1016/J.BIORTECH.2015.09.088

Abdulrazaq Y et al (2020) Classification, potential routes and risk of emerging pollutants/contaminant. Emerg Contam. https://doi.org/10.5772/INTECHOPEN.94447

Agrahari R, Bayar B, Abubackar HN, Giri BS, Rene ER, Rani R (2022) Advances in the development of electrode materials for improving the reactor kinetics in microbial fuel cells. Chemosphere 290:133184. https://doi.org/10.1016/J.CHEMOSPHERE.2021.133184

Air pollution and your health. Accessed 6 Dec 2023. [Online]. Available: https://www.niehs.nih.gov/health/topics/agents/air-pollution/index.cfm

Air pollution facts, causes and the effects of pollutants in the air. Accessed 6 Dec 2023. [Online]. Available: https://www.nrdc.org/stories/air-pollution-everything-you-need-know

Antunes E et al (2021) Application of biochar for emerging contaminant mitigation. Adv Chem Pollut Environ Manag Protect 7:65–91. https://doi.org/10.1016/bs.apmp.2021.08.003

Badwal SPS, Giddey SS, Munnings C, Bhatt AI, Hollenkamp AF (2014) Emerging electrochemical energy conversion and storage technologies. Front Chem 2:79. https://doi.org/10.3389/fchem.2014.00079

Bala S et al (2022) Recent strategies for bioremediation of emerging pollutants: a review for a green and sustainable environment. Toxics 10(8). https://doi.org/10.3390/TOXICS10080484

Bexfield LM, Toccalino PL, Belitz K, Foreman WT, Furlong ET (2019) Hormones and pharmaceuticals in groundwater used as a source of drinking water across the United States. Environ Sci Technol 53(6):2950–2960. https://doi.org/10.1021/ACS.EST.8B05592

Briffa J, Sinagra E, Blundell R (2020) Heavy metal pollution in the environment and their toxicological effects on humans. Heliyon 6(9). https://doi.org/10.1016/J.HELIYON.2020.E04691

Cao Z, Yan W, Ding M, Yuan Y (2022) Construction of microbial consortia for microbial degradation of complex compounds. Front Bioeng Biotechnol 10:1051233. https://doi.org/10.3389/FBIOE.2022.1051233/BIBTEX

Chandel H, Wang B, Verma ML (2022) Nanomaterial-based fluorescent biosensors for monitoring environmental pollutants. Encyclopedia Sens Biosens 1-4, First Edition:742–754. https://doi.org/10.1016/B978-0-12-822548-6.00146-1

Chaturvedi V, Verma P (2016) Microbial fuel cell: a green approach for the utilization of waste for the generation of bioelectricity. Bioresour Bioprocess 3(1):1–14. https://doi.org/10.1186/S40643-016-0116-6

Contaminants of emerging concern. Accessed 6 Dec 2023. [Online]. Available: https://portal.ct.gov/DEEP/Remediation%2D%2DSite-Clean-Up/Contaminants-of-Emerging-Concern/Contaminants-of-Emerging-Concern

Contaminants of emerging concern - Wikipedia. Accessed 6 Dec 2023. [Online]. Available: https://en.wikipedia.org/wiki/Contaminants_of_emerging_concern

Cristina P, Silveira Brandão C Emerging pollutants a brief overview

Dealing with emerging contaminants: types and specific. Accessed 6 Dec 2023. [Online]. Available: https://www.cascade-env.com/resources/blogs/dealing-with-emerging-contaminants/

Edwards B, Jaswal R, Pathak A, Chauhan A (2022) Current trends in algal biotechnology for the generation of sustainable biobased products. In: Microbial biodegradation and bioremediation:

techniques and case studies for environmental pollution, pp 213–239. https://doi.org/10.1016/B978-0-323-85455-9.00031-X

Emerging pollutants in wastewater: an increasing threat. Accessed 6 Dec 2023. [Online]. Available: https://www.unep.org/events/unep-event/emerging-pollutants-wastewater-increasing-threat

Evidence rising: the emerging pollutants poisoning our environment. Accessed 6 Dec 2023. [Online]. Available: https://www.unep.org/news-and-stories/story/evidence-rising-emerging-pollutants-poisoning-our-environment

Ezziat L, Elabed A, Ibnsouda S, El Abed S (2019) Challenges of microbial fuel cell architecture on heavy metal recovery and removal from wastewater. Front Energy Res 7:410850. https://doi.org/10.3389/FENRG.2019.00001/BIBTEX

Fatehbasharzad P, Aliasghari S, Shaterzadeh Tabrizi I, Khan JA, Boczkaj G (2022) Microbial fuel cell applications for removal of petroleum hydrocarbon pollutants: a review. Water Resour Ind 28:100178. https://doi.org/10.1016/J.WRI.2022.100178

Hamed MS, Majdi HS, Hasan BO (2020) Effect of electrode material and hydrodynamics on the produced current in double chamber microbial fuel cells. ACS Omega 5(18):10339–10348. https://doi.org/10.1021/ACSOMEGA.9B04451/ASSET/IMAGES/LARGE/AO9B04451_0007.JPEG

Huang X et al (2021) Role of electrode materials on performance and microbial characteristics in the constructed wetland coupled microbial fuel cell (CW-MFC): a review. J Clean Prod 301:126951. https://doi.org/10.1016/J.JCLEPRO.2021.126951

Kausar A, Ahmad I, Zhao T, Maaza M, Bocchetta P (2023) Green nanocomposite electrodes/electrolytes for microbial fuel cells—cutting-edge technology. J Compos Sci 7(4):166. https://doi.org/10.3390/JCS7040166

Koleva R, Peeva G, Yemendzhiev H, Nenov V (2022) Potential use of microbial fuel cell technology in wastewater treatment. PRO 10(3):486. https://doi.org/10.3390/PR10030486

Koo B, Jung SP (2021) Improvement of air cathode performance in microbial fuel cells by using catalysts made by binding metal-organic framework and activated carbon through ultrasonication and solution precipitation. Chem Eng J 424:130388. https://doi.org/10.1016/J.CEJ.2021.130388

Kumar SS et al (2019) Microbial fuel cells (MFCs) for bioelectrochemical treatment of different wastewater streams. Fuel 254:115526. https://doi.org/10.1016/J.FUEL.2019.05.109

Lei M et al (2015) Overview of emerging contaminants and associated human health effects. Biomed Res Int 2015. https://doi.org/10.1155/2015/404796

Li F et al (2019) Engineering microbial consortia for high-performance cellulosic hydrolyzates-fed microbial fuel cells. Front Microbiol 10:439965. https://doi.org/10.3389/FMICB.2019.00409/BIBTEX

Malik S et al (2023) A perspective review on microbial fuel cells in treatment and product recovery from wastewater. Water 15(2):316. https://doi.org/10.3390/W15020316

Manisalidis I, Stavropoulou E, Stavropoulos A, Bezirtzoglou E (2020) Environmental and health impacts of air pollution: a review. Front Public Health 8:14. https://doi.org/10.3389/FPUBH.2020.00014

N. O. and A. A. US Department of Commerce, "Nonpoint Source Pollution, NOS Education Offering"

Naddeo V (2020) Development of environmental biotechnology and control of emerging biological contaminants: the grand challenge for a sustainable future. Water Environ Res 92(9):1246–1248. https://doi.org/10.1002/wer.1439

Naha A, Debroy R, Sharma D, Shah MP, Nath S (2023) Microbial fuel cell: a state-of-the-art and revolutionizing technology for efficient energy recovery. Clean Circ Bioecon 5:100050. https://doi.org/10.1016/J.CLCB.2023.100050

Naik S, Eswari JS (2021) Experimental and validation with neural network time series model of microbial fuel cell bio-sensor for phenol detection. J Environ Manag 290:112594. https://doi.org/10.1016/J.JENVMAN.2021.112594

Nawaz A et al (2022) Microbial fuel cells: insight into simultaneous wastewater treatment and bioelectricity generation. Process Saf Environ Prot 161:357–373. https://doi.org/10.1016/J.PSEP.2022.03.039

Obileke KC, Onyeaka H, Meyer EL, Nwokolo N (2021) Microbial fuel cells, a renewable energy technology for bio-electricity generation: a mini-review. Electrochem Commun 125:107003. https://doi.org/10.1016/J.ELECOM.2021.107003

Palanisamy G, Jung HY, Sadhasivam T, Kurkuri MD, Kim SC, Roh SH (2019) A comprehensive review on microbial fuel cell technologies: processes, utilization, and advanced developments in electrodes and membranes. J Clean Prod 221:598–621. https://doi.org/10.1016/J.JCLEPRO.2019.02.172

Pereira LC et al (2015) A perspective on the potential risks of emerging contaminants to human and environmental health. Environ Sci Pollut Res Int 22(18):13800–13823. https://doi.org/10.1007/S11356-015-4896-6

Priya AK et al (2022) Advancements on sustainable microbial fuel cells and their future prospects: a review. Environ Res 210:112930. https://doi.org/10.1016/J.ENVRES.2022.112930

Rafeeq H et al (2023) Genetically engineered microorganisms for environmental remediation. Chemosphere 310:136751. https://doi.org/10.1016/J.CHEMOSPHERE.2022.136751

Roy H et al (2023) Microbial fuel cell construction features and application for sustainable wastewater treatment. Membranes (Basel) 13(5). https://doi.org/10.3390/MEMBRANES13050490/S1

Sanchez W, Egea E (2018) Health and environmental risks associated with emerging pollutants and novel green processes. Environ Sci Pollut Res 25(7):6085–6086. https://doi.org/10.1007/S11356-018-1372-0/METRICS

Sauvé S, Desrosiers M (2014) A review of what is an emerging contaminant. Chem Cent J 8(1):15. https://doi.org/10.1186/1752-153X-8-15

Savla N, Anand R, Pandit S, Prasad R (2020) Utilization of nanomaterials as anode modifiers for improving microbial fuel cells performance. J Renew Mater 8(12):1581–1604. https://doi.org/10.32604/JRM.2020.011803

Selvasembian R et al (2022) Recent progress in microbial fuel cells for industrial effluent treatment and energy generation: fundamentals to scale-up application and challenges. Bioresour Technol 346:126462. https://doi.org/10.1016/J.BIORTECH.2021.126462

Sen Thapa B et al (2022) Application of Microbial Fuel Cell (MFC) for pharmaceutical wastewater treatment: an overview and future perspectives. Sustain For 14(14):8379. https://doi.org/10.3390/SU14148379

Shaffer RM et al (2021) Fine particulate matter and dementia incidence in the adult changes in thought study. Environ Health Perspect 129(8). https://doi.org/10.1289/EHP9018

Sivasankar P, Poongodi S, Seedevi P, Sivakumar M, Murugan T, Loganathan S (2019) Bioremediation of wastewater through a quorum sensing triggered MFC: a sustainable measure for waste to energy concept. J Environ Manag 237:84–93. https://doi.org/10.1016/J.JENVMAN.2019.01.075

Suresh R, Rajendran S, Kumar PS, Dutta K, Vo DVN (2022) Current advances in microbial fuel cell technology toward removal of organic contaminants—a review. Chemosphere 287:132186. https://doi.org/10.1016/J.CHEMOSPHERE.2021.132186

Tan WH et al (2021) Microbial fuel cell technology—a critical review on scale-up issues. PRO 9(6):985. https://doi.org/10.3390/PR9060985

Thulluru LP, Raj R, Chowdhury S, Ghangrekar MM (2023) Microbial fuel cell as an approach for bioremediation of emerging contaminants. In: Management and mitigation of emerging pollutants, pp 221–248. https://doi.org/10.1007/978-3-031-41005-5_8

Tsekouras GJ et al (2022) Microbial fuel cell for wastewater treatment as power plant in smart grids: utopia or reality? Front Energy Res 10:843768. https://doi.org/10.3389/FENRG.2022.843768/BIBTEX

Venkatramanan V, Shah S, Prasad R (2021) A critical review on microbial fuel cells technology: perspectives on wastewater treatment. Open Biotechnol J 15(1):131–141. https://doi.org/10.2174/1874070702115010131

Vinayak V, Khan MJ, Varjani S, Saratale GD, Saratale RG, Bhatia SK (2021) Microbial fuel cells for remediation of environmental pollutants and value addition: special focus on coupling diatom microbial fuel cells with photocatalytic and photoelectric fuel cells. J Biotechnol 338:5–19. https://doi.org/10.1016/J.JBIOTEC.2021.07.003

Vishwanathan AS (2021) Microbial fuel cells: a comprehensive review for beginners. 3 Biotech 11(5):248. https://doi.org/10.1007/S13205-021-02802-Y
Wang J et al (2022) A review of recent advances in microbial fuel cells: preparation, operation, and application. Biotech 11(4):44. https://doi.org/10.3390/BIOTECH11040044
Where does air pollution come from? - Air (U.S. National Park Service). Accessed 6 Dec 2023. [Online]. Available: https://www.nps.gov/subjects/air/sources.htm
Wu Q, Jiao S, Ma M, Peng S (2020) Microbial fuel cell system: a promising technology for pollutant removal and environmental remediation. Environ Sci Pollut Res Int 27(7):6749–6764. https://doi.org/10.1007/S11356-020-07745-0
Yadav D et al (2021) Environmental and health impacts of contaminants of emerging concerns: recent treatment challenges and approaches. Chemosphere 272:129492. https://doi.org/10.1016/J.CHEMOSPHERE.2020.129492
Zhou M, Wang H, Hassett DJ, Gu T (2013) Recent advances in microbial fuel cells (MFCs) and microbial electrolysis cells (MECs) for wastewater treatment, bioenergy and bioproducts. J Chem Technol Biotechnol 88(4):508–518. https://doi.org/10.1002/JCTB.4004
Zhou A, Ángel M, Zavala L, Cámara Gutiérrez IC (2023) Effects of external resistance, new electrode material, and catholyte type on the energy generation and performance of dual- chamber microbial fuel cells. Fermentation 9(4):344. https://doi.org/10.3390/FERMENTATION9040344
Zhu Q et al (2022) Recent advances on the development of functional materials in microbial fuel cells: from fundamentals to challenges and outlooks. Energy Environ Mater 5(2):401–426. https://doi.org/10.1002/EEM2.12173

Chapter 22
Removal of Emerging Contaminants Microplastics by Using Biochar Technology, a Review

Debarati Roy and Suchismita Das

Abstract In recent years, living world is severely threatened by the imminent issues of emergent pollutants. Microplastics (MPs) are one of the major emergent contaminants that possess properties toxic to any organism such as high rigidity, durable characteristics, and extensive resistance in water and soil. MPs transfer through a variety of ecological networks mainly via human activity. Due to this reason, numerous technologies have been originated in order to eradicate MPs, but biochar adsorption process is receiving more limelight because of its widespread availability, efficiency, low energy requirement, eco-friendly technology, and cost-effectiveness in removing MPs from the environment. Biochar functions as bioremediation technique, derived from the pyrolysis of various organic residues. Therefore, the chapter mainly focuses on the various aspects of MPs and their removal methods especially via biochar technology.

Keywords Emergent · Microplastics · Contaminants · Biochar · Adsorption · Pyrolysis

22.1 Introduction

Biochar is a carbon-based porous substance formed in the course of thermochemical breakdown of feedstock (biomass) under 900 °C in the absence of oxygen (Godlewska et al. 2017). Any organic waste material, such as sewage sludge, forest, and agriculture residues, saw dust, manures, algae, and organic municipal solid wastes, and also their compost forms are suitable as biochar's feedstock (Liu et al. 2020). Changes in the physical or chemical attributes and structures of biochar are

D. Roy · S. Das (✉)
Aquatic Toxicology and Remediation Laboratory, Department of Life Science and Bioinformatics, Assam University, Silchar, India
e-mail: suchismita.das@aus.ac.in

S. Dey, S. Bhattacharya (eds.), *Biotechnological Interventions in the Removal of Emerging Pollutants*, Interdisciplinary Biotechnological Advances,
https://doi.org/10.1007/978-981-97-9922-0_22

closely associated with the temperature of pyrolysis (Tomczyk et al. 2020). Biochar technology is gaining a lot of attention due to its ability to remove various pollutants as well as microplastics (MPs).

After the industrial revolution, the increase in population and urbanization forced the world to look after a new type of material for mass productions. The first synthetic polymer was created in the year 1869 by John Wesley Hyatt as a replacement of ivory in the production of billiard balls (Gilbert 2016). Prior to the plastic age, most things were made from natural elements such as wood, stones, sands, bones, metals etc., but because of the cost and rarity of these elements the production capacity was quite limited (Abuwatfa et al. 2021). Plastics entered the field as game changer while eliminating most of humanity's production problems.

Plastics are polymer-based processable materials that are usually treated with different chemical compounds for making them suitable for use. By 2050, it is predicted that there will be about 12,000 metric ton of plastic glut in junkyard and the ecosystem (Geyer et al. 2017). Due to UV radiation exposure and climate change, plastics will break down into tiny particles known as MPs. MPs can be defined as 5 mm long fragments of plastic bits, and once they began damaging the environment it is quite challenging to remove them because of their size. In every region of the world, including Antarctica MPs are found (Stapleton and Hai 2023). MPs of the following categories are frequently found in large quantities in the environment such as polythene (PE), polypropylene (PP), acrylic, polystyrene (PS), polyester (PES), and polyamides (PA) (Jiang 2018).

Since they have the ability to transmit hazardous pollutants as well as toxic pollutants, scientist and public concern over MP contamination have increased dramatically over time. Aquatic habitat unknowingly consumes MPs, which may then enter the food chain and reach the human body. They are entering into the atmosphere by a variety of human activities, from sewage systems, houses and businesses (Kumar et al. 2023). Another significant source of MPs in aquatic bodies is wastewater. Their influence appears on the soil particles stability, fundamental units in the soil structure, which changes the soil environment and its overall density. It may also interfere with nutrient transfer within a plant by obstructing cell walls or links between cells. The increased absorption of MPs by dietary value may offer substantial health hazards to people (Dad et al. 2023).

There have been several methods created to reduce MPs including biological process namely membrane bioreactor and conventional activated sludge, physical methods include ultrafiltration, membrane technologies, adsorption and magnetic separation and chemical methods cover treatments for chemical agglomeration, advanced oxidation, and coagulation. Yet, there are also drawbacks to technology, including high installation and upkeep costs, pH dependency, high energy usage, membrane fouling, and scaling. As a result, these technologies are not practical from a financial standpoint and cannot be widely used in underdeveloped nations. Hence, biochar adsorption technology proves to be more successful and promising approach in removing MPs from environment. The following chapter presents an overall basics of MPs and role of biochar technology in its extermination.

22.2 Microplastics: A Threat to the Ecosystem

As a result of extensive plastic usage, MPs are practically found in all medium. The amount of plastic garbage generated worldwide in the past years has reached astounding 6.3×10^9 tons (Geyer et al. 2017). Recently, COVID-19 pandemic abruptly came out leading to a glaring shortage of waste disposal worldwide. The overuse of protective equipment kit and face masks during this period also increased the pollution. According to estimates, between 0.15 and 0.39 thousand tons of disposable trash will wind up in seas in the upcoming period (Chowdhury et al. 2021).

MPs have a detrimental effect on the biosphere and animal health. It reduces the conductivity of water, porosity of the soil, and microbial growth after they are integrated into the soil. The progress of plants is directly impacted by changes in the chemical and physical qualities of the soil caused by MPs (Zhang and Liu 2018). MP contamination of agroecosystems is a global problem; research has shown that greater quantities of MPs are detected in mulched farmlands than in unoccupied regions. Coated fertilizers and mulch residues are two major sources of MPs in agricultural soil (Ramasamy et al. 2021). The prevailing causes of MPs in plants are reduced photosynthesis, germinating rate, impaired length of roots and shoots in plant and genotoxic and free radical damage (Dad et al. 2023). The MPs size is highly relative to its ability to be absorbed by plants. In addition to soil and water, MPs have been found in the air, which is an underappreciated yet possible route of exposure through the respiratory and oral systems. Urban dust, abrasion from rubber tire wear, furniture, synthetic textile clothes and building materials are the main causes of MPs in the air (Amato-Lourenço et al. 2020).

Plastic materials include substances like antimicrobial agents and nanoparticles that are hazardous to variety of species. For example, researchers reported that it causes some adverse effects on aquatic organism such as inhibited energy storage, decrease feeding capacity, damaged gill tissues and organisms DNA. In terrestrial organism's MPs can enter via three pathways—surface, ingestion, and tissues. When soil fauna including consume MPs, they may experience a range of harmful consequences, including metabolic problems, histopathological damage, oxidative stress, dysbiosis of the intestinal microbiota, DNA damage, reproductive toxicity, genotoxicity, neurotoxicity, or effects on blood and hemolymph. They can also act as bio transform enzymes (Chang et al. 2022). Therefore, studies in this area need to be conducted and monitored as individuals are thought to consume between 39,000 and 52,000 MP particles annually (Stapleton and Hai 2023).

22.3 Origin, Substances, and Transportation of Microplastics

22.3.1 *Origin of Microplastics (Types and How They Release)*

In the year 2004, marine biologist Prof. Richard Thompson first coined the term microplastics (MPs) to characterize the formation of minute fragments of plastic in marine sediments and the water column of European seas (Thompson et al. 2004). According to Galloway et al. (2013), MPs are categorized into two forms on the basis of their origin: primary (made to be of microscopic dimensions), and secondary MPs (resulting from degradation and fragmentation processes in the environment). In recent times, MPs are divided into three types depending on their source-primary and secondary MPs and nanoplastics. Primary MPs are formed in the MPs (<5 mm) size range, and are used in the production of numerous household and personal goods as basic ingredients. The purposeful production of primary MPs is utilized in the air blasting industry or to make toiletries like cosmetics or toothpaste. The creation of secondary MPs is a result of chemical, physical, biological, thermal and UV radiation impacts on large raw plastic waste. The two main sources of secondary MPs are lost plastic fishing nets and non-biodegradable water bottles due to their deterrable structures. Contrary to them, nano-plastics are tiny pieces of plastics with two dimension that are less than 100 nm (<100 nm). The breakdown of plastic products, such as expanding polystyrene with an enhanced mechanical abrasion mechanism, and the fragmentation of synthetic fibers during laundry are the two processes behind the creation of nano-plastics (Ramasamy et al. 2021).

22.3.2 *Microplastic Substances*

Many chemical substances like phthalates, polybrominated diphenyl ethers, and bisphenol A, are present in MPs, which intensifies some targeted plastic properties such as flame retardants, colors, plasticizers, and antioxidants. The majority of MP polymers contain these plastic additives and they are readily discharged into the environment. Due to their hormone disrupting nature they are quite hazardous to human health. It has been documented that biological processes are a key factor for the accumulation of these substances in human bodies (Padervand et al. 2020).

22.3.3 *Transportation of Microplastics*

Freshwaters and terrestrial habitats are considered to be the primary suppliers of MPs, whereas the sea and ocean are thought to be the main consumers. A number of biological processes, such as accumulation, biofouling, breakdown, flocculation, egestion and adsorption in the environment, occur during MPs movement (Kumar et al. 2021). They are distributed throughout the soil via a number of techniques, including harvesting, bioturbation, wet-dry cycles, and soil management techniques. According to Born and Brüll (2022), UV light and increased temperatures can lead to the degradation of plastics in the crust, even tiny MPs (less than 5 mm). It has been determined that plastic pollution represents the negative aspect of aquatic environment modernization. Numerous direct and indirect places exposure, like dumping site degradation, surface runoffs, agricultural drainage activities, sewage effluents, rainwater, home activities, etc., opens ways for plastics to enter the aquatic environment (Horton and Dixon 2018; Kumar et al. 2021). It can enter the atmosphere through a variety of sources, such as fabrics made of artificial materials, erosion of tires, household items, burning waste, garbage dumps, sewage sludge, abrasive powders, construction products, 3D printer items, and resurrection of polymer fragments in urban dust. As the movements of MPs are stern at a greater scale, forecasting and imitating its transfer patterns can be quite challenging (Amato-Lourenço et al. 2020).

22.4 Harmful Impacts of Microplastics: Chemical, Physical, and Microorganisms Carried by Microplastics

Nowadays, MP contamination is a threat to the entire human race as it has detrimental effects on aquatic habitat, animal and human health. Moreover, MPs serve as effective transporters for poisons, hazardous compounds, pathogens, and pollutants as a result of their hydrophobic nature, porous properties and substantial ratio of surface area to volume. Three key aspects contribute to their toxicity: their chemical and physical structures, as well as their ability to transport microbes (Abuwatfa et al. 2021). Polymers, oligomers, and chemical supplements that escape from the plastic during prolonged rubbing are the sources of chemical toxicity of MPs (Thompson et al. 2004). These plastic leftovers may be absorbed by human bodies through a variety of routes once their concentration hits a certain point (Padervand et al. 2020). A toxicological investigation implemented by Nobre et al. (2015) to examine the MPs exposure impact in sea urchins (*Lytechinus variegatus*) growth and formation during their embryonic stage. The tests replicated the release of chemical additives into the water from pure and beach-stranded MPs. Exposure to unprocessed MPs increased the possibilities of creating abnormal embryo up to 66.5%. MP concentrations, size, form, density, and texture are the main cause of physical impact. According to Ory et al. (2017), blue-colored polyethylene MP particles were ingested by 80% of planktivorous fishes because they resembled their copepod prey's size and color. This experiment is an example of physical toxicity of MPs. In the year 2016, Inga V. Kirstein discovered that Pathogenic *Vibrio parahaemolyticus* bacterial strains were adsorbed on the surfaces of twelve distinct kinds of MP particles, such as polyethylene, polypropylene, and polystyrene. Unprocessed chemical additions, the binding and movement of contaminants and microorganisms as well as their release into the environment and life itself are the main causes of MP toxicity. It has also been shown through research that virus-contaminated MPs might be even more hazardous than regular ones.

22.5 Decomposition Process of Microplastics

Plastics are synthetic materials composed of small organic molecules. It takes hundreds of years to breakdown entirely into the environment. Throughout the whole deterioration process, plastic pieces changes into macro to micro size range, this deterioration process is neither uniform nor linear. Fragmentation of plastics caused by various degradation mechanisms, these degradation mechanisms are categorized into four main degradation categories—chemical, photo, biological, and thermal degradation.

Chemical degradation primarily encompasses hydrolytic and oxidative degradation. Hydrolysis is the breakdown of polymer chains by water, whereas oxidation process can be initiated by light or heat. The main variables that influence the effectiveness of hydrolysis are pH, water activity, temperature, and time (Lucas et al. 2008). In oxidative degradation oxygen is the powerful agent that can cause degradation; by changing their structure, generating free radicals and also atmospheric oxygen can attack plastic covalent bonds. Similarly, Photodegradation is the process where plastics transform into MPs through exposure to ultraviolet (UV) light which can eventually convert plastic pieces into smaller fragments in the environment. Biodegradation of plastics occur due to microorganisms such as bacteria and fungi. Microorganisms transform MPs into carbon dioxide, water and other by-products through various processes like oxidative reactions, biofilm formation, and enzymatic mechanisms. Although the entire degradation process may be time consuming, these activities subdue the plastic polymers, leading to the formation of tinier plastic pieces. Thermal degradation of plastics occurs when plastics are dealt with high temperatures. This process also breaks down the polymer chains, leading to the formation of MPs (Liu et al. 2022) (Fig. 22.1).

22.6 Removal Methods

Emerging contaminants are artificial and natural substances found in environment that has a detrimental effect on human and environmental health such as hormones, MPs, medicines, surfactants, and personal care items because of their uncertain future (Jagadeesh and Sundaram 2023). Among these, MPs are destroying land, aquatic as well as atmospheric biota along their negative influence on intuitive systems make it imperative to remove in order to maintain agreeable environments. As a result, researchers have been looking into a number of effective removal

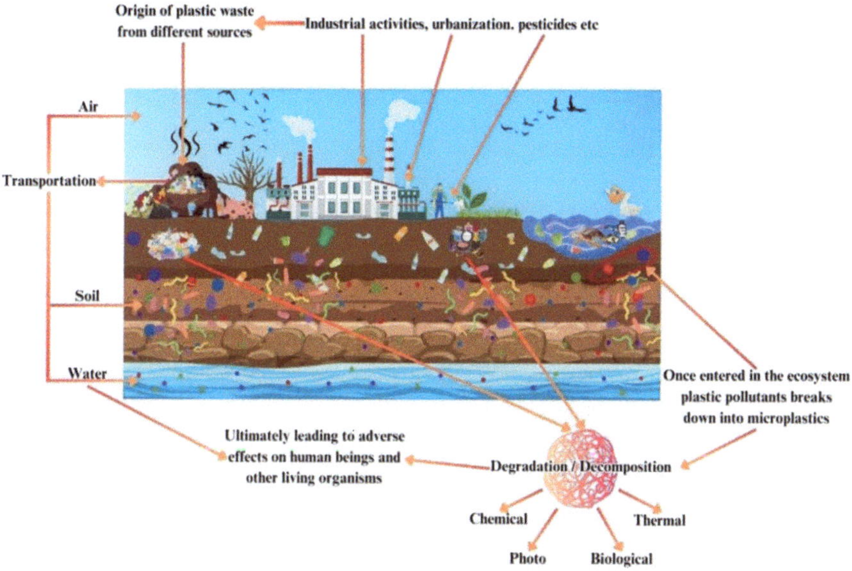

Fig. 22.1 Origin of plastic waste and breaks down of plastic waste into MPs, transportation, and decomposition process of MPs into the ecosystem

techniques for cleaning out MPs. MP removal techniques are divided into three categories—biological, chemical, and physical methods.

22.6.1 Biological Methods

In natural environments plastics are biologically degrade into CO_2, water and other simpler forms. These biodegradations happen due to the numerous microorganisms (especially bacteria) that are present in soil. Now, there are lots of research coming to light on MPs' removal by microorganisms, for example, researchers found that types of bacteria of *Enterobacter asburiae* and *Bacillus* sp. isolated from the insects, i.e., *Plodia interpunctella,* guts molder polyethylene MPs (Jiang 2018; Yang et al. 2017). However, they take long period of time to break the MP particles in minute pieces and hence not effective in comparison to other methods. Apart from the microorganisms there are some aquatic species, which have a critical contribution on the degradation process. Giant clams of red sea (Arossa et al. 2019), corals (Martin et al. 2019) and Krills of Antarctica (Dawson et al. 2018) represent this type of contribution in the decomposition of MPs by soaking it through ingestion, retention, or adhesion process. The other biologically intended techniques are sewage sludge methods and membrane bioreactors. Sludge from sewage systems utilization is a prominent origin of MPs contamination in the natural biota. In 2020, Chen and his coworkers revealed a hyper-thermophilic compost technique (HTC) can remove sewage sludge-originate MPs in complete scale and in laboratory settings for natural breakdown (Chen et al. 2020). Several reports suggested that both the methods work in effective manner in environment remediation from the MPs with the highest efficiency level of 98.3% up to 99% removal (Lares et al. 2018; Talvitie et al. 2017).

22.6.2 Chemical Methods

MP pieces are cleaned using chemical methods which refers to the use of chemicals or solutions that are designed to segregate MP pieces, and agglomerate them for the further process that is sedimentation or filtering. Coagulation and flocculation are main processes behind the sedimentation or filtration processes as it makes it easier for removal from the infested surroundings. Shahi and colleagues in 2020 have documented that deletion of MPs is contingent upon the MPs framework, with elongated and rough MPs exhibiting a higher rate of removal compared to elongated and smooth MPs, followed by spherical and rough MPs, and finally, spherical and smooth MPs. Concentration, pH, coagulation kind, and medium chemical composition were acknowledged to have an impact on efficiency (Padervand et al. 2020). In another experiment, Herbort et al. (2018) used a new technique sol–gel-induced agglomeration reaction to eliminate MPs through alkoxy-silyl elements as an adhesive particle to form agglomerates of MPs. Inevitably, the outcomes of this experiment suggested that agglomeration is also a successful approach for the removal of MPs.

22.6.3 Physical Methods

As per lab studies the most popular physical technique for removing MPs is filtration. Other physical techniques such as sedimentation, advanced filtration, dynamic membranes and adsorption. Compared to plastics, MPs cannot be eliminated by traditional sedimentation method. A study by Simon et al. (2019) examined the efficiency of a disc filter as a removal technique to mitigate the MP particles from treated wastewater. He reported that the disc filter with 18 μm can effectively remove 89.7% of the MPs that are larger than 10 μm. In other experiment, Sundbæk et al. (2018) evaluated the sorption of fluorescent polystyrene (PS) MPs to the edible seaweed (*Fucus vesiculosus)* and also found the amount of wash-off potential of adsorbed PS particles with an applicable industrial technique. According to this study, washing significantly reduces the PS microplastic particles by 94.5% along with demonstration of their sorption to *F. vesiculosus.* The findings of this work helped to clarify the working interaction behind the direct connection of MP with microalgae where they are most likely "trapped" in alginate of the microalga. Multiple investigations have revealed that physical procedures such as precipitation, filtration, coagulation, and degradation play a vital role in the removal of MPs from water treatment systems around 60–90%. There are other sophisticated adsorption technologies, such as the three-dimensional-reduced graphene oxide (3-D RGO) sheets (Yuan et al. 2020a), chitin and graphene oxides, superhydrophobic non-fluorinated aluminum surface, colloid gas coagulative band, and magnetic polyoxometalate-supported ionic liquid phases (Badola et al. 2022). Since these adsorption techniques require sophisticated equipment, high energy, high installation costs, and are also time consuming for the researchers, so they prefer biochar adsorption technology, which is the most effective, low energy and time-consuming technology at a reasonable price for the removal of pollutants. In the process of adsorption, the substance which gets adsorbed on the surface is called adsorbate, which can remain in several forms like solid, liquid, and gas. The substance on whose surface the whole process can occur is known as adsorbent such as charcoal, biochar, silica gel, etc. Biochar technology gained its popularity in the field of pollutant removal due to its accessibility, flexibility, potential to manage waste by converting it into biomass and adaptability over a broad spectrum of uses. For small size spherical polystyrene microbeads (10 μm), activated biochar has successfully attained a removal effectiveness of 95% (Wang et al. 2020). It is possible to extract 100% of the polyethylene fibers using pine and spruce biochar (Siipola et al. 2020) (Table 22.1).

Table 22.1 Pros and cons of commonly used techniques for removing MPs

Categories	Concepts	Procedure	Pros	Cons	References
Biological	Bio-degradation	Membrane bioreactor	Potentially removes 79–99.9% for large microplastics. Powerful, adaptable, cost-effective	Size and shape affect removal efficiency. Aggregation of microbes might be a problem for the procedure	Lares et al. (2018), Talvitie et al. (2017), Yuan et al. (2020b)
		Conventional activated sludge	Cost-effective, resilient, and adaptable; capable of being successfully upscaled ((98.3% removal efficiency)	Need extended periods of retention. Requires large space for processing and disposal with excessive costs	Gurung et al. (2016), Lares et al. (2018)
Chemical	Fluctuation and coagulation	Fluctuation and coagulation with different substances	Simple operating requirements with adaptable conditions. High percentage of elimination up to 70–>99%	The use of chemicals is necessary for producing byproducts and secondary pollutants, making it suitable only for a limited size range of microparticles	Herbort et al. (2018), Shahi et al. (2020)

(continued)

Table 22.1 (continued)

Categories	Concepts	Procedure	Pros	Cons	References
Physical	Filtration	Rapid Sand Filter	Budget-friendly and eliminates every MP	Ineffective in small-sized MPs	Talvitie et al. (2017)
		Ultrafiltration	Removal efficiency of more than 90% was tested in a combination of traditional wastewater treatment plants (WWTP)	Large particles cause clogging of filter pores, necessitating regular backflushing	Pramanik et al. (2021)
		Disc filter	Excellent elimination effectiveness of up to 98.5%	The clogging of filter pores by large particles necessitates frequent maintenance	Talvitie et al. (2017)
Physical	Adsorption	Microalgae Marine	Removal efficiency up to 94.5%. Electrostatic charge and other surface features of microplastics can be used to adjust selectivity	Unrecyclable method. The photosynthesis and growth activities of the microalgae are inhibited by the adhesion of microplastics to the algae	Lagarde et al. (2016), Zhang et al. (2017)
		Biochar	Biochar filters are easy to create. 95–100% elimination effectiveness is high cheaper	Tested exclusively for specific types of MPs Lower MP sizes result in decreased removal efficiency	Siipola et al. (2020), Wang et al. (2020)

22.7 Identification Methods

MPs can be examined by their physical and chemical properties (Jiang 2018). In regard to the physical characterization, scanning electron microscopy (SEM) and scanning electron microscopy-energy dispersive X-ray spectroscopy (SEM-EDX) fall under the category of imagination methods and facilitate the visualization of particle morphology (Gyre et al. 2013). Moreover, techniques such as Fourier-transform infrared spectroscopy (FTIR) and Raman spectroscopy are employed for the purpose of detecting and characterizing chemical properties of MPs, thereby allowing for molecular identification. Comprehending the dispersion and influence of MPs in various environmental settings is made possible by the use of these methodologies.

22.8 Biochar Technology

A carbonaceous substance "Biochar" is a solid, stable, porous, and reasonably priced adsorbent that is extensively utilized for the removal of pollutants from environments. It is found that Biochar research originated from the terra preta ("dark earth") soils of the Amazon Basin, which are rich in fertility and carbon (Jagadeesh and Sundaram 2023).

22.8.1 Synthesis of Biochar

Production of Biochar from different feedstock first undergoes through different pre-treatment procedures such as physical, chemical and biological pre-treatments. Physical pre-treatments include cleaning, drying, crushing/grinding and sieving. In chemical pre-treatments some chemicals are applied to the biomass to change their properties such as broad porous surface structure, functional groups, and acidic or basic nature etc. Biological pretreatments mainly include engineered biochar-based materials such as bacterial treatments. Bacterial treatments mainly include anaerobic digestion which converts normal surface area to larger specific surface area (Xiang et al. 2020).

After that using thermochemical conversion methods biomass can be converted into biochar. These methods include pyrolysis (slow and fast pyrolysis), gasification, hydrothermal carbonization (HTC) methods and microwave-assisted pyrolysis (MAP). Among this four methods pyrolysis was found to be the most effective method for producing high amount of biochar. Slow pyrolysis is a slow and continuous process where temperature of pyrolysis is between 300 and 800 °C (Liu et al. 2015). While fast pyrolysis is performed at high temperature between the 400 and 650 °C range at a heating rate around 100–1000 °C/s with the vapor residence time of ~0.5–10 seconds. In the gasification process biomass are oxidized at 800 °C and mainly utilized to produce fuel gases (Oliver et al. 2013). Microwave-assisted pyrolysis (MAP) is a viable process for producing bioenergetics products such as bio-oil, bio-gas, and biochar at 400–500 W temperature range within 1–10 min time period. In other way, hydrothermal carbonization is a process of converting feedstock into biochar at temperature between 120 and 260 °C with moist condition (without pre-drying/when water is present) (Fang et al. 2018). For producing biochar, with the efficient feedstock properties and high carbon output, hydrothermal carbonization and slow pyrolysis procedures are among the most significant thermochemical conversion methods. However, compared to HTC, the slow pyrolysis method exhibited a significantly larger capacity for storing carbon due to its greater stability (Praveen et al. 2022).

22.8.2 Components Impacting the Production of Biochar

Three steps were involved in the manufacturing process of biochar: initial, primary, and final pyrolysis for the creation of organic-rich minerals (Lee et al. 2017). In these three steps several factors are involved in the production of biochar like heating rate, temperature, feedstock materials and particle size of feedstock. A temperature rise has an inverse relationship with the yield of biochar formation, and the molecular arrangement and qualities of biochar rely on the heating rate of the pyrolysis (Jindo et al. 2014; Tomczyk et al. 2020). The product materials such as biofuel or bio-oil and syngas are indirectly impacted by the heating rate during the high pyrolysis temperature. Increased temperatures result in lower amount of biochar with higher syngas generation and disorganized porous arrangement, while slow pyrolysis temperature along with average rate of heating produces biochar adsorbent with optimal characteristics (Tomczyk et al. 2020). Combustion temperatures encourage the biochar generation while minimizing losses from optimal ingredients. High yield of biochar also has a connection with organic and inorganic components present in the biomass. For example, Shariff et al. (2016) noted that the production of biochar is influenced by differences in the lignocellulosic component of the feedstock. In the case of feedstock type size of the feedstock performed an important part in the production of biochar. At the time of pyrolysis process, transmission of heat and weight depends on the size of the particles. Depending on feedstock type and pyrolysis process the feedstock particle size can be determined.

22.8.3 Biochar Properties

The physicochemical characteristics of biochar are highly influenced by the nature of the raw material used, specifically the preparation process, the feedstock's particle size, and the duration of the raw material combustion process with the fluctuation of temperatures. At the time of pyrolysis process the resulting heat creates porous structures in biochar surface, which is also a physical property of biochar. The contaminants are more likely to adsorb on its surface because of its porosity and stable molecular structure. Chemical properties of biochar include elemental makeup, cation/anion transfer capacity, pH, moiety, ash concentration, and hydrophobic nature (Di Chen et al. 2020). Hydrophobicity generally used to measure the amount of organic-rich matter and also significantly increases the binding capability. The main factors influencing the chemical characteristics of biochar are its higher specific surface area (SSA), pore's size, immovable substances with improved binding capability, and functional groups including methyl, carbonyl, hydroxyl and carboxyl groups (Jagadeesh and Sundaram 2023).

22.9 Removal of Microplastics Using Biochar Adsorption

In order to successfully remove of MPs, the main factors effecting the nature of adsorption are the surface area, surface texture, and porosity of the adsorbent. Among all the adsorption methods biochar technology provides this type of adsorption nature. Biochar is a dependable and relatively inexpensive substance that is obtained by the pyrolysis process of biomass without oxygen. In the process of pyrolysis, moisture and volatile components are released and only the pure carbon form of biochar is produced. According to Chen et al. (2012), the rising pyrolytic temperatures have a direct impact on the level of carbonization, improving the biochar's sorption capacity and other physical and chemical characteristics. Recently, the following benefits of biochar over other adsorbents have led researchers to choose it as a form of adsorbent—(i) broad, depth and porous arrangements surface area which enhance the adsorption capacity, (ii) it reduces greenhouse gases emission and carbon sequestration, (iii) its beneficial for soil and act as a fertilizer that increases the crop productivity, (iv) remove organic and inorganic contaminants from water, (v) reduce the storage of waste production, and also (vi) cost effective and recyclable nature. Consequently, biochar gains significance as a potential remedy for several worldwide issues such as soil degradation, pollutants removal, and climate change.

The growing interest in using biochar to remove MPs has been brought to light by recent research endeavors. Siipola et al. (2020) used gradual pyrolysis technique at a temperature of 475 °C to create activated biochar of pine as well as spruce bark. The biochar was subjected to steam activation at 800 °C, for producing larger pores to improve its adsorption properties. Two gram each of spherical/smooth/cylindrical polyethylene (PE) particles along with fibers were used to imitate MP particles of different sizes and shapes. The fleece shirt fibers were almost 100% retained and cylindrical pieces of PE were fully conserved by steam generating biochar. The study proposed the direct interaction on biochar components was responsible to eliminate the bigger fragments of plastics, notwithstanding the lack of clarity around the retention mechanisms. The elimination of the smallest MP particles may benefit from increased meso- and macropore contents, according to the results.

In another experiment, Wang et al. (2020) looked at the strength of combining maize straw and hardwood feedstock as compost form of biochar to relegate MP spheres (of 10 μm) at three various certain temperatures (i.e., 300 °C, 400 °C, and 500 °C). In order to ameliorate of wastewater treatment facilities for eliminating microbeads, the examination included biochar in sand filtering apparatus. The outcomes showed that, in contrast to an effectualness of 60–80%, utilizing the unchanged sand filtering apparatus, the removal efficiency attained 95%. The adsorption of the MP spheres in biochar surface was discovered to be controlled by three ways: "Stucked," "Trapped," and "Entangled." The immobilization processes known as "Trapped" and "Entangled" provide the biochar a higher effectiveness in removing compared to sand filters of a comparable size. While Tong et al. (2020a) used a simple precipitation process to create biochar with modification of magnetite

(Fe_3O_4) biochar at ambient temperature. The use of magnetite biochar and normal biochar is essential for trapping plastic bits in sand-column. While both materials show decreased particles movement and improved retention qualities, magnetite biochar works better. Its special qualities change the dynamics of suspension, increasing the adsorption capacity of porous media and enabling more particles deposition. Minimal biochar and magnetite biochar release is observed, proving their stability. Their potential to reduce the amount of plastic in terrestrial regions is highlighted by effective recuperation and desorption processes.

According to Tong et al. (2020b), the coagulation of plastic particles and induced biochar steric repulsion led to increased cellulose-derived biochar transport for small-sized plastic particles such as 20 nm nanoplastics and 0.2 μm microplastics in pH 6 solutions containing 5 and 25 mM NaCl. Nevertheless, in a 5 mM NaCl solution, the biochar transport of large-sized particles (~2 μm MPs) was found to be enhanced, but in a 25 mM NaCl solution, one observed decreased transport. A coagulation binding formed between the MPs and the biochar, resulting in biochar being considerably placed for every plastic particle for both low/high NaCl concentrations. However, Singh et al. (2021) altered the surface of biochar by incorporating iron nanoparticles (NPs) at 550 and 850 °C. Utilizing biochar's magnetic and surface properties, they effectively removed various nanoparticle sizes. Batch adsorption analysis showed minimal influence of temperature, pH and other ions on nanoparticle removal, highlighting the adsorbent's robustness. Although raw biochar only had a removal effectiveness of 75%, iron-modified biochar obtained an almost 100% removal efficiency. This work established a standard for addressing NPs in an aquatic environment with strong evidence of electrostatic interaction between NPs and biochar composites. Moreover, its minimal electrical consumption and minimal operating expenses make it a potential removal approach in the sector.

While conducting their study, Abdoul Magid et al. (2021) made the observation that the pyrolyzed biomass derived from corncob exhibited a considerable capacity for adsorbing polystyrene nanoplastics when subjected to high temperatures. However, in comparison to biochar derived from corncob, the oxidized biochar obtained from corncob displayed a higher degree of adsorption, surpassing 90%. This enhanced adsorption can be attributed to the presence of hydroxyl groups that facilitate the process of polystyrene adsorption. The capacity of oxidized corncob biochar to adsorb polystyrene increased with the rise in pyrolyzing temperature. Various factors performed an essential part in this sorption process, including H-bonding, hydrophobicity electrostatic interactions, pore-filling, and specific surface area. Based on the analysis of the hydrogen-bonds and hydrophobicity it has been determined that the presence of ester or hydroxyl groups may be accountable for the notable reduction observed.

Another recent study by Wang et al. (2021) delved into the exploration of the elimination of MPs through the utilization of Magnesium and Zinc modification charcoal adsorbents. Removal efficiencies reached 98.8%, 99.5%, and 94.8% for different adsorbents, with promising recycling potential explored through heat treatment. Moreover, this investigation also delved with the potential reprocess capabilities of the sorbents, which involved subjecting them to a combined heat treatment in order to renew the adsorbents and simultaneously destroy the adsorbed MPs. This approach not only highlights the effectiveness of these modified magnetic charcoal adsorbents, but also emphasizes the importance of finding sustainable methods for the management and removal of MPs in our environment. In a similar vein, Wang et al. (2022b) have shown that polystyrene MPs have been maintained at different cationic strength (Na+, Ca_2+) and humic acid concentrations containing biochar of peanut shell (PSB) and modified MgO PSB (MgO-PSB). According to the findings, there was a clear decrease in the transit of MP colloids in porous medium without HA when PSB or MgO-PSB were added to quartz sand. Increased cationic strength lowers the negative charges on the surfaces of both raw and MgO-modified peanut shell biochar, which in turn limits the transport of MPs.

Wang et al. (2022c) also observed a considerable retention of polystyrene MPs (PSMPs) using raw and Fe_3O_4-modified biochar in saturated column studies, taking into account the effects of ions or humic acids in sandy porous media. According to the findings, PSMPs can be prevented from moving through porous media without humic acid (HA) by adding biochar or Fe_3O_4 biochar. Fe_3O_4 biochar, on the other hand, was more successful at preventing PSMPs from moving through the process of electrostatic adsorption and complexity, with an optimal retention effectiveness of 92.36%. Wang et al. (2022a) discovered that activated sludge waste biochar efficiently eliminates polyethylene nanoplastics compared to anaerobic granular sludge, with a difference of 11.1–23.9%, but the precise mechanisms of micro(nano) plastic adsorption using biochar from sewage sludge have not been studied.

In addition to all these numerous investigations have explored to examine the capacity of biochar of various compositions to retain MPs within river or wastewater samples. Built wetlands are artificial that are intentionally designed to mimic the characteristics of natural wetlands. These engineered ecosystems have gained widespread popularity and have provided a diverse range of ecological benefits, such as the efficient cycling of nutrients, the provision of habitats for various organisms, and the preservation of water resources. Studies on the interaction processes between MPs and biochar in saturated sand columns have not received comprehensive scrutiny so far, thus necessitating further investigation to fully comprehend their specific mechanisms (Table 22.2).

Table 22.2 A review of research on the use of biochar to remove microplastics under different environmental condition

Feedstock properties of Biochar/Biochar Feedstock Properties	Production condition and operating setup	Environmental factors	Kind of microplastics removed	Removal capabilities and efficiency	Interaction mechanism	References
Steam Activated pine and spruce bark feedstock biochar	Pyrolyzed at 475 °C temperature and steam-activated at 800 °C	Temperature = 25 °C	Polyethylene (PE) microbeads (10 μm), cylindrical and smooth PE pieces (2–3 mm), and fleece shirt fibers	˜100%	Physical adsorption with biochar particles	Siipola et al. (2020)
Corn straw and hardwood feedstock biochar	Pyrolyzed at 500 °C temperature	pH = 7.5	Polystyrene microplastic spheres (10 μm diameter)	>95%	Stuck, trapped and entangled	Wang et al. (2020)
Biochar/Magnetic biochar (Fe3O4-biochar)	Pyrolyzed at 400 °C for 6 h.	Temperature = 25 °C	Three different sized plastic particles (0.02 mm nano-plastic particles, and 0.2 mm and 2 mm micro-plastic particles)	>74%	Ionic strength	Tong et al. (2020a)
Cellulose-derived biochar	Pyrolyzed at 400 °C	Ionic strength 5 mM NaCl solution, pH = 6	20 nm nanoplastics and 0.2 μm microplastics	3% 32%	Hetero aggregation	Tong et al. (2020b)
Iron modified biochar derived from prosopisjuliflora	Iron (Fe3O4) modification pyrolyzed at 550 °C and 850 °C	Temperature = 25 °C and pH = 5.5 and	Nano-plastics (diameter: 30 nm and 1000 nm)	˜100%; at 850 °C temp. 206.46 mg/g 225.11 mg/g At 550 °C temp 290.20 mg/g	The creation of surface complexity and electrostatic attraction	Singh et al. (2021)
Corncob raw and oxidized biochar	Pyrolyzed at 500, 700 and 900 °C, HNO3-H2SO4 was used as oxidizing agent	pH = 5	Polystyrene Nanoplastics(1000 nm)	>90%	Hydrophobic interaction and H- bonding	Abdoul Magid et al. (2021)

Magnetic Mg/Zn modified biochar	Pyrolyzed at 550 °C with Mg/Zn modification	Temperature = 25 °C	Polystyrene microplastic spheres (1 μm)	98.75% for Mg-MBC 99.46% for Zn-MBC.	Chemical bonding and electrostatic attraction	Wang et al. (2021)
Peanut shell Biochar and MgO-peanut shell biochar	Pyrolyzed at 500 °C and MgCl2.H2O was used for modification	Humic acid solution, Na+ and Ca2+ solutions; Ionic strength 10–100 mM	Polystyrene microplastics	59.1% 75.5%	Electrostatic repulsion, steric hindrance, and ionic competition between microplastics and humic acids	Wang et al. (2022b)

22.10 Adsorption Mechanism of Microplastics on Biochar

The adsorption was greatly influenced by the hydrophobicity, H-bonding, electrostatic interactions, pore volume and specific surface area. When a polar medium, like water, is present, two nonpolar substances will aggregate in hydrophobic interactions. Since several polymers, including PE, PS, PP, and PET (polyethylene terephthalate), are inherently hydrophobic, hydrophobicity can play a major role in MPs that serve as vectors for a variety of pollutants (Torres et al. 2021). The process via which two molecules are attracted or rejected by other molecules that are negatively or positively charged is known as electrostatic interaction. The surface of MPs is negatively charged in low pH environments, and in contrast, biochar possesses positive charge; therefore, the adsorption capacity will increase. Wang et al. (2021) reported that due to the chemical interaction and electrostatic attraction within the biochar and MPs in aquatic settings, where the Magnesium and Zinc modification biochar had greatest removal rates of 98.75% and 99.64%, respectively. The sorption of substances can be significantly affected by pore filling interactions, which in turn depends on the characteristics of the polymer. Among the various types of polymers such as polyvinyl chloride (PVC), polystyrene (PS), and polyamide (PA), pore-filling interactions exert the most significant influence on sorption capacity (Hüffer and Hofmann 2016). Vander Waals interactions and π–π interactions are other types of interaction (Stapleton and Hai 2023). Some aliphatic polymers, including polyethylene, exhibit this type of interaction. Polystyrene and other aromatic polymers exhibit this type of interaction (Fig. 22.2).

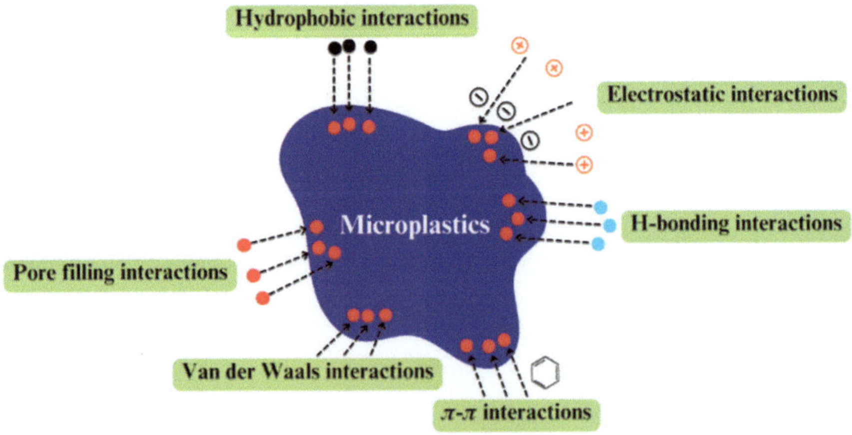

Fig. 22.2 Adsorption mechanisms of microplastics on biochar

22.11 Conclusions and Future Perspectives

The global ecosystem is facing the adverse effects of the growing use and demand of plastic items. The widespread presence of these MPs particles in soil, water and air has far-reaching repercussions. Its bioaccumulation and biomagnification disturb marine ecosystems, affecting both aquatic life and humans who rely on seafood. The consequences are seen in terrestrial ecosystems as well, as soil contamination throws off important nutrient cycles. It also has an impact on crop output, and there are worries regarding MPs adsorption by plants, that might have an impact on people's wellness when the plants are consumed. Moreover, the hazards to living things are increased by the capacity of MPs to transport harmful compounds. As mentioned before, they are extremely dangerous since their poisonous effects of chemical and physical structure, and capacity to absorb and transfer poisons via microorganisms. These negative effects are harmful to animals, humans and aquatic life equally. Because of their persistent nature, MPs require extensive study, cutting-edge cleanup techniques, and international campaigns to reduce plastic use. Research in this field is limited due to conducting long term-controlled tests. The mitigation of MPs detrimental impacts is essential for maintaining ecosystems, safeguarding animals, and promoting community well-being. In order to ensure a healthier and more resilient world for present and future generations, it is imperative that we address the MP problem as we move toward sustainable practices.

However, this prolonged battle with MPs has given us many allies and one of the most trustworthy among them is biochar. This remarkable technology provides us a tremendous edge against MPs. It has gained more and more popularity as a growing technique among researchers for eliminating pollutants due to its low-cost adsorption systems developed from different biomass, such as saw dust, agricultural leftovers, forest residues and municipal solid waste, etc. The process of biochar adsorption is potentially cost-effective, making it an attractive option for large scale implementation. Biochar's attractiveness as a removal method due to its a range of intriguing qualities that renders it highly attractive for the efficient elimination of MPs. One such challenge lies in the limited number of experimental studies that have been conducted on this particular subject matter. It is evident that more experimental researches are required to fully comprehend the potential of biochar as a solution to the MPs predicament. Therefore, in order to formulate effective strategies that diminish the utilization of plastics on a global scale, a comprehensive evaluation of the potential hazards associated with MPs must be conducted.

There are some areas that will have a substantial influence in the future for research:

(a) Biochar as a Remediation Agent

The consideration of biochar as a remediation agent for MPs is predicted to be further explored. It will be indispensable to decipher its efficiency in the process of adsorption and immobilization of this particles in different environmental conditions.

(b) Cooperative Methods

To improve the elimination efficacy of MPs, scientists may investigate synergistic ways that merge biochar with other materials. This may include creating composite materials that incorporate the advantages of biochar with other well-proven remediation methods.

(c) Biochar for Waste Water Treatment

Eventual outlooks may focus on using biochar into water treatment systems to mitigate MP pollution in marine environment. Biochar's utility in eradicating MPs from water sources as well as its application in purification networks may be examined.

(d) Agricultural Applications and Soil Health

Researchers will likely pay more attention on the effects of biochar regarding soil health and plant growth with MPs may lead to the development of sustainable farming techniques.

(e) Extended-Duration Environmental Effect Research

The long-term environmental consequences and possible adverse effects of biochar will be crucial as it becomes more popular. Implementation that is responsible and successful will be guided by an assessment of its sustainability and compatibility with various ecosystems.

It will be essential to raise public knowledge of the advantages and uses of biochar in reducing the pollution caused by MPs. Education campaigns may be quite effective in spreading awareness of and acceptance for biochar as a workable alternative.

In summary, research on MPs, with a focus on biochar specifically, is positioned to investigate creative and comprehensive solutions. Through the integration of scientific discoveries with pragmatic implementations and the consideration of ecological, agricultural, and social aspects, investigators want to provide a valuable and resilient contribution to the resolution of the issues raised by MPs. The of abundance of MPs in ground, particularly in farming regions, as well as the variables that influence them, have to be elucidated. Further research is required to determine the influence of MPs in soil microbes.

References

Abdoul Magid ASI, Islam MS, Chen Y, Weng L, Li J, Ma J, Li Y (2021) Enhanced adsorption of polystyrene nanoplastics (PSNPs) onto oxidized corncob biochar with high pyrolysis temperature. Sci Total Environ 784. https://doi.org/10.1016/j.scitotenv.2021.147115

Abuwatfa WH, Al-Muqbel D, Al-Othman A, Halalsheh N, Tawalbeh M (2021) Insights into the removal of microplastics from water using biochar in the era of COVID-19: a mini review. Case Stud Chem Environ Eng 4:100151. https://doi.org/10.1016/j.cscee.2021.100151

Amato-Lourenço LF, dos Santos Galvão L, de Weger LA, Hiemstra PS, Vijver MG, Mauad T (2020) An emerging class of air pollutants: potential effects of microplastics to respiratory human health? Sci Total Environ 749:141676. https://doi.org/10.1016/j.scitotenv.2020.141676

Arossa S, Martin C, Rossbach S, Duarte CM (2019) Microplastic removal by Red Sea giant clam (Tridacna maxima). Environ Pollut 252:1257–1266. https://doi.org/10.1016/j.envpol.2019.05.149

Badola N, Bahuguna A, Sasson Y, Chauhan JS (2022) Microplastics removal strategies: a step toward finding the solution. Front Environ Sci Eng 16(1):1–18. https://doi.org/10.1007/s11783-021-1441-3

Born MP, Brüll C (2022) From model to nature — a review on the transferability of marine (micro-) plastic fragmentation studies. Sci Total Environ 811. https://doi.org/10.1016/j.scitotenv.2021.151389

Chang X, Fang Y, Wang Y, Wang F, Shang L, Zhong R (2022) Microplastic pollution in soils, plants, and animals : A review of distributions, effects and potential mechanisms. Sci Total Environ 850:157857. https://doi.org/10.1016/j.scitotenv.2022.157857

Chen Z, Chen B, Chiou CT (2012) Fast and slow rates of naphthalene sorption to biochars produced at different temperatures. Environ Sci Technol 46(20):11104–11111. https://doi.org/10.1021/es302345e

Chen Z, Zhao W, Xing R, Xie S, Yang X, Cui P, Lü J, Liao H, Yu Z, Wang S, Zhou S (2020) Enhanced in situ biodegradation of microplastics in sewage sludge using hyperthermophilic composting technology. J Hazard Mater 384:121271. https://doi.org/10.1016/j.jhazmat.2019.121271

Chowdhury H, Chowdhury T, Sait SM (2021) Estimating marine plastic pollution from COVID-19 face masks in coastal regions. Mar Pollut Bull 168(March):112419. https://doi.org/10.1016/j.marpolbul.2021.112419

Dad FP, Khan W, Ud D, Kirkham MB, Bolan N, Tanveer M (2023) Microplastics: a review of their impacts on different life forms and their removal methods. Environ Sci Pollut Res 30(37):86632–86655. https://doi.org/10.1007/s11356-023-28513-w

Dawson AL, Kawaguchi S, King CK, Townsend KA, King R, Huston WM, Bengtson Nash SM (2018) Turning microplastics into nanoplastics through digestive fragmentation by Antarctic krill. Nat Commun 9(1):1–8. https://doi.org/10.1038/s41467-018-03465-9

Di Chen Y, Wang R, Duan X, Wang S, Ren N, Qi, Ho SH (2020) Production, properties, and catalytic applications of sludge derived biochar for environmental remediation. Water Res 187:116390. https://doi.org/10.1016/j.watres.2020.116390

Fang J, Zhan L, Ok YS, Gao B (2018) Minireview of potential applications of hydrochar derived from hydrothermal carbonization of biomass. J Ind Eng Chem 57:15–21. https://doi.org/10.1016/j.jiec.2017.08.026

Galloway T, Cole M, Lewis C (2013) ORE open research Exeter. J Clean Prod:0–48

Geyer R, Jambeck JR, Law KL (2017) Production, use, and fate of all plastics ever made. Sci Adv 3(7):25–29. https://doi.org/10.1126/sciadv.1700782

Gilbert M (2016) Plastics materials: introduction and historical development. In: Brydson's plastics materials: eighth edition, p 2–18. https://doi.org/10.1016/B978-0-323-35824-8.00001-3

Godlewska P, Schmidt HP, Ok YS, Oleszczuk P (2017) Biochar for composting improvement and contaminants reduction. A review. Bioresour Technol 246(July):193–202. https://doi.org/10.1016/j.biortech.2017.07.095

Gurung K, Ncibi MC, Fontmorin JM (2016) Incorporating submerged MBR in conventional activated sludge process for municipal wastewater treatment: a feasibility and performance assessment. J Membrane Sci Technol 6(3). https://doi.org/10.4172/2155-9589.1000158

Gyre S, Eriksen M, Cummins A, Maximenko N, Thiel M, Lattin G, Wilson S, Hafner J, Zellers A, Rifman S (2013) Plastic pollution in the south pacific. Plast Eng 69(5):38–44. https://doi.org/10.1002/j.1941-9635.2013.tb01006.x

Herbort AF, Sturm MT, Fiedler S, Abkai G, Schuhen K (2018) Alkoxy-silyl induced agglomeration: a new approach for the sustainable removal of microplastic from aquatic systems. J Polym Environ 26(11):4258–4270. https://doi.org/10.1007/s10924-018-1287-3

Horton AA, Dixon SJ (2018) Microplastics: an introduction to environmental transport processes. Wiley Interdiscip Rev Water 5(2):1–10. https://doi.org/10.1002/WAT2.1268

Hüffer T, Hofmann T (2016) Sorption of non-polar organic compounds by micro- sized plastic particles in aqueous solution. Environ Pollut 214:194–201. https://doi.org/10.1016/j.envpol.2016.04.018

Jagadeesh N, Sundaram B (2023) Adsorption of pollutants from wastewater by biochar: a review. J Hazard Mater Adv 9. https://doi.org/10.1016/j.hazadv.2022.100226

Jiang JQ (2018) Occurrence of microplastics and its pollution in the environment: a review. Sustain Prod Consump 13:16–23. https://doi.org/10.1016/j.spc.2017.11.003

Jindo K, Mizumoto H, Sawada Y, Sanchez-Monedero MA, Sonoki T (2014) Physical and chemical characterization of biochars derived from different agricultural residues. Biogeosciences 11(23):6613–6621. https://doi.org/10.5194/bg-11-6613-2014

Kirstein IV et al (2016) Dangerous hitchhikers? Evidence for potentially pathogenic Vibrio spp. on microplastic particles. Mar Environ Res 120:1–8. https://doi.org/10.1016/J.MARENVRES.2016.07.004

Kumar R, Sharma P, Manna C, Jain M (2021) Abundance, interaction, ingestion, ecological concerns, and mitigation policies of microplastic pollution in riverine ecosystem: a review. Sci Total Environ 782:146695. https://doi.org/10.1016/j.scitotenv.2021.146695

Kumar R, Verma A, Rakib MRJ, Gupta PK, Sharma P, Garg A, Girard P, Aminabhavi TM (2023) Adsorptive behavior of micro(nano)plastics through biochar: co-existence, consequences, and challenges in contaminated ecosystems. Sci Total Environ 856(September 2022):159097. https://doi.org/10.1016/j.scitotenv.2022.159097

Lagarde F, Olivier O, Zanella M, Daniel P, Hiard S, Caruso A (2016) Microplastic interactions with freshwater microalgae: hetero-aggregation and changes in plastic density appear strongly dependent on polymer type. Environ Pollut 215:331–339. https://doi.org/10.1016/j.envpol.2016.05.006

Lares M, Ncibi MC, Sillanpää M, Sillanpää M (2018) Occurrence, identification and removal of microplastic particles and fibers in conventional activated sludge process and advanced MBR technology. Water Res 133:236–246. https://doi.org/10.1016/j.watres.2018.01.049

Lee XJ, Lee LY, Gan S, Thangalazhy-Gopakumar S, Ng HK (2017) Biochar potential evaluation of palm oil wastes through slow pyrolysis: thermochemical characterization and pyrolytic kinetic studies. Bioresour Technol 236:155–163. https://doi.org/10.1016/j.biortech.2017.03.105

Liu W, Jiang H, Yu H (2015) Development of biochar-based functional materials: toward a sustainable platform carbon material. Chem Rev. https://doi.org/10.1021/acs.chemrev.5b00195

Liu J, Huang S, Chen K, Wang T, Mei M, Li J (2020) Preparation of biochar from food waste digestate: pyrolysis behavior and product properties. Bioresour Technol 302(December 2019):122841. https://doi.org/10.1016/j.biortech.2020.122841

Liu L, Xu M, Ye Y, Zhang B (2022) On the degradation of (micro) plastics: degradation methods, influencing factors, environmental impacts. Sci Total Environ 806:151312

Lucas N, Bienaime C, Belloy C et al (2008) Polymer biodegradation: mechanisms and estimation techniques - a review. Chemosphere 73(4):429–442

Martin C, Corona E, Mahadik GA, Duarte CM (2019) Adhesion to coral surface as a potential sink for marine microplastics. Environ Pollut 255:113281. https://doi.org/10.1016/j.envpol.2019.113281

Nobre CR, Santana MFM, Maluf A, Cortez FS, Cesar A, Pereira CDS, Turra A (2015) Assessment of microplastic toxicity to embryonic development of the sea urchin Lytechinus variegatus (Echinodermata: Echinoidea). Mar Pollut Bull 92(1–2):99–104. https://doi.org/10.1016/j.marpolbul.2014.12.050

Oliver DP, Pan YF, Anderson JS, Lin TF, Kookana RS, Douglas GB, Wendling LA (2013) Sorption of pesticides by a mineral sand mining by-product, neutralised used acid (NUA). Sci Total Environ 442:255–262. https://doi.org/10.1016/j.scitotenv.2012.10.020

Ory NC, Sobral P, Ferreira JL, Thiel M (2017) Amberstripe scad Decapterus muroadsi (Carangidae) fish ingest blue microplastics resembling their copepod prey along the coast of Rapa Nui

(Easter Island) in the South Pacific subtropical gyre. Sci Total Environ 586:430–437. https://doi.org/10.1016/j.scitotenv.2017.01.175

Padervand M, Lichtfouse E, Robert D, Wang C (2020) Removal of microplastics from the environment. A review. Environ Chem Lett 18(3):807–828. https://doi.org/10.1007/s10311-020-00983-1

Pramanik BK, Pramanik SK, Monira S (2021) Understanding the fragmentation of microplastics into nano-plastics and removal of nano/microplastics from wastewater using membrane, air flotation and nano-ferrofluid processes. Chemosphere 282(May):131053. https://doi.org/10.1016/j.chemosphere.2021.131053

Praveen S, Jegan J, Bhagavathi Pushpa T, Gokulan R, Bulgariu L (2022) Biochar for removal of dyes in contaminated water: an overview. Biochar 4(1):1–16. https://doi.org/10.1007/s42773-022-00131-8

Ramasamy R, Murugan M, Nadu T (2021) Microplastics: distribution, isolation, detection, and effects on Flora and Fauna—a mini review. All rights reserved WJES © 2014 available online at : www.environmentaljournals.org. 11(4):1–8

Shahi NK, Maeng M, Kim D, Dockko S (2020) Removal behavior of microplastics using alum coagulant and its enhancement using polyamine-coated sand. Process Saf Environ Prot 141:9–17. https://doi.org/10.1016/j.psep.2020.05.020

Shariff A, Mohamad Aziz NS, Md Saleh N, Ruzali NSI (2016) The effect of feedstock type and slow pyrolysis temperature on biochar yield from coconut wastes. Int J Chem Mol Eng 10(12):1410–1414. https://waset.org/Publication/the-effect-of-feedstock-type-and-slow-pyrolysistemperature-on-biochar-yield-from-coconut-wastes/10005789

Siipola V, Pflugmacher S, Romar H, Wendling L, Koukkari P (2020) Low- cost biochar adsorbents for water purification including microplastics removal. Appl Sci (Switzerland) 10(3). https://doi.org/10.3390/app10030788

Simon M, Vianello A, Vollertsen J (2019) Removal of > 10 μm microplastic particles from treated wastewater by a disc filter. Water (Switzerland) 11(9). https://doi.org/10.3390/w11091935

Singh N, Khandelwal N, Ganie ZA, Tiwari E, Darbha GK (2021) Eco- friendly magnetic biochar: an effective trap for nanoplastics of varying surface functionality and size in the aqueous environment. Chem Eng J 418(March):129405. https://doi.org/10.1016/j.cej.2021.129405

Stapleton MJ, Hai FI (2023) Microplastics as an emerging contaminant of concern to our environment: a brief overview of the sources and implications. Bioengineered 14(1):2244754. https://doi.org/10.1080/21655979.2023.2244754

Sundbæk KB, Koch IDW, Villaro CG, Rasmussen NS, Holdt SL, Hartmann NB (2018) Sorption of fluorescent polystyrene microplastic particles to edible seaweed Fucus vesiculosus. J Appl Phycol 30(5):2923–2927. https://doi.org/10.1007/s10811-018-1472-8

Talvitie J, Mikola A, Koistinen A, Setälä O (2017) Solutions to microplastic pollution – removal of microplastics from wastewater effluent with advanced wastewater treatment technologies. Water Res 123:401–407. https://doi.org/10.1016/j.watres.2017.07.005

Thompson RC, Olson Y, Mitchell RP, Davis A, Rowland SJ, John AWG, McGonigle D, Russell AE (2004) Lost at sea: where is all the plastic? Science 304(5672):838. https://doi.org/10.1126/science.1094559

Tomczyk A, Sokołowska Z, Boguta P (2020) Biochar physicochemical properties: pyrolysis temperature and feedstock kind effects. Rev Environ Sci Biotechnol 19(1):191–215. https://doi.org/10.1007/s11157-020-09523-3

Tong M, He L, Rong H, Li M, Kim H (2020a) Transport behaviors of plastic particles in saturated quartz sand without and with biochar/Fe3O4-biochar amendment. Water Res 169:115284. https://doi.org/10.1016/j.watres.2019.115284

Tong M, Li T, Li M, He L, Ma Z (2020b) Cotransport and deposition of biochar with different sized-plastic particles in saturated porous media. Sci Total Environ 713:136387. https://doi.org/10.1016/j.scitotenv.2019.136387

Torres FG, Dioses-Salinas DC, Pizarro-Ortega CI, De-la-Torre GE (2021) Sorption of chemical contaminants on degradable and non-degradable microplastics: recent progress and research trends. Sci Total Environ 757:143875. https://doi.org/10.1016/j.scitotenv.2020.143875

Wang Z, Sedighi M, Lea-langton A (2020) Filtration of microplastic spheres by biochar: removal efficiency and immobilisation mechanisms. Water Res 116165. https://doi.org/10.1016/j.watres.2020.116165

Wang J, Sun C, Huang QX, Chi Y, Yan JH (2021) Adsorption and thermal degradation of microplastics from aqueous solutions by Mg/Zn modified magnetic biochars. J Hazard Mater 419(April):126486. https://doi.org/10.1016/j.jhazmat.2021.126486

Wang C, Wei W, Zhang Y-T, Ni B-J (2022a) Evaluating the role of biochar in mitigating the inhibition of polyethylene nanoplastics on anaerobic granular sludge. Water Res 221:118855. https://doi.org/10.1016/j.watres.2022.118855

Wang X, Dan Y, Diao Y, Liu F, Wang H, Sang W (2022b) Transport and retention of microplastics in saturated porous media with peanut shell biochar (PSB) and MgO-PSB amendment: co-effects of cations and humic acid. Environ Pollut 305(December 2021):119307. https://doi.org/10.1016/j.envpol.2022.119307

Wang X, Dan Y, Diao Y, Liu F, Wang H, Sang W, Zhang Y (2022c) Transport characteristics of polystyrene microplastics in saturated porous media with biochar/Fe3O4-biochar under various chemical conditions. Sci Total Environ 847(June):157576. https://doi.org/10.1016/j.scitotenv.2022.157576

Xiang W, Zhang X, Chen J, Zou W, He F, Hu X, Tsang DCW, Ok YS, Gao B (2020) Biochar technology in wastewater treatment: a critical review. Chemosphere 252:126539. https://doi.org/10.1016/j.chemosphere.2020.126539

Yang X, Wen P, Yang Y (2017) Plastic biodegradation by in vitro environmental microorganisms and in vivo gut microorganisms of insects. Front Microbiol 13:1001750

Yuan F, Yue L, Zhao H, Wu H (2020a) Study on the adsorption of polystyrene microplastics by three-dimensional reduced graphene oxide. Water Sci Technol 81(10):2163–2175. https://doi.org/10.2166/wst.2020.269

Yuan J, Ma J, Sun Y, Zhou T, Zhao Y, Yu F (2020b) Microbial degradation and other environmental aspects of microplastics/plastics. Sci Total Environ 715:136968. https://doi.org/10.1016/j.scitotenv.2020.136968

Zhang GS, Liu YF (2018) The distribution of microplastics in soil aggregate fractions in southwestern China. Sci Total Environ 642:12–20. https://doi.org/10.1016/j.scitotenv.2018.06.004

Zhang C, Chen X, Wang J, Tan L (2017) Toxic effects of microplastic on marine microalgae Skeletonema costatum: interactions between microplastic and algae. Environ Pollut 220:1282–1288. https://doi.org/10.1016/j.envpol.2016.11.005

Chapter 23
Diatoms: An Important Biomonitoring and Bioindicator Tool of Aquatic Ecosystems with Potential in Novel Nanomaterial Synthesis and Downstream Applications in Pollutant Removal

Rohan Kr Biswas, Arghadeep Das, and Avik Kumar Choudhury

Abstract Health of natural aquatic ecosystems has become a cause of concern in recent times due to uncontrolled anthropogenic influences. So, the importance of monitoring the health of aquatic ecosystems has become a cause of concern like never before. However, monitoring of water through physical or chemical parameters is time consuming and not cheap. To tide over this issue, bioindicators have come to the forefront due to their high responsiveness to environmental protuberances. Diatoms are one of the most abundant forms of photosynthetic protists in aquatic environment that is well recognized as a biomonitoring tool by international organizations. Diatoms are nanocontainers that can remove different pollutants like trace metals, polymers and drugs which are harmful for living organisms. Thus, the presence of diatoms in natural aquatic habitats makes the water bodies less toxic and more habitable for other aquatic organisms. Even though diatoms are tolerant algal group, overexposure to pollutants may lead to morphological and physiological changes in diatoms.

Thus, observation of morphometric changes and community composition of diatoms in aquatic habitats can be an excellent biomonitoring tool to observe the health of natural aquatic ecosystems against pollutant exposure. Diatoms have also been well documented as an excellent framework for production of novel metal nanoparticles and nanoconjugates under experimental conditions. These diatoms are excellent biomining tool for green synthesis of novel nanoparticles. In the process of nanoscale particle synthesis, diatom can convert novel metals into nontoxic or less toxic forms by conjugation with different functional groups on diatom cells. The other important aspect of diatom-derived nanoparticle is their ability in pollutant

R. K. Biswas · A. Das · A. K. Choudhury (✉)
Phycology Ecology and Remote Sensing Lab (PEARL), Department of Botany, Ramakrishna Mission Vivekananda Centenary College, Kolkata, India
e-mail: avik.choudhury@rkmvccrahara.org

S. Dey, S. Bhattacharya (eds.), *Biotechnological Interventions in the Removal of Emerging Pollutants*, Interdisciplinary Biotechnological Advances,
https://doi.org/10.1007/978-981-97-9922-0_23

removal. The biogenic nanoparticles have recently been well established as resource for successful removal of pollutants and as surfactants. Thus, the present work will aim to highlight the applicability of diatoms both as a biomonitoring tool as well as an efficient entity in production of novel metal nanomaterials for removal of pollutants both under experimental and in situ conditions.

Keywords Diatoms · Aquatic habitats · Biomonitoring · Nanomaterial · Pollutant

23.1 Introduction

Aquatic habitats include both lentic (lacustrine or stagnant) and lotic (riverine or flowing) ecosystems based upon the nature of water flow, light penetrance and nutrient availability. Many of these habitats represent unique biodiversity and ecosystem that are often designated as "ecoregions." Natural ecoregions are affected by several natural physical factors (temperature, pH, salinity, conductivity) and sensitivity of water masses to nutrient enrichment (Painting et al. 2007). Ecologically, biological communities are differently affected by the above-mentioned physical attributes that concomitantly influence biological productivity and diversity. Other climatological factors (warming events, seasonal precipitation, episodic cyclones) and localized changes (CO_2 enrichment, terrestrial and agricultural runoffs, anthropogenic influences) further affect natural biological communities in different ecoregions (Tyson et al. 2002). Even though habitat characterization using abiotic variables has been a traditional method, in recent times as climatic alterations are increasingly becoming erratic, such an approach often gives incorrect assessment of habitats. As natural biotic communities are highly responsive to climate driven changes in habitat conditions, so they can be excellent indicators and monitoring tools. Thus, biomonitoring of the health of ecosystem using biotic proxies and assessment of pollution status of habitat using a combination of bioindicators and natural abiotic parameters is the future direction of study in ecosystem and environmental science. Among bioindicators, phytoplankton has been recommended as a highly efficient bioindicator by various international agencies like United Nations' Environmental Programme (UNEP), Marine Strategic Framework Directives (MSFD) and similar other organizations. However, the community composition and the dominance of phytoplankton in any aquatic habitat depends upon the nature of the habitat.

The nature of habitat mainly pertaining to their stagnated or free flowing nature determines the type of phytoplankton community that will be abundantly present which can possibly work as a proxy for characterization of that habitat. Studies have suggested that free flowing lotic habitats tend to be dominated by diatoms and dinoflagellates with some representative of other classes like Cyanobacteria, Chlorophyta and members of Heterokontophyta.

However, lentic habitat photosynthetic protist community is less diverse with dominance of members of Chlorophyceae and Euglenophyceae with some members of Bacillariophyceae. Since diatoms are abundant both lotic and lentic habitats, the present work considers applicability of these unicellular protists as a potential bioindicator tool for habitat characterization. Thus, an idea about the different types of natural aquatic habitats will be central in understanding the kind of phytoplankton that may possibly provide us with information about the dynamics of that habitat.

23.2 Different Types of Aquatic Ecosystem

Lotic and lentic aquatic ecosystems acquire distinct physical and chemical characters from terrestrial ecosystems. According to the amount of incoming light absorbance by water and photosynthetic organism different zones of an aquatic ecosystem can be marked as photic zone and aphotic zone. These zones together constitute the pelagic zone, beneath which is the abyssal zone with no light penetration. Thus, based upon the physical and chemical characters as well as biotic communities, different aquatic ecosystems of the pelagic zone can be classified that gives us an idea about the various forms of lotic aquatic ecosystems that constitutes the global water cycle (Reece et al. 2017).

23.2.1 Lake Ecosystem

Lake ecosystems are one of the major lentic systems that includes water bodies from ponds to several square meter lakes, depending on the incoming light and thermal stratification. In summer the surface layer or the epilimnion with high light incidence has typical homogeneous mixing of wind. In the subsurface layers there is a sudden temperature drop (Hairston Jr and Fussmann 2002), causing a thermocline that represent a homogeneous cooler water zone called the hypolimnion. Shallow, high latitude and altitude water bodies do not show thermal stratification although incidences of winter stratification seldom occur in high altitudes. In terms of chemical environment, high oxygen and low nutrient availability and bottom sediment disposition is characteristic of oligotrophic lakes and vice versa in eutrophic lakes.

23.2.2 Wetland Ecosystem

According to the Ramsar convention article 1.1, wetlands are defined as "areas of marsh, fen, peatland or water, whether natural or artificial, permanent or temporary, with water that is static or flowing, fresh, brackish or salt, including areas of marine water the depth of which at low tide does not exceed six metres." Major wetland types are as follows (Gell et al. 2023):

- Marine (coastal wetlands including coastal lagoons, rocky shores, seagrass beds and coral reefs)
- estuarine (including deltas, tidal marshes and mudflats and mangrove swamps)
- lacustrine (wetlands associated with lakes)
- riverine (wetlands along rivers and streams)
- palustrine (meaning "marshy"—marshes, swamps and bogs)

Wetlands are termed as nature's kidney having the ability to filter dissolved nutrients, which supports rich floral and faunal diverse groups of organisms.

23.2.3 Streams and Rivers

The primary physical factors for river and stream assessment is the velocity and amount of the flowing water. At the source of a river the water is clearer, colder, turbulence is higher. As it flows downstream, turbidity increases, turbulence decreases,other tributaries may join which carry more sediments. Salinity and other dissolved nutrients are low and oxygen concentration is high at the source. At source, phytoplankton and aquatic plants are dominant, and in downstream, fish communities and invertebrates inhabit (Reece et al. 2017).

23.2.4 Estuaries

Estuaries represent a transition area where the river and sea merge. A distinct ecosystem is characterized by typical salinity profile lower than sea but much higher than rest of the river from source. Several well adapted halotolerant plant species survive and phytoplankton belonging to classes like cyanobacteria, Chlorophyceae, Euglenophyceae, Bacillariophyceae, and Dinophyceae dominate these habitats (Pal and Choudhury 2014). These areas experience inflow of denser seawater through the channels from sea that mixes with low density river water, resulting in a significant increase in density of this ecosystem (Reece et al. 2017).

23.2.5 Intertidal Zones

They occur in any space where the ocean meets land; these areas remain in submerged conditions generally bearing rocky and sandy geological features. These areas have high oxygen and nutrient concentrations with nutrient replenishment being achieved through tidal variations. Several alga genera attached on the rocky surface are common but rare in sandy intertidal zones. Upper and lower intertidal zones show distinct heterotrophic diversity, where animals are adapted to remain attached in the hard substratum (Reece et al. 2017). The intertidal zone can be classified into four basic types, namely (https://oceanservice.noaa.gov/facts/intertidal-zone):

- Spray zone
- High intertidal zone
- Middle intertidal zones
- Low intertidal zones

23.2.6 Oceanic Pelagic Zone

In the oceanic zone, the photic zone is deeper as compared to the coastal zone. Water mixing by winds is a constant phenomenon with high oxygen level and low nutrient concentrations as compared to the coasts. Average depth is about 4000 m, and the deepest point could be more than 10,000 m. Phytoplankton and other photosynthetic bacteria are major photosynthesis carrying organisms. And these are basic food source for other organism found here like zooplanktons, worms, copepods, shrimps, jellyfish, etc. (Reece et al. 2017).

23.2.7 Marine Benthic Zone

This zone is represented by the sea floor with negligible or no sunlight. Temperature is nearly about 3 °C and the organisms here are adapted in this cold high-pressure environment. Enough oxygen supports diverse group of organisms like seaweeds and filamentous algae found in shallow benthic area to run their photosynthesis. The hydrothermal vents support organisms such as chemoautotrophic bacteria. Beyond the photic zone organisms depend on the above-mentioned members for nutrition (Reece et al. 2017).

23.3 Importance of Aquatic Ecosystems in Carbon Sequestration and Blue Economy

Aquatic ecosystems are most diverse and productive of all the ecosystem (Poff et al. 2002). Many people living near aquatic ecosystems are dependent economically on aquatic ecosystems (Korsgaard and Schou 2010). Many freshwater aquatic ecosystems are the source of drinking water and act as natural water purifiers, filter pollutants and sediments that provide clean water for human consumption (Hammer and Bastian 2020). Agriculture and fisheries are directly depended on aquatic ecosystems (Blann et al. 2009).

Aquatic ecosystems fill up the groundwater storage (Danielopol et al. 2003). Aquatic ecosystems release most of the oxygen in the environment, mainly contributed phytoplankton. By photosynthetic events aquatic ecosystems fix carbon di oxide (CO_2), sequester CO_2 and participate in climate regulation (Santos et al. 2022). Aquatic ecosystem supports the largest food web where the dominant primary producers are phytoplankton (Potapov et al. 2019). The other important aspect of aquatic ecosystem is the huge biodiversity of both flora and fauna that it houses. It expectedly includes both autotrophic and heterotrophic biotic communities and the mutual interdependence of both these communities decides the yield and productivity of natural aquatic ecosystems. Drawdown of carbon by autotrophic communities like phytoplankton for photosynthesis in one hand reduces carbon footprints as well as provide resources for sustainable blue economy. Thus, decrease in carbon sequestration efficacy of natural photosynthetic protist communities in aquatic ecosystems would concomitantly reduce availability of carbon stocks for blue economy (Fig. 23.1).

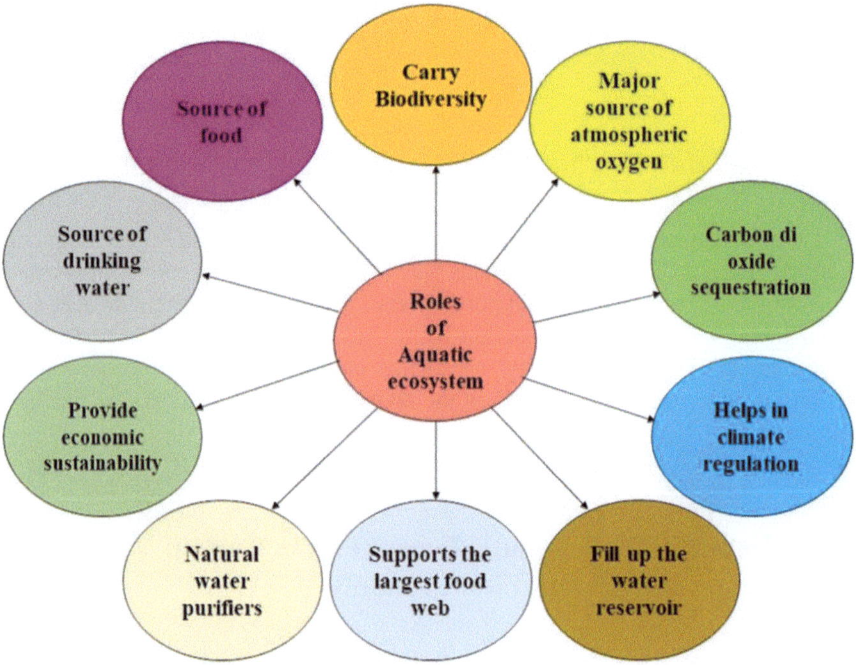

Fig. 23.1 Multifarious roles of natural aquatic ecosystem and their importance in sustenance of the health of planet Earth

23.4 Impacts and Implications of Pollutants in Aquatic Ecosystems

"Pollution is defined as the production and or introduction by man, directly or indirectly substances or energy into the environment, resulting in deleterious effects to living resources, including human beings or interfere with amenities and other uses of the environment" (Don-Pedro 1990). Natural aquatic habitats are impacted by both natural and chemical pollutants from point and nonpoint sources that affect the health of the concerned ecosystem. Natural pollutants are primarily contributed by human population growth that accounts for the amount, types and ecological implications of pollutant loads in coastal, estuarine and marine watersheds (Nixon 1995). Anthropogenic influences and increase in contamination by pollutants in the aquatic environment is very much interconnected, that results in negative consequences on wellbeing of the environment in totality. Population growth in recent times has contributed maximally in enhancing anthropogenic influences. UNESCO reported six-fold rise of freshwater consumption in past 100 years that has impacted water quality negatively for human use as well (Xu et al. 2022). Deterioration of coastal ecosystems due to natural pollutants like anthropogenic nutrient enrichment have become a cause of concern in the recent years (Ocean Studies Board and National Research Council 2000). From a biological perspective the impacts of nutrient enrichment have led to eutrophication (Nixon 1995), habitat degradation (Diaz and Solow 1999), and changes in function and structure of food webs (Jackson 2001). The alterations in food web dynamics due to anthropogenic nutrient loadings have led to a decrease in primary consumer population because of loss of palatability of primary producers, thereby enhancing the producer stock in aquatic ecosystems (Jackson 2001). The impacts of these changes have been multifarious and resulted in altered nutrient cycling, biodiversity, and ecosystem health (Ocean Studies Board and National Research Council 2000).

These habitats are also heavily influenced by chemical contamination (like microfibers, heavy metals, plastics, detergent) released from industries and household (Bhat et al. 2017). One of the major chemical pollutants are Oxides of Nitrogen (NOxs) and Oxides of Sulfur (SOxs) which are the major air pollutants responsible for acid rain. When these NOxs and SOxs mix with water through acid rain they alter the pH of the habitat through formation of sulfuric acid and nitric acid both of which are potent acids if precipitated in high quantities in aquatic ecosystems. Nearly 80% of untreated wastes from different municipal areas from all over the world are discharged in aquatic habitats increasing the deterioration of health of that aquatic body. Industrial wastewater is often laden with high quantities of minerals and ions like Calcium (Ca^{2+}), Magnesium (Mg^{2+}), Iron (Fe^{2+}/Fe^{3+}) which render them as hard water. These ions often bind together to form insoluble salts that easily precipitates and causes "scaling" of biotic communities. As aquatic ecosystems are regarded as nature's kidney, they are highly potent to take up the huge amount of heavy metals contribution by industries. In recent times metal toxicity has become a concern for aquatic ecosystem. When the concentration of metals reaches more than the safe recommended limit, it then termed as metal toxicity. Metal toxicity harms every lifeform directly or indirectly, mainly through bio-magnification (Baby et al. 2010).

Implications of such increase in both natural and chemical pollutants may have indirect multifarious effects pertaining to socio-economic wellbeing of people dependent on natural aquatic habitats. A major economic component dependent on natural aquatic habitats is fisheries. Such pollution of habitats would affect inland aquaculture and fisheries as benthic organisms are most affected by such chemical pollution of natural habitats (Martínez-Córdova and Enriquez-Ocaña 2007). Consequential habitat degradation would affect numerous organisms that may result in significant fluctuations in species diversity and biodiversity. Thus, to minimize habitat degradation due to natural and chemical pollutants, it is important to monitor these habitats so that the impacts of pollutants on natural communities of the ecosystems is easily observable and understandable. Based upon such monitoring efforts, severity of pollutants can be easily assessed, which in turn will be beneficial to envisage the threats of pollution and their impacts on the ecosystem functioning.

23.5 Biomonitoring of Aquatic Ecosystem and Importance of Bioindicators

Furley et al. (2018) put forward the priorities which are to be considered for attaining environmental sustainability like contaminants and their metabolites identification, prediction of effects of individual and hybrid contaminants, improvement of environmental management and regulatory tools. Monitoring program suing bioindicators as proxies meets all these areas of priority. Various living organisms serve as bioindicators and provide us with information on the amount of pollutants and help us to better assess their effects on the ecosystem. Some of the bioindicators that are presently operative are as follows:

- Alteration in growth and reproduction pattern is regarded as an indicator of environmental stress in a community (Orfanidis et al. 2007).
- It is reported that aquatic organisms show responses to contaminants even when they are in small amounts in water or sediments (Wijeyaratne and Pathiratne 2006).
- Bioindicators are used for identification of pollutants in their natural environment as well as under laboratory conditions.
- As bioindicators aquatic organisms are used once and again due to their comprehensive exposure to environmental variations.
- Not only the presence or absence of indicator species but also various advanced techniques like Saprobic index, Saprobic valency, indicative weight, trent biotic index, score system and Hilsenhoff Biotic Index are used in biomonitoring to get clearer vision about the concern.

- There are some group of organisms that are frequently used as bioindicators such as diatoms, macroinvertebrates, gastropods, fishes, zooplankton, phytoplankton and macrophytes for the monitoring of aquatic ecosystem.
- Benthic macroinvertebrates have a major role in biomonitoring and serve as good indicators for evaluation of ecological conditions.
- There are some important biomonitoring techniques in freshwater ecosystem, such as "The diversity indices, Biotic indices, Multimetric approaches, Multivariate approaches, Indices of Biological Integrity (IBI), Functional Approaches and macroinvertebrate trait-based approaches."
- Different genera of phytoplankton belonging to Chlorophyceae, Bacillariophyceae, Cyanobacteria and Euglenophyceae also act as good bioindicators and can provide the reader with both long term and short-term reports of various environmental diversifications.
- Several biomonitoring indices based on algae are used like "Palmer's Algal Index; Chlorophycean Index; Nygaard's phytoplankton indices (Myxophycean Index, Bacillariophycean Index, Euglenophycean Index and Compound Coefficient Index)."

23.6 Diatoms

Diatoms are chromophytic protists and are the most important primary producers (Biswas and Choudhury 2021) that contribute for more than 20% of global productivity. Lee (2018) classified algal classes into four groups based on endosymbiotic theory and placed diatoms (Bacillariophyceae) in the fourth group (eukaryotic algae with chloroplast surrounded by two membranes of chloroplast endoplasmic reticulum) under the class Heterokontophyta (two flagella—anterior tinsel and posterior whiplash; contains chlorophyll a and c, fucoxanthin; chrysolaminarin usually the storage product). Morphologically diatoms can be grouped as centric and pennate based on their plane of symmetry. Pennate diatoms may be symmetrical along the apical axis, the transapical axis or both whereas centric diatoms are mostly radially symmetrical (Biswas and Choudhury 2021). The presence of girdle bands is another important character for valve morphology studies as it represents overlapping parts of the epitheca and the hypotheca. Raphe system is the other major morphometric character that help to classify pennate diatoms. Thus, based on frustule morphology and raphe structure, diatoms can be classified as below, as reported by the North American Association for diatoms (https://diatoms.org/morphology) (Biswas and Choudhury 2021) (Table 23.1).

Diatom cell wall is unique as it is noncellulosic and is made up of silica (Schröfel et al. 2011). Highly specific Silicic Acid Transporter (SIT), present in diatom frustules transport silicic acid into the cell from the habitat (Lee 2018; Biswas and Choudhury 2021). Silica in natural aquatic habitat remains as metasilicic acid, which autopolymerizes at pH greater than 9. Once silica polymerizes, it can then form siloxane crystals, which is then assimilated in the diatom frustule. Frustules

Table 23.1 The classification of diatoms based on valve morphology and patterns of raphe system as recommended by the US Association of Diatoms (https://diatoms.org/morphology) (Spaulding et al. 2021; Biswas and Choudhury 2021)

Types	Morphological characteristics			Examples
	Valve morphology	Raphe system	Other details	
Centric	Radial	Absent	Cells may possess rimoportulae and fultoportulae	*Actinocyclus*, *Cyclotella*, *Chaetoceros*, *Brevisira*, etc.
Araphid	Bilaterally symmetrical	Absent	Cells may possess rimoportulae	*Diatoma*, *Ctenophora*, *Asterionella*, *Fragillaria*, etc.
Eunotioid	Bilaterally symmetrical; asymmetrical to the apical axis	Short	Cells may possess two or more rimoportulae	*Actinella*, *Amphicampa*, *Eunotia*, *Peronia*, etc.
Symmetrical biraphid	Bilaterally symmetrical; symmetrical to both apical and transapical axis	Well developed	Cells may be highly motile	*Cavinula*, *Amicula*, *Caloneis*, *Fallacia*, etc.
Monoraphid	Bilaterally symmetrical	Present on one valve only	Heterovalvar ornamentation	*Cocconeis*, *Eucocconeis*, *Platessa*, *Achananthes*, etc.
Asymmetrical biraphid	Asymmetrical to apical axis or transapical axis or both	Well developed	Some genera secrete mucilaginous stalk through apical pore fields, some secrete mucilaginous tubes	*Amphola*, *Cymbella*, *Gomphonema*, *Gomphosinica*, etc.
Epithemioid	Bilaterally symmetrical; asymmetrical to apical axis	Well developed	Raphe enclosed within a canal. Raphe margin is near to valve margin.	*Epithemia*, *Rhopalodia*, etc.
Nitzschoid	Bilaterally symmetrical; symmetrical to both apical and transapical axes	Well developed	Raphe enclosed within a canal and may be raised onto a keel	*Cylindrotheca*, *Denticula*, *Nitzschia*, *Pseudonitzschia*, etc.
Surirelloid	Bilaterally symmetrical	Extremely well developed	Raphe enclosed within a canal and positioned around the entire valve margin and raised onto a keel	*Surirella*, *Iconella*, *Entomoneis*, *Camohylodiscus*, etc.

are formed by biomineralization of silica through formation of Silica Deposition Vesicles (SDVs) by the process of biofabrication (Fischer et al. 2016). The diatom frustule is a nanopore structure with different levels of sieving properties through progressive decrease in pore size from foramen to cribellum. It contains different functional groups like amino (–NH_2), silanol (Si–O–Si), carboxyl (–COOH), aldehyde (–CHO), ketone (=CO), esters, etc. These groups are responsible for changing the charge properties of frustules, which ultimately helps in bioaccumulation of metal nanoparticles (NP) and in detoxification of toxic metal NP (Biswas and Choudhury 2021). Frustulin protein, which is found all over the cell wall of diatom, have protective function rather than biogenesis. Unlike frustulin, other proteins like silaffins, silacidins, pleuralins, and cingulins are specialized silica forming proteins that form the basic scaffold for deposition patterns of amorphous silica. These proteins play the major role in determining the species-specific morphometry of the frustule of the diatom cell.

Physiologically, diatoms are present from freshwater to marine habitats with ubiquitous distribution. Pennate forms dominate freshwater conditions whereas centric diatoms are abundantly present in estuarine and marine habitats, although exceptions exist. Marine to estuarine forms is euryhaline, whereas freshwater forms tend to be stenohaline. They can be present from oligotrophic to eutrophic conditions based upon the efficacy of nutrient utilization. Even though they are abundant in tropical to temperate regions, thermophilic and psychrophilic are also reported from contrasting habitats like tropical oceans and the Arctic fjords. However, they are highly sensitive to even slightest alteration in habitat conditions and can be well used as a marker for habitat characterization (Reynolds et al. 2002; Padisák et al. 2009) (Table 23.2).

Table 23.2 Habitat characterization and corresponding codon specification for different diatom taxa used as indicator species for specific habitat characters

Codon	Habitat characteristics	Selected representative taxa
A	Clear, deep, base poor lakes with species showing sensitivity towards pH variations	*Rhizosolenia* spp., *Thalassiosira* spp.
B	Mesotrophic conditions with phytoplankton species being sensitive to stratification of water column	*Aulacoseira islandica*, *Stephanodiscus neostrea*, *Cyclotella bodanica*
C	Eutrophic habitat with species sensitive to the onset of stratification	*Aulacoseira ambigua*, *Stephanodiscus*
D	Shallow turbid habitat that includes both lotic and lentic systems	*Synedra ulna*, *Nitzschia acicularis*
MP	Frequently stirred up inorganically turbid shallow lake	*Pleurosigma* spp., *Cyclotella* spp.
P	Habitat conditions is well mixed with light availability but tends to be more eutrophic	*Fragilaria cortonensis*, *Aulacoseira granulate*
WS	Ponds with high organic matter resulting in nonacidic environments	*Synura* spp. (planktonic forms)
X1	Shallow, with eutrophic to hypereutrophic conditions of the habitat	*Manoraphidium contortum*, *Monoraphidium* spp.

Modified from Reynolds et al. (2002); Padisák et al. (2009)

23.7 Diatom as Biomonitoring and Bioindicator Tool and Metal Toxicity Indicator

The adaptive capacity of diatom is better than any other algal species (Biswas and Choudhury 2021). Not only had the extensive evolutionary variations of metabolic pathway of diatoms made them a comparatively more productive species in algal groups, but also their siliceous cell wall made them more tolerant. Different functional groups in diatom frustules can convert toxic substances into nontoxic forms. The response of diatom to any disturbances from their natural habitable environment is very rapid as these have a short life cycle (Biswas and Choudhury 2021). These protists respond rapidly to the changing aquatic environment brought about by changes in pH, salinity, inorganic nutrients, organic matters, and metal toxicity (Solak et al. 2020). The siliceous cell wall of diatom provides it a superior adaptation over other algal members, which may not survive the environment where diatom species can survive. In comparison of other algal members, diatoms have larger vacuoles.

Therefore, these can store more quantity of nutrients than other algal genera, which help these survive even in a low nutrient environment (Falkowski et al. 2004). Finding only diatom without any other algal species can be an indication of water pollution. Though diatoms have great adaptations to different pollutants, especially metal pollutants, the frustule structure gets deformed and the physiological characters of diatoms get affected. Apart from the higher tolerance capacity of diatoms, their wide distribution in aquatic environments and also the autecology of most of the diatom species make them a highly efficient tool for biomonitoring. Different diatom indices are used to assess the health of different aquatic ecosystems although they can be area specific.

Due to ever increasing need for monitoring of aquatic ecosystem, new and better methods are necessary [Urban Wastewater Treatment Directive (Taylor 1991)]. Trophic Diatom Index (TDI), was mainly developed to assess the trophic status of UK Rivers because of the correlation of individual taxa to a single variable (molybdate reactive phosphorus or "orthophosphate"). After extensive work from different lotic habitats, TDI was proposed that considered the relationship between individual diatom taxon and orthophosphate (Kelly and Whitton 1995). Each of the taxa was assigned a sensitivity value between 1 and 5 based on the concentration of orthophosphate at which the taxa were recorded to be most abundant Table 23.3. The TDI was calculated as:

$$\mathrm{TDI} = \frac{\Sigma a_j.v_j.i_j}{\Sigma a_j.v_j}$$

where TDI = Trophic Diatom Index, a_j = abundance (proportion) of species j in sample, v_j = isndicator value (1–3) and i_j = pollution sensitivity of (1–5) of species j. Based on the calculated values, TDI ranges from 1 (very low nutrient

Table 23.3 Taxon sensitivity and indicator value of selected phytoplankton taxa using TDI (Kelly and Whitton 1995) as reference criterion

S. no.	Taxon	Sensitivity	Indicator value
1	*Asterionella japonica*	3	1
2	*Aulacosira granulata*	0	0
3	*Cyclotella meneghiniana*	5	2
4	*Cyclotella striata*	5	2
5	*Diploneis interrupta*	1	1
6	*Diploneis weissflogii*	1	1
7	*Fragilaria* sp.	2	1
8	*Gyrosigma scalproides*	5	2
9	*Gyrosigma spenceri*	5	2
10	*Gyrosigma beaufortsanam*	5	2
11	*Melosira* sp.	4	2
12	*Navicula paradoxa*	4	1
13	*Navicula mutica*	4	1
14	*Navicula minima*	4	1
15	*Navicula obtusa*	4	1
16	*Navicula pacifica*	4	1
17	*Navicula* sp.	4	1
18	*Nitzschia amphibia*	4	1
19	*Nitzschia delicatissima*	4	1
20	*Nitzschia ignorata*	4	1
21	*Nitzschia punctata*	4	1
22	*Nitzschia peregrina*	4	1
23	*Nitzschia sigmoidea*	4	1
24	*Stephanodiscus* sp.	5	3
25	*Surirella fastuosa*	3	1
26	*Thalassiosira decipiens*	4	1

concentrations) to 5 (very high nutrient concentrations), which represent oligotrophic to hypereutrophic conditions.

Based upon the proportionate contribution of each diatom taxa tolerant of organic pollution to the total phytoplankton population, a list to interpret the level of organic pollution was further recommended. It was recommended that if the population of organic pollution resistant taxa was less than 20% of total population, then the habitat was almost free of organic pollution whereas a contribution of more than 61% suggested heavy contamination of organic pollutants in the habitat (Vollenweider 1992). Several other diatom-based indices were developed between 1975 and 2015 including the Sládeček index [SLA, the Czech Republic (Sládeček 1986)], the Eutrophication Pollution Index using Diatoms [EPI-D, Italy (Dell'Uomo 2004)], the Schiefele–Schreiner Diatom Index [SHE, Germany (Schiefele and Schreiner 1991)], the standardized Biological Diatom Index [BDI, France (Lenoir and Coste 1996)] and the Generic Diatom Index [GDI, France (Rumeau and Coste 1988)], although the scope of application of these indices was more localized and not successfully implemented for other geographical regions due to the differences in the flora and environmental conditions of the habitat. Thus, Shen et al. (2018) suggested that instead of using region specific formulae, a combination of similar formulae may work as a more accurate descriptor of the pollution level and trophic status of the habitat. However, the method of calculation and application remained similar in most cases where diatom dominance (Mcnaughton Index) was calculated, and parameters like SPI (Specific Pollution Sensitivity Index, France) and GDI were determined by modification of formulae as proposed by Zellnka and Marvan (1961). The specific formulae used are (McNaughton 1967):

$$\text{Mcnaughton Index}\left(Y\right) = n_i \,/\, N \times f_i \times 100\%$$

where N is the number of all benthic diatoms in the samples, n_i is the number of benthic diatoms i, f_i is the frequency of the species i appearing at each sampling point.

$$\text{SPI} = \frac{\sum_{j=1}^{n} A_j V_j I_j}{A_j V_j};\quad \text{GDI} = \frac{\sum_{j=1}^{n} A_j V_j S_j}{A_j V_j}$$

where, A_j/A_i = relative abundance of the species j/i, V_j/V_i = indicative value ($1 \le V_j/V_i \le 3$), and I_j/S_i = pollution sensitivity ($1 \le I_j/S_i \le 5$).

The values range from 1 to 20, where SPI/GDI $\geq$ 16 indicate zero pollution or low eutrophication, 13.5 $\leq$ SPI/GDI < 16 indicate moderate eutrophication, 11 $\leq$ SPI/GDI < 13.5 indicate moderate pollution or heavy eutrophication, 7 $\leq$ SPI/GDI < 11 indicate high pollution, and SPI/GDI < 7 very heavy pollution (Shen et al. 2018).

Although diatom is tolerant to heavy metal toxicity, the morphological and physiological properties of diatom are affected due to enhanced exposure to heavy metals. Excessive amount of metal accumulation by diatom leads to abnormal cellular and biochemical processes. The growth rate of these protists are stunned cell division is stopped, and ultimately the cell dies (Biswas and Choudhury 2021). Even though different metals may be associated with different enzymes, variable amounts of exposure of these metals cause negative impact on diatom metabolic and cellular processes (Biswas and Choudhury 2021). However, the tolerance levels of different diatom species to different metals and different concentration of metals are different (Morin et al. 2012). The diatom with more tolerance can tolerate a higher concentration of a particular metal rather than a less tolerant diatom species, which can tolerate a lower concentration of metal of a particular type.

There are some important physiological parameters of diatoms that can be considered when it is used as a biomonitoring tool, such as the size of diatom community, alteration of diatom frustule structure, and diatom morphology. Pollutant exposure, especially metal pollutants, to the diatom cells alters the metabolic processes of photosynthesis and respiration. Both vegetative and sexual reproduction gets affected and ultimately causes death of the cell (Morin et al. 2012). The reduction of cell size of diatoms can be a protective measure taken by the protists to reduce the rate of bioaccumulation, bioabsorption, and bioadsorption because these processes are positively correlated to surface area to volume ratio (SA:V), which is why minimizing the size of cell helps diatom to survive for a greater extent of pollution (Morin et al. 2012; Biswas and Choudhury 2021). As diatoms are protists, single-celled unicelluar organisms, the means of vegetative reproduction is simply by mitotic cell division. The vegetative reproduction, which is the dominant reproductive process of diatoms, continues until it reaches the limiting size, after which it cannot divide again. But in the water body, which is affected by metal toxicity, diatom cells minimize their size to reduce the exposure of metal pollutants to it. That reduces the number of vegetative reproductions in diatoms which ultimately reduces the biomass of diatoms (Morin et al. 2012; Biswas and Choudhury 2021). The three-dimensional architecture of diatoms is altered due to exposure to metal stress. This change of architecture is favorable to some species. These species become tolerant and/or resistant to metal stress as the changed three-dimensional architecture helps these species to not only increase their size, but also these can perform the metabolic activities almost normally. Due to this, in a metal stress condition these tolerant species survives while the less tolerant and species are gets eliminated (Morin et al. 2012; Biswas and Choudhury 2021). The major structural alteration which are seen in diatoms due to metal stress are deformed outline of cells and alteration in striation pattern (Falasco et al. 2009).

Falasco et al. (2009) pointed out seven types of alteration in diatoms due to metals stress. Type 1 is described as deformity in valve outline, seen in *Surirella angusta* in Charest River, Canada; type 2 is alteration in striation pattern, seen in *Caloneis bacillum* in Nant d'Avril River, Switzerland; type 3 is change of size and shape, seen in *Planothidium frequentissimum* in Riou Mort River, France; type 4 is modification in raphe system, seen in *Gomphonema micropus* in Riou-Mort River, France; type 5 is modification in raphe canal system, seen in *Nitzschia dissipata* in Osor River, Spain; type 6 is cells forming colonies with unusual arrangements and type 7 is mixed type of alteration in valve architecture, both seen in *Gomphonema truncatum* in Deûle River, France. By observing the alteration of diatom morphology and by studying the physiological properties of diatom genera, biomonitoring of water bodies can be done. Thus, observing the diversity of diatoms in a water body can give an idea about the health of the water body.

23.8 Metal Particles Removal and Biosynthesis of Metal Nanostructures by Diatom

In recent times the other interesting application of diatom frustule is as absorbent of toxic metals from wastewater due to their high surface: volume ratio, porous and nonreactive nature. The process of removal of heavy metal from wastewater by physical methods like filtration, oxidation-reduction, sedimentation, media filtration and membrane filtration are often tedious and expensive (Anjum 2017). The efficacy of diatom surpasses all these hindrances and studies have revealed that salinity directly regulate surface potential and surface: volume ratio and sulfur-containing groups in the cell wall that alters the efficiency of diatom nanostructures in removal of heavy metal toxicity (Ma et al. 2019). However, teratological changes in diatom architecture, which can be both physical and chemical are indicative of metal exposure. Teratology is basically nonadaptive phenotypic abnormalities on valve morphometry. Thus, increase in teratological abnormalities further allow us to envisage the effects of metal toxicity on diatom cells (Table 23.4).

Diatom is a great tool for biosynthesis of metal nanopariticles (MNP) due its porous cell wall. Due to the nanopore structure of diatom frustules, it is best suited for the synthesis of metal nanoparticles (MNP). The different groups present in the diatom frustules, as carboxyl group (−COO−), hydroxide group (OH−), phosphoryl group (PO_{43-}), etc., create a negative environment that attracts the positively charged metal particles toward the diatom cell (Ding et al. 2017; Hedayatkhah et al. 2018). The metal particles accumulated outside of the diatom cell wall by the process of bioadsorption and internalized inside the cell by bioabsorption (Biswas and Choudhury 2021). The extracellular accumulation of metal particles is a passive process that depends on pH, temperature, SA:V ratio, contact time, biomass concentration, metal concentration, etc. (Ding et al. 2017). The bioabsorption process, on the other hand, is an active process and faster than bioadsorption. The pumps present in the diatom cell wall use ATP to influx the metal particles. By enzymatic action, the toxic forms of metals converted into less toxic or nontoxic forms (Marella et al. 2020). Metal particles gets accumulated inside the cell and form MNP. Gold

Table 23.4 The teratological properties of different pennate diatom taxa on their exposures to different levels of metal pollutant exposure

Diatom taxa	Teratology	Pollutant exposure
Nitzschia fonticola	The raphe channel is deformed	Copper and zinc
Amphora pediculus	Outline of frustule is distorted	Cadmium
Nitzschia amphibian	Outline of frustule is distorted	Zinc, copper, mercury, lead, cadmium, arsenic
Eunotia sp.	Distorted valve morphology with twist in the middle	Copper
Eunotia spp.	Bent or dialated frustule with loss of symmetry	Copper, iron, zinc, nickel
Phaeodactylum tricornutum	Frustules with abnormal boundary or with a wrinkled shape	Copper and zinc
Gomphonema pseudoaugur	Middle portion of the frustule is dilated	Selenium and lead
Brachysira microcephala	Frustule with a distorted outline	Zinc, cadmium, lead
Discostella pseudostelligera	Striation patterns are deformed at the side of the valve	Heavy metals
Fragilaria nanoides	Central area of the frustule is abnormal	Copper

Modified from Khan et al. (2021)

(Au) is a precious metal and toxic beyond a safer limit. It is evident from nature and laboratory that pennate diatoms bioaccumulate gold particles by both bioadsorption and bioabsorption processes (Chakraborty et al. 2006; Biswas and Choudhury 2021). By accumulating metal particles, diatoms not only plays an important role in the removal of metal particles but also participate in the biosynthesis of important metal nanoparticles and nanoconjugate biosynthesis. Studying the diatom frustules can give an idea about metal toxicity in an aquatic ecosystem.

23.9 Conclusion

Thus, diatoms as a dominant autotrophic component in aquatic ecosystems not only contribute to primary productivity but participates well as a biomonitoring agent. The physiological adaptations as well as responsiveness to variable habitat conditions make them an excellent tool for monitoring of habitats especially in context to trophic status. A series of reference criteria and estimation tools are available from different references which may have significant applications in our systems as well. In India, monitoring of aquatic habitats has largely been in reference to human use often referred as "water quality." The Central Pollution Control Board (CPCB) proposed different reference guidelines although none considered the ecological perspective, rather focused on water quality. As mentioned in this chapter, shifts in diatom community composition and morphometric alterations in diatom frustule architecture can be much robust tools to monitor the trophic status of the habitat rather than physicochemical parameters and biotic indicators like CFU counts. The other important aspect of metal tolerance and metal bioaccumulation by diatoms make them unique tool for both monitoring of metal toxicity under in situ conditions and MNP generation under in vitro conditions. Such applications can be easily achieved as they do not overlap but will have major scientific implications in the future. In this era of climate change, we propose that diatoms are the most accurate descriptor of environmental protuberances and validating it as a biomonitoring tool will be a significant boost for field-based research from coastal habitats in the Indian diaspora. However, the other contrasting aspect of diatoms in forming MNPs can be exploited through lab-based research which has immense potential in areas like semiconductor preparation, drug delivery, MNP supplementation and formation of nanoconjugates. If properly utilized, these diatoms can well be the future for biomining of novel metal nanoparticles and nanoconjugates as a cheap yet reliable alternative to chemically synthesized MNPs. As a concluding statement, it can be said that even though diatoms are submicroscopic entities, they are crucial to us and can be an excellent sustainable alternative for the future.

Acknowledgments Rohan Kr Biswas acknowledges CSIR-HRDG for the Junior Research Fellowship (JRF) in this work.

References

Anjum A (2017) Adsorption technology for removal of toxic pollutants. In: Sustainable heavy metal remediation. Springer, pp 25–80

Baby J, Raj JS, Biby ET, Sankarganesh P, Jeevitha MV, Ajisha SU, Rajan SS (2010) Toxic effect of heavy metals on aquatic environment. Int J Biol Chem Sci 4(4):939–952. https://doi.org/10.4314/ijbcs.v4i4.62976

Bhat SA, Singh J, Vig AP (2017) Amelioration and degradation of pressmud and bagasse wastes using vermitechnology. Bioresour Technol 243:1097–1104

Biswas RK, Choudhury AK (2021) Diatoms: miniscule biological entities with immense importance in synthesis of targeted novel bioparticles and biomonitoring. J Biosci 46(4):102. https://doi.org/10.1007/s12038-021-00222-x

Blann KL, Anderson JL, Sands GR, Vondracek B (2009) Effects of agricultural drainage on aquatic ecosystems: a review. Crit Rev Environ Sci Technol 39(11):909–1001. https://doi.org/10.1080/10643380801977966

Chakraborty N, Pal R, Ramaswami A, Nayak D, Lahiri S (2006) Diatom: a potential bio-accumulator of gold. J Radioanal Nucl Chem 270:645–649. https://doi.org/10.1007/s10967-006-0475-0

Danielopol DL, Griebler C, Gunatilaka A, Notenboom J (2003) Present state and future prospects for groundwater ecosystems. Environ Conserv 30(2):104–130

Dell'Uomo A (2004) L'indice diatomico di eutrofizzazione/polluzione (EPI-D) nel monitoraggio delle acque correnti. Linee Guida, pp 1–101

Diaz RJ, Solow A (1999) Ecological and economic consequences of hypoxia: topic 2 report for the integrated assessment on hypoxia in the Gulf of Mexico, NOAA Coastal Ocean Program decision analysis series no. 16. NOAA Coastal Ocean Program, Silver Spring, pp 1–45

Ding T, Yang M, Zhang J, Yang B, Lin K, Li J, Gan J (2017) Toxicity, degradation and metabolic fate of ibuprofen on freshwater diatom Navicula sp. J Hazard Mater 330:127–134. https://doi.org/10.1016/j.jhazmat.2017.02.004

Don-Pedro KN (1990) May pesticide pollution-biological resources for control and management. In: Proceedings of the conference on pesticide pollution detection and management at the University of Agriculture, Abeokuta, Nigeria

Falasco E, Bona F, Ginepro M, Hlúbiková D, Hoffmann L, Ector L (2009) Morphological abnormalities of diatom silica walls in relation to heavy metal contamination and artificial growth conditions. Water SA 35(5):595–606. https://doi.org/10.4314/wsa.v35i5.49185

Falkowski PG, Katz ME, Knoll AH, Quigg A, Raven JA, Schofield O, Taylor FJR (2004) The evolution of modern eukaryotic phytoplankton. Science 305(5682):354–360

Fischer C, Adam M, Mueller AC, Sperling E, Wustmann M, van Pée KH, Kaskel S, Brunner E (2016) Gold nanoparticle-decorated diatom biosilica: a favorable catalyst for the oxidation of d-glucose. ACS Omega 1(6):1253–1261. https://doi.org/10.1021/acsomega.6b00406

Furley TH, Brodeur J, Silva de Assis HC, Carriquiriborde P, Chagas KR, Corrales J, Denadai M, Fuchs J, Mascarenhas R, Miglioranza KS, Miguez Carames DM (2018) Toward sustainable environmental quality: identifying priority research questions for Latin America. Integr Environ Assess Manag 14(3):344–357

Gell PA, Finlayson CM, Davidson NC (2023) An introduction to the Ramsar convention on wetlands. In: Ramsar wetlands. Elsevier, pp 1–36

Hairston NG Jr, Fussmann GF (2002) Lake ecosystems. In: Encyclopedia of life sciences. Macmillan Publishers Ltd. https://doi.org/10.1038/npg.els.0003191

Hammer DA, Bastian RK (2020) Wetlands ecosystems: natural water purifiers? In: Constructed wetlands for wastewater treatment. CRC Press, pp 5–19

Hedayatkhah A, Cretoiu MS, Emtiazi G, Stal LJ, Bolhuis H (2018) Bioremediation of chromium contaminated water by diatoms with concomitant lipid accumulation for biofuel production. J Environ Manag 227:313–320. https://doi.org/10.1016/j.jenvman.2018.09.011

Jackson GA (2001) Effect of coagulation on a model planktonic food web. Deep-Sea Res I Oceanogr Res Pap 48(1):95–123

Kelly MG, Whitton BA (1995) The trophic diatom index: a new index for monitoring eutrophication in rivers. J Appl Phycol 7:433–444

Khan MJ et al (2021) Diatom microalgae as smart nanocontainers for biosensing wastewater pollutants: recent trends and innovations. Bioengineered 12(2):9531–9549

Korsgaard L, Schou JS (2010) Economic valuation of aquatic ecosystem services in developing countries. Water Policy 12(1):20–31

Lee RE (2018) Phycology. Cambridge University Press

Lenoir A, Coste M (1996) Development of a practical diatom index of overall water quality applicable to the French National Water Board Network. In: International symposium, Volksbildungsheim Grilhof Vill, AUT, 17–19 September 1995. Universität Innsbruck, pp 29–43

Ma J et al (2019) Salinity-dependent nanostructures and composition of cell surface and its relation to Cd toxicity in an estuarine diatom. Chemosphere 215:807–814

Marella TK, Saxena A, Tiwari A (2020) Diatom mediated heavy metal remediation: a review. Bioresour Technol 305:123068

Martínez-Córdova LR, Enriquez-Ocaña F (2007) Study of the benthic fauna in a discharge lagoon of a shrimp faro with special emphasis on polychaeta. J Biol Sci 7:12–17. https://doi.org/10.3844/ojbsci.2007.12.17

McNaughton SJ (1967) Relationships among functional properties of Californian grassland. Nature 216(5111):168–169

Morin S, Cordonier A, Lavoie I, Arini A, Blanco S, Duong TT, Tornés E, Bonet B, Corcoll N, Faggiano L, Laviale M (2012) Consistency in diatom response to metal-contaminated environments. In: Emerging and priority pollutants in rivers: bringing science into river management plans. Springer, Berlin, pp 117–146. https://doi.org/10.1007/978-3-642-25722-3_5

Nixon SW (1995) Coastal marine eutrophication: a definition, social causes, and future concerns. Ophelia 41(1):199–219

Ocean Studies Board and National Research Council (2000) Clean coastal waters: understanding and reducing the effects of nutrient pollution. Committee on the Causes and Management of Eutrophication, Ocean Studies Board, Water Science and Technology Board, National Academies Press., ISBN: 0-309-51615-3, pp 1–428

Orfanidis S, Papathanasiou V, Sabetta L, Pinna M, Gigi V, Gounaris S, Tsiagga E, Nakou K, Theodosiou TH (2007) Benthic macrophyte communities as bioindicators of transitional and coastal waters: relevant approaches and tools. Transit Water Bull 1(3):45–49. https://doi.org/10.1285/i1825229Xv1n3p45

Padisák J, Crossetti LO, Naselli-Flores L (2009) Use and misuse in the application of the phytoplankton functional classification: a critical review with updates. Hydrobiology 621(1):1–19

Painting SJ, Devlin MJ, Malcolm SJ, Parker ER, Mills DK, Mills C, Tett P, Wither A, Burt J, Jones R, Winpenny K (2007) Assessing the impact of nutrient enrichment in estuaries: susceptibility to eutrophication. Mar Pollut Bull 55(1–6):74–90. https://doi.org/10.1016/j.marpolbul.2006.08.020

Pal R, Choudhury AK (2014) An introduction to phytoplankton: diversity and ecology (No. 14781). Springer, New Delhi. https://doi.org/10.1007/978-81-322-1838-8_2

Poff NL, Brinson MM, Day JW (2002) Aquatic ecosystems and global climate change. Pew Centre on Global Climate Change, Arlington, 44, pp 1–36

Potapov AM, Brose U, Scheu S, Tiunov AV (2019) Trophic position of consumers and size structure of food webs across aquatic and terrestrial ecosystems. Am Nat 194(6):823–839. https://doi.org/10.1086/705811

Reece JB, Urry LA, Cain ML (2017) Campbell biology. Pearson

Reynolds CS, Huszar V, Kruk C, Naselli-Flores L, Melo S (2002) Towards a functional classification of the freshwater phytoplankton. J Plankton Res 24(5):417–428

Rumeau A, Coste M (1988) Introduction into the systematics of freshwater diatoms. For a useful generic diatomic index. Bull Fr Peche Piscic 309:1–69

Santos IR, Hatje V, Serrano O, Bastviken D, Krause-Jensen D (2022) Carbon sequestration in aquatic ecosystems: recent advances and challenges. Limnol Oceanogr 67:S1–S5. https://doi.org/10.1002/lno.12268

Schiefele S, Schreiner C (1991) Use of diatoms for monitoring nutrient enrichment acidification and impact salts in Germany and Austria. In: Whitton BA, Rott E, Friedrich G (eds) Use of algae for monitoring rivers. Institut für Botanik, Universität Innsbruck

Schröfel A, Kratošová G, Bohunická M, Dobročka E, Vávra I (2011) Biosynthesis of gold nanoparticles using diatoms—silica-gold and EPS-gold bionanocomposite formation. J Nanopart Res 13:3207–3216. https://doi.org/10.1007/s11051-011-0221-6

Shen R, Ren H, Yu P, You Q, Pang W, Wang Q (2018) Benthic diatoms of the Ying River (Huaihe River Basin, China) and their application in water trophic status assessment. Water 10(8):1013

Sládeček V (1986) Diatoms as indicators of organic pollution. Acta Hydrochim Hydrobiol 14(5):555–566

Solak CN, Peszek Ł, Yilmaz E, Ergül HA, Kayal M, Ekmekçi F, Várbíró G, Yüce AM, Canli O, Binici MS, Ács É (2020) Use of diatoms in monitoring the Sakarya river basin, Turkey. Water 12(3):703. https://doi.org/10.3390/w12030703

Spaulding SA et al (2021) Diatoms.org: supporting taxonomists, connecting communities. Diatom Res 36(4):291–304

Taylor P (1991) The European Community and the state: assumptions, theories and propositions. Rev Int Stud 17(2):109–125

Tyson PD, Lee-Thorp JA, Holmgren K, Thackeray JF (2002) Changing gradients of climate change in southern Africa during the past millennium: implications for population movements. Clim Chang 52:129–135

Vollenweider RA (1992) Coastal marine eutrophication: principles and control. In: Marine coastal eutrophication. Elsevier, pp 1–20

Wijeyaratne WMDN, Pathiratne A (2006) Acetylcholinesterase inhibition and gill lesions in Rasbora caverii, an indigenous fish inhabiting rice field associated waterbodies in Sri Lanka. Ecotoxicology 15:609–619. https://doi.org/10.1007/s10646-006-0101-5

Xu X, Yang H, Li C (2022) Theoretical model and actual characteristics of air pollution affecting health cost: a review. Int J Environ Res Public Health 19(6):3532. https://doi.org/10.3390/ijerph19063532

Zellnka M, Marvan P (1961) Zur Prazisierung der biologischen Klassifikation der Reinheit fliessender Gewiisser. Arch Hydrobiol 57:389–407

Chapter 24
Engineered Nanomaterials and Associated Threats in the Environment Risk Assessment Strategies

Abhishek Kumar

Abstract Given the potential ecological and human health implications, assessing the risks associated with engineered nanomaterials in the environment is a critical endeavor. The complex nature of ENMs, characterized by their unique physicochemical properties and diverse applications, necessitates comprehensive risk assessment strategies. This chapter delves into the multifaceted challenges ENMs pose and explores strategies for evaluating their environmental impact. It reviews established methodologies for detecting and quantifying ENMs in natural environments, emphasizing the importance of robust pretreatment techniques and advanced detection technologies. Furthermore, it examines the intricate mechanisms governing the fate, transport, and transformation of ENMs, drawing insights from laboratory studies. Understanding the toxicological effects of ENMs at genetic, cellular, individual, and ecosystem levels is crucial, and this chapter provides a preliminary understanding of ENM-induced toxicity. However, it also highlights the complexity of extrapolating laboratory findings to real-world scenarios, where multiple environmental factors and diverse organisms come into play. To address this challenge, a holistic approach and the utilization of mesocosmic model ecosystems are proposed for systematic ecological health research on ENMs. By amalgamating these insights, this chapter aims to advance risk assessment strategies, providing a comprehensive framework for evaluating the threats associated with ENMs in the environment.

Keywords Engineered nanomaterials (ENMs) · Environmental risk assessment · Nanoparticle toxicity · Detection methods · Mesocosmic model ecosystems

A. Kumar (✉)
Department of Chemical Engineering, National Institute of Technology Rourkela, Rourkela, Odisha, India

S. Dey, S. Bhattacharya (eds.), *Biotechnological Interventions in the Removal of Emerging Pollutants*, Interdisciplinary Biotechnological Advances,
https://doi.org/10.1007/978-981-97-9922-0_24

24.1 Introduction

The concept of environmental health was initially proposed by the European members of the World Health Organization (WHO) in 1993 (WHO 1993). They defined it as the field of study that aims to identify hazardous agents and their associated environmental sources and risks in food, water, soil, and other environmental mediums. Additionally, environmental health seeks to evaluate and control the detrimental impacts of these agents on human and ecological health. Environmental health research aims to accomplish the following:

(a) Identify critical pollutants in the environment that present risks to human health.
(b) Utilize the integration of environmental and genetic methodologies to clarify the mechanisms through which environmental pollutants cause toxicity and their effects on human health.
(c) Efficiently manage the health risks associated with pollutants in the environment.
(d) Promote a comprehensive "One Health" paradigm that encompasses all aspects of health. The detrimental effects that pollutants have on human and environmental health are often irreversible.

According to the WHO's global assessment report, contaminants are responsible for 24% of human diseases and 23% of deaths. Each day, pregnant women are exposed to more than 43 distinct categories of pollutants that are primarily cytotoxic and genotoxic. It poses a threat to future generations. When environmental pollutants, which are frequently produced inadvertently by human activities, accumulate to a specific concentration over an extended period and result in pollution incidents, they become a matter of concern. Concerns regarding environmental contaminants first surfaced during the 1960s, when the extensive application of dichlorodiphenyltrichloroethane (DDT) for the eradication of malaria resulted in its accumulation in the environment. The detrimental inadvertent toxic consequences of DDT, as detailed in Rachel Carson's seminal work Silent Spring (1962), caused the extinction of numerous species of animals, insects, fish, and even humans, thereby causing ecological disruptions and biodiversity loss (Prüss-Ustün et al. 2017; Rani and Dhok 2023).

Notwithstanding the lack of DDT application for more than five decades, the repercussions endure. An additional noteworthy illustration is mercury (Hg), a metallic additive frequently incorporated into batteries, personal care items, medical supplements, and thermometers (Bose-O'Reilly et al. 2017). Although mercury has been utilized for millennia, its investigation into environmental health did not begin until 1956, in response to the outbreak of "Minamata disease," which was distinguished by neuromuscular impairment, renal dysfunction, and immunotoxicity in organisms. Mercury pollution has the potential to cause mortality rates reaching as high as 34%. The dated philosophy of "immediate mitigation after pollution" has been a contributing factor to the postponement of research and the proliferation of pollution-related incidents. It emphasizes the necessity of conducting environmental health research on novel substances or compounds before they are extensively utilized in order to safeguard environmental sustainability (Louis et al. 2006; Zeitz et al. 2002).

Emerging contaminants are a group of compounds that are either recently identified or causing new worries because of their toxicity, persistence, and ability to build up in the environment, unlike traditional pollutants such as DDT and mercury (Arman et al. 2021; Khan et al. 2022). These compounds lack efficient management measures, have little known environmental characteristics, and their toxicological and health concerns are not thoroughly investigated. Commonly acknowledged emerging contaminants include persistent organic pollutants such as organochlorine insecticides and polybrominated diphenyl ethers, endocrine disruptors like perfluorinated chemicals, personal care products, engineered nanomaterials, antibiotics, and resistance genes. Identifying and assessing the environmental health hazards of new pollutants is difficult since they are often present in low quantities, making them hard to detect and evaluate accurately (Ashesh et al. 2022; Panda et al. 2023). Once introduced into the environment, these pollutants go through various biogeochemical processes such as transfer, transformation, and accumulation within the food chain. This results in modifications to their characteristics, making them harder to detect and causing changes in their environmental impact. For example, perfluorooctanoic acids, an emerging pollutant, may attach to soil or dust particles, decreasing their movement and availability. Additionally, the transfer of resistance genes across bacteria worsens their environmental health hazards. Furthermore, the ageing of microplastics in the environment may lead to the simultaneous movement of existing pollutants and the liberation of intrinsic components, which increases the complexity of environmental concerns linked to microplastics (Guo et al. 2019; Pal and Sen 2023a, b).

Laboratory investigations have shown the harmful health impacts of new pollutants, such as neurological problems, cytotoxicity, and genotoxicity (Bose-O'Reilly et al. 2017; Genchi et al. 2022; Pašukoniene et al. 2014). These results are often obtained from trials carried out in controlled settings with elevated levels of contaminants and individual species, which may not accurately represent the actual environmental hazards. During the last 20 years, there has been increasing worry and official measures taken against certain kinds of new pollutants, resulting in their ban. However, ENMs have continued to be a primary area of interest. ENMs, when introduced into the environment by human activities, have reactive surfaces that promote transformation yet make identification difficult. The varied characteristics shown by various ENMs complicate their evaluation of environmental health (Mudumba et al. 2023; Rivero et al. 2019).

Since 2003, well-known scientific publications have emphasized the need to deal with the environmental health impacts of ENMs. Nanoscale titanium dioxide (nTiO2), a frequently used ENMs, has been recognized as a carcinogen. It has led to stricter regulatory measures and more extensive studies on the environmental health and safety of ENMs. ENMs were deemed significant and added to the United States Environmental Protection Agency's list of emerging pollutants in 2010 (Parfomak and Folger 2010). Considerable advancements have been achieved in comprehending the environmental health impacts of ENMs in the last 20 years despite the difficulties encountered. The knowledge acquired from ENM research may be a helpful asset in dealing with environmental health problems related to other emerging pollutants, especially in their early stages of investigation. Understanding the environmental destiny of Emerging Contaminants, such as ENMs, is essential due to unknown events, little toxicological data, and intricate environmental behaviors (Mohanty and Das 2023).

This chapter aims to explore how basic research on emerging contaminants' environmental health might be undertaken to generate objective data for accurate and objective environmental health risk assessments. This study intends to use the vast literature on ENMs and their environmental health effects to provide insights into the environmental health studies of other emerging pollutants, thereby establishing a basis for comprehending their behaviors in natural settings. A thorough grasp of the environmental geochemical mechanisms that control emerging pollutants is crucial for guiding their scientific management in the future.

24.2 Environmental Health Implications of ENMs: Progress and Challenges

Nanotechnology has infiltrated several areas, such as the chemical and pharmaceutical industries, medicine, agriculture, and environmental research (He et al. 2019). It is a summit of competitive scientific inquiry on a global scale. The current worldwide yearly output of ENMs is around 2.4 million tons, which presents environmental health hazards at every stage, including manufacture, use, and disposal. In the last 20 years, there has been notable progress in comprehending the environmental health effects of ENMs, with a specific emphasis on four primary areas.

(a) Identifying environmental sources and exposure hazards of ENMs and developing detection methods to determine relevant environmental concentrations.
(b) An analysis of the ambient behaviors exhibited by ENMs, including their transfer, transformation, and the factors that influence them.
(c) Exhaustive research has been conducted on the toxicity mechanisms of ENMs, focusing primarily on the genetic, cellular, and individual levels, while ecotoxicity and human toxicity have received less attention.
(d) Environmental risk assessment and management of ENMs strategies are to be developed.

Notwithstanding these efforts, the environmental health risk assessment of ENMs faces obstacles arising from significant deficiencies in prior research. An example of this is the insufficiency of trace analysis methodologies for environmental nanomaterials in laboratory studies, which has led to the establishment of elevated exposure concentrations and consequently hindered the accurate assessment of genuine environmental hazards. In addition, complex behaviors are exhibited by ENMs upon passage into the environment; however, the biogeochemical processes that influence these behaviors, which are essential for objective risk assessment, remain largely unknown.

To date, the majority of laboratory-based toxicity assessments of ENMs have concentrated on genetic, cellular, and individual levels, while their ecotoxicological and human health effects have received scant attention. However, the purpose of this chapter is to provide a summary of fundamental research discoveries obtained from investigating the environmental health hazards of ENMs. These findings have the potential to guide environmental health investigations concerning other emerging contaminants (Romero-Franco et al. 2017).

24.2.1 Environmental Occurrence and Detection of ENMs

It is of the utmost importance to comprehend the origins and prevalence of ENMs in natural settings in order to assess their environmental health risks with precision. This segment delves into the diverse origins and pathways by which ENMs reach the environment, the difficulties that arise in their characterization and detection, and the approaches utilized to distinguish them from naturally occurring nanomaterials (Lowry et al. 2010).

24.2.1.1 Sources and Routes of Entry

ENMs are introduced into the environment via natural processes as well as anthropogenic activities:

(i) Anthropogenic Sources

These encompass industrial operations such as production, consumption, and refuse management. The numerous industries in which ENMs are utilized, including healthcare, textiles, food packaging, and automotive, all contribute to their pervasive distribution.

Notable examples include:

- Metal ENMs: Silver (nAg0), iron (nFe0)
- Metal oxide ENMs: Titanium dioxide (nTiO2), zinc oxide (nZnO), cerium dioxide (nCeO2), iron oxide (nFe2O3)
- Carbon-based ENMs: Carbon black, graphene, plastic nanomaterials

(ii) Natural Sources

Natural occurrences, including forest fires, volcanic eruptions, mineral weathering, sandstorms, dust deposition, and sea surges, also produce nanoscale particles.

Breakdown Processes:

ENMs may be produced via the following mechanisms when larger particles disintegrate:

(a) Mechanical wearing: Nanoscale particles can be generated by means of frictional forces operating upon more significant substances.
(b) Chemical corrosion: Chemical reactions, such as oxidation, can break down more extensive materials into nanoscale components.
(c) Biodegradation: Biological processes, including microbial degradation, can lead to the fragmentation of larger particles into nanoscale counterparts.
(d) Solar photolysis: Exposure to sunlight can degrade certain materials, leading to the formation of nanoscale particles.

(iii) Challenges in Characterization and Detection

The detection and characterization of ENMs in natural environments present formidable obstacles:

(a) Complex compositions: Frequently, the intricate compositions of natural nanomaterials resemble those of engineered ENMs.
(b) Low concentrations: ENMs are typically present in natural environments at low concentrations, which complicates their detection and quantification.
(c) Dynamic behaviors: The detection of ENMs is further complicated by their dynamic behaviors in the environment, which include aggregation, dispersion, and transformation.
(d) Matrix Effects: Detection and quantification of ENMs may be inhibited by the existence of complex environmental matrices, including soil, sediment, and water.

Differentiation Strategies

Distinguishing between natural and ENMs requires careful consideration of their production processes and structural characteristics:

(a) Analytical methods: ENMs are characterized and quantified utilizing an array of analytical techniques, such as scanning electron microscopy (SEM), transmission electron microscopy (TEM), atomic force microscopy-infrared spectroscopy (AFM-IR), and single-particle inductively coupled plasma mass spectrometry (Sp-ICP-MS).
(b) Chemical fingerprinting: By employing chemical analysis methods like mass spectrometry (MS), one can discern particular impurities or modifiers that are linked to engineered ENMs.
(c) Isotopic analysis: Isotopic fingerprinting methods can distinguish artificial nanomaterials from naturally occurring ones by analyzing their isotopic signatures.

It is critical to detect and characterize ENMs in natural environments in order to assess their potential dangers and comprehend their environmental fate. Notwithstanding the obstacles entailed, progress in analytical methodologies and differentiation tactics presents auspicious prospects for enhancing our capacity to identify, define, and measure ENMs in natural surroundings (Minz and Mishra

2023; Sahoo and Mishra 2023; Santhosh et al. 2024; Santhosh and Sarkar 2022; United Nations Environment Programme 2009).

24.2.2 Environmental Dynamics of ENMs

ENMs exhibit a highly complex ability to traverse various environmental matrices, including air, soil, and water, which results in their gradual accumulation and subsequent modification of physical and chemical properties. ENMs' environmental behaviors, which include transformation and transfer, are affected by an infinite number of biotic and abiotic factors. The authentic environmental health implications of ENMs are determined by the attributes (morphology, surface charge, and aggregate structure) that are intricately influenced by these factors that define the environmental distribution of ENMs. Additionally, the detection of ENMs and the unbiased evaluation of their environmental dangers are complicated by these factors. The biogeochemical processes that regulate ENMs have attracted considerable interest owing to their critical significance and intrinsic intricacies. For an objective evaluation of the environmental hazards posed by these emerging contaminants, it is essential to establish a standardized framework for conducting environmental behavior studies on ENMs. This framework should obtain data that accurately represent natural environmental conditions (Joo and Zhao 2017; Pal and Sen 2024).

24.2.2.1 ENMs Transfer

It is crucial to comprehend the transfer mechanisms of ENMs in order to identify their movement within environmental systems. As ENMs traverse the air, water, and soil, their distribution and potential impacts are influenced by processes that occur within these compartments. This concise synopsis examines the primary routes by which ENMs migrate between matrices in the environment (Gottschalk et al. 2013).

ENMs Transfer in Atmospheric Environments

It is crucial to comprehend the environmental behavior and potential health hazards posed by ENMs by examining their motion within atmospheric systems. Particles with diameters smaller than 100 nm, specifically ENMs, make a substantial contribution to the transportation of particulate matter in the atmosphere. An annual discharge of nanomaterials into the atmosphere occurs in significant quantities, varying from 1×10^{12} to 10×10^{12} g, via forest fires, mining operations, industrial processes, and transportation emissions. ENMs have a comparatively brief residence time in the atmosphere prior to their condensation and subsequent deposition into terrestrial or aquatic environments via the force of gravity. Potential health hazards may result from cutaneous contact or inhalation of airborne ENMs that come into contact with humans. Significantly, specific ENMs, including plastic nanomaterials, demonstrate a reduced density in comparison to ambient dust. This characteristic empowers them to traverse horizontal circulation over long distances. The detection of plastic nanomaterials in remote regions such as Mount Everest and Antarctica can be attributed to this phenomenon.

Environmental elements such as light, humidity, coexisting particles, and other contaminants interact with ENMs rather than facilitating their migration as individual particles in the atmosphere. As a result, it is difficult to model the migration patterns of ENMs and to forecast their fate in the atmosphere. The simultaneous occurrence of vertical and horizontal migrations further complicates sample collection and detection efforts. The obstacles mentioned above impede the progress of developing precise models to evaluate the spatial and temporal distribution of ENM concentrations in atmospheric environments. It underscores the imperative for ongoing investigation in this field (Gardea-Torresdey et al. 2014; John et al. 2017).

ENM Transfer in Terrestrial Environments

ENMs breach terrestrial ecosystems via various routes, such as atmospheric deposition, surface discharge, and sediment application, thereby engaging in complex biogeochemical cycles. A multitude of factors impact their conduct within the soil, including microorganisms, dissolved ions, pH, and redox conditions, in addition to soil organic matter (OM). The presence of humic acids in the pore water of soil has a substantial impact on the stability and dispersion of ENMs. ENMs are coated, which enhances their stability and bioavailability, particularly in soils with low ion concentrations, by increasing steric hindrance between particles. Nevertheless, the aggregation of ENM can be facilitated through the bridging of particles or the compression of double layers in the presence of ions (Ma et al. 2021).

Due to increased gravitational effects, ENMs can be adsorbed to hydrophilic natural organic matter (NOM), which facilitates their deposition in soil. An elevated pH and soil characteristics such as a high organic matter concentration and cation exchange capacity can promote the migration of cations and electron nematodes. However, the presence of clay minerals can impede this process. ENM characteristics, such as particulate size, shape, and surface charge, as well as soil properties, impact the transfer of ENMs in terrestrial environments. Additionally, biotic factors, including the activities of soil organisms, are vital in ENM migration. Soil organisms are dominated by microorganisms, specifically bacteria, which strongly influence ENM migration. By means of their burrowing and excretion mechanisms, earthworms can facilitate substantial downward migration of ENMs within the soil (Yu et al. 2018).

Concerns exist regarding plant–root interactions with ENMs in agricultural soil. Although adsorbent ENMs may have restricted mobility on root surfaces, they are capable of migrating downward with root growth. ENM properties can be altered by root exudates, which can impact the processes of dissolution and agglomeration. Along the food chain, the absorption of ENMs by plant roots poses hazards to human health. Despite advancements in research, laboratory-based investigations into ENM transmission in soil might not comprehensively capture the intricacies of interactions that occur in natural environments. Additionally, in situ soil environments merit additional research in order better to comprehend the environmental fate and dangers of ENMs (Deng et al. 2014; Panda and Gangawane 2023).

Transfer of ENMs in Aquatic Environments

The movement of ENMs in aquatic environments is a multifaceted phenomenon that occurs via biological, vertical, and horizontal migrations. Particles originating from terrestrial and atmospheric sources accumulate in aquatic ecosystems, functioning as substantial repositories for ENMs through mechanisms such as surface discharge and deposition. Aquatic environments exhibit longer migration distances and greater bioavailability in comparison to terrestrial environments, thereby intensifying the environmental health hazards associated with ENMs. Nevertheless, comprehending the migration of ENMs in aquatic environments is a complex task, as a multitude of interconnected environmental factors impact it (Cross et al. 2015; Williams et al. 2019).

(a) *Vertical migration*: Vertical migration of ENMs occurs when they are transported in water columns in an upward or downward direction as a result of resuspension mechanisms propelled by hydrodynamics, bioturbation, or gravitational sedimentation. A multitude of environmental factors, such as organisms, natural organic matter (NOMs), minerals, pH, and ionic strength, intricately govern the dynamics of ENM migration. For example, shear forces in water can increase the toxicity of particular ENMs by promoting their resuspension. This results in an extended residence time for the ENMs. Minerals promote sedimentation, while NOMs facilitate ENM suspension; the interaction outcomes are dependent on the composition and properties of the minerals.

(b) *Horizontal migration*: The predominant mechanism by which long-distance ENMs are transported in natural water environments is horizontal migration. Despite this, horizontal migration research is still in its infancy in comparison to that of vertical migration. Examining this process presents difficulties due to its infrequent manifestation in microcosmic systems, where vertical migration is more prevalent.

(c) *Biological activities*: Vessel and horizontal ENM migration are both substantially impacted by biological processes. Co-migration between suspended ENMs and aquatic organisms can occur through mechanisms including adsorption-desorption or uptake-excretion. Additionally, biological macromolecules can be released by organisms, forming ecological coronas on ENM surfaces and influencing their migration indirectly. ENMs, despite being predominantly indigestible, amass in diverse biological organs throughout trophic levels, thereby inducing biomagnification effects in aquatic food chains.

It is essential to comprehend the transfer of ENMs within aquatic environments in order to analyze their environmental distribution and associated exposure risks comprehensively. Although numerous studies have examined the environmental factors that influence ENM migration, there is a scarcity of research that investigates the intricate interplay among multiple coexisting factors. Considering the established sampling and detection techniques for ENMs in aquatic environments, it is necessary to conduct additional research in natural water environments to determine how environmental factors mediate ENM transfer and subsequent changes in environmental hazards.

24.2.2.2 ENM Transformation in the Environment

The process of transforming ENMs entails modifications to their physical and chemical characteristics, as opposed to a simple transfer that primarily impacts the environmental bioavailability of ENMs. These transformations transpire via a multitude of mechanisms, encompassing both biological and physicochemical processes (e.g., degradation and sulfidation, dissolution, and photochemical reactions). These modifications have a substantial effect on the persistence and toxicity of ENMs in the environment (Kansara et al. 2022; Nowack et al. 2012).

Physicochemical Transformation
Environmental factors, including temperature, dissolved oxygen, magnetic interference, and light, all exert an influence on the physicochemical changes of ENMs.

- *Fragmentation*: When large plastic particulates are physically fragmented, plastic NMs and associated toxic additives are released, thereby increasing their toxicity.
- *Light exposure*: Light exposure can oxidize nanosilver (nAg^0) to more bioavailable Ag^+ ions under visible light, increasing their toxicity, while ultraviolet light can reduce Ag^+ back to less toxic Ag^0.
- *Dissolved oxygen*: Higher concentrations of dissolved oxygen facilitate the liberation of metal ions, which in turn influences the redox reactions of metal ENMs.
- *Coexisting ions*: Enzymes that coexist with phosphate and carbonate can impede the release of toxic ions.
- *Temperature*: Transformation processes are influenced by temperature variations, whereby elevated temperatures facilitate the liberation of metal ions from ENMs.

Biological Transformation
In many places, microbes break down carbon-based ENMs and help metal ENMs go through redox changes. This is called biotransformation.

- *Fungi*: Fullerol can be broken down by fungi like white-rot fungi, which use it in their lipid metabolism.
- *Bacteria*: mushrooms, such as white-rot mushrooms, can break down fullerol and use it in their fatty metabolism.

- *Animal and plant interaction*: Plants and animals are also capable of contributing to ENM biotransformation. For example, nAg0 can be ingested by giant fleas and oxidize to Ag+ in their intestines, thereby increasing its toxicity. ENMs are capable of undergoing oxidation or reduction reactions in the phyllosphere or rhizosphere of plants, which modify their toxicity.

Complexity and Challenges

The combined processes of biotransformation and physicochemical transformation of ENMs in natural settings create difficulties in evaluating their availability to organisms, toxicity, and impact on the environment.

- *Interactions*: The precise impact of these changes on total toxicity and their collaborative processes are yet unknown.
- *Research need*: To fully understand the relative roles of each transformation process and how they interact with each other, more study is needed. It is because the environmental transformation processes of ENMs are very complicated.

24.2.3 Environmental and Human Health Impacts of ENMs

In order to assess the environmental health risks that ENMs present, toxicological investigations furnish instrumental foundational data. Commencing in the 2000s, these investigations have unveiled a multitude of toxic impacts of distinct ENMs on a variety of organisms, which are frequently inconsistent. Notwithstanding a decade-long endeavor, the establishment of a robust and all-encompassing risk assessment model continues to be arduous on account of constraints in experimental design and the comparability of data.

Positive environmental impacts associated with ENMs, including those in nanofertilizers and agricultural technologies, have been observed to have emerged since 2016. This phenomenon emphasizes the potential advantages of nanotechnology in the business sector and inspires consumer trust. However, it also raises concerns regarding the associated dangers to human and environmental health and the escalating introduction of ENMs into the environment.

Although ENMs provide notable benefits such as widespread consumer acceptance and industrial applications, their environmental health implications transcend mere toxicity to affect individual organisms. These incorporate more extensive risks to both ecological systems and human welfare. In spite of this, the evaluation of the biological toxicity of ENMs at concentrations pertinent to the environment continues to be difficult, impeding the development of exhaustive risk assessments for emerging contaminants such as ENMs. This review provides a critical synthesis of almost 20 years of research pertaining to the environmental and human health impacts of ENMs. It identifies significant voids in knowledge that must be filled in order to formulate effective management strategies. However, it does recognize possible biases that may have influenced the review process, including the prevalence of in vitro experiments and the probability of publication bias in the available literature (Pietroiusti et al. 2018; Romero-Franco et al. 2017).

24.2.3.1 ENM Toxicity and Individual Impact

Toxicity issues arise in ENMs due to their nanoscale dimensions, increased bioavailability, stable biokinetics, and distinctive surface toxicological characteristics. According to research, ENMs cause genotoxicity, immunotoxicity, cytotoxicity, and neurotoxicity, which manifest in organisms as physical damage, inflammation, and metabolic disorders. The phytotoxicity of ENMs, which inhibited plant photosynthesis, was identified in 2011. Common toxicity mechanisms include oxidative stress, which is caused by the interaction of O_2 and ENMs that generate reactive oxygen species (ROS). Through the oxidation of lipids, proteins, and nucleic acids, ROS can cause cell injury. In addition, ENMs are capable of initiating apoptosis and programmed cell death by targeting the mitochondrial system. A reduction in cell proliferation and inhibition of nitrogen metabolism characterizes cytotoxicity. Enzymatic activities implicated in lipid and amino acid metabolism are disrupted at the individual level by ENM-induced metabolic disorders, which have detrimental effects on the growth of organisms such as freshwater fish. Proteins and fatty acids function as biomarkers to assess metabolic alterations that occur in response to stress. Intestinal tissues are especially susceptible to ENMs because ingestion is the principal route of exposure for organisms. The immunotoxic effects of this ingestion are demonstrated by the downregulation of functional genes involved in apoptosis and signaling. Environmental nanomaterials frequently induce liver infections and other forms of chronic toxicity in addition to immunotoxicity.

The exposure concentration of ENMs is intricately linked to their toxicity. According to meta-analyses, the toxic effects of ENMs are predominantly observed at high concentrations, while low concentrations may promote the growth of organisms. To illustrate, the induction of DNA damage in tobacco plants by nTiO2 was limited to concentrations exceeding 319 mg/L. In contrast, the development of algae and zooplankton was stimulated by 10 µg/L of nAg0, whereas phytoplankton photosynthesis was inhibited at concentrations exceeding 500 µg/L. Similarly, modest concentrations of carbon nanotubes, nMgO, and nFe2O3 promoted the growth of algae.

Notwithstanding the relatively low concentrations of ENMs in the environment, the majority of toxicology studies utilize elevated levels of exposure, which may lead to an overestimation of environmental hazards. ENMs interact with other pollutants upon introduction into the environment, resulting in mutual toxicity. ENMs are capable of altering the metabolic pathways of pollutants and increasing their bioaccessibility to organisms. Metabolic disorders and structural damage are among the synergistic deleterious effects between ENMs and pollutants, as demonstrated by research involving model organisms. In contrast, by aggregating and decreasing pollutant bioavailability or by enhancing organismal defense mechanisms, ENMs can mitigate pollutant toxicity.

As an illustration, it was discovered that nTiO2 inhibits ATP synthase expression and slows metabolism, thereby mitigating the neurotoxicity and genotoxicity caused by coexisting pollutants such as phoxim. In general, the interaction between ENMs and environmental pollutants can result in a variety of unanticipated impacts on ecosystem dynamics and organismal health.

ENMs are subject to complex environmental modifications that result in changes to their characteristics and subsequent toxicity. Significantly, the toxicity of ENMs differs between aquatic and soil environments, with soils exhibiting diminished toxicity as a result of interactions with soil particles and heterogeneous agglomeration with minerals. By means of steric hindrance, soil organic matter (OM) functions as an inherent barrier, reducing the bioavailability of ENM. On the contrary, NOMs can either alleviate or worsen the toxicity of ENMs in aquatic environments through the promotion of dispersion or the facilitation of toxic ion release.

Because exposure scenarios are simplified in laboratory studies, ENM toxicity in natural settings is frequently misrepresented. Moreover, the transmission of ENM between environmental media is made possible by interconnected air-soil-water ecosystems, which increases exposure pathways. An example of this is the probable mechanism by which plants become harmed by ENMs: concurrent exposure through roots (soil) and leaves/fruits (atmosphere). Biological elements, including moisture and aerosol particles, have an impact on the concentration and properties of ENMs, thereby influencing atmospheric toxic risks despite their brief residence periods. When atmospheric exposure pathways are disregarded in toxicology studies, the environmental dangers associated with ENMs are underestimated (Pietroiusti et al. 2018).

24.2.3.2 Ecotoxicity of ENMs

ENMs induce a range of reactions and exposure hazards in organisms with distinct characteristics. In order to evaluate the ecotoxicity of ENM, alterations in community diversity, species composition, biogeochemical processes, and energy transfer within an ecosystem are assessed. Despite the acknowledged significance of investigating the ecotoxicity of ENMs, extensive research on this subject did not commence until 2014, with a specific emphasis on the material and energy cycles that occur within ecosystems.

Sensitive microorganisms exposed to ENMs exhibit diminished biomass and diversity in soil when exposed to concentrations of nAg^0 and nCuO that are environmentally pertinent. The impairment of biogeochemical processes is demonstrated by the reduction in the abundance of functional genes that are implicated in ammunition transformation. Primary producers, including marine microalgae, are also affected by ENMs, which impede photosynthesis and diminish the oxygen supply capacity in the oceans. Additionally, plastic nanoparticles reduce ocean productivity by interfering with carbon sequestration and biodiversity (Kumar and Gangawane 2022a, b).

Alterations in primary producers and decomposers in terrestrial ecosystems are the primary focus of research, while the effects on consumer biodiversity are primarily disregarded. Furthermore, there are substantial concerns regarding the transfer, bioaccumulation, and biomagnification of ENMs throughout food chains. For instance, the accumulation of ENMs in organisms such as D. magna leads to oxidative stress and genotoxicity. Mesocosmic experiments and advanced ecological technologies, including environmental DNA metabarcoding and multi-omics, are critical for comprehending ecosystem alterations caused by ENMs. Real-world ecosystem changes in biodiversity and functional diversity can be elucidated by means of these instruments (Kumar et al. 2023b).

Moreover, biological evolution must be taken into account, given that ENMs have the potential to induce genotoxicity and aid in the adaptation of organisms. However, from the perspective of global ecosystems, the effects of ENMs on ecological function and evolution remain poorly understood, necessitating additional research and the disclosure of genetic information (Handy et al. 2012; Jahan et al. 2017).

24.2.3.3 Human Health Impacts of ENMs

The environmental consequences of ENMs inevitably have implications for human health. Notwithstanding the notable advancements in nanotechnology that have enabled substantial progress in domains such as apparel, cosmetics, and medical treatment, direct exposure to ENMs presents hazards to human health. Therefore, it is imperative to comprehend the impacts of ENMs on human health in order to advance nanotechnology in a manner that inspires public confidence (Kumar et al. 2022a; Pietroiusti et al. 2018; Romero-Franco et al. 2017).

- *Inhalation exposure*: The principal mode of ENM exposure is through inhalation; on an annual basis, humans inhale around 55,000 particles of nTiO2. As a result, conditions such as chronic bronchitis, asthma, and pulmonary fibrosis manifest, rendering the respiratory system more vulnerable! Notably, the presence of numerous ENMs in human lungs has sparked concerns regarding the potential health hazards they may pose.
- *Skin absorption*: Skin absorption of ENMs is another route by which they can penetrate the body; indoor concentrations can reach 59.4 grains/m^3, while outdoor concentrations range from 0.3 to 1.5 grains/m^3. The utilization of functional apparel and personal care products additionally contributes to skin exposure. An example of this is the association between exposure to nCuO from paint and coating and atypical development in children. Plastic NMs are capable of penetrating cells, causing cellular toxicity and impairing immune processes, according to in vitro research.
- *Ingestion*: ENMs are predominantly ingested by humans via water and a variety of dietary sources. Consequently, the gastrointestinal tract becomes an ENM target. It is crucial to comprehend the bioavailability and transformation mechanisms

of ENMs within the digestive tract in order to evaluate the potential health hazards they pose. In gastric fluid, nAg0 particles, for instance, can transform into more bioavailable and toxic forms. In addition, metal ions released from metal oxide ENMs are capable of causing adverse effects on the digestive tract.

- *Organ toxicity*: It is the interaction between epithelial cells and ENMs that dictates their toxicity and transport efficacy to organs, including the liver, kidney, heart, and nervous system. ENMs are capable of interfering with intestinal processes, modifying the structures of microbial communities, and causing harm to reproductive organs. Moreover, they have the potential to cause neurological disorders, oxidative DNA damage, and even affect the health of future generations.
- *Research challenges*: The majority of research on the health effects of ENMs is derived from laboratory experiments or predictions made by machine learning. These studies, nevertheless, frequently disregard environmental factors that affect the properties and toxicity of ENMs. Furthermore, the presence of diverse cell types, culture systems, and exposure pathways further complicates the attempt to establish a unified understanding of ENM toxicity. In order to fill these voids in knowledge, dynamic human environmental exposome technology must be utilized to monitor biological responses to ENMs systematically.

24.3 Advances and Challenges in ENM Environmental Health Research

In the last 20 years, there has been substantial progress in researching the environmental health effects of ENMs. Advanced pretreatment methods and detection methodologies have been developed, specifically for inorganic ENMs, enabling the qualitative and quantitative study of ENM presence in natural environments. Laboratory studies have systematically examined the processes that control the transformation and movement of ENMs across various environmental mediums. These studies provide vital insights into how ENMs behave and what happens to them outside of controlled settings (Kumar and Gangawane 2023a, b; Zickgraf et al. 2023).

Advancements have been made in comprehending the processes of ENM-induced toxicity in individual species and their transmission via essential food chains, which is crucial for a thorough evaluation of their environmental health effects. Several scientific challenges are yet unanswered. Accurately assessing the amounts of exposure to environmentally non-friendly materials ENM and their presence in natural settings is a crucial obstacle in detecting the health concerns connected with them. Secondly, the process of incorporating extensive data from laboratory experiments into applicable risk evaluations for natural settings is a significant difficulty. Further research is needed to investigate the relationships between environmental elements (such as oxygen, light, temperature) and genetic factors (including organismal genetics and evolution) in determining the environmental health consequences of ENM (Kumar et al. 2023a; Lewis et al. 2019; Maharramov et al. 2019).

Future studies should prioritize improving our understanding of ENM transmission and accumulation across food chains and generations. It is crucial to clarify how ENMs function as transporters for simultaneous contaminants or diseases, resulting in combined toxicity. It is crucial to set hazard levels for ENMs in natural settings to provide guidelines for assessing and managing their environmental health risks.

Other developing pollutants provide similar scientific hurdles in environmental health studies. Contaminants such as ENMs often exist at low concentrations in the environment and go through intricate transfer and transformation processes in natural environments. Furthermore, their cumulative and environmentally lasting characteristics need in-depth research on how they are passed down food chains and their overall effects on environmental health and biological evolution. Applying knowledge gained from research on environmentally novel materials may help direct the investigation of other newly identified pollutants, aiding in the creation of a thorough plan for environmental health research in this field (Shetty et al. 2023; Wang and Liu 2022).

24.3.1 Emerging Contaminants: Detection Challenges and Research Strategies

Emerging toxins are characterized by their low ambient concentrations and unclear presence in the environment. It is crucial to accurately identify the origins and presence of pollutants in various environmental contexts to assess their environmental health concerns and plan research efforts effectively. The toxicity of emerging contaminants, such as ENMs, depends on their structural characteristics and exposure levels, which are affected by complex environmental conditions (Geissen et al. 2015).

Detecting organic emerging pollutants involves identifying their chemical composition by analyzing specific structural characteristics, often done using high-resolution mass spectrometry (MS) or nuclear magnetic resonance (NMR) technology. Detecting these contaminants is difficult since they are present in very small amounts and may be mixed with other pollutants or environmental chemicals such as natural organic matter (NOMs). Therefore, it is essential to create effective pretreatment techniques to remove and separate new pollutants from complex environmental samples and concentrate them to detectable concentrations (Kumar et al. 2022b).

An effective research method includes combining several pretreatment methods, such as structural modification, selective extraction, and functional membrane filtration, customized to the specific pollutants and environmental conditions. Improving the precision of instrument detection is a crucial research direction for identifying new pollutants. Moreover, the lack of established criteria for these pollutants poses a difficulty. Utilizing mass spectral fingerprinting, database matching, and deep machine learning may help identify the structures in such circumstances. When direct concentration determination is complex, using effect-directed screening models combined with environmental economics may help identify emerging toxins in natural habitats. This comprehensive plan includes technology improvements and novel methodologies to identify emerging pollutants and solve their numerous difficulties correctly (Vu et al. 2023).

24.3.2 Advancing Environmental Health Assessment of Emerging Contaminants

Contemporary studies on emerging toxins sometimes include high-dose exposure despite their usual low ambient concentrations. Laboratory studies are valuable for replicating environmental circumstances but are limited by controlled settings, restricted response space, short exposure periods, and simplified biological systems. Moreover, the enduring tendency to accumulate several new pollutants provides continuous ecological and health hazards, making it challenging to estimate toxicity accurately in the face of genetic changes and environmental development.

To fill these information gaps, using mesocosmic model ecosystems is suggested to understand the environmental impacts of new pollutants in natural environments. Mesocosmic ecosystems combine a variety of natural environmental elements and living organisms, including typical biological organisms, across long periods. It allows for unbiased assessments of large-scale biogeochemical processes, biological reactions, and ecological impacts of new pollutants, overcoming the constraints of conventional lab tests and avoiding problems linked to on-site field studies.

Creating a microcosmic ecosystem involves addressing issues, including maintaining ecological stability, setting up an independent food web, and closely monitoring all input and output variables. It involves integrating biological aspects, such as interactions among various trophic levels and environmental characteristics, including dissolved oxygen, temperature, and pH. Thoroughly monitoring many factors, such as environmental pollutant behaviors, organism activities, and ecosystem growth, allows for a detailed evaluation of mass movement and ecological concerns related to new contaminants (Murnyak et al. 2011).

24.3.3 Enhancing Environmental Risk Assessment of Emerging Contaminants

Research on evaluating the environmental dangers of ENMs has been conducted for more than 10 years. However, a complete theoretical and methodological framework for this evaluation is still lacking. This gap underscores the need for more research on control mechanisms for ENMs.

Systematic evaluation and management of environmental health risks for emerging contaminants like ENMs encounter obstacles owing to limited data on their origins, emissions, occurrence, biological processes, health impacts, and danger thresholds. Discrepancies in exposure doses, durations, and experimental variables across various research lead to inconsistencies and hinders the comparability of findings. A comprehensive technique, known as the "whole-chain research methodology," is suggested to tackle these difficulties.

This method includes collecting data on the origin, path, environmental levels, and release rate of new pollutants via analyzing environmental samples and using models. The data guide the development of laboratory research on the environmental health impacts and biogeochemical processes of new pollutants. Continuous mesocosm research tracks evolving environmental conditions, aiding in the advancement of analytical tools and comprehension of diverse health impacts in various natural settings. Health effects and mechanism studies provide toxicity thresholds that are used to evaluate the danger posed by new pollutants.

Consistent environmental health studies on new pollutants enable data to be gathered in a large-scale system for analysis, leading to further research and the development of worldwide scientific management and control plans. An integrated management strategy is recommended to mitigate risks associated with pollutants throughout their life cycle, from raw materials to disposal (Lui et al. 2021).

Interdisciplinary cooperation among academics from many fields, such as environmental science, materials science, geochemistry, biology, and medical science, is crucial due to the intricate nature of environmental health impacts related to new pollutants. Extensive efforts are necessary to enhance scientific knowledge of biogeochemical processes and health hazards, guaranteeing environmental sustainability. Developing research paradigms and benchmarking methodologies for newly identified pollutants may provide scientific backing for risk mitigation and control initiatives (Hong et al. 2018).

24.4 Conclusion

It is crucial to comprehend the environmental health effects of new pollutants, including ENMs, in order to protect human and ecological health. Despite progress in understanding the origins, behaviors, and toxicity mechanisms of ENMs, issues remain in accurately assessing and managing risks because of the complexities of environmental dynamics and transformation processes. Continuous research is essential for creating standardized frameworks, enhancing detection methods, and increasing our knowledge of the environmental destiny of ENMs. Utilizing knowledge from ENM research may help develop solutions for environmental health issues caused by new contaminants, promoting sustainable environmental management and worldwide health protection.

Assessing the environmental and human health implications of ENMs is crucial, yet there are still significant obstacles to overcome. Various toxicological studies have shown different impacts on organisms, but creating a complete risk assessment model is difficult due to limitations in experiments and challenges in comparing data. ENMs provide advantages like nanofertilizers in agriculture, yet there are ongoing worries about their possible hazards to human and environmental health. The concerns go beyond the toxicity of individual organisms to include larger ecological systems and human well-being. Challenges in evaluating ENM toxicity at levels found in the environment hinder thorough risk evaluations. It is essential to fill these knowledge gaps to develop successful management methods. However, biases such as the predominance of in vitro experiments and possible publishing biases might impact study results. Future research should focus on investigating the toxicity mechanisms of ENMs, their interactions with other contaminants, and their environmental destiny in various ecosystems. Research is crucial for creating risk assessment frameworks and implementing effective management methods to safeguard environmental and human health from emerging pollutants such as ENMs.

References

Arman NZ, Salmiati S, Aris A, Salim MR, Nazifa TH, Muhamad MS, Marpongahtun M (2021) A review on emerging pollutants in the water environment: existences, health effects and treatment processes. *Water* 13(22):3258. https://doi.org/10.3390/w13223258

Ashesh A, Singh S, Linthoingambi Devi N, Chandra Yadav I (2022) Organochlorine pesticides in multi-environmental matrices of India: a comprehensive review on characteristics, occurrence, and analytical methods. Microchem J 177:107306. https://doi.org/10.1016/j.microc.2022.107306

Bose-O'Reilly S, Bernaudat L, Siebert U, Roider G, Nowak D, Drasch G (2017) Signs and symptoms of mercury-exposed gold miners. Int J Occup Med Environ Health 30(2):249–269. https://doi.org/10.13075/ijomeh.1896.00715

Cross RK, Tyler C, Galloway TS (2015) Transformations that affect fate, form and bioavailability of inorganic nanoparticles in aquatic sediments. Environ Chem 12(6):627. https://doi.org/10.1071/EN14273

Deng Y, White JC, Xing B (2014) Interactions between engineered nanomaterials and agricultural crops: implications for food safety. J Zheijang Univ Sci A 15(8):552–572. https://doi.org/10.1631/jzus.A1400165

Gardea-Torresdey JL, Rico CM, White JC (2014) Trophic transfer, transformation, and impact of engineered nanomaterials in terrestrial environments. Environ Sci Technol 48(5):2526–2540. https://doi.org/10.1021/es4050665

Geissen V, Mol H, Klumpp E, Umlauf G, Nadal M, van der Ploeg M, van de Zee SEATM, Ritsema CJ (2015) Emerging pollutants in the environment: a challenge for water resource management. Int Soil Water Conserv Res 3(1):57–65. https://doi.org/10.1016/j.iswcr.2015.03.002

Genchi G, Lauria G, Catalano A, Carocci A, Sinicropi MS (2022) Arsenic: a review on a great health issue worldwide. Appl Sci 12(12):6184. https://doi.org/10.3390/app12126184

Gottschalk F, Kost E, Nowack B (2013) Engineered nanomaterials in water and soils: a risk quantification based on probabilistic exposure and effect modeling. Environ Toxicol Chem 32(6):1278–1287. https://doi.org/10.1002/etc.2177

Guo W, Pan B, Sakkiah S, Yavas G, Ge W, Zou W, Tong W, Hong H (2019) Persistent organic pollutants in food: contamination sources, health effects and detection methods. Int J Environ Res Public Health 16(22):4361. https://doi.org/10.3390/ijerph16224361

Handy RD, Cornelis G, Fernandes T, Tsyusko O, Decho A, Sabo-Attwood T, Metcalfe C, Steevens JA, Klaine SJ, Koelmans AA, Horne N (2012) Ecotoxicity test methods for engineered nanomaterials: practical experiences and recommendations from the bench. Environ Toxicol Chem 31(1):15–31. https://doi.org/10.1002/etc.706

He X, Deng H, Hwang H (2019) The current application of nanotechnology in food and agriculture. J Food Drug Anal 27(1):1–21. https://doi.org/10.1016/j.jfda.2018.12.002

Hong L, Ye ZS, Ling R (2018) Environmental risk assessment of emerging contaminants using degradation data. J Agric Biol Environ Stat 23(3):390–409. https://doi.org/10.1007/s13253-018-0326-9

Jahan S, Yusoff IB, Alias YB, Bakar AFBA (2017) Reviews of the toxicity behavior of five potential engineered nanomaterials (ENMs) into the aquatic ecosystem. Toxicol Rep 4:211–220. https://doi.org/10.1016/j.toxrep.2017.04.001

John A, Küpper M, Manders-Groot A, Debray B, Lacome J-M, Kuhlbusch T (2017) Emissions and possible environmental implication of engineered nanomaterials (ENMs) in the atmosphere. *Atmosphere* 8(5):84. https://doi.org/10.3390/atmos8050084

Joo SH, Zhao D (2017) Environmental dynamics of metal oxide nanoparticles in heterogeneous systems: a review. J Hazard Mater 322:29–47. https://doi.org/10.1016/j.jhazmat.2016.02.068

Kansara K, Bolan S, Radhakrishnan D, Palanisami T, Al-Muhtaseb AH, Bolan N, Vinu A, Kumar A, Karakoti A (2022) A critical review on the role of abiotic factors on the transformation, environmental identity and toxicity of engineered nanomaterials in aquatic environment. Environ Pollut 296:118726. https://doi.org/10.1016/j.envpol.2021.118726

Khan S, Naushad M, Govarthanan M, Iqbal J, Alfadul SM (2022) Emerging contaminants of high concern for the environment: current trends and future research. Environ Res 207:112609. https://doi.org/10.1016/j.envres.2021.112609

Kumar A, Gangawane KM (2022a) Effect of precipitating agents on the magnetic and structural properties of the synthesized ferrimagnetic nanoparticles by co-precipitation method. Powder Technol 401:117298. https://doi.org/10.1016/j.powtec.2022.117298

Kumar A, Gangawane KM (2022b) Synthesis and effect on the surface morphology & magnetic properties of ferrimagnetic nanoparticles by different wet chemical synthesis methods. Powder Technol 410:117867. https://doi.org/10.1016/j.powtec.2022.117867

Kumar A, Gangawane KM (2023a) Advanced computational approaches for water treatment. In: Advanced computational approaches for water treatment. CRC Press, pp 1–12. https://doi.org/10.1201/9781003325147-1

Kumar A, Gangawane KM (2023b) Molecular dynamics for the membrane process. In: Advanced computational approaches for water treatment. CRC Press, pp 13–38. https://doi.org/10.1201/9781003325147-2

Kumar A, Gangawane KM, Ramanjaneyulu B (2022a) Ferrofluids for waste-water treatment. In: International conference on chemical, bio and environmental engineering. Springer, Cham, pp 723–744. https://link.springer.com/chapter/10.1007/978-3-030-96554-9_48

Kumar R, Vuppaladadiyam AK, Antunes E, Whelan A, Fearon R, Sheehan M, Reeves L (2022b) Emerging contaminants in biosolids: presence, fate and analytical techniques. Emerg Contam 8:162–194. https://doi.org/10.1016/j.emcon.2022.03.004

Kumar A, Jena HM, Gangawane KM (2023a) Effect of ferrimagnetic nanoparticles as an oxygen absorber on the shelf life of tomatoes. In: 9th international food convention (IFCoN)

Kumar A, Jena HM, Gangawane KM (2023b) Variations in magnetic behavior of ferrimagnetic nanoparticles with temperature. In: 75th CHEMCON 2023

Lewis RW, Bertsch PM, McNear DH (2019) Nanotoxicity of engineered nanomaterials (ENMs) to environmentally relevant beneficial soil bacteria – a critical review. Nanotoxicology 13(3):392–428. https://doi.org/10.1080/17435390.2018.1530391

Louis GB, Damstra T, Díaz-Barriga F, Faustman E, Hass U, Kavlock R, Kimmel C, Kimmel G, Krishnan K, Luderer U, Sheldon L (2006) Principles for evaluating health risks in children associated with exposure to chemicals. Environ Health Crit 237:1–327

Lowry GV, Hotze EM, Bernhardt ES, Dionysiou DD, Pedersen JA, Wiesner MR, Xing B (2010) Environmental occurrences, behavior, fate, and ecological effects of nanomaterials: an introduction to the special series. J Environ Qual 39(6):1867–1874. https://doi.org/10.2134/jeq2010.0297

Lui LM, Majumder ELW, Smith HJ, Carlson HK, von Netzer F, Fields MW, Stahl DA, Zhou J, Hazen TC, Baliga NS, Adams PD, Arkin AP (2021) Mechanism across scales: a holistic modeling framework integrating laboratory and field studies for microbial ecology. Front Microbiol 12:642422. https://doi.org/10.3389/fmicb.2021.642422

Ma B, Stirling E, Liu Y, Zhao K, Zhou J, Singh BK, Tang C, Dahlgren RA, Xu J (2021) Soil biogeochemical cycle couplings inferred from a function-taxon network. Research 2021:7102769. https://doi.org/10.34133/2021/7102769

Maharramov AM, Hasanova UA, Suleymanova IA, Osmanova GE, Hajiyeva NE (2019) The engineered nanoparticles in food chain: potential toxicity and effects. SN Appl Sci 1(11):1–25. https://doi.org/10.1007/s42452-019-1412-5

Minz P, Mishra S (2023) Jackfruit peel derived ZnCl2-impregnated activated carbon: optimization, characterization, and application in dye removal. Biomass Convers Biorefinery:0123456789. https://doi.org/10.1007/s13399-023-04592-0

Mohanty B, Das A (2023) Heavy metals in agricultural cultivated products irrigated with wastewater in India: a review. AQUA Water Infrastruct Ecosyst Soc 72(6):851–867. https://doi.org/10.2166/aqua.2023.122

Mudumba T, Stimpson B, Jingo S, Montgomery RA (2023) The implications of global oil exploration for the conservation of terrestrial wildlife. Environ Chall 11:100710. https://doi.org/10.1016/j.envc.2023.100710

Murnyak G, Vandenberg J, Yaroschak PJ, Williams L, Prabhakaran K, Hinz J (2011) Emerging contaminants: presentations at the 2009 toxicology and risk assessment conference. Toxicol Appl Pharmacol 254(2):167–169. https://doi.org/10.1016/j.taap.2010.10.021

Nowack B, Ranville JF, Diamond S, Gallego-Urrea JA, Metcalfe C, Rose J, Horne N, Koelmans AA, Klaine SJ (2012) Potential scenarios for nanomaterial release and subsequent alteration in the environment. Environ Toxicol Chem 31(1):50–59. https://doi.org/10.1002/etc.726

Pal D, Sen S (2023a) Emerging petroleum pollutants and their adverse effects on the environment. In: Behera ID, Das AP (eds) Impact of petroleum waste on environmental pollution and its sustainable management through circular economy. Environmental science and engineering. Springer Nature, Cham, pp 103–137. https://doi.org/10.1007/978-3-031-48220-5_5

Pal D, Sen S (2023b) In-depth coverage of petroleum waste sources, characteristics, environmental impact, and sustainable remediation process. In: Behera ID, Das AP (eds) Impact of petroleum waste on environmental pollution and its sustainable management through circular economy. Springer Nature, Cham, pp 1–38. https://doi.org/10.1007/978-3-031-48220-5_1

Pal D, Sen S (2024) Optimal synthesis of dolochar derived faujasite zeolite X for highly effective Cd(II) removal. Environ Res 240:117494. https://doi.org/10.1016/j.envres.2023.117494

Panda D, Gangawane KM (2023) Development of superhydrophobic hybrid silica-cellulose aerogel as promising thermal insulation and sound absorption. Int J Mater Res 114(4–5):405–417. https://doi.org/10.1515/ijmr-2022-0066

Panda D, Sharma S, Gangawane KM (2023) Synthesis of novel biocatalyst from organic waste protonated with acid treatment for hydrogen production. Int J Hydrog Energy 48(63):24242–24254. https://doi.org/10.1016/j.ijhydene.2023.03.117

Parfomak PW, Folger P (2010) Carbon dioxide (CO2) pipelines for carbon sequestration: emerging policy issues. Carbon Capture and Greenhouse Gases, pp 95–114

Pašukoniene V, Mlynska A, Steponkiene S, Poderys V, Matulionyte M, Karabanovas V, Statkute U, Purviniene R, Kraśko JA, Jagminas A, Kurtinaitiene M, Strioga M, Rotomskis R (2014) Accumulation and biological effects of cobalt ferrite nanoparticles in human pancreatic and ovarian cancer cells. Medicina (Lithuania) 50(4):237–244. https://doi.org/10.1016/j.medici.2014.09.009

Pietroiusti A, Stockmann-Juvala H, Lucaroni F, Savolainen K (2018) Nanomaterial exposure, toxicity, and impact on human health. WIREs Nanomed Nanobiotechnol 10(5):e1513. https://doi.org/10.1002/wnan.1513

Prüss-Ustün A, Wolf J, Corvalán C, Neville T, Bos R, Neira M (2017) Diseases due to uanhealthy environments: an updated estimate of the global burden of disease attributable to environmental determinants of health. J Public Health 39(3):464–475. https://doi.org/10.1093/pubmed/fdw085

Rani P, Dhok A (2023) Effects of pollution on pregnancy and infants. Cureus 15(1):1–7. https://doi.org/10.7759/cureus.33906

Rivero M, Marín-Barba M, Gutiérrez L, Lozano-Velasco E, Wheeler GN, Sánchez-Marcos J, Muñoz-Bonilla A, Morris CJ, Ruiz A (2019) Toxicity and biodegradation of zinc ferrite nanoparticles in Xenopus laevis. J Nanopart Res 21(8):1–30. https://doi.org/10.1007/s11051-019-4631-1

Romero-Franco M, Godwin HA, Bilal M, Cohen Y (2017) Needs and challenges for assessing the environmental impacts of engineered nanomaterials (ENMs). Beilstein J Nanotechnol 8(1):989–1014. https://doi.org/10.3762/bjnano.8.101

Sahoo A, Mishra S (2023) Synthesis of microporous-activated carbon via direct H3PO4 and KOH activation of Pterospermum acerifolium fruits: process optimization and characterization. Biomass Convers Biorefinery:0123456789. https://doi.org/10.1007/s13399-023-05072-1

Santhosh R, Sarkar P (2022) Jackfruit seed starch/tamarind kernel xyloglucan/zinc oxide nanoparticles-based composite films: preparation, characterization, and application on tomato (Solanum lycopersicum) fruits. Food Hydrocoll 133:107917. https://doi.org/10.1016/j.foodhyd.2022.107917

Santhosh R, Babu DM, Thakur R, Nath D, Hoque M, Gaikwad KK, Ahmed J, Sarkar P (2024) Effect of atmospheric cold plasma treatment on structural, thermal, and mechanical properties of pea protein isolate edible films. Sustain Chem Pharm 37:101398. https://doi.org/10.1016/j.scp.2023.101398

Shetty SS, Deepthi D, Harshitha S, Sonkusare S, Naik PB, Kumari NS, Madhyastha H (2023) Environmental pollutants and their effects on human health. Heliyon 9(9):e19496. https://doi.org/10.1016/j.heliyon.2023.e19496

United Nations Environment Programme (2009) Developing integrated solid waste management plan-training manual, vol 4. United Nations Environment Programme (UNEP)

Vu KA, Nguyen TA, Nguyen TP (2023) Advances in pretreatment methods for free nucleic acid removal in wastewater samples: enhancing accuracy in pathogenic detection and future directions. Appl Microbiol 4(1):1–15. https://doi.org/10.3390/applmicrobiol4010001

Wang T, Liu W (2022) Emerging investigator series: metal nanoparticles in freshwater: transformation, bioavailability and effects on invertebrates. Environ Sci Nano 9(7):2237–2263. https://doi.org/10.1039/D2EN00052K

WHO (1993) A report of a World Health Organization consultation on health and environment in preparation for 2nd European conference on environment and health. World Health Organization, Sofia

Williams RJ, Harrison S, Keller V, Kuenen J, Lofts S, Praetorius A, Svendsen C, Vermeulen LC, van Wijnen J (2019) Models for assessing engineered nanomaterial fate and behaviour in the aquatic environment. Curr Opin Environ Sustain 36:105–115. https://doi.org/10.1016/j.cosust.2018.11.002

Yu S, Liu J, Yin Y, Shen M (2018) Interactions between engineered nanoparticles and dissolved organic matter: a review on mechanisms and environmental effects. J Environ Sci 63:198–217. https://doi.org/10.1016/j.jes.2017.06.021

Zeitz P, Orr MF, Kaye WE (2002) Public health consequences to mercury spills: hazardous substances emergency events surveillance system, 1993-1998. Environ Health Perspect 110(2):129–132. https://doi.org/10.1289/ehp.02110129

Zickgraf FM, Murali A, Landsiedel R (2023) Engineered nanomaterials and the microbiome: implications for human health. Curr Opin Toxicol 35:100429. https://doi.org/10.1016/j.cotox.2023.100429

Chapter 25
Ecotoxicological Footprint of Agricultural Nanoparticles: Balancing Productivity with Environmental Safety

Saswati Bhattacharya and Anirban Kundu

Abstract Given the rapid population growth and alarming shift in climatic patterns, there is an increasing demand for food alongside a diminishing per capita land availability. Stochastic fluctuations in the environment have rendered plants vulnerable to both abiotic and biotic stressors, ultimately leading to a decline in productivity. To address crop losses resulting from abiotic stresses or pathogen/pest infestations, various strategies have been implemented. Among these, the application of nanoparticles (NPs) has gained widespread popularity in contemporary agricultural practices. Besides, ongoing research has also highlighted the potential harm that NPs may pose to nontarget organisms and their associated environmental threats. Upon release into the environment, NPs experience a complex interplay involving transformation, aggregation, and subsequent dis-aggregation, which determines their ecotoxicological potential. Leaching of NPs infiltrates the soil, compromising the integrity of the surroundings, polluting adjacent aquatic bodies, and thereby adversely affecting both terrestrial and aquatic ecosystems. These NPs undergo biomagnification within the food chain, consequently posing health threats to primary, secondary, and tertiary consumers. The metal and metal-oxide NPs have also been documented for their ability to alter the natural growth and proliferation of rhizospheric microorganisms. This chapter, therefore, provides a comprehensive overview of the various NPs employed within the realm of agriculture, along with an exploration of their ecotoxicological threats and possible regulatory approaches.

Keywords Nanotechnology · Agriculture · Environmental risk · Ecotoxicology · Pollution · Regulatory strategies

---

S. Bhattacharya
Department of Botany, Dr. A.P.J. Abdul Kalam Government College, Kolkata, West Bengal, India

A. Kundu (✉)
Plant Genomics and Bioinformatics Laboratory, P.G. Department of Botany, Ramakrishna Mission Vivekananda Centenary College (Autonomous), Kolkata, West Bengal, India

S. Dey, S. Bhattacharya (eds.), *Biotechnological Interventions in the Removal of Emerging Pollutants*, Interdisciplinary Biotechnological Advances,
https://doi.org/10.1007/978-981-97-9922-0_25

25.1 Introduction

In agriculture, nanotechnology involves the application of nanoscale materials and/or technologies to address the various challenges to improve farming practices (Bhattacharya et al. 2023). As the demand for food escalates, contemporary agricultural practices necessitate advanced technologies like the application of nanomaterials (NMs)/nanoparticles (NPs) to boost crop yields while working within the confines of existing resources. In the past, farmers striving for agricultural excellence employed various techniques aimed at improving crop yields. These methods included application of high-quality fertilizers, experimenting with genetic modifications, and implementing breeding programs to develop varieties with greater yield, stress tolerance, and resistance to pests and pathogens (Laskar et al. 2023). While chemical fertilizers can initially improve crop productivity, their long-term application results in irreversible depletion of soil fertility (Sheteiwy et al. 2021). The use of NPs is therefore regarded as an effective strategy impacting sustainable agriculture and precision farming to enhance crop yields, while minimizing the application of fertilizers, pesticides, and herbicides (Bhattacharya et al. 2023). In this context, NPs have demonstrated remarkable superiority for their greater surface area, enhanced reactivity, diverse functionalities, customizable nature, improved transport properties, exceptional stability, and high adsorption capacity compared to their bulk or dissolved equivalents (Ansari et al. 2020; Bhattacharya and Saha 2023). Therefore, nanotechnology unfurls a new avenue that holds the potential to promote sustainable growth in agriculture while reducing the negative impacts on crop yields.

The rapid advancements in the field of nanotechnology have led to various adverse effects on the ecosystem due to the introduction of NPs/NMs in the food chain. The increasing manufacture of synthetic NPs and their widespread application in agricultural settings are anticipated to increase public's exposure to these particles in the coming years (Xiong et al. 2017). Given that, NPs/NMs can quickly spread to water, air, and soil ecosystems, their toxic and harmful effects can impact different facets of our environment (Kwak and An 2015). Bioaccumulation and biomagnification of these leached NPs pose threats to all the stakeholders of the environment. Therefore, a legitimate concern persists regarding their potential release into the environment. This chapter presents an organized overview, exploring the intricacies of NPs/NMs in the realm of agriculture, considering their origins, fate, and impacts, and shedding light on their interactions with organisms within the multifaceted aquatic and terrestrial ecosystems. This review also aims to provide a better understanding of the interplay between NPs, agriculture and environment, and propose future research directions in nanoecotoxicology to formulate advance recommendations in this field.

25.2 Application of Nanoparticles in Agriculture

25.2.1 Nanofertilizers

The use of nanofertilizers has been instrumental in delivering essential nutrients to crop plants, thereby enhancing growth and increasing productivity (Liu and Lal 2015). A variety of NP formulations containing carbon nanotubes (CNT), Si, Zn, Fe, Ag, and Mo have been applied in diverse agricultural settings, impacting plant growth by stimulating shoot and root elongation, augmenting biomass, chlorophyll content, and accelerating seed germination at specific concentrations (Chhipa 2017). Nanofertilizers also offer promising solution to issues such as eutrophication while improving nutrient utilization efficiency (Shukla et al. 2019). They can be categorized as nanocomposite fertilizers, controlled-release fertilizers, or controlled-loss fertilizers, utilizing integrated nanodevices to deliver various macro- and micronutrients within optimal parameters (Lateef et al. 2016). Engineered NPs encapsulate or carry essential nutrients like nitrogen, phosphorus, and potassium, protecting them from leaching or volatilization and providing controlled release over time. This targeted delivery tailored for controlled release to plant roots promotes efficient absorption and utilization, reduces nutrient loss, regulates rhizospheric microflora, and enhances crop productivity and growth (Wang et al. 2020). Chhipa (2017) classified nanofertilizers into three groups: derived from macronutrients, derived from micronutrients, and those derived from fertilizers. In the first two categories, NPs serve as nutrients, while in the third category, they act as supplements. Compared to traditional phosphorus fertilizers, hydroxyapatite NPs containing calcium and phosphorus significantly increased seed yield in *Glycine max* (Xiu-mei et al. 2005). Similarly, ZnO-NPs have been reported to enhance mineral absorption in cucumber seedlings (Zhao et al. 2014). Soybean exposed to CeO_2 NPs at 1 g/kg caused an accumulation of Cu and P and a decline in Ca content in the pods (Peralta-Videa et al. 2014). Furthermore, ZnO-NPs have significantly increased biomass production, shoot elongation, root development, chlorophyll content, protein content, and phosphatase enzyme activity across various plant species, including *Vigna radiata*, *Brassica napus*, *Cucumis sativus*, *Cicer arietinum*, cluster bean, and *Raphanus sativus* (Lin and Xing 2007; Mahajan et al. 2011; Zhao et al. 2014; Raliya and Tarafdar 2013).

25.2.2 Nanofungicides

NPs are increasingly being explored as potential alternatives to traditional fungicides due to their unique properties and potential benefits. Numerous metal NPs and nanoformulations have been proposed to combat fungal pathogen infections owing to their enhanced reactivity and targeted delivery. Ag-NPs, for instance, have shown broad-spectrum activity against various fungal pathogens such as *Pythium*, *Botrytis*,

Trichoderma, *Sphaerotheca*, *Phoma*, *Rhizoctonia*, and *Fusarium* among others (Park et al. 2006). Similarly, Cu-NPs have also demonstrated efficacy against a wide range of fungal pathogens (Hazarika et al. 2022). In a comparative study between commercial fungicides and Cu-NPs against *Phytophthora infestans*-infected tomatoes, Cu-NPs exhibited superior efficacy, particularly at lower concentrations Furthermore, Cu(II) complexed with chitosan nanogels has shown inhibitory effects against *Fusarium graminearum* (Brunel et al. 2013). Additionally, ZnO-NPs and MgO-NPs have been demonstrated to regulate the growth and proliferation of fungi such as *Fusarium oxysporum*, *Rhizopus stolonifer*, and *Mucor plumbeus*, by invoking systemic defense mechanisms (Shukla et al. 2024). Bioengineered Ag-NPs have exhibited antifungal properties against *Bipolaris sorokiniana*, the causal pathogen of blotch disease in wheat (Mishra et al. 2014). According to Singh et al. (2021), Ag-NPs hold promise as substitutes for conventional chemical fungicides in potato cultivation against *Phytophthora infestans*. Furthermore, amalgamated Si-Ag-NPs have shown formidable antifungal attributes against *Colletotrichum gloeosporioides*, *Rhizoctonia solani*, and *Botrytis cinerea* (Manikandan and Sathiyabama 2016).

25.2.3 Nanobactericides

NPs are increasingly studied for their potential as bactericides, offering advantages in combating bacterial infections. For instance, Cu-NPs have been effective against pathogens like *Xanthomonas oryzae* and *X. campestris*, responsible for rice blast and leaf spot disease, and have shown antimicrobial activity against various food spoilage bacteria, including *E. coli*. Atiq et al. (2023) investigated the antimicrobial activity of Cu and Ag-NPs from *Moringa oleifera* against chili leaf spot caused by *Xanthomonas campestris* pv. *vesicatoria*. In another study, silver and gold NPs synthesized from Brazilian cherry fruit extract exhibited effectiveness against gram-negative and gram-positive bacteria as well as *Candida albicans* due to essential functional groups. Fructose-stabilized Ag-NPs developed by Dzimitrowicz et al. (2018) demonstrated efficacy against bacterial phytopathogens like *Clavibacter michiganensis*, *Dickeya solani*, *Ralstonia solanacearum*, *Erwinia amylovora*, and *X. campestris* pv. *campestris* (Kalaba et al. 2021). Furthermore, Ag-NPs have shown antibacterial activity against plant pathogens such as *Citrobacter freundii* and *Erwinia cacticida* (Aravinthan et al. 2015).

25.2.4 Nanoinsecticides/Nanopesticides

In conventional agriculture, insects and pests pose a significant threat, causing considerable damage to crop productivity and thereby affecting agricultural yields. Recently, the integration of nano-formulated pesticides with metallic or polymer NPs has emerged as a frontier in the pesticide industry, offering potential solutions to address crop loss due to pest infestation. Nanoencapsulation and formulation of insecticides present numerous advantages, including enhanced thermal stability, solubility, permeability, crystallinity, stiffness, and biodegradability, all of which are vital for promoting sustainable agriculture (Dwivedi et al. 2016). Notably, Ag-NPs derived from *Plumeria rubra* latex have demonstrated significant larvicidal activity against *Anopheles stephensi* and *Aedes aegypti* mosquitoes (Patil et al. 2012). Furthermore, SiO2 nanospheres have been found to improve the permeation of pesticides into plant cells and across crops, enhancing their efficiency, particularly against pests with chewing or sucking feeding behavior. The integration of Si-NP has also shown promise in altering the fertility of *Spodoptera littoralis*, leading to a decline in insect populations and subsequent reduction in crop loss (El-bendary and El-Helaly 2013). Si-NPs have exhibited efficacy against various insect pests, including coconut mites, whiteflies, mustard weevils, and rice weevils, with their insecticidal action attributed to physiological absorption in insect cuticle lipids, ultimately resulting in insect mortality (Barik et al. 2008).

25.2.5 Nanoherbicides

Herbicides, whether produced synthetically or derived from natural sources, are formulated to combat invasive weeds by preventing their spread or causing them to perish outside their native habitats (Jiang et al. 2021). NPs are increasingly being investigated as promising substitutes for traditional herbicides, offering potential benefits through the development of NM-based agrochemical formulations or nano-sized preparations. Conventional herbicides often target only the aboveground parts of weeds, leaving underground rhizomes or tubers untouched, leading to potential regrowth. To overcome this limitation, researchers are exploring the development of herbicide molecules encapsulated with NPs, aiming to target receptors found in weed roots and potentially enhancing their effectiveness. Metal-based NPs, such as Cu-NPs, have demonstrated promise as herbicides due to their ability to disrupt cellular processes and inhibit weed proliferation (Pérez-Hernández et al. 2022). Additionally, Cu-NPs offer advantages such as reduced environmental impact and decreased likelihood of weed resistance development compared to conventional herbicides. Similarly, ZnO-NPs have shown herbicidal activity against various weed species. Carbon-based NPs, including CNTs and graphene, have shown their potential as herbicides by disrupting weed cell membranes or interfering with essential metabolic processes, resulting in weed suppression. Research into the development and application of NPs as herbicides continues to grow, to provide sustainable solutions for weed management in agriculture (Alghuthaymi et al. 2020).

25.2.6 Nanopriming-Enhanced Seed Germination

Amidst the challenges posed by climate change, seed dormancy has emerged as a critical obstacle to sustainable agriculture. Nanotechnology offers a potential solution by enhancing crop productivity and stress tolerance (Lowry et al. 2019). Nanopriming, in particular, stands out as a promising technique for boosting seed germination, seedling vigor, and overall plant growth. This process involves seed treatment with NMs such as NPs or nanocomposites at extremely low concentrations, which can modify seed metabolism, germination signaling pathways, and seedling establishment. Moreover, nanopriming plays a role in safeguarding seeds during storage, synchronizing seed germination, and conferring resistance to environmental stresses. Research has shown that $Mg(OH)_2$ NP can stimulate maize seed germination both in vivo and in vitro (Shinde et al. 2020). Similarly, in watermelon, seed priming with Ag-NP has demonstrated significant enhancements in fruit yield, quality, germination rate, and growth rate (Mahakham et al. 2017). Moreover, nanopriming using CNTs demonstrate notable improvements in both seed germination and seedling growth rates.

25.2.7 Soil Texture Improvement

Soil texture improvement has become a focal point in agricultural research, with NPs attracting increasing interest due to their small size and extensive surface area, allowing them to interact with soil particles and organic matter. This interaction influences soil properties, thereby affecting soil texture significantly. NPs promote the formation of larger aggregates, serving as binding agents and enhancing soil structure, porosity, and water holding ability (Abbas et al. 2020). This proves particularly advantageous in soils with poor structure or low organic matter content, facilitating plant root growth and nutrient uptake. Moreover, NPs, such as nanoclays or nanosilica, increase soil water retention capacity by reducing evaporation and improving water holding, crucial for maintaining optimal soil moisture levels during dry periods, thereby promoting plant growth and reducing irrigation needs. Additionally, specific NPs can adsorb and release essential nutrients like nitrogen, phosphorus, and potassium, thus enhancing nutrient availability while reducing leaching. Furthermore, NPs can influence soil pH by adsorbing or releasing H^+ ions, optimizing conditions for plant growth. Their application also aids in soil erosion mitigation by improving stability and cohesion. When applied to erosion-prone areas, NPs reduce the detachment and transport of soil particles by wind or water erosion, thus preserving soil fertility and preventing loss of arable land. However, considering potential environmental implications and long-term effects is essential before widespread NP application, given the complexity of interactions between NPs and soil ecosystems, necessitating further research.

25.3 Fate of Agricultural NPs in the Environment

The introduction of nanotechnology in agriculture marks a new era for farming practices. Since then, there is a growing concern about what happens when NPs get released into our environment. Upon their release in the atmosphere, they are subjected to various transformation processes that include homo- or hetero-agglomeration, sedimentation/dissolution, and adsorption or undergo various chemical and biological transformations. These alterations collectively alter their mobility and bioavailability. This is a crucial aspect to consider when assessing the environmental risks associated with the agricultural NPs.

25.3.1 *Physical Transformation of NPs*

The behavior and path of agricultural NPs undergo significant changes due to physical transformations, especially processes like clumping together (agglomeration) and sedimentation/deposition. In the natural environment, only a meager fraction of NPs exist as individual particles out of the total pool. This happens because NPs have a lot of surface energy, which comes from their high specific surface area (Rodrigues et al. 2016). When NPs are released into the environment, they stick together either among themselves (homo-agglomeration) or with both organic and inorganic particles (hetero-agglomeration) (Lei et al. 2018). In the environment, agglomeration of NPs with natural colloids is more common than homo-agglomeration owing to the presence of trace amounts of NPs. Aquatic and terrestrial environments host a diverse array of natural colloid particles like clay minerals, iron, and manganese oxides, organic matters, and extracellular polymeric substances (EPS). The interaction between natural colloids and NPs is crucial in influencing the stability and agglomeration dynamics of NPs in the environment. Agglomeration in natural systems occurs due to differences in precipitation rates, Brownian motion, and fluid motion (Abbas et al. 2020). The agglomeration process leads to gradual increase in the size of particle, indicating a decrease in the concentration of NPs in soil or water suspensions. This reduction in NP concentration facilitates the deposition of larger agglomerated particles (Mitrano and Nowack 2017). Additionally, agglomerated NPs exhibit reduced surface area to volume ratio with fewer accessible active sites compared to primary particles, resulting in their decreased reactivity and bioavailability. For instance, when Fe_2O_3, TiO_2, and Ag-NPs are introduced into aquatic ecosystems, they tend to aggregate into micrometer-sized clusters. This aggregation reduces their mobility and limits exposure to vulnerable aquatic organisms (Chekli et al. 2015). The interaction between natural colloids and NPs plays a crucial role in regulating stability and its agglomeration dynamics in the environment. According to Chen et al. (2019), the aggregation energy barrier is determined by the combined effect of van der Waals forces and electrostatic repulsion between NPs. These minute particles, originally in constant molecular motion, experience collisions driven by Brownian motion, but undergo agglomeration when equilibrium between van der Waals force and repulsive electrostatic force has been reached. In the microalgae *Isochrysis galbana*, toxicity of TiO_2-NPs is ameliorated by the combined effect of agglomeration process and its subsequent deposition (Hu et al. 2018). At an exposure concentrations of 50 and 100 mg L^{-1}, TiO_2-NPs agglomerate on the surface of *Microcystis aeruginosa*, making a coating effect that impedes algal growth (Liu et al. 2017). NP agglomeration is influenced by factors such as ionic strength, pH, NP type, concentration, and physicochemical properties (Abbas et al. 2020). Higher concentrations of NPs, such as silver and zero-valent iron (nZVI), accelerate the formation of large, stable clusters (Zhang et al. 2011). Moreover, NP characteristics like size, shape, and chemical composition also impact their tendency to agglomerate. pH plays a pivotal role in determining stability and surface charge, with minimal agglomeration observed near the isoelectric point (Lin et al. 2010). Coatings on NPs can also modify their surface charge, thus influencing agglomeration and toxicity. For instance, citrate-coated Ag-NPs exhibited greater toxicity toward zebrafish after a 96-hour exposure compared to their PVP-coated counterparts (Liu et al. 2019).

Additionally, stability, fate, and toxicity of NPs are also governed by processes such as sulfidation, photochemical reactions, biodegradation, and biomodifications. Some NPs, such as Ag, Cu, and Zn, exhibit a strong tendency to form complexes with sulfide ligands in the soil, creating relatively insoluble compounds that decrease their bioavailability and toxicity (Lv et al. 2012). However, contrasting response was observed during the sulfidation of CuO-NPs, where their bioavailability is increased by its solubilization and release of environmentally harmful Cu^{2+} ions. Photochemically induced reactions like photo-oxidation and reduction, photolysis, etc., not only affect the NP coating and its oxidation state but also hinder their environmental persistence (Mitrano and Nowack 2017). Meanwhile, biomodification and biodegradation mediated by plants and microbes modify the surface properties of NPs, potentially leading to the complete degradation of carbon-based nanomaterials.

25.3.2 Chemical Transformation of NPs

Besides physical transformations, NPs applied in agro-industries interact with organic or inorganic particles in the environments and undergo a plethora of chemical transformations that includes such as oxidation–reduction (redox), dissolution, adsorption, complexation, and degradation (Abbas et al. 2020).

25.3.2.1 Oxidation–Reduction (Redox) Reactions

Redox reactions are primarily characterized by transfer of electrons between the two reactants. One of them undergoes oxidation, losing electrons, while the other undergoes reduction, gaining those electrons. In the agricultural field, these reactions are governed by environmental parameters such as pH and the presence of electron donors or acceptors (Nowack et al. 2012). Agricultural soils are commonly identified as oxidizing environments, while in the subsurface soils oxygen availability is limited and may result in the reduction of metal-based NPs (Chen 2018). Redox reactions, involving the transfer of electrons, profoundly impact the functionality, stability and interactions of NPs within the agricultural systems. Oxidation–reduction processes are common in metal-based NPs, such as Ag, Cu, Fe, Zn, and nZVI due to their redox-active nature. It was observed that oxidation of agricultural NPs can be accelerated under conditions of low pH and high O_2 levels, as seen in aerated soils with low saturation in the field (Levard et al. 2012). Under such conditions, metal NPs may undergo oxidation to form metal oxides or hydroxides, altering their chemical properties and reactivity. Additionally, carbon-based NPs, such as graphene or CNTs, can also undergo oxidation–reduction reactions. Oxidative treatments may introduce functional groups onto the NP surface, modifying their physicochemical properties and enhancing their interactions with soil constituents.

25.3.2.2 Dissolution of NPs

Dissolution of agricultural NPs refers to the myriad of processes through which NPs are solubilized or disintegrated typically within the agricultural soil or plant systems. The solubility or dissolution rate of agricultural NPs are predominantly governed by their intrinsic properties, including chemical composition, shape, size, surface area and chemistry, crystalline phase, and more. Ma et al. (2012) demonstrated that smaller Ag-NPs exhibit greater solubility that could be correlated only with their particle size. Considering their solubility, NPs can be categorized into highly soluble, such as Ag, FeO, Cu/CuO, Zn/ZnO; moderately soluble NPs such as TiO_2 and CeO_2, while some are typically insoluble, like CB, CNTs, graphene, and fullerenes (Singh et al. 2023). The extent NPs dissolution varies significantly depending on environmental conditions. For instance, Tangaa et al. (2016) demonstrated that Ag, ZnO, and CuO NPs show high solubility (1–80%) differing in various substratum, while CeO_2 (20 nm) exhibit a much lower solubility rate, with only 3% observed in artificial soil (Cornelis et al. 2013).

25.3.2.3 Adsorption of NPs

Agricultural NPs interacts and adhere to the surfaces of soil particles, organic matter, or plant roots in agricultural environments. This phenomenon of NP adsorption plays a significant role altering surface properties, transport, solubility, toxicity, and bioavailability of NPs in the plant-soil system. Majority of the productive soils contain substantial amounts of organic matter, which significantly influences the behavior of metal-based NPs within the soil matrix (Yu et al. 2021). Functional groups present in the dissolved organic matter (DOM), such as thiol, carboxyl, hydroxyl, ketones, and aldehyde have the ability to chelate metallic NPs, resulting in a coating of DOM on the NP surfaces impacting their aggregation, sedimentation, and deposition. Additionally, DOM plays a crucial role in the chemical transformations of NPs, including oxidation, reduction, dissolution, and their adsorption capacity for other compounds and ions (Yu et al. 2021). Chekli et al. (2013) demonstrated that humic acid, acting as a surface modifier, enhances the stability of Fe_2O_3-NPs by improving their electrostatic stability. The interaction between Fe_2O_3-NPs and DOM was evidenced by the formation of small agglomerates held together by hydrogen bonds, which could be reverted to their initial size.

25.3.2.4 Photochemical Transformation of NPs

Photochemically induced NP transformation refers to the chemical reactions that occur in agricultural NPs under the influence of light. This process involves various photochemical reactions such as photo-oxidation, photo-reduction, photolysis, and photo-catalysis. The bandgap energy of NPs and the wavelength of incident light determine their photochemical transformation potential. When the bandgap energy of NPs is lower than the energy of incident photons, photo-activation occurs, leading to the generation of reactive oxygen species (ROS) and other reactive intermediates. These photochemical reactions can alter the coating, oxidation state, and reactivity of NPs, influencing their behavior and persistence in the environment. NPs with photocatalytic properties, such as titanium dioxide (TiO_2) and zinc oxide (ZnO), are widely used in various agricultural applications. However, in the agricultural field, exposure to sunlight can enhance the toxicity of photoactive NPs, leading to adverse effects on organisms and ecosystems. Toxicity of photoactive NPs varies significantly under light conditions compared to darkness. For example, the toxicity of TiO_2 NPs to *E. coli* decreased drastically from 583 to 1.68 mg L^{-1} when exposed to sunlight for 30 minutes (Dasari et al. 2013). Even at exposure concentrations below 1 mg L^{-1}, ZnO-NPs caused cytotoxicity in freshwater algae *Scenedesmus obliquus*, resulting in viability loss, membrane damage, and ROS generation (Bhuvaneshwari et al. 2015). Therefore, understanding the photochemically induced transformation of NPs is crucial for assessing their environmental impact and ensuring the safe use of NMs.

25.3.2.5 Chemical Complexation and Degradation

In agricultural soils, NPs such as CuO or ZnO can form complexes with DOM, such as humic substances or plant exudates, as well as soil minerals, influencing their retention and transport in the soil. Over time, agricultural NPs may also undergo degradation processes, resulting in their breakdown into smaller, less reactive species. Metal-based NPs like Fe_2O_3 or TiO_2 can degrade through redox reactions or photochemical processes (Abdel-Wahab et al. 2017). Microorganisms present in agricultural soils can also contribute to NP degradation through enzymatic reactions or microbial metabolism. This degradation reduces the persistence and potential environmental risks associated with agricultural NPs.

25.3.3 Biological Transformation of NPs

Biological transformations of agricultural NPs involve an array of processes that include biodegradation, bioaccumulation, and bioavailability within the agricultural environment. The extent and importance of these processes are reliant upon the interactions of NPs with extracellular enzymes, extracellular polymeric substances (EPS), plants, and microorganisms. Biodegradation of agricultural NPs by microorganisms or enzymes present in the soil or aquatic settings is a significant process of NP detoxification. Microorganisms can metabolize NPs, using them as carbon or energy sources, thereby breaking them down into smaller constituents or converting them into harmless by-products. This biodegradation process is pivotal in determining the destiny and longevity of NPs in agricultural soils. For instance, it has been demonstrated that soil bacteria can degrade Ag NPs, reducing their antimicrobial properties and mitigating their potential environmental toxicities. NPs can accumulate within living organisms, mostly in plants, and bacteria, after being absorbed from soil, water, or air. This accumulation often occurs in plant tissues, such as roots, stems, leaves, and fruits, affecting their growth and physiology. Bioaccumulation can additionally affect the movement of NPs through the food chain, potentially influencing the health of ecosystems. Studies have shown that TiO_2-NPs applied to soil can accumulate in various tissues of lettuce and wheat (Hu et al. 2020). Furthermore, organisms inhabiting agricultural soil can uptake NPs from their surroundings and disperse them within their cells or microbial populations. Research by Zhang et al. (2020) revealed the internalization and distribution of CNTs within soil bacteria, impacting microbial growth and metabolic functions. Additionally, earthworms can uptake ZnO-NPs from soil, influencing their growth, reproduction, and the processes within the soil ecosystem (Rajput et al. 2018). NPs taken up by plant roots may also undergo biotransformation within plant cells, resulting in changes to their chemical composition or structure. For example, Ag NPs has been observed to transform into silver sulfide (Ag_2S) NPs within plant tissues, influencing their bioaccumulation and phytotoxicity (Wang et al. 2015).

25.4 Nanotoxicology

Nanotoxicology is the scientific discipline dedicated to thoroughly investigating and understanding the potential toxicity of NPs (Fig. 25.1). Despite the multitude of in vivo and in vitro studies conducted, our understanding of the toxicological impact of NMs within the realms of agriculture, food, and the environment remains incomplete. It is acknowledged that not all NPs exhibit adverse effects, as their toxicity is contingent upon various factors such as size, aggregation, crystallinity, composition, and surface properties. Nonetheless, there exists a significant correlation between NPs and toxicity. This association is largely attributed to the widespread and indiscriminate use of NPs, as well as their improper disposal, which poses unprecedented health and environmental risks (Hu et al. 2019). Moreover, the absence of clear and precise risk management frameworks, along with the failure to establish unified regulatory frameworks in nanosafety, has exacerbated the issue. Therefore, thorough investigations into their effects on human health and the environment have become imperative through nanotoxicity studies.

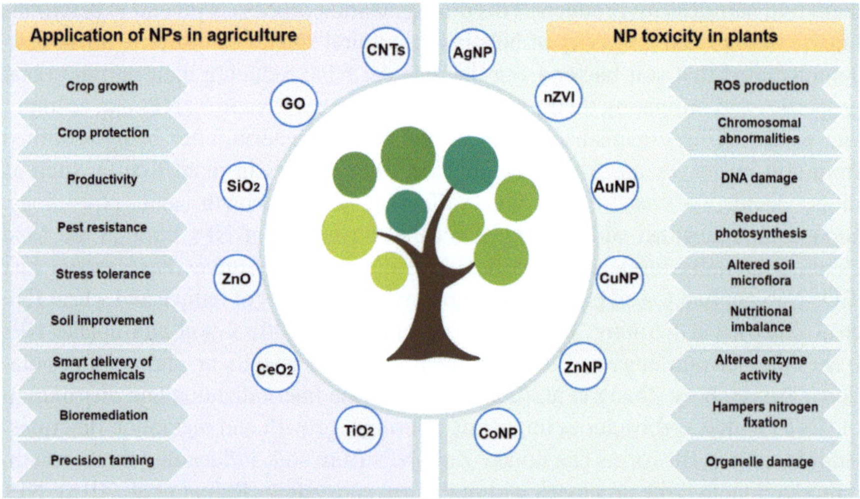

Fig. 25.1 A schematic diagram illustrating the diverse applications of NPs in modern agricultural practices, alongside their potential toxic effects on plants: NPs like AgNPs, AuNPs, ZnNPs, CuNPs, CoNPs, nZVI, CNTs, GO, SiO2-NPs, Zno-NPs, TiO2-NPs, and CeO2-NPs are widely utilized for bolstering crop productivity and may also pose toxicity risks due to their indiscriminate usage and improper disposal

25.4.1 Mechanism of NP Toxicity

NPs exhibit physicochemical reactivity that can lead to irreversible harm to cells and organelles. Due to their small size, they are capable of translocating into the circulatory or lymphatic systems and various body tissues/organs upon internalization. Depending on their composition and size, certain NMs may generate elevated levels of reactive oxygen species (ROS) or free radicals. Processes such as depletion in intracellular glutathione levels and reduction in mitochondrial membrane potentials are significant contributors to the toxicity of metal NPs (Teow et al. 2011). Silver, gold, magnetite, CuO, and TiO2-NPs have been evaluated for their toxicity using various in vivo as well as in vitro studies. Different NPs have demonstrated distinct toxic properties and mechanisms. The primary mechanism underlying nanotoxicity is often the generation of ROS, leading to DNA strand breakage, oxidative modifications of DNA and proteins, the production of free radicals, changes in gene expression levels compromising cellular defense mechanisms (Ghosh et al. 2016) and cellular death.

In NP toxicology, size and concentration are crucial characteristics. Nanoscale NPs with increased surface area, facilitate greater uptake and interaction with plants, thereby augmenting the risk of toxicity (Nel et al. 2006). It has been demonstrated that NPs smaller than 5 nm can easily traverse plant cell wall, while those under 20 nm can easily translocate through plasmodesmata, indicating a higher likelihood of uptake and toxicity even at minute concentrations (Ma et al. 2011; Wang et al. 2013). In contrast, larger NPs like graphene oxide (GO) experience hindered translocation from roots to shoots, reducing their bioaccumulation potential (Rani et al. 2023). Besides size, concentration of NPs also significantly influences their toxicity. Zhao et al. (2013) reported that ZnO and CeO_2 when administered at higher concentrations (800 mg/kg) demonstrated phytotoxicity in cucumber, which was not observed at lower concentrations (400 mg/kg). Contrasting reports in peanut plants illustrates that ZnO-NPs when applied at a concentration of 1000 ppm stimulate germination rate, initiating chlorophyll synthesis and enhancing seedling vigor (Prasad et al. 2012). However, NP application could significantly elevate expression of genes involved in carotenoid biosynthesis pathway and may contribute to reduction in photosynthetic efficiency. These NPs impart toxicity by triggering the excessive formation of ROS, primarily by ionizing biological macromolecules, and disrupting the electron transport chain of mitochondria and chloroplasts (Yan and Chen 2019). Additionally, physicochemical reactivity of metallic NPs influence carbon fixation processes through photoinhibition, generating more superoxide radicals. Excessive generation of ROS disrupts cellular integrity through DNA damage, protein modifications, and lipid peroxidation (Wang et al. 2020).

Furthermore, oxidation of zero-valent Ag-NPs can generate ionic silver (Ag^+) within the cell, promoting significant production of reactive oxygen species (ROS) and causing phytotoxicity (Montes et al. 2017). Additionally, Ag^+ hinders photosynthesis by competing with Cu^+ in plastocyanin (Pc), a copper-binding protein in the thylakoid lumen of chloroplasts. Iron-based Fe-NPs are increasingly acknowledged for their agricultural significance. Studies indicate that Fe-NPs exhibit a slower rate of translocation to aerial conduits (Tripathi et al. 2017) but modulate antioxidant enzyme activity. Exposure to $CoFe_2O_4$ NPs resulted in inhibited catalase (CAT) activity across all treatments (López-Moreno et al. 2016).

Similar instances of phytotoxicity have also been noted for Ni- and Cu-NPs, with increases in ROS levels, chromosomal abnormalities and disruptions in the cell cycle. Furthermore, these NPs have been reported to induce DNA damage and hinder plant growth in annual and perennial ryegrass, and in *Raphanus sativus* (Atha et al. 2012). In addition to their toxic effects, NP induced ROS generation has a beneficial role in mitigating biotic and abiotic stresses. This occurs through the direct or indirect activation of oxidative enzymatic pathways, maintaining an equilibrium between ROS production and scavenging (Faisal et al. 2013; Bhattacharya and Kundu 2020). Zhao et al. (2022) reported adverse effects of graphene oxide (GO) on the physiology of *Trifolium repens*. Their study demonstrated that GO negatively influenced seedling development, nutrient absorption, and photosynthetic factors. Exposure to GO resulted in oxidative damage and electrolyte leakage in seedlings, which can be attributed to decreased levels of antioxidant enzymes and increased MDA levels. Moreover, the investigation revealed that GO induced the highest levels of ROS at concentrations of 0.2% and 0.4% (Zhao et al. 2022). Parallel findings were also noted for fullerene, another carbon NP, which was observed to block the root pore structure, compress endothelial cells, and induce internal wall injuries (He et al. 2021). Fullerene uptake by plant cells led to increased ROS production and accumulation, triggering reactions with unsaturated fatty acids in the cell membrane and altering the ratio of saturated vs. unsaturated fatty acids.

Recent research suggests that NPs negatively influence phytohormone signaling. NP exposure has been associated with a concomitant decrease in the levels of auxins, cytokinins, and salicylic acid, resulting in severe hormonal imbalance in plants thereby impacting overall metabolism (Vankova et al. 2017). Additionally, elevated concentrations of NPs can significantly impair photosynthesis, potentially causing suppressed plant growth or even plant death. The genotoxic effects of NPs primarily occur through mechanisms involving direct interaction with chromosomes during the cell cycle. This interaction can result in binding to DNA, inhibiting processes such as replication or transcription, or inducing chromosomal loss or breakage. Additionally, NPs can indirectly induce genotoxicity by generating reactive oxygen species (ROS) and releasing toxic elements that interfere with proteins essential for DNA replication, transcription, or repair. Irreversible alterations in the chemical structure of purines and pyrimidines may lead to improper pairing during replication, potentially causing lethal mutations (Thongkumkoon et al. 2014). Moreover, NPs can target protein kinases associated with cell cycle events, disrupting cell cycle checkpoint functions (Huang et al. 2009). Smaller NPs exhibit higher reactivity and greater genotoxicity; for example, Ag-NPs were found to induce greater DNA damage in freshwater crustaceans (*Daphnia*) and zebrafish embryos (Park and Choi 2010).

25.4.2 Effect on Soil Microflora

The soil microflora is of utmost importance not only because it nurtures crops but also serves as the foundation for agricultural sustainability. The intentional or unintentional introduction of NPs into the natural environment poses a potential threat to these soil-borne beneficial microbial communities and also interferes with the mycorrhizal colonization process (Rajput et al. 2018). In paddy fields, TiO_2 and CuO-NPs elicit a decreased enzymatic activity and diminution of soil microbial flora that ultimately leads to an altered community structure (Xu et al. 2015). The altered activity of the bacterial enzymes (invertase, phosphatase, urease, and catalase) and shifts in their community structure impede the nitrogen fixation capability decreasing soil fertility. According to Chavan et al. (2020), plant growth-promoting bacteria (PGPR), spanning into three distinct categories, the nitrogen fixers, the phosphate solubilizers, and the biofilm producers, encountered significant inhibition upon direct contact with TiO_2-NPs. Furthermore, TiO_2-NPs appeared to exert greater inhibitory effects on Gram-positive bacteria, indicating the potential significance of bacterial surface characteristics in the toxicity of TiO_2 NPs. Cr_2O_3, CuO, and Ni-NPs at higher concentrations reduce the activity of dehydrogenase enzyme among soil-borne bacteria (Jośko et al. 2014). The interaction between ZnO-NPs and rhizospheric microbiota damages bacterial cell surfaces and morphological alterations within rhizobial cells. Furthermore, ZnO-NPs agitate the process of root nodulation, hamper nitrogen fixation, and initiate premature senescence of nodules (Fan et al. 2014). Leached Zn^+ NPs in the soil are reported to exert oxidative stress on *Campylobacter jejuni* and *Staphylococcus aureus* (Xie et al. 2011; Premanathan et al. 2011). The pH and the organic matter content of soil are the sentinel factors that affect the impact of ions released from NP on microorganisms (Reidy et al. 2013). In acidic soil, the toxicity of ZnO-NPs toward *Folsomia candida* was observed to escalate (Waalewijn-Kool et al. 2013). CeO_2-NPs are reported to cause membrane damage and stall the growth in beneficial soil bacteria like Bacillus subtilis, and *E. coli* (Krishnamoorthy et al. 2014). In contemporary agriculture, NPs are used as ubiquitous tool for augmenting soil fertility and facilitating the bioremediation of pollutants. Consequently, future endeavors must prioritize the crafting of NPs with minimal toxicity toward microorganisms. Furthermore, it is imperative to explore soil management strategies to ensure conditions that deactivate NPs and ameliorate their detrimental impact on microorganisms (Kumar et al. 2018).

25.5 Regulatory Strategies to Minimize Environmental Hazards

The indiscriminate application of NPs in agro-nanotechnologies raises concerns about their direct release into the environment and subsequent incorporation into the food chain. The existing challenges related to the use of agricultural NPs arise from the potential passage of NMs through the food chain, underscoring the need for prompt safety assessment. In this context, plants play a crucial role in mediating the transfer of NPs to various trophic levels and ultimately into the food web, given their widespread consumption by organisms at lower trophic levels, animals, and humans (López-Moreno et al. 2018). The concept of One Health, as advocated by World Health Organization (WHO), Food and Agriculture Organization (FAO), and the World Organization for Animal Health (WOAH), plays a pivotal role in raising awareness of its impact. Specifically, the excessive and improper use of agrochemicals significantly contributes to the migration of NPs into the food chain (Okeke et al. 2022). Addressing nanotoxicity and the associated risks to human health, therefore, necessitates the development of an intelligent strategy for evaluating and mitigating the risks of agricultural-NPs. Improper disposal of nanowaste can cause environmental contamination, and increased NP exposure underscore the importance of disposal methods. Guidelines adopted by the government of various countries emphasize compliance with regulations governing NP production, use, and disposal, ensuring adherence to occupational health and safety standards and environmental regulations. They also promote public awareness and education about the risks of NP exposure and the importance of responsible handling and disposal practices for all stakeholders. Some of the safety measures for handling, waste disposal, and cleaning protocols for NPs to consider are as follows:

Recycling: Where feasible, NPs can be recycled or reused in agricultural or industrial processes to minimize waste generation.

Landfilling: NPs can be disposed of in designated landfills equipped to handle hazardous materials. However, precautions must be taken to prevent leaching into soil and groundwater.

Incineration: Thermal treatment through incineration can be an option for certain NPs, but it requires specialized facilities to ensure complete combustion and proper management of resulting emissions.

Neutralization: Some NPs may be modified to less hazardous compounds through chemical treatments or neutralization processes before disposal.

Controlled release: In some cases, NPs may be incorporated into controlled-release formulations for agricultural use, reducing the risk of environmental exposure.

Regulatory compliance: It is essential to adhere to local regulations and guidelines governing the disposal of hazardous NMs, to minimize environmental risks and ensure compliance with safety standards.

25.6 Conclusion and Future Perspective

In conclusion, this chapter has highlighted the emerging field of agricultural nanotechnology, highlighting the potential risks associated with the use of NPs in agricultural practices. Through a comprehensive review of current literature, various factors have been explored contributing to nanotoxicity in agriculture, including NP properties, exposure routes, and their interactions with the organisms and ecosystem. It is evident that while NPs offer promising benefits such as enhanced crop productivity, stress management, precision farming and soil remediation, their potential adverse effects on environmental and human health cannot be overlooked. The variability in toxicity across different NP types, concentrations, and environmental conditions emphasizes the complexity of this phenomenon. Moving forward, it is imperative to continue research efforts aimed at better understanding of the nanotoxicity mechanisms in agriculture and developing effective risk assessment strategies. Furthermore, proactive measures such as regulatory frameworks, green synthesis of NPs, and environmentally sustainable NP design should be implemented to mitigate potential risks and ensure the safe and responsible use of nanotechnology in agriculture. By addressing these challenges, one can harness the potential of nanotechnology to revolutionize agriculture while safeguarding the health of our planet and its inhabitants for future generations to come.

References

Abbas Q, Yousaf B, Ali MU et al (2020) Transformation pathways and fate of engineered nanoparticles (ENPs) in distinct interactive environmental compartments: a review. Environ Int 138:105646

Abdel-Wahab A-M, Al-Shirbini A-S, Mohamed O, Nasr O (2017) Photocatalytic degradation of paracetamol over magnetic flower-like TiO2/Fe2O3 core-shell nanostructures. J Photochem Photobiol A Chem 347:186–198

Alghuthaymi M, Mostafa M, Abd-Elsalam K A (2020) Carbon nanotubes: an efficient sorbent for herbicide sensing and remediation. In Carbon Nanomaterials for Agri-Food and Environmental Applications. Elsevier, pp. 429-457

Ansari M, Shahzadi K, Ahmed S (2020) Nanotechnology: a breakthrough in agronomy. In: Javad S (ed) Nanoagronomy. Springer International Publishing, Cham, pp 1–21

Aravinthan A, Govarthanan M, Selvam K et al (2015) Sunroot mediated synthesis and characterization of silver nanoparticles and evaluation of its antibacterial and rat splenocyte cytotoxic effects. Int J Nanomedicine 10:1977–1983. https://doi.org/10.2147/IJN.S79106

Atha DH, Wang H, Petersen EJ et al (2012) Copper oxide nanoparticle mediated DNA damage in terrestrial plant models. Environ Sci Technol 46:1819–1827. https://doi.org/10.1021/es202660k

Atiq M, Fatima T, Rajput NA et al (2023) *Moringa olifera* plant based copper and silver nano particles and its antibacterial activity towards leaf spot of chilli caused by *Xanthomonas campestris* pv. *vesicatoria*. Int J Phytopathol 12:63–72. https://doi.org/10.33687/phytopath.012.01.4306

Barik TK, Sahu B, Swain V (2008) Nanosilica—from medicine to pest control. Parasitol Res 103:253–258. https://doi.org/10.1007/s00436-008-0975-7

Bhattacharya S, Kundu A (2020) Sugars and sugar polyols in overcoming environmental stresses. In: Roychoudhury A, Tripathi DK (eds) Protective chemical agents in the amelioration of plant abiotic stress, 1st edn. Wiley, pp 71–101. https://doi.org/10.1002/9781119552154.ch4

Bhattacharya S, Saha J (2023) Green synthesis and application of biogenic nanomaterials as a blueprint in mitigation of abiotic stress in crop plants: a conceptual review. In: Das J, Halder S (eds) Advancement of GI-science and sustainable agriculture: a multi-dimensional approach. Springer Nature, Cham, pp 155–182. https://doi.org/10.1007/978-3-031-36825-7_11

Bhattacharya S, Gupta S, Saha J (2023) Nanoparticles regulate redox metabolism in plants during abiotic stress within hormetic boundaries. Funct Plant Biol 50:850–869. https://doi.org/10.1071/FP23068

Bhuvaneshwari M, Iswarya V, Archanaa S et al (2015) Cytotoxicity of ZnO NPs towards fresh water algae Scenedesmus obliquus at low exposure concentrations in UV-C, visible and dark conditions. Aquat Toxicol 162:29–38

Brunel F, El Gueddari NE, Moerschbacher BM (2013) Complexation of copper(II) with chitosan nanogels: toward control of microbial growth. Carbohydr Polym 92:1348–1356. https://doi.org/10.1016/j.carbpol.2012.10.025

Chavan S, Sarangdhar V, Nadanathangam V (2020) Toxicological effects of TiO2 nanoparticles on plant growth promoting soil bacteria. Emerg Contam 6:87–92

Chekli L, Zhao YX, Tijing LD et al (2015) Aggregation behaviour of engineered nanoparticles in natural waters: characterising aggregate structure using on-line laser light scattering. J Hazard Mater 284:190–200

Chen H (2018) Metal based nanoparticles in agricultural system: behavior, transport, and interaction with plants. Chem Speciat Bioavailab 30:123–134. https://doi.org/10.1080/09542299.2018.1520050

Chen X, Zhu Y, Yang K et al (2019) Nanoparticle TiO2 size and rutile content impact bioconcentration and biomagnification from algae to daphnia. Environ Pollut 247:421–430

Chekli L, Phuntsho S, Roy M et al (2013) Characterisation of Fe-oxide nanoparticles coated with humic acid and Suwannee River natural organic matter. Sci Total Environ 461:19-27

Chhipa H (2017) Nanofertilizers and nanopesticides for agriculture. Environ Chem Lett 15:15–22. https://doi.org/10.1007/s10311-016-0600-4

Cornelis G, Pang L, Doolette C et al (2013) Transport of silver nanoparticles in saturated columns of natural soils. Sci Total Environ 463:120–130

Dasari TP, Pathakoti K, Hwang H-M (2013) Determination of the mechanism of photoinduced toxicity of selected metal oxide nanoparticles (ZnO, CuO, Co 3O4 and TiO2) to *E. coli* bacteria. J Environ Sci 25:882–888

Dwivedi S, Saquib Q, Al-Khedhairy AA, Musarrat J (2016) Understanding the role of nanomaterials in agriculture. In: Microbial inoculants in sustainable agricultural productivity. Vol. 2, Functional applications. Springer, New Delhi, pp 271–288

Dzimitrowicz A, Motyka-Pomagruk A, Cyganowski P et al (2018) Antibacterial activity of fructose-stabilized silver nanoparticles produced by direct current atmospheric pressure glow discharge towards quarantine pests. Nanomaterials 8:751. https://doi.org/10.3390/nano8100751

El-Bendary HM, El-Helaly AA (2013) First record nanotechnology in agricultural: silica nanoparticles a potential new insecticide for pest control. Appl Sci Rep 4:241–246

Faisal M, Saquib Q, Alatar AA et al (2013) Phytotoxic hazards of NiO-nanoparticles in tomato: a study on mechanism of cell death. Hazard Mater 250:318–332

Fan R, Huang YC, Grusak MA et al (2014) Effects of nano-TiO2 on the agronomically-relevant Rhizobium–legume symbiosis. Sci Total Environ 466:503–512

Ghosh M, Jana A, Sinha S et al (2016) Effects of ZnO nanoparticles in plants: cytotoxicity, genotoxicity, deregulation of antioxidant defenses, and cell-cycle arrest. Mutat Res Genet Toxicol Environ Mutagen 807:25–32

Hazarika A, Yadav M, Yadav DK, Yadav HS (2022) An overview of the role of nanoparticles in sustainable agriculture. Biocatal Agric Biotechnol 43:102399. https://doi.org/10.1016/j.bcab.2022.102399

He A, Jiang J, Ding J, Sheng GD (2021) Blocking effect of fullerene nanoparticles (nC60) on the plant cell structure and its phytotoxicity. Chemosphere 278:130474

Hu J, Wang J, Liu S et al (2018) Effect of TiO2 nanoparticle aggregation on marine microalgae *Isochrysis galbana*. J Environ Sci 66:208–215

Hu D, Shen M, Zhang Y et al (2019) Microplastics and nanoplastics: would they affect global biodiversity change? Environ Sci Pollut Res 26:19997–20002. https://doi.org/10.1007/s11356-019-05414-5

Hu J, Wu X, Wu F et al (2020) TiO2 nanoparticle exposure on lettuce (*Lactuca sativa* L.): dose-dependent deterioration of nutritional quality. Environ Sci Nano 7:501–513

Huang S, Chueh PJ, Lin Y-W et al (2009) Disturbed mitotic progression and genome segregation are involved in cell transformation mediated by nano-TiO2 long-term exposure. Toxicol Appl Pharmacol 241:182–194

Jiang M, Song Y, Kanwar MK et al (2021) Phytonanotechnology applications in modern agriculture. J Nanobiotechnol 19:430. https://doi.org/10.1186/s12951-021-01176-w

Jośko I, Oleszczuk P, Futa B (2014) The effect of inorganic nanoparticles (ZnO, Cr2O3, CuO and Ni) and their bulk counterparts on enzyme activities in different soils. Geoderma 232:528–537

Kalaba MH, Moghannem SA, El-Hawary AS et al (2021) Green synthesized ZnO nanoparticles mediated by *Streptomyces plicatus*: characterizations, antimicrobial and nematicidal activities and cytogenetic effects. Plants 10:1760. https://doi.org/10.3390/plants10091760

Krishnamoorthy K, Veerapandian M, Zhang L-H et al (2014) Surface chemistry of cerium oxide nanocubes: toxicity against pathogenic bacteria and their mechanistic study. J Ind Eng Chem 20:3513–3517

Kumar P, Burman U, Kaul RK (2018) Ecological risks of nanoparticles: effect on soil microorganisms. In: Nanomaterials in plants, algae, and microorganisms. Elsevier, pp 429–452

Kwak JI, An Y-J (2015) A review of the ecotoxicological effects of nanowires. Int J Environ Sci Technol 12:1163–1172. https://doi.org/10.1007/s13762-014-0727-4

Laskar P, Hazra A, Pal A, Kundu A (2023) Deciphering the role of alternative splicing as modulators of defense response in the MYMIV – *Vigna mungo* pathosystem. Physiol Plant 175:e13922. https://doi.org/10.1111/ppl.13922

Lateef A, Nazir R, Jamil N et al (2016) Synthesis and characterization of zeolite based nanocomposite: an environment friendly slow release fertilizer. Microporous Mesoporous Mater 232:174–183. https://doi.org/10.1016/j.micromeso.2016.06.020

Lei C, Sun Y, Tsang DC, Lin D (2018) Environmental transformations and ecological effects of iron-based nanoparticles. Environ Pollut 232:10–30

Levard C, Hotze EM, Lowry GV, Brown GE (2012) Environmental transformations of silver nanoparticles: impact on stability and toxicity. Environ Sci Technol 46:6900–6914. https://doi.org/10.1021/es2037405

Lin D, Xing B (2007) Phytotoxicity of nanoparticles: inhibition of seed germination and root growth. Environ Pollut 150:243–250. https://doi.org/10.1016/j.envpol.2007.01.016

Lin D, Tian X, Wu F, Xing B (2010) Fate and transport of engineered nanomaterials in the environment. J Environ Qual 39:1896–1908. https://doi.org/10.2134/jeq2009.0423

Liu R, Lal R (2015) Potentials of engineered nanoparticles as fertilizers for increasing agronomic productions. Sci Total Environ 514:131–139. https://doi.org/10.1016/j.scitotenv.2015.01.104

Liu X, Li J, Huang Y et al (2017) Adsorption, aggregation, and deposition behaviours of carbon dots on minerals. Environ Sci Technol 51:6156–6164. https://doi.org/10.1021/acs.est.6b06558

Liu H, Wang X, Wu Y et al (2019) Toxicity responses of different organs of zebrafish (*Danio rerio*) to silver nanoparticles with different particle sizes and surface coatings. Environ Pollut 246:414–422

López-Moreno ML, Avilés LL, Pérez NG et al (2016) Effect of cobalt ferrite (CoFe2O4) nanoparticles on the growth and development of *Lycopersicon lycopersicum* (tomato plants). Sci Total Environ 550:45–52

López-Moreno ML, Cassé C, Correa-Torres SN (2018) Engineered NanoMaterials interactions with living plants: benefits, hazards and regulatory policies. Curr Opin Environ Sci Health 6:36–41

Lowry GV, Avellan A, Gilbertson LM (2019) Opportunities and challenges for nanotechnology in the agri-tech revolution. Nat Nanotechnol 14:517–522

Lv J, Zhang S, Luo L et al (2012) Dissolution and microstructural transformation of ZnO nanoparticles under the influence of phosphate. Environ Sci Technol 46:7215–7221. https://doi.org/10.1021/es301027a

Ma X, Wu Y, Jin S et al (2011) Gold nanoparticles induce autophagosome accumulation through size-dependent nanoparticle uptake and lysosome impairment. ACS Nano 5:8629–8639. https://doi.org/10.1021/nn202155y

Ma R, Levard C, Marinakos SM et al (2012) Size-controlled dissolution of organic-coated silver nanoparticles. Environ Sci Technol 46:752–759. https://doi.org/10.1021/es201686j

Mahajan P, Dhoke SK, Khanna AS (2011) Effect of nano-ZnO particle suspension on growth of mung (Vigna radiata) and gram (Cicer arietinum) seedlings using plant agar method. J Nanotechnol 2011:e696535. https://doi.org/10.1155/2011/696535

Mahakham W, Sarmah AK, Maensiri S, Theerakulpisut P (2017) Nanopriming technology for enhancing germination and starch metabolism of aged rice seeds using phytosynthesized silver nanoparticles. Sci Rep 7:8263

Manikandan A, Sathiyabama M (2016) Preparation of Chitosan nanoparticles and its effect on detached rice leaves infected with *Pyricularia grisea*. Int J Biol Macromol 84:58–61. https://doi.org/10.1016/j.ijbiomac.2015.11.083

Mishra S, Singh BR, Singh A et al (2014) Biofabricated silver nanoparticles act as a strong fungicide against Bipolaris sorokiniana causing spot blotch disease in wheat. PLoS One 9:e97881. https://doi.org/10.1371/journal.pone.0097881

Mitrano DM, Nowack B (2017) The need for a life-cycle based aging paradigm for nanomaterials: importance of real-world test systems to identify realistic particle transformations. Nanotechnology 28:072001

Montes A, Bisson MA, Gardella JA Jr, Aga DS (2017) Uptake and transformations of engineered nanomaterials: critical responses observed in terrestrial plants and the model plant *Arabidopsis* thaliana. Sci Total Environ 607:1497–1516

Nel A, Xia T, Mädler L, Li N (2006) Toxic potential of materials at the nanolevel. Science 311:622–627. https://doi.org/10.1126/science.1114397

Nowack B, Ranville JF, Diamond S et al (2012) Potential scenarios for nanomaterial release and subsequent alteration in the environment. Environ Toxicol Chem 31:50–59. https://doi.org/10.1002/etc.726

Okeke ES, Ezeorba TPC, Mao G et al (2022) Nano-enabled agrochemicals/materials: potential human health impact, risk assessment, management strategies and future prospects. Environ Pollut 295:118722

Park HJ, Kim SH, Kim HJ et al (2006) A new composition of nanosized silica-silver for control of various plant diseases. Plant Pathol J 22:295-302

Park S, Choi J (2010) Geno-and ecotoxicity evaluation of silver nanoparticles in freshwater crustacean *Daphnia magna*. Environ Eng Res 15:23–27

Patil CD, Patil SV, Borase HP et al (2012) Larvicidal activity of silver nanoparticles synthesized using Plumeria rubra plant latex against *Aedes aegypti* and *Anopheles stephensi*. Parasitol Res 110:1815–1822. https://doi.org/10.1007/s00436-011-2704-x

Peralta-Videa JR, Hernandez-Viezcas JA, Zhao L et al (2014) Cerium dioxide and zinc oxide nanoparticles alter the nutritional value of soil cultivated soybean plants. Plant Physiol Biochem 80:128–135. https://doi.org/10.1016/j.plaphy.2014.03.028

Pérez-Hernández H, García-Mayagoitia S, Torres-Gómez PA et al (2022) Ecological effects of copper NPs: advantages and drawbacks regarding current and potential applications. In: Copper nanostructures: next-generation of agrochemicals for sustainable agroecosystems. Elsevier, pp 719–750

Prasad TNVKV, Sudhakar P, Sreenivasulu Y et al (2012) Effect of nanoscale zinc oxide particles on the germination, growth and yield of peanut. J Plant Nutr 35:905–927. https://doi.org/10.1080/01904167.2012.663443

Premanathan M, Karthikeyan K, Jeyasubramanian K, Manivannan G (2011) Selective toxicity of ZnO nanoparticles toward Gram-positive bacteria and cancer cells by apoptosis through lipid peroxidation. Nanomedicine 7:184–192

Rajput VD, Minkina T, Sushkova S et al (2018) Effect of nanoparticles on crops and soil microbial communities. J Soils Sediments 18:2179–2187. https://doi.org/10.1007/s11368-017-1793-2

Raliya R, Tarafdar JC (2013) ZnO nanoparticle biosynthesis and its effect on phosphorous-mobilizing enzyme secretion and gum contents in clusterbean (*Cyamopsis tetragonoloba* L.). Agric Res 2:48–57. https://doi.org/10.1007/s40003-012-0049-z

Rani S, Kumari N, Sharma V (2023) Uptake, translocation, transformation and physiological effects of nanoparticles in plants. Arch Agron Soil Sci 69:1579–1599. https://doi.org/10.1080/03650340.2022.2103549

Reidy B, Haase A, Luch A et al (2013) Mechanisms of silver nanoparticle release, transformation and toxicity: a critical review of current knowledge and recommendations for future studies and applications. Materials 6:2295–2350

Rodrigues SM, Trindade T, Duarte AC et al (2016) A framework to measure the availability of engineered nanoparticles in soils: trends in soil tests and analytical tools. TrAC Trends Anal Chem 75:129–140

Sheteiwy MS, Shaghaleh H, Hamoud YA et al (2021) Zinc oxide nanoparticles: potential effects on soil properties, crop production, food processing, and food quality. Environ Sci Pollut Res 28:36942–36966. https://doi.org/10.1007/s11356-021-14542-w

Shinde S, Paralikar P, Ingle AP, Rai M (2020) Promotion of seed germination and seedling growth of Zea mays by magnesium hydroxide nanoparticles synthesized by the filtrate from *Aspergillus niger*. Arab J Chem 13:3172–3182

Shukla P, Chaurasia P, Younis K et al (2019) Nanotechnology in sustainable agriculture: studies from seed priming to post-harvest management. Nanotechnol Environ Eng 4:11. https://doi.org/10.1007/s41204-019-0058-2

Shukla K, Mishra V, Singh J et al (2024) Nanotechnology in sustainable agriculture: a double-edged sword. J Sci Food Agric 104:5675–5688. https://doi.org/10.1002/jsfa.13342

Singh RP, Handa R, Manchanda G (2021) Nanoparticles in sustainable agriculture: an emerging opportunity. J Control Release 329:1234–1248. https://doi.org/10.1016/j.jconrel.2020.10.051

Singh A, Singh V, Arora M et al (2023) Nanomaterials for renewable energy. In: Metal oxide–based carbon nanocomposites for environmental remediation and safety. CRC Press, Boca Raton, pp 83–113

Tangaa SR, Selck H, Winther-Nielsen M, Khan FR (2016) Trophic transfer of metal-based nanoparticles in aquatic environments: a review and recommendations for future research focus. Environ Sci Nano 3:966–981

Teow Y, Asharani PV, Hande MP, Valiyaveettil S (2011) Health impact and safety of engineered nanomaterials. Chem Commun 47:7025–7038

Thongkumkoon P, Sangwijit K, Chaiwong C et al (2014) Direct nanomaterial-DNA contact effects on DNA and mutation induction. Toxicol Lett 226:90–97

Tripathi D, Ahmad P, Sharma S et al (2017) Nanomaterials in plants, algae, and microorganisms. Elsevier

Vankova R, Landa P, Podlipna R et al (2017) ZnO nanoparticle effects on hormonal pools in *Arabidopsis thaliana*. Sci Total Environ 593:535–542

Waalewijn-Kool PL, Ortiz MD, Lofts S, Van Gestel CAM (2013) The effect of pH on the toxicity of zinc oxide nanoparticles to *Folsomia candida* in amended field soil. Environ Toxicol Chem 32:2349–2355. https://doi.org/10.1002/etc.2302

Wang W-N, Tarafdar JC, Biswas P (2013) Nanoparticle synthesis and delivery by an aerosol route for watermelon plant foliar uptake. J Nanopart Res 15:1417. https://doi.org/10.1007/s11051-013-1417-8

Wang P, Menzies NW, Lombi E et al (2015) Silver sulfide nanoparticles (Ag2S-NPs) are taken up by plants and are phytotoxic. Nanotoxicology 9:1041–1049. https://doi.org/10.3109/17435390.2014.999139

Wang W, Liu J, Ren Y et al (2020) Phytotoxicity assessment of copper oxide nanoparticles on the germination, early seedling growth, and physiological responses in *Oryza sativa* L. Bull Environ Contam Toxicol 104:770–777. https://doi.org/10.1007/s00128-020-02850-9

Xie Y, He Y, Irwin PL et al (2011) Antibacterial activity and mechanism of action of zinc oxide nanoparticles against *Campylobacter jejuni*. Appl Environ Microbiol 77:2325–2331. https://doi.org/10.1128/AEM.02149-10

Xiong T, Dumat C, Dappe V et al (2017) Copper oxide nanoparticle foliar uptake, phytotoxicity, and consequences for sustainable urban agriculture. Environ Sci Technol 51:5242–5251. https://doi.org/10.1021/acs.est.6b05546

Xiu-mei LIU, Fu-dao Z, Shu-qing Z et al (2005) Responses of peanut to nano-calcium carbonate. J Plant Nutr Fert 11:385–389. https://doi.org/10.11674/zwyf.2005.0317

Xu C, Peng C, Sun L et al (2015) Distinctive effects of TiO2 and CuO nanoparticles on soil microbes and their community structures in flooded paddy soil. Soil Biol Biochem 86:24–33

Yan A, Chen Z (2019) Impacts of silver nanoparticles on plants: a focus on the phytotoxicity and underlying mechanism. Int J Mol Sci 20:1003

Yu J, Wang D, Geetha N et al (2021) Current trends and challenges in the synthesis and applications of chitosan-based nanocomposites for plants: a review. Carbohydr Polym 261:117904

Zhang W, Yao Y, Li K et al (2011) Influence of dissolved oxygen on aggregation kinetics of citrate-coated silver nanoparticles. Environ Pollut 159:3757–3762

Zhang M, Tao S, Wang X (2020) Interactions between organic pollutants and carbon nanomaterials and the associated impact on microbial availability and degradation in soil: a review. Environ Sci Nano 7:2486–2508

Zhao L, Sun Y, Hernandez-Viezcas JA et al (2013) Influence of CeO2 and ZnO nanoparticles on cucumber physiological markers and bioaccumulation of Ce and Zn: a life cycle study. J Agric Food Chem 61:11945–11951. https://doi.org/10.1021/jf404328e

Zhao L, Peralta-Videa JR, Rico CM et al (2014) CeO2 and ZnO nanoparticles change the nutritional qualities of cucumber (*Cucumis sativus*). J Agric Food Chem 62:2752–2759. https://doi.org/10.1021/jf405476u

Zhao S, Zhu X, Mou M, Wang Z, Duo L (2022) Assessment of graphene oxide toxicity on the growth and nutrient levels of white clover (*Trifolium repens* L.). Ecotoxicol Environ Saf 234:113399. https://doi.org/10.1016/j.ecoenv.2022.113399

Chapter 26
Ecotoxicity Potential of Carbon and Metal Nanoparticles: Delving into the Darker Realm

Bidisha Ghosh, Subhasis Sarkar, and Santanu Paul

Abstract The contribution of nanotechnology has grown significantly since the last century, and numerous reports highlighted its potential in various arenas. Being economically feasible, nanotechnology finds its generous application in diverse domains including medicine, bioimaging, biosensing, electrical devices and in remediating pollutants from environment. However, in contrast, there had been significant arguments over its stability and ecotoxicity which need to be pondered for its long-term affliction over human health. The harsh chemicals and other conditions, including high temperature and large amount of water and heat, were required for the synthesis of the nanoparticles. Subsequently, in turn, huge amounts of robust chemicals, heat, and toxic by-products were released into the environment as well. These result in bioaccumulation and transfer of nanoparticles into different trophic levels which consequently can lead to major health hazards including immune-modulatory defects, oxidative stress, and mortality. Hence, the ill-fated impact of metal nanoparticles and carbon-based nanomaterials on human health, flora, and fauna is the major module to be discussed in this chapter.

The original version of the chapter has been revised. A correction to this chapter can be found at https://doi.org/10.1007/978-981-97-9922-0_27

B. Ghosh
Department of Biotechnology, School of Life Sciences, Swami Vivekananda University, Bara Kanthalia, Barrackpore, West Bengal, India
e-mail: subhasiss@svu.ac.in

S. Sarkar (✉)
Department of Microbiology, School of Life Sciences, Swami Vivekananda University, Bara Kanthalia, Barrackpore, West Bengal, India
e-mail: subhasiss@svu.ac.in

S. Paul (✉)
Department of Biology, Indian Institute of Science Education and Research, Tirupati, India
e-mail: santanupaul@labs.iisertirupati.ac.in

S. Dey, S. Bhattacharya (eds.), *Biotechnological Interventions in the Removal of Emerging Pollutants*, Interdisciplinary Biotechnological Advances,
https://doi.org/10.1007/978-981-97-9922-0_26

Keywords Bioaccumulation · Ecotoxicity · Immunomodulatory defects · Nanomaterials · Oxidative stress · Trophic levels

26.1 Introduction

26.1.1 The Active Protagonist: Metal Nanoparticles and Carbon Nanoparticles

The field of nanotechnology and nanoparticles has given impetus to different branches of science and technology in the recent years. Particles with at least one dimension smaller than 100 nm are categorized as nanoparticles (Kreyling et al. 2006). Applications of nanotechnology can be found in diverse fields such as packaging (Espitia et al. 2012), agriculture (Singh et al. 2021), food processing (Chen et al. 2023), pharmaceuticals (Marszałł 2011), new materials development, and environmental protection. The use of nanoemulsion pesticides in agriculture (Feng et al. 2018), the use of nanoceramic devices (Guduru and Dixit 2024) for large surface areas in the food production sector, the use of merged nanoparticles into packages or nanosensors for intensive care and detection in the maintenance sector—all demonstrate how nanomaterials may be practical in most stages of the food production process. The use of nanoproducts has increased at a rapid pace as a result of accelerated industrialization. In a survey, it has been observed that engineered nanoparticles are largely been used by consumers, and in another survey, it has been observed that the global consumers for engineered nanoproducts is summed to a figure of about 11.5 million tons with the largest production with silver (Ag) nanoparticles (Ghobashy et al. 2021).

The food industry uses metal nanoparticles for storage, packing, and nanosensor manufacturing. It also uses united active nanoparticles for migration properties, nanopore filters for purification, nanoencapsulated additives, and nanosized food and nutrients as supplements. These are the most commonly used nanoparticles in the food industry. There are numerous research on the toxicological characteristics of nanoparticles. Numerous investigations have demonstrated that a portion of them are able to go past the different boundaries that keep living things safe. Once the inhaled nanoparticles have gotten past all of the digestive and respiratory defense systems, they may find their way into the circulation. Afterward, they are dispersed throughout the different organs and gather at specific locations. Just as they can get past cell barriers, they can go down the olfactory nerves and enter the brain directly (Zhang et al. 2016). The use of metal- and carbon-based nanomaterials is highly popular due to their unprecedented advancements and potential for diverse applications. Metal-based nanomaterials that typically consist of noble metals, primarily gold (Au), silver (Ag), and platinum (Pt), have always been efficient candidates in a plethora of industrial and medicinal applications. Among them, Au nanoparticle finds their application in the field of sensing, photonic, catalytic and huge therapeutic modalities in unison with Ag nanoparticles as well, which are used in various household applications like water filters to several antimicrobials (Shepard et al. 2020), electronic devices (Chen et al. 2009) to several antimicrobials (Durán et al. 2016). The philanthropic use of Ag nanoparticles finds its application in their impregnation in bandages, medical adhesives, and in prosthetics such as catheters (Gorup et al. 2017). Pt nanoparticles, on the other hand, have found various uses in heterogeneous catalysis, fuel cell applications, sensing of different analytes in samples, and as a dopant material in different nanomaterials (Kim et al. 2005) bringing forth enhanced activity.

Metal nanoparticles categorize the most popular nanomaterials like oxides of titanium (TiO_2), copper (CuO_x), and Zinc (ZnO). These nanomaterials are surface modified with the help of various organic and biological ligands and are been used for various purposes in modern day life. Mention should be made of TiO_2 nanoparticles that are synthesized with the help of various stabilizers and capping molecules that finds its use in varied usage. Being a wide gap semiconductor nanomaterial (Reddy et al. 2003), TiO_2 are used in remediating notable notorious pollutant in aqueous and in terrestrial ecosystems (Vijayaraj et al. 2018), in cosmetic products (Dreno et al. 2019), in various antibacterial treatment strategies (Vargas and Rodríguez-Páez 2017) with their concomitant advantages in cancer antidotes (Chu et al. 2019). Copper and zinc being essentials nutrients to the cellular system are also harnessed in a large number of biological applications in the form of their elemental states as well as their oxides. A lot of literature reports are available that clearly hints at CuO and ZnO nanoparticles, their industrial as well as their biomedical application. Elemental Copper (Cu) and CuO_x oxide nanoparticles have been prepared through various experimental routes and have been surface modified. The nanoparticles in nascent colloidal solution or in their solid state and in impregnated condition in films are used as fillers (Hassan et al. 2014), dielectric materials in nanodevices (Yan et al. 2020), catalysis (Astruc 2020), in removal of heavy metals (Simeonidis et al. 2018), and in drug delivery (Mitchell et al. 2021).

Recently, carbon nanomaterials have been widely hailed for their extensive properties in the field of nanotechnology. It has been predicted that by 2050 the energy requirement of the world will have to be doubled and thus carbon nanomaterials are found to be the apt participants of energy storage materials (Dai et al. 2012). Carbon nanostructures, predominately used in the form of carbon dots (CDs), have been utilized for a vast array of purposes with the introduction of defects through hetero atom doping with nitrogen, boron, phosphorous, and sulfur (Table 26.1).

Carbon nanotubes (CNTs) and carbon nanofibers (CNFs) are widely used for their vast array of applications. CNTs have high thermal conductivity and when embedded into matrix they augment wear and increase mechanical properties like stiffness (Garg et al. 2021). CNTs may be single or double walled depending on the temperature of synthesis, as multiwalled CNTs are synthesized at low temperatures (De Jong and Geus 2000). CNFs along with fullerenes are known to be close relatives of carbon nanotubes, can be of varied shapes. Unlike CNTs that have stack of carbon structures with graphene walls, the arrangement of CNFs is brought about by stacking the cones with large surface to volume ratio (Yadav et al. 2020). The

Table 26.1 Type of carbon dots, dopant material, and applications

Carbon source	Dopant material	Applications	References
Citric acid	Nitrogen	Solar cell	Dai et al. (2012)
Carbonized L-DOPA	Nitrogen	Live cell imaging	Wang et al. (2016)
Chionanthus retusus extract	Nitrogen	Sensing of Fe^{3+} ion	Park et al. (2016)
Acetic acid	Boron	Sensing potassium sorbate, Vit B12	Jia et al. (2019)
Catechol	Boron	Sensing intracellular Mg^{2+} in cells	Sadhanala et al. (2021)
Trisodium citrate	Phosphorous	Iron sensor	Kalaiyarasan et al. (2020)
Thiomalic acid	Sulfur	Bioimaging	Chandra et al. (2013)
Sucrose	Sulfur	Fe^{3+} in acidic environment	Naik et al. (2018)

Table 26.2 Recent advances in carbon nanomaterials, their mode of synthesis and applications

Carbon nanostructure	Synthesis route	Applications	References
CNT	Pyrolytic synthetic strategy	Hydrogen electrocatalysis	Li et al. (2024)
SmSe@CNT	Hydrothermal route	Water splitting	Alenad et al. (2024)
SWCNF	Electrospinning	Luteolin sensing	Zhang et al. (2023)
ZnO@CNF	Biosynthesis	Waste water treatment	Arabkhani et al. (2023)
CNF	Enzymatic hydrolysis	Electrode for supercapacitors	Dong et al. (2023)

widest recent applications of carbon nanotubes, carbon dots and carbon fibers encapsulates its use in optoelectronic devices, in energy harvesting devices in catalysis and in various medicinal applications which have been summarized in Table 26.2.

26.1.2 Opening the Pandora Box: A Question to the Benevolence of Metal and Carbon-Based Nanomaterials

Nanoparticles are highly mobile in the all realms of the ecosystem and this property is being brought about by the property of its stability. This relates to the fact that once destabilized in the solution, a nanoparticle will be much more prone to aggregation (Gimbert et al. 2007). In contrast to the different mineral nanoparticles of known importance metal nanoparticles are seen to accumulate more within the ecosystem and thus arises the predicament in terms of their degradability (Rana and

Kalaichelvan 2013). It has been reported that nanoparticles in contact with biological cells and molecules induces particle aggregation, enzymatic degradation and opsonization (Kolosnjaj-Tabi et al. 2016). The results of toxicity are observed at the cellular, organ or at the protein levels. This brings forth queries regarding their bio toxicity and propensity to induce an inflammatory response. Metal nanoparticles most notably Ag, TiO_2 nanoparticles are seen to afflict the central nervous system through the release of proinflammatory cytokines by the activation of glial cells resulting into "neuroinflammation" (Wu and Tang 2018). Similar scenarios are observed with the exposure of ZnO nanoparticles where marked increase in the TNFα and IL-1β and subsequent decrease in glutathione brings about immunomodulatory defects (Senapati et al. 2015). Concentrations about 200 μg/mL of Carbon dots has resulted into alarming consequences of yolk sac edema, spinal cord flexure, and death of zebra fish (Liu et al. 2020).

26.1.3 Assessment of Nanoparticle Toxicity: Fate of the Nanoparticles Within the Living System

The toxicity assessment imposed by nanoparticle can be done in various organisms. The following section shows different classes of nanoparticles used on experimental models and toxic effects imposed by nanoparticles. Toxicity assessment can be achieved by in vitro and in vivo methods.

26.2 In Vitro Assessment Methods

In order to assess the toxicity of nanoparticle, in vitro assay has been an important method. This assessment provides cost-effective, rapid solution with the least ethical concerns. In context, numerous studies get recognized including proliferation assay, apoptosis assay, necrosis assay, oxidative stress assay, and DNA damage assay.

26.2.1 Proliferation Assays

To study cellular metabolism, metabolically active cells are important. Studying in vitro toxicity of nanoparticle requires MTT assay (Sayes et al. 2007). This technique has got attention across the globe as it requires less time, easy to reproduce and involving less manipulation of cells (Marshall et al. 1995). The test is requiring tetrazolium salt with addition of culture media additives. In this assay, formazan is produced (Molinari et al. 2003). Another method of cell proliferation is [^{3}H] thymidine incorporation. However, this technique exhibit toxicity and thus avoided as its expensive (Hussain et al. 2006). Beside this, the assessment of cellular redox potential using alamarBlue is a potential method in compared to MTT assay which involves simple way of sample preparation (Punshon et al. 2005). Clonogenic assay is conducted by counting the proliferating cells by visual inspection after the exposure of nanoparticle to the cells (Casey et al. 2007).

26.2.2 Apoptosis Assay

Toxicity assessment of nanoparticle is achieved by apoptosis assay as apoptosis is one of the potential biomarker. The generation and accumulation of free radical inside the cell is recognized as the cause of apoptosis and DNA damage (Ryter et al. 2007; Li and Osborne 2008). Oxidative stress inside the cell culture environment, is one of the causes (Ryter et al. 2007). Scientific records show that the nanoparticle exposure is responsible for apoptosis. To study apoptosis, exposure of nanoparticle in vitro is carried out in embryonic mouse model (Ahamed et al. 2008). In context, caspase 3 and caspase 9 are considered as biomarker are examined in *Drosophila melanogaster* model system. In parallel, the upregulation of p53 gene was also found to be happened after the exposure of silver nanoparticle in *D. melanogaster.* Cell death involving apoptosis is mediated via the upregulation of genes (Liang and Clarke 2001; Sherr 2004). Also, a few methods like Comet assay (Jin et al. 2007), TdT-mediated dUTP-biotin nick end labeling (TUNEL) assay may be considered for inspecting the morphological changes during apoptosis. The endonuclease cleavage products of apoptosis is observed using DNA laddering technique (Wyllie 1980). The cell size and DNA fragmentation process is confirmed if induction of apoptosis is happened. In that case, agarose gel electrophoresis pinpoint on apoptosis and helps in distinguishing between apoptosis and necrosis (Kerr et al. 1972; Suman et al. 2012).

The fragmentation of genomic DNA during necrosis of cells can be observed via electrophoresis and the ladder like electrophoretic pattern in the gel reveal the apoptotic internucleosomal DNA fragmentation (Abdel-Khalek 2016). In this context, annexin-V and propidium iodide (PI) are considered as potential cell death marker in toxicity assessment. The foundation of the assessment lies in the interaction between annexin-V and phosphatidylserine and indicting the externalization of the plasma membrane by increased fluorescence. The underlying mechanism of externalization of plasma membrane is induced by caspase-dependent pathway. PI staining is relevant in determining the late stage of apoptosis as it is able to stain the nucleus only while the membrane is lost. The exposure of silica nanoparticle on human HepG2 hepatoma cells is determined by morphological changes in nucleus (Lu et al. 2011). The induction of apoptosis can be also be monitored when HeLa cell lines are treated with gold nanoparticle using Annexin V/PI (Baharara et al. 2016).

Beside this, comet assay (single-cell gel electrophoresis) is a significant tool for detection of the mutagenic ability of any test sample (Kisin et al. 2007). This technique is aimed at detecting the strand breakage in DNA molecule (Bajpayee et al. 2013; Kumar et al. 2013) and also in monitoring the extent of oxidative DNA damage (Kumar et al. 2013; Dusinska and Collins 2008; Pavanello and Clonfero 2000). The basic principle of this method lies in monitoring the migration of damaged DNA fragments outside of the cell, while intact DNA molecule remained inside the cell nucleus. DNA damage is assessed in terms of the size and shape of the tail and the distribution of DNA within the comet (Fairbairn et al. 1995; Lemay and Wood 1999; Malyapa et al. 1998). Here, the cell lysis process removed protein and therefore, damaged DNA is allowed to move from nucleus. Staining with the florescent dye, specific to the DNA molecule is used. The level of fluorescence in head, tail, and tail length is analyzed in the gel (Henderson et al. 1998; Collins et al. 1995; Singh and Stephens 1997). The wide usage of comet assay in determining the toxicity of zinc oxide nanoparticle at 25 mg Z/L concentration on *D. tertiolecta* revealed the 55% nuclei damage (Schiavo et al. 2016). Comet assay can also be used to measure the toxicity imposed by SiO_2 nanoparticles on *D. tertiolecta* at 125 mg/L which is an indication of genotoxicity after 72 hours (Handy et al. 2012). Another experimental approach, TUNEL assay is one of the potential methods for understanding DNA damage in situ (Gavrieli et al. 1992). This assay is a method for staining cells undergoing apoptosis and internucleosomal DNA fragmentation (Arends et al. 1990; Bortner et al. 1995; Loo 2002). The underlying principle of TUNEL assay lies in the ability of enzyme terminal deoxynucleotidyl transferase to incorporate labeled dUTP into free 3′-hydroxyl termini, the result of DNA fragmentation (Gavrieli et al. 1992). Other than apoptosis study, DNA damage can also be monitored in nonapoptotic cells including necrotic cell death, induced by the action of toxicants (Ansari et al. 1993). Also, the TUNEL assay is used to study the DNA repair in cells (Kanoh et al. 1999). The understanding of pancreatic islet function of Goto Kakizaki rats can be estimated of TUNEL assay while insulin loaded selenium nanoparticle (25 IU/kg) is exposed to the pancreatic β-cells daily for 2 weeks and the result showed the decrease in apoptotic cells from 17.6% to 2.3% after the treatment regime (Deng et al. 2017).

Carbon nanoparticle exhibits toxicity and biological response of the cells (Donaldson et al. 2006; Pulskamp et al. 2007). One of the known carbon-based nanomaterials is fullerenes. Their ubiquitous existence in our environment released from fuel combustion. The distribution of nonfunctionalized fullerones C60 are prevalent in all sot of tissues and in long term exposure, the accumulation is noticed in important organs, the liver, kidney, bones, and spleen (Qingnuan et al. 2002; Cagle et al. 1999). In vitro studies revealed that the fullerenes may exert genotoxicity. The underlying mechanism of genotoxicity lies in the breakage of DNA strands, chromosomal damage, and micronucleus formation. This is achieved after exposure of fullerenes (1 ng/mL) to Chinese hamster ovary cells, human epidermoid-like carcinoma cells and human embryonic kidney cells (HEK293) for 80 days (Dhawan et al. 2006; Niwa and Iwai 2006). In contrast, another study revealed that the fullerenes exhibit no significant effect on DNA strand breakage as evidenced by comet assay (Jacobsen et al. 2008). Carbon NPs cannot be used extensively due to the lack of comprehensive evaluation of toxicity results.

26.2.3 Necrosis Assay

Necrosis is evaluated in terms of the integrity of membrane and widely used in determining the viability of cells. The integrity of membrane is assessed with the help of a dye Neutral Red and Trypan Blue (Monteiro-Riviere et al. 2005; Huang et al. 2004). There is an urgent need for a reliable, rapid, and reproducible quantitative in vitro assay for screening of nanoparticle. The neutral red is a dye which readily diffuses across the plasma membrane. It accumulates inside the lysosome and is bale to bind by electrostatic hydrophobic interactions (Bitensky 1963; Nemes et al. 1979). Alterations of the cell surface lead to cause fragile lysosomal membrane. These changes take place in exposure of xenobiotics or nanoparticle (Nemes et al. 1979; Lüllmann-Rauch 1979). In turn, the uptake and binding of Neutral Red gets reduced. The difference between dead and live cell can be obtained by standard methods (Borenfreund and Puerner 1985). The stability of endosome-lysosome can be measured by neutral red assay which is evidenced by decreasing trend after the exposure of silver nanoparticle (30–81%). The live and dead cell can be distinguished by Trypan blue as it stains the dead cell by entering inside while remain outside of the live cell. Trypan blue exclusion method is widely used as the dye enters dead cells and stay outside the living cells (Strober 2001). This test confirms the stability of cell membrane. The exposure of zinc nanoparticle give rise to cytotoxicity at 369 μm and higher (Kononenko et al. 2017).

26.2.4 Oxidative Stress Assay

Nanoparticles are responsible for generation of reactive oxygen species (ROS) and reactive nitrogen species (RNS) at cellular level (Magder 2006). Therefore, the determination of ROS and NOS has become so important. The determination of ROS and NOS involving the reaction of 2,2,6,6-tetramethylpiperidine (TEMP) with O_2^- stable radical and this can be detected using X-band electron paramagnetic resonance (EPR) (Zang et al. 1995). This particular method is still expensive and therefore, fluorescent probe molecules have been recognized as an alternative and cost effective way (Gomes et al. 2005). Though there is limitation since the florescence molecules are inefficient to react with a variety of reactive species which ultimately produces ambiguous results (Halliwell and Whiteman 2004). Toward resolving this issue, the use of 2′,7′-dichlorofluorescein diacetate (DCFDA), a non-fluorescent probe, is being explored. This DCFDA is reactive to HO·, RO·, ROO·, and H_2O_2 (Wagner et al. 2007). The estimation of oxidative stress is carried out by lipid peroxidation C11-BIODIPY assay and TBA assay for malondialdehyde

(Fantel 1996). A couple of assays are relevant in this contrast which include Amplex Red assay, to determine lipid hydro peroxide's measurement, antioxidant depletion measurement using 5,5′-dithiobis-(2-nitrobenzoic acid) (DTNB) and superoxide dismutase (SOD) activity by Nitro blue tetrazolium assay (Hussain et al. 2006).

26.3 In Vitro Toxicity of Nanoparticles

26.3.1 Cell Viability and Lethality

Carbon nanoparticles, or CNTs, are the most widely employed type of nanoparticle when evaluating the lethality and viability of cells. Because of their special qualities, they are widely employed in chemical, industrial, and biological applications (Guo et al. 2012; Sathyanarayana and Hübner 2013). Both single-walled and multi-walled carbon nanotubes (SW- and MW-CNTs) are produced from them (Madani et al. 2013). Studies on a variety of bacteria have shown that CNTs have antimicrobial qualities because of the mechanical harm that the nanotubes produce (Amarnath et al. 2012). Functionalized CNTs have been shown to influence soil bacterial diversity (Kerfahi et al. 2015), according to a recent study. The results of the toxicity tests conducted on a fish (*Oryzias latipes*), a microcrustacean (*Daphnia magna*), and freshwater microalgae (*Raphidocelis subcapitata* and *Chlorella vulgaris*) showed that the algae's growth was inhibited with effective SW-CNT concentrations of 30.96 mg/mL and 29.99 m/L, respectively (Sohn et al. 2015). Rat mesenchymal stem cells, human macrophages, murine macrophages, and human hepatocellular carcinoma cells have all been shown to be hazardous to the iron oxide-based nanoparticles. For 2 hours, iron oxide nanoparticles were hazardous to mouse macrophage cells at concentrations between 25 and 200 μg/mL. One of the effects that the study saw was a decline in cell viability (Naqvi et al. 2009).

A different study found that treating murine macrophage cells with 0.1 mg/mL iron oxide nanoparticles for 7 days reduced the cells' viability (Jeng and Swanson 2006). Cell viability was found to have decreased in another toxicity investigation that used rat mesenchymal stem cells at 0.1 mg/mL for 2 days (Delcroix et al. 2009). The outcomes revealed a possible decline in cell viability (Park et al. 2010).

26.3.2 Effects on Cell Lines

On a variety of cell lines, the toxicity of nanoparticles was assessed. Researchers have studied the effects of self-assembled nanotubes (SW-CNTs) on human cell lines, such as human HEL 293 cells, human macrophage cells, human HEK cells, human A549 cells and human epithelial-like Hela cells (Yehia et al. 2007; Davoren et al. 2007). To assess the toxicity, lung fibroblast cells were subjected to CNT treatment in a study (Kisin et al. 2007). Another study found that 72 hours of exposure to 250–500 μg/mL of SW-CNTs caused oxidative response and inflammatory response-induced membrane damage in A549 cells (Choi et al. 2009). In vitro reduction of inflammatory mediators such as IL-6, IL-8, and MCP-1 was documented in another study (Herzog et al. 2009). Human epidermal keratinocytes were also used to assess the effects of multiwall carbon nanotubes (Monteiro-Riviere et al. 2005).

Pro-inflammatory effects that are aided by NF-κB and ROS have been proposed as the mechanism of toxicity caused by multiwall carbon nanotubes (Ye et al. 2009). Numerous toxicological effects of MW-CNTs, including as oxidative stress, DNA damage, and apoptosis in mammalian cell lines, have been observed in vitro investigations. In human aortic endothelial cells, other effects include actin filament integrity and the distribution of VE-cadherin (Cveticanin et al. 2009; Patlolla et al. 2010a, b; Ravichandran et al. 2013; Reddy et al. 2010).

In MRC-5 human lung fibroblasts, gold nanoparticles (AuNPs) promoted autophagy together with oxidative stress (Li et al. 2010; Tarantola et al. 2008). In one study, the cytotoxicity of semiconductor and metal nanoparticles on animal cells was demonstrated using the motility of the cells. The investigation makes use of electrical cell substrate impedance analysis's potential as an excellent tool for estimating the in vitro cytotoxicity of gold nanorods and quantum dots. Dark field microscopy and fluorescence were used to validate the procedure (Tarantola et al. 2008). The study examined the toxicity of starch-coated silver nanoparticles using human lung fibroblast cells (IMR-90) and human glioblastoma cells (U251). DNA damage and ATP levels decreased in a dose-dependent manner as a result of the study. According to the study, cell cycle arrest in the G2/M phase is the result of silver nanoparticle interaction and deposition on DNA, which causes DNA damage (AshaRani et al. 2008). A cytotoxic investigation conducted on NIH3T3 fibroblast cells revealed that silver nanoparticles caused mitochondria-dependent apoptosis linked to reactive oxygen species and JNK activation (Hsin et al. 2008). The human hematoma cell line HepG2 was used to examine the toxicity of silver nanoparticles using DNA microarray analysis, viability assays and the micronucleus test (Kawata et al. 2009). A study revealed that silver nanoparticles were harmful to HeLa cells. The genes for oxidative stress, mt-2A, and ho-1 were upregulated as a result of exposure (Miura and Shinohara 2009). The outcomes of treating E. coli with three different kinds of silver nanoparticles revealed impaired risk gene replication function (Yang et al. 2009). According to the study, CdTe quantum dots were shown to be cytotoxic, causing significant lysosomal damage as well as ROS generation. The findings indicated that the release of IL-1β, IL-6, and IL-8 was improved by quantum dots coated with carboxylic acid. It was found that the main factor influencing immunotoxicity and cytotoxicity in HEKs is surface coating.

The human brain microvascular endothelial cells (HBMVECs) were treated with aluminum nanoparticles in the size range of 1–10 μM for a full day. According to the study, the treatment led to a decline in cell viability, oxidative stress, and mitochondrial function (Chen et al. 2004). To investigate the harmful effects, 10–400 μg/mL of aluminum nanoparticles were applied to mammalian cells. According to the reported data, there was no discernible harm on cell viability during the investigation (Radziun et al. 2011). In an additional investigation, the vitality of cells was assessed in relation to the interaction between human bone marrow-derived mesenchymal stem cells (HMSC) and aluminum nanoparticles within the 25–40 μg/mL range. The study's findings showed that the viability of the cells had decreased (Alshatwi et al. 2012). After 2 hours of treatment, aluminum nanoparticles at a concentration of 0–5000 μg/mL were implicated in DNA damage, according to a study done on mammalian cell lines (Dhawan et al. 2006).

26.4 In Vivo Toxicity Assessment Methods

Toxicity assessment can be done in animal models such as mice and rat. This is in vivo toxicity assessment method. Biodistribution, clearance, hematology, serum chemistry, and histopathology are popular methods of toxicity assessment (Kim et al. 2001). The excretion and metabolism of nanoparticle at various time points after the exposure is assessed (Li et al. 2001). In vivo toxicity assessment can also be conducted by the examination of changes in serum chemistry and also the changes of cell type after the exposure of nanoparticle (Baker et al. 2008). Nanoparticle toxicity is also assessed by histopathology of cell while tissue and organ is exposed to nanoparticle (Lei et al. 2008). The examination oriented to histopathology is conducted in several tissues such as lung, eyes, brain, liver, kidneys, heart, and spleen after the exposure of nanoparticles (Baker et al. 2008; Zhu et al. 2008). The advancement of technology gives rise to the another assessment technique known to us as micro-electrochemistry and microfluidics (Ewing et al. 1983).

26.4.1 In Vivo Toxicity of Nanoparticles

The dosage, duration, and exposure circumstances of nanoparticles all affect their toxicity. In a study, 1 hour of exposure to 1–10 mm size SWCNTs was given to 3T3 cells. Twenty percent of the treated cells were viable (Tkachenko et al. 2004). In a different study, SWCNTs were applied to immortalized epidermal keratinocytes with 80% consistency for 2, 4, 6, and 8 hours at concentrations ranging from 0.06 to 0.24 mg/mL. After 4 hours, the treatment reduced vitality, and when exposed to nanoparticles at a concentration of 0.24 mg/mL, viability was around 65% (Shukla et al. 2005). Following 24–120 hours of treatment with SWCNTs at concentrations between 0.7812 and 200 μg/mL, it was found that the cytotoxicity was dose- and time-dependent, resulting in G1 cell cycle arrest in 43.5% of the treated cells after day 1 (Pernodet et al. 2006). However, when SWCNTs in the 0–7.3 μg/mL range were applied to mouse peritoneal macrophage-like cells for 4, 8, 12, and 18 hours, the cells consumed the nanoparticles without experiencing any harmful effects (Su et al. 2007). The study found that guinea pig alveolar macrophages treated with

SWCNTs at 1.41–226 μg/cm² for 3 hours showed necrosis at 3.06 μg/cm² and cytotoxicity at 0.38 μg/cm² (Gupta and Wells 2004).

The cytotoxicity of SWCNTs was shown to be dose-dependent, and treatment of rat alveolar epithelial cells (NR8383) and human alveolar epithelial cells (A549) at 5–100 μg/mL for 24–96 hours resulted in a 60–80% drop in cell number (Hu et al. 2006). The strongest negative effect of SWCNTs was seen when HEKs were treated during 24–120 hours at 0.8–100 μg/mL. After being treated for 1, 3, and 5 days, the study's findings showed 79%, 50%, and 31% viability at 100 μg/mL, respectively. A variety of cell lines were used to assess the toxicity of MWCNTs. A study showing ~73% vitality at 0.4 mg/mL of MWCNTs with 80% consistency at 0.1–0.4 mg/mL for 1, 4, 8, 12, 24, and 48 hours was conducted on HEK cells. It was shown that the toxicity was dose dependent when human skin fibroblasts with 70% confluency were exposed to pure MWCNTs at concentrations of 0.06–0.6 mg/mL for 24 and 48 hours (William et al. 2006). HeLa and 3T3/NIH cells exposed to gold nanoparticles for 3 hours at 150 pm saw a 20% and 5% reduction in cell viability, respectively (Hoshino et al. 2004). Gold nanoparticles with a particle size range of 10–100 μm were exposed to macrophage cells for a duration of 24–72 hours in a different investigation. The research shown that within 72 hours, the cell viability was reduced to 85% by the 100 μm nanoparticles (Guo et al. 2007). Human dermal fibroblasts were employed in a study to evaluate the toxicity of gold nanoparticles at a concentration of 0–0.8 mg/mL for a period of 2–6 days. The study's findings showed that the cell area and density decreased in a dose-dependent manner (Lovrić et al. 2005). Studying the cytotoxic effects also involved the usage of gold nanoshells. In a study, Vero cells exposed to gold nanoshells for 6 and 24 hours at concentrations between 0.001 and 200 μg/mL showed a reduction in cell viability (Kim et al. 2009).

Fe_3O_4 nanoparticles' exposure to human fibroblast cells at 0–1000 μg/mL for 24 hours resulted in 25–50% reduction in cell viability (Kim et al. 2009). In another study conducted on mouse macrophages, Fe_3O_4 nanoparticles were exposed at 0.2 mg/mL for 1 and 4 days. The study demonstrated that cytotoxicity was dose dependent. When human breast carcinoma SK-BR-3 cells were exposed to Fe_3O_4 nanoparticles at 10–400 nm for 1–48 hours, the study resulted in 91% viability (Kim et al. 2010).

26.4.2 Effects on Organ Systems

The way that CNTs are administered seems to affect their effect in in vivo models. Numerous study teams have discovered that lung cancer, emphysema, bronchitis, and asthma may arise from breathing in nanoparticles. The intestinal system can allow nanoparticles to enter, which can cause Crohn's disease and colon cancer. Moreover, it has been found that heart illness and blood clots may arise from the circulatory system's exposure to the nanoparticles (Madani et al. 2013; Buzea et al. 2007). A few studies evaluating the in vivo toxicity of carbon nanotubes toward animals used guinea pigs, mice, and rats as their study subjects. On guinea pigs, the effects of soot-containing carbon nanotubes containing Co/Ni were noted (Huczko et al. 2001). Many researchers studied the effects of metal-containing single-walled carbon nanotubes in mice and rats (Lam et al. 2004, 2006; Shvedova et al. 2005, Warheit et al. 2004). Acute lung toxicity of intratracheally injected SW-CNTs in rats was assessed in one of the aforementioned investigations (Warheit et al. 2004). In this investigation, transitory inflammatory responses, multifocal granulomas, and cell damage were noted. A different study found that there was an increase in proinflammatory cytokines, macrophage and lymphocyte infiltration, early neutrophil accumulation, and fibrogenic transforming growth factor (Shvedova et al. 2005).

A rat investigation was also done to assess the impact of MW-CNTs (Muller et al. 2005). Rats were given MW-CNTs intratracheally during the trial. The investigation assessed the fibrotic responses, lung persistence and inflammation from a histological and biochemical perspective. Collagen-rich granulomas were identified in the bronchial lumen, indicating the presence of pulmonary lesions. Additionally, stimulation for TNF-α production was noted. It has also been noted that carbon nanotubes are toxic to aquatic species like rainbow trout (Smith et al. 2007). For 10 days, rainbow trout in the study received 0.1–0.5 mg/L of SWCNT treatment. According to the study, the SW-CNTs are a respiratory toxicant that causes cell cycle abnormalities and neurotoxicity. Additionally, MW-CNTs have been linked to chromosomal abnormalities and increased micronuclei frequency, as well as the induction of allergy responses in mice, the activation of cyclooxygenase enzymes through the inhibition of systemic immune function in the spleen, and changes in hepatic gene expression (Ji et al. 2009; Mitchell et al. 2009; Nygaard et al. 2009; Park et al. 2009; Patlolla et al. 2010a, b). Additional effects of MW-CNTs include apoptosis, phenotypic abnormalities, bacterial toxicity and the development of aberrant spinal cords in zebrafish embryos (Asharani et al. 2008; Cheng et al. 2009; Kang et al. 2008).

Gold nanoparticle exposure in vivo has been linked to sperm head and tail penetration ability, acute inflammation and apoptosis in the liver, and bioaccumulation in other organs (Cho et al. 2009; Lasagna-Reeves et al. 2010; Wiwanitkit et al. 2009). PEG-coated gold nanoparticles measuring 13 nm were employed in a study to monitor liver damage. Particles were found to collect in the spleen and liver for up to 7 days following injection. The liver experienced inflammation and apoptosis as a result of these nanoparticles. The investigation additionally revealed the existence of PEG-coated gold nanoparticles in liver lysosomes and splenic macrophages (Cho et al. 2009). According to in vivo research, exposure to silver nanoparticles causes astrocyte swelling, disruption of the blood–brain barrier, oxidative stress brought on by free radicals, changes in gene expression, and neuronal degeneration (Rahman et al. 2009).

Through subcutaneous injection, the distribution and accumulation of silver nanoparticles in rats were studied. According to the findings, the liver, spleen, kidney, brain, and lung all had particles (Tang et al. 2009). Rats were used to study the permeability of the blood–brain barrier and the capacity to develop edema through intraperitoneal, intracerebral, and intravenous injection. The findings indicated that by disrupting the blood–brain barrier in vivo, nanoparticles could cause edema development in the brain (Sharma et al. 2010). Silver nanoparticles (25 nm) were used in a study to assess the effects of the material on the expression of certain genes in the various brain areas of mice. According to the findings, free radical production by the silver nanoparticles may have caused neurotoxicity (Rahman et al. 2009). The findings showed that the kidney, gills, and liver all absorbed silver nanoparticles.

Through subcutaneous injection, the distribution and accumulation of silver nanoparticles in rats were studied. According to the findings, the liver, spleen, kidney, brain, and lung all had particles (Tang et al. 2009). Rats were used to study the permeability of the blood–brain barrier and the capacity to develop edema through intraperitoneal, intracerebral, and intravenous injection. The findings revealed that by disrupting the blood–brain barrier in vivo, nanoparticles could cause edema development in the brain (Sharma et al. 2010). Silver nanoparticles (25 nm) were used in a study to assess the effects of the material on the expression of certain genes in the various brain areas of mice. According to the findings, free radical production by the silver nanoparticles may have caused neurotoxicity (Rahman et al. 2009).

26.5 Conclusion

The rapidly expanding science of nanotechnology will enable the creation of materials with novel features. Based on the evidence that is now available, it appears that certain insoluble nanoparticles have the ability to withstand various barriers, travel throughout the body, and gather in many organs. Nanoparticles can be found throughout the body, even inside cells, while toxic effects have previously been shown at the pulmonary, cardiac, reproductive, renal, cutaneous, and cellular levels. These particles provide a large contact surface per mass unit due to their minuscule size. Additionally, the charge at the membrane surface determines the cytotoxicity that nanoparticles have on cells. The electrostatic interaction of nanoparticles with membranes and their interaction with the cytoplasm may be the cause of nanotoxicity. Therefore, current article has been able to discuss and forecast the NP toxicity in addition to exploring ways to more precisely characterize the molecular interactions with NPs at the cellular level.

Acknowledgments The authors are highly thankful to Swami Vivekananda University and IISER Tirupati for constant support and encouragement to complete this study.

Conflict of Interest The authors declare that they have no conflict of interest.

References

Abdel-Khalek AA (2016) Comparative evaluation of genotoxic effects induced by CuO bulk and nano-particles in *Nile tilapia, Oreochromis niloticus*. Water Air Soil Pollut 227(1):35

Ahamed M et al (2008) DNA damage response to different surface chemistry of silver nanoparticles in mammalian cells. Toxicol Appl Pharmacol 233(3):404–410

Alenad AM, Aman S, Ahmad N, Rashid AR, Abid AG, Manzoor S et al (2024) Facile synthesis of SmSe2 over multiwalled carbon nanotubes for efficient water-splitting applications. J Korean Ceram Soc 61(1):44–54

Alshatwi AA et al (2012) Al_2O_3 nanoparticles induce mitochondria-mediated cell death and upregulate the expression of signaling genes in human mesenchymal stem cells. J Biochem Mol Toxicol 26(11):469–476

Amarnath S, Hussain MA, Nanjundiah V, Sood AK (2012) β-Galactosidase leakage from Escherichia coli points to mechanical damages likely cause of carbon nanotube toxicity. Soft Nanosci Lett 2(3):41–45

Ansari B et al (1993) In situ end-labelling detects DNA strand breaks in apoptosis and other physiological and pathological states. J Pathol 170(1):1–8

Arabkhani P, Saeedi N, Sadeghi H, Nouripour-Sisakht S, Gharaghani M, Asfaram A (2023) Plant extracts-mediated green synthesis of zinc oxide/carbon nanofiber nanocomposites with highly efficient photocatalytic and antimicrobial properties for wastewater treatment. J Water Process Eng 54:104020

Arends M, Morris R, Wyllie A (1990) Apoptosis. The role of the endonuclease. Am J Pathol 136(3):593

AshaRani P et al (2008) Cytotoxicity and genotoxicity of silver nanoparticles in human cells. ACS Nano 3(2):279–290

Astruc D (2020) Introduction: nanoparticles in catalysis. Chem Rev 120(2):461–463

Baharara J et al (2016) Induction of apoptosis by green synthesized gold nanoparticles through activation of caspase-3 and 9 in human cervical cancer cells. Avicenna J Med Biotechnol 8(2):75

Bajpayee M, Kumar A, Dhawan A (2013) The comet assay: assessment of in vitro and in vivo DNA damage. In: Dhawan A, Bajpayee M (eds) Genotoxicity assessment, Methods in molecular biology (methods and protocols), vol 1044. Humana Press, Totowa, pp 325–345

Baker GL et al (2008) Inhalation toxicity and lung toxicokinetics of C60 fullerene nanoparticles and microparticles. Toxicol Sci 101(1):122–131

Bitensky L (1963) The reversible activation of lysosomes in normal cells and the effects of pathological conditions. In: Ciba foundation symposium—lysosomes. Wiley Online Library, pp 362–383

Borenfreund E, Puerner JA (1985) Toxicity determined in vitro by morphological alterations and neutral red absorption. Toxicol Lett 24(2–3):119–124

Bortner CD, Oldenburg NB, Cidlowski JA (1995) The role of DNA fragmentation in apoptosis. Trends Cell Biol 5(1):21–26

Buzea C, Pacheco II, Robbie K (2007) Nanomaterials and nanoparticles: sources and toxicity. Biointerphases 2(4):MR7–MR71

Cagle DW, Kennel SJ, Mirzadeh S, Alford JM, Wilson LJ (1999) In vivo studies of fullerene-based materials using endohedral metallofullerene radiotracers. Proc Natl Acad Sci USA 96(9):5182–5187

Casey A et al (2007) Spectroscopic analysis confirms the interactions between single walled carbon nanotubes and various dyes commonly used to assess cytotoxicity. Carbon 45(7):1425–1432

Chandra S, Patra P, Pathan SH, Roy S, Mitra S, Layek A et al (2013) Luminescent S-doped carbon dots: an emergent architecture for multimodal applications. J Mater Chem B 1(18):2375–2382

Chen Y et al (2004) Comparing study of the effect of nanosized silicon dioxide and microsized silicon dioxide on fibrogenesis in rats. Toxicol Ind Health 20(1–5):21–27

Chen D, Qiao X, Qiu X, Chen J (2009) Synthesis and electrical properties of uniform silver nanoparticles for electronic applications. J Mater Sci 44:1076–1081

Chen J, Guo Y, Zhang X, Liu J, Gong P, Su Z et al (2023) Emerging nanoparticles in food: sources, application, and safety. J Agric Food Chem 71(8):3564–3582

Cheng C, Müller KH, Koziol KK, Skepper JN, Midgley PA, Welland ME, Porter AE (2009) Toxicity and imaging of multi-walled carbon nanotubes in human macrophage cells. Biomaterials 30(25):4152-4160

Cho W-S et al (2009) Acute toxicity and pharmacokinetics of 13 nm-sized PEG-coated gold nanoparticles. Toxicol Appl Pharmacol 236(1):16–24

Choi S-J, Oh J-M, Choy J-H (2009) Toxicological effects of inorganic nanoparticles on human lung cancer A549 cells. J Inorg Biochem 103(3):463–471

Chu X, Mao L, Johnson O, Li K, Phan J, Yin Q et al (2019) Exploration of TiO2 nanoparticle mediated microdynamic therapy on cancer treatment. Nanomedicine 18:272–281

Collins AR, Ai-Guo M, Duthie SJ (1995) The kinetics of repair of oxidative DNA damage (strand breaks and oxidised pyrimidines) in human cells. Mutat Res 336(1):69–77

Cveticanin J et al (2009) Using carbon nanotubes to induce micronuclei and double strand breaks of the DNA in human cells. Nanotechnology 21(1):015102

Dai L, Chang DW, Baek JB, Lu W (2012) Carbon nanomaterials for advanced energy conversion and storage. Small 8(8):1130–1166

Davoren M et al (2007) In vitro toxicity evaluation of single walled carbon nanotubes on human A549 lung cells. Toxicol In Vitro 21(3):438–448

De Jong KP, Geus JW (2000) Carbon nanofibers: catalytic synthesis and applications. Catal Rev 42(4):481–510

Delcroix GJ-R et al (2009) Mesenchymal and neural stem cells labeled with HEDP-coated SPIO nanoparticles: in vitro characterization and migration potential in rat brain. Brain Res 1255:18–31

Deng W et al (2017) Selenium nanoparticles as versatile carriers for oral delivery of insulin: insight into the synergic antidiabetic effect and mechanism. Nanomedicine 13(6):1965–1974. https://doi.org/10.1016/j.nano.2017.05.002

Dhawan A, Taurozzi JS, Pandey AK, Shan W, Miller SM, Hashsham SA, Tarabara VV (2006) Stable colloidal dispersions of C60 fullerenes in water: evidence for genotoxicity. Environ Sci Technol 40(23):7394–7401

Donaldson K, Aitken R, Tran L, Stone V, Duffin R, Forrest G, Alexander A (2006) Carbon nanotubes: a review of their properties in relation to pulmonary toxicology and workplace safety. Toxicol Sci 92(1):5–22

Dong Y, Wang W, Wang W, Ma D, Ma S, Wang C et al (2023) Synthesis of activated carbon nanofibers by bio-enzymatic method as electrode material for supercapacitors. Int J Electrochem Sci 18(3):100024

Dreno B, Alexis A, Chuberre B, Marinovich M (2019) Safety of titanium dioxide nanoparticles in cosmetics. J Eur Acad Dermatol Venereol 33:34–46

Durán N, Durán M, De Jesus MB, Seabra AB, Fávaro WJ, Nakazato G (2016) Silver nanoparticles: a new view on mechanistic aspects on antimicrobial activity. Nanomedicine 12(3):789–799

Dusinska M, Collins AR (2008) The comet assay in human biomonitoring: gene–environment interactions. Mutagenesis 23(3):191–205

Espitia PJP, Soares NDFF, Coimbra JSDR, de Andrade NJ, Cruz RS, Medeiros EAA (2012) Zinc oxide nanoparticles: synthesis, antimicrobial activity and food packaging applications. Food Bioprocess Technol 5:1447–1464

Ewing AG, Bigelow JC, Wightman RM (1983) Direct in vivo monitoring of dopamine released from two striatal compartments in the rat. Science 221(4606):169–171

Fairbairn DW, Olive PL, O'Neill KL (1995) The comet assay: a comprehensive review. Mutat Res 339(1):37–59

Fantel AG (1996) Reactive oxygen species in developmental toxicity: review and hypothesis. Teratology 53(3):196–217

Feng J, Zhang QI, Liu QI, Zhu Z, McClements DJ, Jafari SM (2018) Application of nanoemulsions in formulation of pesticides. In: Nanoemulsions. Academic Press, pp 379–413

Garg A, Chalak HD, Belarbi MO, Zenkour AM, Sahoo R (2021) Estimation of carbon nanotubes and their applications as reinforcing composite materials–an engineering review. Compos Struct 272:114234

Gavrieli Y, Sherman Y, Ben-Sasson SA (1992) Identification of programmed cell death in situ via specific labelling of nuclear DNA fragmentation. J Cell Biol 119:493

Ghobashy MM, Elkodous MA, Shabaka SH, Younis SA, Alshangiti DM, Madani M et al (2021) An overview of methods for production and detection of silver nanoparticles, with emphasis on their fate and toxicological effects on human, soil, and aquatic environment. Nanotechnol Rev 10(1):954–977

Gimbert LJ, Hamon RE, Casey PS, Worsfold PJ (2007) Partitioning and stability of engineered ZnO nanoparticles in soil suspensions using flow field-flow fractionation. Environ Chem 4(1):8–10

Gomes A, Fernandes E, Lima JL (2005) Fluorescence probes used for detection of reactive oxygen species. J Biochem Biophys Methods 65(2):45–80

Gorup LF, Neto FNS, Kubo AM, Souza JAS, Fernandes RA, Fernandes GL et al (2017) Nanostructured functional materials: silver nanoparticles in polymer for the generation of antimicrobial characteristics. In: Recent advances in complex functional materials: from design to application. Springer, Cham, pp 271–292

Guduru RK, Dixit U (2024) Nanoceramics: food and bioprocessing industry. In: Industrial applications of nanoceramics. Elsevier, pp 51–62

Guo G et al (2007) Probing the cytotoxicity of CdSe quantum dots with surface modification. Mater Lett 61(8):1641–1644

Guo NL et al (2012) Multiwalled carbon nanotube-induced gene signatures in the mouse lung: potential predictive value for human lung cancer risk and prognosis. J Toxicol Environ Health A 75(18):1129–1153

Gupta AK, Wells S (2004) Surface-modified superparamagnetic nanoparticles for drug delivery: preparation, characterization, and cytotoxicity studies. IEEE Trans Nanobiosci 3(1):66–73

Halliwell B, Whiteman M (2004) Measuring reactive species and oxidative damage in vivo and in cell culture: how should you do it and what do the results mean? Br J Pharmacol 142(2):231–255

Handy RD et al (2012) Practical considerations for conducting ecotoxicity test methods with manufactured nanomaterials: what have we learnt so far? Ecotoxicology 21(4):933–972

Hassan TA, Rangari VK, Jeelani S (2014) Value-added biopolymer nanocomposites from waste eggshell-based CaCO3 nanoparticles as fillers. ACS Sustain Chem Eng 2(4):706–717

Henderson L et al (1998) The ability of the comet assay to discriminate between genotoxins and cytotoxins. Mutagenesis 13(1):89–94

Herzog E et al (2009) SWCNT suppress inflammatory mediator responses in human lung epithelium in vitro. Toxicol Appl Pharmacol 234(3):378–390

Hoshino A et al (2004) Physicochemical properties and cellular toxicity of nanocrystal quantum dots depend on their surface modification. Nano Lett 4(11):2163–2170

Hsin Y-H et al (2008) The apoptotic effect of nanosilver is mediated by a ROS-and JNK-dependent mechanism involving the mitochondrial pathway in NIH3T3 cells. Toxicol Lett 179(3):130–139

Hu F et al (2006) Cellular response to magnetic nanoparticles "PEGylated" via surface-initiated atom transfer radical polymerization. Biomacromolecules 7(3):809–816

Huang M, Khor E, Lim L-Y (2004) Uptake and cytotoxicity of chitosan molecules and nanoparticles: effects of molecular weight and degree of deacetylation. Pharm Res 21(2):344–353

Huczko A et al (2001) Physiological testing of carbon nanotubes: are they asbestos-like? Fuller Sci Technol 9(2):251–254

Hussain SM et al (2006) The interaction of manganese nanoparticles with PC-12 cells induces dopamine depletion. Toxicol Sci 92(2):456–463

Jacobsen NR, Pojana G, White P, Moller P, Cohn CA, Korsholm KS, Vogel U, Marcomini A, Loft S, Wallin H (2008) Genotoxicity, cytotoxicity, and reactive oxygen species induced by single-walled carbon nanotubes and C(60) fullerenes in the FE1-Mutatrade markMouse lung epithelial cells. Environ Mol Mutagen 49(6):476–487

Jeng HA, Swanson J (2006) Toxicity of metal oxide nanoparticles in mammalian cells. J Environ Sci Health A 41(12):2699–2711

Ji Z et al (2009) The hepatotoxicity of multi-walled carbon nanotubes in mice. Nanotechnology 20(44):445101

Jia Y, Hu Y, Li Y, Zeng Q, Jiang X, Cheng Z (2019) Boron doped carbon dots as a multifunctional fluorescent probe for sorbate and vitamin B12. Microchim Acta 186:1–10

Jin Y et al (2007) Toxicity of luminescent silica nanoparticles to living cells. Chem Res Toxicol 20(8):1126–1133

Kalaiyarasan G, Joseph J, Kumar P (2020) Phosphorus-doped carbon quantum dots as fluorometric probes for iron detection. ACS Omega 5(35):22278–22288

Kanoh M et al (1999) Significance of myocytes with positive DNA in situ nick end-labeling (TUNEL) in hearts with dilated cardiomyopathy. Circulation 99(21):2757–2764

Kang S, Mauter MS, Elimelech M (2008) Physicochemical determinants of multiwalled carbon nanotube bacterial cytotoxicity. Environ Sci Technol 42:7528–7534

Kawata K, Osawa M, Okabe S (2009) In vitro toxicity of silver nanoparticles at noncytotoxic doses to HepG2 human hepatoma cells. Environ Sci Technol 43(15):6046–6051

Kerfahi D et al (2015) Effects of functionalized and raw multi-walled carbon nanotubes on soil bacterial community composition. PLoS One 10(3):e0123042

Kerr JF, Wyllie AH, Currie AR (1972) Apoptosis: a basic biological phenomenon with wide-ranging implications in tissue kinetics. Br J Cancer 26(4):239

Kim SC et al (2001) In vivo evaluation of polymeric micellar paclitaxel formulation: toxicity and efficacy. J Control Release 72(1):191–202

Kim S, Hwang SJ, Choi W (2005) Visible light active platinum-ion-doped TiO2 photocatalyst. J Phys Chem B 109(51):24260–24267

Kim Y-J et al (2009) Genotoxicity of aluminum oxide (Al_2O_3) nanoparticle in mammalian cell lines. Mol Cell Toxicol 5(2):172–178

Kim JS et al (2010) Determination of cytotoxicity attributed to multiwall carbon nanotubes (MWCNT) in normal human embryonic lung cell (WI-38) line. J Toxicol Environ Health A 73(21–22):1521–1529

Kisin ER et al (2007) Single-walled carbon nanotubes: geno-and cytotoxic effects in lung fibroblast V79 cells. J Toxicol Environ Health A 70(24):2071–2079

Kolosnjaj-Tabi J, Lartigue L, Javed Y, Luciani N, Pellegrino T, Wilhelm C et al (2016) Biotransformations of magnetic nanoparticles in the body. Nano Today 11(3):280–284

Kononenko V, Repar N, Marušič N, Drašler B, Romih T, Hočevar S, Drobne D (2017) Comparative in vitro genotoxicity study of ZnO nanoparticles, ZnO macroparticles and ZnCl2 to MDCK kidney cells: size matters. Toxicol in Vitro 40:256-263

Kreyling WG, Semmler-Behnke M, Möller W (2006) Health implications of nanoparticles. J Nanopart Res 8:543–562

Kumar A, Sharma V, Dhawan A (2013) Methods for detection of oxidative stress and genotoxicity of engineered nanoparticles. In: Armstrong D, Bharali D (eds) Oxidative stress and nanotechnology, Methods in molecular biology (methods and protocols), vol 1028. Humana Press, Totowa, pp 231–246

Lam CW, James JT, McCluskey R, Hunter RL (2004) Pulmonary toxicity of single-wall carbon nanotubes in mice 7 and 90 days after intratracheal instillation. Toxicol Sci 77(1):126-134Lam C-W et al (2006) A review of carbon nanotube toxicity and assessment of potential occupational and environmental health risks. Crit Rev Toxicol 36(3):189–217

Lasagna-Reeves C et al (2010) Bioaccumulation and toxicity of gold nanoparticles after repeated administration in mice. Biochem Biophys Res Commun 393(4):649–655

Lei R et al (2008) Integrated metabolomic analysis of the nano-sized copper particle-induced hepatotoxicity and nephrotoxicity in rats: a rapid in vivo screening method for nanotoxicity. Toxicol Appl Pharmacol 232(2):292–301

Lemay M, Wood KA (1999) Detection of DNA damage and identification of UV-induced photoproducts using the CometAssay [TM] kit. Biotechniques 27:846–848

Li G-Y, Osborne NN (2008) Oxidative-induced apoptosis to an immortalized ganglion cell line is caspase independent but involves the activation of poly (ADP-ribose) polymerase and apoptosis-inducing factor. Brain Res 1188:35–43

Li Y-P et al (2001) PEGylated PLGA nanoparticles as protein carriers: synthesis, preparation and biodistribution in rats. J Control Release 71(2):203–211

Li JJ et al (2010) Autophagy and oxidative stress associated with gold nanoparticles. Biomaterials 31(23):5996–6003

Li W, Liu K, Feng S, Xiao Y, Zhang L, Mao J et al (2024) Well-defined Ni3N nanoparticles armored in hollow carbon nanotube shell for high-efficiency bifunctional hydrogen electrocatalysis. J Colloid Interface Sci 655:726–735

Liang SH, Clarke MF (2001) Regulation of p53 localization. FEBS J 268(10):2779–2783

Liu W, Huang G, Su X, Li S, Wang Q, Zhao Y et al (2020) Zebrafish: a promising model for evaluating the toxicity of carbon dot-based nanomaterials. ACS Appl Mater Interfaces 12(43):49012–49020

Loo DT (2002) TUNEL assay: an overview of techniques. In: Didenko VV (ed) Situ detection of DNA damage: methods and protocols. Humana Press, New York, pp 21–30

Lovrić J et al (2005) Unmodified cadmium telluride quantum dots induce reactive oxygen species formation leading to multiple organelle damage and cell death. Chem Biol 12(11):1227–1234

Lu X et al (2011) In vitro cytotoxicity and induction of apoptosis by silica nanoparticles in human HepG2 hepatoma cells. Int J Nanomed 6:1889

Lüllmann-Rauch R (1979) Drug-induced lysosomal storage disorders. Front Biol 48:49

Madani SY, Mandel A, Seifalian AM (2013) A concise review of carbon nanotube's toxicology. Nano Rev 4:21521–21534

Magder S (2006) Reactive oxygen species: toxic molecules or spark of life? Crit Care 10(1):208

Malyapa RS et al (1998) Detection of DNA damage by the alkaline comet assay after exposure to low-dose gamma radiation. Radiat Res 149(4):396–400

Marshall NJ, Goodwin CJ, Holt SJ (1995) A critical assessment of the use of microculture tetrazolium assays to measure cell growth and function. Growth Regul 5(2):69–84

Marszałł MP (2011) Application of magnetic nanoparticles in pharmaceutical sciences. Pharm Res 28:480–483

Mitchell L et al (2009) Mechanisms for how inhaled multiwalled carbon nanotubes suppress systemic immune function in mice. Nat Nanotechnol 4(7):451–456

Mitchell MJ, Billingsley MM, Haley RM, Wechsler ME, Peppas NA, Langer R (2021) Engineering precision nanoparticles for drug delivery. Nat Rev Drug Discov 20(2):101–124

Miura N, Shinohara Y (2009) Cytotoxic effect and apoptosis induction by silver nanoparticles in HeLa cells. Biochem Biophys Res Commun 390(3):733–737

Molinari BL et al (2003) Cell-based quantitative evaluation of the MTT assay. Anal Quant Cytol Histol 25(5):254–262

Monteiro-Riviere NA et al (2005) Multi-walled carbon nanotube interactions with human epidermal keratinocytes. Toxicol Lett 155(3):377–384

Muller J et al (2005) Respiratory toxicity of multi-wall carbon nanotubes. Toxicol Appl Pharmacol 207(3):221–231

Naik VM, Gunjal DB, Gore AH, Pawar SP, Mahanwar ST, Anbhule PV, Kolekar GB (2018) Quick and low cost synthesis of sulphur doped carbon dots by simple acidic carbonization of sucrose for the detection of Fe3+ ions in highly acidic environment. Diam Relat Mater 88:262–268

Naqvi S et al (2009) Concentration-dependent toxicity of iron oxide nanoparticles mediated by increased oxidative stress. Int J Nanomed 5:983–989

Nemes Z et al (1979) The pharmacological relevance of vital staining with neutral red. Experientia 35(11):1475–1476

Niwa Y, Iwai N (2006) Genotoxicity in cell lines induced by chronic exposure to water-soluble fullerenes using micronucleus test. Environ Health Prev Med 11(6):292–297

Nygaard UC, Hansen JS, Samuelsen M, Alberg T, Marioara CD, Løvik M (2009) Single-walled and multi-walled carbon nanotubes promote allergic immune responses in mice. Toxicol Sci 109(1):113–123

Park E-J et al (2009) Pro-inflammatory and potential allergic responses resulting from B cell activation in mice treated with multi-walled carbon nanotubes by intratracheal instillation. Toxicology 259(3):113–121

Park Y-H et al (2010) Assessment of dermal toxicity of nanosilica using cultured keratinocytes, a human skin equivalent model and an invivo model. Toxicology 267(1):178–181

Park Y, Yoo J, Lim B, Kwon W, Rhee SW (2016) Improving the functionality of carbon nanodots: doping and surface functionalization. J Mater Chem A 4(30):11582–11603

Patlolla A, Patlolla B, Tchounwou P (2010a) Evaluation of cell viability, DNA damage, and cell death in normal human dermal fibroblast cells induced by functionalized multiwalled carbon nanotube. Mol Cell Biochem 338(1–2):225–232

Patlolla AK et al (2010b) Comparative study of the clastogenicity of functionalized and nonfunctionalized multiwalled carbon nanotubes in bone marrow cells of Swiss-Webster mice. Environ Toxicol 25(6):608–621

Pavanello S, Clonfero E (2000) Biological indicators of genotoxic risk and metabolic polymorphisms. Mutat Res Rev Mutat Res 463(3):285–308

Pernodet N et al (2006) Adverse effects of citrate/gold nanoparticles on human dermal fibroblasts. Small 2(6):766–773

Pulskamp K, Diabaté S, Krug HF (2007) Carbon nanotubes show no sign of acute toxicity but induce intracellular reactive oxygen species in dependence on contaminants. Toxicol Lett 168(1):58–74

Punshon G et al (2005) Interactions between endothelial cells and a poly(carbonate-silsesquioxane-bridge-urea)urethane. Biomaterials 26(32):6271–6279

Qingnuan L, Yan X, Xiaodong Z, Ruili L, Qieqie D, Xiaoguang S, Shaoliang C, Wenxin L (2002) Preparation of (99m) Tc-C (60)(OH)(x) and its biodistribution studies. Nucl Med Biol 29(6):707–710

Radziun E et al (2011) Assessment of the cytotoxicity of aluminium oxide nanoparticles on selected mammalian cells. Toxicol In Vitro 25(8):1694–1700

Rahman M et al (2009) Expression of genes related to oxidative stress in the mouse brain after exposure to silver-25 nanoparticles. Toxicol Lett 187(1):15–21

Rana S, Kalaichelvan P (2013) Ecotoxicity of nanoparticles. ISRN Toxicol 2013:574648

Ravichandran P et al (2013) Induction of apoptosis in rat lung epithelial cells by multiwalled carbon nanotubes. J Biochem Mol Toxicol 27(5):286

Reddy KM, Manorama SV, Reddy AR (2003) Bandgap studies on anatase titanium dioxide nanoparticles. Mater Chem Phys 78(1):239–245

Reddy ARN et al (2010) Multi wall carbon nanotubes induce oxidative stress and cytotoxicity in human embryonic kidney (HEK293) cells. Toxicology 272(1):11–16

Ryter SW et al (2007) Mechanisms of cell death in oxidative stress. Antioxid Redox Signal 9(1):49–89

Sadhanala HK, Pagidi S, Gedanken A (2021) High quantum yield boron-doped carbon dots: a ratiometric fluorescent probe for highly selective and sensitive detection of Mg 2+ ions. J Mater Chem C 9(5):1632–1640

Sathyanarayana S, Hübner C (2013) Thermoplastic nanocomposites with carbon nanotubes. In: Njuguna J (ed) Structural nanocomposites. Engineering materials. Springer, Berlin, pp 19–60

Sayes CM, Reed KL, Warheit DB (2007) Assessing toxicity of fine and nanoparticles: comparing in vitro measurements to in vivo pulmonary toxicity profiles. Toxicol Sci 97(1):163–180

Schiavo S et al (2016) Genotoxic and cytotoxic effects of ZnO nanoparticles for *Dunaliella tertiolecta* and comparison with SiO_2 and TiO_2 effects at population growth inhibition levels. Sci Total Environ 550:619–627

Senapati VA, Kumar A, Gupta GS, Pandey AK, Dhawan A (2015) ZnO nanoparticles induced inflammatory response and genotoxicity in human blood cells: a mechanistic approach. Food Chem Toxicol 85:61–70
Sharma HS et al (2010) Influence of nanoparticles on blood–brain barrier permeability and brain edema formation in rats. In: Brain edema XIV. Springer, Berlin, pp 359–364
Shepard ZJ, Lux EM, Oyanedel-Craver VA (2020) Performance of silver nanoparticle-impregnated ovoid ceramic water filters. Environ Sci Nano 7(6):1772–1780
Sherr CJ (2004) Principles of tumor suppression. Cell 116(2):235–246
Shukla R et al (2005) Biocompatibility of gold nanoparticles and their endocytotic fate inside the cellular compartment: a microscopic overview. Langmuir 21(23):10644–10654
Shvedova AA et al (2005) Unusual inflammatory and fibrogenic pulmonary responses to single-walled carbon nanotubes in mice. Am J Physiol Lung Cell Mol Physiol 289(5):L698–L708
Simeonidis K, Martinez-Boubeta C, Zamora-Perez P, Rivera-Gil P, Kaprara E, Kokkinos E, Mitrakas M (2018) Nanoparticles for heavy metal removal from drinking water. Environ Nanotechnol 1:75–124
Singh N, Stephens R (1997) Microgel electrophoresis: sensitivity, mechanisms, and DNA electrostretching. Mutat Res 383(2):167–175
Singh RP, Handa R, Manchanda G (2021) Nanoparticles in sustainable agriculture: an emerging opportunity. J Control Release 329:1234–1248
Smith CJ, Shaw BJ, Handy RD (2007) Toxicity of single walled carbon nanotubes to rainbow trout, (*Oncorhynchus mykiss*): respiratory toxicity, organ pathologies, and other physiological effects. Aquat Toxicol 82(2):94–109
Sohn EK, Chung YS, Johari SA, Kim TG, Kim JK, Lee JH, Lee YH, Kang SW, Yu IJ (2015) Acute toxicity comparison of single-walled carbon nanotubes in various freshwater organisms. Biomed Res Int 2015:323090. https://doi.org/10.1155/2015/323090
Strober W (2001) Trypan blue exclusion test of cell viability. In: Current protocols in immunology. Wiley, pp A3.B.1–A3.B.3
Su C-H et al (2007) Nanoshell magnetic resonance imaging contrast agents. In: 2007 Digest of papers. Microprocesses and nanotechnology. IEEE
Suman S, Pandey A, Chandna S (2012) An improved non-enzymatic "DNA ladder assay" for more sensitive and early detection of apoptosis. Cytotechnology 64(1):9–14
Tang J et al (2009) Distribution, translocation and accumulation of silver nanoparticles in rats. J Nanosci Nanotechnol 9(8):4924–4932
Tarantola M et al (2008) Cytotoxicity of metal and semiconductor nanoparticles indicated by cellular micromotility. ACS Nano 3(1):213–222
Tkachenko AG et al (2004) Cellular trajectories of peptide-modified gold particle complexes: comparison of nuclear localization signals and peptide transduction domains. Bioconjug Chem 15(3):482–490
Vargas MA, Rodríguez-Páez JE (2017) Amorphous TiO2 nanoparticles: synthesis and antibacterial capacity. J Non-Cryst Solids 459:192–205
Vijayaraj V, Liné C, Cadarsi S, Salvagnac C, Baqué D, Elger A et al (2018) Transfer and ecotoxicity of titanium dioxide nanoparticles in terrestrial and aquatic ecosystems: a microcosm study. Environ Sci Technol 52(21):12757–12764
Wagner AJ et al (2007) Cellular interaction of different forms of aluminum nanoparticles in rat alveolar macrophages. J Phys Chem B 111(25):7353–7359
Wang H, Sun P, Cong S, Wu J, Gao L, Wang Y et al (2016) Nitrogen-doped carbon dots for "green" quantum dot solar cells. Nanoscale Res Lett 11:1–6
Warheit DB et al (2004) Comparative pulmonary toxicity assessment of single-wall carbon nanotubes in rats. Toxicol Sci 77(1):117–125
William WY et al (2006) Aqueous dispersion of monodisperse magnetic iron oxide nanocrystals through phase transfer. Nanotechnology 17(17):4483
Wiwanitkit V, Sereemaspun A, Rojanathanes R (2009) Effect of gold nanoparticles on spermatozoa: the first world report. Fertil Steril 91(1):e7–e8

Wu T, Tang M (2018) The inflammatory response to silver and titanium dioxide nanoparticles in the central nervous system. Nanomedicine 13(2):233–249

Wyllie AH (1980) Glucocorticoid-induced thymocyte apoptosis is associated with endogenous endonuclease activation. Nature 284(5756):555–556

Yadav D, Amini F, Ehrmann A (2020) Recent advances in carbon nanofibers and their applications–a review. Eur Polym J 138:109963

Yan J, Liu X, Ma C, Huang Y, Yang G (2020) All-dielectric materials and related nanophotonic applications. Mater Sci Eng R Rep 141:100563

Yang W, Shen C, Ji Q, An H, Wang J, Liu Q, Zhang Z (2009) Food storage material silver nanoparticles interfere with DNA replication fidelity and bind with DNA. Nanotechnology 20(8):085102

Ye S-F et al (2009) ROS and NF-κB are involved in upregulation of IL-8 in A549 cells exposed to multi-walled carbon nanotubes. Biochem Biophys Res Commun 379(2):643–648

Yehia HN et al (2007) Single-walled carbon nanotube interactions with HeLa cells. J Nanobiotechnol 5(1):1

Zang LY, Stone K, Pryor WA (1995) Detection of free radicals in aqueous extracts of cigarette tar by electron spin resonance. Free Radic Biol Med 19(2):161-167

Zhang TT, Li W, Meng G, Wang P, Liao W (2016) Strategies for transporting nanoparticles across the blood–brain barrier. Biomater Sci 4(2):219–229

Zhang L, Qin D, Feng J, Tang T, Cheng H (2023) Rapid quantitative detection of luteolin using an electrochemical sensor based on electrospinning of carbon nanofibers doped with single-walled carsbon nanoangles. Anal Methods 15(25):3073–3083

Zhu M-T et al (2008) Comparative study of pulmonary responses to nano-and submicron-sized ferric oxide in rats. Toxicology 247(2):102–111

Correction to: Ecotoxicity Potential of Carbon and Metal Nanoparticles: Delving into the Darker Realm

Bidisha Ghosh, Subhasis Sarkar, and Santanu Paul

Correction to: Chapter 26 in: S. Dey, S. Bhattacharya (eds.), *Biotechnological Interventions in the Removal of Emerging Pollutants*, Interdisciplinary Biotechnological Advances, https://doi.org/10.1007/978-981-97-9922-0_26

The book is published with two departments for the corresponding author of Chapter 26. The affiliation of Dr. Subhasis Sarkar is updated with "Department of Microbiology, School of Life Sciences, Swami Vivekananda University, Bara Kanthalia, Barrakpore, West Bengal, India." The first affiliation "Department of Biotechnology, School of Life Sciences, Swami Vivekananda University, Bara Kanthalia, Barrakpore, West Bengal, India of Dr. Subhasis Sarkar" has been removed.

The updated version of this chapter can be found at
https://doi.org/10.1007/978-981-97-9922-0_26

S. Dey, S. Bhattacharya (eds.), *Biotechnological Interventions in the Removal of Emerging Pollutants*, Interdisciplinary Biotechnological Advances,
https://doi.org/10.1007/978-981-97-9922-0_27

Index

S. Dey, S. Bhattacharya (eds.), *Biotechnological Interventions in the Removal of Emerging Pollutants*, Interdisciplinary Biotechnological Advances,
https://doi.org/10.1007/978-981-97-9922-0

MIX
Papier aus verantwortungsvollen Quellen
Paper from responsible sources
FSC® C105338

If you have any concerns about our products,
you can contact us on
ProductSafety@springernature.com

In case Publisher is established outside the EU,
the EU authorized representative is:
Springer Nature Customer Service Center GmbH
Europaplatz 3, 69115 Heidelberg, Germany

Printed by Libri Plureos GmbH
in Hamburg, Germany